Generalized Ordinary Differential Equations in Abstract Spaces and Applications

Generalized Ordinary Differential Equations in Abstract Spaces and Applications

Edited by

Everaldo M. Bonotto
Universidade de São Paulo
São Carlos, SP, Brazil

Márcia Federson
Universidade de São Paulo
São Carlos, SP, Brazil

Jaqueline G. Mesquita
Universidade de Brasília
Brasília, DF, Brazil

This edition first published 2021

Registered Office
John Wiley & Sons, Inc., 111 River Street, Hoboken, NJ 07030, USA

Editorial Office
111 River Street, Hoboken, NJ 07030, USA

For details of our global editorial offices, customer services, and more information about Wiley products visit us at www.wiley.com.

Wiley also publishes its books in a variety of electronic formats and by print-on-demand. Some content that appears in standard print versions of this book may not be available in other formats.

Library of Congress Cataloging-in-Publication Data applied for

ISBN: 9781119654933

Cover design by Wiley
Cover images: © piranka/E+/Getty Images

Set in 9.5/12.5pt STIXTwoText by Straive, Chennai, India

10 9 8 7 6 5 4 3 2 1

To God, Vanda, Humberto,
and Filipe, with deep gratitude.
—S. M. Afonso

To my family.
—F. G. de Andrade

To my family,
Teresinha, José and Tiago.
—F. Andrade da Silva

To God.
To my parents,
Geraldo and Zulmira.
To my brother, Daniel.
—M. Ap. Silva

To my wife Fabiana
and my son Arthur.
—E. M. Bonotto

To my family.
—R. Collegari

To God, Yahweh.
—M. Federson

To my parents,
Lourival and Helena.
—M. Frasson

To God, my family,
and my hometown Sabanalarga,
with all my love.
—R. Grau

To my husband, Luís.
—J. G. Mesquita

To God, for his infinite
love, which reached me and
changed my life forever.
—M. C. Mesquita

To God. To my family.
—P. H. Tacuri

To my wife, Michelle.
—E. Toon

and we dedicate this book to the memories of our friends
Luciene P. Gimenes Arantes & Štefan Schwabik.

Contents

List of Contributors

Suzete M. Afonso
Departamento de
Matemática - Instituto de Geociências
e Ciências Exatas
Universidade Estadual Paulista "Júlio
de Mesquita Filho" (UNESP)
Rio Claro–SP, Brazil

Fernando G. Andrade
Colégio Técnico de Bom Jesus
Universidade Federal do Piauí
Bom Jesus–PI, Brazil

Fernanda Andrade da Silva
Departamento de Matemática
Instituto de Ciências Matemáticas e de
Computação (ICMC)
Universidade de São Paulo
São Carlos–SP, Brazil

Marielle Ap. Silva
Departamento de Matemática
Instituto de Ciências Matemáticas e de
Computação (ICMC)
Universidade de São Paulo
São Carlos–SP, Brazil

Everaldo M. Bonotto
Departamento de Matemática
Aplicada e Estatística
Instituto de Ciências Matemáticas e de
Computação (ICMC)
Universidade de São Paulo
São Carlos–SP, Brazil

Rodolfo Collegari
Faculdade de Matemática
Universidade Federal de Uberlândia
Uberlândia–MG, Brazil

Márcia Federson
Departamento de Matemática
Instituto de Ciências Matemáticas e de
Computação (ICMC)
Universidade de São Paulo
São Carlos–SP, Brazil

Miguel V. S. Frasson
Departamento de Matemática
Aplicada e Estatística
Instituto de Ciências Matemáticas e de
Computação (ICMC)
Universidade de São Paulo
São Carlos–SP, Brazil

Luciene P. Gimenes (in memorian)
Departamento de Matemática – Centro de Ciências Exatas
Universidade Estadual de Maringá
Maringá-PR, Brazil

Rogelio Grau
Departamento de Matemáticas y Estadística
División de Ciencias Básicas
Universidad del Norte
Barranquilla, Colombia

Jaqueline G. Mesquita
Departamento de Matemática
Instituto de Ciências Exatas
Universidade de Brasília
Brasília-DF, Brazil

Maria Carolina Mesquita
Departamento de Matemática
Instituto de Ciências Matemáticas e de Computação (ICMC)
Universidade de São Paulo
São Carlos-SP, Brazil

Patricia H. Tacuri
Departamento de Matemática – Centro de Ciências Exatas
Universidade Estadual de Maringá
Maringá-PR, Brazil

Eduard Toon
Departamento de Matemática
Instituto de Ciências Exatas
Universidade Federal de Juiz de Fora
Juiz de Fora-MG, Brazil

Foreword

Since the origins of the calculus, the development of ordinary differential equations has always both influenced and followed the development of the integral calculus. Newton's theory of fluxions was instrumental for solving the equations of mechanics, Leibniz and the Bernoulli brothers explored the differential equations solvable by quadrature, Euler developed similar numerical methods for computing definite integrals and approximate solutions of differential equations, that Cauchy converted into existence results for the integral of continuous functions and the solutions of initial value problems.

Following this distinguished line, Jaroslav Kurzweil solved in 1957 the delicate question of finding optimal conditions for the continuous dependence on parameters of solutions of ordinary differential systems. To this aim, he introduced a new concept of integral generalizing Ward's extension of the Perron integral, and showed how to define it through a technically minor but basic modification of the definition of the Riemann integral. It happened that, under this type of convergence, the limit of a sequence of ordinary differential equations could be a more general object, a generalized differential equations, defined in terms of the Kurzweil integral. As it often happens in mathematics, a similar integral was introduced independently in 1961 by Ralph Henstock during his investigations on the Ward integral.

This new theory, besides providing a simple, elegant, and pedagogical approach to the classical concepts of Lebesgue and Perron integrals, shed a new light on many classical questions in differential equations like stability and the averaging method. The theory was actively developed in Praha, around Kurzweil, by Jiří Jarník, Štefan Schwabik, Milán Tvrdy, Ivo Vrkoč, Dana Fraňková, and others. Their results are beautifully described in the monographs *Generalized Ordinary Differential Equations* of Schwabik and *Generalized Ordinary Differential Equations* of Kurzweil.

Contributions came also from many other countries. In particular, inspired by the fruitful visits of Márcia Federson in Praha and the enthusiastic lectures of Štefan Schwabik in São Carlos, a school was created and developed at the University of São Paulo and spread into other Brazilian institutions. Those researches gave a great impetus to the theory of generalized ordinary differential equations from the fundamental theory to many new important applications.

Because of its definition through suitable limits of Riemann sums, the Kurzweil–Henstock integral directly applies to functions with values in Banach spaces. This is particularly emphasized by the monograph *Topics in Banach Space Integration* of Schwabik and Ye GuoJu. It was therefore a natural and fruitful idea to consider generalized ordinary differential equations in Banach spaces, not for just the sake of generality, but also because of their specific applications, and in particular their use in reinterpreting the concept of functional-differential equation. This is the viewpoint adopted in the present substantial monograph, whose red wire is to show that many types of evolution equations in Banach space can be treated in an unified and more general way as a special case of the Kurzweil generalized differential equations.

The general ideas are introduced and motivated through an elegant treatment of the measure functional differential equations, where in the integrated form of the differential equation, the Lebesgue measure is replaced by some Stieltjes one. The approach covers differential equations with impulses and the dynamic equations on time scales. The generalized differential equations in Banach spaces are then introduced and developed in a systematic way. Most classical problems of the theory of ordinary differential equations extending to this new and general setting. This includes the existence, continuation, and continuous dependence of solutions, the linear equations, the Lyapunov stability, the periodic and bounded solutions, the averaging method, some control theory, the dichotomy theory, questions of topological dynamics, and the measure neutral functional differential equations.

It is not surprising that the richness and wideness of the content of this substantial monograph is the result of the intensive team activity of 14 contributors: S.M. Afonso, F. Andrade da Silva, M. Ap. Silva, E.M. Bonotto, R. Collegari, F.G. de Andrade, M. Federson, M. Frasson, L.P. Gimenes, R. Grau, J.G. Mesquita, M.C. Mesquita, P.H. Tacuri, and E. Toon. Most of them have already published contributions to the area and other ones got a PhD recently in this direction. Each chapter is authored by a selected subgroup of the team, but it appears that a joint final reading took place for the final product. Each chapter provides a state-of-the-art of its title, and many classes of readers will find in this book a renewed picture of their favorite area of expertise and an inspiration for further research.

In a world where, even for scientific research, competition is too often praised as a necessary driving force, this monograph shows that the true answer is better found in an open, enthusiastic, and unselfish cooperation. In this way, the book not only magnifies the work of Jaroslav Kurzweil and of Štefan Schwabik but also their wonderful spirit.

October 2020

Jean Mawhin
Louvain-la-Neuve, Belgium

Preface

It is well known that the remarkable theory of generalized ordinary differential equations (we write generalized ODEs, for short) was born in Czech Republic in the year 1957 with the brilliant paper [147] by Professor Jaroslav Kurzweil. In Brazil, the theory of generalized ODEs was introduced by Professor Štefan Schwabik during his visit to the Universidade de São Paulo, in the city of São Paulo, in 1989.

Nevertheless, it was only in 2002 that the theory really started to be developed here. The article by Professors Márcia Federson and Plácido Táboas, published in the *Journal of Differential Equations* in 2003 (see [92]), was the first Brazilian publication on the subject. Now, 18 years later, members of the Brazilian research group on *Functional Differential Equations and Nonabsolute Integration* decided to gather the results obtained over these years in order to produce a comprehensive literature about our results developed so far, regarding the theory of generalized ODEs in abstract spaces.

Originally, this monograph was thought to be organized by professors Márcia Federson, Everaldo M. Bonotto, and Jaqueline G. Mesquita, with the contribution of the following authors: Suzete M. Afonso, Fernando G. Andrade, Fernanda Andrade da Silva, Marielle Ap. Silva, Rodolfo Collegari, Miguel Frasson, Luciene P. Gimenes, Rogelio Grau, Maria Carolina Mesquita, Patricia H. Tacuri, and Eduard Toon. However, after a while, it became a production of us all, with contributions of everyone to all chapters and to the uniformity, coherence, language, and interrelationship of the results. We, then, present this carefully crafted work to disseminate the theories involved here, especially those on Kurzweil–Henstock nonabsolute integration and on generalized ODEs.

In the introductory chapter, named *Preliminaries*, we brought together two main issues that permeate this book. The first one concerns the spaces where the functions within the right-hand sides of differential or integral equations live. The other one concerns the theory of nonabsolute vector-valued integrals in the senses of J. Kurzweil and R. Henstock. Sections 1.1 and 1.2 are devoted to properties of the space of regulated functions and the space of functions of bounded $\mathcal{B}$-variation. Among the main results of Section 1.1, we mention a characterization, based on [96, 97], of relatively compact sets of the space of regulated functions. Section 1.2 deals with properties of functions of bounded $\mathcal{B}$-variation where the Helly's choice principle for abstract spaces is a spotlight. The book [127] is the main reference for this section. The third section is devoted to nonabsolute vector-valued integrals. The basis of this theory is presented here and results specialized for Perron–Stieltjes integrals are included. We highlight substitution formulas and an integration by parts formula coming from [212]. Other important references to this section are [72, 73, 210].

The second chapter is devoted to the integral as defined by Jaroslav Kurzweil in [147]. We compiled some historical data on how the idea of the integral came about. Highlights of this chapter include the Saks–Henstock lemma, the Hake-type theorem, and the change of variables theorem. We end this chapter with a brief history of the Kapitza pendulum equation whose solution is highly oscillating and, therefore, suitable for being treated via Kurzweil–Henstock nonabsolute integration theory. An important reference to Chapter 2 is [209].

Before entering the theory of generalized ODEs, we take a trip through the theory of measure functional differential equations (we write measure FDEs for short). Then, the third chapter appears as an embracing collection of results on measure FDEs for Banach space-valued functions. In particular, we investigate equations of the form

$$y(t) = y(t_0) + \int_{t_0}^{t} f(y_s, s)\, dg(s), \quad t \in [t_0, t_0 + \sigma],$$

where y_s is a memory function and the integral on the right-hand side is in the sense of Perron–Stieltjes. We show that these equations encompass not only impulsive functional dynamic equations on time scales but also impulsive measure FDEs. Examples illustrating the relations between any two of these equations are also included. References [85, 86] feature as the foundation for this relations. Among other topics covered by Chapter 3, we mention averaging principles, covering the periodic and nonperiodic cases, and results on continuous dependence of solutions on time scales. References [21, 82, 178] are crucial here.

In Chapter 4, we enter the theory of generalized ODE itself. We begin by recalling the concept of a nonautonomous generalized ODE of the form

$$\frac{dx}{d\tau} = DF(x, t),$$

where F takes a pair (x, t) of a regulated function x and a time t to a regulated function. The main reference to this chapter is [209]. Measure FDEs in the integral form described above feature in Chapter 4 as supporting actors, because now their solutions can be related to solutions of the generalized ODEs, whose right-hand sides involve functions which look like

$$F(x,t)(\vartheta) = \begin{cases} 0, & t_0 - r \le \vartheta \le t_0, \\ \displaystyle\int_{t_0}^{\vartheta} f(x_s, s)\, dg(s), & t_0 \le \vartheta \le t \le t_0 + \sigma, \\ \displaystyle\int_{t_0}^{t} f(x_s, s)\, dg(s), & t \le \vartheta \le t_0 + \sigma. \end{cases}$$

This characteristic of generalized ODEs plays an important role in the entire manuscript, since it allows one to translate results from generalized ODEs to measure FDEs.

Chapter 5, based on [78], brings together the foundations of the theory of generalized ODEs. Section 5.1 concerns local existence and uniqueness of a solution of a nonautonomous generalized ODE with applications to measure FDEs and functional dynamic equations on time scales. Second 5.2 is devoted to results on prolongation of solutions of generalized ODEs, measure differential equations, and dynamic equations on time scales.

Chapter 6 deals with a very important class of differential equations, the class of linear generalized ODEs. The origins of linear generalized ODEs goes back to the papers [209–211]. Here, we recall the notion of fundamental operator associated with a linear generalized ODE for Banach space-valued functions and we travel on the same road as the authors of [45] to obtain a variation-of-constants formula for a linear perturbed generalized ODE. Concerning applications, we extend the class of equations to include linear measure FDEs.

After linear generalized ODEs are investigated, we move to results on continuous dependence of solutions on parameters. This is the core of Chapter 7 which is based on [4, 95, 96, 177]. Given a family of generalized ODEs, we present sufficient conditions so that the family of their corresponding solutions converge uniformly, on compact sets, to the solution of the limiting generalized ODE. We also prove that given a generalized ODE and its solution $x_0 : [a, b] \to X$, where X is a Banach space, one can obtain a family of generalized ODEs whose solutions converge uniformly to x_0 on $[a, b]$.

As we mentioned before, many types of differential equations can be regarded as generalized ODEs. This fact allows us to derive stability results for these equations through the relations between the solutions of a certain equations and the solutions of a generalized ODEs. At the present time, the stability theory for generalized ODEs is undergoing a remarkable development. Recent results in this respect are contained in [3, 7, 80, 89, 90] and are gathered in Chapter 8. We also show the effectiveness of Lyapunov's Direct Method to obtain several stability results, in addition to proving converse Lyapunov theorems for some types of stability. The types of stability explore here are variational stability, Lyapunov stability, regular stability, and many relations permeating these concepts.

The existence of periodic solutions to any kind of equation is also of great interest, especially in applications. Chapter 9 is devoted to this matter in the framework of generalized ODEs, whose results are also specified to measure differential equations and impulsive differential equations. Section 9.1 brings together a result which provides conditions for the solutions of a linear generalized ODE taking values in $\mathbb{R}^n$ to be periodic and a result which relates periodic solutions of linear nonhomogeneous generalized ODEs to periodic solutions of linear homogeneous generalized ODEs. Still in this section, a characterization of the fundamental matrix of periodic linear generalized ODEs is established. This is the analogue of the Floquet theorem for generalized ODEs involving finite dimensional space-valued functions. In Section 9.2, inspired in an approach by Jean Mawhin to treat periodic boundary value problem (we write periodic BVP for short), we introduce the concept of a (θ, T)-periodic solution for a nonlinear homogeneous generalized ODE in Banach spaces, where $T > 0$ and $\theta > 0$. A result that ensures a correspondence between solutions of a (θ, T)-periodic BVP and (θ, T)-periodic solutions of a nonlinear homogeneous generalized ODE is the spotlight here. Then, the existence of a (θ, T)-periodic solution is guaranteed.

Averaging methods are used to investigate the solutions of a nonautonomous differential equations by means of the solutions of an "*averaged*" autonomous equation. In Chapter 10, we present a periodic averaging principle as well as a nonperiodic one for generalized ODEs. The main references to this chapter are [83, 178].

Chapter 11 is designed to provide the reader with a systematic account of recent developments in the boundedness theory for generalized ODEs. The results of this chapter were borrowed from the articles [2, 79].

Chapter 12 is devoted to the control theory in the setting of abstract generalized ODEs. In its first section, we introduce concepts of observability, exact controllability, and approximate controllability, and we give necessary and sufficient conditions for a system of generalized ODEs to be exactly controllable, approximately controllable, or observable. In Section 12.2, we apply the results to classical ODEs.

The study of exponential dichotomy for linear generalized ODEs of type

$$\frac{dx}{d\tau} = D[A(t)x]$$

is the heartwood of Chapter 13, where sufficient conditions for the existence of exponential dichotomies are obtained, as well as conditions for the existence of bounded solutions for the nonhomogeneous equation

$$\frac{dx}{d\tau} = D[A(t)x + f(t)].$$

Using the relations between the solutions of generalized ODEs and the solutions of other types of equations, we translate our results to measure differential equations and impulsive differential equations. The main reference for this chapter is [29].

The aim of Chapter 14 is to bring together the theory of semidynamical systems generated by generalized ODEs. We show the existence of a local semidynamical system generated by a nonautonomous generalized ODE of the form

$$\frac{dx}{d\tau} = DF(x, t),$$

where F belongs to a compact class of right-hand sides. We construct an impulsive semidynamical system associated with a generalized ODE subject to *external* impulse effects. For this class of impulsive systems, we present a LaSalle's invariance principle-type result. Still in this chapter, we present some topological properties for impulsive semidynamical systems as minimality and recurrence. The main reference here is [4].

Chapter 15 is intended for applications of the theory developed in some of the previous chapters to a class of more general functional differential equations, namely, measure FDE of neutral type. In Section 15.1, some historical notes ranging from the beginnings of the term *equation*, passing through "*functional differential equation*," and reaching *functional differential equation of neutral type* are put together. Then, we present a correspondence between solutions of a measure FDE of neutral type with finite delays and solutions of a generalized ODEs. Results on existence and uniqueness of a solution as well as continuous dependence of solutions on parameters based on [76] are also explored.

We end this preface by expressing our immense gratitude to professors Jaroslav Kurzweil, Štefan Schwabik (*in memorian*) and Milan Tvrdý for welcoming several members of our research group at the Institute of Mathematics of the Academy of Sciences of the Czech Republic so many times, for the countably many good advices and talks, and for the corrections of proofs and theorems during all these years.

October 2020

Everaldo M. Bonotto
Márcia Federson
Jaqueline G. Mesquita
São Carlos, SP, Brazil

1

Preliminaries

Everaldo M. Bonotto[1], Rodolfo Collegari[2], Márcia Federson[3], Jaqueline G. Mesquita[4], and Eduard Toon[5]

[1] *Departamento de Matemática Aplicada e Estatística, Instituto de Ciências Matemáticas e de Computação (ICMC), Universidade de São Paulo, São Carlos, SP, Brazil*
[2] *Faculdade de Matemática, Universidade Federal de Uberlândia, Uberlândia, MG, Brazil*
[3] *Departamento de Matemática, Instituto de Ciências Matemáticas e de Computação (ICMC), Universidade de São Paulo, São Carlos, SP, Brazil*
[4] *Departamento de Matemática, Instituto de Ciências Exatas, Universidade de Brasília, Brasília, DF, Brazil*
[5] *Departamento de Matemática, Instituto de Ciências Exatas, Universidade Federal de Juiz de Fora, Juiz de Fora, MG, Brazil*

This preliminary chapter is devoted to two pillars of the theory of generalized ordinary differential equations for which we use the short form *"generalized ODEs"*. One of these pillars concerns the spaces in which the solutions of a generalized ODE are generally placed. The other pillar concerns the theory of nonabsolute integration, due to Jaroslav Kurzweil and Ralph Henstock, for integrands taking values in Banach spaces. As a matter of fact, such integration theory permeates the entire book. It (the theory of non absolute integration) is within the heartwood of the theory of generalized ODEs, appearing (the same theory of nonabsolute integration) in the integral form of a very special case of nonautonomous generalized ODEs, namely (now we mention the name of the special case of generalized ODEs), measure functional differential equations.

The solutions of a Cauchy problem for a generalized ODE, with right-hand side in a class of functions introduced by J. Kurzweil in [147–149], usually belong to a certain space of functions of bounded variation (see Lemma 4.9). However, since functions of bounded variation are also regulated functions in the sense described by Jean Dieudonné and, more generally, by the group Nicolas Bourbaki, and because the space of regulated functions is more adequate for dealing with discontinuous functions appearing naturally in Stieltjes-type integrals, it is important to present a substantial content about this space. Thus, the first section of this chapter describes the main properties of the space of regulated functions

Generalized Ordinary Differential Equations in Abstract Spaces and Applications, First Edition.
Edited by Everaldo M. Bonotto, Márcia Federson, and Jaqueline G. Mesquita.

with the icing of the cake being a characterization of its relatively compact subsets due to D. Franková.

Regarding functions of bounded variation, which are known to be of bounded semivariation and, hence, of bounded $\mathcal{B}$-variation, we present, in the second section of this chapter, a coherent overview of functions of bounded $\mathcal{B}$-variation over bilinear triples. Among the results involving functions of bounded variation, the theorem of Helly (or the Helly's choice principle for Banach space-valued function) due to C. S. Hönig is a spotlight. On the other hand, functions of bounded semivariation appear, for instance, in the integration by parts formula for Kurzweil and Henstock integrals of Stieltjes-type.

In the third section of this chapter, we describe the second pillar and main background of the theory of generalized ODEs, namely, the framework of vector-valued nonabsolute integrals of Kurzweil and Henstock. Here, we call the reader's attention to the fact that we refer to Kurzweil vector integrals as Perron–Stieltjes integrals so that, when a more general definition of the Kurzweil integral is presented in Chapter 2, the reader will not be confused. One of the highlights of the third section is, then, the integration by parts formula for Perron–Stieltjes integrals.

An extra section called *"Appendix,"* which can be skipped in a first reading of the book, concerns other types of gauge-based integrals which use the interesting idea of Edward James McShane. The well-known Bochner–Lebesgue integral comes into the scene and an equivalent definition of it as the limit of Riemannian-type sums comes up.

1.1 Regulated Functions

Regulated functions appear in the works by J. Dieudonné [58, p. 139] and N. Bourbaki [32, p. II.4]. The *raison d'être* of regulated functions lies on the fact that every regulated function $f : [0,T] \subset \mathbb{R} \to \mathbb{R}^n$ has a primitive, that is, there exists a continuous function $F : [0,T] \subset \mathbb{R} \to \mathbb{R}^n$ such that $\frac{dF}{dt}(t) = f(t)$ almost everywhere in $[0,T]$, in the sense of the Lebesgue measure. The interested reader may want to check this fact as described, for instance, by the group N. Bourbaki in [32, Corollaire I, p. II.6].

1.1.1 Basic Properties

Let X be a Banach space with norm $\|\cdot\|$. Here, we describe regulated functions $f : [a,b] \to X$, where $[a,b]$, with $a < b$, is a compact interval of the real line $\mathbb{R}$.

Definition 1.1: A function $f : [a,b] \to X$ is called *regulated*, if the lateral limits

$$\lim_{s \to t^-} f(s) = f(t^-), \quad t \in (a,b], \qquad \text{and} \qquad \lim_{s \to t^+} f(s) = f(t^+), \quad t \in [a,b),$$

exist. The space of all regulated functions $f : [a, b] \to X$ will be denoted by $G([a, b], X)$.

We denote the subspace of all continuous functions $f : [a, b] \to X$ by $C([a, b], X)$ and, by $G^-([a, b], X)$, we mean the subspace of regulated functions $f : [a, b] \to X$ which are left-continuous on $(a, b]$. Then, the following inclusions clearly hold

$$C([a, b], X) \subset G^-([a, b], X) \subset G([a, b], X).$$

Remark 1.2: Let $O \subset X$. By $G([a, b], O)$, we mean the set of all elements $f \in G([a, b], X)$ for which $f(t) \in O$ for every $t \in [a, b]$. Thus, it is clear that $G([a, b], O) \subset G([a, b], X)$ and the range of $f \in G([a, b], O)$ belongs to O. Note that, for a given $t \in [a, b]$, $f(t^+)$ and $f(t^-)$ do not necessarily belong to O.

Any finite set $d = \{t_0, t_1, \ldots, t_m\}$ of points in the closed interval $[a, b]$ such that

$$a = t_0 < t_1 < \cdots < t_m = b$$

is called a *division* of $[a, b]$. We write simply $d = (t_i)$. Given a division $d = (t_i)$ of $[a, b]$, its elements are usually denoted by $t_0, t_1, \ldots, t_{|d|}$, where, from now on, $|d|$ denotes the number of intervals in which $[a, b]$ is divided through the division d. The set of all divisions $d = (t_i)$ of $[a, b]$ is denoted by $D_{[a,b]}$.

Definition 1.3: A function $f : [a, b] \to X$ is called a *step function*, if there is a division $d = (t_i) \in D_{[a,b]}$ such that for each $i = 1, \ldots, |d|$, $f(t) = c_i \in X$, for all $t \in (t_{i-1}, t_i)$. We denote by $E([a, b], X)$ the set of all step functions $f : [a, b] \to X$.

It is clear that $E([a, b], X) \subset G([a, b], X)$. Moreover, we have the following important result which is a general version of the result presented in [127, Theorem I.3.1].

Theorem 1.4: *Let $O \subset X$ and consider a function $f : [a, b] \to O$. The assertions below are equivalent:*

(i) *$f : [a, b] \to O$ is the uniform limit of step functions $\{f_n\}_{n\in\mathbb{N}}$, with $f_n : [a, b] \to O$;*
(ii) *$f \in G([a, b], O)$;*
(iii) *given $\epsilon > 0$, there exists a division $d : a = t_0 < t_1 < t_2 < \cdots < t_{|d|} = b$ such that*

$$\sup\{\| f(t) - f(s) \| : \ t, s \in (t_{j-1}, t_j), \ j = 1, \ldots, |d|\} < \epsilon.$$

Proof. We will prove (i) $\Rightarrow$ (ii), (ii) $\Rightarrow$ (iii) and, then, (iii) $\Rightarrow$ (i).

(i) ⇒ (ii) Note that $f(t) \in O$ for all $t \in [a,b]$. We need to show that $f \in G([a,b],X)$, see Remark 1.2. Let $t \in [a,b)$. We will only prove that $\lim_{s\to t^+} f(s)$ exists, because the existence of $\lim_{s\to t^-} f(s)$ follows analogously. Consider a sequence $\{t_n\}_{n\in\mathbb{N}}$ in $[a,b]$ such that $t_n \searrow t$, that is, $t_n \geqslant t$, for every $n \in \mathbb{N}$, and t_n converges to t as $n \to \infty$. Consider the sequence $\{f_n\}_{n\in\mathbb{N}}$ of step functions from $[a,b]$ to O such that $f_n \to f$ uniformly as $n \to \infty$. Then, given $\epsilon > 0$, there exists $k \in \mathbb{N}$ such that $\| f(t) - f_k(t) \| < \epsilon/4$, for all $t \in [a,b]$. In addition, since f_k is a step function, there exists $N_0 \in \mathbb{N}$ such that $\| f_k(t_n) - f_k(t^+) \| < \epsilon/4$, for all $n \geqslant N_0$. Therefore, for $n, m \geqslant N_0$, we have

$$\begin{aligned}\| f(t_n) - f(t_m) \| &\leqslant \| f(t_n) - f_k(t_n) \| + \| f_k(t_n) - f_k(t^+) \| \\ &\quad + \| f_k(t^+) - f_k(t_m) \| + \| f_k(t_m) - f(t_m) \| < \epsilon.\end{aligned}$$

Then, once X is a Banach space, $\lim_{s\to t^+} f(s)$ exists.

(ii) ⇒ (iii) Let $\epsilon > 0$ be given. Since $f \in G([a,b],O)$, it follows that $f \in G([a,b],X)$ (see Remark 1.2). Thus, for every $t \in (a,b)$, there exists $\delta_t > 0$ such that

$$\sup_{w,s\in(t-\delta_t,t)} \| f(w) - f(s) \| < \epsilon \quad \text{and} \quad \sup_{w,s\in(t,t+\delta_t)} \| f(w) - f(s) \| < \epsilon.$$

Similarly, there are $\delta_a, \delta_b > 0$ such that

$$\sup_{w,s\in(a,a+\delta_a)} \| f(w) - f(s) \| < \epsilon \quad \text{and} \quad \sup_{w,s\in(b-\delta_b,b)} \| f(w) - f(s) \| < \epsilon.$$

Notice that the set A of intervals $\{[a,a+\delta_a),(t-\delta_t,t+\delta_t),(b-\delta_b,b] : t \in (a,b)\}$ is an open cover of the interval $[a,b]$ and, hence, there is a division $d = (t_i)$ of $[a,b]$, with $i = 1,2,\dots,|d|$, such that $\{[a,a+\delta_a),(t_1-\delta_{t_1},t_1+\delta_{t_1}),\dots,(b-\delta_b,b]\}$ is a finite subcover of A for $[a,b]$ and, moreover,

$$\sup \{\| f(t) - f(s) \| : t,s \in (t_{i-1},t_i),\ i = 1,\dots,|d|\} < \epsilon.$$

(iii) ⇒ (i). Given $n \in \mathbb{N}$, let $d_n = (t_i)$, $i = 1,2,\dots,|d_n|$, be a division of $[a,b]$ such that

$$\sup \{\| f(t) - f(s) \| : t,s \in (t_{i-1},t_i),\ i = 1,2,\dots,|d_n|\} < \frac{1}{n}$$

and $\tau_i \in (t_{i-1},t_i)$, $i = 1,2,\dots,|d_n|$. Define

$$f_n(t) = \sum_{i=1}^{|d_n|} f(\tau_i)\chi_{(t_{i-1},t_i)}(t) + \sum_{i=0}^{|d_n|} f(t_i)\chi_{t_i}(t),$$

where χ_B denotes the characteristic function of a measurable set $B \subset \mathbb{R}$. Note that $f_n(t) \in O$ for all $t \in [a,b]$ and all $n \in \mathbb{N}$. Moreover, $\{f_n\}_{n\in\mathbb{N}}$ is a sequence of step functions which converge uniformly to f, as $n \to \infty$. □

It is a consequence of Theorem 1.4 (with $O = X$) that the closure of $E([a,b],X)$ is $G([a,b],X)$. Therefore, $G([a,b],X)$ is a Banach space when equipped with the usual supremum norm

$$\| f \|_\infty = \sup_{t \in [a,b]} \| f(t) \| .$$

See also [127, Theorem I.3.6].

If $B([a, b], X)$ denotes the Banach space of bounded functions from $[a, b]$ to X, equipped with the supremum norm, then the inclusion

$$G([a, b], X) \subset B([a, b], X)$$

follows from Theorem 1.4, items (i) and (ii), taking the limit of step functions which are constant on each subinterval of continuity.

Recently, D. Franková established a fourth assertion equivalent to those assertions of Theorem 1.4 in the case where $O = X$. See [97, Theorem 2.3]. One can note, however, that such result also holds for any open set $O \subset X$. This is the content of the next lemma.

Lemma 1.5: *Let $O \subset X$ and $f : [a, b] \to O$ be a function. Then the assertions of Theorem 1.4 are also equivalent to the following assertion:*

(iv) *for every $\epsilon > 0$, there is a division $a = t_0 < t_1 < \cdots < t_{|d|} = b$ such that*

$$\| f(t'') - f(t') \| < \epsilon, \text{ for all } t_{i-1} < t' < t'' < t_i \text{ and } i = 1, 2, \ldots, |d|.$$

Proof. Note that condition (iii) from Theorem 1.4 implies condition (iv). Now, assume that condition (iv) holds. Given $\epsilon > 0$, there is a division $a = t_0 < t_1 < \cdots < t_{|d|} = b$ such that $\| f(t'') - f(t') \| < \epsilon$, for all $t_{i-1} < t' < t'' < t_i$ and $i = 1, 2, \ldots, |d|$. According to [97, Theorem 2.3], take $\tau_i \in (t_{i-1}, t_i)$ and consider a step function $h : [a, b] \to O$ given by

$$h(t) = \begin{cases} f(\tau_i), & t \in (t_{i-1}, t_i), \quad i = 1, 2, \ldots, |d|, \\ f(t_i), & t = t_i, \quad i = 0, 1, \ldots, |d|. \end{cases}$$

Hence, $\sup_{t \in [a,b]} \| h(t) - f(t) \| \leqslant \epsilon$. □

The next result, borrowed from [7, Lemma 2.3], specifies the supremum of a function $f \in G([a, b], X)$.

Proposition 1.6: *Let $f \in G([a, b], X)$. Then $\sup_{t \in [a,b]} \| f(t) \| = c$, where either $c = \| f(\sigma) \|$, for some $\sigma \in [a, b]$, or $c = \| f(\sigma^-) \|$, for some $\sigma \in (a, b]$, or $c = \| f(\sigma^+) \|$ for some $\sigma \in [a, b)$.*

Proof. Let $c = \sup_{t \in [a,b]} \| f(t) \|$. Since $G([a, b], X) \subset B([a, b], X)$, $c < \infty$. By the definition of the supremum, for all $n \in \mathbb{N}$, one can choose $x_n \in [a, b]$ such that $0 \leqslant c - \| f(x_n) \| < \frac{1}{n}$ which implies

$$\lim_{n \to \infty} \| f(x_n) \| = c.$$

Since $\{x_n\}_{n\in\mathbb{N}} \subset [a,b]$, there exists a subsequence $\{x_{n'}\}_{n'\in\mathbb{N}} \subset \{x_n\}_{n\in\mathbb{N}}$ such that $x_{n'} \to \sigma \in [a,b]$ as $n' \to \infty$. Since f is regulated, $c = \lim_{n\to\infty} \| f(x'_n) \|$ belongs to $\{\| f(\sigma) \|, \| f(\sigma^-) \|, \| f(\sigma^+) \|\}$ and the proof is complete. □

The composition of regulated functions may not be a regulated function as shown by the next example proposed by Dieudonné as an exercise. See, for instance, [58, Problem 2, p. 140].

Example 1.7: Consider, for instance, functions $f, g : [0,1] \to \mathbb{R}$ given by $f(t) = t \sin \frac{1}{t}$, for $t \in (0,1]$, $f(0) = 0$ and $g(t) = \operatorname{sgn} t$, that is, g is the sign function. Both f and g belong to $G([0,1],\mathbb{R})$. However, the composition $g \circ f$ does not.

The next result, borrowed from [209, Theorem 10.11], gives us an interesting property of left-continuous regulated functions. Such result will be used in Chapters 8 and 11. We state it here without any proof.

Proposition 1.8: *Let $f, g \in G^-([a,b],\mathbb{R})$. If for every $t \in [a,b)$, there exists $\delta(t) > 0$ such that for every $\eta \in (0, \delta(t))$, we have $f(t+\eta) - f(t) \leqslant g(t+\eta) - g(t)$, then*

$$f(s) - f(a) \leqslant g(s) - g(a), \quad s \in [a,b].$$

We end this first section by introducing some notation for certain spaces of regulated functions defined on unbounded intervals of the real line $\mathbb{R}$. Given $t_0 \in \mathbb{R}$, we denote by $G([t_0,\infty),X)$ the space of regulated functions from $[t_0,\infty)$ to X. In order to obtain a Banach space, we can intersect the space $G([t_0,\infty),X)$ with the space $B([t_0,\infty),X)$ of bounded functions from $[t_0,\infty)$ to X, in which case, we write $BG([t_0,\infty),X)$ and equip such space with the supremum norm,

$$f \in BG([t_0,\infty),X) \mapsto \| f \|_\infty = \sup_{t\in[t_0,\infty)} \| f(t) \| \in \mathbb{R}_+,$$

where $\mathbb{R}_+ = [0,\infty)$. Alternatively, we can consider a subspace $G_0([t_0,\infty),X)$ of $G([t_0,\infty),X)$ formed by all functions $f : [t_0,\infty) \to X$ such that

$$\sup_{t\in[t_0,\infty)} e^{-(t-t_0)} \| f(t) \| < \infty.$$

The next result shows that $G_0([t_0,\infty),X)$ is a Banach space with respect to a special norm. This result, whose proof follows ideas similar to those of [124, 220], will be largely used in. Chapters 5 and 8

Proposition 1.9: *The space $G_0([t_0,\infty),X)$, equipped with the norm*

$$\| f \|_{[t_0,\infty)} = \sup_{t\in[t_0,\infty)} e^{-(t-t_0)} \| f(t) \|, \quad f \in G_0([t_0,\infty),X),$$

is a Banach space.

Proof. Let $T : G_0([t_0, \infty), X) \to BG([t_0, \infty), X)$ be the linear mapping defined by

$$(Ty)(t) = e^{-(t-t_0)}y(t),$$

for all $y \in G_0([t_0, \infty), X)$ and $t \in [t_0, \infty)$.

Claim. T is an isometric isomorphism. Indeed, T is an isometry because

$$\| T(y)\|_{BG([t_0,\infty),X)} = \sup_{t\in[t_0,\infty)} \| (Ty)(t) \| = \sup_{t\in[t_0,\infty)} \| y(t) \| e^{-(t-t_0)} = \| y\|_{G_0([t_0,\infty),X)},$$

for all $y \in G_0([t_0, \infty), X)$. Moreover, if $y \in BG([t_0, \infty), X)$, then $u : [t_0, \infty) \to X$ defined by

$$u(t) = e^{t-t_0}y(t), \quad \text{for all } t \in [t_0, \infty)$$

is such that $Ty = u$ and $u \in G_0([t_0, \infty), X)$, since

$$\sup_{s\in[t_0,\infty)} \| u(s) \| e^{-(s-t_0)} = \sup_{s\in[t_0,\infty)} \| y(s) \| e^{s-t_0}e^{-(s-t_0)} = \sup_{s\in[t_0,\infty)} \| y(s) \| < \infty.$$

Therefore, T is onto and the *Claim* is proved.

Once T is an isometric isomorphism and $BG([t_0, \infty), X)$ is a Banach space, we conclude that $G_0([t_0, \infty), X)$ is also a Banach space. □

1.1.2 Equiregulated Sets

In this subsection, our goal is to investigate important properties of equiregulated sets. In addition to [97], the reference [40] also deals with a characterization of subsets of equiregulated functions.

Definition 1.10: A set $\mathcal{A} \subset G([a, b], X)$ is called *equiregulated*, if it has the following property: for every $\epsilon > 0$ and every $\sigma \in [a, b]$, there exists $\delta > 0$ such that

(i) if $f \in \mathcal{A}$, $t' \in [a, b]$ and $\sigma - \delta < t' < \sigma$, then $\| f(\sigma^-) - f(t') \| < \epsilon$;
(ii) if $f \in \mathcal{A}$, $t'' \in [a, b]$ and $\sigma < t'' < \sigma + \delta$, then $\| f(t'') - f(\sigma^+) \| < \epsilon$.

The next result, which can be found in [97, Proposition 3.2], gives a characterization of equiregulated functions taking values in a Banach space.

Theorem 1.11: *A set $\mathcal{A} \subset G([a, b], X)$ is equiregulated if and only if for every $\epsilon > 0$, there is a division $d : a = t_0 < t_1 < \cdots < t_{|d|} = b$ of $[a, b]$ such that*

$$\| f(t') - f(t) \| \leqslant \epsilon, \tag{1.1}$$

for every $f \in \mathcal{A}$ and $[t, t'] \subset (t_{j-1}, t_j)$, for $j = 1, 2, \dots, |d|$.

Proof. (⇒) Let $\epsilon > 0$ be given and let B be the set of all $\gamma \in (a, b]$ such that there is a division $d' \in D_{[a,\gamma]}$, that is, $d' : a = t_0 < t_1 < \cdots < t_{|d'|} = \gamma$ for which (1.1) holds with $|d'|$ instead of $|d|$. Since $\mathcal{A}$ is equiregulated, there is $\delta_1 \in (0, b - a]$ such that $\| f(t) - f(a^+) \| \leqslant \frac{\epsilon}{2}$, for every $f \in \mathcal{A}$ and $t \in (a, a + \delta_1)$. Let $a_1 = a + \delta_1$ and $a = t_0 < t_1 = a_1$. Thus, for $[t, t'] \subset (a, a_1)$ and $f \in \mathcal{A}$, we have

$$\| f(t) - f(t') \| \leqslant \| f(t) - f(a^+) \| + \| f(t') - f(a^+) \| \leqslant \epsilon.$$

Hence, $a_1 \in B$.

Let $\tilde{c}$ be the supremum of the set B. Since $f \in \mathcal{A}, f$ is regulated. Thus, there exists $\delta > 0$ such that $\| f(\tilde{c}^-) - f(t) \| \leqslant \frac{\epsilon}{2}$ for every $f \in \mathcal{A}$ and $t \in (\tilde{c} - \delta, \tilde{c}) \cap [a, b]$. Take $c \in B \cap (\tilde{c} - \delta, \tilde{c})$ and a division $d'' \in D_{[a,c]}$, say, $d'' : a = t_0 < t_1 < \cdots < t_{|d''|} = c$, such that (1.1) holds with $|d''|$ instead of $|d|$. Denote $t_{|d''|+1} = \tilde{c}$. Then, for $[t, t'] \subset (t_{|d''|}, t_{|d''|+1})$ and $f \in \mathcal{A}$, we have

$$\| f(t) - f(t') \| \leqslant \| f(t) - f(\tilde{c}^-) \| + \| f(t') - f(\tilde{c}^-) \| \leqslant \epsilon,$$

which implies $\tilde{c} \in B$. Thus, we have two possibilities: either $\tilde{c} = b$ or $\tilde{c} < b$. In the first case, the proof is finished. In the second case, one can use a similar argument as the one we used before in order to find $e \in (\tilde{c}, b]$ such that $e \in B$, and this contradicts the fact that $\tilde{c} = \sup B$. Thus, $\tilde{c} = b$, and we finish the proof of the sufficient condition.

(⇐) Now, we prove the necessary condition. Given $\epsilon > 0$, there exists a division $d' \in D_{[a,b]}$, say, $d' : a = t_0 < t_1 < \cdots < t_{|d'|} = b$ such that the inequality (1.1) is fulfilled, for every $f \in \mathcal{A}$ and every $[t, t'] \subset (t_{j-1}, t_j)$, with $j = 1, 2, \ldots, |d'|$. Then, for every $j = 1, 2, \ldots, |d'|$, take $\tau_j \in (t_{j-1}, t_j)$ and $\delta > 0$ such that $(\tau_j - \delta, \tau_j + \delta) \subset (t_{j-1}, t_j)$. Thus, (1.1) is satisfied, for all $t, t' \in (\tau_j - \delta, \tau_j + \delta)$. In particular, if either $t = \tau_j$ and $t' \in (\tau_j - \delta, \tau_j]$, or $t = \tau_j$ and $t' \in [\tau_j, \tau_j + \delta)$, then the inequality (1.1) holds. Thus, $\mathcal{A}$ is equiregulated. □

The next result describes an interesting property of equiregulated sets of $G([a, b], X)$. Such result can be found in [97, Proposition 3.8].

Theorem 1.12: *Assume that a set $\mathcal{A} \subset G([a, b], X)$ is equiregulated and, for any $t \in [a, b]$, there is a number γ_t such that*

$$\| f(t) - f(t^-) \| \leqslant \gamma_t, \; t \in (a, b], \quad \text{and} \quad \| f(t^+) - f(t) \| \leqslant \gamma_t, \; t \in [a, b).$$

Then, there is a constant $K > 0$ such that, for every $f \in \mathcal{A}$,

$$\| f(t) - f(a) \| \leqslant K, \quad t \in [a, b].$$

Proof. Take B as the set of all numbers $\tau \in (a, b]$ fulfilling the condition that there exists a positive number K_τ for which we have $\| f(t) - f(a) \| \leqslant K_\tau$, for every $f \in \mathcal{A}$

and every $t \in [a, \tau]$. Since $\mathcal{A}$ is an equiregulated set, there exists $\delta > 0$ such that $\| f(t) - f(a^+) \| \leqslant 1$, for every $f \in \mathcal{A}$ and every $t \in (a, a + \delta]$. From this fact and the hypotheses, we can infer that for every $f \in \mathcal{A}$ and every $t \in (a, a + \delta]$, we have

$$\| f(t) - f(a) \| \leqslant \| f(t) - f(a^+) \| + \| f(a^+) - f(a) \| \leqslant 1 + \gamma_a = K_{(a+\delta)}.$$

Then, $(a, a + \delta] \subset B$.

Let $\tau_0 = \sup B$. The equiregulatedness of $\mathcal{A}$ implies that there exists $\delta' > 0$ such that $\| f(t) - f(\tau_0^-) \| \leqslant 1$, for every $f \in \mathcal{A}$ and $t \in [\tau_0 - \delta', \tau_0)$. Take $\tau \in B \cap [\tau_0 - \delta', \tau_0)$. Thus, for every $f \in \mathcal{A}$,

$$\| f(\tau_0^-) - f(a) \| \leqslant \| f(\tau_0^-) - f(\tau) \| + \| f(\tau) - f(a) \| \leqslant 1 + K_\tau, \quad t \in (\tau, \tau_0),$$

which, together with the hypotheses, yield

$$\| f(\tau_0) - f(a) \| \leqslant \| f(\tau_0) - f(\tau_0^-) \| + \| f(\tau_0^-) - f(a) \| \leqslant \gamma_{\tau_0} + 1 + K_\tau.$$

Hence, $\tau_0 \in B$.

Let $\tau_0 < b$. Since $\mathcal{A}$ is equiregulated, there exists $\delta'' > 0$ such that, for every $f \in \mathcal{A}$, $\| f(t) - f(\tau_0^+) \| \leqslant 1$, for all $t \in (\tau_0, \tau_0 + \delta'']$. Therefore, for every $f \in \mathcal{A}$, we have

$$\begin{aligned}\| f(t) - f(a) \| &\leqslant \| f(t) - f(\tau_0^+) \| + \| f(\tau_0^+) - f(\tau_0) \| + \| f(\tau_0) - f(a) \| \\ &\leqslant 1 + \gamma_{\tau_0} + K_{\tau_0} = K_{(\tau_0+\delta'')},\end{aligned}$$

for $t \in (\tau_0, \tau_0 + \delta'']$, where $K_{\tau_0} = \gamma_{\tau_0} + 1 + K_\tau$. Note that $\tau_0 + \delta'' \in B$ which contradicts the fact that $\tau_0 = \sup B$. Hence, $\tau_0 = b$ and the statement follows. □

1.1.3 Uniform Convergence

This subsection brings a few results borrowed from [177]. In particular, Lemma 1.13 describes an interesting and useful property of equiregulated converging sequences of Banach space-valued functions and it is used later in the proof of a version of Arzelà–Ascoli theorem for Banach space-valued regulated functions.

Lemma 1.13: *Let $\{f_k\}_{k\in\mathbb{N}}$ be a sequence of functions from $[a, b]$ to X. If the sequence $\{f_k\}_{k\in\mathbb{N}}$ converges pointwisely to f_0 and is equiregulated, then it converges uniformly to f_0.*

Proof. By hypothesis, the sequence of functions $\{f_k\}_{k\in\mathbb{N}}$ is equiregulated. Then, Theorem 1.11 yields that, for every $\epsilon > 0$, there is a division $d = (t_i) \in D_{[a,b]}$ for which

$$\| f_k(t) - f_k(s) \| < \frac{\epsilon}{3},$$

for every $k \in \mathbb{N}$ and $t_{i-1} < s < t < t_i$, $i = 1, 2, \dots, |d|$.

Take $\tau_i \in (t_{i-1}, t_i)$. Because the sequence $\{f_k\}_{k\in\mathbb{N}}$ converges pointwisely to f_0, we have $f_k(t_i) \to f_0(t_i)$ and also $f_k(\tau_i) \to f_0(\tau_i)$, for $i = 1, 2, \dots, |d|$. Thus, for every $\epsilon > 0$, there is $k_0 \in \mathbb{N}$ such that, whenever $k > k_0$, we have $\| f_k(t_i) - f_0(t_i) \| < \frac{\epsilon}{3}$ and $\| f_k(\tau_i) - f_0(\tau_i) \| < \frac{\epsilon}{3}$, for $i = 1, 2, \dots, |d|$.

Take an arbitrary $t \in [a, b]$. Then, either $t = t_i$ for some i, or $t \in (t_{i-1}, t_i)$ for some i. In the former case, $\| f_k(t) - f_0(t) \| < \frac{\epsilon}{3}$. The other case yields

$$\| f_k(t) - f_0(t) \| \leqslant \| f_k(t) - f_k(\tau_i) \| + \| f_k(\tau_i) - f_0(\tau_i) \| + \| f_0(\tau_i) - f_0(t) \| < \epsilon.$$

Then, $\| f_k - f_0 \|_\infty < \epsilon$ and, therefore, $f_k \to f_0$ uniformly on $[a, b]$. □

Lemma 1.14: *Let $\{f_k\}_{k\in\mathbb{N}}$ be a sequence in $G([a, b], X)$. The following assertions hold:*

(i) *if the sequence of functions f_k converges uniformly to f_0 as $k \to \infty$ on $[a, b]$, then $f_k(t^-) \to f_0(t^-)$, for $t \in (a, b]$, and $f_k(t^+) \to f_0(t^+)$, for $t \in [a, b)$;*

(ii) *if the sequence of functions f_k converges pointwisely to f_0 as $k \to \infty$ on $[a, b]$ and $f_k(t^-) \to f_0(t^-)$, for $t \in (a, b]$, and $f_k(t^+) \to f_0(t^+)$, for $t \in [a, b)$, where $f_0 \in G([a, b], X)$, then the sequence f_k converges uniformly to f_0 as $k \to \infty$.*

Proof. We start by proving (i). By hypothesis, the sequence $\{f_k\}_{k\in\mathbb{N}}$ converges uniformly to f_0. Then, Moore–Osgood theorem (see, e.g., [19]) implies

$$\lim_{k\to\infty}\lim_{s\to t^-} f_k(s) = \lim_{s\to t^-}\lim_{k\to\infty} f_k(s), \quad t \in (a, b].$$

Therefore, $f_k(t^-) \to f_0(t^-)$, for $t \in (a, b]$. In a similar way, one can show that $f_k(t^+) \to f_0(t^+)$, for every $t \in [a, b)$.

Now, we prove (ii). It suffices to show that $\{f_k : [a, b] \to X : k \in \mathbb{N}\}$ is an equiregulated set. Indeed, by Lemma 1.13, f_k converges uniformly to f_0.

Since the function f_0 is regulated, its lateral limits exist. Then, for every $t_0 \in [a, b]$ and every $\epsilon > 0$, there is $\delta > 0$ such that

$$\| f_0({t_0}^-) - f_0(t) \| < \epsilon, \text{for } t_0 - \delta \leqslant t < t_0,$$
$$\| f_0(t) - f_0({t_0}^+) \| < \epsilon, \text{for } t_0 < t \leqslant t_0 + \delta,$$

for every $t \in [a, b]$.

But the hypotheses say that we can find $n_0 \in \mathbb{N}$ such that, for $n \geqslant n_0$, we have

$$\| f_n(t_0 - \delta) - f_0(t_0 - \delta) \| < \epsilon, \ \| f_n(t_0) - f_0(t_0) \| < \epsilon, \ \| f_n(t_0^+) - f_0(t_0^+) \| < \epsilon,$$
$$\| f_n(t_0 + \delta) - f_0(t_0 + \delta) \| < \epsilon \ \text{ and } \ \| f_n(t_0^-) - f_0(t_0^-) \| < \epsilon.$$

When $t \in [a, b]$ satisfies $t_0 - \delta \leqslant t < t_0$, we have, for every $n \geqslant n_0$,

$$\| f_n(t_0^-) - f_n(t) \| \leqslant \| f_n(t_0^-) - f_0(t_0^-) \| + \| f_0(t_0^-) - f_0(t) \| + \| f_0(t) - f_n(t) \| < 3\epsilon.$$

On the other hand, when $t \in [a, b]$ satisfies $t_0 < t \leqslant t_0 + \delta$, we get, for $n \geqslant n_0$,

$$\| f_n(t) - f_n(t_0^+) \| \leqslant \| f_n(t) - f_0(t) \| + \| f_0(t) - f_0(t_0^+) \| + \| f_n(t_0^+) - f_0(t_0^+) \| < 3\epsilon.$$

But this yields the fact that $\{f_n\}_{n\in\mathbb{N}}$ is an equiregulated sequence, and the proof is complete. □

The next lemma guarantees that, if a sequence of functions $\{f_k\}_{k\in\mathbb{N}}$ is bounded by an equiregulated sequence of functions, then $\{f_k\}_{k\in\mathbb{N}}$ is also equiregulated.

Lemma 1.15: *Let $\{f_k\}_{k\in\mathbb{N}}$ be a sequence of functions in $G([a, b], X)$. Suppose, for each $k \in \mathbb{N}$, the function f_k satisfies*

$$\| f_k(s_2) - f_k(s_1) \| \leqslant |h_k(s_2) - h_k(s_1)|, \tag{1.2}$$

for every $s_1, s_2 \in [a, b]$, where $h_k : [a, b] \to \mathbb{R}$ for each $k \in \mathbb{N}$ and the sequence $\{h_k\}_{k\in\mathbb{N}}$ is equiregulated. Then, the sequence $\{f_k\}_{k\in\mathbb{N}}$ is equiregulated.

Proof. Let $\epsilon > 0$ be given. Since the sequence $\{h_k\}_{k\in\mathbb{N}}$ is equiregulated, it follows from Theorem 1.11 that there is a division $d = (t_i) \in D_{[a,b]}$ such that $|h_k(t') - h_k(t)| \leqslant \epsilon$, for every $k \in \mathbb{N}$ and $[t, t'] \subset (t_{j-1}, t_j)$, for $j = 1, 2, \dots, |d|$. Thus, by (1.2), $\| f_k(t') - f_k(t) \| \leqslant \epsilon$, for every $k \in \mathbb{N}$ and every interval $[t, t'] \subset (t_{j-1}, t_j)$, with $j = 1, 2, \dots, |d|$. Finally, Theorem 1.11 ensures the fact that the sequence $\{f_k\}_{k\in\mathbb{N}}$ is equiregulated. □

A clear outcome of Lemmas 1.13 and 1.15 follows below:

Corollary 1.16: *Let $\{f_k\}_{k\in\mathbb{N}}$ be a sequence of functions from $[a, b]$ to X and suppose the function f_k satisfies condition (1.2) for every $k \in \mathbb{N}$ and $s_1, s_2 \in [a, b]$, where $h_k : [a, b] \to \mathbb{R}$ and the sequence $\{h_k\}_{k\in\mathbb{N}}$ is equiregulated. If the sequence $\{f_k\}_{k\in\mathbb{N}}$ converges pointwisely to a function f_0, then it also converges uniformly to f_0.*

1.1.4 Relatively Compact Sets

In this subsection, we investigate an extension of the Arzelà–Ascoli theorem for regulated functions taking values in a general Banach space X with norm $\| \cdot \|$.

Unlike the finite dimensional case, when we consider functions taking values in X, the relatively compactness of a set $\mathcal{A} \subset G([a, b], X)$ does not come out as a consequence of the equiregulatedness of the set $\mathcal{A}$ and the boundedness of the set $\{f(t) : f \in \mathcal{A}\} \subset X$, for each $t \in [a, b]$. In the following lines, we present an example, borrowed from [177] which illustrates this fact.

Example 1.17: Let $Z \subset X$ be bounded. Suppose Z not relatively compact in X. Thus, given an arbitrary $\epsilon > 0$, there is a sequence of functions $\{z_n\}_{n\in\mathbb{N}}$ in Z for which

$$\| z_n \| \leqslant K \quad \text{and} \quad \| z_n - z_m \| \geqslant \epsilon,$$

for all $n \neq m$ and for some constant $K > 0$. Hence, the set

$$B = \{y_n : [0,1] \to X : y_n(t) = tz_n,\ n \in \mathbb{N}\}$$

is bounded, once $\{z_n\}_{n\in\mathbb{N}}$ is bounded. Moreover, B is equiregulated and $\{y_n(0)\}_{n\in\mathbb{N}}$ is relatively compact in X. On the other hand, B is not relatively compact in $G([0,1],X)$.

At this point, it is important to say that, in order to guarantee that a set $\mathcal{A} \subset G([a,b],X)$ is relatively compact, one needs an additional condition. It is clear that, if one assumes that, for each $t \in [a,b]$, the set $\{f(t) : f \in \mathcal{A}\}$ is relatively compact in X, then $\mathcal{A}$ becomes relatively compact in $G([a,b],X)$. This is precisely what the next result says, and we refer to it as the Arzelà–Ascoli theorem for Banach space-valued regulated functions. Such important result can be found in [97] and [177] as well.

Theorem 1.18: *Suppose $\mathcal{A} \subset G([a,b],X)$ is equiregulated and, for every $t \in [a,b]$, $\{f(t) : f \in \mathcal{A}\}$ is relatively compact in X. Then, $\mathcal{A}$ is relatively compact in $G([a,b],X)$.*

Proof. Take a sequence of functions $\{f_n\}_{n\in\mathbb{N}} \subset \mathcal{A}$. The set $\mathcal{A}$ is equiregulated by hypothesis. Then, for every $\epsilon > 0$, there exists a division $d = (t_i) \in D_{[a,b]}$ fulfilling

$$\| f_n(t') - f_n(t) \| < \frac{\epsilon}{4},$$

for every $n \in \mathbb{N}$ and $t_{i-1} < t < t' < t_i$, $i = 1,2,\dots,|d|$.

On the other hand, the sets $\{f_n(t_i)\}_{n\in\mathbb{N}}$ and $\{f_n(\tau_i)\}_{n\in\mathbb{N}}$ are relatively compact in X for every $i = 1,2,\dots,|d|$, where $t_{i-1} < \tau_i < t_i$. Thus, there is a subsequence of indexes $\{n_k\}_{k\in\mathbb{N}} \subset \mathbb{N}$, with $n_{k+1} > n_k$, for which $\{f_{n_k}(t_i)\}_{k\in\mathbb{N}}$ and $\{f_{n_k}(\tau_i)\}_{k\in\mathbb{N}}$ are also relatively compact sets of X, for every i.

This last statement implies that there exist $\{y_i : i = 0,1,2,\dots,|d|\} \subset X$ and $\{z_i : i = 1,2,\dots,|d|\} \subset X$ satisfying

$$y_i = \lim_{k\to\infty} f_{n_k}(t_i) \quad \text{and} \quad z_i = \lim_{k\to\infty} f_{n_k}(\tau_i).$$

Thus, there exists $N \in \mathbb{N}$ such that

$$\left\| f_{n_k}(t_i) - y_i \right\| < \frac{\epsilon}{4} \quad \text{and} \quad \left\| f_{n_k}(\tau_i) - z_i \right\| < \frac{\epsilon}{4},$$

provided $k > N$. In particular, for $q > k$, we have

$$\| f_{n_q}(t_i) - y_i \| < \frac{\epsilon}{4} \quad \text{and} \quad \| f_{n_q}(\tau_i) - z_i \| < \frac{\epsilon}{4},$$

for $i = 1, 2, \dots, |d|$.

Take $t \in [a, b]$ and consider $q \in \mathbb{N}$ such that $q > k$. Therefore, either $t = t_i$, for some $i \in \{1, 2, \dots, |d|\}$, in which case, we have

$$\| f_{n_k}(t) - f_{n_q}(t) \| \leqslant \| f_{n_k}(t_i) - y_i \| + \| f_{n_q}(t_i) - y_i \| < \frac{\epsilon}{2},$$

or $t \in (t_{i-1}, t_i)$, for some $i \in \{1, 2, \dots, |d|\}$, in which case, we have

$$\begin{aligned} \| f_{n_k}(t) - f_{n_q}(t) \| &\leqslant \| f_{n_k}(t) - f_{n_k}(\tau_i) \| + \| f_{n_k}(\tau_i) - f_{n_q}(\tau_i) \| + \| f_{n_q}(t) - f_{n_q}(\tau_i) \| \\ &\leqslant \| f_{n_k}(t) - f_{n_k}(\tau_i) \| + \| f_{n_q}(t) - f_{n_q}(\tau_i) \| + \| f_{n_k}(\tau_i) - z_i \| \\ &\quad + \| f_{n_q}(\tau_i) - z_i \| < \frac{\epsilon}{4} + \frac{\epsilon}{4} + \frac{\epsilon}{4} + \frac{\epsilon}{4} = \epsilon. \end{aligned}$$

Hence, for every $t \in [a, b]$, $\{f_{n_k}(t)\}_{k \in \mathbb{N}} \subset X$ satisfies the Cauchy condition. Due to the fact that X is a complete space and because $\{f_{n_k}(t)\}_{k \in \mathbb{N}}$ is a Cauchy sequence, the limit $\lim_{k \to \infty} f_{n_k}(t)$ exists.

We conclude by considering $f_0(t) = \lim_{k \to \infty} f_{n_k}(t)$. Then, $f_{n_k} \to f_0$ on $[a, b]$, by Lemma 1.13. Hence, f_0 is the uniform limit of the subsequence $\{f_{n_k}\}$ in $G([a, b], X)$. Finally, any sequence $\{f_n\}_{n \in \mathbb{N}} \subset \mathcal{A}$ admits a converging subsequence which, in turn, implies that $\mathcal{A}$ is a relatively compact set, and the proof is finished. □

We end this subsection by mentioning an Arzelà–Ascoli-type theorem for regulated functions taking values in $\mathbb{R}^n$. A slightly different version of it can be found in [96].

Corollary 1.19: *The following conditions are equivalent:*

(i) *a set $\mathcal{A} \subset G([a, b], \mathbb{R}^n)$ is relatively compact;*
(ii) *the set $\{f(a) : f \in \mathcal{A}\}$ is bounded, and there are an increasing continuous function $\eta : [0, \infty) \to [0, \infty)$, with $\eta(0) = 0$, and a nondecreasing function $v : [a, b] \to \mathbb{R}$ such that, for every $f \in \mathcal{A}$,*

$$|f(t_2) - f(t_1)| \leqslant \eta(v(t_2) - v(t_1)),$$

for $a \leqslant t_1 \leqslant t_2 \leqslant b$;
(iii) *$\mathcal{A}$ is equiregulated, and for every $t \in [a, b]$, the set $\{f(t) : f \in \mathcal{A}\}$ is bounded.*

We point out in [96, Theorem 2.17], item (ii), it is required that v is an increasing function. However, it is not difficult to see that if u is a nondecreasing function, then taking $v(t) = u(t) + t$ yields v is an increasing function. Therefore, Corollary 1.19 follows as an immediate consequence of [96, Theorem 2.17].

1.2 Functions of Bounded $\mathcal{B}$-Variation

The concept of a function of bounded $\mathcal{B}$-variation generalizes the concepts of a function of semivariation and of a function of bounded variation, as we will see in the sequel.

Definition 1.20: A *bilinear triple* (we write BT) is a set of three vector spaces E, F, and G, where F and G are normed spaces with a bilinear mapping $\mathcal{B} : E \times F \to G$. For $x \in E$ and $y \in F$, we write $xy = \mathcal{B}(x, y)$, and we denote the BT by $(E, F, G)_{\mathcal{B}}$ or simply by (E, F, G). A *topological* BT is a BT (E, F, G), where E is also a normed space and $\mathcal{B}$ is continuous.

IfE and F are normed spaces, then we denote by $L(E, F)$ the space of all linear continuous functions from E to F. We write $E' = L(E, \mathbb{R})$ and $L(E) = L(E, E)$, where $\mathbb{R}$ denotes the real line. Next, we present examples, borrowed from [127], of bilinear triples.

Example 1.21: Let X, Y, and Z denote Banach spaces. The following are BT:

(i) $E = L(X, Y), F = L(Z, X), G = L(Z, Y)$, and $\mathcal{B}(v, u) = v \circ u$;
(ii) $E = L(X, Y), F = X, G = Y$, and $\mathcal{B}(u, x) = u(x)$;
(iii) $E = Y, F = Y', G = \mathbb{R}$, and $\mathcal{B}(y, y') = \langle y, y' \rangle$;
(iv) $E = G = Y, F = \mathbb{R}$ and $\mathcal{B}(y, \lambda) = \lambda y$.

Given a BT $(E, F, G)_{\mathcal{B}}$, we define, for every $x \in E$, a norm

$$\| x \|_{\mathcal{B}} = \sup \{ \| \mathcal{B}(x, y) \| : \| y \| \leqslant 1 \}$$

and we set $E_{\mathcal{B}} = \{ x \in E : \| x \| < \infty \}$. Whenever the space $E_{\mathcal{B}}$ is endowed with the norm $\| \cdot \|_{\mathcal{B}}$, we say that the topological BT $(E_{\mathcal{B}}, F, G)$ is associated with the BT (E, F, G).

Let E be a vector space and Γ_E be a set of seminorms defined on E such that $p_1, \dots, p_m \in \Gamma_E$ implies $\sup [p_1, \dots, p_m] \in \Gamma_E$. Then, Γ_E defines a topology on E, and the sets $V_{p,\epsilon} = \{ x \in E : p(x) < \epsilon \}$, $p \in \Gamma_E$, $\epsilon > 0$, form a basis of neighborhoods of 0. The sets $x_0 + V_{p,\epsilon}$ form a basis of the neighborhood of $x_0 \in E$. Moreover, when endowed with this topology, E is called a *locally convex space* (see [127], p. 3, 4).

Example 1.22: Every normed or seminormed space E is a locally convex space.

For other examples of locally convex spaces, we refer to [110].

Definition 1.23: Given a BT $(E, F, G)_{\mathcal{B}}$, and a function $\alpha : [a, b] \to E$, for every division $d = (t_i) \in D_{[a,b]}$, we define

$$SB_d(\alpha) = SB_{[a,b],d}(\alpha) = \sup \left\{ \left\| \sum_{i=1}^{|d|} \left[\alpha(t_i) - \alpha(t_{i-1})\right] y_i \right\| : y_i \in F, \|y_i\| \leqslant 1 \right\} \quad \text{and}$$

$$SB(\alpha) = SB_{[a,b]}(\alpha) = \sup \left\{ SB_d(\alpha) : d \in D_{[a,b]} \right\}.$$

Then, $SB(\alpha)$ is the $\mathcal{B}$-variation of α on $[a, b]$. We say that α is a *function of bounded $\mathcal{B}$-variation*, whenever $SB(\alpha) < \infty$. In this case, we write $\alpha \in SB([a, b], E)$.

The following properties are not difficult to prove. See, e.g. [127, 4.1 and 4.2].

(SB1) $SB([a, b], E)$ is a vector space and the mapping $\alpha \in SB([a, b], E) \mapsto SB(\alpha) \in \mathbb{R}_+$ is a seminorm.
(SB2) Given $\alpha \in SB([a, b], E)$, the function $t \in [a, b] \mapsto SB_{[a,t]}(\alpha) \in \mathbb{R}_+$ is increasing.
(SB3) Given $\alpha \in SB([a, b], E)$ and $c \in (a, b)$, $SB_{[a,b]}(\alpha) \leqslant SB_{[a,c]}(\alpha) + SB_{[c,b]}(\alpha)$.

Definition 1.24: Consider the BT $(L(X, Y), X, Y)$. Then, instead of $SB(\alpha)$ and $SB([a, b], L(X, Y))$, we write simply $SV(\alpha)$ and $SV([a, b], L(X, Y))$, respectively. Hence,

$$SV([a, b], L(X, Y)) = SB([a, b], L(X, Y))$$

and we call any element of $SV(\left[a, b\right], L(X, Y))$ a *function of bounded semivariation.*

Definition 1.25: Given a function $\alpha : [a, b] \to E$, E a normed space, and a division $d = (t_i) \in D_{[a,b]}$, we define

$$\text{var}_d(\alpha) = \sum_{i=1}^{|d|} \|\alpha(t_i) - \alpha(t_{i-1})\|$$

and the variation of α is given by

$$\text{var}(\alpha) = \text{var}_a^b(\alpha) = \sup \left\{ \text{var}_d(\alpha) : d \in D_{[a,b]} \right\}.$$

If $\text{var}(\alpha) < \infty$, then α is called a *function of bounded variation*, in which case, we write $\alpha \in BV([a, b], E)$.

It is not difficult to prove that $BV([a, b], L(E, F)) \subset SV([a, b], L(E, F))$ and $SV([a, b], E') = BV([a, b], E')$.

Moreover (see [127, Corollary I.3.4]), $BV([a, b], X) \subset G([a, b], X)$.

The space $BV([a, b], X)$ is complete when equipped with the variation norm, $\|\cdot\|_{BV}$, given by

$$\| f \|_{BV} = \| f(a) \| + \text{var}_a^b(f),$$

for $f \in BV([a,b],X)$. When there is no room for misunderstanding, we may use the notation $\|\cdot\|$ instead of $\|\cdot\|_{BV}$.

Remark 1.26: Consider a BT (E,F,G). The definition of variation of a function $\alpha : [a,b] \to E$, where E is a normed space, can also be considered as a particular case of the $\mathcal{B}$-variation of α in two different ways.

(i) Let $E = F'$, $G = \mathbb{R}$ or $G = \mathbb{C}$ and $\mathcal{B}(x',x) = \langle x, x'\rangle$. By the definition of the norm in $E = F'$, we have

$$\begin{aligned} \mathrm{var}_d(\alpha) &= \sum_{i=1}^{|d|} \|\alpha(t_i) - \alpha(t_{i-1})\| \\ &= \sup \left\{ \left| \sum_{i=1}^{|d|} \langle x_i, \alpha(t_i) - \alpha(t_{i-1})\rangle \right| : x_i \in F, \; \| x_i \| \leqslant 1 \right\} = SB_d(\alpha). \end{aligned}$$

Thus, when we consider the BT $(Y', Y, \mathbb{R})$, we write $BV(\alpha)$ and $BV([a,b],Y')$ instead of $SB(\alpha)$ and $SB([a,b],Y')$, respectively.

(ii) Let $F = E'$, $G = \mathbb{R}$ or $G = \mathbb{C}$ and $\mathcal{B}(x,x') = \langle x, x'\rangle$. By the Hahn–Banach Theorem, we have

$$\|\alpha(t_i) - \alpha(t_{i-1})\| = \sup \left\{ |\langle \alpha(t_i) - \alpha(t_{i-1}), x_i'\rangle| : x_i' \in E', \; \| x_i' \| \leqslant 1 \right\}$$

and, hence,

$$\begin{aligned} \mathrm{var}_d(\alpha) &= \sum_{i=1}^{|d|} \|\alpha(t_i) - \alpha(t_{i-1})\| \\ &= \sup \left\{ \left| \sum_{i=1}^{|d|} \langle \alpha(t_i) - \alpha(t_{i-1}), x_i'\rangle \right| : x_i' \in E', \; \| x_i' \| \leqslant 1 \right\} = SB_d(\alpha). \end{aligned}$$

Definition 1.27: Given $c \in [a,b]$, we define the spaces

$$BV_c([a,b],X) = \{f \in BV([a,b],X) : f(c) = 0\} \quad \text{and}$$
$$SV_c([a,b],L(X,Y)) = \{\alpha \in SV([a,b],L(X,Y)) : \alpha(c) = 0\}.$$

Such spaces are complete when endowed, respectively, with the norm given by the variation $\mathrm{var}_a^b(f)$ and the norm given by the semivariation

$$SV(\alpha) = \sup_{d \in D_{[a,b]}} SV_d(\alpha),$$

where

$$SV_d(\alpha) = \sup_{\|x_i\| \leqslant 1} \left\| \sum_{i=1}^{|d|} [\alpha(t_i) - \alpha(t_{i-1})] \, x_i \right\|,$$

and $d : a = t_0 < t_1 < \cdots < t_{|d|} = b$ is a division of $[a,b]$.

The following properties are not difficult to prove:

(V1) Every $\alpha \in BV([a,b],E)$ is bounded and $\| \alpha(t) \| \leqslant \| \alpha(a) \| + \text{var}_a^t(\alpha), t \in [a,b]$.
(V2) Given $\alpha \in BV([a,b],E)$ and $c \in (a,b)$, we have $\text{var}_a^b(\alpha) = \text{var}_a^c(\alpha) + \text{var}_c^b(\alpha)$.

Remark 1.28: Note that property (V1) above implies $\| \alpha\|_\infty \leqslant \| \alpha\|_{BV}$ for all $\alpha \in BV([a,b],X)$.

For more details about the spaces in Definition 1.27, the reader may want to consult [127]. The next results are borrowed from [126]. We include the proofs here since this reference is not easily available. Lemmas 1.29 and 1.30 below are, respectively, Theorems I.2.7 and I.2.8 from [126].

Lemma 1.29: *Let $\alpha \in BV([a,b],X)$. Then,*

(i) *For all $t \in (a,b]$, there exists $\alpha(t^-) = \lim_{\epsilon\to 0^+} \alpha(t-\epsilon)$.*
(ii) *For all $t \in [a,b)$, there exists $\alpha(t^+) = \lim_{\epsilon\to 0^+} \alpha(t+\epsilon)$.*

Proof. We only prove item (i), because item (ii) follows analogously. Consider an increasing sequence $\{t_n\}_{n\in\mathbb{N}}$ in $[a,t)$ converging to t. Then,

$$\sum_{i=1}^{n} \| \alpha(t_i) - \alpha(t_{i-1}) \| \leqslant \text{var}_a^t(\alpha),$$

for all $n \in \mathbb{N}$. Therefore, we have $\sum_{i=1}^{\infty} \| \alpha(t_i) - \alpha(t_{i-1}) \| \leqslant \text{var}_a^t(\alpha)$ and, hence, $\sum_{i=j}^{\infty} \| \alpha(t_i) - \alpha(t_{i-1}) \| \to 0$, as $j \to \infty$. Thus, $\{\alpha(t_n)\}_{n\in\mathbb{N}}$ is a Cauchy sequence, since for any given $\epsilon > 0$, we have

$$\| \alpha(t_m) - \alpha(t_n) \| \leqslant \sum_{i=n+1}^{m} \| \alpha(t_i) - \alpha(t_{i-1}) \| \leqslant \epsilon,$$

for sufficiently large m, n. Finally, note that the limit $\alpha(t^-)$ of $\{\alpha(t_n)\}_{n\in\mathbb{N}}$ is independent of the choice of $\{t_n\}_{n\in\mathbb{N}}$, and we finish the proof. □

It comes from Lemma 1.29 that all functions $x : [a,b] \to X$ of bounded variation are also regulated functions (see, e.g. [127, Corollary 3.4]) which, in turn, are Darboux integrable [127, Theorem 3.6].

Lemma 1.30: *Let $\alpha \in BV([a,b],X)$. For every $t \in [a,b]$, let $v(t) = \text{var}_a^t(\alpha)$. Then,*

(i) $v(t^+) - v(t) = \| \alpha(t^+) - \alpha(t) \|, t \in [a,b)$;
(ii) $v(t) - v(t^-) = \| \alpha(t) - \alpha(t^-) \|, t \in (a,b]$.

Proof. By property (SB2), v is increasing and, hence, $v(t^+)$ and $v(t^-)$ exist. By Lemma 1.29, $\alpha(t^+)$ and $\alpha(t^-)$ also exist. We prove (i). The proof of (ii) follows analogously.

Suppose $s > t$. Then, property (V2) implies $\mathrm{var}_a^s(\alpha) = \mathrm{var}_a^t(\alpha) + \mathrm{var}_t^s(\alpha)$. Thus,

$$\| \alpha(s) - \alpha(t) \| \leqslant \mathrm{var}_t^s(\alpha) = \mathrm{var}_a^s(\alpha) - \mathrm{var}_a^t(\alpha)$$

and, hence, $\| \alpha(t^+) - \alpha(t) \| \leqslant v(t^+) - v(t)$.

Conversely, for any given $d \in D_{[a,t]}$, let $v_d(t) = \mathrm{var}_d(\alpha)$. Then for every $\epsilon > 0$, there exists $\delta > 0$ such that $v(t+\sigma) - v(t^+) \leqslant \epsilon$ and $\| \alpha(t+\sigma) - \alpha(t^+) \| \leqslant \epsilon$, and there exists a division $d : a = t_0 < t_1 < \cdots < t_n = t < t_{n+1} = t + \sigma$ such that

$$v(t+\sigma) - v_d(t+\sigma) \leqslant \epsilon, \quad \text{for all } 0 < \sigma \leqslant \delta.$$

Then,

$$\begin{aligned} v(t+\sigma) - v(t) &\leqslant v_d(t+\sigma) + \epsilon - v_d(t) = \| \alpha(t+\sigma) - \alpha(t) \| + \epsilon \\ &\leqslant \| \alpha(t^+) - \alpha(t) \| + 2\epsilon \end{aligned}$$

and, hence, $v(t^+) - v(t) \leqslant \| \alpha(t^+) - \alpha(t) \|$ which completes the proof. □

Using the fact that $\| \alpha(t) \| \leqslant \| \alpha(a) \| + \mathrm{var}_a^t(\alpha)$, the following corollary follows immediately from Lemma 1.30.

Corollary 1.31: *Let $\alpha \in BV([a,b],X)$. Then the sets*

$$\{t \in [a,b) : \| \alpha(t^+) - \alpha(t) \| \geqslant \epsilon\} \quad \textit{and} \quad \{t \in (a,b] : \| \alpha(t) - \alpha(t^-) \| \geqslant \epsilon\}$$

are finite for every $\epsilon > 0$.

Thus, we have the next result which can be found in [126, Proposition I.2.10].

Proposition 1.32: *Let $\alpha \in BV([a,b],X)$. Then the set of points of discontinuity of α is countable.*

Let us define

$$BV_a^+([a,b],X) = \{\alpha \in BV_a([a,b],X) : \alpha(t^+) = \alpha(t),\ t \in (a,b)\}.$$

A proof that $BV_a^+([a,b],X)$ equipped with the variation norm, $\| \cdot \|_{BV}$, is complete can be found in [126, Theorem I.2.11]. We reproduce it in the next theorem.

Theorem 1.33: *$BV_a^+([a,b],X)$, equipped with the variation norm, is a Banach space.*

Proof. We know that $BV_a([a,b],X)$, with the variation norm, is a Banach space. Let $\{f_n\}_{n\in\mathbb{N}}$ be a sequence in $BV_a^+([a,b],X)$ converging to $f \in BV_a([a,b],X)$ in the variation norm. Then, since for every $n \in \mathbb{N}$ and every $t \in [a,b)$, $f_n(t) = f(t^+)$, we obtain

$$\| f(t) - f(t^+) \| \leqslant \| f(t) - f_n(t) \| + \| f_n(t^+) - f(t^+) \| \leqslant 2\,\mathrm{var}_a^b(f - f_n),$$

which tends to zero as $n \to \infty$. Hence, $f \in BV_a^+([a,b],X)$. □

We end this section with the Helly's choice principle for Banach space-valued functions due to C. S. Hönig. See [127, Theorem I.5.8].

Theorem 1.34 (Theorem of Helly): *Let (E,F,G) be a BT and consider a sequence $\{\alpha_n\}_{n\in\mathbb{N}}$ of elements of $SB([a,b],E)$, with $SB(\alpha_n) \leqslant M$, for all $n \in \mathbb{N}$, and such that there exists $\alpha : [a,b] \to E$, with $\alpha_n(t)y \to \alpha(t)y$ for all $t \in [a,b]$ and all $y \in F$. Then, $\alpha \in SB([a,b],E)$ and $SB(\alpha) \leqslant M$. Moreover, if $\alpha_n \in BV([a,b],E)$, with $var_a^b(\alpha_n) \leqslant M$, for all $n \in \mathbb{N}$, then $\alpha \in BV([a,b],E)$ and $BV(\alpha) \leqslant M$.*

Proof. Consider a division $d : a = t_0 < t_1 < \cdots < t_{|d|} = b$ and let $y_i \in F$, with $\| y_i \| \leqslant 1$, for $i = 1,2,\ldots,|d|$. Then, for all $n \in \mathbb{N}$, we have

$$\left\| \sum_{i=1}^{|d|} \left[\alpha(t_i) - \alpha(t_{i-1})\right] y_i \right\| \leqslant \left\| \sum_{i=1}^{|d|} \left[\alpha_n(t_i) - \alpha_n(t_{i-1})\right] y_i \right\| + \left\| \sum_{i=1}^{|d|} \left[\alpha_n(t_i) - \alpha(t_i)\right] y_i - \left[\alpha_n(t_{i-1}) - \alpha(t_{i-1})\right] y_i \right\|,$$

where the first member on the right-hand side of the inequality is smaller than M, since $SB(\alpha_n) \leqslant M$. Moreover, by hypothesis, given $\epsilon > 0$, there is $N > 0$ such that, for all $n \geqslant N$, $n \in \mathbb{N}$, $\left\| \left[\alpha_n(t_i) - \alpha(t_i)\right] y_i \right\| \leqslant \frac{\epsilon}{2|d|}$, for $i = i, i-1$. Hence, for all $n \geqslant N$,

$$\left\| \sum_{i=1}^{|d|} \left[\alpha(t_i) - \alpha(t_{i-1})\right] y_i \right\| \leqslant M + \epsilon$$

and we conclude the proof of the first part. The second part follows analogously□

For more details about functions of bounded variation, the reader may want to consult [68], for instance.

1.3 Kurzweil and Henstock Vector Integrals

Throughout this section, we consider functions $\alpha : [a,b] \to L(X,Y)$ and $f : [a,b] \to X$, where X and Y are Banach spaces.

1.3.1 Definitions

We start by recalling some definitions of vector integrals in the sense of J. Kurzweil and R. Henstock. At first, we need some auxiliary concepts, namely tagged division, gauge, and δ-fine tagged division.

Definition 1.35: Let $[a, b]$ be a compact interval.

(i) Any set of point-interval pairs $(\xi_i, [t_{i-1}, t_i])$ such that $d = (t_i) \in D_{[a,b]}$ and $\xi_i \in [t_{i-1}, t_i]$ for $i = 1, 2, \dots, |d|$, is called a *tagged division* of $[a, b]$. In this case, we write $d = (\xi_i, [t_{i-1}, t_i]) \in TD_{[a,b]}$, where $TD_{[a,b]}$ denotes the set of all tagged divisions of $[a, b]$.

(ii) Any subset of a tagged division of $[a, b]$ is a *tagged partial division* of $[a, b]$ and, in this case, we write $d \in TPD_{[a,b]}$.

(iii) A gauge on a set $A \subset [a, b]$ is any function $\delta : A \to (0, \infty)$. Given a gauge δ on $[a, b]$, we say that $d = (\xi_i, [t_{i-1}, t_i]) \in TPD_{[a,b]}$ is a *δ-fine tagged partial division*, whenever $[t_{i-1}, t_i] \subset \{t \in [a, b] : |t - \xi_i| < \delta(\xi_i)\}$ for $i = 1, 2, \dots, |d|$, that is,

$$\xi_i \in [t_{i-1}, t_i] \subset (\xi_i - \delta(\xi_i), \xi_i + \delta(\xi_i)),$$

whenever $i = 1, 2, \dots, |d|$.

Before presenting the definition of any integral based on Riemannian sums concerning δ-fine tagged divisions of an interval $[a, b]$, we bring up an important result which guarantees the existence of a δ-fine tagged division for a given gauge δ. This result is known as Cousin Lemma, and a proof of it can be found in [120, Theorem 4.1].

Lemma 1.36 (Cousin Lemma): *Given a gauge δ of $[a, b]$, there exists a δ-fine tagged division of $[a, b]$.*

Definition 1.37: We say that α is *Kurzweil f-integrable* (or *Kurzweil integrable with respect to f*), if there exists $I \in Y$ such that for every $\epsilon > 0$, there is a gauge δ on $[a, b]$ such that for every δ-fine $d = (\xi_i, [t_{i-1}, t_i]) \in TD_{[a,b]}$,

$$\left\| \sum_{i=1}^{|d|} \alpha(\xi_i) \left[f(t_i) - f(t_{i-1})\right] - I \right\| < \epsilon.$$

In this case, we write $I = \int_a^b \alpha(t)\, df(t)$ and $\alpha \in K_f([a, b], L(X, Y))$.

Analogously, we define the Kurzweil integral of $f : [a, b] \to X$ with respect to a function $\alpha : [a, b] \to L(X, Y)$.

Definition 1.38: We say that f is *Kurzweil α– integrable* (or *Kurzweil integrable with respect to α*), if there exists $I \in Y$ such that given $\epsilon > 0$, there is a gauge δ on $[a, b]$ such that

$$\left\| \sum_{i=1}^{|d|} \left[\alpha(t_i) - \alpha(t_{i-1}) \right] f(\xi_i) - I \right\| < \epsilon,$$

whenever $d = (\xi_i, [t_{i-1}, t_i]) \in TD_{[a,b]}$ is δ-fine. In this case, we write $I = \int_a^b d\alpha(t) f(t)$ and $f \in K^\alpha(\left[a, b\right], X)$.

Suppose the Kurzweil vector integral $\int_a^b \alpha(t)\, df(t)$ exists. Then, we define

$$\int_b^a \alpha(t)\, df(t) = -\int_a^b \alpha(t)\, df(t).$$

An analogous consideration holds for the Kurzweil vector integral $\int_a^b d\alpha(t) f(t)$.

If the gauge δ in the definition of $\alpha \in K_f([a, b], L(X, Y))$ is a constant function, then we obtain the Riemann–Stieltjes integral $\int_a^b \alpha(t)\, df(t)$, and we write $\alpha \in R_f([a, b], L(X, Y))$. Similarly, when we consider only constant gauges δ in the definition of $f \in K^\alpha(\left[a, b\right], X)$, we obtain the Riemann–Stieltjes integral $\int_a^b d\alpha(t) f(t)$, and we write $f \in R^\alpha([a, b], X)$. Hence, if we denote by $R_f([a, b], L(X, Y))$ the set of all functions $\alpha : [a, b] \to L(X, Y)$ which are Riemann integrable with respect to $f : [a, b] \to X$ and by $R^\alpha([a, b], X)$ the set of all functions $f : [a, b] \to X$ which are Riemann integrable with respect to $\alpha : [a, b] \to L(X, Y)$, then we have

$$R_f([a, b], L(X, Y)) \subset K_f([a, b], L(X, Y)) \quad \text{and}$$
$$R^\alpha([a, b], X) \subset K^\alpha([a, b], X).$$

The next very important remark concerns the terminology we adopt from now on in this book concerning Kurzweil vector integrals given by Definitions 1.37 and 1.38.

Remark 1.39: We refer to the vector integrals from Definitions 1.37 and 1.38, namely,

$$\int_a^b \alpha(t)\, df(t) \quad \text{and} \quad \int_a^b d\alpha(t) f(t),$$

where $f : [a, b] \to X$ and $\alpha : [a, b] \to L(X, Y)$, as Perron–Stieltjes integrals. For the particular case where $\alpha : [a, b] \to L(X)$ is the identity in $L(X)$ and $f : [a, b] \to X$, we refer to the integral

$$\int_a^b f(t)\, dt$$

simply as a Perron integral. Our choice to use this terminology is due to the fact that, in Chapter 2, we deal with a more general definition of the Kurzweil integral which encompasses all integrals presented here. Moreover, since the same notation for the integrals

$$\int_a^b \alpha(t)\, df(t), \quad \int_a^b d\alpha(t) f(t), \quad \text{and} \quad \int_a^b f(t)\, dt$$

are used for Riemann–Stieltjes integrals, we will specify which integral we are dealing with whenever there is possibility for an ambiguous interpretation.

The vector integral of Henstock, which we define in the sequel, is more restrictive than the Kurzweil vector integral for integrands taking values in infinite dimensional Banach spaces.

Again, consider functions $f : [a, b] \to X$ and $\alpha : [a, b] \to L(X, Y)$.

Definition 1.40: We say that α is *Henstock f-integrable* (or *Henstock variationally integrable with respect to f*), if there exists a function $A_f : [a, b] \to Y$ (called the *associate function* of α) such that for every $\epsilon > 0$, there is a gauge δ on $[a, b]$ such that for every δ-fine $d = (\xi_i, [t_{i-1}, t_i]) \in TD_{[a,b]}$,

$$\sum_{i=1}^{|d|} \left\| \alpha(\xi_i) \left[f(t_i) - f(t_{i-1}) \right] - \left[A_f(t_i) - A_f(t_{i-1}) \right] \right\| < \epsilon.$$

We write $\alpha \in H_f([a, b], L(X, Y))$ in this case.

In an analogous way, we define the Henstock α-integrability of $f : [a, b] \to X$.

Definition 1.41: We say that f is *Henstock α-integrable* (or *Henstock variationally integrable with respect to α*), if there exists a function $F_\alpha : [a, b] \to Y$ (called the *associate function* of f) such that for every $\epsilon > 0$, there is a gauge δ on $[a, b]$ such that for every δ-fine $d = (\xi_i, [t_{i-1}, t_i]) \in TD_{[a,b]}$,

$$\sum_{i=1}^{|d|} \left\| [\alpha(t_i) - \alpha(t_{i-1})] f(\xi_i) - \left[F_\alpha(t_i) - F_\alpha(t_{i-1}) \right] \right\| < \epsilon.$$

We write $f \in H_\alpha([a, b], X)$ in this case.

Next, we define indefinite vector integrals.

Definition 1.42 (Indefinite Vector Integrals):

(i) Given $f : [a, b] \to X$ and $\alpha \in K_f([a, b], L(X, Y))$, we define the *indefinite integral* $\tilde{\alpha}_f : [a, b] \to Y$ of α with respect to f by

$$\tilde{\alpha}_f(t) = \int_a^t \alpha(s)\, df(s), \quad t \in [a,b].$$

If, in addition, $\alpha \in H_f([a,b], L(X,Y))$, then $\tilde{\alpha}_f(t) = A_f(t) - A_f(a)$, $t \in [a,b]$.

(ii) Given $\alpha : [a,b] \to L(X,Y)$ and $f \in K^\alpha([a,b],X)$, we define the *indefinite integral* $\tilde{f}^\alpha : [a,b] \to Y$ of f with respect to α by

$$\tilde{f}^\alpha(t) = \int_a^t d\alpha(s) f(s), \quad t \in [a,b].$$

If, moreover, $f \in H^\alpha([a,b],X)$, then $\tilde{f}^\alpha(t) = F^\alpha(t) - F^\alpha(a)$, $t \in [a,b]$.

Note that Definition 1.42 yields the inclusions

$$H_f([a,b], L(X,Y)) \subset K_f([a,b], L(X,Y)) \quad \text{and}$$
$$H^\alpha([a,b],X) \subset K^\alpha([a,b],X).$$

If in item (ii) of Definition 1.42, we consider the particular case, where $X = Y$, and for every $t \in [a,b]$, $\alpha(t)$ is the identity in X, then instead of $K^\alpha([a,b],X)$, $R^\alpha([a,b],X)$, $H^\alpha([a,b],X)$ and $\tilde{f}^\alpha$, we write simply

$$K([a,b],X), \; R([a,b],X), \; H([a,b],X) \text{ and } \tilde{f}$$

respectively, where

$$\tilde{f}(t) = \int_a^t f(s)\, ds, \quad t \in [a,b].$$

If in item (i) of Definition 1.42, we have $X = Y = \mathbb{R}$, then one can identify the isomorphic spaces $L(\mathbb{R},\mathbb{R})$ and $\mathbb{R}$ and, hence, the spaces $K_f([a,b], L(\mathbb{R}))$, $K_f([a,b],\mathbb{R})$, $H_f([a,b], L(\mathbb{R}))$, and $H_f([a,b],\mathbb{R})$ can also be identified, because one has

$$K_f([a,b],\mathbb{R}) = H_f([a,b],\mathbb{R}).$$

Indeed. It is clear that $H_f([a,b],\mathbb{R}) \subset K_f([a,b],\mathbb{R})$. In order to prove that $K_f([a,b],\mathbb{R}) \subset H_f([a,b],\mathbb{R})$, it is enough to write the usual Riemannian-type sum as two sums (with the positive and negative parts of the sum):

$$\begin{aligned}
&\sum_{i=1}^{|d|} \left|\alpha(\tau_i)[f(t_i) - f(t_{i-1})] - [\tilde{\alpha}(t_i) - \tilde{\alpha}(t_{i-1})]\right| \\
&= \sum_{i=1}^{|d|} \left\{\alpha(\tau_i)[f(t_i) - f(t_{i-1})] - [\tilde{\alpha}(t_i) - \tilde{\alpha}(t_{i-1})]\right\}_+ \\
&+ \sum_{i=1}^{|d|} \left\{\alpha(\tau_i)[f(t_i) - f(t_{i-1})] - [\tilde{\alpha}(t_i) - \tilde{\alpha}(t_{i-1})]\right\}_- \\
&\leqslant \epsilon + \epsilon = 2\epsilon,
\end{aligned}$$

for every δ-fine $d = (\tau_i, [t_{i-1}, t_i]) \in TD_{[a,b]}$ corresponding to a given $\epsilon > 0$.

As we already mentioned, we will consider a more general definition of the Kurzweil integral in Chapter 2. Thus, in the remaining of this chapter, we refer to the integrals

$$\int_a^t \alpha(s)\, df(s), \quad \text{and} \quad \int_a^t d\alpha(s) f(s), \quad t \in [a,b],$$

as Perron–Stieltjes integrals, where $\alpha : [a,b] \to L(X,Y)$ and $f : [a,b] \to X$.

As it should be expected, the above integrals are linear and additive over nonoverlapping intervals. These facts will be put aside for a while, because in Chapter 2 they will be proved for the more general form of the Kurzweil integral. In the meantime, we present a simple example of a function which is Riemann improper integrable (and, hence, also Perron integrable, due to Theorem 2.9), but it is not Lebesgue integrable (because it is not absolutely integrable).

Example 1.43: Let $f : [0,\infty) \to \mathbb{R}$ be given by $f(t) = \frac{\sin t}{t}$, for $t \in (0,\infty)$, and $f(0) = L$, for some $L \in \mathbb{R}$. Then, it is not difficult to prove that $\lim_{t\to\infty} \int_0^t f(s)\, ds$ exists, but $\int_0^\infty \left|\frac{\sin s}{s}\right| ds = \infty$, once

$$\int_0^\infty \left|\frac{\sin s}{s}\right| ds \geqslant \sum_{k=1}^{n} \int_{(k-1)\pi}^{k\pi} \left|\frac{\sin s}{s}\right| ds \geqslant \sum_{k=1}^{n} \frac{1}{k\pi} \int_{(k-1)\pi}^{k\pi} |\sin s|\, ds = \frac{2}{\pi} \sum_{k=1}^{n} \frac{1}{k}.$$

Another example is also needed at this point. Borrowed from [73, example 2.1], the example below exhibits a function $f \in R([a,b],X) \setminus H([a,b],X)$ (that is, f belongs to $R([a,b],X)$, but not to $H([a,b],X)$), satisfying $\tilde{f} = 0$. However, $f(t) \neq 0$ for almost every $t \in [a,b]$. Thus, such a function is also an element of $K([a,b],X)$ (because it belongs to $R([a,b],X)$) which does not belong to $H([a,b],X)$, showing that, in the infinite dimensional-valued case, $H([a,b],X)$ may be a proper subset of $K([a,b],X)$.

Example 1.44: Let $I \subset \mathbb{R}$ be an arbitrary set and let E be a normed space. A family $\{x_i\}_{i\in I}$ of elements of E is *summable* with sum $x \in E$ (we write $\sum_{i\in I} x_i = x$), if for every $\epsilon > 0$, there is a finite subset $F_\epsilon \subset I$ such that for every finite subset $F \subset I$ with $F \supset F_\epsilon$, $\| x - \sum_{i\in F} x_i \| < \epsilon$.

Let $l_2(I)$ denote the set of all families $\{x_i\}_{i\in I}$, $x_i \in \mathbb{R}$, such that the family $\left\{|x_i|^2\right\}_{i\in I}$ is summable, that is,

$$l_2(I) = \left\{ x = \{x_i\}_{i\in I},\ x_i \in \mathbb{R} : \sum_{i\in I} |x_i|^2 < \infty \right\}.$$

It is known that the expression $\langle x, y\rangle = \sum_{i\in I} x_i y_i$ defines an inner product and $l_2(I)$, equipped with the norm $\| x\|_2 = (\sum_{i\in I} |x_i|^2)^{1/2}$ is a Hilbert space. As a consequence of the Basis Theorem, since $l_2(I)$ is a Hilbert space, $\{e_i\}_{i\in I}$ is a maximal orthonormal system for $l_2(I)$, that is, $e_i(j) = \langle e_i, e_j\rangle = \delta_{ij}$ and δ_{ij} stands for the Kronecker delta (see [128, Theorem 4.6], item 6, p. 61)).

In what follows, we will use the *the Bessel equality* given as

$$\| x \|_2^2 = \sum_{i\in I} |\langle x_i, e_i\rangle|^2 = \sum_{i\in I} |x_i|^2, \quad x \in l_2(I).$$

Let $[a,b]$ be a nondegenerate closed interval of $\mathbb{R}$ and $X = l_2([a,b])$ be equipped with the norm

$$x \mapsto \|x\|_2 = \left(\sum_{i\in[a,b]} |x_i|^2 \right)^{1/2}.$$

Consider a function $f : [a,b] \to X$ given by $f(t) = e_t$, $t \in [a,b]$. Given $\epsilon > 0$, there exists $\delta > 0$, with $\delta^{\frac{1}{2}} < \frac{\epsilon}{(b-a)^{\frac{1}{2}}}$, such that for every $(\frac{\delta}{2})$-fine $d = (\xi_j, [t_{j-1}, t_j]) \in TD_{[a,b]}$,

$$\left\| \sum_{j=1}^{|d|} f(\xi_j)(t_j - t_{j-1}) - 0 \right\|_2 = \left\| \sum_{j=1}^{|d|} e_{\xi_j}(t_j - t_{j-1}) \right\|_2$$

$$= \left[\sum_{j=1}^{|d|} |t_j - t_{j-1}|^2 \right]^{\frac{1}{2}} < \delta^{\frac{1}{2}} \left[\sum_{j=1}^{|d|} (t_j - t_{j-1}) \right]^{\frac{1}{2}} < \epsilon,$$

where we applied the Bessel equality. Thus, $f \in R([a,b],X) \subset K([a,b],X)$ and $\tilde{f} = 0$, since $\int_a^t f(s)\,ds = 0$ for every $t \in [a,b]$. On the other hand,

$$\sum_{j=1}^{|d|} \left\| f(\xi_j)(t_j - t_{j-1}) - 0 \right\|_2 = b - a$$

for every $(\xi_j, [t_{j-1}, t_j]) \in TD_{[a,b]}$. Hence, $f \notin H([a,b],X)$.

1.3.2 Basic Properties

The first result we present in this subsection is known as the Saks–Henstock lemma, and it is useful in many situations. For a proof of it, see [210, Lemma 16], for instance. Similar results hold if we replace $K_f([a,b], L(X,Y))$ by $R_f([a,b], L(X,Y))$ and also $K^\alpha([a,b],X)$ by $R^\alpha([a,b],X)$.

Lemma 1.45 (Saks–Henstock Lemma): *The following assertions hold.*

(i) *Let $\alpha : [a,b] \to L(X,Y)$ and $f \in K^\alpha([a,b],X)$. Given $\epsilon > 0$, let δ be a gauge on $[a,b]$ such that for every δ-fine $d = (\xi_i, [t_{i-1}, t_i]) \in TD_{[a,b]}$,*

$$\left\| \sum_{i=1}^{|d|} [\alpha(t_i) - \alpha(t_{i-1})] f(\xi_i) - \int_a^b d\alpha(t) f(t) \right\| < \epsilon.$$

Then, for every δ-fine $d' = (\eta_j, [s_{j-1}, s_j]) \in TPD_{[a,b]}$, we have

$$\left\| \sum_{j=1}^{|d'|} \left[[\alpha(s_j) - \alpha(s_{j-1})] f(\eta_j) - \int_{s_{j-1}}^{s_j} d\alpha(t) f(t) \right] \right\| \leqslant \epsilon.$$

(ii) *Let $f : [a,b] \to X$ and $\alpha \in K_f([a,b], L(X,Y))$. Given $\epsilon > 0$, let δ be a gauge on $[a,b]$ such that for every δ-fine $d = (\xi_i, [t_{i-1}, t_i]) \in TD_{[a,b]}$,*

$$\left\| \sum_{i=1}^{|d|} \alpha(\xi_i) \left[f(t_i) - f(t_{i-1}) \right] - \int_a^b \alpha(t)\, df(t) \right\| < \epsilon.$$

Then, for every δ-fine $d' = (\eta_j, [s_{j-1}, s_j]) \in TPD_{[a,b]}$, we have

$$\left\| \sum_{j=1}^{|d'|} \left[\alpha(\eta_j) \left[f(s_j) - f(s_{j-1}) \right] - \int_{s_{j-1}}^{s_j} \alpha(t)\, df(t) \right] \right\| \leqslant \epsilon.$$

Now, we define some important sets of functions.

Definition 1.46: Let $C^{\sigma}([a,b], L(X,Y))$ be the set of all functions $\alpha : [a,b] \to L(X,Y)$ which are *weakly continuous*, that is, for every $x \in X$, the function $t \in [a,b] \mapsto \alpha(t)x \in Y$ is continuous, and we denote by $G^{\sigma}([a,b], L(X,Y))$ the set of all *weakly regulated* functions $\alpha : [a,b] \to L(X,Y)$, that is, for every $x \in X$, the function $t \in [a,b] \mapsto \alpha(t)x \in Y$ is r egulated.

Given $\alpha \in G^{\sigma}([a,b], L(X,Y))$ and $x \in X$, let us define

$$\alpha(t\hat{+})x = (\alpha x)(t^+) = \lim_{\rho \to 0^+} (\alpha x)(t + \rho),\ t \in [a,b)$$

$$\alpha(t\hat{-})x = (\alpha x)(t^-) = \lim_{\rho \to 0^+} (\alpha x)(t - \rho),\ t \in (a,b].$$

By the Banach–Steinhaus theorem, the limits $\alpha(t\hat{+})$ and $\alpha(t\hat{-})$ exist and belong to $L(X,Y)$. Then, by the Uniform Boundedness Principle, $G^{\sigma}([a,b], L(X,Y))$ is a Banach space when equipped with the usual supremum norm. It is also clear that

$$C^{\sigma}([a,b], L(X,Y)) \subset G^{\sigma}([a,b], L(X,Y)).$$

The next result concerns the existence of Perron–Stieltjes integrals. A proof of its item (i) can be found in [210, Theorem 15]. A proof of item (ii) follows similarly as the proof of item (i). See also [212, Proposition 7].

Theorem 1.47: *The following assertions hold.*

(i) *If $\alpha \in G([a,b], L(X,Y))$ and $f \in BV([a,b], X)$, then $\alpha \in K_f([a,b], L(X,Y))$.*
(ii) *If $\alpha \in SV([a,b], L(X,Y)) \cap G^{\sigma}([a,b], L(X,Y))$ and $f \in G([a,b], X)$, then we have $f \in K^{\alpha}([a,b], X)$.*

The following consequence of Theorem 1.47 will be used later in many chapters. The inequalities follow after some calculations. See, for instance, [210, Proposition 10].

Corollary 1.48: *The following assertions hold.*

(i) *If $\alpha \in G([a,b], L(X))$ and $f \in BV([a,b], X)$, then the Perron–Stieltjes integral $\int_a^b \alpha(s)\, df(s)$ exists, and we have*

$$\left\| \int_a^b \alpha(s)\, df(s) \right\| \leqslant \int_a^b \| \alpha(s) \| \, d[var_a^s(f)] \leqslant \| \alpha \|_\infty var_a^b(f).$$

Similarly, if $f \in G([a,b], X)$ and $g : [a,b] \to \mathbb{R}$ is nondecreasing, then

$$\left\| \int_a^b f(s)\, dg(s) \right\| \leqslant \| f \|_\infty [g(b) - g(a)].$$

(ii) *If $\alpha \in BV([a,b], L(X))$ and $f \in G([a,b], X)$, then the Perron–Stieltjes integral $\int_a^b d\alpha(s) f(s)$ exists, and we have*

$$\left\| \int_a^b d\alpha(s) f(s) \right\| \leqslant \int_a^b d[var_a^s(\alpha)] \, \| f(s) \| \leqslant var_a^b(\alpha) \, \| f \|_\infty .$$

The next result, borrowed from [74, Theorem 1.2], gives us conditions for indefinite Perron–Stieltjes integrals to be regulated functions.

Theorem 1.49: *The following assertions hold:*

(i) *if $\alpha \in G^\sigma([a,b], L(X,Y))$ and $f \in K^\alpha([a,b], X)$, then $\tilde{f}^\alpha \in G([a,b], Y)$;*
(ii) *if $f \in G([a,b], X)$ and $\alpha \in K_f([a,b], L(X,Y))$, then $\tilde{\alpha}_f \in G([a,b], Y)$.*

Proof. We prove (i). Item (ii) follows similarly. For item (i), it is enough to show that

$$\tilde{f}^\alpha(\xi^+) - \tilde{f}^\alpha(\xi) = \left[\alpha(\xi\hat{+}) - \alpha(\xi)\right] f(\xi), \quad \xi \in [a,b),$$

because, in this case, the equality

$$\tilde{f}^\alpha(\xi) - \tilde{f}^\alpha(\xi^-) = [\alpha(\xi) - \alpha(\xi\hat{-})] f(\xi), \quad \xi \in (a,b]$$

follows in an analogous way. By hypothesis, $f \in K^\alpha([a,b], X)$. Hence, given $\epsilon > 0$, there is a gauge δ on $[a,b]$ such that for every δ-fine division $d = (\xi_i, [t_{i-1}, t_i]) \in TD_{[a,b]}$,

$$\left\| \sum_{i=1}^{|d|} \left[\alpha(t_i) - \alpha(t_{i-1})\right] f(\xi_i) - \int_a^b d\alpha(t) f(t) \right\| < \frac{\epsilon}{2}.$$

Now, let $\xi \in [a,b)$. Since $\alpha \in G^{\sigma}([a,b], L(X,Y))$, there exists $(\alpha x)(\xi^+)$, for every $x \in X$. In particular, there exists $\mu > 0$ such that

$$\left\| \left[\alpha(\xi+\rho) - \alpha(\xi\hat{+})\right] f(\xi) \right\| < \frac{\epsilon}{2}, \text{ for } 0 < \rho < \mu.$$

If $\delta(\xi) < \mu$ and $0 < \rho < \delta(\xi)$, then by the Saks–Henstock lemma (Lemma 1.45)

$$\left\| [\alpha(\xi+\rho) - \alpha(\xi)] f(\xi) - \int_{\xi}^{\xi+\rho} d\alpha(t) f(t) \right\| \leqslant \frac{\epsilon}{2}.$$

Thus,

$$\begin{aligned}
&\left\| \tilde{f}^{\alpha}(\xi^+) - \tilde{f}^{\alpha}(\xi) - \left[\alpha(\xi\hat{+}) - \alpha(\xi)\right] f(\xi) \right\| \\
&= \left\| \int_{\xi}^{\xi+\rho} d\alpha(t) f(t) - \left[\alpha(\xi\hat{+}) - \alpha(\xi)\right] f(\xi) \right\| \\
&\leqslant \left\| \int_{\xi}^{\xi+\rho} d\alpha(t) f(t) - [\alpha(\xi+\rho) - \alpha(\xi)] f(\xi) \right\| \\
&+ \left\| [\alpha(\xi+\rho) - \alpha(\xi)] f(\xi) - \left[\alpha(\xi\hat{+}) - \alpha(\xi)\right] f(\xi) \right\| < \epsilon
\end{aligned}$$

and the proof is complete. □

With Theorem 1.49 at hand, the next corollary follows immediately.

Corollary 1.50: *The following statements hold:*

(i) *If* $\alpha \in C^{\sigma}([a,b], L(X,Y))$ *and* $f \in K^{\alpha}([a,b], X)$, *then* $\tilde{f}^{\alpha} \in C([a,b], Y)$.
(ii) *If* $f \in C([a,b], X)$ *and* $\alpha \in K_f([a,b], L(X,Y))$, *then* $\tilde{\alpha}_f \in C([a,b], Y)$.

Next, we state a uniform convergence theorem for Perron–Stieltjes vector integrals. A proof of such result can be found in [210, Theorem 11].

Theorem 1.51: *Let* $\alpha \in SV([a,b], L(X,Y))$ *and* $f, f_n \in G([a,b], X)$, $n \in \mathbb{N}$, *be such that the Perron–Stieltjes integral* $\int_a^b d\alpha(t) f_n(t)$ *exists for every* $n \in \mathbb{N}$ *and* $f_n \to f$ *uniformly in* $[a,b]$. *Then,* $\int_a^b d\alpha(t) f(t)$ *exists and*

$$\int_a^b d\alpha(t) f(t) = \lim_{n\to\infty} \int_a^b d\alpha(t) f_n(t).$$

We finish this subsection by presenting a Grönwall-type inequality for Perron–Stieltjes integrals. For a proof of it, we refer to [209, Corollary 1.43].

Theorem 1.52 (Grönwall Inequality): *Let $g : [a,b] \to [0,\infty)$ be a nondecreasing left-continuous function, $k > 0$ and $\ell \geqslant 0$. Assume that $f : [a,b] \to [0,\infty)$ is bounded and satisfies*

$$f(t) \leqslant k + \ell \int_a^t f(s)\, dg(s), \quad t \in [a,b].$$

Then,

$$f(t) \leqslant k e^{\ell [g(t)-g(a)]}, \quad t \in [a,b].$$

Other properties of Perron–Stieltjes integrals can be found in Chapter 2, where they appear within the consequences of the main results presented there.

1.3.3 Integration by Parts and Substitution Formulas

The first result of this section is an Integration by Parts Formula for Riemann–Stieltjes integrals. It is a particular consequence of Proposition 1.70 presented in the end of this section. A proof of it can be found in [126, Theorem II.1.1].

Theorem 1.53 (Integration by Parts): *Let (E,F,G) be a BT. Suppose*

(i) *either $\alpha \in SB([a,b],E)$ and $f \in C([a,b],F)$;*
(ii) *or $\alpha \in C([a,b],E)$ and $f \in BV([a,b],F)$.*

Then, $\alpha \in R_f([a,b],E)$ and $f \in R^{\alpha}([a,b],F)$, that is, the Riemann–Stieltjes integrals $\int_a^b d\alpha(t)f(t)$ and $\int_a^b \alpha(t)\, df(t)$ exist, and moreover,

$$\int_a^b d\alpha(t)f(t) = \alpha(b)f(b) - \alpha(a)f(a) - \int_a^b \alpha(t)\, df(t).$$

Next, we state a result which is not difficult to prove using the definitions involved in the statement. See [72, Theorem 5]. Recall that the indefinite integral of a function f in $K([a,b],X)$ is denoted by (see Definition 1.42)

$$\tilde{f}(t) = \int_a^t f(s)\, ds, \ t \in [a,b].$$

Theorem 1.54: *Suppose $f \in H([a,b],X)$ and $\alpha \in K_{\tilde{f}}([a,b],L(X,Y))$ is bounded. Then $\alpha f \in K([a,b],Y)$ and*

$$\int_a^b \alpha(t)f(t)\, dt = \int_a^b \alpha(t) d\tilde{f}(t). \tag{1.3}$$

If, in addition, $\alpha \in H_{\tilde{f}}([a,b],L(X,Y))$, then $\alpha f \in H([a,b],Y)$.

By Theorem 1.47, the Perron–Stieltjes integral $\int_a^b \alpha(t)d\tilde{f}(t)$ exists and the next corollary follows.

Corollary 1.55: *Let $\alpha \in G([a,b], L(X,Y))$ and $f \in H([a,b],X)$ be such that $\tilde{f} \in BV([a,b],X)$. Then, $\alpha f \in K([a,b],Y)$ and (1.3) holds.*

A second corollary of Theorem 1.54 follows by the fact that Riemann–Stieltjes integrals are special cases of Perron–Stieltjes integrals. Then, it suffices to apply Theorems 1.49 and 1.53.

Corollary 1.56: *Suppose the following conditions hold:*

(i) *either $\alpha \in C([a,b], L(X,Y))$ and $f \in H([a,b],X)$, with $\tilde{f} \in BV([a,b],X)$;*
(ii) *or $f \in H([a,b],X)$ and $\alpha \in SV([a,b], L(X,Y))$.*

Then, $\alpha f \in K([a,b],Y)$, equality (1.3) holds, and we have

$$\int_a^b \alpha(t)d\tilde{f}(t) = \alpha(b)\tilde{f}(b) - \alpha(a)\tilde{f}(a) - \int_a^b d\alpha(t)\tilde{f}(t). \tag{1.4}$$

The next theorem is due to C. S. Hönig (see [129]), and it concerns multipliers for Perron–Stieltjes integrals.

Theorem 1.57: *Suppose $f \in K([a,b],X)$ and $\alpha \in SV([a,b], L(X,Y))$. Then, $\alpha f \in K([a,b],Y)$ and Eqs. (1.3) and (1.4) hold.*

Since $H([a,b],X) \subset K([a,b],X)$ and $BV([a,b], L(X,Y)) \subset SV([a,b], L(X,Y))$, it is immediate that if $f \in H([a,b],X)$ and $\alpha \in BV([a,b], L(X,Y))$, then $\alpha f \in K([a,b],Y)$. As a matter of fact, the next result gives us information about the multipliers for the Henstock vector integral. See [72, Theorem 7].

Theorem 1.58: *Assume that $f \in H([a,b],X)$ and $\alpha \in BV([a,b], L(X,Y))$. Then, $\alpha f \in H([a,b],Y)$ and equalities (1.3) and (1.4) hold.*

Proof. Since $f \in H([a,b],X)$, $\tilde{f}$ is continuous by Theorem 1.49. Thus, given $\epsilon > 0$, there exists $\delta^* > 0$ such that

$$\omega(\tilde{f}, [c,d]) = \sup \{ \| \tilde{f}(t) - \tilde{f}(s) \| : t,s \in [c,d] \} < \epsilon,$$

whenever $0 < d - c < \delta^*$, where $[c,d] \subset [a,b]$. Moreover, there is a gauge δ on $[a,b]$, with $\delta(t) < \frac{\delta^*}{2}$ for $t \in [a,b]$, such that for every δ-fine $d = (\xi_i, [t_{i-1}, t_i]) \in TD_{[a,b]}$,

$$\sum_{i=1}^{|d|} \left\| f(\xi_i)(t_i - t_{i-1}) - \int_{t_{i-1}}^{t_i} f(t)\, dt \right\| < \epsilon.$$

Thus,

$$\begin{aligned}
&\sum_{i=1}^{|d|}\left\|\alpha(\xi_i)f(\xi_i)(t_i-t_{i-1})-\int_{t_{i-1}}^{t_i}\alpha(t)f(t)\,dt\right\| \\
&\leqslant \sum_{i=1}^{|d|}\left\|\alpha(\xi_i)\left[f(\xi_i)(t_i-t_{i-1})-\int_{t_{i-1}}^{t_i}f(t)\,dt\right]\right\| + \sum_{i=1}^{|d|}\left\|\int_{t_{i-1}}^{t_i}\left[\alpha(t)-\alpha(\xi_i)\right]f(t)\,dt\right\| \\
&< \|\alpha\|_\infty\epsilon + \sum_{i=1}^{|d|}\left\|\int_{t_{i-1}}^{t_i}\left[\alpha(t)-\alpha(\xi_i)\right]f(t)\,dt\right\|.
\end{aligned}$$

But by Corollary 1.56, item (ii), $\alpha f \in K([a,b],Y)$ and

$$\int_a^b \alpha(t)f(t)\,dt = \int_a^b \alpha(t)d\tilde{f}(t) = \alpha(b)\tilde{f}(b) - \alpha(a)\tilde{f}(a) - \int_a^b d\alpha(t)\tilde{f}(t)$$

and a similar formula also holds for every subinterval contained in $[a,b]$. Hence, for $\beta_{t_i} = \left[\alpha(t_i)-\alpha(\xi_i)\right]\tilde{f}(t_i)$ and $\beta_{t_{i-1}} = \left[\alpha(t_{i-1})-\alpha(\xi_i)\right]\tilde{f}(t_{i-1})$, we have

$$\begin{aligned}
&\sum_{i=1}^{|d|}\left\|\int_{t_{i-1}}^{t_i}\left[\alpha(t)-\alpha(\xi_i)\right]f(t)\,dt\right\| = \sum_{i=1}^{|d|}\left\|\beta_{t_i}-\beta_{t_{i-1}}-\int_{t_{i-1}}^{t_i}d\alpha(t)\tilde{f}(t)\right\| \\
&= \sum_{i=1}^{|d|}\left\|\beta_{t_i}-\int_{\xi_i}^{t_i}d\alpha(t)\tilde{f}(t)-\beta_{t_{i-1}}-\int_{t_{i-1}}^{\xi_i}d\alpha(t)\tilde{f}(t)\right\| \\
&= \sum_{i=1}^{|d|}\left\|\int_{\xi_i}^{t_i}d\alpha(t)\left[\tilde{f}(t_i)-\tilde{f}(t)\right]+\int_{t_{i-1}}^{\xi_i}d\alpha(t)\left[\tilde{f}(t_{i-1})-\tilde{f}(t)\right]\right\| \leqslant \epsilon\,\mathrm{var}_a^b(\alpha),
\end{aligned}$$

since for every $t \in \left[t_{i-1},t_i\right]$, we have

$$\begin{aligned}
&\left\|\tilde{f}(t_i)-\tilde{f}(t)\right\| \leqslant \sup\{\|\tilde{f}(t)-\tilde{f}(s)\| : t,s\in[t_{i-1},t_i]\} \quad \text{and} \\
&\left\|\tilde{f}(t_{i-1})-\tilde{f}(t)\right\| \leqslant \sup\{\|\tilde{f}(t)-\tilde{f}(s)\| : t,s\in[t_{i-1},t_i]\}.
\end{aligned}$$

The proof is then complete. □

A proof of the next result, borrowed from [72, Theorem 8], follows from the definitions of the integrals.

Theorem 1.59: *Let $\alpha \in H([a,b],L(X,Y))$ and $f \in K^{\tilde{\alpha}}([a,b],X)$. If f is bounded, then $\alpha f \in K(\left[a,b\right],Y)$ and*

$$\int_a^b \alpha(t)f(t)\,dt = \int_a^b d\tilde{\alpha}(t)f(t). \tag{1.5}$$

If, in addition, $f \in H^{\tilde{\alpha}}([a,b],X)$, then $\alpha f \in H([a,b],Y)$.

Corollary 1.60: *Suppose $\alpha \in H([a,b],L(X,Y))$ with $\tilde{\alpha} \in SV([a,b],L(X,Y))$ and $f \in G(\left[a,b\right],X)$. Then, $\alpha f \in K([a,b],Y)$ and (1.5) holds.*

Proof. By Theorem 1.49, $\tilde{\alpha} \in C([a,b], L(X,Y))$. Then, the result follows from Theorem 1.47, since $C([a,b], L(X,Y)) \subset G^{\sigma}([a,b], L(X,Y))$. □

The next corollaries follow from Theorems 1.49 and 1.53.

Corollary 1.61: *Suppose $\alpha \in H([a,b], L(X,Y))$ with $\tilde{\alpha} \in SV([a,b], L(X,Y))$ and $f \in C([a,b], X)$. Then, $\alpha f \in K([a,b], Y)$, and we have*

$$\int_a^b \alpha(t)f(t)\,dt = \int_a^b d\tilde{\alpha}(t)f(t) \tag{1.6}$$

and the following integration by parts formula holds

$$\int_a^b d\tilde{\alpha}(t)f(t) = \tilde{\alpha}(b)f(b) - \tilde{\alpha}(a)f(a) - \int_a^b \tilde{\alpha}(t)\,df(t). \tag{1.7}$$

Corollary 1.62: *Consider functions $\alpha \in H([a,b], L(X,Y))$ and $f \in BV([a,b], X)$. Then, $\alpha f \in K([a,b], Y)$ and equalities (1.6) and (1.7) hold.*

The next two theorems generalize Corollary 1.62. For their proofs, the reader may want to consult [72].

Theorem 1.63: *Consider $f \in BV([a,b], X)$. If $\alpha \in K([a,b], L(X,Y))$ (respectively, $\alpha \in H([a,b], L(X,Y))$), then $\alpha f \in K([a,b], Y)$ (respectively, $\alpha f \in H([a,b], Y)$) and both (1.6) and (1.7) hold.*

The next result is a Substitution Formula for Perron–Stieltjes integrals. A similar result holds for Riemann–Stieltjes integrals. For a proof of it, see [72, Theorem 11].

Theorem 1.64: *Consider functions $\alpha \in SV([a,b], L(X,Y))$, $f : [a,b] \to Z$, and $\beta \in K_f([a,b], L(Z,X))$. Let*

$$g(t) = \tilde{\beta}_f(t) = \int_a^t \beta(s)\,df(s), \quad t \in [a,b].$$

Then, $\alpha \in K_g([a,b], L(X,Y))$ if and only if $\alpha\beta \in K_f([a,b], L(Z,Y))$, in which case, we have

$$\int_a^b \alpha(t)\beta(t)\,df(t) = \int_a^b \alpha(t)\,dg(t) = \int_a^b \alpha(t)d\left[\int_a^t \beta(s)\,df(s)\right] \quad \text{and} \tag{1.8}$$

$$\left\| \int_a^b \alpha(t)\beta(t)\,df(t) \right\| \leqslant \left[SV_{[a,b]}(\alpha) + \|\alpha(a)\|\right] \|g\|_{\infty}. \tag{1.9}$$

Using Theorem 1.53, one can prove the next corollary. See [72, Corollary 8]. From now on, X, Y, Z, and W are Banach spaces.

Corollary 1.65: *Consider functions $\alpha \in SV([a,b], L(X,Y))$, $f \in C([a,b], W)$, and $\beta \in K_f([a,b], L(W,X))$, and define*

$$g(t) = \tilde{\beta}_f(t) = \int_a^t \beta(s)\, df(s), \quad t \in [a,b].$$

Then, $\alpha\beta \in K_f([a,b], L(W,Y))$ and equality (1.8) and inequality (1.9) hold.

Another substitution formula for Perron–Stieltjes integrals is presented next. Its proof uses a very nice trick provided by Professor C. S. Hönig while advising M. Federson's Master Thesis. Such result is borrowed from [72, Theorem 12].

Theorem 1.66: *Consider functions $\gamma : [a,b] \to L(W,Y)$, $\alpha \in K^\gamma([a,b], L(X,W))$, $f \in BV([a,b], X)$ and $\beta = \tilde{\alpha}^\gamma : [a,b] \to L(X,Y)$, that is,*

$$\beta(t) = \int_a^t d\gamma(s)\alpha(s), \quad t \in [a,b].$$

Then, $f \in K^\beta([a,b], X)$, if and only if $\alpha f \in K^\gamma([a,b], Y)$ and

$$\int_a^t d\gamma(t)\alpha(t)f(t) = \int_a^t d\beta(t)f(t). \tag{1.10}$$

Proof. Since $\alpha \in K^\gamma([a,b], L(X,W))$, given $\epsilon > 0$, there exists a gauge δ on $[a,b]$ such that, for every δ-fine $d = (\xi_i, [t_{i-1}, t_i]) \in TD_{[a,b]}$,

$$\left\| \sum_{i=1}^{|d|} \left\{ \left[\gamma(t_i) - \gamma(t_{i-1})\right] \alpha(\xi_i) - \int_{t_i-1}^{t_i} d\gamma(t)\alpha(t) \right\} \right\| < \epsilon.$$

Taking approximated sums for $\int_a^b d\gamma(t)\alpha(t)f(t)$ and $\int_a^b d\beta(t)f(t)$, we obtain

$$\left\| \sum_{i=1}^{|d|} \left[\gamma(t_i) - \gamma(t_{i-1})\right] \alpha(\xi_i)f(\xi_i) - \sum_{i=1}^{|d|} \left[\beta(t_i) - \beta(t_{i-1})\right] f(\xi_i) \right\|$$
$$= \left\| \sum_{i=1}^{|d|} \left\{ \left[\gamma(t_i) - \gamma(t_{i-1})\right] \alpha(\xi_i) - \int_{t_{i-1}}^{t_i} d\gamma(t)\alpha(t) \right\} f(\xi_i) \right\|.$$

But, if $\gamma_i \in L(X,Y)$ and $x_i \in X$, then

$$\sum_{i=1}^{n} \gamma_i x_i = \left(\sum_{i=1}^{n} \gamma_i \right) x_0 + \sum_{j=1}^{n} \left(\sum_{i=j}^{n} \gamma_i \right) (x_j - x_{j-1}), \; n \in \mathbb{N}.$$

Hence, taking $\gamma_i = \left[\gamma(t_i) - \gamma(t_{i-1})\right] \alpha(\xi_i) - \int_{t_i-1}^{t_i} d\gamma(t)\alpha(t)$, $x_i = f(\xi_i)$, $x_0 = f(a)$ and $n = |d|$, we obtain

$$I \leqslant \left\| \sum_{i=1}^{|d|} \left\{ \left[\gamma(t_i) - \gamma(t_{i-1})\right] \alpha(\xi_i) - \int_{t_i-1}^{t_i} d\gamma(t)\alpha(t) \right\} \right\| \|f(a)\|$$

$$+ \sum_{j=1}^{|d|} \left\| \sum_{i=j}^{|d|} \gamma_i \right\| \|f(t_i) - f(t_{i-1})\| < \epsilon \|f(a)\| + \epsilon \mathrm{var}(f),$$

where we applied the Saks–Henstock lemma (Lemma 1.45) to obtain

$$\left\| \sum_{i=j}^{|d|} \gamma_i \right\| = \left\| \sum_{i=j}^{|d|} \left\{ \left[\gamma(t_i) - \gamma(t_{i-1})\right] \alpha(\xi_i) - \int_{t_i-1}^{t_i} d\gamma(t)\alpha(t) \right\} \right\| \leqslant \epsilon,$$

for every $j = 1, 2, \ldots, |d|$. □

A proof of the next proposition follows similarly as the proof of Theorem 1.66.

Proposition 1.67: *Let J be any interval of the real line and $a, b \in J$, with $a < b$. Consider functions $f : J \to L(X)$ and $v : J \to \mathbb{R}$ of locally bounded variation. Assume that $\alpha : J \to L(X)$ is locally Perron–Stieltjes integrable with respect to v, that is, the Perron–Stieltjes integral $\int_I \alpha(t)\, dv(t)$ exists, for every compact interval $I \subset J$. Assume, further, that $\beta : J \to L(X)$, defined by*

$$\beta(t) = \int_a^t \alpha(s)\, dv(s), \quad t \in J,$$

is also of locally bounded variation. Then, the Perron–Stieltjes integrals $\int_a^b d\beta(r) f(r)$ and $\int_a^b \alpha(r) f(r)\, dv(r)$ exist and

$$\int_a^b d\beta(r) f(r) = \int_a^b \alpha(r) f(r)\, dv(r). \tag{1.11}$$

Yet another substitution formula for Perron–Stieltjes integrals, borrowed from [72, Theorem 11], is brought up here and, again, another interesting trick provided by Professor Hönig is used in its proof. Such substitution formula will be used in Chapter 3 in order to guarantee the existence of some Perron–Stieltjes integrals. As a matter of fact, the corollary following Theorem 1.68 will do the job.

Theorem 1.68: *Consider functions $f : [a, b] \to X$, $\alpha \in K_f([a, b], L(X, W))$, $\beta = \tilde{\alpha}_f : [a, b] \to W$, that is,*

$$\beta(t) = \int_a^t \alpha(s)\, df(s), \quad \textit{for every } t \in [a, b]$$

and assume that $\gamma \in SV([a,b], L(W,Y))$. Thus, $\gamma \in K_\beta([a,b], L(W,Y))$ if and only if $\gamma\,\alpha \in K_f([a,b], L(X,Y))$, in which case, we have

$$\int_a^b \gamma(t)\alpha(t)\, df(t) = \int_a^b \gamma(t) d\beta(t). \tag{1.12}$$

Proof. By hypothesis, $\alpha \in K_f([a,b], L(X,W))$. Therefore, for every $\epsilon > 0$, there is a gauge δ of $[a,b]$ such that for every δ-fine $d = (\xi_i, t_i) \in TD_{[a,b]}$, we have

$$\left\| \sum_{i=1}^{|d|} \left\{ \alpha(\xi_i)\left[f(t_i) - f(t_{i-1})\right] - \int_{t_{i-1}}^{t_i} \alpha(t)\, df(t) \right\} \right\| < \epsilon.$$

Taking approximated Riemannian-type sums for the integrals $\int_a^b \gamma(t)\alpha(t)\, df(t)$ and $\int_a^b \gamma(t) d\beta(t)$, we obtain

$$\left\| \sum_{i=1}^{|d|} \gamma(\xi_i)\alpha(\xi_i)\left[f(t_i) - f(t_{i-1})\right] - \sum_{i=1}^{|d|} \gamma(\xi_i)\left[\beta(t_i) - \beta(t_{i-1})\right] \right\|$$
$$= \left\| \sum_{i=1}^{|d|} \gamma(\xi_i) \left\{ \alpha(\xi_i)\left[f(t_i) - f(t_{i-1})\right] - \int_{ti-1}^{t_i} \alpha(t)\, df(t) \right\} \right\| = I.$$

On the other hand, when $\gamma_i \in L(X,Y)$ and $x_i \in X$, we have

$$\sum_{i=i}^{n} \gamma_i x_i = \sum_{j=1}^{n} (\gamma_j - \gamma_{j-1}) \left(\sum_{i=j}^{n} x_i \right) + \gamma_0 \left(\sum_{i=j}^{n} x_i \right), \qquad n \in \mathbb{N}.$$

Then, taking $x_i = \alpha(\xi_i)\left[f(t_i) - f(t_{i-1})\right] - \int_{ti-1}^{t_i} \alpha(t)\, df(t)$, $\gamma_i = \gamma(\xi_i)$, $\gamma_0 = \gamma(a)$, and $n = |d|$, we get

$$I = \left\| \sum_{j=1}^{|d|} \left[\gamma(\xi_j) - \gamma(\xi_{j-1})\right] \left(\sum_{i=j}^{|d|} x_i \right) + \gamma_0 \left(\sum_{i=j}^{|d|} x_i \right) \right\| \leqslant SV(\gamma)\,\epsilon + \| \gamma(a) \|\, \epsilon,$$

because the Saks-Henstock lemma (Lemma 1.45) yields $\left\| \sum_{i=j}^{|d|} x_i \right\| \leqslant \epsilon$, for every $j \in \{1, 2, \ldots, |d|\}$. □

Corollary 1.69: *Consider functions $f : [a,b] \to X$, $\alpha \in K_f([a,b], L(X,W))$, $\beta = \tilde{\alpha}_f \in BV([a,b], W)$, and $\gamma \in G([a,b], L(W,Y)) \cap SV([a,b], L(W,Y))$. Then, we have $\gamma \in K_\beta([a,b], L(W,Y))$, $\gamma\,\alpha \in K_f(a,b], L(X,Y))$, and Eq. (1.12) holds.*

Proof. Theorem 1.47, item (i), yields $\gamma \in K_g([a,b], L(W,Y))$. Then, the statement follows from Theorem 1.68. □

The next result gives us an integration by parts formula for Perron–Stieltjes integrals. A proof of it can be found in [212, Theorem 13].

Proposition 1.70: *Suppose $\alpha \in G([a,b], L(X))$ and $f \in BV([a,b], X)$ or $\alpha \in SV([a,b], L(X))$ and $f \in G([a,b], X)$. Then, the Perron–Stieltjes integrals $\int_a^b d\alpha(t)f(t)$ and $\int_a^b \alpha(t)\, df(t)$ exist, and the following equality holds:*

$$\int_a^b d\alpha(t)f(t) + \int_a^b \alpha(t)\, df(t) = \alpha(b)f(b) - \alpha(a)f(a)$$
$$- \sum_{a \leqslant \tau < b} \Delta^+\alpha(\tau)\Delta^+ f(\tau) + \sum_{a \leqslant \tau < b} \Delta^-\alpha(\tau)\Delta^- f(\tau),$$

where $\Delta^+\alpha(\tau) = \alpha(\tau^+) - \alpha(\tau)$, $\Delta^-\alpha(\tau) = \alpha(\tau) - \alpha(\tau^-)$, $\Delta^+ f(\tau) = f(\tau^+) - f(\tau)$, and $\Delta^- f(\tau) = f(\tau) - f(\tau^-)$.

As an immediate consequence of the previous proposition, we have the following result.

Corollary 1.71: *If $f \in G([a,b], X)$ and $g : [a,b] \to \mathbb{R}$ is a nondecreasing function, then the integral $\int_a^b f(t)\, dg(t)$ exists.*

We end this subsection by presenting a result, borrowed from [172] and [179, Theorem 5.4.5], which gives us a change of variable formula for Perron–Stieltjes integrals.

Theorem 1.72: *Suppose $\phi : [a,b] \to [\alpha, \beta]$ is increasing and maps $[a,b]$ onto $[\alpha, \beta]$ and consider functions $g, h : [\alpha, \beta] \to \mathbb{R}$. Then, both integrals*

$$\int_\alpha^\beta g(\tau) dh(\tau), \quad \int_a^b g(\phi(s)) d[h(\phi(s))]$$

exists, whenever one of the integrals exists, in which case, we have

$$\int_\alpha^\beta g(\tau)\, dh(\tau) = \int_a^b g(\phi(s)) d[h(\phi(s))].$$

1.3.4 The Fundamental Theorem of Calculus

The first result we present in this section is the Fundamental Theorem of Calculus for the variational Henstock integral. The proof follows standard steps (see [172], p. 43, for instance) adapted to Banach space-valued functions.

Theorem 1.73 (Fundamental Theorem of Calculus): *Suppose $F : [a,b] \to X$ is a function such that there exists the derivative $F'(t) = f(t)$, for every $t \in [a,b]$. Then, $f \in H([a,b], X)$ and*

$$\int_a^t f(s)\, ds = F(t) - F(a), \quad t \in [a,b].$$

Next, we give an example, borrowed from [73], of a Banach space-valued function $f : [0, 1] \to X$ which is Riemann–McShane integrable (see the appendix to this chapter for this definition). However, f is not variationally Henstock integrable, nor it is integrable in the sense of Bochner–Lebesgue.

Example 1.74: Let $X = G^-([0,1], \mathbb{R}) = \{f \in G([0,1], \mathbb{R}) : f \text{ is left-continuous}\}$ and consider the function $f : [0,1] \to X$ given by $f(t) = \chi_{[t,1]}$, where χ_A denotes the characteristic function of a measurable set $A \subset [0,1]$.

Since $f \in SV([0,1], L(\mathbb{R}, X))$ (see Definition 1.24) and the function $\phi(t) = t$, $t \in [0,1]$, is an element of $C([0,1], \mathbb{R})$, the abstract Riemann–Stieltjes integral, $\int_0^1 df(t)\phi(t)$, exists (see [127, Theorem 4.6], p. 24). Moreover, the Riemann–Stieltjes integral, $\int_0^1 f(t)d\phi(t)$, also exists and the integration by parts formula

$$\int_0^1 f(t)\,dt = \int_0^1 f(t)d\phi(t) = (f(t)\cdot t)|_0^1 - \int_0^1 df(t)\phi(t)$$

holds (see Theorem 1.53). Hence, $f \in R([0,1], X) \subset K([0,1], X)$.

Consider the indefinite integral $\tilde{f}(t) = \int_0^t f(r)\,dr$, $t \in [0,1]$, of f. Then,

$$\left(\int_0^t f(r)\,dr\right)(s) = \left(\int_0^t \chi_{[r,1]}\,dr\right)(s) = \int_0^t \chi_{[r,1]}(s)\,dr = \int_0^{t\wedge s} dr = t \wedge s$$

and, hence,

$$\tilde{f}(t)(s) = t \wedge s = \inf\{t, s\}.$$

Thus, $\tilde{f}$ is absolutely continuous.

On the other hand, $\tilde{f}$ is nowhere differentiable (see [73], Example 3.1). Then, the Lebesgue theorem implies $f \notin \mathcal{L}_1([0,1], X)$, where by $\mathcal{L}_1([0,1], X)$, we denote the space of functions from $[0,1]$ to X which are Lebesgue integrable with finite integral. See the appendix of this chapter. As a matter of fact, the Fundamental Theorem of Calculus for the Henstock integral (see Theorem 1.73) yields that $f \notin H([0,1], X)$. Optionally, one can verify that $f \notin H([0,1], X)$ simply by noticing that

$$\left\| f(\xi_i)(t_i - t_{i-1}) - \int_{t_{i-1}}^{t_i} f(t)\,dt \right\| \geqslant \frac{1}{2}(t_i - t_{i-1}),$$

for every $(\xi_i, [t_{i-1}, t_i]) \in TD_{[0,1]}$.

Claim. $f \in RMS([0,1], X)$, that is, f is Riemann–McShane integrable (see the appendix of this chapter).

Is is sufficient to prove that, given $\epsilon > 0$, we can find $\delta > 0$ such that for every δ-fine $d = (\xi_i, [t_{i-1}, t_i]) \in STD_{[0,1]}$ (the reader may want to check the notation $STD_{[a,b]}$ in the appendix of this chapter),

$$\left\| \tilde{f}(1) - \sum_{i=1}^{|d|} f(\xi_i)(t_i - t_{i-1}) \right\| < \epsilon.$$

Consider $0 < \delta < \epsilon$ and take a δ-fine $d = (\xi_i, [t_{i-1}, t_i]) \in TPD_{[0,1]}$.

- If $\xi_i \leqslant s$ and $t_i < \xi_i + \delta$, then $t_i < s + \delta$. Therefore, $\sum_{\xi_i \leqslant s}(t_i - t_{i-1}) < s + \delta$ and, hence,

$$-s + \sum_{\xi_i \leqslant s}(t_i - t_{i-1}) < \delta.$$

- If $\xi_j > s$ and $t_{j-1} > \xi_j - \delta$, then $t_{j-1} > s - \delta$. Therefore,

$$0 \leqslant \sum_{\xi_j > s}(t_j - t_{j-1}) < 1 - (s - \delta) = \sum_{i=1}^{|d|}(t_i - t_{i-1}) - s + \delta$$

and we obtain

$$s - \sum_{\xi_i \leqslant s}(t_i - t_{i-1}) < \delta.$$

Finally, we get

$$\begin{aligned}\left\| \tilde{f}(1) - \sum_{i=1}^{|d|} f(\xi_i)(t_i - t_{i-1}) \right\|_\infty &= \sup_{0 \leqslant s \leqslant 1} \left| \tilde{f}(1)(s) - \sum_{i=1}^{|d|} f(\xi_i)(s)(t_i - t_{i-1}) \right| \\ &= \sup_{0 \leqslant s \leqslant 1} \left| s - \sum_{\xi_i \leqslant s}(t_i - t_{i-1}) \right| < \delta < \epsilon\end{aligned}$$

and the *Claim* is proved.

A less restrict version of the Fundamental Theorem of Calculus is stated next. A proof of it follows as in [108, Theorem 9.6].

Theorem 1.75 (Fundamental Theorem of Calculus): *Suppose $F : [a, b] \to X$ is a continuous function such that there exists the derivative $F'(t) = f(t)$, for nearly everywhere on $[a, b]$ (i.e. except for a countable subset of $[a, b]$). Then, $f \in H([a, b], X)$ and*

$$\int_a^t f(s)\, ds = F(t) - F(a), \quad t \in [a, b].$$

Now, we present a class of functions $f : [a, b] \to X$, laying between absolute continuous and continuous functions, for which we can obtain a version of the Fundamental Theorem of Calculus for Henstock vector integrals. Let m denote the Lebesgue measure.

Definition 1.76: A function $f : [a, b] \to X$ satisfies the *strong Lusin condition*, and we write $f \in SL([a, b], X)$, if given $\epsilon > 0$ and $B \subset [a, b]$ with $m(B) = 0$, then there is a gauge δ on B such that for every δ-fine $d = (\xi_i, [t_{i-1}, t_i]) \in TPD_{[a,b]}$ with $\xi_i \in B$ for all $i = 1, 2, \ldots, |d|$, we have $\sum_{i=1}^{|d|} \|f(t_i) - f(t_{i-1})\| < \epsilon$.

If we denote by $AC([a,b], X)$ the space of all absolutely continuous functions from $[a,b]$ to X, then it is not difficult to prove that

$$AC([a,b], X) \subset SL([a,b], X) \subset C([a,b], X).$$

In $SL([a,b], X)$, we consider the usual supremum norm, $\| \cdot \|_\infty$, induced by $C([a,b], X)$.

The next two versions of the Fundamental Theorem of Calculus for Henstock vector integrals, as described in Definition 1.41, are borrowed from [70, Theorems 1 and 2]. We use the term *almost everywhere* in the sense of the Lebesgue measure m.

Theorem 1.77: *If $f \in SL([a,b], X)$ and $A \in SL([a,b], Y)$ are both differentiable and $\alpha : [a,b] \to L(X, Y)$ is such that $A'(t) = \alpha(t) f'(t)$ for almost every $t \in [a,b]$, then $\alpha \in H_f([a,b], L(X, Y))$ and $A = \tilde{\alpha}_f$, that is,*

$$\int_a^t \alpha(s) f'(s)\, ds = \int_a^t \alpha(s)\, df(s), \quad t \in [a,b].$$

Theorem 1.78: *If $f \in SL([a,b], X)$ is differentiable and $\alpha \in H_f([a,b], L(X, Y))$ is bounded, then $\tilde{\alpha}_f \in SL([a,b], Y)$, and there exists the derivative $(\tilde{\alpha}_f)'(t) = \alpha(t) f'(t)$ for almost every $t \in [a,b]$, that is,*

$$\frac{d}{dt}\left[\int_a^t \alpha(s)\, df(s)\right] = \alpha(t) f'(t), \quad \textit{almost everywhere in } [a,b].$$

Corollary 1.79: *Suppose $f \in SL([a,b], X)$ is differentiable and nonconstant on any nondegenerate subinterval of $[a,b]$ and $\alpha \in H_f([a,b], L(X, Y))$ is bounded and such that $\tilde{\alpha}_f = 0$. Then, $\alpha = 0$ almost everywhere in $[a,b]$.*

From Corollary 1.50, we know that if $f \in C([a,b], X)$ and $\alpha \in K_f([a,b], L(X, Y))$, then $\tilde{\alpha}_f \in C([a,b], Y)$. For the Henstock vector integral, we have the following analogue whose proof can be found in [70, Theorem 7].

Theorem 1.80: *If $f \in SL([a,b], X)$ and $\alpha \in H_f([a,b], L(X, Y))$, then we have $\tilde{\alpha}_f \in SL([a,b], Y)$.*

The next result is borrowed from [70, Theorem 5]. We reproduce its proof here.

Theorem 1.81: *Suppose $f \in SL([a,b],X)$ and $\alpha : [a,b] \to L(X,Y)$ is such that $\alpha = 0$ almost everywhere on $[a,b]$. Then, $\alpha \in H_f([a,b],L(X,Y))$ and $\tilde{\alpha}_f = 0$, that is,*

$$\tilde{\alpha}_f(t) = \int_a^t \alpha(s)\, df(s) = 0, \text{ for every } t \in [a,b].$$

Proof. Consider the sets

$$E = \{t \in \mathbb{R} : \alpha(t) \neq 0\} \quad \text{and} \quad E_n = \{t \in E : n-1 < \| \alpha(t) \| \leqslant n\}, \\ \text{for each } n \in \mathbb{N}.$$

By hypothesis, $m(E) = 0$, where m denotes the Lebesgue measure. Hence, $m(E_n) = 0$ for every $n \in \mathbb{N}$. In addition, $f \in SL([a,b],X)$. Then, for every $n \in \mathbb{N}$ and every $\epsilon > 0$, there exists a gauge δ_n on E_n such that, for every δ_n-fine tagged partial division $d = (\xi_{n_i}, [t_{n_i-1}, t_{n_i}]) \in TPD_{[a,b]}$, with $\xi_{n_i} \in E_n$, for $i = 1, 2, \dots, |d|$, we have

$$\sum_{i=1}^{|d|} \| f(t_{n_i}) - f(t_{n_i-1}) \| < \frac{\epsilon}{n2^n}.$$

Consider a gauge δ of $[a,b]$ such that $\delta(\xi) = \delta_n(\xi)$, whenever $\xi \in E_n$, and $\delta(\xi)$ can assume any value in $(0,\infty)$, otherwise. Then, for every δ-fine $d = (\xi_i, [t_{i-1}, t_i]) \in TD_{[a,b]}$, we have

$$\begin{aligned} \sum_{i=1}^{|d|} \| \alpha(\xi_i)[f(t_i) - f(t_{i-1})] \| &\leqslant \sum_{n \in \mathbb{N}} \sum_{\xi_i \in E_n} \| \alpha(\xi_i) \| \, \| f(t_i) - f(t_{i-1}) \| \\ &\leqslant \sum_{n \in \mathbb{N}} n \sum_{\xi_i \in E_n} \| f(t_i) - f(t_{i-1}) \| < \epsilon \end{aligned}$$

and we complete the proof. □

Given a function $f : [a,b] \to X$, since $H_f([a,b],L(X,Y)) \subset K_f([a,b],L(X,Y))$, Theorem 1.81 holds for $K_f([a,b],L(X,Y))$ instead of $H_f([a,b],L(X,Y))$. Then, next proposition follows easily (see, also, [70, Corollary after Theorem 5]).

Proposition 1.82: *Suppose $f \in SL([a,b],X)$ and $\alpha \in K_f([a,b],L(X,Y))$. Assume, in addition, that $\beta : [a,b] \to L(X,Y)$ is such that $\beta = \alpha$ almost everywhere in $[a,b]$. Then, $\beta \in K_f([a,b],L(X,Y))$ and $\tilde{\beta}_f(t) = \tilde{\alpha}_f(t)$, for every $t \in [a,b]$. If, moreover, $\alpha \in H_f([a,b],L(X,Y))$, then $\beta \in H_f([a,b],L(X,Y))$.*

In view of Proposition 1.82, we can define equivalence classes of nonabsolute vector integrable functions.

Definition 1.83: Let us assume that $f \in SL([a,b],X)$. Two functions $\beta, \alpha \in K_f([a,b],L(X,Y))$ are *equivalent* if and only if their indefinite integrals coincide, that is, $\tilde{\beta}_f = \tilde{\alpha}_f$. By $\mathbf{K}_f([a,b], L(X,Y))$ and $\mathbf{H}_f([a,b], L(X,Y))$ we mean, respectively, the spaces of equivalence classes of functions of $K_f([a,b], L(X,Y))$ and of $H_f([a,b], L(X,Y))$, and we endow these spaces with an Alexiewicz-type norm

$$\|\alpha\|_{A,f} = \left\|\tilde{\alpha}_f\right\|_\infty = \sup\left\{\left\|\int_a^t \alpha(s)\,df(s)\right\| : t \in [a,b]\right\}.$$

From Example 1.74, we know that although $g, f \in R([a,b],X) \subset K([a,b],X)$ may belong to the same equivalence class, that is, $\int_a^t g(s)\,ds = \int_a^t f(s)\,ds$ for all $t \in [a,b]$, one cannot conclude that $g = f$ almost everywhere in $[a,b]$. It is known, however, that the space of all equivalence classes of real-valued Perron integrable functions $f : [a,b] \to \mathbb{R}$, equipped with the usual Alexiewicz norm (see [5]) given by

$$\| f\|_A = \| \tilde{f}\|_\infty = \sup_{t\in[a,b]} \left|\int_a^t f(s)\,ds\right|$$

is noncomplete (see [24], for instance). The same applies to Banach space-valued functions. Let us denote by $\mathbf{K}([a,b],X)$ the space of all *equivalence classes* of functions $f \in K([a,b],X)$, equipped with the Alexiewicz norm $\| f\|_A = \| \tilde{f}\|_\infty$. The space $\mathbf{K}([a,b],X)$ is noncomplete (see [129]). However, $\mathbf{K}([a,b],X)$ is ultrabornological (see [105]) and, therefore, barrelled. In particular, good functional analytic properties hold, such as the Banach–Steinhaus theorem and the Uniform Boundedness Principle (see, for instance, [142]). The same applies to the space $\mathbf{H}([a,b],X)$ of equivalence classes of functions $f \in H([a,b],X)$, endowed with the Alexiewicz norm $\| f\|_A = \| \tilde{f}\|_\infty$.

The next example, borrowed from [73], exhibits a Cauchy sequence of Henstock integrable functions which is not convergent in the usual Alexiewicz norm, $\| \cdot\|_A$.

Example 1.84: Consider functions $f_n : [0,1] \to l_2(\mathbb{N} \times \mathbb{N})$, $n \in \mathbb{N}$ defined by $f_n = \sum_{i=1}^n g_i$, where $g_i(t) = e_{ij}$, whenever $\frac{j-1}{2^i} \leqslant t < \frac{j}{2^i}$, $j = 1, 2, \dots, 2^i$, and $g_i(t) = 0$ otherwise.

Hence,

$$\|g_1\|_A = \sup_{0\leqslant t\leqslant 1}\left\|\int_0^t g_1\right\|_2 = \left\|\int_0^1 g_1\right\|_2 = \left\|\frac{1}{2}e_{11} + \frac{1}{2}e_{12}\right\|_2 = \left(\frac{1}{2}\right)^{\frac{1}{2}}.$$

Then,

$$\int_0^1 g_2 = \int_0^{\frac{1}{4}} e_{21} + \int_{\frac{1}{4}}^{\frac{1}{2}} e_{22} + \int_{\frac{1}{2}}^{\frac{3}{4}} e_{23} + \int_{\frac{3}{4}}^1 e_{24} = \frac{1}{4}(e_{21} + e_{22} + e_{23} + e_{24})$$

and, hence,

$$\|g_2\|_A = \sup_{0\leqslant t\leqslant 1}\left\|\int_0^t g_2\right\|_2 = \left\|\int_0^1 g_2\right\|_2 = \left(4\frac{1}{4^2}\right)^{\frac{1}{2}} = \left(\frac{1}{4}\right)^{\frac{1}{2}}.$$

By induction, one can show that

$$\|g_i\|_A = \left\|\sum_{j=1}^{2^i}\int_{\frac{j-1}{2^i}}^{\frac{j}{2^i}} e_{ij}\right\|_2 = \left[2^i\left(\frac{1}{2^i}\right)^2\right]^{\frac{1}{2}} = \frac{1}{2^{\frac{i}{2}}},$$

for every $i \in \mathbb{N}$. Then,

$$\|f_n - f_m\|_A = \left\|\sum_{i=n+1}^{m} g_i\right\|_A \leqslant \sum_{i=n+1}^{m}\frac{1}{2^{\frac{i}{2}}}$$

which goes to zero for sufficiently large $n, m \in \mathbb{N}$, with $n > m$. Thus, $\{f_n\}_{n\in\mathbb{N}}$ is a $\|\cdot\|_A$-Cauchy sequence. On the other hand,

$$\|f_n(t)\|_2 = \|g_1(t) + g_2(t) + \cdots + g_n(t)\|_2 = \sqrt{n},$$

for every $t \in [0,1]$. Hence, there is no function $f(t) \in l_2(\mathbb{N}\times\mathbb{N})$, with $t \in [0,1]$, such that $\lim_{n\to\infty}\|f_n - f\|_A = 0$.

The next result follows from Theorem 1.80. A proof of it can be found in [75, Theorem 5].

Theorem 1.85: *Suppose $f \in SL([a,b],X)$ is nonconstant on any nondegenerate subinterval of $[a,b]$. Then, the mapping*

$$\alpha \in \mathbf{H}_f([a,b],L(X,Y)) \mapsto \tilde{\alpha}_f \in C_a([a,b],X)$$

is an isometry, that is $\left\|\tilde{\alpha}_f\right\|_\infty = \|\alpha\|_{A,f}$ onto a dense subspace of $C_a([a,b],X)$.

The next result, known as *straddle Lemma*, will be useful to prove that the space $G([a,b],L(X,Y))$ of regulated functions from $[a,b]$ to $L(X,Y)$ is dense in $\mathbf{K}_f([a,b],L(X,Y))$ in the Alexiewicz norm $\|\cdot\|_{A,f}$. For a proof of the straddle Lemma, the reader may want to consult [130, 3.4] or [119].

Lemma 1.86 (Straddle Lemma): *Suppose $f, F : [a,b] \to X$ are functions such that F is differentiable, with $F'(\xi) = f(\xi)$, for all $\xi \in [a,b]$. Then, given $\epsilon > 0$, there exists $\delta(\xi) > 0$ such that*

$$\|F(t) - F(s) - f(\xi)(t-s)\| < \epsilon(t-s),$$

whenever $\xi - \delta(\xi) < s < \xi < t < \xi + \delta(\xi)$.

The next result is adapted from [75, Theorem 8].

Proposition 1.87: *Suppose $f \in SL([a,b],X)$ is differentiable and nonconstant on any nondegenerate subinterval of $[a,b]$. Then, the Banach space $G([a,b],L(X,Y))$ is dense in $\mathbf{K}_f([a,b],L(X,Y))$ under the Alexiewicz norm $\|\cdot\|_{A,f}$.*

Proof. Assume that $\alpha \in \mathbf{K}_f([a,b],L(X,Y))$ and let $\epsilon > 0$ be given. We need to find a function $\beta \in G([a,b],L(X,Y))$ such that $\|\beta - \alpha\|_{A,f} < \epsilon$, or equivalently,

$$\left\| \int_a^t \beta(s)\, df(s) - \int_a^t \alpha(s)\, df(s) \right\| < \epsilon, \quad t \in [a,b]. \tag{1.13}$$

By Corollary 1.50, $\tilde{\alpha}_f \in C_a([a,b],Y) = \{x \in C([a,b],Y) : x(a) = 0\}$. Let us denote by $C_a^{(1)}([a,b],Y)$ the subspace of $C_a([a,b],Y)$ of functions which are differentiable with continuous derivative. Hence, there is a function $h \in C_a^{(1)}([a,b],Y)$ such that

$$\left\| h - \tilde{\alpha}_f \right\|_\infty < \epsilon. \tag{1.14}$$

Let $\beta : [a,b] \to L(X,Y)$ be defined by $\beta(t)x = h'(t)$, for all $x \in X$ such that $x \neq 0$, and by $\beta(t)0 = 0$. In particular, $\beta(t)f'(t) = h'(t)$ whenever $f'(t) \neq 0$. Therefore, $\beta(t)f'(t) = h'(t)$ for almost every $t \in [a,b]$, since $f : [a,b] \to X$ is differentiable and nonconstant on any nondegenerate subinterval of $[a,b]$. Hence, $\beta \in G([a,b],L(X,Y))$. Then, the Riemann integral $\int_a^b \beta(s)f'(s)\, ds$ exists and

$$\int_a^t \beta(s)f'(s)\, ds = \int_a^t h'(s)\, ds = h(t), \quad t \in [a,b], \tag{1.15}$$

where we applied the Fundamental Theorem of Calculus for the Riemann integral in order to obtain the last equality. Thus, replacing (1.15) in (1.14), we obtain

$$\left\| \int_a^t \beta(s)f'(s)\, ds - \int_a^t \alpha(s)\, df(s) \right\| < \epsilon, \quad t \in [a,b]. \tag{1.16}$$

Now, in view of (1.13), it remains to prove that the Perron–Stieltjes integral $\int_a^b \beta(s)\, df(s)$ exists and

$$\int_a^t \beta(s)\, df(s) = \int_a^t \beta(s)f'(s)\, ds, \quad t \in [a,b]. \tag{1.17}$$

Let δ_1 be the gauge on $[a,b]$ from the definition of $\int_a^b \beta(s)f'(s)\, ds$. Take $t \in [a,b]$, and for every $\xi \in [a,t]$, let $\delta_2(\xi) > 0$ be such that if $\xi - \delta_2(\xi) < s < \xi < u < \xi + \delta_2(\xi)$, then, by the Straddle Lemma (Lemma 1.86), we have

$$\|f(u) - f(s) - f'(\xi)(u-s)\| < \epsilon(u-s). \tag{1.18}$$

Fix $t \in [a, b]$. We now define a gauge δ on $[a, t]$ by $\delta(\xi) = \min\{\delta_1(\xi), \delta_2(\xi)\}$, for every $\xi \in [a, t]$. Hence, for every δ-fine $d = (\xi_i, [t_{i-1}, t_i]) \in TD_{[a,t]}$, we have

$$\begin{aligned}
&\left\| \sum_{i=1}^{|d|} \beta(\xi_i)\left[f(t_i) - f(t_{i-1})\right] - \int_a^t \beta(s) f'(s)\, ds \right\| \\
&\leqslant \left\| \sum_{i=1}^{|d|} \beta(\xi_i)\left[f(t_i) - f(t_{i-1})\right] - \sum_{i=1}^{|d|} \beta(\xi_i) f'(t_i)(t_i - t_{i-1}) \right\| \\
&\quad + \left\| \sum_{i=1}^{|d|} \beta(\xi_i) f'(t_i)(t_i - t_{i-1}) - \int_a^t \beta(s) f'(s)\, ds \right\| \\
&< \|\beta\| \sum_{i=1}^{|d|} \|f(t_i) - f(t_{i-1}) - f'(t_i)(t_i - t_{i-1})\| + \epsilon \\
&< \|\beta\| \sum_{i=1}^{|d|} \epsilon(t_i - t_{i-1}) + \epsilon = \|\beta\|\, \epsilon(t-a) + \epsilon,
\end{aligned}$$

by (1.18) and by the Riemann integrability of $\beta(\cdot) f'(\cdot)$. Finally, (1.13) follows from (1.16) and (1.17) and the proof is complete. □

1.3.5 A Convergence Theorem

As the last result of this introductory chapter, we mention a convergence theorem for Perron–Stieltjes integrals. Such result is used in Chapter 3. A proof of it can be found in [[180], Theorem 2.2].

Theorem 1.88: *Consider functions $f, f_n \in G([a, b], X)$ and $\alpha, \alpha_n \in BV([a, b], L(X))$, for $n \in \mathbb{N}$. Suppose*

$$\lim_{n\to\infty} \| f_n - f\|_\infty = 0, \quad \lim_{n\to\infty} \| \alpha_n - \alpha\|_\infty = 0 \quad \textit{and} \quad \sup_{n\in\mathbb{N}} var_a^b(\alpha_n) < \infty.$$

Then

$$\lim_{n\to\infty} \left(\sup_{t\in[a,b]} \left\| \int_a^t d\alpha_n(s) f_n(s) - \int_a^t d\alpha(s) f(s) \right\| \right) = 0.$$

Appendix 1.A: The McShane Integral

The integrals introduced by J. Kurzweil [152] and independently by R. Henstock [118] in the late 1950s are equivalent to the restricted Denjoy integral and the Perron integral for integrands taking values in $\mathbb{R}$ (see [108], for instance). In particular, the definitions of the so-called "Kurzweil–Henstock" integrals are based on Riemannian sums, and are therefore easy to deal with even by undergraduate students. Not only that, but the Kurzweil–Henstock–Denjoy–Perron integral encompasses the integrals of Newton, Riemann, and Lebesgue.

In 1969, E. J. McShane (see [173, 174]) showed that a small change in the subdivision process of the domain of integration within the Kurzweil–Henstock (or Perron) integral leads to the Lebesgue integral. This is a very nice finding, since now the Lebesgue integral can be taught by presenting its Riemannian definition straightforwardly and, then, obtaining immediately some very interesting properties such as the linearity of the Lebesgue integral which comes directly from the fact that the Riemann sum can be split into two sums. The monotone convergence theorem for the Lebesgue integral is another example of a result which is naturally obtained from its equivalent definition due to McShane.

The Kurzweil integral and the variational Henstock integral can be extended to Banach space-valued functions as well as to the evaluation of integrands over unbounded intervals. The extension of the McShane integral, proposed by R. A. Gordon (see [107]) to Banach space-valued functions, gives a more general integral than that of Bochner–Lebesgue. As a matter of fact, the idea of McShane into the definition due to Kurzweil enlarges the class of Bochner–Lebesgue integrals.

On the other hand, when the idea of McShane is employed in the variational Henstock integral, one gets precisely the Bochner–Lebesgue integral. This interesting fact was proved by W. Congxin and Y. Xiabo in [47] and, independently, by C. S. Hönig in [131]. Later, L. Di Piazza and K. Musal generalized this result (see [55]). We clarify here that unlike the proof by Congxin and Xiabo, based on the Fréchet differentiability of the Bochner–Lebesgue integral, Hönig's idea to prove the equivalence between the Bochner–Lebesgue integral and the integral we refer to as *Henstock–McShane integral* uses the fact that the indefinite integral of a Henstock–McShane integrable function is itself a function of bounded variation and the fact that absolute Henstock integrable functions are also functions of bounded variation. In this way, the proof provided in [131] becomes simpler. We reproduce it in the next lines, since reference [131] is not easily available. We also refer to [73] for some details.

Definition 1.89: We say that a function $f : [a,b] \to X$ is *Bochner–Lebesgue integrable* (we write $f \in \mathcal{L}_1([a,b],X)$), if there exists a sequence $\{f_n\}_{n\in\mathbb{N}}$ of simple functions, $f_n : [a,b] \to X$, $n \in \mathbb{N}$, such that

(i) $f_n \to f$ almost everywhere (i.e. $\lim_{n\to\infty} \|f_n(t) - f(t)\| = 0$ for almost every $t \in [a,b]$), and

(ii) $\lim_{n,m\to\infty} \int_a^b \|f_n(t) - f_m(t)\| \, dt = 0.$

With the notation of Definition 1.89, we define

$$\int_a^b f(t)\,dt = \lim_{n\to\infty} \int_a^b f_n(t)\,dt \quad \text{and} \quad \|f\|_1 = \int_a^b \|f(t)\|\,dt.$$

Then, the space of all equivalence classes of Bochner–Lebesgue integrable functions, equipped with the norm $\|f\|_1$, is complete.

The next definition can be found in [239], for instance.

Definition 1.90: We say that a function $f : [a, b] \to X$ is *measurable*, whenever there is a sequence of simple functions $f_n : [a, b] \to X$ such that $f_n \to f$ almost everywhere. When this is the case,

$$f \in \mathcal{L}_1([a,b],X) \quad \text{if and only if} \quad \int_a^b \|f(t)\|\,dt < \infty. \tag{1.A.1}$$

Again, we explicit the "name" of the integral we are dealing with, whenever we believe there is room for ambiguity.

As we mentioned earlier, when only real-valued functions are considered, the Lebesgue integral is equivalent to a modified version of the Kurzweil–Henstock (or Perron) integral called *McShane integral.* The idea of slightly modifying the definition of the Kurzweil–Henstock integral is due to E. J. McShane [173, 174]. Instead of taking tagged divisions of an interval $[a, b]$, McShane considered what we call *semitagged divisions*, that is,

$$a = t_0 < t_1 \ldots < t_{|d|} = b$$

is a division of $[a, b]$ and, to each subinterval $[t_{i-1}, t_i]$, with $i = 1, 2, \ldots, |d|$, we associate a point $\xi_i \in [a, b]$ called "*tag*" of the subinterval $[t_{i-1}, t_i]$. We denote such semitagged division by $d = (\xi_i, [t_{i-1}, t_i])$ and, by $STD_{[a,b]}$, we mean the set of all semitagged divisions of the interval $[a, b]$. But what is the difference between a semitagged division and a tagged division? Well, in a semitagged division $(\xi_i, [t_{i-1}, t_i]) \in STD_{[a,b]}$, it is not required that a tag ξ_i belongs to its associated subinterval $[t_{i-1}, t_i]$. In fact, neither the subintervals need to contain their corresponding tags. Nevertheless, likewise for tagged divisions, given a gauge δ of $[a, b]$, in order for a semitagged division $(\xi_i, [t_{i-1}, t_i]) \in STD_{[a,b]}$ to be δ-fine, we need to require that

$$\left[t_{i-1}, t_i\right] \subset \left\{t \in [a, b] : \left|t - \xi_i\right| < \delta(\xi_i)\right\} \quad \text{for all } i = 1, 2, \ldots$$

This simple modification provides an elegant characterization of the Lebesgue integral through Riemann sums (see [174]).

Let us denote by $KMS([a, b], \mathbb{R})$ the space of all real-valued Kurzweil–McShane integrable functions $f : [a, b] \to \mathbb{R}$, that is, $f \in KMS([a, b], \mathbb{R})$ is integrable in the sense of Kurzweil with the modification of McShane. Formally, we have the next definition which can be extended straightforwardly to Banach space-valued functions.

Definition 1.91: We say that $f : [a,b] \to \mathbb{R}$ is *Kurzweil-McShane integrable*, and we write $f \in KMS([a,b],\mathbb{R})$ if and only if there exists $I \in \mathbb{R}$ such that for every $\epsilon > 0$, there is a gauge δ on $[a,b]$ such that

$$\left| I - \sum_{i=1}^{|d|} f(\xi_i)(t_i - t_{i-1}) \right| < \epsilon,$$

whenever $d = (\xi_i, [t_{i-1}, t_i]) \in STD_{[a,b]}$ is δ-fine. We denote the Kurzweil-McShane integral of a function $f : [a,b] \to \mathbb{R}$ by $(KMS) \int_a^b f(t)\, dt$.

The following inclusions hold

$$R([a,b],\mathbb{R}) \subset \mathcal{L}_1([a,b],\mathbb{R}) = KMS([a,b],\mathbb{R}) \subset K([a,b],\mathbb{R}) = H([a,b],\mathbb{R}).$$

Moreover, $K([a,b],\mathbb{R}) \setminus \mathcal{L}_1([a,b],\mathbb{R}) \neq \emptyset$ as one can note by the next classical example.

Example 1.92: Let $F : [0,1] \to \mathbb{R}$ be defined by $F(t) = t^2 \sin(t^{-2})$, if $0 < t \leqslant 1$, and $F(0) = 0$, and consider $f = F'$. Since f is Riemann improper integrable, $f \in K([a,b],\mathbb{R}) = H([a,b],\mathbb{R})$, because the Kurzweil–Henstock (or Perron) integral contains its improper integrals (see Theorem 2.9, [158], or [213]). However, $f \notin \mathcal{L}_1([a,b],\mathbb{R})$, since f is not absolutely integrable (see also [227]).

Example 1.92 tells us that the elements of $K([a,b],\mathbb{R}) = H([a,b],\mathbb{R})$ are not absolutely integrable.

When McShane's idea is applied to Kurzweil and Henstock vector integrals, the story changes. In fact, the modification of McShane applied to the Kurzweil vector integral originates an integral which encompasses the Bochner–Lebesgue integral (see Example 1.74). On the other hand, when McShane's idea is used to modify the variational Henstock integral, we obtain exactly the Bochner–Lebesgue integral (see [47 and 131]). Thus, if $HMS([a,b],X)$ denotes the space of Henstock–McShane integrable functions $f : [a,b] \to X$, that is, $f \in HMS([a,b],X)$ is integrable in the sense of Henstock with the modification of McShane, then

$$HMS([a,b],X) = \mathcal{L}_1([a,b],X).$$

We will prove this equality in the sequel. Furthermore, $HMS([a,b],X) \subset H([a,b],X)$, $KMS([a,b],X) \subset K([a,b],X)$, and $RMS([a,b],X) \subset R([a,b],X)$, where we use the notation $KMS([a,b],X)$ and $RMS([a,b],X)$ to denote, respectively, the spaces of Kurzweil–McShane and Riemann–McShane integrable functions from $[a,b]$ to X. For other interesting results, the reader may want to consult [55].

Our aim in the remaining of this chapter is to show that the integrals of Bochner–Lebesgue and Henstock–McShane coincide. See [132, Theorem 10.4]. The next results are due to C. S. Hönig. They belong to a brochure of a series of lectures Professor Hönig gave in Rio de Janeiro in 1993. We include the proofs here, once the brochure is in Portuguese.

Lemma 1.93: *Let $\{f_n\}_{n\in\mathbb{N}}$ be a sequence in $KMS([a,b],X)$ and $f : [a,b] \to X$ be a function. Suppose there exists*

$$\lim_{n\to\infty} (KMS) \int_a^b \|f_n(t) - f(t)\| \, dt = 0.$$

Then, $f \in KMS([a,b],X)$ and

$$\lim_{n\to\infty} (KMS) \int_a^b f_n(t)\, dt = (KMS) \int_a^b f(t)\, dt.$$

Proof. Given $\epsilon > 0$, take n_ϵ such that for $m, n \geqslant n_\epsilon$,

$$(KMS) \int_a^b \|f_n(t) - f_m(t)\| \, dt < \epsilon$$

and take a gauge δ on $[a,b]$ such that for every δ-fine $d = (\xi_i, [t_{i-1}, t_i]) \in STD_{[a,b]}$,

$$\sum_{i=1}^{|d|} \left\| f_{n_\epsilon}(\xi_i) - f(\xi_i) \right\| (t_i - t_{i-1}) < \epsilon. \tag{1.A.2}$$

The limit $I = \lim_{n\to\infty} (KMS) \int_a^b f_n(t)\, dt$ exists, since for $m, n \geqslant n_\epsilon$,

$$\begin{aligned} &\left\| (KMS) \int_a^b f_n(t)\, dt - (KMS) \int_a^b f_m(t)\, dt \right\| \\ &\leqslant (KMS) \int_a^b \|f_n(t) - f(t)\| \, dt + (KMS) \int_a^b \|f(t) - f_m(t)\| \, dt \leqslant 2\epsilon. \end{aligned}$$

Hence, if $I_n = (KMS) \int_a^b f_n(t)\, dt$, then

$$\begin{aligned} &\left\| \sum_{i=1}^{|d|} f(\xi_i)(t_i - t_{i-1}) - I \right\| \\ &\leqslant \sum_{i=1}^{|d|} \left\| f(\xi_i) - f_{n_\epsilon}(\xi_i) \right\| (t_i - t_{i-1}) + \left\| \sum_{i=1}^{|d|} f_{n_\epsilon}(\xi_i)(t_i - t_{i-1}) - I_{n_\epsilon} \right\| + \left\| I_{n_\epsilon} - I \right\|. \end{aligned}$$

Thus, the first summand on the right-hand side of the last inequality is smaller than ϵ by (1.A.2), the third summand is smaller than ϵ by the definition of n_ϵ and, if we refine the gauge δ, we may suppose, by the definition of I_{n_ϵ}, that the second summand is smaller than ϵ, and the proof is complete. □

We show next that Lemma 1.93 remains valid if we replace *KMS* by *HMS*, that is, if instead of the space $KMS([a, b], X)$ of Kurzweil–McSchane integrable functions, we consider the space $HMS([a, b], X)$ of Henstock–McSchane integrable functions.

Lemma 1.94: *Let $\{f_n\}_{n\in\mathbb{N}}$ be a sequence in $HMS([a, b], X)$ and $f : [a, b] \to X$ be a function. If $\lim_{n\to\infty} \int_a^b \|f_n(t) - f(t)\|\, dt = 0$, then $f \in HMS([a, b], X)$ and*

$$\lim_{n\to\infty} (KMS) \int_a^b f_n(t)\, dt = (KMS) \int_a^b f(t)\, dt.$$

Proof. By Lemma 1.93, $f \in KMS([a, b], X)$, and we have the convergence of the integrals. It remains to prove that $f \in HMS([a, b], X)$, that is, for every $\epsilon > 0$, there exists a gauge δ on $[a, b]$ such that for every δ-fine $d = (\xi_i, [t_{i-1}, t_i]) \in TPD_{[a,b]}$,

$$\sum_{i=1}^{|d|} \left\| (KMS) \int_{t_{i-1}}^{t_i} f(t)\, dt - f(\xi_i)(t_i - t_{i-1}) \right\| \leqslant \epsilon.$$

However,

$$\sum_{i=1}^{|d|} \left\| (KMS) \int_{t_{i-1}}^{t_i} f(t)\, dt - f(\xi_i)(t_i - t_{i-1}) \right\| \leqslant \sum_{i=1}^{|d|} \left\| (KMS) \int_{t_{i-1}}^{t_i} \left[f(t) - f_n(t)\right] dt \right\|$$
$$+ \sum_{i=1}^{|d|} \left\| (KMS) \int_{t_{i-1}}^{t_i} f_n(t)\, dt - f_n(\xi_i)(t_i - t_{i-1}) \right\| + \sum_{i=1}^{|d|} \|f_n(\xi_i) - f(\xi_i)\| \,(t_i - t_{i-1}).$$

Since $\int_a^b \|f_n(t) - f(t)\|\, dt \to 0$ as n tends to infinity, there exists $n_\epsilon > 0$ such that the first summand in the last inequality is smaller than $\frac{\epsilon}{3}$ for all $n \geqslant n_\epsilon$. Choose an $n \geqslant n_\epsilon$. Then, we can take δ such that the third summand is smaller than $\frac{\epsilon}{3}$, because it approaches $\int_a^b \|f_n(t) - f(t)\|\, dt$. In addition, once $f_n \in HMS([a, b], X)$, we can refine δ so that the second summand becomes smaller than $\frac{\epsilon}{3}$, and we finish the proof. □

For a proof of the next lemma, it is enough to adapt the proof found in [107, Theorem 16] for the case of Banach space-valued functions.

Lemma 1.95: $\mathcal{L}_1([a, b], X) \subset KMS([a, b], X)$.

Now, we are able to prove the next inclusion.

Theorem 1.96: $\mathcal{L}_1([a, b], X) \subset HMS([a, b], X)$.

Proof. By Lemma 1.95, $\mathcal{L}_1([a, b], X) \subset KMS([a, b], X)$. Then, following the steps of the proof of Lemma 1.95 and using Lemma 1.94, we obtain the desired result. □

For the next result, which says that the indefinite integral of any function of $HMS([a,b],X)$ belongs to $BV([a,b],X)$, we employ a trick based on the fact that if $g = f$ almost everywhere, then $g \in HMS([a,b],X)$ and $\tilde{g} = \tilde{f}$, that is, the indefinite integrals of f and g coincide. This fact follows by a straightforward adaptation of [108, Theorem 9.10] for Banach space-valued functions (see also [70]). Thus, if we change a function $f \in HMS([a,b],X)$ on a set of Lebesgue measure zero, its indefinite integral does not change. Therefore, we consider, for instance, that f vanishes at such points.

Lemma 1.97: *If $f \in HMS([a,b],X)$, then $\tilde{f} \in BV([a,b],X)$.*

Proof. It is enough to show that every $\xi \in [a,b]$ has a neighborhood where $\tilde{f}$ is of bounded variation. By hypothesis, given $\epsilon > 0$, there exists a gauge δ on $[a,b]$ such that for every δ-fine semitagged division $d = (\xi_i, [t_{i-1}, t_i])$ of $[a,b]$,

$$\sum_{i=1}^{|d|} \left\| \tilde{f}(t_i) - \tilde{f}(t_{i-1}) - f(\xi_i)(t_i - t_{i-1}) \right\| < \epsilon. \tag{1.A.3}$$

Let $s_0 < s_1 < \cdots < s_m$ be any division of $[\xi - \delta(\xi), \xi + \delta(\xi)]$. If we take $\xi_j = \xi$ for $j = 1, 2, \ldots, m$, then the point-interval pair $(\xi_j, [s_{j-1}, s_j])$ is a δ-fine tagged partial division of $[\xi - \delta(\xi), \xi + \delta(\xi)]$ and, therefore, from (1.A.3) and since we can assume, without loss of generality (see comments in the paragraph before the statement), that $f(\xi_j) = f(\xi) = 0$ for $j = 1, 2, \ldots, m$, we have

$$\sum_{j=1}^{m} \left\| \tilde{f}(s_j) - \tilde{f}(s_{j-1}) \right\| < \epsilon$$

and the proof is complete. □

Lemma 1.98: *Suppose $f \in H([a,b],X)$. The following properties are equivalent:*

(i) f is absolutely integrable;
(ii) $\tilde{f} \in BV([a,b],X)$.

Proof. (i)$\Rightarrow$(ii). Suppose f is absolutely integrable. Since the variation of $\tilde{f}$, $\mathrm{var}_a^b(\tilde{f})$, is given by

$$\mathrm{var}_a^b(\tilde{f}) = \sup \left\{ \sum_{i=1}^{|d|} \left\| \tilde{f}(t_i) - \tilde{f}(t_{i-1}) \right\| \; : \; d = (t_i) \in D_{[a,b]} \right\}$$

we have

$$\sum_{i=1}^{|d|} \left\| \tilde{f}(t_i) - \tilde{f}(t_{i-1}) \right\| = \sum_{i=1}^{|d|} \left\| \int_{t_{i-1}}^{t_i} f(t)\, dt \right\| \leqslant \sum_{i=1}^{|d|} \int_{t_{i-1}}^{t_i} \|f(t)\|\, dt = \int_a^b \|f(t)\|\, dt.$$

(ii) $\Rightarrow$ (i). Suppose $\tilde{f} \in BV([a,b],X)$. We prove that the integral $\int_a^b \|f(t)\|\, dt$ exists and $\int_a^b \|f(t)\|\, dt = \mathrm{var}_a^b(\tilde{f})$. Given $\epsilon > 0$, we need to find a gauge δ on $[a,b]$ such that

$$\left| \sum_{i=1}^{|d|} \|f(\xi_i)\| (t_i - t_{i-1}) - \mathrm{var}_a^b(\tilde{f}) \right| < \epsilon,$$

whenever $d = (\xi_i, [t_{i-1}, t_i]) \in TD_{[a,b]}$ is δ-fine. However,

$$\begin{aligned} &\left| \sum_{i=1}^{|d|} \|f(\xi_i)\| (t_i - t_{i-1}) - \mathrm{var}_a^b(\tilde{f}) \right| \\ &\leqslant \sum_{i=1}^{|d|} \left| \|f(\xi_i)\| (t_i - t_{i-1}) - \left\| \int_{t_{i-1}}^{t_i} f(t)\, dt \right\| \right| + \left| \sum_{i=1}^{|d|} \left\| \int_{t_{i-1}}^{t_i} f(t)\, dt \right\| - \mathrm{var}_a^b(\tilde{f}) \right| \\ &\leqslant \sum_{i=1}^{|d|} \left\| f(\xi_i)(t_i - t_{i-1}) - \int_{t_{i-1}}^{t_i} f(t)\, dt \right\| + \left| \sum_{i=1}^{|d|} \left\| \tilde{f}(t_i) - \tilde{f}(t_{i-1}) \right\| - \mathrm{var}_a^b(\tilde{f}) \right|. \end{aligned} \tag{1.A.4}$$

By the definition of $\mathrm{var}_a^b(\tilde{f})$, we may take $(t_i) \in D_{[a,b]}$ such that the last summand in (1.A.4) is smaller than $\frac{\epsilon}{2}$. Because $f \in H([a,b],X)$, we may take a gauge δ such that for every δ-fine $(\xi_i, [t_{i-1}, t_i]) \in TD_{[a,b]}$, the first summand in (1.A.4) is also smaller than $\frac{\epsilon}{2}$ (and we may suppose that the points chosen for the second summand are the points of the δ-fine tagged division $(\xi_i, [t_{i-1}, t_i])$). □

The next result is a consequence of the fact that $HMS([a,b],X) \subset H([a,b],X)$ and Lemmas 1.97 and 1.98. A proof of it can be found in [[132], Theorem 10.3].

Corollary 1.99: *All functions of $HMS([a,b],X)$ are absolutely.integrable*

The reader can find a proof of the next lemma in [35, Theorem 9].

Lemma 1.100: *All functions of $H([a,b],X)$ are.measurable*

Finally, we can prove the following inclusion.

Theorem 1.101: $HMS([a,b],X) \subset \mathcal{L}_1([a,b],X)$.

Proof. The result follows from the facts that all functions of $H([a,b],X)$ and, hence, of $HMS([a,b],X)$ are measurable (Lemma 1.100), and all functions of $HMS([a,b],X)$ are absolutely integrable by Corollary 1.99. □

As we mentioned before, the inclusion $\mathcal{L}_1([a,b],X) \subset KMS([a,b],X)$ always holds. However, when X is an infinite dimensional Banach space, then for sure $KMS([a,b],X) \setminus \mathcal{L}_1([a,b],X) \neq \emptyset$, as shown by the next result due to C. S. Hönig (personal communication by him to his students in 1990 at the University of São Paulo) and presented in [73].

Proposition 1.102 (Hönig): *If X is an infinite dimensional Banach space, then there exists $f \in KMS([a,b],X) \setminus \mathcal{L}_1([a,b],X)$.*

Proof. Let $\dim X$ denote the dimension of X. If $\dim X = \infty$, then the Theorem of Dvoretsky–Rogers (see [60] and also [57]) implies there exists a sequence $\{x_n\}_{n\in\mathbb{N}}$ in X which is summable but not absolutely summable. Thus, if we define a function $f : [1,\infty] \to X$ by $f(t) = x_n$, for $n \leqslant t < n+1$, then $(KMS) \int_a^b f(t)\,dt = \sum_{n\in\mathbb{N}} x_n \in X$, whenever the integral exists. However, $f \notin \mathcal{L}_1([a,b],X)$, since

$$\int_a^b \|f(t)\|\,dt = \|x_1\| + \|x_2\| + \|x_3\| + \cdots = \infty.$$

and this completes the proof. □

In the next example, borrowed from [73, Example 3.4], we exhibit a Banach space-valued function which is integrable in the variational Henstock sense and also in the sense of Kurzweil–McShane. Nevertheless, it is not absolutely integrable.

Example 1.103: Let $f : [0,1] \to l_2(\mathbb{N})$ be given by

$$f(t) = \frac{2^i}{i} e_i, \text{ for } \frac{1}{2^i} \leqslant t < \frac{1}{2^{i-1}} \text{ and } i \in \mathbb{N}.$$

Then,

$$\int_{\frac{1}{2^i}}^{\frac{1}{2^{i-1}}} \frac{2^i}{i} e_i\,dt = \frac{1}{i} e_i$$

which is summable in $l_2(\mathbb{N})$. Since the Henstock integral contains its improper integrals (and the same applies to the Kurzweil integral), we have $f \in H([0,1], l_2(\mathbb{N}))$. However, $f \notin \mathcal{L}_1([0,1], l_2(\mathbb{N}))$ because the sequence $\left\{\frac{1}{i} e_i\right\}_{i\in\mathbb{N}}$ is non-summable in $\mathcal{L}_1([0,1], l_2(\mathbb{N}))$. By the Monotone Convergence Theorem for the Kurzweil–McShane integral (which follows the ideas of [71] with obvious adaptations), $f \in KMS([0,1], l_2(\mathbb{N}))$. But $f \notin RMS([0,1], l_2(\mathbb{N}))$, since f is not bounded, where by $RMS([a,b],X)$ we denote the space of Riemann–McShane integrable functions from $[a,b]$ to X.

2

The Kurzweil Integral

Everaldo M. Bonotto[1], *Rodolfo Collegari*[2], *Márcia Federson*[3], *and Jaqueline G. Mesquita*[4]

[1]*Departamento de Matemática Aplicada e Estatística, Instituto de Ciências Matemáticas e de Computação (ICMC), Universidade de São Paulo, São Carlos, SP, Brazil*
[2]*Faculdade de Matemática, Universidade Federal de Uberlândia, Uberlândia, MG, Brazil*
[3]*Departamento de Matemática, Instituto de Ciências Matemáticas e de Computação (ICMC), Universidade de São Paulo, São Carlos, SP, Brazil*
[4]*Departamento de Matemática, Instituto de Ciências Exatas, Universidade de Brasília, Brasília, DF, Brazil*

This chapter is devoted to the theory of integration introduced by Jaroslav Kurzweil in the form presented in his articles dated 1957, 1958, 1959, and 1962, and so on. See [147–151]. Although very short, this chapter brings the heart of the theory of generalized ordinary differential equations (ODEs) which is precisely the Kurzweil integration theory, presented here in a concise form which includes its most fundamental properties – those usually expected that a "*good integral*" will fulfill.

As pointed out by Š. Schwabik in [209] (see also [136]), it is known that in a paper published in the early 1950s, I.I. Gichman observed that the nonperiodic averaging method for ODEs proposed by Bogolyubov through certain approximations lies heavily on the continuous dependence of the solutions on a parameter (see [103] and also [144], for instance). In 1955, M.A. Krasnosel'skiĭ and S.G. Krein (see [143]), also while investigating averaging methods for ODEs, established a result on the "*continuity*" of integrals whose integrands $f_k(x, t)$, $k \in \mathbb{N}_0$, are also right-hand sides of certain nonautonomous ODEs. Then, Krasnosel'skiĭ and Krein required that these integrands are equicontinuous and uniformly bounded in order to obtain continuous dependence on a parameter.

In 1957, two papers by J. Kurzweil appeared concomitantly referring to one another: one is a solo article in English and Russian, and the other is coauthored by Z. Vorel and appears only in Russian (see [147, 154]). In [154], Kurzweil and Vorel presented a more general form of the following result:

Generalized Ordinary Differential Equations in Abstract Spaces and Applications, First Edition.
Edited by Everaldo M. Bonotto, Márcia Federson, and Jaqueline G. Mesquita.

Let $f_k : \mathcal{O} \times [0, T] \to \mathbb{R}^n, k \in \mathbb{N}_0$, be a sequence of functions and $\mathcal{O}$ be an open subset of the real line $\mathbb{R}$. Let x_k be a solution of the differential equation

$$\dot{x} = f_k(x, t), \quad x(0) = 0$$

and let x_0 be uniquely defined on $[0, T]$, $T > 0$. If

$$F_k(x, t) = \int_0^t f_k(x, s)ds \to \int_0^t f_0(x, s)ds = F_0(x, t)$$

uniformly with $k \to \infty$ and if the functions $f_k(x, t)$, $k \in \mathbb{N}_0$, are equicontinuous in x for fixed t, then for sufficiently large k the solutions x_k are defined on $[0, T]$ and $x_k(t) \to x_0(t)$, with $k \to \infty$ uniformly on $[0, T]$.

Then, in [147], referring to the above result, Kurzweil mentioned that, unlike that stated by Krasnosel'skiĭ and Krein, the authors of [154] did not require uniform boundedness on the integrands $f_k(x, t)$, $k \in \mathbb{N}_0$.

Still concerning the continuous dependence result of [154], it is within the best possible results according to Z. Artstein in [8] and as quoted in [136]. On the other hand, it thrust a question on how to relate the solutions of ODEs with right-hand sides $f_k(x, t)$, $k \in \mathbb{N}_0$ ($\mathbb{N}_0 = \mathbb{N} \cup \{0\}$), to the indefinite integrals

$$F_k(x, t) = \int_0^t f_k(x, s)ds, \quad k \in \mathbb{N}_0.$$

This issue was solved by Kurzweil in [147] when he introduced the concept of generalized ODEs whose framework is the scope of the book. In this chapter, as we said at the beginning, we focus our attention on the theory of integration discovered in 1957 in the paper [147], where Kurzweil introduced the concept of what he called generalized Perron integral, which we call Kurzweil integral, which appears on the right-hand side of generalized ODEs.

Other references for the Kurzweil integration theory are [153, 179, 209].

2.1 The Main Background

2.1.1 Definition and Compatibility

With the terminologies and notations of Chapter 1, we present the definition of a Kurzweil integrable function U defined on a square $[a, b] \times [a, b]$, with $-\infty < a \leqslant b < \infty$, taking values in an arbitrary Banach space X, endowed with a norm denoted by $\|\cdot\|$.

Definition 2.1: Let X be a Banach space. A function $U : [a, b] \times [a, b] \to X$ is called *Kurzweil integrable* on $[a, b]$ if there is an element $I \in X$ such that for each

$\epsilon > 0$, there is a gauge δ on $[a,b]$ so that

$$\left\| \sum_{i=1}^{|d|} [U(\tau_i, t_i) - U(\tau_i, t_{i-1})] - I \right\| < \epsilon$$

for every δ-fine tagged division $d = (\tau_i, [t_{i-1}, t_i])$, $i = 1, 2, \ldots, |d|$, of $[a,b]$. In this case, I is called the *Kurzweil integral* of U over $[a,b]$ and will be denoted by $\int_a^b DU(\tau, t)$.

Sometimes we refer to the Riemann-type sum $\sum_{i=1}^{|d|} [U(\tau_i, t_i) - U(\tau_i, t_{i-1})]$ simply by $S(U, d)$. When the Kurzweil integral $\int_a^b DU(\tau, t)$ exists, we define $\int_b^a DU(\tau, t) = -\int_a^b DU(\tau, t)$ and $\int_a^a DU(\tau, t) = 0$.

At this point, the reader may have paid attention to the fact that Definition 2.1 only makes sense if, for a given gauge δ defined on $[a,b]$, there exists at least one δ-fine tagged division d of $[a,b]$. This is assured by a result, stated below, known as Cousin Lemma (or Compatibility Theorem) which can be proved by application of the Heine–Borel Theorem. See, e.g. [172, Section S1.8].

Lemma 2.2 (Cousin Lemma): *Given a gauge δ on $[a,b]$, there is a δ-fine tagged division $d = (\tau_i, [t_{i-1}, t_i])$, $i = 1, 2, \ldots, |d|$, of $[a,b]$.*

In order to extend the Kurzweil integral to unbounded intervals of $\mathbb{R}$, we need to define δ-neighborhoods of $-\infty$ and ∞. Consider the extended real line $\overline{\mathbb{R}} = \mathbb{R} \cup \{-\infty\} \cup \{\infty\}$ with the operations: $0 \cdot (\pm\infty) = 0 = (\pm\infty) \cdot 0$, $x + (\pm\infty) = \pm\infty = (\pm\infty) + x$ for $x \in \mathbb{R}$, $x \cdot (\pm\infty) = \pm\infty = (\pm\infty) \cdot x$ for $x > 0$, and $x \cdot (\pm\infty) = \mp\infty = (\pm\infty) \cdot x$ for $x < 0$. Take $\delta(-\infty) > 0$ and $\delta(\infty) > 0$ and consider the neighborhoods

$$\left[-\infty, -\frac{1}{\delta(-\infty)}\right) \quad \text{and} \quad \left(\frac{1}{\delta(\infty)}, \infty\right].$$

Following the ideas of [12], for instance, consider $U : \overline{\mathbb{R}} \times \overline{\mathbb{R}} \to X$ and set $U(-\infty, -\infty) = U(-\infty, t) = U(\infty, \infty) = U(\infty, t) = 0$ for all $t \in \mathbb{R}$. Then, given a gauge δ on $\overline{\mathbb{R}}$, that is, a function $\delta : \overline{\mathbb{R}} \to (0, \infty)$, consider a δ-fine tagged division of $\overline{\mathbb{R}}$, say,

$$d = \{(\tau_0, [-\infty, t_0]), (\tau_1, [t_0, t_1]), \ldots, (\tau_{|d|}, [t_{|d|-1} t_{|d|}]), (\tau_{|d|+1}, [t_{|d|}, \infty])\}, \tag{2.1}$$

with

$$\tau_i \in [t_{i-1}, t_i] \subset (\tau_i - \delta(\tau_i), \tau_i + \delta(\tau_i)), \quad i = 1, \ldots, n,$$

$$[-\infty, t_0] \subset \left[-\infty, -\frac{1}{\delta(-\infty)}\right) \quad \text{and} \quad [t_{|d|-1}, \infty] \subset \left(\frac{1}{\delta(\infty)}, \infty\right].$$

This definition of δ-fineness forces the tags τ_0 and $\tau_{|d|+1}$ to satisfy $\tau_0 = -\infty$ and $\tau_{|d|+1} = \infty$. Consequently, the corresponding Riemann-type sum becomes

$$S(U,d) = \sum_{i=1}^{|d|} [U(\tau_i, t_i) - U(\tau_i, t_{i-1})],$$

since $U(-\infty, t_0) - U(-\infty, -\infty) = 0$ and $U(\infty, \infty) - U(\infty, t_{|d|}) = 0$. In this manner, the results involving the Kurzweil integral of a function $U : [a,b] \times [a,b] \to X$ can be extended or easily adapted to the case where $U : J \times J \to X$, with J being any unbounded interval of the real line $\mathbb{R}$ (i.e. $J = (-\infty, \infty)$, $J = [c, \infty)$ or $J = (-\infty, c]$).

2.1.2 Special Integrals

In the following lines, we specify some particular cases of the Kurzweil integral.

In Definition 2.1, when we consider a BT $(L(X,Y), X, Y)$, where X and Y are Banach spaces, $\alpha : [a,b] \to L(X,Y)$, $f : [a,b] \to X$, and $U : [a,b] \times [a,b] \to Y$ are functions such that $U(\tau, t) = \alpha(\tau)f(t)$, we obtain the Kurzweil vector integral as presented in Definition 1.37, that is,

$$\int_a^b DU(\tau, t) = \int_a^b \alpha(s) df(s).$$

Similarly, when $U(\tau, t) = \alpha(t)f(\tau)$, we obtain the integral as presented in Definition 1.38, that is,

$$\int_a^b DU(\tau, t) = \int_a^b d\alpha(s) f(s).$$

In particular, when $f : [a,b] \to \mathbb{R}^n$, $g : [a,b] \to \mathbb{R}$, and $U(\tau, t) = f(\tau)g(t)$, we have

$$\int_a^b DU(\tau, t) = \int_a^b f(s) dg(s)$$

and, here, $f \in K_g([a,b], \mathbb{R}^n) = H_g([a,b], \mathbb{R}^n)$ (see comments before Example 1.44). When $f : [a,b] \to \mathbb{R}^n$ and $U(\tau, t) = f(\tau)t$, we have

$$\int_a^b DU(\tau, t) = \int_a^b f(s) ds$$

and, in this case, $f \in K([a,b], \mathbb{R}^n) = H([a,b], \mathbb{R}^n)$.

We refer to the integrals

$$\int_a^b f(s) ds \quad \text{and} \quad \int_a^b f(s) dg(s),$$

respectively, as Perron and Perron–Stieltjes integrals. See Remark 1.39. We recall that, in the finite dimensional case, the Perron–Stieltjes and the Kurzweil–Henstock–Stieltjes integrals coincide. See for instance [120, 158].

We end this section with an observation concerning other possibilities of more general integrals as, for instance, path integrals. Let $[a, b]$ be a compact interval of $\mathbb{R}$ and $\mathbb{R}^{[a,b]}$ be the set of all functions from $[a, b]$ to $\mathbb{R}$. Consider a function f defined for values $\tau \in \mathbb{R}^{[a,b]}$, that is, $f : \mathbb{R}^{[a,b]} \to \mathbb{R}$. Let $I(\mathbb{R}^{[a,b]})$ denote the set of all cylindrical intervals of $\mathbb{R}^{[a,b]}$ (see, e.g. [183]) and consider a function μ of cylindrical intervals J of $\mathbb{R}^{[a,b]}$, that is, $\mu : I(\mathbb{R}^{[a,b]}) \to \mathbb{R}$. Then, consider Riemann-type sums of the form

$$\sum f(\tau)\mu(J), \quad \tau \in J,$$

over τ-dependent divisions of the function space $\mathbb{R}^{[a,b]}$. Sometimes, these sums approximate a path-type integral denoted by

$$\int f(\tau)\mu(I).$$

More generally, one can consider a function h of point-interval pairs (τ, J) and, in particular, such function h can be given as

$$h(\tau, J) = f(\tau)\mu(J).$$

An integral of type $\int h(\tau, J)$ is called Henstock (path) integral. In this respect, the reader may want to consult [121, 122, 181–183].

Coming to Definition 2.1, it is possible to consider $U : [a, b] \times [a, b] \to X$ as a particular case of a function h of point-interval pairs (τ, J), where $\tau \in [a, b]$ and J are subintervals of $[a, b]$. Take, for instance $J = [c, d] \subset [a, b]$ and

$$h(\tau, J) = U(\tau, d) - U(\tau, c).$$

Then, the Riemann-type sum

$$\sum_{i=1}^{|d|} h(\tau_i, J_i) = \sum_{i=1}^{|d|} [U(\tau_i, t_i) - U(\tau_i, t_{i-1})],$$

where $J_i = [t_{i-1}, t_i]$ and $d = (\tau_i, [t_{i-1}, t_i]) \in TD_{[a,b]}$, may approximate an integral of type

$$\int_{[a,b]} h(\tau, J) = \int_a^b DU(\tau, t),$$

whenever the latter exists.

2.2 Basic Properties

Similar to the Riemann and Lebesgue integrals, the Kurzweil integral presented in Definition 2.1 has the usual properties of uniqueness, linearity, and additivity with respect to adjacent intervals, integrability on subintervals, among other properties. Here, we state those results we need, some of which are without proofs. Those proofs that we omit can be easily adapted from [209] to the case where the integrands take values in a general Banach space.

Our first result concerns the linearity of the Kurzweil integral whose proof can be made following [209, Theorem 1.9].

Theorem 2.3 (Linearity): *If $U, V : [a,b] \times [a,b] \to X$ are Kurzweil integrable functions and $c_1, c_2 \in \mathbb{R}$, then $c_1 U + c_2 V$ is also Kurzweil integrable and*

$$\int_a^b D\left[c_1 U(\tau,t) + c_2 V(\tau,t)\right] = c_1 \int_a^b DU(\tau,t) + c_2 \int_a^b DV(\tau,t).$$

Proof. Let $c_1, c_2 \in \mathbb{R}$ be constants and d be an arbitrary tagged division of $[a,b]$ given by $d = (\tau_j, [t_{j-1}, t_j])$, $j = 1, 2, \ldots, |d|$, and let $S(U,d)$ and $S(V,d)$ be the Riemann sums corresponding to the functions $U, V : [a,b] \times [a,b] \to X$, respectively. Then,

$$\begin{aligned}
S(c_1 U + c_2 V, d) &= \sum_{j=1}^{|d|} [(c_1 U + c_2 V)(\tau_j, t_j) - (c_1 U + c_2 V)(\tau_j, t_{j-1})] \\
&= \sum_{j=1}^{|d|} [(c_1 U)(\tau_j, t_j) + (c_2 V)(\tau_j, t_j) - (c_1 U)(\tau_j, t_{j-1}) - (c_2 V)(\tau_j, t_{j-1})] \\
&= \sum_{j=1}^{|d|} [(c_1 U)(\tau_j, t_j) - (c_1 U)(\tau_j, t_{j-1})] + \sum_{j=1}^{|d|} [(c_2 V)(\tau_j, t_j) - (c_2 V)(\tau_j, t_{j-1})] \\
&= c_1 S(U,d) + c_2 S(V,d),
\end{aligned}$$

which completes the proof after appropriate choice of gauges. □

The next lemma is known as Cauchy Criterion for the Kurzweil integral. Its proof follows as in [209, Theorem 1.7], and we omit it here.

Theorem 2.4 (Cauchy Criterion): *A function $U : [a,b] \times [a,b] \to X$ is Kurzweil integrable over $[a,b]$, if and only if for every $\epsilon > 0$, there exists a gauge δ on $[a,b]$ such that*

$$\| S(U, d_1) - S(U, d_2) \| < \epsilon$$

for every δ-fine tagged divisions d_1, d_2 of $[a,b]$.

The next result concerns integrability on subintervals of $[a,b]$. It generalizes [209, Theorem 1.10] for the case of Banach space-valued integrands. The proof follows similarly.

Theorem 2.5 (Integrability on Subintervals): *Suppose $U : [a,b] \times [a,b] \to X$ is Kurzweil integrable over $[a,b]$. Then, given $[c,d] \subset [a,b]$, U is also Kurzweil integrable over $[c,d]$.*

Proof. Let $\epsilon > 0$ be given. By the Cauchy Criterion (Theorem 2.4), since the integral $\int_a^b DU(\tau,t)$ exists, there is a gauge δ on $[a,b]$ such that

$$\| S(U,d_1) - S(U,d_2) \| < \epsilon \tag{2.2}$$

for every δ-fine tagged divisions d_1 and d_2 of $[a,b]$.

Let $\widetilde{d_1}$ and $\widetilde{d_2}$ be arbitrary δ-fine tagged divisions of $[c,d]$ and assume that $a < c < d < b$. Consider, also, δ-fine tagged divisions d_L of $[a,c]$ and d_R of $[d,b]$ which are guaranteed by the Cousin Lemma (Lemma 2.2). Consider

$$\widetilde{d_1} = (\tau_i, [t_{i-1}, t_i]), \quad i = 1, 2, \dots, |\widetilde{d_1}|,$$

$$d_L = \left(\tau_j^L, \left[t_{j-1}^L, t_j^L\right]\right), \quad j = 1, 2, \dots, |d_L|,$$

$$d_R = \left(\tau_k^R, \left[t_{k-1}^R, t_k^R\right]\right), \quad k = 1, 2, \dots, |d_R|,$$

and define a δ-fine tagged division d_1 of $[a,b]$ from the union of the δ-fine tagged divisions d_L, $\widetilde{d_1}$, and d_R which implies that d_1 is a δ-fine tagged division of $[a,b]$.

Analogously, the union of the δ-fine tagged divisions d_L, $\widetilde{d_2}$, and d_R generate a δ-fine tagged division of $[a,b]$ which we will denote by d_2. Therefore, by (2.2)

$$\| S(U,\widetilde{d_1}) - S(U,\widetilde{d_2}) \| = \| S(U,d_1) - S(U,d_2) \| < \epsilon,$$

which yields, by Cauchy Criterion, that the integral $\int_c^d DU(\tau,t)$ exists. □

The additivity of the integral with respect to subjacent intervals is described in the next result. A proof of it can be carried out as in [209, Theorem 1.11] with obvious adaptations.

Theorem 2.6 (Additivity on Adjacent Intervals): *Let $c \in (a,b)$ and $U : [a,b] \times [a,b] \to X$ be a function such that $U \in K([a,c],X)$ and $U \in K([c,b],X)$. Then $U \in K([a,b],X)$ and*

$$\int_a^b DU(\tau,t) = \int_a^c DU(\tau,t) + \int_c^b DU(\tau,t).$$

The next result is known as Saks–Henstock Lemma. See [209, Lemma 1.13], for instance.

Lemma 2.7 (Saks–Henstock): *Let $U : [a,b] \times [a,b] \to X$ be Kurzweil integrable over $[a,b]$. Given $\epsilon > 0$, let δ be a gauge on $[a,b]$ such that*

$$\left\| \sum_{i=1}^{|d|} [U(\tau_i, t_i) - U(\tau_i, t_{i-1})] - \int_a^b DU(\tau, t) \right\| < \epsilon \tag{2.3}$$

for every δ-fine tagged division $d = (\tau_i, [t_{i-1}, t_i])$, $i = 1, 2, \ldots, |d|$, of $[a, b]$. If

$$a \leqslant \beta_1 \leqslant \xi_1 \leqslant \gamma_1 \leqslant \beta_2 \leqslant \xi_2 \leqslant \gamma_2 \leqslant \cdots \leqslant \beta_m \leqslant \xi_m \leqslant \gamma_m \leqslant b$$

represents a δ-fine tagged partial division $\tilde{d} = (\xi_j, [\beta_j, \gamma_j])$ of $[a, b]$, with $j = 1, 2, \ldots, m$, then

$$\left\| \sum_{j=1}^{m} \left[U(\xi_j, \beta_j) - U(\xi_j, \gamma_j) - \int_{\beta_j}^{\gamma_j} DU(\tau, t) \right] \right\| \leqslant \epsilon. \tag{2.4}$$

Proof. By hypothesis, $\beta_j \leqslant \gamma_j$ for $j = 1, 2, \ldots, m$. We can suppose, without loss of generality, that $\beta_j < \gamma_j$ for every $j = 1, 2, \ldots, m$, and set $\gamma_0 = a$ and $\beta_{m+1} = b$.

If $\gamma_j < \beta_{j+1}$ for some $j = 0, 1, \ldots, m$, then the existence of the integral $\int_a^b DU(\tau, t)$ together with Theorem 2.5 imply the existence of the integral $\int_{\gamma_j}^{\beta_{j+1}} DU(\tau, t)$. Thus, given $\eta > 0$, there exists a gauge δ_j on $[\gamma_j, \beta_{j+1}]$ such that

$$\left\| S(U, d^j) - \int_{\gamma_j}^{\beta_{j+1}} DU(\tau, t) \right\| < \frac{\eta}{m+1}$$

for every δ_j-fine tagged division d^j of $[\gamma_j, \beta_{j+1}]$. Note that we can also take δ_j as a refinement of δ, that is, δ_j satisfies $\delta_j(\tau) < \delta(\tau)$ for every $\tau \in [\gamma_j, \beta_{j+1}]$.

On the other hand, if $\gamma_j = \beta_{j+1}$, then $S(U, d^j) = 0$.

Since the union of $\tilde{d}$ and all d^j, $j = 1, 2, \ldots, m$, forms a δ-fine tagged division d of $[a, b]$ whose corresponding Riemann sum is given by

$$\sum_{j=1}^{m} [U(\xi_j, \gamma_j) - U(\xi_j, \beta_j)] + \sum_{j=1}^{m} S(U, d^j),$$

then, by (2.3), we obtain

$$\left\| \sum_{j=1}^{m} [U(\xi_j, \gamma_j) - U(\xi_j, \beta_j)] + \sum_{j=1}^{m} S(U, d^j) - \int_a^b DU(\tau, t) \right\| < \epsilon.$$

Note that $\int_a^b DU(\tau, t) = \sum_{j=1}^m \int_{\beta_j}^{\gamma_j} DU(\tau, t) + \sum_{j=0}^m \int_{\gamma_j}^{\beta_{j+1}} DU(\tau, t)$. Hence,

$$\begin{aligned} &\left\| \sum_{j=1}^{m} [U(\xi_j, \beta_j) - U(\xi_j, \gamma_j) - \int_{\beta_j}^{\gamma_j} DU(\tau, t)] \right\| \\ &= \left\| \sum_{j=1}^{m} [U(\xi_j, \beta_j) - U(\xi_j, \gamma_j)] + \sum_{j=0}^{m} \int_{\gamma_j}^{\beta_{j+1}} DU(\tau, t) - \int_a^b DU(\tau, t) \right\| \end{aligned}$$

$$\leqslant \left\| \sum_{j=1}^{m} [U(\xi_j, \beta_j) - U(\xi_j, \gamma_j)] - \int_a^b DU(\tau, t) + \sum_{j=0}^{m} S(U, d^j) \right\|$$
$$+ \left\| \sum_{j=0}^{m} S(U, d^j) - \sum_{j=0}^{m} \int_{\gamma_j}^{\beta_{j+1}} DU(\tau, t) \right\|$$
$$\leqslant \left\| \sum_{j=1}^{m} [U(\xi_j, \beta_j) - U(\xi_j, \gamma_j)] - \int_a^b DU(\tau, t) + \sum_{j=0}^{m} S(U, d^j) \right\|$$
$$+ \sum_{j=0}^{m} \left\| S(U, d^j) - \int_{\gamma_j}^{\beta_{j+1}} DU(\tau, t) \right\| < \epsilon + (m+1)\frac{\eta}{m+1} = \epsilon + \eta,$$

which holds for every $\eta > 0$. Therefore, (2.4) is satisfied, and we complete the proof. □

As a consequence of the Saks–Henstock Lemma, we have the next result.

Corollary 2.8: *Let $U : [a, b] \times [a, b] \to X$ be Kurzweil integrable over $[a, b]$. Given $\epsilon > 0$, there exists a gauge δ on $[a, b]$ such that, if $[\gamma, v] \subset [a, b]$, then*

(i) $(v - \gamma) < \delta(\gamma)$ *implies* $\left\| U(\gamma, v) - U(\gamma, \gamma) - \int_\gamma^v DU(\tau, t) \right\| < \epsilon,$

(ii) $(v - \gamma) < \delta(v)$ *implies* $\left\| U(v, v) - U(v, \gamma) - \int_\gamma^v DU(\tau, t) \right\| < \epsilon.$

Proof. Let $\epsilon > 0$ be given. Since $U : [a, b] \times [a, b] \to X$ is Kurzweil integrable over $[a, b]$, there exists a gauge δ on $[a, b]$ such that (2.3) holds for every δ-fine tagged division $d = (\tau_i, [t_{i-1}, t_i])$, $i = 1, 2, \dots, |d|$, of $[a, b]$. For item (i), let $[\gamma, v] \subset [a, b]$ be such that $(v - \gamma) < \delta(\gamma)$. Then $(\gamma, [\gamma, v])$ is a δ-fine tagged partial division of $[a, b]$. For item (ii), note that if $[\gamma, v] \subset [a, b]$ and $(v - \gamma) < \delta(v)$, then $(v, [\gamma, v])$ is also a δ-fine tagged partial division of $[a, b]$. Then, the result follows easily from the Saks–Henstock Lemma. □

The next theorem concerns the Cauchy extension for the Kurzweil integral. Its statement says that the Kurzweil integral is invariant under Cauchy extensions. The theorem is also known as being a Hake-type theorem for the Kurzweil integral. A proof of it follows as in [209, Theorem 1.14] with obvious changes for the case of Banach space-valued functions. The reader may want to consult the results from [213] as well.

Theorem 2.9 (Cauchy Extension): *If $U : [a,b] \times [a,b] \to X$ is a function such that for every $c \in [a,b)$, U is integrable on $[a,c]$ and the limit*

$$\lim_{c \to b^-} \left[\int_a^c DU(\tau,t) - U(b,c) + U(b,b) \right] = I \in X$$

exists, then the function U is integrable on $[a,b]$ and it satisfies $\int_a^b DU(\tau,t) = I$. Similarly, if the function U is integrable on $[c,b]$ for every $c \in (a,b]$ and the limit

$$\lim_{c \to a^+} \left[\int_c^b DU(\tau,t) + U(a,c) - U(a,a) \right] = I \in X$$

exists, then the function U is integrable on $[a,b]$, and we have $\int_a^b DU(\tau,t) = I$.

In particular, we have the following Hake-type theorem for Perron–Stieltjes integrals.

Corollary 2.10: *Consider a pair of functions $f : [a,b] \to X$ and $g : [a,b] \to \mathbb{R}$.*

(i) *Suppose the Perron–Stieltjes integral $\int_a^s f(t)dg(t)$ exists, for every $s \in [a,b)$, and*

$$\lim_{s \to b^-} \left[\int_a^s f(t)dg(t) + f(b)[g(b) - g(s)] \right] = I.$$

Then $\int_a^b f(t)dg(t) = I$.

(ii) *Suppose the Perron–Stieltjes integral $\int_s^b f(t)dg(t)$ exists, for every $s \in (a,b]$, and*

$$\lim_{s \to a^+} \left[\int_s^b f(t)dg(t) + f(a)(g(s) - g(a)) \right] = I.$$

Then $\int_a^b f(t)dg(t) = I$.

Proof. We prove item (i) and leave item (ii) to the reader, since it follows similarly.

Consider a function $U : [a,b] \times [a,b] \to X$ defined by $U(\tau,t) = f(\tau)g(t)$. Then U is Kurzweil integrable over $[a,s)$ for all $s \in [a,b)$, since, by hypothesis, the Perron–Stieltjes integral $\int_a^s f(t)dg(t)$ exists for every $s \in [a,b)$. Then, by the Hake-type Theorem for the Kurzweil integral (Theorem 2.9), U is Kurzweil integrable over $[a,b]$. Then, clearly, we obtain

$$\begin{aligned}
&\lim_{s \to b^-} \int_a^s DU(\tau,s) - U(b,s) + U(b,b) \\
&= \lim_{s \to b^-} \int_a^s f(t)dg(t) + f(b)[g(b) - g(s)] = I.
\end{aligned}$$

Hence, by Theorem 2.9, $\int_a^b f(t)dg(t) = \int_a^b DU(\tau,t) = I$. □

The next example, borrowed from [44], illustrates the use of the Hake Theorem for Perron integrals (see Corollary 2.9).

Example 2.11: Let $f : [0,1] \to \mathbb{R}$ be a function given by $f(t) = 2t \sin\left(\frac{2}{t^2}\right) - \frac{2}{t}\cos\left(\frac{2}{t^2}\right)$, if $t \in (0,1]$, and $f(0) = 0$. Note that f is a highly oscillating function which is not absolutely integrable over $[0,1]$, that is, f is a function of unbounded variation and, hence, it is not Lebesgue integrable over $[0,1]$. Since the improper Riemann integral of f exists, using a Hake-type theorem for the Perron integral (which coincides with the Kurzweil–Henstock integral), it follows that the Perron integral of f also exists and it has the same value as the improper Riemann integral of f.

If $U : [a,b] \times [a,b] \to X$ is a Kurzweil integrable function, its indefinite integral $\int_a^s DU(\tau,t)$, $s \in [a,b]$, is not always continuous. The next result describes such a behavior. In particular, it implies that the indefinite integral of U is continuous at $c \in [a,b]$, if and only if $U(c,\cdot) : [a,b] \to X$ is continuous at c. Its proof, which we include here, follows as in [209, Theorem 1.16].

Theorem 2.12: *Let $U : [a,b] \times [a,b] \to X$ be Kurzweil integrable over $[a,b]$ and $c \in [a,b]$. Then*

$$\lim_{s \to c} \left[\int_a^s DU(\tau,t) - U(c,s) + U(c,c) \right] = \int_a^c DU(\tau,t).$$

Proof. Let $\epsilon > 0$ be given. Since U is integrable on $[a,b]$, there exists a gauge δ on $[a,b]$ such that the inequality

$$\|S(U,d) - I\| = \left\| \sum_{j=1}^{|d|} [U(\tau_j,t_j) - U(\tau_j,t_{j-1})] - I \right\| < \epsilon$$

holds for every δ-fine division $d = (\tau_j, [t_{j-1},t_j]), j = 1,2,\dots,|d|$, of $[a,b]$.

Now, take an arbitrary $c \in (a,b]$. If $s \in (c-\delta(c), c+\delta(c)) \cap [a,b]$, then by the Corollary 2.8, we have

$$\left\| U(c,s) - U(c,c) - \int_c^s DU(\tau,t) \right\| < \epsilon.$$

Therefore,

$$\left\| \int_a^s DU(\tau,t) - U(c,s) + U(c,c) - \int_a^c DU(\tau,t) \right\|$$
$$= \left\| \int_c^s DU(\tau,t) - U(c,s) + U(c,c) \right\| < \epsilon,$$

and the result follows. □

Remark 2.13: Analogously as in Theorem 2.12, one can prove that, for $c \in [a, b]$,

$$\lim_{s \to c} \left[\int_s^b DU(\tau, t) + U(c, s) - U(c, c) \right] = \int_c^b DU(\tau, t).$$

The next result, borrowed from [179, Corollary 6.5.5], will be essential to prove some correspondence between equations presented in Chapters 3 and 4.

Corollary 2.14: *Let $f : [a, b] \to X$ and $g : [a, b] \to \mathbb{R}$ be functions such that g is regulated and the Perron–Stieltjes integral $\int_a^b f(t)dg(t)$ exists. Then the functions*

$$h(t) = \int_a^t f(s)dg(s) \quad \text{and} \quad k(t) = \int_t^b f(s)dg(s)$$

are regulated on $[a, b]$ and satisfy

$$h(t^+) = h(t) + f(t)\Delta^+ g(t) \quad \text{and} \quad k(t^+) = k(t) - f(t)\Delta^+ g(t), \qquad t \in [a, b),$$
$$h(t^-) = h(t) - f(t)\Delta^- g(t) \quad \text{and} \quad k(t^-) = k(t) + f(t)\Delta^- g(t), \qquad t \in (a, b],$$

where

$$\Delta^+ g(t) = g(t^+) - g(t) \quad \text{and} \quad \Delta^- g(t) = g(t) - g(t^-).$$

Proof. We prove the first equality. The others will follow analogously.

Consider a function $U : [a, b] \times [a, b] \to X$ defined by $U(\tau, t) = f(\tau)g(t)$. Then $\int_a^b DU(\tau, t) = \int_a^b f(t)dg(t)$ which exists by hypothesis. Hence, U is Kurzweil integrable over $[a, b]$. Thus, by Theorem 2.12, we have

$$\lim_{s \to t^+} \left[\int_a^s DU(\tau, u) - U(t, s) + U(t, t) \right] = \int_a^t DU(\tau, u),$$

for $t \in [a, b)$, which implies

$$\lim_{s \to t^+} \left[\int_a^s f(u)dg(u) - f(t)g(s) + f(t)g(t) \right] = \int_a^t f(u)dg(u),$$

hence,

$$\lim_{s \to t^+} \left[\int_a^s f(u)dg(u) \right] = \int_a^t f(u)dg(u) + f(t)[g(t^+) - g(t)].$$

Therefore, $h(t^+) = h(t) + f(t)\Delta^+ g(t)$ and the proof is complete. □

The next theorem is a generalization of [209, Theorem 1.35] for the case of Banach space-valued integrands. The proof follows straightforward as in [209].

Theorem 2.15: *Let $U : [a,b] \times [a,b] \to X$ be Kurzweil integrable over $[a,b]$. If $V : [a,b] \times [a,b] \to \mathbb{R}$ is Kurzweil integrable and if there exists a gauge function θ on $[a,b]$ such that*

$$|t-\tau|\|U(\tau,t)-U(\tau,\tau)\| \leqslant (t-\tau)[V(\tau,t)-V(\tau,\tau)]$$

for every $t \in (\tau-\theta(\tau), \tau+\theta(\tau))$, then

$$\left\|\int_a^b DU(\tau,t)\right\| \leqslant \int_a^b DV(\tau,t).$$

Proof. Fix an arbitrary $\epsilon > 0$. Owing to the fact that both Kurzweil integrals $\int_a^b DU(\tau,t)$ and $\int_a^b DV(\tau,t)$ exist, one can find a gauge δ on $[a,b]$, with $\delta(s) \leqslant \theta(s)$ for all $s \in [a,b]$, such that for every δ-fine tagged division $d = (\tau_j, [t_{j-1}, t_j])$ of $[a,b]$, with $j = 1, 2, \ldots, |d|$, the following inequalities hold:

$$\left\|\sum_{j=1}^{|d|}[U(\tau_j,t_j)-U(\tau_j,t_{j-1})] - \int_a^b DU(\tau,t)\right\| < \epsilon \quad \text{and} \tag{2.5}$$

$$\left|\sum_{j=1}^{|d|}[V(\tau_j,t_j)-V(\tau_j,t_{j-1})] - \int_a^b DV(\tau,t)\right| < \epsilon. \tag{2.6}$$

By hypothesis, for each $i = 1, 2, \ldots, |d|$, we have

$$(t_i-\tau_i)\|U(\tau_i,t_i)-U(\tau_i,\tau_i)\| \leqslant (t_i-\tau_i)[V(\tau_i,t_i)-V(\tau_i,\tau_i)],$$

since $t_i \geqslant \tau_i$. Thus, $\|U(\tau_i,t_i)-U(\tau_i,\tau_i)\| \leqslant V(\tau_i,t_i)-V(\tau_i,\tau_i)$, for $i = 1, 2, \ldots, |d|$ and, analogously, $\|U(\tau_i,t_{i-1})-U(\tau_i,\tau_i)\| \leqslant V(\tau_i,\tau_i)-V(\tau_i,t_{i-1})$, for $i = 1, 2, \ldots, |d|$. Hence, for each $i = 1, 2, \ldots, |d|$, we have

$$\begin{aligned}\|U(\tau_i,t_i)-U(\tau_i,t_{i-1})\| &\leqslant \|U(\tau_i,t_i)-U(\tau_i,\tau_i)\| + \|U(\tau_i,\tau_i)-U(\tau_i,t_{i-1})\| \\ &\leqslant V(\tau_i,t_i)-V(\tau_i,\tau_i)+V(\tau_i,\tau_i)-V(\tau_i,t_{i-1}) \\ &= V(\tau_i,t_i)-V(\tau_i,t_{i-1}).\end{aligned}$$

Then, by (2.5) and (2.6), we obtain

$$\begin{aligned}\left\|\int_a^b DU(\tau,t)\right\| \leqslant &\left\|\sum_{j=1}^{|d|}[U(\tau_j,t_j)-U(\tau_j,t_{j-1})] - \int_a^b DU(\tau,t)\right\| \\ &+ \left\|\sum_{j=1}^{|d|}[U(\tau_j,t_j)-U(\tau_j,t_{j-1})]\right\|\end{aligned}$$

$$< \epsilon + \sum_{j=1}^{|d|} [V(\tau_j, t_j) - V(\tau_j, t_{j-1})] - \int_a^b DV(\tau, t) + \int_a^b DV(\tau, t)$$

$$< 2\epsilon + \int_a^b DV(\tau, t).$$

Since ϵ is arbitrary, the result follows. □

The following result follows as in [209, Corollary 1.36] by using $V(\tau, t) = f(\tau)g(t)$ in Theorem 2.15.

Corollary 2.16: *Suppose $U : [a, b] \times [a, b] \to X$ is Kurzweil integrable over $[a, b]$, and $f : [a, b] \to \mathbb{R}$ and $g : [a, b] \to \mathbb{R}$ are functions such that the Perron–Stieltjes integral $\int_a^b f(s)dg(s)$ exists. If there is a gauge δ on $[a, b]$ such that*

$$|t - \tau| \|U(\tau, t) - U(\tau, \tau)\| \leqslant (t - \tau)f(\tau)[g(t) - g(\tau)]$$

for every $t \in (\tau - \delta(\tau), \tau + \delta(\tau))$, then

$$\left\| \int_a^b DU(\tau, t) \right\| \leqslant \int_a^b f(s)dg(s).$$

Corollary 2.17: *Let $f_1, f_2 : [a, b] \to \mathbb{R}$ be Perron–Stieltjes integrable functions with respect to a nondecreasing function $g : [a, b] \to \mathbb{R}$ such that $f_1(t) \leqslant f_2(t)$, for $t \in [a, b]$. Then,*

$$\int_a^b f_1(t)dg(t) \leqslant \int_a^b f_2(t)dg(t).$$

The final result of this chapter concerns the change of variables in the Kurzweil integral. A proof of it can be carried out as the proof of [209, Theorem 1.18] with obvious adaptation for the Banach space-valued case.

Theorem 2.18: *Suppose $-\infty < c < d < \infty$ and $\phi : [c, d] \to \mathbb{R}$ is a continuous function which is strictly increasing on $[c, d]$. Let $U : [\phi(c), \phi(d)] \times [\phi(c), \phi(d)] \to X$ be given. If one of the Kurzweil integrals $\int_{\phi(c)}^{\phi(d)} DU(\tau, t)$ or $\int_c^d DU(\phi(\sigma), \phi(s))$ exists, then the other one also exists and*

$$\int_{\phi(c)}^{\phi(d)} DU(\tau, t) = \int_c^d DU(\phi(\sigma), \phi(s)).$$

Remark 2.19: In the case of Perron–Stieltjes integrals, the change of variable formula presented in Theorem 1.72 is obtained as a corollary from Theorem 2.18.

Remark 2.20: Suppose in the statements of Theorems 1.72 and 2.18, $\phi : [c, d] \to \mathbb{R}$ is a continuous function which is strictly decreasing on $[c, d]$ instead. Then, analogous results hold with

$$\int_{\phi(c)}^{\phi(d)} DU(\tau, t) = -\int_{c}^{d} DU(\phi(\sigma), \phi(s)) \quad \text{and}$$

$$\int_{\phi(c)}^{\phi(d)} f(t)dg(t) = -\int_{c}^{d} f(\phi(t))dg(\phi(t)).$$

2.3 Notes on Kapitza Pendulum

The aim of this small section is not to develop the theory around Kapitza pendulum, but rather to present a short historical background and its relation with the Kurzweil integral.

A common rigid flat pendulum, whose pivot is forced to oscillate along the vertical line, has only one stable point, that is, vertically downwards. A well-known fascinating feature of classical mechanics is the dynamic stabilization of the inverted pendulum or Kapitza pendulum. When the frequency and amplitude of these vibrations are large enough, the inverted pendulum shows no tendency to stabilize. Indeed, Kapitza pendulum is stabilized by minimizing the potential energy.

This phenomenon of dynamic stabilization was originally mentioned by Andrew Stephenson in a series of studies in the early 1900s (see [224, 225]). Stephenson was the first to state that the upward equilibrium position could be linearly stable by an appropriate choice of parameters. In 1951, this extraordinary behavior of the pendulum, whose suspension point oscillated in the vertical direction with great frequency and small amplitude, was rediscovered, explained physically and investigated experimentally in detail by Pjotr Kapitza (see [139, 140]). In [140], Kapitza says:

> *Demonstration of the phenomenon of the oscillation of the turned pendulum is very effective: rapid and fine displacements caused by the vibrations are not perceptible visually, and so the behavior of the pendulum in the turned position produces an unexpected impression on the observer.*

The equation derived by Kapitza in 1951 [140] for the inverted pendulum is given by

$$\ddot{\Theta} = \left(gL^{-1} - AL^{-1}\omega^2 \sin \omega t\right) \sin \Theta,$$

where $g, L, \omega, A \in \mathbb{R}_+$, with L being the length of the pendulum and g being the gravitational constant. Using a method of averaging, taking a large frequency ω

and a small amplitude A, Kapitza substituted the solution of the above equation by the solution of

$$\ddot{\Theta} = \left(gL^{-1} - \frac{1}{2}A^2L^{-2}\omega^2 \cos\Theta\right) \sin\Theta.$$

Since then, the words "*averaging method*" started to be used. A few years later, the same equations were investigated by N.N. Bogolyubov and Yu.A. Mitropol'skiĭ in [20], also using an averaged method.

In 1965, J. Jarník (see [134]) considered the approach described by Kurzweil in [147, 148], by means of the Kurzweil integral, to obtain a result on the existence of a unique stable periodic solution of a second-order differential equation of which Kapitza equations is a particular case. Roughly speaking, the main assumptions on the rapidly oscillating term

$$\varphi(\Theta, t, \omega) = AL^{-1}\omega^2 \sin \omega t \sin \Theta,$$

which is periodic in t, are that the second primitive function of φ with respect to t goes to zero as $\omega \to \infty$ and the first primitive function of φ is bounded independently of ω.

For those readers who are used to the class of right-hand sides of generalized ODEs described in Chapter 4, the paper [134] is very nice to read. The interested reader may want to consult the second chapter of the book [153] by Kurzweil. See also [135].

With respect to applications, vibrational stabilization is a very relevant research field. Back there in 1951, Kapitza mentioned

> *The concept of the vibrational moment can be applied to any body, be it a colloidal particle or a molecule. If a force equivalent to those applied to the body is not passing through the body's centre of gravity, then in their vibration there arises a vibrational moment tending to set the body in such a position in which its centre of gravity would be on the axis of the oscillation. Since the nature of the vibrational moment has so far escaped notice of theoretical physics, one has neither experimentally sought for the aligning effect on colloidal and molecular particles, which in the case of a particle's asymmetric shape could be caused, for example, by the application of ultrasound vibrations or oscillations of an electric nature. It is interesting to note that the anisotropy of the amplitudes of thermal vibrations of molecules, which takes place in a crystal lattice, will produce by itself no vibrational moment, since by virtue of the law of uniform thermal-energy distribution the mean kinetic energy of the molecular vibrations will be the same in all directions.*

Recent applications include investigations of mechanisms for capturing elementary particles in Quantum Mechanics, such as the Paul trap (see [104]), and studies in Finance on the behavior of the market around a new equilibrium price (see [125]).

We conclude this chapter calling the reader's attention to the fact that the main feature of the Kurzweil integral is to handle well highly oscillating functions as shown by the Kapitza equation for the inverted pendulum.

3

Measure Functional Differential Equations

Everaldo M. Bonotto[1], Márcia Federson[2], Miguel V. S. Frasson[1], Rogelio Grau[3], and Jaqueline G. Mesquita[4]

[1]*Departamento de Matemática Aplicada e Estatística, Instituto de Ciências Matemáticas e de Computação (ICMC), Universidade de São Paulo, São Carlos, SP, Brazil*
[2]*Departamento de Matemática, Instituto de Ciências Matemáticas e de Computação (ICMC), Universidade de São Paulo, São Carlos, SP, Brazil*
[3]*Departamento de Matemáticas y Estadística, División de Ciencias Básicas, Universidad del Norte, Barranquilla, Colombia*
[4]*Departamento de Matemática, Instituto de Ciências Exatas, Universidade de Brasília, Brasília, DF, Brazil*

Let X be a Banach space equipped with norm $\|\cdot\|$. Consider the following *functional differential equation* (we write FDE, for short)

$$\dot{x}(t) = f(x_t, t), \quad t \in [t_0, t_0 + \sigma], \tag{3.1}$$

where $\sigma > 0$, $t_0 \in \mathbb{R}$, $B \subset X$ is an open set, $f : P \times [t_0, t_0 + \sigma] \to X$ is a function, where $P \subset G([-r, 0], B)$ is an open set and $r > 0$. Consider the *memory function*, $x_t : [-r, 0] \to X$, which describes the past events, for each $t \in \mathbb{R}$, happening on $[t - r, t]$. This memory function is commonly defined by

$$x_t(\theta) = x(t + \theta), \quad \theta \in [-r, 0],$$

for each $t \in [t_0, t_0 + \sigma]$. The integral form of (3.1) is given by

$$x(t) = x(t_0) + \int_{t_0}^{t} f(x_s, s)ds, \quad t \in [t_0, t_0 + \sigma], \tag{3.2}$$

where the integral on the right-hand side can be considered, for instance, in the sense of Riemann, Lebesgue, or Perron (see Chapter 1).

This type of equation describes processes which admit a lapse of time between cause and effect, being very important to bring a more realistic description of certain phenomena such as administration and ingestion of medicine and its effect into the body of the patient, evolution of diseases with incubation period, period

Generalized Ordinary Differential Equations in Abstract Spaces and Applications, First Edition.
Edited by Everaldo M. Bonotto, Márcia Federson, and Jaqueline G. Mesquita.

of gestation/pregnancy, among others. It is also possible to consider a more general class of functional differential equations such as FDEs with infinite delays, FDEs with time-dependent delays, and FDEs with state-dependent delays. Good references on functional and delay differential equations are [56, 115, 146] and the references therein. We also refer to Chapter 15 which is dedicated to FDEs of neutral type. In particular, the introduction of Chapter 15 provides a very good account on the history of FDEs and, in particular, of neutral FDEs.

The study of *measure differential equations* (we write MDEs for short) goes back to the 1960s with the pioneers works from W. W. Schmaedeke [206] in control theory, followed, among others, by the contributions of P. C. Das and R. R. Sharma in [54], R. R. Sharma in [215, 216], and S. G. Pandit and S. G. Deo in [194]. This type of equation plays an important role in applications, since they model naturally problems involving jumps and, therefore, such type of equations present functions having discontinuities.

In 2012, the authors of [85] introduced a new class of differential equations called *measure functional differential equations* for which we use the short form *measure FDEs*. It is known that measure FDEs encompass other types of equations such as

- impulsive functional differential equations (we write impulsive FDEs for short);
- functional dynamic equations on time scales;
- impulsive functional dynamic equations on time scales.

We describe these facts carefully in the present chapter.

Here, we consider a measure FDE in its integral form as below

$$x(t) = x(t_0) + \int_{t_0}^{t} f(x_s, s)dg(s), \quad t \in [t_0, t_0 + \sigma], \tag{3.3}$$

where $\sigma > 0$ and $t_0 \in \mathbb{R}$ are given, $\Omega \subset G([-r, 0], B)$ is any open set, with $B \subset X$ an open set and $r > 0$, and $f : \Omega \times [t_0, t_0 + \sigma] \to X$ is a function. Besides, $g : [t_0, t_0 + \sigma] \to \mathbb{R}$ is a nondecreasing function and $x_s : [-r, 0] \to B$ is a function called "*memory function*" or "*history function*" given as in (3.2). This memory function describes the history of events in the lapse time between cause and effect. The integral on the right-hand side of (3.3) can be understood in the sense of Riemann–Stieltjes, Lebesgue–Stieltjes, or Perron–Stieltjes, for instance. But here we consider it in the sense of Perron–Stieltjes due to its generality.

Clearly, when $g(s) = s$, Eq. (3.3) becomes the usual FDE described by (3.1). However, it is possible to define the function $g : [t_0, t_0 + \sigma] \to \mathbb{R}$ in another way so that different types of equations are included in the setting of Eq. (3.3). As we mentioned above, this is covered in this chapter. Moreover, one can also consider Eq. (3.3) in a more general setting, taking infinite delays, time-dependent delays and state-dependent delays into account. See, for instance, [100, 116, 220].

Even though measure FDEs are the core of this chapter, we are also interested in investigating equations without delays, that is, *measure differential equations*, which we refer to simply as MDEs and which we consider in its integral form

$$x(t) = x(t_0) + \int_{t_0}^{t} f(x(s), s)dg(s), \quad t \in [t_0, t_0 + \sigma], \tag{3.4}$$

where $B \subset X$ is open, $t_0 \in \mathbb{R}$ and $\sigma > 0$ are given, $g : [t_0, t_0 + \sigma] \to \mathbb{R}$ is a nondecreasing function, and $f : B \times [t_0, t_0 + \sigma] \to X$ is Perron–Stieltjes integrable with respect to g.

Considering the theory presented in [85] for the n-dimensional case, an interesting approach is to tackle analogous ideas, but now for case of Banach space-valued functions. In [29], for instance, the authors established results on dichotomies and boundedness of solutions for abstract MDEs and impulsive differential equations (we write IDEs) , using similar hypotheses to those employed in [85]. Thus, the investigation of this type of equations has been shown to be important to applications. Then, in this chapter, we investigate the following types of equations: measure FDEs, impulsive FDEs, functional dynamic equations on time scales, impulsive functional dynamic equations on time scales, all of which involving Banach space-valued functions.

This chapter is based on the articles [21, 82, 84–86, 178]. In the sequel, we describe how we organized it. In Section 3.1, we define what is a solution of an initial value problem (we write IVP for short) for a measure FDE of the form (3.3) and, in particular, we come with a definition of a solution of an IVP concerning equation (3.4). In addition, we include a lemma which says that the norm of the memory function of a regulated function is also regulated. In Section 3.2, we describe a relation between the solutions of impulsive measure FDEs and the solutions of measure FDEs. In order to prove this result, we require conditions on the Perron–Stieltjes integral of a function $f : [a, b] \to X$ instead of considering conditions on the integrand f itself. This "*slight*" different way of tacking integral equations allows us to consider integrands which behave as "*badly*" as the Perron–Stieltjes integral can handle, and it is a very well-known fact that Perron–Stieltjes integrals cope well with many jumps and highly oscillating functions (see Chapter 1). Still in Section 3.2, we present an example of a function whose indefinite integral satisfies a Carathéodory-type condition, showing that indeed our conditions are more general than those found in the literature for classical differential equations.

Section 3.3 is based mostly on the papers [85, 86]. In this section, we recall the basic concepts and properties of the theory of dynamic equations on time scales and we prove that functional dynamic equations on time scales can be regarded as measure FDEs. We also present a correspondence between the solutions of impulsive functional dynamic equations on time scales and the solutions of impulsive measure FDEs. This result yields the fact that measure FDEs are, in addition, a useful tool to explore impulsive functional dynamic equations on time scales.

Section 3.4 is dedicated to averaging principles for measure FDEs, as well as for impulsive measure FDEs and impulsive functional dynamic equations on time scales. The main references used in this section are [82, 84–86, 178]. At first, we present periodic averaging principles for various types of equations and, then, we investigate nonperiodic averaging principles. The results concerning averaging principles are very important, since they allow us to understand, for instance, the asymptotic behavior of the solutions of nonlinear systems involving parameters, by using the approach of the solutions of "*averaged*" autonomous equations, which are easier to deal with.

In the last section of this chapter, namely Section 3.5, we investigate continuous dependence on time scales for impulsive functional dynamic equations on time scales. See [21], for instance. The main goal here is to provide sufficient conditions which ensure that the sequence of unique solutions of the impulsive functional dynamic equation on time scales

$$\begin{cases} x(t) = x(t_0) + \displaystyle\int_{t_0}^{t} f(x_s, s)\Delta s + \sum_{\substack{k\in\{1,\dots,m\}\\ t_k<t}} I_k(x(t_k)), & t \in [t_0, t_0+\sigma]_{\mathbb{T}_n}, \\ x(t) = \phi(t), & t \in [t_0 - r, t_0]_{\mathbb{T}_n}, \end{cases}$$

converges uniformly to the unique solution of the impulsive functional dynamic equation on time scales

$$\begin{cases} x(t) = x(t_0) + \displaystyle\int_{t_0}^{t} f(x_s, s)\Delta s + \sum_{\substack{k\in\{1,\dots,m\}\\ t_k<t}} I_k(x(t_k)), & t \in [t_0, t_0+\sigma]_{\mathbb{T}}, \\ x(t) = \phi(t), & t \in [t_0 - r, t_0]_{\mathbb{T}}, \end{cases}$$

whenever $d_H(\mathbb{T}_n, \mathbb{T}) \to 0$ as $n \to \infty$, where $d_H(\mathbb{T}_n, \mathbb{T})$ denotes the Hausdorff metric (see [21] for details). To achieve the results for impulsive functional dynamic equations on time scales, the strategy is to employ the correspondence between the solutions of these equations and the solutions of impulsive measure FDEs, as well as the correspondence between the solutions of impulsive measure FDEs and the solutions of measure FDEs (see [21] for details). These results on continuous dependence of solutions of dynamic equations on variable time scales have several applications in numerical analysis.

3.1 Measure FDEs

In this section, our goal is to place the very basis of measure FDEs given by

$$\begin{cases} x(t) = x(t_0) + \displaystyle\int_{t_0}^{t} f(x_s, s)dg(s), & t \in [t_0, t_0+\sigma], \\ x_{t_0} = \phi, \end{cases} \tag{3.5}$$

where $t_0 \in \mathbb{R}$, $\sigma > 0$, $B \subset X$ is open, $P = G([-r, 0], B)$, $\phi \in P$, $f : P \times [t_0, t_0 + \sigma] \to X$ is a function, and $g : [t_0, t_0 + \sigma] \to \mathbb{R}$ is a nondecreasing function. The integral on the right-hand side of (3.5) is considered in the sense of Perron–Stieltjes. We start by presenting a concept of solution for this type of equation.

Definition 3.1: Let $B \subset X$ be open and $P = G([-r, 0], B)$. We say that a function $x : [t_0 - r, t_0 + \sigma] \to X$ is a *solution of the measure FDE* (3.5), with initial condition $x_{t_0} = \phi$, if it satisfies the following conditions:

(i) $x(t) = \phi(t - t_0)$ for all $t \in [t_0 - r, t_0]$;
(ii) $x \in G([t_0 - r, t_0 + \sigma], B)$ and $(x_t, t) \in P \times [t_0, t_0 + \sigma]$;
(iii) the Perron–Stieltjes integral $\int_{t_0}^{t_0+\sigma} f(x_s, s)dg(s)$ exists;
(iv) the equality

$$x(t) = x(t_0) + \int_{t_0}^{t} f(x_s, s)dg(s)$$

holds for all $t \in [t_0, t_0 + \sigma]$.

We are also interested in the particular case of these equations, the so-called *measure differential equations* (without delays). In many chapters of this book, the reader may find several applications for this type of equations.

In an analogous way, a measure differential equation (MDE) is given by

$$\begin{cases} x(t) = x(t_0) + \int_{t_0}^{t} f(x(s), s)dg(s), & t \in [t_0, t_0 + \sigma], \\ x(t_0) = x_0, \end{cases} \tag{3.6}$$

where $t_0 \in \mathbb{R}$, $\sigma > 0$, $B \subset X$ is open, $f : B \times [t_0, t_0 + \sigma] \to X$ is a function, and $g : [t_0, t_0 + \sigma] \to \mathbb{R}$ is a nondecreasing function. In addition, the integral in the right-hand side of (3.6) is understood in the sense of Perron–Stieltjes. In particular, we present a definition of a solution of the initial value problem (3.6).

Definition 3.2: Let $B \subset X$ be open. We say that a function $x : [t_0, t_0 + \sigma] \to X$ is a *solution of the measure differential equation* (3.6), with initial condition $x(t_0) = x_0$, if it satisfies the following conditions:

(i) $x(t_0) = x_0 \in B$;
(ii) $x \in G([t_0, t_0 + \sigma], B)$ and $(x(t), t) \in B \times [t_0, t_0 + \sigma]$;
(iii) the Perron–Stieltjes integral $\int_{t_0}^{t_0+\sigma} f(x(s), s)dg(s)$ exists;
(iv) the equality

$$x(t) = x_0 + \int_{t_0}^{t} f(x(s), s)dg(s)$$

holds for all $t \in [t_0, t_0 + \sigma]$.

Recall that any norm in the Banach space X is denoted by $\|\cdot\|$. We finish this section by presenting an important property of the memory function y_s, $s \in [t_0, t_0 + \sigma]$, coming from a regulated function $y : [t_0 - r, t_0 + \sigma] \to X$. This fact, borrowed from [85], is used in the entire book.

Lemma 3.3: *Let $y : [t_0 - r, t_0 + \sigma] \to X$ be a regulated function. Then, the norm of the memory function $[t_0, t_0 + \sigma] \ni s \mapsto \|y_s\|_\infty \in \mathbb{R}_+$, where $\|y_s\|_\infty = \sup_{\theta \in [-r,0]} \|y_s(\theta)\|$, is regulated on $[t_0, t_0 + \sigma]$.*

Proof. We need to prove that $\lim_{s \to s_0^-} \|y_s\|_\infty$ exists for every $s_0 \in (t_0, t_0 + \sigma]$, and $\lim_{s \to s_0^+} \|y_s\|_\infty$ exists for every $s_0 \in [t_0, t_0 + \sigma)$. We will only prove the existence of the first limit, since the second one follows analogously.

Let $s_0 \in (t_0, t_0 + \sigma]$. By hypothesis, y is regulated. Thus, y satisfies the Cauchy condition at $s_0 - r$ and s_0, which means that, given $\epsilon > 0$, there exists $\delta \in (0, s_0 - t_0)$ such that the following inequalities

$$\|y(u) - y(v)\| < \epsilon, \quad u, v \in (s_0 - r - \delta, s_0 - r), \quad \text{and} \tag{3.7}$$

$$\|y(u) - y(v)\| < \epsilon, \quad u, v \in (s_0 - \delta, s_0) \tag{3.8}$$

hold. Let s_1, s_2 be such that $s_0 - \delta < s_1 < s_2 < s_0$. If $s_2 - r \geqslant s_1$, then (3.7) yields

$$\|y(s)\| < \|y(s_2 - r)\| + \epsilon \leqslant \|y_{s_2}\|_\infty + \epsilon$$

for every $s \in [s_1 - r, s_2 - r]$. If, on the other hand, $s_2 - r < s_1$, then $\|y(s)\| \leqslant \|y_{s_2}\|_\infty$ for every $s \in [s_2 - r, s_1]$, whence

$$\|y(s)\| < \|y_{s_2}\|_\infty + \epsilon, \quad \text{for every } s \in [s_1 - r, s_1].$$

In any case, $\|y_{s_1}\|_\infty \leqslant \|y_{s_2}\|_\infty + \epsilon$.

Finally, (3.8) yields $\|y_{s_2}\|_\infty \leqslant \|y_{s_1}\|_\infty + \epsilon$ and hence,

$$\left| \|y_{s_1}\|_\infty - \|y_{s_2}\|_\infty \right| \leqslant \epsilon, \quad \text{for every} \quad s_1, s_2 \in (s_0 - \delta, s_0),$$

which ensures the existence of $\lim_{s \to s_0^-} \|y_s\|_\infty$ for $s_0 \in (t_0, t_0 + \sigma]$. □

3.2 Impulsive Measure FDEs

In this section, our goal is to investigate the impulsive measure FDEs and to show how to relate these equations to measure FDEs. In fact, we are going to prove that a certain type of impulsive measure FDEs can be regarded as measure FDEs. The main reference for this section is [86].

Let $\{t_k\}_{k=1}^m$ be moments of impulse effects which we assume to form an increasing sequence, that is, $t_0 \leqslant t_1 < \cdots < t_m < t_0 + \sigma$. Set $J_0 = [t_0, t_1]$ and $J_k = (t_k, t_{k+1}]$, for $k \in \{1, \dots, m-1\}$, and $J_m = (t_m, t_0 + \sigma]$. We consider the impulsive measure FDE given by

$$\begin{cases} x(v) - x(u) = \int_u^v f(x_s, s)dg(s), & u, v \in J_k, \quad k \in \{0, \dots, m\}, \\ \Delta^+ x(t_k) = I_k(x(t_k)), & k \in \{1, \dots, m\}, \\ x_{t_0} = \phi, \end{cases} \tag{3.9}$$

where $\Delta^+ x(t_k) = x(t_k^+) - x(t_k)$, $t_0 \in \mathbb{R}$, $\sigma > 0$, $B \subset X$ is open, $P = G([-r, 0], B)$, $f : P \times [t_0, t_0 + \sigma] \to X$ is a function, $g : [t_0, \infty) \to \mathbb{R}$ is a left-continuous and nondecreasing function, and for each $k \in \{1, \dots, m\}$, $I_k : B \to X$ is an impulse operator.

Let $x : [t_0 - r, t_0 + \sigma] \to X$ be a solution of the system (3.9). In order to $x(t) \in B$ for all $t \in [t_0 - r, t_0 + \sigma]$, we need to impose that $x(t_k^+) \in B$ for every $k \in \{1, \dots, m\}$, so that, after the action of the impulse operator at time t_k, the solution remains in B. Thus, we shall assume that $I + I_k : B \to B$ for every $k \in \{1, \dots, m\}$, where $I : B \to B$ is the identity operator.

We start by recalling that, from the properties of Perron–Stieltjes integrals, given $u, v \in J_k$, we have

$$\int_u^v f(x_s, s)dg(s) = \int_u^v f(x_s, s)d\tilde{g}(s),$$

where $\tilde{g} : [t_0, t_0 + \sigma] \to \mathbb{R}$ is such $\tilde{g} - g$ is constant on each interval J_k. Due to this fact, we can assume, without loss of generality, that g satisfies $\Delta^+ g(t_k) = 0$, for every $k \in \{1, \dots, m\}$. This assumption plays an important role here, since it implies that g is continuous at the times of impulse effects. This means that the function g and the impulse operators I_k do not "jump" at the same points, having discontinuities at different times. From this fact and by Corollary 2.14, the function

$$[t_0, t_0 + \sigma] \ni t \mapsto \int_{t_0}^t f(x_s, s)dg(s)$$

is continuous at $t_1, \dots, t_m$. Thus, we are talking about the following equivalent formulation of problem (3.9)

$$\begin{cases} x(t) = x(t_0) + \int_{t_0}^t f(x_s, s)dg(s) + \sum_{\substack{k \in \{1, \dots, m\} \\ t_k < t}} I_k(x(t_k)), & t \in [t_0, t_0 + \sigma], \\ x_{t_0} = \phi. \end{cases}$$

Recall the definition of the left-continuous Heaviside function, $H_T : [t_0, \infty) \to \mathbb{R}$, concentrated at time $T \in (t_0, \infty)$

$$H_T(t) = \begin{cases} 0, & t_0 \leqslant t \leqslant T, \\ 1, & t > T. \end{cases}$$

Thus, for $t \in [t_0, t_0 + \sigma]$, we can rewrite the previous initial value problem in the form

$$\begin{cases} x(t) = x(t_0) + \int_{t_0}^{t} f(x_s, s)dg(s) + \sum_{k=1}^{m} I_k(x(t_k))H_{t_k}(t), \quad t \in [t_0, t_0 + \sigma], & (3.10) \\ x_{t_0} = \phi. & (3.11) \end{cases}$$

Clearly, when $g(s) = s$, the initial value problem (3.10) and (3.11) is equivalent to the classic impulsive FDE (see [88, 91, 217] and the references therein) described by

$$\begin{cases} x(t) = x(t_0) + \displaystyle\int_{t_0}^{t} f(x_s, s)ds + \sum_{k=1}^{m} I_k(x(t_k))H_{t_k}(t), \quad t \in [t_0, t_0 + \sigma], \\ x_{t_0} = \phi. \end{cases} \tag{3.12}$$

In what follows, we present a definition of a solution of problem (3.10) and (3.11) and, then, an auxiliary result, borrowed from [86], which is essential to achieve the main results of this section, since it tells us how to turn a Perron–Stieltjes integral plus a sum carrying some impulses back into another Perron–Stieltjes integral. This feature will be used later to transform an impulsive system into a measure-type equation.

Definition 3.4: Let $B \subset X$ be open and $P = G([-r, 0], B)$. We say that a function $x : [t_0 - r, t_0 + \sigma] \to X$ is a *solution of the impulsive measure FDE* (3.10) and (3.11), if it satisfies the following conditions:

(i) $x(t) = \phi(t - t_0)$ for all $t \in [t_0 - r, t_0]$;
(ii) $x \in G([t_0 - r, t_0 + \sigma], B)$ and $(x_t, t) \in P \times [t_0, t_0 + \sigma]$;
(iii) the Perron–Stieltjes integral $\int_{t_0}^{t_0+\sigma} f(x_s, s)dg(s)$ exists;
(iv) the equality (3.10) holds for all $t \in [t_0, t_0 + \sigma]$.

Lemma 3.5: *Let $m \in \mathbb{N}$. Suppose for each $k \in \{1, \dots, m\}$, $t_k \in [t_0, t_0 + \sigma]$, $t_0 \leqslant t_1 < t_2 < \cdots < t_m < t_0 + \sigma$, and $g : [t_0, t_0 + \sigma] \to \mathbb{R}$ is regulated, left-continuous on $[t_0, t_0 + \sigma]$, and continuous at t_k for each $k \in \{1, \dots, m\}$. Let $f : [t_0, t_0 + \sigma] \to X$ be a function and consider $\tilde{f} : [t_0, t_0 + \sigma] \to X$ such that $\tilde{f}(t) = f(t)$, for every $t \in [t_0, t_0 + \sigma] \setminus \{t_1, \dots, t_m\}$, and take $\tilde{g} : [t_0, t_0 + \sigma] \to \mathbb{R}$ such that $\tilde{g} - g$ is constant on each of the intervals $[t_0, t_1], (t_1, t_2], \dots, (t_{m-1}, t_m], (t_m, t_0 + \sigma]$. Then, the Perron–Stieltjes integral $\int_{t_0}^{t_0+\sigma} \tilde{f}(s)d\tilde{g}(s)$ exists if and only if the Perron–Stieltjes integral $\int_{t_0}^{t_0+\sigma} f(s)dg(s)$ exists. In this case,*

$$\int_{t_0}^{t_0+\sigma} \tilde{f}(s)d\tilde{g}(s) = \int_{t_0}^{t_0+\sigma} f(s)dg(s) + \sum_{\substack{k \in \{1,\dots,m\} \\ t_k < t_0+\sigma}} \tilde{f}(t_k)\Delta^+\tilde{g}(t_k).$$

Proof. Since $\tilde{g} - g$ is constant on each interval $[t_0, t_1], (t_1, t_2], \ldots, (t_{m-1}, t_m], (t_m, t_0 + \sigma]$, by the properties of the Perron–Stieltjes integral, we have the following equalities

$$\int_{t_0}^{t_1} \tilde{f}(s)d(\tilde{g} - g)(s) = 0, \quad \lim_{\tau \to t_m^+} \int_{\tau}^{t_0+\sigma} \tilde{f}(s)d(\tilde{g} - g)(s) = 0 \tag{3.13}$$

and, for each $k \in \{1, \ldots, m-1\}$, we have

$$\lim_{\tau \to t_k^+} \int_{\tau}^{t_{k+1}} \tilde{f}(s)d(\tilde{g} - g)(s) = 0. \tag{3.14}$$

Then, Corollary 2.14 and (3.14) yield

$$\begin{aligned}\int_{t_k}^{t_{k+1}} \tilde{f}(s)d(\tilde{g} - g)(s) &= \lim_{\tau \to t_k^+} \int_{\tau}^{t_{k+1}} \tilde{f}(s)d(\tilde{g} - g)(s) + \lim_{\tau \to t_k^+} \int_{t_k}^{\tau} \tilde{f}(s)d(\tilde{g} - g)(s) \\ &= \lim_{\tau \to t_k^+} \int_{\tau}^{t_{k+1}} \tilde{f}(s)d(\tilde{g} - g)(s) + \tilde{f}(t_k)\Delta^+(\tilde{g} - g)(t_k) \\ &= \tilde{f}(t_k)\Delta^+\tilde{g}(t_k),\end{aligned} \tag{3.15}$$

for every $k \in \{1, \ldots, m-1\}$.

If $t_m = t_0 + \sigma$, then $\int_{t_m}^{t_0+\sigma} \tilde{f}(s)d(\tilde{g} - g)(s) = 0$. Otherwise, we get

$$\begin{aligned}\int_{t_m}^{t_0+\sigma} \tilde{f}(s)d(\tilde{g} - g)(s) &= \lim_{\tau \to t_m^+} \int_{\tau}^{t_0+\sigma} \tilde{f}(s)d(\tilde{g} - g)(s) + \tilde{f}(t_m)\Delta^+(\tilde{g} - g)(t_m) \\ &= \tilde{f}(t_m)\Delta^+\tilde{g}(t_m).\end{aligned} \tag{3.16}$$

Combining (3.14), (3.15) and (3.16), we derive that the Perron–Stieltjes integral $\int_{t_0}^{t_0+\sigma} \tilde{f}(s)d(\tilde{g} - g)(s)$ exists and

$$\int_{t_0}^{t_0+\sigma} \tilde{f}(s)d(\tilde{g} - g) = \sum_{\substack{k \in \{1, \ldots, m\} \\ t_k < t_0+\sigma}} \tilde{f}(t_k)\Delta^+\tilde{g}(t_k). \tag{3.17}$$

Then, using Corollaries 2.10 and 2.14, together with the continuity of g at each t_k and the definition of $\tilde{f}$, we obtain

$$\int_{t_0}^{t_1} \tilde{f}(s)dg(s) = \lim_{\tau \to t_1^-} \int_{t_0}^{\tau} \tilde{f}(s)dg(s) = \lim_{\tau \to t_1^-} \int_{t_0}^{\tau} f(s)dg(s) = \int_{t_0}^{t_1} f(s)dg(s),$$

$$\int_{t_k}^{t_{k+1}} \tilde{f}(s)dg(s) = \lim_{\substack{\sigma \to t_k^+ \\ \tau \to t_{k+1}^-}} \int_{\sigma}^{\tau} \tilde{f}(s)dg(s) = \lim_{\substack{\sigma \to t_k^+ \\ \tau \to t_{k+1}^-}} \int_{\sigma}^{\tau} f(s)dg(s) = \int_{t_k}^{t_{k+1}} f(s)dg(s),$$

for every $k \in \{1, \ldots, m-1\}$, and also

$$\begin{aligned}\int_{t_m}^{t_0+\sigma} \tilde{f}(s)dg(s) &= \lim_{\tau \to t_m^+} \int_{\tau}^{t_0+\sigma} \tilde{f}(s)dg(s) \\ &= \lim_{\tau \to t_m^+} \int_{\tau}^{t_0+\sigma} f(s)dg(s) = \int_{t_m}^{t_0+\sigma} f(s)dg(s).\end{aligned}$$

In the equalities above, the integrals on the left-hand sides exist if and only if those on the right-hand sides exist. Consequently, the Perron–Stieltjes integral $\int_{t_0}^{t_0+\sigma} \widetilde{f}(s)dg(s)$ exists if and only if the Perron–Stieltjes integral $\int_{t_0}^{t_0+\sigma} f(s)dg(s)$ exists and they have the same value. Moreover,

$$\int_{t_0}^{t_0+\sigma} \widetilde{f}(s)d\widetilde{g}(s) = \int_{t_0}^{t_0+\sigma} \widetilde{f}(s)dg(s) + \int_{t_0}^{t_0+\sigma} \widetilde{f}(s)d(\widetilde{g} - g)(s)$$

$$= \int_{t_0}^{t_0+\sigma} f(s)dg(s) + \sum_{\substack{k\in\{1,\dots,m\}\\ t_k<t_0+\sigma}} \widetilde{f}(t_k)\Delta^+\widetilde{g}(t_k),$$

which implies the result. □

Next, we prove a result which establishes a relation between the solutions of a measure FDEs and the solutions of an impulsive measure FDE. Such result is borrowed from [86].

Theorem 3.6: *Let $m \in \mathbb{N}$. Assume that, for each $k \in \{1, \dots, m\}$, $t_k \in [t_0, t_0 + \sigma]$, with $t_0 \leqslant t_1 < t_2 < \cdots < t_m < t_0 + \sigma$, and $I_k : B \to X$ is an impulse operator such that $I + I_k : B \to B$, where $B \subset X$ is an open set and $I : B \to B$ is the identity operator. Suppose $P = G([-r, 0], B)$, $f : P \times [t_0, t_0 + \sigma] \to X$ is Perron–Stieltjes integrable with respect to a function $g \in G^-([t_0, t_0 + \sigma], \mathbb{R})$ which is continuous at each t_k, for $k \in \{1, \dots, m\}$. For every $y \in P$, define*

$$\widetilde{f}(y, t) = \begin{cases} f(y, t), & t \in [t_0, t_0 + \sigma] \setminus \{t_1, \dots, t_m\}, \\ I_k(y(0)), & t = t_k,\ k \in \{1, \dots, m\}. \end{cases}$$

Moreover, for all $k \in \{1, \dots, m-1\}$, let $c_k \in \mathbb{R}_+$ be a constant with $c_k \leqslant c_{k+1}$, and define $\widetilde{g} : [t_0, t_0 + \sigma] \to \mathbb{R}$ by

$$\widetilde{g}(t) = \begin{cases} g(t), & t \in [t_0, t_1], \\ g(t) + c_k, & t \in (t_k, t_{k+1}],\ k \in \{1, \dots, m-1\}, \\ g(t) + c_m, & t \in (t_m, t_0 + \sigma], \end{cases}$$

satisfies $\Delta^+\widetilde{g}(t_k) = 1$, for $k \in \{1, \dots, m\}$. Then, $x \in G([t_0 - r, t_0 + \sigma], B)$ is a solution of the impulsive measure FDE

$$\begin{cases} x(t) = x(t_0) + \displaystyle\int_{t_0}^{t} f(x_s, s)dg(s) + \sum_{\substack{k\in\{1,\dots,m\}\\ t_k<t}} I_k(x(t_k)), & t \in [t_0, t_0 + \sigma], \\ x_{t_0} = \phi, \end{cases} \tag{3.18}$$

if and only if x is a solution of the measure FDE

$$\begin{cases} x(t) = x(t_0) + \displaystyle\int_{t_0}^{t} \widetilde{f}(x_s, s)d\widetilde{g}(s), & t \in [t_0, t_0 + \sigma], \\ x_{t_0} = \phi. \end{cases} \tag{3.19}$$

Proof. Using Lemma 3.5, we obtain

$$\int_{t_0}^{t} \widetilde{f}(x_s, s)d\widetilde{g}(s) = \int_{t_0}^{t} f(x_s, s)dg(s) + \sum_{\substack{k\in\{1,\dots,m\} \\ t_k<t}} \widetilde{f}(x_{t_k}, t_k)\Delta^+\widetilde{g}(t_k).$$

Then, the definitions of the functions $\widetilde{f}$ and $\widetilde{g}$ yield

$$\int_{t_0}^{t} \widetilde{f}(x_s, s)d\widetilde{g}(s) = \int_{t_0}^{t} f(x_s, s)dg(s) + \sum_{\substack{k\in\{1,\dots,m\} \\ t_k<t}} I_k(x(t_k)),$$

proving the result. □

As a direct consequence of Theorem 3.6, we also obtain a relation between the solutions of an impulsive MDE (without delays) and the solutions of an MDE (without delays).

Corollary 3.7: *Let $m \in \mathbb{N}$. Assume that, for each $k \in \{1, \dots, m\}$, $t_k \in [t_0, t_0 + \sigma]$, with $t_0 \leqslant t_1 < t_2 < \cdots < t_m < t_0 + \sigma$, and $I_k : B \to X$ is an impulse operator such that $I + I_k : B \to B$, where $B \subset X$ is an open set and $I : B \to B$ is the identity operator. Suppose $f : B \times [t_0, t_0 + \sigma] \to X$ is Perron–Stieltjes integrable with respect to a function $g \in G^-([t_0, t_0 + \sigma], \mathbb{R})$ which is continuous at each t_k, $k \in \{1, \dots, m\}$. Define an auxiliary function $\widetilde{f} : B \times [t_0, t_0 + \sigma] \to X$ by*

$$\widetilde{f}(z, t) = \begin{cases} f(z, t), & t \in [t_0, t_0 + \sigma] \setminus \{t_1, \dots, t_m\}, \\ I_k(z), & t = t_k,\ k \in \{1, \dots, m\}. \end{cases}$$

Moreover, for all $k \in \{1, \dots, m-1\}$, let $c_k \in \mathbb{R}_+$ be a constant with $c_k \leqslant c_{k+1}$, and define $\widetilde{g} : [t_0, t_0 + \sigma] \to \mathbb{R}$ by

$$\widetilde{g}(t) = \begin{cases} g(t), & t \in [t_0, t_1], \\ g(t) + c_k, & t \in (t_k, t_{k+1}],\ k \in \{1, \dots, m-1\}, \\ g(t) + c_m, & t \in (t_m, t_0 + \sigma], \end{cases}$$

satisfies $\Delta^+\widetilde{g}(t_k) = 1$, for all $k \in \{1, \dots, m\}$. Then, $x \in G([t_0, t_0 + \sigma], B)$ is a solution of the impulsive MDE

$$\begin{cases} x(t) = x(t_0) + \displaystyle\int_{t_0}^{t} f(x(s), s)dg(s) + \sum_{\substack{k\in\{1,\dots,m\} \\ t_k<t}} I_k(x(t_k)), & t \in [t_0, t_0 + \sigma], \\ x(t_0) = x_0, \end{cases} \tag{3.20}$$

if and only if $x \in G([t_0, t_0+\sigma], B)$ *is a solution of the MDE*

$$\begin{cases} x(t) = x(t_0) + \displaystyle\int_{t_0}^{t} \widetilde{f}(x(s), s)d\widetilde{g}(s), & t \in [t_0, t_0+\sigma], \\ x(t_0) = x_0. \end{cases} \tag{3.21}$$

Remark 3.8: Notice that if $\widetilde{g} : [t_0, t_0+\sigma] \to \mathbb{R}$ is defined by

$$\widetilde{g}(t) = \begin{cases} g(t), & t \in [t_0, t_1], \\ g(t) + c_k, & t \in (t_k, t_{k+1}],\ k \in \{1, \dots, m-1\}, \\ g(t) + c_m, & t \in (t_m, t_0+\sigma], \end{cases}$$

then the condition $\Delta^+\widetilde{g}(t_k) = 1, k \in \{1, \dots, m\}$, which appears in Theorem 3.6 and Corollary 3.7, is automatically satisfied, since $\Delta^+\widetilde{g}(t_k) = \widetilde{g}(t_k^+) - \widetilde{g}(t_k) = k+1-k = 1$ and $0 \leqslant c_k \leqslant c_{k+1}$, for $k \in \{1, \dots, m-1\}$.

The next result shows us how to relate certain conditions on impulsive measure FDEs with their analogues on measure FDEs. The version presented here is more general than that found in [86] and its proof is inspired in the results contained in [6] for this specific case.

Lemma 3.9: *Let* $m \in \mathbb{N}$. *Suppose, for each* $k \in \{1, \dots, m\}$, $t_k \in [t_0, t_0+\sigma]$ *and* $t_0 \leqslant t_1 < t_2 < \cdots < t_m < t_0+\sigma$. *Consider* $B \subset X$ *an open set and impulse operators* $I_k : B \to X$ *such that* $I + I_k : B \to B$, $k \in \{1, \dots, m\}$, *where* $I : B \to B$ *is the identity operator. Take* $P = G([-r, 0], B)$, $O = G([t_0 - r, t_0+\sigma], B)$, *and assume that* $g : [t_0, t_0+\sigma] \to \mathbb{R}$ *is a left-continuous and nondecreasing function. Let* $f : P \times [t_0, t_0+\sigma] \to X$ *be an arbitrary function and define* $\widetilde{f} : P \times [t_0, t_0+\sigma] \to X$ *by*

$$\widetilde{f}(y, \tau) = \begin{cases} f(y, \tau), & \tau \in [t_0, t_0+\sigma] \setminus \{t_1, \dots, t_m\}, \\ I_k(y(0)), & \tau = t_k, k \in \{1, \dots, m\}, \end{cases}$$

and $\widetilde{g} : [t_0, t_0+\sigma] \to \mathbb{R}$ *by*

$$\widetilde{g}(\tau) = \begin{cases} g(\tau), & \tau \in [t_0, t_1], \\ g(\tau) + c_k, & \tau \in (t_k, t_{k+1}],\ k \in \{1, \dots, m-1\}, \\ g(\tau) + c_m, & \tau \in (t_m, t_0+\sigma], \end{cases}$$

where $\Delta^+\widetilde{g}(t_k) = 1$, $k \in \{1, \dots, m\}$, *and* $0 \leqslant c_k \leqslant c_{k+1}$ *for each* $k \in \{1, \dots, m-1\}$. *Then, the following statements are valid:*

(i) *the function* $\widetilde{g}$ *is nondecreasing and left-continuous;*

(ii) *given* $u_1, u_2 \in [t_0, t_0 + \sigma]$ *and* $x \in O$, *if the Perron–Stieltjes integral* $\int_{u_1}^{u_2} f(x_s, s)dg(s)$ *exists, then the Perron–Stieltjes integral* $\int_{u_1}^{u_2} \widetilde{f}(x_s, s)d\widetilde{g}(s)$ *also exists;*

(iii) *if there exist a Perron–Stieltjes integrable function* $M_1 : [t_0, t_0 + \sigma] \to \mathbb{R}$ *with respect to g such that*

$$\left\| \int_{u_1}^{u_2} f(x_s, s)dg(s) \right\| \leqslant \int_{u_1}^{u_2} M_1(s)dg(s), \tag{3.22}$$

for all $x \in O$ *and all* $t_0 \leqslant u_1 \leqslant u_2 \leqslant t_0 + \sigma$, *and a constant* $M_2 > 0$ *such that* $\|I_k(z)\| \leqslant M_2$, *for* $k \in \{1, \dots, m\}$ *and* $z \in B$, *then there exists a Perron–Stieltjes integrable function* $M : [t_0, t_0 + \sigma] \to \mathbb{R}$ *with respect to* $\widetilde{g}$ *satisfying*

$$\left\| \int_{u_1}^{u_2} \widetilde{f}(x_s, s)d\widetilde{g}(s) \right\| \leqslant \int_{u_1}^{u_2} M(s)d\widetilde{g}(s),$$

for all $x \in O$ *and all* $t_0 \leqslant u_1 \leqslant u_2 \leqslant t_0 + \sigma$;

(iv) *if there exist a Perron–Stieltjes integrable function* $L_1 : [t_0, t_0 + \sigma] \to \mathbb{R}$ *with respect to g such that*

$$\left\| \int_{u_1}^{u_2} [f(x_s, s) - f(z_s, s)]\, dg(s) \right\| \leqslant \int_{u_1}^{u_2} L_1(s)\|x_s - z_s\|_\infty \, dg(s),$$

for all $x, z \in O$ *and* $t_0 \leqslant u_1 \leqslant u_2 \leqslant t_0 + \sigma$ *and a constant* $L_2 > 0$ *such that*

$$\|I_k(w) - I_k(y)\| \leqslant L_2 \, \|w - y\| \,,$$

for $k \in \{1, \dots, m\}$ *and* $w, y \in B$, *then there exists a Perron–Stieltjes integrable function* $L : [t_0, t_0 + \sigma] \to \mathbb{R}$ *with respect to* $\widetilde{g}$ *satisfying*

$$\left\| \int_{u_1}^{u_2} \left[\widetilde{f}(x_s, s) - \widetilde{f}(z_s, s)\right] \, d\widetilde{g}(s) \right\| \leqslant \int_{u_1}^{u_2} L(s)\|x_s - z_s\|_\infty \, d\widetilde{g}(s),$$

for all $x, z \in O$ *and all* $t_0 \leqslant u_1 \leqslant u_2 \leqslant t_0 + \sigma$.

Proof. Before starting the proof itself, we would like to highlight that the integral on the right-hand side of the inequality in (iv) exists by Corollary 1.69, since the Perron–Stieltjes integral $\int_{t_0}^{t_0+\sigma} L_1(s)dg(s)$ exists (by assumption), the function $s \in [t_0, t_0 + \sigma] \mapsto \|x_s - z_s\|$ is regulated for all $x, z \in O$ and the facts that $L(\mathbb{R})$ and $\mathbb{R}$ are isomorphic and $SV([t_0, t_0 + \sigma], L(\mathbb{R})) = BV([t_0, t_0 + \sigma], \mathbb{R})$ (hence, the spaces $G([t_0, t_0 + \sigma], \mathbb{R})$ and $G([t_0, t_0 + \sigma], L(\mathbb{R})) = G([t_0, t_0 + \sigma], L(\mathbb{R})) \cap SV([t_0, t_0 + \sigma], L(\mathbb{R}))$ are also isomorphic).

Since g is a nondecreasing and left-continuous function, by definition, $\widetilde{g}$ inherits the same properties, proving (i).

Combining (i), the hypotheses on $\widetilde{f}$, and Lemma 3.5, we obtain (ii).

In order to prove (iii), consider $x \in O$ and $t_0 \leqslant u_1 \leqslant u_2 \leqslant t_0 + \sigma$. According to Lemma 3.5, we have

$$\int_{u_1}^{u_2} \widetilde{f}(x_s, s)d\widetilde{g}(s) = \int_{u_1}^{u_2} f(x_s, s)dg(s) + \sum_{\substack{k \in \{1,\dots,m\} \\ u_1 \leqslant t_k < u_2}} \widetilde{f}(x_{t_k}, t_k)\Delta^+\widetilde{g}(t_k)$$

$$= \int_{u_1}^{u_2} f(x_s, s)dg(s) + \sum_{\substack{k \in \{1,\dots,m\} \\ u_1 \leqslant t_k < u_2}} I_k(x(t_k))\Delta^+\widetilde{g}(t_k).$$

Therefore,

$$\left\| \int_{u_1}^{u_2} \widetilde{f}(x_s, s)d\widetilde{g}(s) \right\| \leqslant \int_{u_1}^{u_2} M_1(s)dg(s) + \sum_{\substack{k \in \{1,\dots,m\} \\ u_1 \leqslant t_k < u_2}} M_2\Delta^+\widetilde{g}(t_k). \tag{3.23}$$

Notice that, by definition, $\widetilde{g}(v) - \widetilde{g}(u) \geqslant g(v) - g(u)$, whenever $t_0 \leqslant u \leqslant v \leqslant t_0 + \sigma$. Then, (3.23) and the properties of the Perron–Stieltjes integral yield

$$\int_{u_1}^{u_2} M_1(s)dg(s) \leqslant \int_{u_1}^{u_2} M_1(s)d\widetilde{g}(s) \leqslant \int_{u_1}^{u_2} \widetilde{M}(s)d\widetilde{g}(s), \tag{3.24}$$

where $\widetilde{M}(s) = 1 + M_2 + M_1(s)$, for all $s \in [t_0, t_0 + \sigma]$.

Define a function $h : [t_0, t_0 + \sigma] \to \mathbb{R}$ by

$$h(t) = \int_{t_0}^{t} \widetilde{M}(s)d\widetilde{g}(s), \quad \text{for } t \in [t_0, t_0 + \sigma].$$

Claim. h is nondecreasing.

Indeed, let $s_1, s_2 \in [t_0, t_0 + \sigma]$, with $s_2 \geqslant s_1$. Then,

$$\int_{t_0}^{s_2} M_1(s)dg(s) - \int_{t_0}^{s_1} M_1(s)dg(s) = \int_{s_1}^{s_2} M_1(s)dg(s) \geqslant 0,$$

where the inequality follows from (3.22). This fact implies that $t \mapsto \int_{t_0}^{t} M_1(s)dg(s)$ is a nondecreasing function. Thus, since M_2 is a positive constant and g is nondecreasing, it follows that h is a nondecreasing function and the *Claim* holds.

By Corollary 2.14, $\Delta^+h(t_k) = \widetilde{M}(t_k)\Delta^+\widetilde{g}(t_k)$, for $k \in \{1, \dots, m\}$. Then,

$$\sum_{\substack{k \in \{1,\dots,m\} \\ u_1 \leqslant t_k < u_2}} M_2\Delta^+\widetilde{g}(t_k) \leqslant \sum_{\substack{k \in \{1,\dots,m\} \\ u_1 \leqslant t_k < u_2}} \widetilde{M}(t_k)\Delta^+\widetilde{g}(t_k) = \sum_{\substack{k \in \{1,\dots,m\} \\ u_1 \leqslant t_k < u_2}} \Delta^+h(t_k)$$

$$\leqslant h(u_2) - h(u_1) = \int_{u_1}^{u_2} \widetilde{M}(s)d\widetilde{g}(s). \tag{3.25}$$

By (3.23), (3.24), and (3.25), we obtain

$$\left\| \int_{u_1}^{u_2} \widetilde{f}(x_s, s)d\widetilde{g}(s) \right\| \leqslant 2 \int_{u_1}^{u_2} \widetilde{M}(s)d\widetilde{g}(s). \tag{3.26}$$

Hence, defining $M(t) = 2\widetilde{M}(t)$, for $t \in [t_0, t_0 + \sigma]$, the proof of (iii) follows.

Now, let $t_0 \leqslant u_1 \leqslant u_2 \leqslant t_0 + \sigma$. By Lemma 3.5, we have

$$\int_{u_1}^{u_2} \left[\widetilde{f}(x_s, s) - \widetilde{f}(z_s, s)\right] d\widetilde{g}(s) = \int_{u_1}^{u_2} [f(x_s, s) - f(z_s, s)]\, dg(s) + \sum_{\substack{k\in\{1,\dots,m\}\\ u_1\leqslant t_k<u_2}} [I_k(x(t_k)) - I_k(z(t_k))]\Delta^+\widetilde{g}(t_k).$$

In consequence,

$$\left\|\int_{u_1}^{u_2} \left[\widetilde{f}(x_s, s) - \widetilde{f}(z_s, s)\right] d\widetilde{g}(s)\right\| \leqslant \int_{u_1}^{u_2} L_1(s)\|x_s - z_s\|_\infty\, dg(s) + \sum_{\substack{k\in\{1,\dots,m\}\\ u_1\leqslant t_k<u_2}} L_2\|x_{t_k} - z_{t_k}\|_\infty \Delta^+\widetilde{g}(t_k).$$

Thus, if $\widetilde{L}(s) = 1 + L_2 + L_1(s)$, for all $s \in [t_0, t_0 + \sigma]$, then we obtain

$$\int_{u_1}^{u_2} L_1(s)\|x_s - z_s\|_\infty\, dg(s) \leqslant \int_{u_1}^{u_2} L_1(s)\|x_s - z_s\|_\infty\, d\widetilde{g}(s) \leqslant \int_{u_1}^{u_2} \widetilde{L}(s)\|x_s - z_s\|_\infty\, d\widetilde{g}(s),$$

where in the first inequality, we use Lemma 3.5.

Defining $\gamma : [t_0, t_0 + \sigma] \to \mathbb{R}$ by

$$\gamma(t) = \int_{t_0}^{t} \widetilde{L}(s)\|x_s - z_s\|_\infty\, d\widetilde{g}(s)$$

and applying a similar reasoning as before, it follows that γ is nondecreasing. Therefore, Corollary 2.14 yields

$$\Delta^+\gamma(t_k) = \widetilde{L}(t_k)\|y_{t_k} - z_{t_k}\|_\infty \Delta^+\widetilde{g}(t_k),$$

for $k \in \{1, \dots, m\}$. Thus,

$$\sum_{\substack{k\in\{1,\dots,m\}\\ u_1\leqslant t_k<u_2}} L_2\|y_{t_k} - z_{t_k}\|_\infty \Delta^+\widetilde{g}(t_k) \leqslant \sum_{\substack{k\in\{1,\dots,m\}\\ u_1\leqslant t_k<u_2}} \widetilde{L}(t_k)\|y_{t_k} - z_{t_k}\|_\infty \Delta^+\widetilde{g}(t_k) = \sum_{\substack{k\in\{1,\dots,m\}\\ u_1\leqslant t_k<u_2}} \Delta^+\gamma(t_k) \leqslant \gamma(u_2) - \gamma(u_1) = \int_{u_1}^{u_2} \widetilde{L}(s)\|x_s - z_s\|_\infty\, d\widetilde{g}(s).$$

Finally, taking $L(t) = 2\widetilde{L}(t)$, for all $t \in [t_0, t_0 + \sigma]$, we get

$$\left\|\int_{u_1}^{u_2} \left[\widetilde{f}(x_s, s) - \widetilde{f}(z_s, s)\right] d\widetilde{g}(s)\right\| \leqslant \int_{u_1}^{u_2} L(s)\|x_s - z_s\|_\infty\, d\widetilde{g}(s),$$

which proves (iv). □

The next example gives us a motivation to require Carathéodory-type conditions on the indefinite integral instead of simply imposing conditions to the integrand. To the integrands, some integrability is required and that is pretty much everything.

Example 3.10: Consider functions $\varphi, \lambda : [0,1] \to \mathbb{R}$ defined by

$$\varphi(t) = \begin{cases} \dfrac{(-1)^{k+1}2^k}{k}, & t \in [d_{k-1}, d_k),\ k \in \mathbb{N}, \\ 0, & t = 1, \end{cases} \tag{3.27}$$

and

$$\lambda(t) = \begin{cases} \dfrac{2^k}{k}, & t \in [d_{k-1}, d_k),\ k \in \mathbb{N}, \\ 0, & t = 1, \end{cases} \tag{3.28}$$

where $d_k = 1 - \frac{1}{2^k}$, $k \in \mathbb{N}_0$. Note that $\lambda(t) = |\varphi(t)|$ for every $t \in [0,1]$. By [12, Example 2.8], the function φ is Riemann improper integrable over $[0,1]$ and

$$I = \int_0^1 \varphi(s)ds = \sum_{k=1}^{\infty} \frac{(-1)^{k+1}}{k}. \tag{3.29}$$

Furthermore, the integral of λ over $[0,1]$ is not finite which means that it is not Perron integrable over $[0,1]$ and, hence, neither Lebesgue integrable. Likewise, φ is not Lebesgue integrable (because it is not absolutely integrable).

Now, define $f : G([-r,0],\mathbb{R}) \times [0,1] \to \mathbb{R}$ by $f(\xi, s) = \varphi(s)$, for every $(\xi, s) \in G([-r,0],\mathbb{R}) \times [0,1]$, where φ is given by (3.27) and $r > 0$. Note that f is independent of the first variable. By (3.29), for every $x \in G([-r,1],\mathbb{R})$, the mapping $s \mapsto f(x_s, s) = \varphi(s)$ is Perron integrable over $[0,1]$ and

$$\left| \int_0^1 f(x_s, s)ds \right| = \left| \int_0^1 \varphi(s)ds \right| = |I| < \int_0^1 M(s)ds = |I| + 1, \tag{3.30}$$

where $M(s) = |I| + 1$ for every $s \in [0,1]$. Note that (3.30) does not imply that $|f(x_s, s)| \leqslant M(s) = |I| + 1$ for all $s \in [0,1]$ and all $x \in G([-r,1],\mathbb{R})$. Otherwise, $|\varphi(s)| = |f(x_s, s)| \leqslant |I| + 1$, for every $s \in [0,1]$ and, hence, φ would be Lebesgue integrable, which is a contradiction.

3.3 Functional Dynamic Equations on Time Scales

The theory of dynamic equations on time scales was introduced by Štefan Hilger in 1988 in his PhD thesis (see [123]). The main advantage behind this theory comes from the fact that it can unify and extend the continuous analysis, when the time scale $\mathbb{T} = \mathbb{R}$, and the discrete analysis, when $\mathbb{T} = \mathbb{Z}$. Moreover, the theory of time scales also encompasses interesting cases such as the scale $\mathbb{T} = q^{\mathbb{N}_0} = \{q^n : n \in \mathbb{N}_0\}$, $q > 1$, which plays an important role in what is called "*quantum calculus.*" It also deals with "*hybrid*" cases involving discrete and continuous parts, having several applications in the study of population dynamics, among others. For more details in this respect, the reader may want to consult [22, 23, 85, 86, 196].

3.3.1 Fundamentals of Time Scales

In this subsection, we present a brief overview of the theory of time scales, bringing the main concepts and the results we need to give an appropriate treatment to the correspondence between measure FDEs and functional dynamic equations on time scales. The main references for this first part are [22, 23, 196].

First of all, we need to have in mind the meaning of the term "*time scale.*" A *time scale* is a closed nonempty subset of $\mathbb{R}$, which is denoted by the symbol $\mathbb{T}$. Examples of time scales are closed intervals, $\mathbb{N}$, $\mathbb{Z}$, $q^{\mathbb{N}_0} = \{q^n : n \in \mathbb{N}_0\}$ for $q > 1$, among others.

Definition 3.11: For every $t \in \mathbb{T}$, we define the *forward jump operator* and the *backward jump operator*, respectively, $\sigma, \rho : \mathbb{T} \to \mathbb{T}$ by

$$\sigma(t) = \inf\{s \in \mathbb{T} : s > t\} \quad \text{and} \quad \rho(t) = \sup\{s \in \mathbb{T} : s < t\}.$$

As in [22], we follow the convention $\inf \emptyset = \sup \mathbb{T}$ and $\sup \emptyset = \inf \mathbb{T}$.

With the above operators at hand, we now recall some standard definitions for the points in a given time scale (see [22, 23]).

Definition 3.12: Consider the backward and forward jump operators ρ and σ. If $\sigma(t) > t$, then t is called a *right-scattered point* and if $t < \sup \mathbb{T}$ and $\sigma(t) = t$, then t is called *right-dense*. On the other hand, if $\rho(t) < t$, then t is said to be *left-scattered*, whereas if $t > \inf \mathbb{T}$ and $\rho(t) = t$, then t is called *left-dense*.

In the sequel, we present the concept of graininess function, which plays an important role in the theory of time scales.

Definition 3.13: The *graininess function* $\mu : \mathbb{T} \to [0, \infty)$ is defined by

$$\mu(t) = \sigma(t) - t.$$

If $t \in \mathbb{T}$ is right-scattered, then $\mu(t) > 0$. On the other hand, if $t \in \mathbb{T}$ is right-dense, then $\mu(t) = 0$. We can also check these properties using the well-known time scales $\mathbb{R}$ and $\mathbb{Z}$. For instance, if $\mathbb{T} = \mathbb{R}$, then $\mu(t) \equiv 0$ for every $t \in \mathbb{R}$. In case $\mathbb{T} = \mathbb{Z}$, then $\mu(t) \equiv 1$ for every $t \in \mathbb{Z}$. It is also important to point out that the graininess function μ is not always constant as these two examples suggest. Indeed, if we consider the time scale $\mathbb{T} = q^{\mathbb{N}_0}$, with $q > 1$, then $\mu(t) = (q-1)t$ for each $t \in q^{\mathbb{N}_0}, q > 1$.

Now, we bring up the analogous concept of a regulated function in the setting of time scales. It coincides with the definition of a regulated function presented in Chapter 1 for the case $\mathbb{T} = \mathbb{R}$, showing consistency.

Definition 3.14: A function $f : \mathbb{T} \to X$ is called *regulated* provided its right-sided limits exist (finite) at all right-dense points in $\mathbb{T}$ and its left-sided limits exist (finite) at all left-dense points in $\mathbb{T}$. The set of all regulated functions $f : \mathbb{T} \to X$ is denoted by $G(\mathbb{T}, X)$.

For each pair of numbers $a, b \in \mathbb{T}$, with $a \leqslant b$, we denote a *time scale interval* by $[a,b]_{\mathbb{T}} = [a,b] \cap \mathbb{T}$. Given a set $B \subset X$, the symbol $G([a,b]_{\mathbb{T}}, B)$ denotes the set of all regulated functions $f : [a,b]_{\mathbb{T}} \to B$.

Definition 3.15: A function $f : \mathbb{T} \to X$ is called *rd-continuous*, if it is regulated on $\mathbb{T}$ and continuous at right-dense points of $\mathbb{T}$. The set of all rd-continuous functions $f : \mathbb{T} \to X$ is denoted by $C_{\mathrm{rd}}(\mathbb{T}, X)$. In an analogous way, one can define an rd-continuous function $f : [a,b]_{\mathbb{T}} \to X$. In this case, we denote by $C_{\mathrm{rd}}([a,b]_{\mathbb{T}}, X)$ the set of all rd-continuous $f : [a,b]_{\mathbb{T}} \to X$.

In sequel, we define a set, denoted by $\mathbb{T}^{\kappa}$, which plays an important role in the definition of derivatives in the framework of time scales – the so-called *delta derivatives*. The set $\mathbb{T}^{\kappa}$ is defined by means of the time scale $\mathbb{T}$ as follows: if $\mathbb{T}$ has a left-scattered maximum m, then $\mathbb{T}^{\kappa} = \mathbb{T} \setminus \{m\}$. Otherwise, $\mathbb{T}^{\kappa} = \mathbb{T}$.

Definition 3.16: For $f : \mathbb{T} \to X$ and $t \in \mathbb{T}^{\kappa}$, we define a function $f^{\Delta} : \mathbb{T}^{\kappa} \to X$ (provided it exists) with the following property: given $\epsilon > 0$, there exists a neighborhood U of t in $\mathbb{T}$ such that the inequality

$$\|f(\sigma(t)) - f(s) - f^{\Delta}(t)[\sigma(t) - s]\| \leqslant \epsilon |\sigma(t) - s|$$

holds for every $s \in U$. In this case, we say that the function f^{Δ} is the *delta-derivative* (or Δ-derivative) of f at t.

Clearly, from the previous definition, if $\mathbb{T} = \mathbb{R}$, then we obtain the usual derivative for the continuous case, showing the consistency of the theory of time scales. For more details, the reader may want to consult [22, 23].

3.3.2 The Perron Δ-integral

In what follows, we introduce the concept of Perron Δ-integrals for Banach spaces, similarly to what is already known from the literature for Kurzweil–Henstock Δ-integrals in the case where $X = \mathbb{R}^n$. As a matter of fact, the notion of Kurzweil–Henstock Δ-integral was introduced by A. Peterson and B. Thompson in [196].

Let $\mathbb{T}$ be a given time scale and consider a time scale interval $[a,b]_{\mathbb{T}}$. We say that $\delta = (\delta_L, \delta_R)$ is a *Δ-gauge* of $[a,b]_{\mathbb{T}}$, whenever $\delta_L(t) > 0$ on $(a,b]_{\mathbb{T}}$, $\delta_R(t) > 0$ on $[a,b)_{\mathbb{T}}$, $\delta_L(a) \geqslant 0$, $\delta_R(b) \geqslant 0$ and $\delta_R(t) \geqslant \mu(t)$, for all $t \in [a,b)_{\mathbb{T}}$. We say that $d_{\mathbb{T}}$ is a *tagged division of* $[a,b]_{\mathbb{T}}$, provided

$$a = s_0 \leqslant \tau_1 \leqslant s_1 \leqslant \cdots \leqslant s_{|d|-1} \leqslant \tau_{|d|} \leqslant s_{|d|} = b$$

with $s_i > s_{i-1}$ and $s_i, \tau_i \in \mathbb{T}$, for $1 \leqslant i \leqslant |d|$. The points τ_i are called *tag points* and s_i are called *end points*. We denote such a tagged division by $d_{\mathbb{T}} = (\tau_i, [s_{i-1}, s_i]_{\mathbb{T}})$, where $[s_{i-1}, s_i]_{\mathbb{T}}$ denotes a usual time scale interval in $d_{\mathbb{T}}$ and τ_i is the *associated tag point* in $[s_{i-1}, s_i]_{\mathbb{T}}$. When δ is a Δ-gauge of $[a,b]_{\mathbb{T}}$, a division $d_{\mathbb{T}}$ is called *δ-fine* if, for $1 \leqslant i \leqslant |d|$, we have

$$\tau_i - \delta_L(\tau_i) \leqslant s_{i-1} < s_i \leqslant \tau_i + \delta_R(\tau_i).$$

Now, we have all the ingredients for the definition of the Δ-integral of a function $f : [a,b]_{\mathbb{T}} \to X$ in the sense of Perron. Owing to the fact that the Kurzweil integral and the variational Henstock integral of a function taking values in an infinite dimensional space may not coincide (see the comments after Definition 1.42 and also Example 1.44) and, in order to be coherent with the terminology of this book, we refer to the Δ-integral of a function $f : [a,b]_{\mathbb{T}} \to X$ given by means of δ-fine tagged divisions of a time scale interval $[a,b]_{\mathbb{T}}$ as a Perron Δ-integral, having in mind that the abstract Perron integral and the Kurzweil integral, as defined in Chapter 1, coincide.

Definition 3.17: A function $f : [a,b]_{\mathbb{T}} \to X$ is *Perron Δ-integrable* over $[a,b]_{\mathbb{T}}$, if there is an element $I \in X$ such that given $\epsilon > 0$, there exists a Δ-gauge δ on $[a,b]_{\mathbb{T}}$ such that

$$\left\| I - \sum_{i=1}^{|d|} f(\tau_i)(s_i - s_{i-1}) \right\| < \epsilon,$$

for all δ-fine tagged division $d_{\mathbb{T}} = (\tau_i, [s_{i-1}, s_i]_{\mathbb{T}})$ of $[a,b]_{\mathbb{T}}$. In this case, we write $I = \int_a^b f(s)\Delta s$ and I is called the *Perron Δ-integral* of f.

Note that, similarly as in the classical definition of the Perron integral in the continuous case, that is, when $\mathbb{T} = \mathbb{R}$ (see Chapter 1), Definition 3.17 only makes sense if, for a given Δ-gauge δ on $[a,b]_{\mathbb{T}}$, one can ensure the existence of at least one δ-fine tagged division $d_{\mathbb{T}}$ of $[a,b]_{\mathbb{T}}$. This is the content of the next result, which is a generalization of the well-known Cousin Lemma (see Lemma 1.36) for a Δ-gauge of a time scale interval. A proof of it can be found in [196, Lemma 1.9].

Lemma 3.18 (Cousin Lemma): *Given a Δ-gauge δ on $[a,b]_{\mathbb{T}}$, there exists a δ-fine tagged division $d_{\mathbb{T}}$ of $[a,b]_{\mathbb{T}}$.*

We finish this subsection by stating a result which says that Perron Δ-integrals are linear and additive over adjacent intervals. Such result can be found in [196] for functions taking values in $\mathbb{R}^n$. But, with obvious adaptations, it also holds for functions taking values in X. Thus, we omit its proof here.

Theorem 3.19: *A function $f : [a,b]_{\mathbb{T}} \to X$ is Perron Δ-integrable over $[a,b]_{\mathbb{T}}$ if and only if f is Perron Δ-integrable over $[a,c]_{\mathbb{T}}$ and $[c,b]_{\mathbb{T}}$. In this case,*

$$\int_a^b f(s)\Delta s = \int_a^c f(s)\Delta s + \int_c^b f(s)\Delta s.$$

Moreover, if $f, g : [a,b]_{\mathbb{T}} \to X$ are Perron Δ-integrable over $[a,b]_{\mathbb{T}}$, then for every $\alpha, \beta \in \mathbb{R}$, $\alpha f + \beta g$ is also Perron Δ-integrable over $[a,b]_{\mathbb{T}}$ and

$$\int_a^b [\alpha f(s) + \beta g(s)]\,\Delta s = \alpha\left(\int_a^b f(s)\Delta s\right) + \beta\left(\int_a^b g(s)\Delta s\right).$$

3.3.3 Perron Δ-integrals and Perron–Stieltjes integrals

In this subsection, our goal is to investigate a relation between Perron Δ-integrals and Perron–Stieltjes integrals. In order to do that, we need to present some extensions of time scales and functions defined in time scales. The main references for this subsection are [86, 219].

We start by defining an *extension* of a given time scale $\mathbb{T}$. We denote the *extended time scale* by $\mathbb{T}^*$. Thus, for an arbitrary time scale $\mathbb{T}$, we define

$$\mathbb{T}^* = \begin{cases} (-\infty, \sup \mathbb{T}], & \text{if } \sup \mathbb{T} < \infty, \\ (-\infty, \infty), & \text{otherwise.} \end{cases}$$

We also need to define an operator $*: \mathbb{T}^* \to \mathbb{T}$ which was introduced by A. Slavík in [219]. Given a time scale $\mathbb{T}$ and a real number $t \leqslant \sup \mathbb{T}$, we define

$$t^* = \inf\{s \in \mathbb{T} : s \geqslant t\}.$$

Due to the fact that $\mathbb{T}$ is a closed set, it is clear that $t^* \in \mathbb{T}$, showing that the operator $*$ is well defined. Notice that t^* may be different from $\sigma(t)$ in many situations. In particular, if $t \in \mathbb{T}$ and t is right-scattered, then $\sigma(t) \neq t^*$, because $t^* = t$ and $\sigma(t) > t$. Take, for instance, $\mathbb{T} = \mathbb{Z}$ and $t = 2$. In this case, $t^* = 2$, while $\sigma(t) = 3$.

Given a function $f : \mathbb{T} \to X$, define its extension $f^* : \mathbb{T}^* \to X$ by

$$f^*(t) = f(t^*), \quad \text{for all } t \in \mathbb{T}^*.$$

Similarly, for a function $f : \mathbb{T} \times X \to X$, we define its extension $f^* : X \times \mathbb{T}^* \to X$ by

$$f^*(x, t) = f(x, t^*), \quad \text{for all } x \in X \text{ and all } t \in \mathbb{T}^*.$$

The next result, borrowed from [219, Lemma 4] and easily adapted to Banach space-valued functions, brings important properties concerning the extension of a function $f : \mathbb{T} \to X$, namely, $f^* : \mathbb{T}^* \to X$. In fact, it shows how f^* inherits the properties of f.

Lemma 3.20: *Let $\mathbb{T}$ be a time scale. If $f : \mathbb{T} \to X$ is a regulated function, then $f^* : \mathbb{T}^* \to X$ is also regulated. If $f : \mathbb{T} \to X$ is left-continuous, then $f^* : \mathbb{T}^* \to X$ is left-continuous on $\mathbb{T}^*$. If $f : \mathbb{T} \to X$ is right-continuous, then $f^* : \mathbb{T}^* \to X$ is right-continuous at right-dense points of $\mathbb{T}$.*

Proof. We start by investigating $\lim_{t \to t_0^-} f^*(t)$, where $t_0 \in \mathbb{T}^*$. Let us consider three cases:

(a) $t_0 \in \mathbb{T}$ and t_0 is left-dense;
(b) $t_0 \in \mathbb{T}$ and t_0 is left-scattered;
(c) $t_0 \notin \mathbb{T}$.

In case (a), we have $\lim_{t \to t_0^-} f^*(t) = \lim_{t \to t_0^-} f(t)$. In case (b), $\lim_{t \to t_0^-} f^*(t) = f(t_0) = f^*(t_0)$. Finally, in case (c), we obtain $\lim_{t \to t_0^-} f^*(t) = f(t_0^*) = f^*(t_0)$.

Now, let us investigate $\lim_{t \to t_0^+} f^*(t)$, for $t_0 \in \mathbb{T}^*$ and $t_0 < \sup \mathbb{T}^*$. This leads us to three cases:

(a′) $t_0 \in \mathbb{T}$ and t_0 is right-dense;
(b′) $t_0 \in \mathbb{T}$ and t_0 is right-scattered;
(c′) $t_0 \notin \mathbb{T}$.

In case (a′), $\lim_{t \to t_0^+} f^*(t) = \lim_{t \to t_0^+} f(t)$. In case (b′), we have $\lim_{t \to t_0^+} f^*(t) = f(\sigma(t_0))$, and, in case (c′), $\lim_{t \to t_0^+} f^*(t) = f(t_0^*) = f^*(t_0)$. □

As an immediate consequence of Lemma 3.20, we obtain the next corollary which describes some properties of the function $g(t) = t^*$, for $t \in [t_0, t_0 + \sigma]$. This result can be found in [78].

Corollary 3.21: *Let $\mathbb{T}$ be a time scale and suppose $t_0, t_0 + \sigma \in \mathbb{T}$. Let the function $g : [t_0, t_0 + \sigma] \to \mathbb{R}$ be given by $g(t) = t^*$, for all $t \in [t_0, t_0 + \sigma]$. Then, g satisfies the following conditions:*

(i) *g is a nondecreasing function;*
(ii) *g is left-continuous on $(t_0, t_0 + \sigma]$.*

Proof. Item (i) is a direct consequence of the definition of g and item (ii) follows immediately from Lemma 3.20. □

The next result can be found in [86], and it describes a very interesting relation between Perron Δ-integrals and Perron–Stieltjes integrals.

Theorem 3.22: *Let $a, b \in \mathbb{T}$, $f : [a, b]_{\mathbb{T}} \to X$ be an arbitrary function and $g : [a, b] \to \mathbb{R}$ be defined by $g(t) = t^*$ for every $t \in [a, b]$. Then, the Perron Δ-integral $\int_a^b f(s)\Delta s$ exists if and only if the Perron–Stieltjes integral $\int_a^b f^*(s)dg(s)$ also exists, in which case, both integrals coincide.*

Proof. Consider an arbitrary tagged division $d = (\tau_i, [s_{i-1}, s_i])$ of $[a, b]$ and set

$$S(d) = \sum_{i=1}^{|d|} f^*(\tau_i)(g(s_i) - g(s_{i-1})) = \sum_{i=1}^{|d|} f(\tau_i^*)(s_i^* - s_{i-1}^*). \tag{3.31}$$

Let us suppose the Perron Δ-integral $\int_a^b f(s)\Delta s$ exists. Then, given an arbitrary $\epsilon > 0$, there is a Δ-gauge $\delta = (\delta_L, \delta_R)$ on $[a, b]_{\mathbb{T}}$ such that

$$\left\| \sum_{i=1}^{|d|} f(\tau_i)(s_i - s_{i-1}) - \int_a^b f(s)\Delta s \right\| < \epsilon$$

for every δ-fine tagged division of $[a, b]_{\mathbb{T}}$. Now, let us define a gauge $\tilde{\delta} : [a, b] \to (0, \infty)$ as follows:

$$\tilde{\delta}(t) = \begin{cases} \min\,(\delta_L(t), \sup\{m : t + m \in [a, b]_{\mathbb{T}}, m \leqslant \delta_R(t)\}), & \text{if } t \in (a, b) \cap \mathbb{T}, \\ \sup\{m : a + m \in [a, b]_{\mathbb{T}}, m \leqslant \delta_R(a)\}, & \text{if } t = a, \\ \delta_L(b), & \text{if } t = b, \\ \frac{1}{2}\inf\{|t - s| : s \in \mathbb{T}\}, & \text{if } t \in [a, b] \setminus \mathbb{T}. \end{cases}$$

Consider an arbitrary $\tilde{\delta}$-fine tagged division $d = (\tau_i, [s_{i-1}, s_i])$ of $[a, b]$. Then, for every $i \in \{1, \ldots, |d|\}$, we have the following two possibilities:

(a) $\tau_i \in \mathbb{T}$;
(b) $[s_{i-1}, s_i] \cap \mathbb{T} = \emptyset$.

Since d is an arbitrary $\widetilde{\delta}$-fine tagged division of $[a, b]$, the division points s_i, $i \in \{0, \dots, m\}$ and tags $\tau_i, i \in \{1, \dots, m\}$, may not belong to $\mathbb{T}$. However, it is possible to construct a $\widetilde{\delta}$–fine tagged division of $[a, b]$ with such property and other important properties.

Claim 1. There exists a division $\widetilde{d}$ satisfying:

(1) all division points and tags belong to $\mathbb{T}$;
(2) $S(d) = S(\widetilde{d})$;
(3) $\widetilde{d}$ is δ-fine.

Let us prove (i). We proceed by induction. Notice that $s_0 = a \in \mathbb{T}$. Now, consider an interval $[s_{i-1}, s_i]$ such that $s_{i-1} \in \mathbb{T}$. Therefore, $[s_{i-1}, s_i] \cap \mathbb{T} \neq \emptyset$, which implies that $\tau_i \in \mathbb{T}$ by case (a).

In case of $s_i \notin \mathbb{T}$, we replace the division point s_i by s_i^*, delete all division points s_j such that belong to (s_i, s_i^*), and all tags τ_j belonging to (s_i, s_i^*). With this operation, the value of the integral sum given by (3.31) remains the same, since the contributions of the intervals $[s_{i-1}, s_i]$ and $[s_{i-1}, s_i^*]$ to the value of the sum are the same, and the contributions of the intervals $[s_{j-1}, s_j]$ contained in (s_i, s_i^*) are zero, since $s_{j-1}^* = s_j^* = s_i^*$.

Therefore, it follows that items (1) and (2) are satisfied. Now, it remains to prove (3) to conclude the proof of the *Claim 1*. Set $K = \sup\left([a, \tau_i + \delta_R(\tau_i)] \cap \mathbb{T}\right)$. Since $\mathbb{T}$ is closed, by definition, it follows that $K \in [a, b]_{\mathbb{T}}$. Since d is $\widetilde{\delta}$-fine, by construction, it follows that

$$s_i \leqslant \tau_i + \widetilde{\delta}(\tau_i) \leqslant \tau_i + \sup\{m : \tau_i + m \in [a, b]_{\mathbb{T}}, m \leqslant \delta_R(\tau_i)\} = K.$$

However, $s_i \notin \mathbb{T}$ and as remarked above, $K \in \mathbb{T}$, then these two facts imply that $s_i^* \leqslant K$, since by the definition s_i^* is the smallest time scale point larger than s_i. Hence, $s_i^* \leqslant K \leqslant \tau_i + \delta_R(\tau_i)$, proving that $\widetilde{d}$ is δ-fine, concluding the *Claim 1*.

Hence, by the *Claim 1*, $\widetilde{d}$ is a δ-fine tagged division of $[a, b]_{\mathbb{T}}$ such that $S(d) = S(\widetilde{d})$. As a consequence, we get

$$\left\| S(d) - \int_a^b f(t)\Delta t \right\| = \left\| S(\widetilde{d}) - \int_a^b f(t)\Delta t \right\| < \epsilon.$$

Hence, we obtain the existence of the Perron–Stieltjes integral $\int_a^b f^*(t)dg(t)$ and also, it equals $\int_a^b f(t)\Delta t$.

Reciprocally, suppose the Perron–Stieltjes integral $\int_a^b f^*(t)dg(t)$ exists. Thus, given an arbitrary $\epsilon > 0$, there is a gauge $\widetilde{\delta} : [a, b] \to (0, \infty)$ such that

$$\left\| \sum_{i=1}^{|d|} f\left(\tau_i^*\right)\left(s_i^* - s_{i-1}^*\right) - \int_a^b f^*(t)dg(t) \right\| < \epsilon$$

for every $\widetilde{\delta}$-fine tagged division d of $[a, b]$. Set a Δ-gauge $\delta = (\delta_L, \delta_R)$ on $[a, b]_{\mathbb{T}}$ as follows: for every $t \in [a, b]_{\mathbb{T}}$, define

$$\delta_L(t) = \widetilde{\delta}(t) \quad \text{and} \quad \delta_R(t) = \max(\widetilde{\delta}(t), \mu(t)). \tag{3.32}$$

Consider an arbitrary δ-fine tagged division $\hat{d}_{\mathbb{T}} = (\tau_i, [s_{i-1}, s_i]_{\mathbb{T}})$ of $[a, b]_{\mathbb{T}}$. By definition, all these points belong to $\mathbb{T}$.

Notice that this δ-fine division does not need to be $\widetilde{\delta}$-fine: for certain values of $i \in \{1, \dots, |\hat{d}_{\mathbb{T}}|\}$, we can have

$$\delta_R(\tau_i) + \tau_i \geqslant s_i > \widetilde{\delta}(\tau_i) + \tau_i.$$

In this case, it follows that $\delta_R(\tau_i) = \mu(\tau_i)$, the point τ_i is right-scattered and $s_i = \sigma(\tau_i)$.

Claim 2. There exists a modified tagged division $\widetilde{d} = (\tau_i, [s_{i-1}, s_i])$ of $[a, b]$ such that is $\widetilde{\delta}$-fine and $S(\hat{d}_{\mathbb{T}}) = S(\widetilde{d})$.

Since $\hat{d}_{\mathbb{T}}$ is δ-fine, we obtain by definition that

$$\tau_i - \widetilde{\delta}(\tau_i) = \tau_i - \delta_L(\tau_i) \leqslant s_{i-1} < s_i \leqslant \tau_i + \delta_R(\tau_i). \tag{3.33}$$

Therefore, let us divide the interval $[s_{i-1}, s_i]$ in two subintervals $[s_{i-1}, \tau_i + \widetilde{\delta}(\tau_i)]$ and $[\tau_i + \widetilde{\delta}(\tau_i), s_i]$. Let us proceed as follows: replace the division point s_i by $\tau_i + \widetilde{\delta}(\tau_i)$ and maintain τ_i as the tag of the interval $[s_{i-1}, \tau_i + \widetilde{\delta}(\tau_i)]$. Then, we obtain by (3.33) that $[s_{i-1}, \tau_i + \widetilde{\delta}(\tau_i)] \subset [\tau_i - \widetilde{\delta}(\tau_i), \tau_i + \widetilde{\delta}(\tau_i)]$. On the other hand, cover $[\tau_i + \widetilde{\delta}(\tau_i), s_i]$ using an arbitrary $\widetilde{\delta}$-fine division. This construction implies that $\widetilde{d}$ is a $\widetilde{\delta}$-fine division of $[a, b]$.

It remains to show the equality $S(\hat{d}_{\mathbb{T}}) = S(\widetilde{d})$ holds in order to conclude *Claim 2* is valid. But this follows immediately from the fact that $u^* = s_i$ for every $u \in (\tau_i, s_i]$ for all $i = 1, \dots, |\hat{d}_{\mathbb{T}}|$. Notice that

$$\begin{aligned} \left\| \sum_{i=1}^{|\hat{d}_{\mathbb{T}}|} f(\tau_i)(s_i - s_{i-1}) - \int_a^b f^*(t)dg(t) \right\| &= \left\| S(\hat{d}_{\mathbb{T}}) - \int_a^b f^*(t)dg(t) \right\| \\ &= \left\| S(\widetilde{d}) - \int_a^b f^*(t)dg(t) \right\| < \epsilon, \end{aligned}$$

whence the Perron Δ-integral $\int_a^b f(t)\Delta t$ exists and equals $\int_a^b f^*(t)dg(t)$. □

Theorem 3.22 shows us that Stieltjes-type integrals can be used to investigate Δ-integrals on time scales. Moreover, a careful examination reveals that it is possible to generalize the previous correspondence for Stieltjes-type of Δ-integrals. In order to do this, one needs to extend the definition of the Perron Δ-integral to a Stieltjes-type integral, say, a Perron–Stieltjes Δ-integral of a function $f : [a, b]_{\mathbb{T}} \to X$ with respect to a function $g : [a, b]_{\mathbb{T}} \to \mathbb{R}$. Let $\int_a^b f(s)\Delta g(s)$ denote such integral, which can be obtained by taking replacing the usual Riemann-type sum in the definition of the Perron Δ-integral by

$$\sum_{i=1}^{|d_{\mathbb{T}}|} f(\tau_i)[g(s_i) - g(s_{i-1})],$$

where $d_{\mathbb{T}} = (\tau_i, [s_{i-1}, s_i]_{\mathbb{T}})$ is a tagged division of $[a, b]_{\mathbb{T}}$. Then, as in the proof of Theorem 3.22, one can show that the resulting Stieltjes-type Δ-integral satisfies

$$\int_a^b f(s)\Delta g(s) = \int_a^b f^*(s)dg^*(s),$$

yielding a more general correspondence than that presented in Theorem 3.22 (see [86, 179]).

In the next two results, we present important properties of the Perron–Stieltjes integral. They are borrowed from [85, 86].

Lemma 3.23: *Let $a, b \in \mathbb{T}$, $a < b$ and consider a function $g : [a, b] \to \mathbb{R}$ given by $g(t) = t^*$ for every $t \in [a, b]$. If $f : [a, b] \to X$ is such that the Perron–Stieltjes integral $\int_a^b f(s)dg(s)$ exists, then for every $c, d \in [a, b]$,*

$$\int_c^d f(s)dg(s) = \int_{c^*}^{d^*} f(s)dg(s).$$

Proof. By definition, the function g is constant on $[c, c^*]$ and on $[d, d^*]$. Therefore, $\int_c^{c^*} f(s)dg(s) = 0$ and $\int_d^{d^*} f(s)dg(s) = 0$ and hence,

$$\int_c^d f(s)dg(s) = \int_c^{c^*} f(s)dg(s) + \int_{c^*}^d f(s)dg(s) + \int_d^{d^*} f(s)dg(s) = \int_{c^*}^{d^*} f(s)dg(s)$$

proving the desired result. □

Theorem 3.24: *Let $\mathbb{T}$ be a time scale and $[a, b] \subset \mathbb{T}^*$. Consider a function $g : [a, b] \to \mathbb{R}$ defined by $g(s) = s^*$, for every $s \in [a, b]$ and functions $f_1, f_2 : [a, b] \to X$ such that $f_1(t) = f_2(t)$ for every $t \in [a, b] \cap \mathbb{T}$. If the Perron–Stieltjes integral $\int_a^b f_1(s)dg(s)$ exists, then the Perron–Stieltjes integral $\int_a^b f_2(s)dg(s)$ also exists and both integrals coincide.*

Proof. Set $I = \int_a^b f_1(s)dg(s)$. Since I exists, given an arbitrary $\epsilon > 0$, there exists a gauge $\delta_1 : [a, b] \to (0, \infty)$ such that

$$\left\| \sum_{i=1}^{|d|} f_1(\tau_i)[g(s_i) - g(s_{i-1})] - I \right\| < \epsilon,$$

for every δ_1-fine tagged division $d = (\tau_i, [s_{i-1}, s_i])$ of $[a, b]$. Now, set

$$\delta_2(t) = \begin{cases} \delta_1(t), & \text{if } t \in [a, b] \cap \mathbb{T}, \\ \min\left(\delta_1(t), \frac{1}{2}\inf\{|t-s| : s \in \mathbb{T}\}\right), & \text{if } t \in [a, b] \setminus \mathbb{T}. \end{cases}$$

By definition, a δ_2-fine tagged division is also δ_1-fine.

Consider an arbitrary δ_2-fine tagged division $\tilde{d} = (\tau_i, [s_{i-1}, s_i])$ of $[a, b]$. For every $i \in \{1, \dots, |\tilde{d}|\}$, there are two possibilities:

(1) $[s_{i-1}, s_i] \cap \mathbb{T} = \emptyset$;
(2) $\tau_i \in \mathbb{T}$.

In case (1), $g(s_{i-1}) = g(s_i)$, which implies

$$f_2(\tau_i)(g(s_i) - g(s_{i-1})) = 0 = f_1(\tau_i)(g(s_i) - g(s_{i-1})),$$

while in case (2), we have

$$f_1(\tau_i) = f_2(\tau_i) \quad \text{and} \quad f_2(\tau_i)(g(s_i) - g(s_{i-1})) = f_1(\tau_i)(g(s_i) - g(s_{i-1})).$$

Combining both cases, we obtain

$$\left\| \sum_{i=1}^{|\tilde{d}|} f_2(\tau_i)(g(s_i) - g(s_{i-1})) - I \right\| = \left\| \sum_{i=1}^{|\tilde{d}|} f_1(\tau_i)(g(s_i) - g(s_{i-1})) - I \right\| < \epsilon.$$

Finally, by the arbitrariness of ϵ, $\int_a^b f_2(s)dg(s) = I$. □

The next result, borrowed from [86, Theorem 4.1], follows from Theorem 3.24.

Theorem 3.25: *Let $f : [a, b]_{\mathbb{T}} \to X$ be a function such that the Perron Δ-integral $\int_a^b f(s)\Delta s$ exists for every $a, b \in \mathbb{T}$, $a < b$. Choose an arbitrary $a \in \mathbb{T}$ and define*

$$F_1(t) = \int_a^t f(s)\Delta s, \ \textit{for } t \in [a, b]_{\mathbb{T}} \quad \textit{and} \quad F_2(t) = \int_a^t f^*(s)dg(s), \ \textit{for } t \in [a, b],$$

where $g(s) = s^*$, for every $s \in [a, b]$. Then, $F_2 = F_1^*$.

Proof. By Theorem 3.22 and Lemma 3.23, we have

$$F_2(t) = \int_a^t f^*(s)dg(s) = \int_a^{t^*} f^*(s)dg(s) = \int_a^{t^*} f(s)\Delta s = F_1(t^*) = F_1^*(t)$$

and the result follows. □

Now, we present an example to illustrate the relation between Perron–Stieltjes integrals and Perron Δ-integrals.

Example 3.26: Let $\mathbb{T} = \mathbb{Z}$ and $a, b \in \mathbb{T}$ with $a < b$. Take $m \in \mathbb{N}$ such that $b = a + m$. Consider a regulated function $f : [a, b]_{\mathbb{T}} \to \mathbb{R}$ and define its extension

$f^* : [a,b]_{\mathbb{T}^*} \to \mathbb{R}$ by $f^*(t) = f(t^*)$. Suppose, in addition, $g : [a,b] \to \mathbb{R}$ is defined by $g(t) = t^*$. By Lemma 3.20, it follows that $f^* : [a,b]_{\mathbb{T}^*} \to \mathbb{R}$ is regulated and by Corollary 3.21, the function g is nondecreasing. Therefore, the Perron–Stieltjes integral $\int_a^b f^*(s)dg(s)$ exists and

$$\int_a^b f^*(s)dg(s) = \int_a^{a+m} f^*(s)dg(s) = \sum_{k=0}^{m-1} \int_{a+k}^{a+k+1} f^*(s)dg(s).$$

Let us show that

$$\sum_{k=0}^{m-1} \int_{a+k}^{a+k+1} f^*(s)dg(s) = \sum_{k=0}^{m-1} f^*(a+k)(g(a+k+1) - g(a+k)).$$

Since $\mathbb{T} = \mathbb{Z}$ and $g(t) = t^*$, for each $k \in \{0,\dots,m-1\}$, the restriction $g|_{(a+k,a+k+1]}$ is constant. In particular,

$$g(s) = g(a+k+1), \quad \text{for every } s \in (a+k, a+k+1]. \tag{3.34}$$

Given $k \in \{0,\dots,m-1\}$, let us prove the equality below

$$\int_{a+k}^{a+k+1} f^*(s)dg(s) = f^*(a+k)(g(a+k+1) - g(a+k)) \tag{3.35}$$

holds. Let $d = (\tau_i, [s_{i-1}, s_i])$ be any tagged division of $[a+k, a+k+1]$. Then, $a+k = s_0 < s_1 < \cdots < s_{|d|-1} < s_{|d|} = a+k+1$. Hence,

$$\begin{aligned}\sum_{i=1}^{|d|} f^*(\tau_i)(g(s_i) - g(s_{i-1})) &= f^*(\tau_1)(g(s_1) - g(s_0)) + \sum_{i=2}^{|d|} f^*(\tau_i)(g(s_i) - g(s_{i-1}))\\ &= f^*(\tau_1)(g(s_1) - g(s_0)) = f^*(\tau_1)(g(s_n) - g(s_0))\\ &= f^*(\tau_1)(g(a+k+1) - g(a+k)),\end{aligned}$$

where the third equality follows by (3.34) and from the facts that $s_1 \in (a+k, a+k+1]$ and $s_{|d|} = a+k+1$.

Now, consider an arbitrary $\epsilon > 0$ and define a gauge $\delta : [a+k, a+k+1] \to (0,\infty)$ fulfilling $0 < \delta(s) < |s-(a+k)|$, for $s \neq a+k$, and $\delta(a+k) = \epsilon$. If $d = (\tau_i, [s_{i-1}, s_1])$ is a δ-fine tagged division of the interval $[a+k, a+k+1]$, then $\tau_1 = a+k$. Thus,

$$\begin{aligned}&\left|\sum_{i=1}^{|d|} f^*(\tau_i)(g(s_i) - g(s_{i-1})) - f^*(a+k)(g(a+k+1) - g(a+k))\right|\\ &= |f^*(a+k)(g(a+k+1) - g(a+k)) - f^*(a+k)(g(a+k+1) - g(a+k))|\\ &= 0 < \epsilon,\end{aligned}$$

which implies that the Perron–Stieltjes integral $\int_{a+k}^{a+k+1} f^*(s)dg(s)$ exists and

$$\int_{a+k}^{a+k+1} f^*(s)dg(s) = f^*(a+k)(g(a+k+1) - g(a+k))$$

proving equality (3.35). Hence, we obtain

$$\begin{aligned}\int_a^b f^*(s)dg(s) &= \sum_{k=0}^{m-1} \int_{a+k}^{a+k+1} f^*(s)dg(s)\\ &= \sum_{k=0}^{m-1} f^*(a+k)(g(a+k+1)-g(a+k))\\ &= \sum_{k=0}^{m-1} f(a+k) = \sum_{k=a}^{a+m-1} f(k) = \sum_{k=1}^{b-1} f(k) = \int_a^b f(s)\Delta s,\end{aligned}$$

showing that the two integrals can be related.

3.3.4 MDEs and Dynamic Equations on Time Scales

In this subsection, our goal is to present a relation between measure differential equations (or simply MDEs) and dynamic equations on time scales. The relation we bring here was first presented by A. Slavík in [219] for functions defined on bounded intervals and taking values in $\mathbb{R}^n$. Then, in [78], the authors extended this correspondence to functions defined on unbounded intervals also taking values in $\mathbb{R}^n$. Here, we extend such relation to Banach space-valued functions defined on bounded intervals. However in Chapter 5, the reader can find a version of this relation for functions defined on unbounded intervals.

Let $\mathbb{T}$ be a time scale such that $t_0, t_0+\sigma \in \mathbb{T}$, where $\sigma > 0$, and consider a dynamic equation on time scales given by

$$x^{\Delta}(t) = f(x(t), t), \quad t \in [t_0, t_0+\sigma]_{\mathbb{T}}, \tag{3.36}$$

where $B \subset X$ is an open set and $f : B \times [t_0, t_0+\sigma]_{\mathbb{T}} \to X$ is a function such that $t \mapsto f(x(t), t)$ is Perron Δ-integrable on $[t_0, t_0+\sigma]_{\mathbb{T}}$. Integrating Eq. (3.36) with respect to t, we obtain

$$x(t) = x(t_0) + \int_{t_0}^{t} f(x(s), s)\Delta s, \quad t \in [t_0, t_0+\sigma]_{\mathbb{T}}, \tag{3.37}$$

where the integral on the right-hand side is the Perron Δ-integral.

Next, we present a concept of a solution of Eq. (3.37) which we refer to as a *dynamic equation on time scales.*

Definition 3.27: We say that a function $x : [t_0, t_0+\sigma]_{\mathbb{T}} \to X$ is a *solution of the dynamic equation on time scales* (3.37), with initial condition $x(s_0) = x_0$, $s_0 \in [t_0, t_0+\sigma]_{\mathbb{T}}$, provided

(i) $x(s_0) = x_0 \in B$;
(ii) $x \in G([t_0, t_0+\sigma]_{\mathbb{T}}, B)$ and $(x(t), t) \in B \times [t_0, t_0+\sigma]_{\mathbb{T}}$;
(iii) the Perron Δ-integral $\int_{t_0}^{t_0+\sigma} f(x(s), s)\Delta s$ exists;
(iv) the equality (3.37) holds for all $t \in [t_0, t_0+\sigma]_{\mathbb{T}}$.

The next result states a correspondence between the solutions of an MDE and the solutions of a dynamic equation on time scales, and it can be found in [78, 179, 219] for $X = \mathbb{R}^n$. We omit its proof here, since it follows as in proof of Theorem 3.30, presented in the next Subsection 3.3.5, with obvious adaptations.

Theorem 3.28: *Let $\mathbb{T}$ be a time scale such that $t_0, t_0 + \sigma \in \mathbb{T}$. Let $B \subset X$ be an open subset and $f : B \times [t_0, t_0 + \sigma]_{\mathbb{T}} \to X$ be a function. Assume that, for every $x \in G([t_0, t_0 + \sigma]_{\mathbb{T}}, B)$, the function $t \mapsto f(x(t), t)$ is Perron Δ-integrable on $[s_1, s_2]_{\mathbb{T}}$, for every $s_1, s_2 \in [t_0, t_0 + \sigma]_{\mathbb{T}}$. Define $g : [t_0, t_0 + \sigma] \to \mathbb{R}$ by $g(s) = s^*$, for every $s \in [t_0, t_0 + \sigma]$. If $x : [t_0, t_0 + \sigma]_{\mathbb{T}} \to B$ is a solution of the following Δ-integral equation on time scales*

$$x(t) = x_0 + \int_{s_0}^{t} f(x(s), s)\Delta s, \quad t \in [t_0, t_0 + \sigma]_{\mathbb{T}}, \tag{3.38}$$

where $x_0 \in B$ and $s_0 \in [t_0, t_0 + \sigma]_{\mathbb{T}}$, then $x^* : [t_0, t_0 + \sigma] \to B$ is a solution of the MDE

$$y(t) = x_0 + \int_{s_0}^{t} f^*(y(s), s)dg(s) = x_0 + \int_{s_0}^{t} f(y(s), s^*)dg(s). \tag{3.39}$$

Conversely, if $y : [t_0, t_0 + \sigma] \to B$ satisfies the MDE (3.39), then it must have the form $y = x^*$, where $x : [t_0, t_0 + \sigma]_{\mathbb{T}} \to B$ is a solution of the dynamic equation on time scales (3.38).

3.3.5 Relations with Measure FDEs

In this subsection, our goal is to show that a functional dynamic equation on time scales of the form

$$\begin{cases} x^{\Delta}(t) = f(x_t^*, t), & t \in [t_0, t_0 + \sigma]_{\mathbb{T}}, \\ x(t) = \phi(t), & t \in [t_0 - r, t_0]_{\mathbb{T}}, \end{cases} \tag{3.40}$$

where $t_0, t_0 + \sigma \in \mathbb{T}, B \subset X$ is open, $O = G([t_0 - r, t_0 + \sigma]_{\mathbb{T}}, B), P = \{x_t^* : x \in O, t \in [t_0, t_0 + \sigma]\}, f : P \times [t_0, t_0 + \sigma]_{\mathbb{T}} \to X$, and $\phi \in G([t_0 - r, t_0]_{\mathbb{T}}, B)$, can be regarded as a measure FDE. This correspondence was firstly presented in [85] for functions taking values in $\mathbb{R}^n$. Here, we extend this relation to the abstract case of Banach space-valued functions.

Before we continue, let us make a few comments. We want to investigate dynamic equations on time scales such that the Δ-derivative of the unknown function $x : \mathbb{T} \to X$ at $t \in \mathbb{T}$ depends on the values of $x(s)$, where $s \in [t - r, t] \cap \mathbb{T}$. But, unlike the classical case, here we have an obstacle to surpass: the function x_t is defined on a subset of $[-r, 0]$. In order to overcome this problem, we consider the function x_t^* instead.

Throughout this section and in the remaining of this chapter, x_t^* stands for $(x^*)_t$. Clearly, x_t^* carries the same information as x_t, but x_t^* is defined on the whole interval $[-r, 0]$. Thus, it seems reasonable to consider functional dynamic equations of the form

$$x^{\Delta}(t) = f(x_t^*, t).$$

Note that the functional dynamic equation on time scales (3.40) can be rewritten in its integral form as

$$\begin{cases} x(t) = x(t_0) + \displaystyle\int_{t_0}^{t} f(x_s^*, s)\Delta s, & t \in [t_0, t_0 + \sigma]_{\mathbb{T}}, \qquad (3.41) \\ x(t) = \phi(t), & t \in [t_0 - r, t_0]_{\mathbb{T}}, \qquad (3.42) \end{cases}$$

where $t_0, t_0 + \sigma \in \mathbb{T}$, $\sigma > 0$, $B \subset X$ is open, $O = G([t_0 - r, t_0 + \sigma]_{\mathbb{T}}, B)$, $P = \{x_t^* : x \in O, t \in [t_0, t_0 + \sigma]\}$, $f : P \times [t_0, t_0 + \sigma]_{\mathbb{T}} \to X$ is a function, and $\phi \in G([t_0 - r, t_0]_{\mathbb{T}}, B)$. What motivates us to deal with the integral form (3.41) instead of the differential form (3.40), is the fact that the hypothesis of our results are required on integrals instead of integrands and this enables our right-hand sides to be "*merely*" integrable.

A concept of a solution of the functional dynamic equation on time scales (3.41) follows next.

Definition 3.29: Let $B \subset X$ be open, $O = G([t_0 - r, t_0 + \sigma]_{\mathbb{T}}, B)$, and $P = \{x_t^* : x \in O, t \in [t_0, t_0 + \sigma]\}$. We say that a function $x : [t_0 - r, t_0 + \sigma]_{\mathbb{T}} \to X$ is a *solution of the functional dynamic equation on time scales* (3.41), provided

(i) $(x_t^*, t) \in P \times [t_0, t_0 + \sigma]_{\mathbb{T}}$;
(ii) the Perron Δ-integral $\int_{t_0}^{t_0+\sigma} f(x_s^*, s)\Delta s$ exists;
(iii) the equalities (3.41) and (3.42) hold for all $t \in [t_0, t_0 + \sigma]_{\mathbb{T}}$.

Now, we are ready to present the main result of this subsection which concerns a relation between the solutions of a measure FDE and the solutions of a functional dynamic equations on time scales.

Theorem 3.30: *Let $[t_0 - r, t_0 + \sigma]_{\mathbb{T}}$ be a time scale interval, with $t_0 \in \mathbb{T}$ and $\sigma > 0$. Let $B \subset X$ be open, $O = G([t_0 - r, t_0 + \sigma]_{\mathbb{T}}, B)$, $P = \{x_t^* : x \in O, t \in [t_0, t_0 + \sigma]\}$, $f : P \times [t_0, t_0 + \sigma]_{\mathbb{T}} \to X$ be a function and $\phi \in G([t_0 - r, t_0]_{\mathbb{T}}, B)$. Assume that for every $z \in P$, the Perron Δ-integral $\int_{u_1}^{u_2} f(z, s)\Delta s$ exists, for all $u_1, u_2 \in [t_0, t_0 + \sigma]_{\mathbb{T}}$. Define $g(s) = s^*$ for every $s \in [t_0, t_0 + \sigma]$. If $x : [t_0 - r, t_0 + \sigma]_{\mathbb{T}} \to B$ is a solution of the functional dynamic equation on time scales (3.41) and (3.42), then $x^* : [t_0 - r, t_0 + \sigma] \to$*

B satisfies

$$\begin{cases} x^*(t) = x^*(t_0) + \int_{t_0}^{t} f(x_s^*, s^*)dg(s), & t \in [t_0, t_0 + \sigma], \\ x_{t_0}^* = \phi_{t_0}^*. \end{cases} \quad (3.43)$$

Conversely, if $y : [t_0 - r, t_0 + \sigma] \to B$ *is a solution of the measure FDE*

$$\begin{cases} y(t) = y(t_0) + \int_{t_0}^{t} f(y_s, s^*)dg(s), & t \in [t_0, t_0 + \sigma], \\ y_{t_0} = \phi_{t_0}^*, \end{cases}$$

then $y = x^*$, *where* $x : [t_0 - r, t_0 + \sigma]_{\mathbb{T}} \to B$ *satisfies* (3.41) *and* (3.42).

Proof. Suppose $x : [t_0 - r, t_0 + \sigma]_{\mathbb{T}} \to B$ is a solution of (3.41) and (3.42). Then,

$$x(t) = x(t_0) + \int_{t_0}^{t} f(x_s^*, s)\Delta s, \quad \text{for } t \in [t_0, t_0 + \sigma]_{\mathbb{T}},$$

and, by Theorem 3.25,

$$x^*(t) = x^*(t_0) + \int_{t_0}^{t} f(x_{s^*}^*, s^*)dg(s), \quad \text{for } t \in [t_0, t_0 + \sigma].$$

Notice that $f(x_{s^*}^*, s^*) = f(x_s^*, s^*)$, for $s \in [t_0, t_0 + \sigma]_{\mathbb{T}}$. Then, Theorem 3.24 implies

$$x^*(t) = x^*(t_0) + \int_{t_0}^{t} f(x_s^*, s^*)dg(s), \quad t \in [t_0, t_0 + \sigma]. \quad (3.44)$$

In addition, for $\theta \in [-r, 0]$, we have

$$x_{t_0}^*(\theta) = x^*(t_0 + \theta) = x((t_0 + \theta)^*) = \phi((t_0 + \theta)^*) = \phi_{t_0}^*(\theta). \quad (3.45)$$

Then, (3.44) and (3.45) yield $x^* : [t_0 - r, t_0 + \sigma] \to B$ is a solution of (3.43).

Now, we assume that y satisfies

$$y(t) = y(t_0) + \int_{t_0}^{t} f(y_s, s^*)dg(s), \quad \text{for } t \in [t_0, t_0 + \sigma].$$

By definition, g is constant on every interval $(\alpha, \beta]$, where $\beta \in \mathbb{T}$ and $\alpha = \sup\{\tau \in \mathbb{T} : \tau < \beta\}$. Then, y inherits the same property and hence, $y = x^*$, for some $x : [t_0 - r, t_0 + \sigma]_{\mathbb{T}} \to B$. Using the same ideas of the first part of the proof, one can show that x satisfies the functional dynamic equation on time scales (3.41) and (3.42). □

Lemma 3.31: *Let* $[t_0, t_0 + \sigma]_{\mathbb{T}}$ *be a time scale interval. If there exists a Perron* Δ*-integrable function* $L : [t_0, t_0 + \sigma]_{\mathbb{T}} \to \mathbb{R}$, *then* $\int_{u_1}^{u_2} L(s)h^*(s)\Delta s$ *exists in the sense of Perron* Δ*-integral for every regulated function* $h^* : [t_0, t_0 + \sigma] \to \mathbb{R}$ *and* $u_1, u_2 \in [t_0, t_0 + \sigma]_{\mathbb{T}}$ *such that* $u_1 \leqslant u_2$.

Proof. Since $\int_{u_1}^{u_2} L(s)\Delta s$ exists for every $u_1, u_2 \in \mathbb{T}$ such that $u_1 \leqslant u_2$, by Theorem 3.22, it follows that $\int_{u_1}^{u_2} L^*(s)dg(s)$ exists in the sense of Perron–Stieltjes for $g(s) = s^*$, where $g : [t_0, t_0 + \sigma] \to \mathbb{R}$. Also, since g is nondecreasing by Corollary 3.21, $\int_{t_0}^{t} L^*(s)dg(s)$ is of bounded variation on $[t_0, t_0 + \sigma]$. Therefore, by Corollary 1.69, since $h^* : [t_0, t_0 + \sigma] \to \mathbb{R}$ is a regulated function, it follows that $\int_{u_1}^{u_2} L^*(s)h^*(s)dg(s)$ exists in the sense of Perron–Stieltjes for each $u_1, u_2 \in [t_0, t_0 + \sigma]$. Applying Theorems 3.22 and 3.25, it follows that $\int_{u_1^*}^{u_2^*} L(s)h^*(s)\Delta s$ exists and

$$\int_{u_1^*}^{u_2^*} L(s)h^*(s)\Delta s = \int_{u_1}^{u_2} L^*(s)h^*(s)dg(s),$$

getting the desired result. □

We finish this section with a crucial result for the application of the correspondence that we presented in Theorem 3.30. The next result provides a way of translating the results from functional dynamic equations on time scales to their analogues in the framework of measure FDEs. The version presented here is more general than that from [86], since our conditions hold for functions instead of constants.

Lemma 3.32: *Let $[t_0 - r, t_0 + \sigma]_{\mathbb{T}}$ be a time scale interval such that $t_0 \in \mathbb{T}$, $B \subset X$ be open, $O = G([t_0 - r, t_0 + \sigma], B)$, $P = G([-r, 0], B)$ and $f : P \times [t_0, t_0 + \sigma]_{\mathbb{T}} \to X$ be an arbitrary function. Define $g(t) = t^*$ and $f^*(z, t) = f(z, t^*)$ for every $z \in P$ and $t \in [t_0, t_0 + \sigma]$. The following statements hold.*

(i) *If the Perron Δ-integral $\int_{t_0}^{t_0+\sigma} f(y_s, s)\Delta s$ exists for every $y \in O$, then the Perron–Stieltjes integral $\int_{t_0}^{t_0+\sigma} f^*(y_s, s)dg(s)$ exists for every $y \in O$.*

(ii) *Suppose there is a Perron Δ-integrable function $M : [t_0, t_0 + \sigma]_{\mathbb{T}} \to \mathbb{R}$ such that, for every $y \in O$ and $u_1, u_2 \in [t_0, t_0 + \sigma]_{\mathbb{T}}$, with $u_1 \leqslant u_2$, we have*

$$\left\| \int_{u_1}^{u_2} f(y_s, s)\Delta s \right\| \leqslant \int_{u_1}^{u_2} M(s)\Delta s.$$

Then, for $t_0 \leqslant u_1 \leqslant u_2 \leqslant t_0 + \sigma$ and $y \in O$, we get

$$\left\| \int_{u_1}^{u_2} f^*(y_s, s)dg(s) \right\| \leqslant \int_{u_1}^{u_2} M^*(s)dg(s).$$

(iii) *Suppose there exists a Perron Δ-integrable function $L : [t_0, t_0 + \sigma]_{\mathbb{T}} \to \mathbb{R}$ such that, for $y, z \in O$ and $u_1, u_2 \in [t_0, t_0 + \sigma]_{\mathbb{T}}$, with $u_1 \leqslant u_2$, we have*

$$\left\| \int_{u_1}^{u_2} \left[f(y_s, s) - f(z_s, s) \right] \Delta s \right\| \leqslant \int_{u_1}^{u_2} L(s)\|y_s - z_s\|_\infty \Delta s.$$

Then, for $t_0 \leqslant u_1 \leqslant u_2 \leqslant t_0 + \sigma$ and $y, z \in O$, we obtain

$$\left\| \int_{u_1}^{u_2} [f^*(y_s, s) - f^*(z_s, s)]\, dg(s) \right\| \leqslant \int_{u_1}^{u_2} L^*(s)\|y_s - z_s\|_\infty\, dg(s).$$

Proof. Let $y \in O$. If the Perron Δ-integral $\int_{t_0}^{t_0+\sigma} f(y_s, s)\Delta s$ exists, then Theorems 3.22 and 3.24 yield

$$\begin{aligned}\int_{t_0}^{t_0+\sigma} f(y_s, s)\Delta s &= \int_{t_0}^{t_0+\sigma} f(y_{s^*}, s^*)dg(s) = \int_{t_0}^{t_0+\sigma} f(y_s, s^*)dg(s)\\ &= \int_{t_0}^{t_0+\sigma} f^*(y_s, s)dg(s),\end{aligned}$$

showing that the last integral also exists which leads to (i).

From Theorems 3.22 and 3.24 and Lemma 3.23, we obtain, for every $y \in O$ and $u_1, u_2 \in [t_0, t_0 + \sigma]_{\mathbb{T}}$, with $u_1 \leqslant u_2$,

$$\begin{aligned}\left\| \int_{u_1}^{u_2} f^*(y_s, s)dg(s) \right\| &= \left\| \int_{u_1}^{u_2} f(y_{s^*}, s^*)dg(s) \right\| = \left\| \int_{u_1^*}^{u_2^*} f(y_{s^*}, s^*)dg(s) \right\|\\ &= \left\| \int_{u_1^*}^{u_2^*} f(y_s, s)\Delta s \right\| \leqslant \int_{u_1^*}^{u_2^*} M(s)\Delta s\\ &= \int_{u_1^*}^{u_2^*} M^*(s)dg(s) = \int_{u_1}^{u_2} M^*(s)dg(s),\end{aligned}$$

which implies (ii).

Now, let us prove item (iii). Notice that the Perron Δ-integral $\int_{u_1}^{u_2} L(s)\|y_s - z_s\|_\infty \Delta s$ exists in the sense of Perron Δ-integral for each $u_2, u_1 \in [t_0, t_0 + \sigma]_{\mathbb{T}}$ by Lemma 3.31, since the function $s \in [t_0, t_0 + \sigma] \mapsto \|y_s - z_s\|$ is regulated for all $y, z \in O$. On the other hand, for every $y, z \in O$ and $u_1, u_2 \in [t_0, t_0 + \sigma]_{\mathbb{T}}$, with $u_1 \leqslant u_2$, we have

$$\begin{aligned}&\left\| \int_{u_1}^{u_2} [f^*(y_s, s) - f^*(z_s, s)]\, dg(s) \right\|\\ &= \left\| \int_{u_1}^{u_2} [f(y_{s^*}, s^*) - f(z_{s^*}, s^*)]\, dg(s) \right\| = \left\| \int_{u_1^*}^{u_2^*} [f(y_{s^*}, s^*) - f(z_{s^*}, s^*)]\, dg(s) \right\|\\ &= \left\| \int_{u_1^*}^{u_2^*} [f(y_s, s) - f(z_s, s)]\, \Delta s \right\| \leqslant \int_{u_1^*}^{u_2^*} L(s)\|y_s - z_s\|_\infty\, \Delta s\\ &= \int_{u_1^*}^{u_2^*} L^*(s)\|y_s - z_s\|_\infty\, dg(s) = \int_{u_1}^{u_2} L^*(s)\|y_s - z_s\|_\infty\, dg(s),\end{aligned}$$

proving (iii). □

3.3.6 Impulsive Functional Dynamic Equations on Time Scales

In this subsection, we focus our attention on functional dynamic equations on time scales subject to impulse effects and their relations with impulsive measure FDEs.

These types of equations have been investigated by many authors (see [14, 15, 37, 109] and the references therein). Consider the following moments of impulses $\{t_k\}_{k=1}^{m} \subset [t_0, t_0+\sigma]_{\mathbb{T}}$ such that $t_0 \leqslant t_1 < \cdots < t_m \leqslant t_0+\sigma$, and the impulses operators $I_k : B \to X$, $B \subset X$ is open, with $k \in \{1, \dots, m\}$. Usually, an impulse condition is described by

$$x\left(t_k^+\right) - x\left(t_k^-\right) = I_k\left(x\left(t_k^-\right)\right), \quad k \in \{1, \dots, m\}. \tag{3.46}$$

When we are dealing with the theory of time scales, the following conventions are commonly adopted:

- $x(t^+) = x(t)$, whenever $t \in \mathbb{T}$ is a right-scattered point;
- $x(t^-) = x(t)$, if $t \in \mathbb{T}$ is a left-scattered point.

Notice that if $x : [t_0 - r, t_0+\sigma]_{\mathbb{T}} \to B$ is a left-continuous function, then the impulsive condition (3.46) can be rewritten as

$$x(t_k^+) - x(t_k) = I_k(x(t_k)), \quad k \in \{1, \dots, m\}. \tag{3.47}$$

As commented in Section 3.2, we shall consider that $I + I_k : B \to B$ for every $k \in \{1, \dots, m\}$, where $I : B \to B$ is the identity operator.

From (3.47), if t_k is right-scattered, then $I_k(x(t_k)) = 0$. Hence, we can assume, without loss of generality, that t_k is right-dense, for every $k \in \{1, \dots, m\}$. This leads us to consider the following impulsive functional dynamic equation on time scales

$$\begin{cases} x^{\Delta}(t) = f\left(x_t^*, t\right), & t \in [t_0, t_0+\sigma]_{\mathbb{T}} \setminus \{t_1, \dots, t_m\}, \\ \Delta^+ x(t_k) = I_k(x(t_k)), & k \in \{1, \dots, m\}, \\ x(t) = \phi(t), & t \in [t_0 - r, t_0]_{\mathbb{T}}, \end{cases} \tag{3.48}$$

where $B \subset X$ is open, $O = G([t_0 - r, t_0+\sigma]_{\mathbb{T}}, B)$, $P = \{x_t^* : x \in O, t \in [t_0, t_0+\sigma]\}$, $f : P \times [t_0, t_0+\sigma]_{\mathbb{T}} \to X$ is a function, $\phi \in G([t_0 - r, t_0]_{\mathbb{T}}, B)$, $t_1, \dots, t_m \in \mathbb{T}$ are right-dense points of impulse effects satisfying

$$t_0 \leqslant t_1 < t_2 < \cdots < t_m < t_0 + \sigma,$$

$I_1, \dots, I_m : B \to X$ are the impulse operators such that $I + I_k : B \to B$ for $k \in \{1, \dots, m\}$, and we are assuming here that the solution is left-continuous. Hence, one can rewrite the solution of the system (3.48) in an integral form given by

$$\begin{cases} x(t) = x(t_0) + \int_{t_0}^{t} f(x_s^*, s)\Delta s + \sum_{\substack{k\in\{1,\dots,m\}\\ t_k<t}} I_k(x(t_k)), & t \in [t_0, t_0+\sigma]_{\mathbb{T}}, \quad (3.49) \\ x(t) = \phi(t), \quad t \in [t_0 - r, t_0]_{\mathbb{T}}, & \quad (3.50) \end{cases}$$

where the Δ-integral is understood in the sense of Perron Δ-integral.

Definition 3.33: Let $B \subset X$ be open, $O = G^-([t_0 - r, t_0 + \sigma]_{\mathbb{T}}, B)$, $P = \{x_t^* : x \in O, t \in [t_0, t_0 + \sigma]\}$. We say that a function $x : [t_0 - r, t_0 + \sigma]_{\mathbb{T}} \to X$ is *a solution of the impulsive functional dynamic equation on time scales* (3.49) and (3.50) on $[t_0 - r, t_0 + \sigma]_{\mathbb{T}}$, if the following conditions are fulfilled:

(i) the Perron Δ-integral $\int_{t_0}^{t_0+\sigma} f(x_s^*, s)\Delta s$ exists;
(ii) $(x_t^*, t) \in P \times [t_0, t_0 + \sigma]_{\mathbb{T}}$;
(iii) the equalities (3.49) and (3.50) hold for all $t \in [t_0, t_0 + \sigma]_{\mathbb{T}}$.

Now, we are ready to present a correspondence, borrowed from [86], between the solutions of an impulsive functional dynamic equation on time scales and the solutions of an impulsive measure FDE.

Theorem 3.34: *Let $[t_0 - r, t_0 + \sigma]_{\mathbb{T}}$ be a time scale interval, with $t_0, t_0 + \sigma \in \mathbb{T}$ and $\sigma > 0$. Consider an open set $B \subset X$, $O = G^-([t_0 - r, t_0 + \sigma]_{\mathbb{T}}, B)$, $P = \{x_t^* : x \in O, t \in [t_0, t_0 + \sigma]\}$, a function $f : P \times [t_0, t_0 + \sigma]_{\mathbb{T}} \to X$ and $\phi \in G([t_0 - r, t_0]_{\mathbb{T}}, B)$. Define a function $g : [t_0, t_0 + \sigma] \to \mathbb{R}$ by $g(s) = s^*$ for every $s \in [t_0, t_0 + \sigma]$. Assume that $t_1, \dots, t_m \in \mathbb{T}$ are right-dense points of impulse effects satisfying $t_0 \leqslant t_1 < t_2 < \cdots < t_m < t_0 + \sigma$, $I_1, \dots, I_m : B \to X$ are the impulse operators such that $I + I_k : B \to B$ for each $k = 1, \dots, m$, where $I : B \to B$ is the identity operator. If $x : [t_0 - r, t_0 + \sigma]_{\mathbb{T}} \to B$ is a solution of the impulsive functional dynamic equation on time scales*

$$\begin{cases} x(t) = x(t_0) + \int_{t_0}^{t} f(x_s^*, s)\Delta s + \sum_{\substack{k\in\{1,\dots,m\}\\ t_k<t}} I_k(x(t_k)), & t \in [t_0, t_0+\sigma]_{\mathbb{T}}, \\ x(t) = \phi(t), & t \in [t_0 - r, t_0]_{\mathbb{T}}, \end{cases} \tag{3.51}$$

then $x^* : [t_0 - r, t_0 + \sigma] \to B$ is a solution of the impulsive measure FDE

$$\begin{cases} y(t) = y(t_0) + \int_{t_0}^{t} f(y_s, s^*)dg(s) + \sum_{\substack{k\in\{1,\dots,m\}\\ t_k<t}} I_k(y(t_k)), & t \in [t_0, t_0+\sigma], \\ y_{t_0} = \phi_{t_0}^*. \end{cases} \tag{3.52}$$

Conversely, if $y : [t_0 - r, t_0 + \sigma] \to B$ *satisfies* (3.52)*, then it has the form* $y = x^*$*, where* $x : [t_0 - r, t_0 + \sigma]_{\mathbb{T}} \to B$ *is a solution of* (3.51).

Proof. Suppose $x : [t_0 - r, t_0 + \sigma]_{\mathbb{T}} \to B$ is a solution of the impulsive functional dynamic equation on time scales (3.51). Using Theorem 3.25, we obtain

$$x^*(t) = x^*(t_0) + \int_{t_0}^{t} f(x^*_{s^*}, s^*)dg(s) + \sum_{\substack{k\in\{1,\ldots,m\}\\ t_k<t^*}} I_k(x(t_k)), \quad t \in [t_0, t_0 + \sigma].$$

Since $t_k \in \mathbb{T}$, for every $k \in \{1, \ldots, m\}$, we have $x(t_k) = x^*(t_k)$ and $t_k < t^*$ if and only if $t_k < t$. The equality $f(x^*_{s^*}, s^*) = f(x^*_s, s^*)$ together with Theorem 3.24 yield

$$x^*(t) = x^*(t_0) + \int_{t_0}^{t} f(x^*_s, s^*)dg(s) + \sum_{\substack{k\in\{1,\ldots,m\}\\ t_k<t}} I_k(x^*(t_k)),$$

for every $t \in [t_0, t_0 + \sigma]_{\mathbb{T}}$. Moreover, for $\theta \in [-r, 0]$, we get

$$x^*_{t_0}(\theta) = x^*(t_0 + \theta) = x((t_0 + \theta)^*) = \phi((t_0 + \theta)^*) = \phi^*_{t_0}(\theta),$$

whence we conclude that x^* is a solution of (3.52).

Now, suppose $y : [t_0 - r, t_0 + \sigma] \to B$ is a solution of the impulsive measure FDE (3.52). If $t \in \mathbb{T}, t_0 < u < t \leqslant t_0 + \sigma$ and $[u, t) \cap \mathbb{T} = \emptyset$, then g is constant on $[u, t]$ and hence, $y(u) = y(t)$. Thus, $y = x^*$, for some $x : [t_0 - r, t_0 + \sigma]_{\mathbb{T}} \to B$. The same argument used in the previous part shows that $x : [t_0 - r, t_0 + \sigma]_{\mathbb{T}} \to B$ is a solution of impulsive functional dynamic equation on time scales (3.51). □

3.4 Averaging Methods

In this section, our goal is to prove averaging principles for measure FDEs and, using the correspondences presented in Subsections 3.3.5 and 3.3.6, we provide averaging principles for impulsive measure FDEs and impulsive functional dynamic equations on time scales. Here, we present both periodic and nonperiodic versions of averaging principles for these equations.

It is known that the right-hand sides of nonautonomous equations which model problems in celestial mechanics may have fast oscillating and slowly oscillating terms. The terms which slowly oscillate indicate the slow evolution of the system of parameters. The terms which are fast oscillating hardly affect the motion and, thus, they can be somehow neglected. The process of omitting the terms which are highly oscillating on the right-hand sides of nonautonomous equation is called "*averaging*". See, also, Chapter 10 on averaging for generalized ordinary differential equations (ODEs).

3.4.1 Periodic Averaging

Let ϵ_0, L, and T be positive numbers and consider the following nonlinear ODE

$$\begin{cases} \dot{x}(t) = \epsilon f(x(t), t) + \epsilon^2 g(x(t), t, \epsilon), \\ x(t_0) = x_0, \end{cases} \tag{3.53}$$

where $\epsilon \in (0, \epsilon_0]$ is a small parameter, $O \subset \mathbb{R}^n$ is open, $x_0 \in O, f : O \times [0, \frac{L}{\epsilon}] \to \mathbb{R}^n$ is T-periodic with respect to the second variable, and $g : O \times [0, \frac{L}{\epsilon}] \times (0, \epsilon_0] \to \mathbb{R}^n$ is a function. The main idea behind periodic averaging principles is to provide sufficient conditions to ensure that the solutions of the nonautonomous equation (3.53) remain close to the solutions of an autonomous ODE of the form

$$\begin{cases} \dot{y}(t) = \epsilon f_0(y(t)), \\ y(t_0) = x_0, \end{cases}$$

where, for every $y \in O$,

$$f_0(y) = \frac{1}{T} \int_{t_0}^{t_0+T} f(y, t)dt.$$

Throughout the years, this type of method for ODEs has been extensively investigated by many authors. See, for instance, [202, 203, 233] and the references therein. The interest in averaging principles lies on the fact that one can investigate asymptotic properties of the solutions of a nonautonomous systems, only by studying the asymptotic behavior of the solutions of the corresponding "*averaged*" autonomous system, which, by the way, is easier to deal with.

Periodic averaging principles for FDEs have also been investigated by many authors. We can mention, for instance, [82, 155, 160–163]. On the other hand, such type of results for measure FDEs are relatively scarce. As far as we are concerned, averaging methods for measure FDEs were first established in the paper [178], where the authors considered a nonautonomous measure FDE and the averaged equation was an autonomous FDE. Thus, it is still an open problem whether it is possible to find an averaged equation which is an autonomous measure FDE.

Here, we investigate periodic averaging principles for measure FDE whose functions take values in a Banach space. We provide conditions under which a solution $x_\epsilon : \left[0, \frac{L}{\epsilon}\right] \to X$ of the initial value problem

$$\begin{cases} x(t) = x(0) + \epsilon \int_0^t f(x_s, s)dh(s) + \epsilon^2 \int_0^t g(x_s, s, \epsilon)dh(s), \\ x_0 = \epsilon\phi, \end{cases} \tag{3.54}$$

can be approximated by a solution $y_\epsilon : \left[0, \frac{L}{\epsilon}\right] \to X$ of the averaged autonomous FDE

$$\begin{cases} y(t) = y(0) + \epsilon \displaystyle\int_0^t f_0(y_s)ds, \\ y_0 = \epsilon\phi, \end{cases} \tag{3.55}$$

where

$$f_0(\varphi) = \frac{1}{T}\int_0^T f(\varphi, s)dh(s), \quad \text{for } \varphi \in O,$$

and, moreover, $\epsilon \in (0, \epsilon_0]$ is a small parameter, $O \subset G([-r, 0], X)$ is an open set, $\phi \in O$, $h : [0, \infty) \to \mathbb{R}$ is a function, $f : O \times [0, \frac{L}{\epsilon}] \to X$ is a T-periodic function with respect to the second variable, and $g : O \times [0, \frac{L}{\epsilon}] \times (0, \epsilon_0] \to X$ is a function.

We also present averaging principles for impulsive measure FDEs and impulsive functional dynamic equations on time scales based on [86].

Let us start by a periodic averaging method for the measure FDE (3.54). Such result is slightly different from the one presented in [178], because our initial condition contains the parameter ϵ, as it also does in (3.54) and (3.55).

Theorem 3.35: *Let ϵ_0, L, and T be positive constants, $B \subset X$ be an open set and $P = G([-r, 0], B)$. Consider a pair of bounded functions $f : P \times [0, \infty) \to X$ and $g : P \times [0, \infty) \times (0, \epsilon_0] \to X$ and a nondecreasing function $h : [0, \infty) \to \mathbb{R}$ such that the following conditions are satisfied:*

(H1) *the Perron–Stieltjes integrals*

$$\int_0^b f(y_s, s)dh(s) \quad \text{and} \quad \int_0^b g(y_s, s, \epsilon)dh(s)$$

exist for every $b > 0$, $y \in G([-r, b], B)$ and $\epsilon \in (0, \epsilon_0]$;

(H2) *f is T-periodic on the second variable;*

(H3) *there exists a constant $\alpha > 0$ such that $h(t + T) - h(t) = \alpha$, for every $t \geqslant 0$;*

(H4) *there exists a constant $C > 0$ such that, for $x, y \in P$ and $t \in [0, \infty)$, we have*

$$\|f(x, t) - f(y, t)\| \leqslant C\|x - y\|_\infty.$$

Define

$$f_0(z) = \frac{1}{T}\int_0^T f(z, s)dh(s), \quad \text{for } z \in P$$

and let $\phi \in P$ be a bounded function. Suppose, in addition, for every $\epsilon \in (0, \epsilon_0]$, x_ϵ : $\left[0, \frac{L}{\epsilon}\right] \to B$ is a solution of the nonautonomous measure FDE

$$\begin{cases} x(t) = x(0) + \epsilon \int_0^t f(x_s, s)dh(s) + \epsilon^2 \int_0^t g(x_s, s, \epsilon)dh(s), \\ x_0 = \epsilon\phi, \end{cases} \tag{3.56}$$

and y_ϵ : $\left[0, \frac{L}{\epsilon}\right] \to B$ is a solution of the autonomous FDE

$$\begin{cases} y(t) = y(0) + \epsilon \int_0^t f_0(y_s)ds, \\ y_0 = \epsilon\phi. \end{cases} \tag{3.57}$$

Then, there exists a constant $J > 0$ such that

$$\|x_\epsilon(t) - y_\epsilon(t)\| \leqslant J\epsilon, \quad \textit{for every } \epsilon \in (0, \epsilon_0] \textit{ and } t \in \left[-r, \frac{L}{\epsilon}\right].$$

Proof. Since $f : P \times [0, \infty) \to X$, $g : P \times [0, \infty) \times (0, \epsilon_0] \to X$, and $\phi \in P$ are bounded functions, we can assume, without loss of generality, that there exists a constant $M > 0$ such that $\|f(z, t)\| \leqslant M$ and $\|g(z, t, \epsilon)\| \leqslant M$ for all $z \in P$, $t \in [0, \infty)$ and all $\epsilon \in (0, \epsilon_0]$, and also $\|\phi\|_\infty \leqslant M$. Then,

$$\|f_0(x)\| = \left\| \frac{1}{T} \int_0^T f(x, s)dh(s) \right\| \leqslant \frac{M}{T}[h(T) - h(0)] = \frac{M\alpha}{T}.$$

Thus, for $\epsilon \in (0, \epsilon_0]$ and $s, t \in [0, \infty)$, with $s \geqslant t$, we consider the following cases:

(i) if $s + \theta, t + \theta < 0$, then

$$\|y_\epsilon(s + \theta) - y_\epsilon(t + \theta)\| = \|\epsilon\phi(s + \theta) - \epsilon\phi(t + \theta)\| \leqslant \epsilon 2M;$$

(ii) if $s + \theta, t + \theta = 0$, then $\|y_\epsilon(s + \theta) - y_\epsilon(t + \theta)\| = 0$;

(iii) if $s + \theta \geqslant 0$ and $t + \theta \leqslant 0$, then

$$\begin{aligned} \|y_\epsilon(s + \theta) - y_\epsilon(t + \theta)\| &\leqslant \left\| \epsilon \int_0^{s+\theta} f_0((y_\epsilon)_\sigma)d\sigma \right\| + \|\epsilon\phi(t + \theta)\| + \|\epsilon\phi(0)\| \\ &\leqslant \frac{\epsilon M(s + \theta)\alpha}{T} + \epsilon 2M \\ &\leqslant \frac{\epsilon M(s + \theta)\alpha}{T} - \frac{\epsilon M(t + \theta)\alpha}{T} + 2\epsilon M \\ &= \frac{\epsilon M(s - t)\alpha}{T} + \epsilon 2M; \end{aligned}$$

(iv) if $s+\theta, t+\theta \geqslant 0$, then

$$\|y_\epsilon(s+\theta) - y_\epsilon(t+\theta)\| = \left\|\epsilon \int_{t+\theta}^{s+\theta} f_0((y_\epsilon)_\sigma)d\sigma\right\| \leqslant \frac{\epsilon M(s-t)\alpha}{T}, \quad \theta \in [-r, 0].$$

Combining all the cases above, we obtain

$$\|(y_\epsilon)_s - (y_\epsilon)_t\|_\infty = \sup_{\theta\in[-r,0]} \|y_\epsilon(s+\theta) - y_\epsilon(t+\theta)\| \leqslant \frac{\epsilon M(s-t)\alpha}{T} + \epsilon 2M, \tag{3.58}$$

for all $s, t \in [0, \infty)$, with $s \geqslant t$.

On the other hand, for every $t \in \left[0, \frac{L}{\epsilon}\right]$, we have

$$\begin{aligned}
\|x_\epsilon(t) - y_\epsilon(t)\| &= \left\|\epsilon \int_0^t f((x_\epsilon)_s, s)dh(s) + \epsilon^2 \int_0^t g((x_\epsilon)_s, s, \epsilon)dh(s)\right.\\
&\quad \left. - \epsilon \int_0^t f_0((y_\epsilon)_s)ds\right\| \leqslant \epsilon \left\|\int_0^t [f((x_\epsilon)_s, s) - f((y_\epsilon)_s, s)]\, dh(s)\right\|\\
&\quad + \epsilon \left\|\int_0^t f((y_\epsilon)_s, s)dh(s) - \int_0^t f_0((y_\epsilon)_s)ds\right\|\\
&\quad + \epsilon^2 \left\|\int_0^t g((x_\epsilon)_s, s, \epsilon)dh(s)\right\|\\
&\leqslant \epsilon \int_0^t C\|(x_\epsilon)_s - (y_\epsilon)_s\|_\infty\, dh(s)\\
&\quad + \epsilon \left\|\int_0^t f((y_\epsilon)_s, s)dh(s) - \int_0^t f_0((y_\epsilon)_s)ds\right\|\\
&\quad + \epsilon^2 M(h(t) - h(0)).
\end{aligned} \tag{3.59}$$

Let us estimate the second term on the right-hand side of (3.59). Given $t \in \left[0, \frac{L}{\epsilon}\right]$, let p be the largest integer such that $pT \leqslant t$. Therefore,

$$\begin{aligned}
&\left\|\int_0^t f((y_\epsilon)_s, s)dh(s) - \int_0^t f_0((y_\epsilon)_s)ds\right\|\\
&\leqslant \sum_{i=1}^p \left\|\int_{(i-1)T}^{iT} [f((y_\epsilon)_s, s) - f((y_\epsilon)_{(i-1)T}, s)]\, dh(s)\right\|\\
&\quad + \sum_{i=1}^p \left\|\int_{(i-1)T}^{iT} f((y_\epsilon)_{(i-1)T}, s)dh(s) - \int_{(i-1)T}^{iT} f_0((y_\epsilon)_{(i-1)T})ds\right\|\\
&\quad + \sum_{i=1}^p \left\|\int_{(i-1)T}^{iT} [f_0((y_\epsilon)_{(i-1)T}) - f_0((y_\epsilon)_s)]\, ds\right\|
\end{aligned}$$

$$+\left\|\int_{pT}^{t} f((y_\epsilon)_s, s)dh(s) - \int_{pT}^{t} f_0((y_\epsilon)_s)ds\right\|.$$

For every $i \in \{1, 2, \dots, p\}$ and every $s \in [(i-1)T, iT]$, (3.58) gives us

$$\|(y_\epsilon)_s - (y_\epsilon)_{(i-1)T}\|_\infty \leqslant \frac{M\epsilon\alpha(s-(i-1)T)}{T} + \epsilon 2M \leqslant M\epsilon(\alpha+2),$$

which, together with the fact that $pT \leqslant \frac{L}{\epsilon}$, imply

$$\sum_{i=1}^{p}\left\|\int_{(i-1)T}^{iT} [f((y_\epsilon)_s, s) - f((y_\epsilon)_{(i-1)T}, s)]\,dh(s)\right\|$$
$$\leqslant \sum_{i=1}^{p} CM\epsilon(\alpha+2)[h(iT) - h((i-1)T)] = CM\epsilon\alpha(\alpha+2)p \leqslant \frac{CML\alpha(\alpha+2)}{T}.$$

On the other hand, for every $z_s, z_t \in P$, with $s, t \geqslant 0$, the definition of f_0 implies

$$\|f_0(z_s) - f_0(z_t)\| \leqslant \frac{1}{T}\left\|\int_0^T [f(z_s,\sigma) - f(z_t,\sigma)]\,dh(\sigma)\right\|$$
$$\leqslant \frac{C}{T}\|z_s - z_t\|_\infty[h(T) - h(0)] = \frac{C}{T}\|z_s - z_t\|_\infty\alpha, \tag{3.60}$$

whence, using the fact that $(y_\epsilon)_s, (y_\epsilon)_{(i-1)T} \in P$ for $s \in [(i-1)T, iT]$, yields

$$\sum_{i=1}^{p}\left\|\int_{(i-1)T}^{iT} [f_0((y_\epsilon)_s) - f_0((y_\epsilon)_{(i-1)T})]\,ds\right\|$$
$$\leqslant \sum_{i=1}^{p}\int_{(i-1)T}^{iT} \left\|f_0((y_\epsilon)_s) - f_0((y_\epsilon)_{(i-1)T})\right\|\,ds$$
$$\leqslant \frac{C}{T}\alpha\sum_{i=1}^{p}\int_{(i-1)T}^{iT} \|(y_\epsilon)_s - (y_\epsilon)_{(i-1)T}\|_\infty\,ds \leqslant \frac{C}{T}\alpha\left[\sum_{i=1}^{p}\epsilon M(\alpha+2)T\right]$$
$$= \epsilon MC\alpha(\alpha+2)p \leqslant \frac{MCL\alpha(\alpha+2)}{T},$$

where the second estimate follows from Corollary 1.48, since $t \mapsto f_0(y_t)$ is a regulated function (because f_0 is continuous by (3.60) and $t \mapsto y_t$ is a regulated function).

By the T-periodicity of f on the second variable and from the definition of f_0, we obtain

$$\sum_{i=1}^{p}\left\|\int_{(i-1)T}^{iT} f((y_\epsilon)_{(i-1)T}, s)dh(s) - \int_{(i-1)T}^{iT} f_0((y_\epsilon)_{(i-1)T})ds\right\|$$
$$= \sum_{i=1}^{p}\left\|\int_0^T f((y_\epsilon)_{(i-1)T}, s)dh(s) - f_0((y_\epsilon)_{(i-1)T})T\right\| = 0.$$

Moreover,

$$\left\| \int_{pT}^{t} f((y_\epsilon)_s, s)dh(s) - \int_{pT}^{t} f_0((y_\epsilon)_s)ds \right\| \leqslant \left\| \int_{pT}^{t} f((y_\epsilon)_s, s)dh(s) \right\|$$

$$+ \int_{pT}^{t} \|f_0((y_\epsilon)_s)\| \; ds \leqslant M[h(t) - h(pT)] + \frac{M\alpha}{T}(t - pT)$$

$$\leqslant M[h((p+1)T) - h(pT)] + \frac{M\alpha T}{T} = 2M\alpha,$$

whence we derive the following estimate

$$\left\| \int_{0}^{t} f((y_\epsilon)_s, s)dh(s) - \int_{0}^{t} f_0((y_\epsilon)_s)ds \right\| \leqslant \frac{2MCL\alpha(\alpha+2)}{T} + 2M\alpha.$$

Set $K = \frac{2MCL\alpha(\alpha+2)}{T} + 2M\alpha$. Then, from (3.59), we obtain

$$\|x_\epsilon(t) - y_\epsilon(t)\| \leqslant \epsilon \int_{0}^{t} C\|(x_\epsilon)_s - (y_\epsilon)_s\|_\infty \, dh(s) + \epsilon K + \epsilon^2 M[h(t) - h(0)]$$

and, defining $\psi(s) = \sup_{\tau \in [0,s]} \|x_\epsilon(\tau) - y_\epsilon(\tau)\|$, we obtain, for every $u \in [0, t]$,

$$\|x_\epsilon(u) - y_\epsilon(u)\| \leqslant \epsilon \int_{0}^{u} C\psi(s)dh(s) + \epsilon K + \epsilon^2 M[h(u) - h(0)]$$

$$\leqslant \epsilon \int_{0}^{t} C\psi(s)dh(s) + \epsilon K + \epsilon^2 M[h(t) - h(0)].$$

Therefore,

$$\psi(t) \leqslant \epsilon \int_{0}^{t} C\psi(s)dh(s) + \epsilon K + \epsilon^2 M[h(t) - h(0)]. \tag{3.61}$$

On the other hand, we have

$$\epsilon[h(t) - h(0)] \leqslant \epsilon[h(L/\epsilon) - h(0)] \leqslant \epsilon[h(\lceil L/(\epsilon T)\rceil \, T) - h(0)]$$

$$\leqslant \epsilon \left\lceil \frac{L}{\epsilon T} \right\rceil \alpha \leqslant \epsilon \left(\frac{L}{\epsilon T} + 1 \right) \alpha \leqslant \left(\frac{L}{T} + \epsilon_0 \right) \alpha, \tag{3.62}$$

where $\lceil q \rceil$ denotes the integer part of the real number q. Then, by (3.61), we obtain

$$\psi(t) \leqslant \epsilon \int_{0}^{t} C\psi(s)dh(s) + \epsilon K + \epsilon M \left(\frac{L}{T} + \epsilon_0 \right) \alpha$$

and hence, the Grönwall inequality (see Theorem 1.52) together with (3.62) yield

$$\psi(t) \leqslant e^{\epsilon C[h(t)-h(0)]} \left(K + M \left(\frac{L}{T} + \epsilon_0 \right) \alpha \right) \epsilon$$

$$\leqslant e^{C\left(\frac{L}{T}+\epsilon_0\right)\alpha} \left(K + M \left(\frac{L}{T} + \epsilon_0 \right) \alpha \right) \epsilon.$$

Define

$$J = e^{C\left(\frac{L}{T}+\epsilon_0\right)\alpha} \left(K + M \left(\frac{L}{T} + \epsilon_0 \alpha \right) \right).$$

Thus, clearly

$$\|x_\epsilon(t) - y_\epsilon(t)\| \leqslant \psi(t) \leqslant J\epsilon,$$

for every $\epsilon \in (0, \epsilon_0]$ and $t \in [0, \frac{L}{\epsilon}]$. Then, the definitions of the initial conditions of both systems imply that, for every $\epsilon \in (0, \epsilon_0]$ and $t \in [-r, 0]$, we have

$$\|x_\epsilon(t) - y_\epsilon(t)\| = 0 \leqslant \psi(t) \leqslant J\epsilon,$$

finishing the proof. □

In the sequel, using the relations presented previously in this chapter between the solutions of a measure FDE and other types of equations (see Theorems 3.6 and 3.34), one can derive periodic averaging principles for impulsive measure FDEs as well as for impulsive functional dynamic equations on time scales. Slightly different versions can be found in [86]. Here, though, similarly as in Theorem 3.35, we consider Banach space-valued functions and a parameter within the initial conditions.

Theorem 3.36: *Assume that $\epsilon_0, L, T > 0$, $P = G([-r, 0], X)$, and there exists $m \in \mathbb{N}$ such that $0 \leqslant t_1 < t_2 < \cdots < t_m < T$. Consider bounded functions $f : P \times [0, \infty) \to X$ and $g : P \times [0, \infty) \times (0, \epsilon_0] \to X$ and a nondecreasing left-continuous function $h : [0, \infty) \to \mathbb{R}$ which is continuous at t_k, for every $k \in \mathbb{N}$. Let $I_k : X \to X$, with $k \in \mathbb{N}$, be bounded and Lipschitz-continuous functions representing the impulse operators. For every integer $k > m$, define t_k and I_k by the recursive formulas $t_k = t_{k-m} + T$ and $I_k = I_{k-m}$. Suppose the following conditions are satisfied:*

(I1) *the Perron–Stieltjes integrals*

$$\int_0^b f(y_s, s)dh(s) \quad \text{and} \quad \int_0^b g(y_s, s, \epsilon)dh(s)$$

exist for every $b > 0$, $y \in G([-r, b], X)$ and $\epsilon \in (0, \epsilon_0]$;

(I2) *f is Lipschitz-continuous with respect to the first variable;*

(I3) *f is T-periodic with respect to the second variable;*

(I4) *there is a constant $\alpha > 0$ for which $h(t + T) - h(t) = \alpha$, for every $t \geqslant 0$;*

(I5) *the Perron–Stieltjes integral*

$$f_0(x) = \frac{1}{T} \int_0^T f(x, s)dh(s)$$

exists for every $x \in P$.

Let

$$I_0(z) = \frac{1}{T} \sum_{k=1}^{m} I_k(z), \quad \text{for all } z \in X$$

and assume that $\phi \in P$ is bounded. Suppose, in addition, for every $\epsilon \in (0, \epsilon_0]$, $x_\epsilon : \left[-r, \frac{L}{\epsilon}\right] \to X$ is a solution of the nonautonomous impulsive measure FDE

$$\begin{cases} x(t) = x(0) + \epsilon \displaystyle\int_0^t f(x_s, s)dh(s) + \epsilon^2 \int_0^t g(x_s, s, \epsilon)dh(s) + \epsilon \sum_{\substack{k\in\mathbb{N}\\ t_k<t}} I_k(x(t_k)), \\ x_0 = \epsilon\phi, \end{cases} \tag{3.63}$$

and $y_\epsilon : \left[-r, \frac{L}{\epsilon}\right] \to X$ is a solution of the initial value problem

$$\begin{cases} y(t) = y(0) + \epsilon \displaystyle\int_0^t [f_0(y_s) + I_0(y(s))]\, ds, \\ y_0 = \epsilon\phi. \end{cases} \tag{3.64}$$

Then, there exists a constant $J > 0$ such that

$$\|x_\epsilon(t) - y_\epsilon(t)\| \leqslant J\epsilon, \quad \text{for all } \epsilon \in (0, \epsilon_0] \text{ and } t \in \left[-r, \frac{L}{\epsilon}\right].$$

Proof. Define a function $\widetilde{h} : [0, \infty) \to \mathbb{R}$ by

$$\widetilde{h}(t) = \begin{cases} h(t), & t \in [0, t_1], \\ h(t) + c_k, & t \in (t_k, t_{k+1}],\ k \in \mathbb{N}, \end{cases}$$

where the sequence $\{c_k\}_{k\in\mathbb{N}}$ is such that $0 \leqslant c_k \leqslant c_{k+1}$, for every $k \in \mathbb{N}$. Suppose $\Delta^+\widetilde{h}(t_k) = 1$, for every $k \in \mathbb{N}$. Since h is a nondecreasing and left-continuous function, the same applies to $\widetilde{h}$. Besides, it is not difficult to see that there exists a constant $\widetilde{\alpha} > 0$ such that

$$\widetilde{h}(t + T) - \widetilde{h}(t) = \widetilde{\alpha}, \quad \text{for all } t \geqslant 0.$$

On the other hand, since $x_\epsilon : [-r, \frac{L}{\epsilon}] \to X$ is a solution of the nonautonomous impulsive measure FDE (3.63), we have

$$x_\epsilon(t) = x_\epsilon(0) + \int_0^t \left[\epsilon f((x_\epsilon)_s, s) + \epsilon^2 g((x_\epsilon)_s, s, \epsilon)\right] dh(s) + \sum_{\substack{k\in\mathbb{N}\\ t_k<t}} \epsilon I_k(x_\epsilon(t_k))$$

for every $\epsilon \in (0, \epsilon_0]$ and $t \in \left[0, \frac{L}{\epsilon}\right]$. Define, for every $y \in P$ and $t \geqslant 0$,

$$F^\epsilon(y, t) = \epsilon\widetilde{f}(y, t) + \epsilon^2\widetilde{g}(y, t, \epsilon),$$

where

$$\widetilde{f}(y, t) = \begin{cases} f(y, t), & t \notin \{t_1, t_2, \dots\}, \\ I_k(y(0)), & t = t_k,\ k \in \mathbb{N}, \end{cases} \quad \text{and}$$

$$\widetilde{g}(y,t,\epsilon)=\begin{cases} g(y,t,\epsilon), & t\notin\{t_1,t_2,\dots\},\\ 0, & t=t_k,\ k\in\mathbb{N}.\end{cases}$$

Then, according to Theorem 3.6, we get

$$x_\epsilon(t)=x_\epsilon(0)+\int_0^t F^\epsilon((x_\epsilon)_s,s)d\widetilde{h}(s),$$

whence, for every $\epsilon\in(0,\epsilon_0]$, the function $x_\epsilon:[-r,L/\epsilon]\to X$ satisfies

$$\begin{cases} x_\epsilon(t)=x_\epsilon(0)+\epsilon\int_0^t \widetilde{f}((x_\epsilon)_s,s)d\widetilde{h}(s)+\epsilon^2\int_0^t \widetilde{g}((x_\epsilon)_s,s,\epsilon)d\widetilde{h}(s),\\ (x_\epsilon)_0=\epsilon\phi.\end{cases}$$

By definition, $\widetilde{f}$ is Lipschitz-continuous with respect to the first variable and T-periodic with respect to the second variable. Then, Lemma 3.5 implies that, for every $x\in P$, we have

$$\begin{aligned}\int_0^T \widetilde{f}(x,s)d\widetilde{h}(s)&=\int_0^T f(x,s)dh(s)+\sum_{k=1}^m \widetilde{f}(x,t_k)\Delta^+\widetilde{h}(t_k)\\ &=\int_0^T f(x,s)dh(s)+\sum_{k=1}^m I_k(x(0)).\end{aligned}$$

Thus, the function $\widetilde{f}_0$ given by

$$\widetilde{f}_0(x)=\frac{1}{T}\int_0^T \widetilde{f}(x,s)d\widetilde{h}(s)$$

fulfills

$$\widetilde{f}_0(x)=\frac{1}{T}\int_0^T f(x,s)dh(s)+\frac{1}{T}\sum_{k=1}^m I_k(x(0))=f_0(x)+I_0(x(0)),\quad x\in P.$$

Finally, by Theorem 3.35, there exists a constant $J>0$ such that

$$\|x_\epsilon(t)-y_\epsilon(t)\|\leqslant J\epsilon,\quad \text{for every } \epsilon\in(0,\epsilon_0] \text{ and } t\in\left[0,\frac{L}{\epsilon}\right].$$

As in the proof of Theorem 3.35, it is clear that

$$\|x_\epsilon(t)-y_\epsilon(t)\|=0\leqslant J\epsilon,\quad \text{for every } \epsilon\in(0,\epsilon_0] \text{ and } t\in[-r,0].$$

Therefore, combining both estimates, we get the desired result. □

Before presenting the last result, let us recall the concept of a T-periodic time scale for a fixed $T>0$. A time scale $\mathbb{T}$ is called a *T-periodic time scale*, whenever

$$t\in\mathbb{T}\quad\text{implies}\quad t+T\in\mathbb{T},\quad\text{and},\quad \mu(t)=\mu(t+T).$$

A periodic averaging principle for impulsive functional dynamic equations on time scales comes next. It slightly differs from the version presented in [86]. Here, though, we consider Banach space-valued functions and a parameter within the initial conditions of both the original and the averaged systems.

Theorem 3.37: *Let $\epsilon_0, T, L, \sigma > 0$. Suppose $\mathbb{T}$ is a T-periodic time scale such that $t_0 \in \mathbb{T}$, and $P = G([-r,0],X)$. Suppose, in addition, there exists $m \in \mathbb{N}$ such that $t_1, \dots, t_m \in \mathbb{T}$ are right-dense points satisfying $t_0 \leqslant t_1 < t_2 < \cdots < t_m < t_0 + T$. For each $k \in \mathbb{N}$, let $I_k : X \to X$ be a bounded Lipschitz-continuous function. For every integer $k > m$, define t_k and I_k by the recursive formulas $t_k = t_{k-m} + T$ and $I_k = I_{k-m}$. Consider bounded functions $f : P \times [t_0, \infty)_{\mathbb{T}} \to X$ and $g : P \times [t_0, \infty)_{\mathbb{T}} \times (0, \epsilon_0] \to X$ fulfilling:*

(TS1) *the Perron Δ-integrals*

$$\int_{t_0}^{b} f(y_s, s)\Delta s \quad \text{and} \quad \int_{t_0}^{b} g(y_s, s, \epsilon)\Delta s$$

exist for every $b > t_0$, $y \in G([t_0 - r, b], X)$ and $\epsilon \in (0, \epsilon_0]$;

(TS2) *f is Lipschitz-continuous with respect to the first variable;*

(TS3) *f is T-periodic with respect to the second variable;*

(TS4) *the Perron Δ-integral*

$$f_0(x) = \frac{1}{T}\int_{t_0}^{t_0+T} f(x,s)\Delta s$$

exists, for every $x \in P$.

Let

$$I_0(z) = \frac{1}{T}\sum_{k=1}^{m} I_k(z), \quad \text{for } z \in X,$$

and $\phi \in G([t_0 - r, t_0]_{\mathbb{T}}, X)$ be bounded. Furthermore, assume that for every $\epsilon \in (0, \epsilon_0]$, $x_\epsilon : \left[t_0 - r, t_0 + \frac{L}{\epsilon}\right]_{\mathbb{T}} \to X$ is a solution of the impulsive functional dynamic equation on time scales

$$\begin{cases} x(t) = x(t_0) + \epsilon \displaystyle\int_{t_0}^{t} f(x_s^*, s)\Delta s + \epsilon^2 \int_{t_0}^{t} g(x_s^*, s, \epsilon)\Delta s + \epsilon \sum_{\substack{k\in\mathbb{N} \\ t_k < t}} I_k(y(t_k)), \\ x(t) = \epsilon\phi(t), \quad t \in [t_0 - r, t_0]_{\mathbb{T}}, \end{cases}$$

and $y_\epsilon : \left[t_0 - r, t_0 + \frac{L}{\epsilon}\right] \to X$ is a solution of the initial value problem

$$\begin{cases} y(t) = y(t_0) + \epsilon \displaystyle\int_{t_0}^{t} [f_0(y_s) + I_0(y(s))]\, ds, \\ y_{t_0} = \epsilon\phi_{t_0}^*. \end{cases}$$

Then, there exists a constant $J > 0$ such that

$$\|x_\epsilon(t) - y_\epsilon(t)\| \leqslant J\epsilon, \quad \textit{for every } \epsilon \in (0, \epsilon_0] \textit{ and } t \in \left[t_0 - r, t_0 + \frac{L}{\epsilon}\right]_{\mathbb{T}}.$$

Proof. We can assume, without loss of generality, that $t_0 = 0$, because if $t_0 \neq 0$, it is enough to consider a shifted problem with the time scale $\widetilde{\mathbb{T}} = \{t - t_0 : t \in \mathbb{T}\}$ and functions $\widetilde{f}(x, t) = f(x, t + t_0)$ and $\widetilde{g}(x, t, \epsilon) = g(x, t + t_0, \epsilon)$, where $\widetilde{f} : P \times [0, \infty)_{\widetilde{\mathbb{T}}} \to X$ and $\widetilde{g} : P \times [0, \infty)_{\widetilde{\mathbb{T}}} \times (0, \epsilon_0] \to X$.

For every $t \in [0, \infty)$, $x \in P$ and $\epsilon \in (0, \epsilon_0]$, consider the following extensions of the functions f and g respectively, given by

$$f^*(x, t) = f(x, t^*) \quad \text{and} \quad g^*(x, t, \epsilon) = g(x, t^*, \epsilon).$$

Define a function $h : [0, \infty) \to \mathbb{R}$ by $h(t) = t^*$, for all $t \in [0, \infty)$. Since $\mathbb{T}$ is T-periodic, it is not difficult to see that $h(t + T) - h(t) = T$, for all $t \geqslant 0$. Then, Theorem 3.25 yields

$$f_0(x) = \frac{1}{T}\int_0^T f(x, s)\Delta s = \frac{1}{T}\int_0^T f^*(x, s)dh(s), \quad \text{for all } x \in P.$$

For every $b > 0$ and every $y \in G([-r, b], X)$, the Perron Δ-integral $\int_0^b f(y_s, s)\Delta s$ exists. Then, from Theorems 3.24 and 3.25, we obtain

$$\int_0^b f(y_s, s)\Delta s = \int_0^b f(y_{s^*}, s^*)dh(s) = \int_0^b f(y_s, s^*)dh(s) = \int_0^b f^*(y_s, s)dh(s),$$

ensuring the existence of the last integral.

By Theorem 3.30, for $\epsilon \in (0, \epsilon_0]$ and $t \in \left[0, \frac{L}{\epsilon}\right]$, $x_\epsilon : [-r, \frac{L}{\epsilon}] \to X$ satisfies

$$\begin{cases} (x_\epsilon)^*(t) &= (x_\epsilon)^*(0) + \epsilon \displaystyle\int_0^t f^*((x_\epsilon)^*_s, s)dh(s) + \epsilon^2 \int_0^t g^*((x_\epsilon)^*_s, s, \epsilon)dh(s) \\ &\quad + \epsilon \displaystyle\sum_{\substack{k \in \mathbb{N} \\ t_k < t}} I_k((x_\epsilon)^*(t_k)), \\ (x_\epsilon)^*_0 &= \epsilon\phi^*. \end{cases}$$

Then, all hypotheses of Theorem 3.36 are fulfilled, and hence, there exists a constant $J > 0$ such that

$$\|(x_\epsilon)^*(t) - y_\epsilon(t)\| \leqslant J\epsilon, \quad \text{for every } \epsilon \in (0, \epsilon_0] \text{ and } t \in \left[0, \frac{L}{\epsilon}\right].$$

Since $(x_\epsilon)^*(t) = x_\epsilon(t)$ for $t \in \left[0, \frac{L}{\epsilon}\right]_{\mathbb{T}}$ and by the initial condition, we obtain

$$\|x_\epsilon(t) - y_\epsilon(t)\| \leqslant J\epsilon, \quad \text{for every } \epsilon \in (0, \epsilon_0] \text{ and } t \in \left[-r, \frac{L}{\epsilon}\right]_{\mathbb{T}},$$

completing the proof. □

3.4.2 Nonperiodic Averaging

In this subsection, our goal is to present nonperiodic averaging principles for measure FDEs, impulsive measure FDEs, and functional dynamic equations on time scales. All the results presented here can be found in [82, 84].

Let $\epsilon_0 > 0$ be given such that $\epsilon \in (0, \epsilon_0]$ and $B \subset X$ be open. We focus our attention on a measure FDE of the form

$$\begin{cases} x(t) = \phi(0) + \epsilon \displaystyle\int_0^t f(x_s, s)dh_1(s) + \epsilon^2 \int_0^t g(x_s, s, \epsilon)dh_2(s), \\ x_0 = \epsilon\phi, \end{cases} \tag{3.65}$$

where $P \subset G([-r, 0], B)$ is open, $r > 0, h_1, h_2 : [0, \infty) \to \mathbb{R}$ are left-continuous and nondecreasing functions, $f : P \times [0, \infty) \to X$ and $g : P \times [0, \infty) \times (0, \epsilon_0] \to X$ are functions, and $\phi \in P$.

In order to obtain an averaging principle for (3.65), we consider the following auxiliary initial value problem

$$\begin{cases} x(t) = \phi(0) + \displaystyle\int_0^t \epsilon f\left(x_{s,\epsilon}, \frac{s}{\epsilon}\right) dh_1\left(\frac{s}{\epsilon}\right) + \int_0^t \epsilon^2 g\left(x_{s,\epsilon}, \frac{s}{\epsilon}, \epsilon\right) dh_2\left(\frac{s}{\epsilon}\right), \ t \in [0, M], \\ x_0 = \epsilon\phi, \end{cases} \tag{3.66}$$

where $x_{t,\epsilon}(\theta) = x(t + \epsilon\theta)$, for $\theta \in [-\frac{r}{\epsilon}, 0]$, $\phi \in P$ and $M > 0$.

Let $\widetilde{P}_\epsilon \subset G([-\frac{r}{\epsilon}, 0], B)$ be an open set and assume that f maps any pair $(\psi, t) \in \widetilde{P}_\epsilon \times [0, \infty)$ into X and that the mapping $t \mapsto f(y_{t,\epsilon}, t)$ is Perron–Stieltjes integrable with respect to h_1, for all $t \in [0, \infty)$. Suppose, in addition, g maps $(\psi, t, \epsilon) \in \widetilde{P}_\epsilon \times [0, \infty) \times (0, \epsilon_0]$ into X and the mapping $t \mapsto g(y_{t,\epsilon}, t, \epsilon)$ is Perron–Stieltjes integrable with respect to h_2, for all $t \in [0, \infty)$. Clearly, if $x_t \in G^-([-r, 0], B)$, then $x_{t,\epsilon} \in G^-\left(\left[-\frac{r}{\epsilon}, 0\right], B\right)$. For details, see [84].

We will refer a few times to the next remark.

Remark 3.38: By a change of variables, we can transform system (3.66) into system (3.65). Indeed, if $x : [-r, M] \to B$ is a solution of Eq. (3.66), then

$$x(t) = \phi(0) + \epsilon \int_0^t f\left(x_{s,\epsilon}, \frac{s}{\epsilon}\right) dh_1\left(\frac{s}{\epsilon}\right) + \epsilon^2 \int_0^t g\left(x_{s,\epsilon}, \frac{s}{\epsilon}, \epsilon\right) dh_2\left(\frac{s}{\epsilon}\right).$$

Define $y(t) = x(\epsilon t)$, for $t \in \left[0, \frac{M}{\epsilon}\right]$, $\psi(s) = \frac{s}{\epsilon}$, for $s \in [0, M]$, and $m(\tau) = f(x_{\epsilon\tau,\epsilon}, \tau)$, for $\tau \in \left[0, \frac{M}{\epsilon}\right]$. Considering $\tau = \psi(s) = \frac{s}{\epsilon}$ in the calculations below

$$\begin{aligned}\int_0^t f\left(x_{s,\epsilon},\frac{s}{\epsilon}\right)dh_1\left(\frac{s}{\epsilon}\right) &= \int_0^t f\left(x_{\epsilon(s/\epsilon),\epsilon},\frac{s}{\epsilon}\right)dh_1\left(\frac{s}{\epsilon}\right)\\ &= \int_0^t m\left(\frac{s}{\epsilon}\right)dh_1\left(\frac{s}{\epsilon}\right) = \int_0^t m(\psi(s))dh_1(\psi(s))\\ &= \int_0^{t/\epsilon} m(\tau)dh_1(\tau) = \int_0^{t/\epsilon} f(x_{\epsilon\tau,\epsilon},\tau)dh_1(\tau)\\ &= \int_0^{t/\epsilon} f(y_\tau,\tau)dh_1(\tau)\end{aligned}$$

we conclude, by Theorem 1.72, that $x(\epsilon t) = y(t)$, for all $t \in \left[0,\frac{M}{\epsilon}\right]$.

Define $n(\tau) = g(x_{\epsilon\tau,\epsilon},\tau,\epsilon)$, for $\tau \in \left[0,\frac{M}{\epsilon}\right]$. Again, using Theorem 1.72, we obtain

$$\begin{aligned}\int_0^t g\left(x_{s,\epsilon},\frac{s}{\epsilon},\epsilon\right)dh_2\left(\frac{s}{\epsilon}\right) &= \int_0^t g\left(x_{\epsilon(s/\epsilon),\epsilon},\frac{s}{\epsilon},\epsilon\right)dh_2\left(\frac{s}{\epsilon}\right)\\ &= \int_0^t n\left(\frac{s}{\epsilon}\right)dh_2\left(\frac{s}{\epsilon}\right) = \int_0^t n(\psi(s))dh_2(\psi(s))\\ &= \int_0^{t/\epsilon} n(\tau)dh_2(\tau) = \int_0^{t/\epsilon} g(x_{\epsilon\tau,\epsilon},\tau,\epsilon)dh_2(\tau)\\ &= \int_0^{t/\epsilon} g(y_\tau,\tau,\epsilon)dh_2(\tau),\end{aligned}$$

since, for every $\tau \in \left[0,\frac{M}{\epsilon}\right]$ and $\theta \in \left[-\frac{r}{\epsilon},0\right]$, we have

$$x_{\epsilon\tau,\epsilon}(\theta) = x(\epsilon(\tau+\theta)) = y(\tau+\theta) = y_\tau(\theta).$$

Therefore, we get

$$\begin{aligned}y(t) - y(0) &= x(\epsilon t) - x(0)\\ &= \epsilon\int_0^{\epsilon t} f\left(x_{s,\epsilon},\frac{s}{\epsilon}\right)dh_1\left(\frac{s}{\epsilon}\right) + \epsilon^2\int_0^{\epsilon t} g\left(x_{s,\epsilon},\frac{s}{\epsilon},\epsilon\right)dh_2\left(\frac{s}{\epsilon}\right)\\ &= \epsilon\int_0^{\epsilon t/\epsilon} f(x_{\epsilon s,\epsilon},s)dh_1(s) + \epsilon^2\int_0^{\epsilon t/\epsilon} g(x_{\epsilon s,\epsilon},s,\epsilon)\,dh_2(s)\\ &= \epsilon\int_0^t f(y_s,s)dh_1(s) + \epsilon^2\int_0^t g(y_s,s,\epsilon)\,dh_2(s),\end{aligned}$$

for $t \in \left[0,\frac{M}{\epsilon}\right]$. Thus, a solution of the measure FDE (3.66) on $[0,M]$ corresponds to a solution of the measure FDE (3.65) on $\left[0,\frac{M}{\epsilon}\right]$ and vice versa. In view of this, we now focus our attention on Eq. (3.66). Then, an averaging principle for Eq. (3.65) will be obtained naturally as a consequence.

Motivated by Remark 3.38, we restrict ourselves to problem (3.66), where $h_1, h_2 : [0, \infty) \to \mathbb{R}$ are left-continuous and nondecreasing functions. Consider open sets $B \subset X$ and $\widetilde{P}_\epsilon \subset G([-\frac{r}{\epsilon}, 0], B)$, and assume that $f : \widetilde{P}_\epsilon \times [0, \infty) \to X$ satisfies the conditions:

(J1) for all $\varphi \in \widetilde{P}_\epsilon$ and all $t \in [0, \infty)$, the Perron–Stieltjes integral

$$\int_0^t f(\varphi, s) dh_1(s)$$

exists;

(J2) there exists a constant $L > 0$ such that, for every $\varphi, \psi \in \widetilde{P}_\epsilon$ and every $u_1, u_2 \in [0, \infty)$, with $u_1 \leqslant u_2$, we have

$$\left\| \int_{u_1}^{u_2} [f(\varphi, s) - f(\psi, s)]\, dh_1(s) \right\| \leqslant L \int_{u_1}^{u_2} \|\varphi - \psi\|_\infty \, dh_1(s).$$

Consider the following assumptions on $g : \widetilde{P}_\epsilon \times [0, \infty) \times (0, \epsilon_0] \to X$:

(J3) the Perron–Stieltjes integral

$$\int_0^t g(\varphi, s, \epsilon) dh_2(s)$$

exists, for every $\varphi \in \widetilde{P}_\epsilon$, $t \in [0, \infty)$ and $\epsilon \in (0, \epsilon_0]$;

(J4) there is a constant $C > 0$ such that for all $\varphi \in \widetilde{P}_\epsilon$, $\epsilon \in (0, \epsilon_0]$ and $u_1, u_2 \in [0, \infty)$, with $u_1 \leqslant u_2$, we have

$$\left\| \int_{u_1}^{u_2} g(\varphi, s, \epsilon) dh_2(s) \right\| \leqslant C \int_{u_1}^{u_2} dh_2(s);$$

(J5) there exists $K > 0$ such that for all $\beta \geqslant 0$, we have

$$\limsup_{T \to \infty} \frac{h_1(T + \beta) - h_1(\beta)}{T} \leqslant K;$$

(J6) there exists $N > 0$ such that for all $\beta \geqslant 0$, we have

$$\limsup_{T \to \infty} \frac{h_2(T + \beta) - h_2(\beta)}{T} \leqslant N;$$

(J7) there exists $\gamma > 0$ such that $\|\phi\|_\infty \leqslant \gamma$.

Suppose, for each $\varphi \in \widetilde{P}_\epsilon$, the limit

$$f_0(\varphi) = \lim_{T \to \infty} \frac{1}{T} \int_0^T f(\varphi, s) dh_1(s) \tag{3.67}$$

exists, where the integral is taken in the sense of Perron–Stieltjes, and consider the averaged FDE

$$\begin{cases} \dot{y} = f_0\left(y_{t,\epsilon}\right), \\ y_0 = \epsilon\phi, \end{cases} \tag{3.68}$$

where $t \in [0, M]$ and f_0 is given by (3.67).

Claim. There is a relation between the solutions of (3.68) and the solutions of the averaged FDE

$$\begin{cases} \dot{y} = \epsilon f_0\left(y_t\right), \\ y_0 = \epsilon \phi, \end{cases} \tag{3.69}$$

where $t \in \left[0, \frac{M}{\epsilon}\right]$ and f_0 is given by (3.67).

Indeed, as in Remark 3.38, denote $\psi(s) = \frac{s}{\epsilon} = \tau$, for $s \in [0, M]$, and $m(\tau) = f_0\left(y_{\epsilon\tau,\epsilon}\right)$, for $\tau \in \left[0, \frac{M}{\epsilon}\right]$. Thus,

$$\begin{aligned} \int_0^t f_0(y_{s,\epsilon})ds &= \int_0^t f_0(y_{\epsilon(s/\epsilon),\epsilon})ds = \int_0^t m(\psi(s))ds \\ &= \epsilon \int_0^{t/\epsilon} m(\tau)d\tau = \epsilon \int_0^{t/\epsilon} f_0(y_{\epsilon\tau,\epsilon})d\tau. \end{aligned}$$

Taking $u(t) = y(t\epsilon)$ for $t \in [0, \frac{M}{\epsilon}]$, we obtain

$$y_{\epsilon\tau,\epsilon}(\theta) = y(\epsilon(\tau + \theta)) = u(\tau + \theta) = u_\tau(\theta),$$

for $\tau \in \left[0, \frac{M}{\epsilon}\right]$ and $\theta \in \left[-\frac{r}{\epsilon}, 0\right]$. Therefore,

$$\begin{aligned} u(t) - u(0) = y(\epsilon t) - y(0) &= \int_0^{\epsilon t} f_0(y_{s,\epsilon})ds \\ &= \epsilon \int_0^{\epsilon(t/\epsilon)} f_0(y_{\epsilon s,\epsilon})ds = \epsilon \int_0^t f_0(u_s)ds, \end{aligned}$$

for every $t \in \left[0, \frac{M}{\epsilon}\right]$ and the *Claim* is proved.

Notice that, if conditions (J2) and (J5) are satisfied, then

$$\begin{aligned} \|f_0(\xi) - f_0(\varphi)\| &= \left\| \lim_{T\to\infty} \frac{1}{T} \int_0^T f(\xi, s)dh_1(s) - \lim_{T\to\infty} \frac{1}{T} \int_0^T f(\varphi, s)dh_1(s) \right\| \\ &\leqslant \lim_{T\to\infty} \frac{1}{T} \int_0^T L\|\xi - \varphi\|_\infty \, dh_1(s) \\ &\leqslant L\|\xi - \varphi\|_\infty \limsup_{T\to\infty} \frac{h_1(T) - h_1(0)}{T} \leqslant LK\|\xi - \varphi\|_\infty. \end{aligned} \tag{3.70}$$

In particular, for every $y_t, y_s \in P$ and every $t, s \in [0, \infty)$, we have

$$\|f_0(y_s) - f_0(y_t)\| \leqslant LK\|y_s - y_t\|_\infty. \tag{3.71}$$

Let $y \in \widetilde{P}_\epsilon$ be a solution of the averaged equation (3.68), where ϕ is bounded by a constant $\gamma > 0$. Let $s, t \in [0, M]$, with $t \leqslant s$, and $\theta \in [-\frac{r}{\epsilon}, 0]$. We consider three cases.

(i) If $s+\epsilon\theta>0$ and $t+\epsilon\theta>0$, then

$$\begin{aligned}
\|y_{s,\epsilon}(\theta)-y_{t,\epsilon}(\theta)\| &= \|y(s+\epsilon\theta)-y(t+\epsilon\theta)\| = \left\|\int_{t+\epsilon\theta}^{s+\epsilon\theta} f_0(y_{\sigma,\epsilon})d\sigma\right\| \\
&\leqslant \int_{t+\epsilon\theta}^{s+\epsilon\theta} \|f_0(y_{\sigma,\epsilon})-f_0(0)\|\, d\sigma + \int_{t+\epsilon\theta}^{s+\epsilon\theta} \|f_0(0)\|\, d\sigma \\
&\leqslant LK\int_{t+\epsilon\theta}^{s+\epsilon\theta} \sup_{\sigma\in[t-r,s]} \|y(\sigma)\|\, d\sigma + (s-t)\|f_0(0)\|,
\end{aligned}$$

where this last inequality follows from (3.71). Therefore,

$$\begin{aligned}
\|y_{s,\epsilon}-y_{t,\epsilon}\|_\infty &= \sup_{\theta\in[-\frac{r}{\epsilon},0]} \|y(s+\epsilon\theta)-y(t+\epsilon\theta)\| \\
&\leqslant LK(s-t)\sup_{\sigma\in[t-r,s]}\|y(\sigma)\| + (s-t)\|f_0(0)\|.
\end{aligned}$$

(ii) If $s+\epsilon\theta\leqslant 0$ and $t+\epsilon\theta\leqslant 0$, then

$$\|y(s+\epsilon\theta)-y(t+\epsilon\theta)\| = \|\epsilon\phi(s+\epsilon\theta)-\epsilon\phi(t+\epsilon\theta)\| \leqslant 2\epsilon\gamma.$$

(iii) If $s+\epsilon\theta>0$ and $t+\epsilon\theta\leqslant 0$, then

$$\begin{aligned}
\|y(s+\epsilon\theta)-y(t+\epsilon\theta)\| &\leqslant \left\|\epsilon\phi(0)+\int_0^{s+\epsilon\theta} f_0(y_{\sigma,\epsilon})d\sigma\right\| + \|\epsilon\phi(t+\epsilon\theta)\| \\
&\leqslant 2\epsilon\gamma + \int_0^{s+\epsilon\theta} \|f_0(y_{\sigma,\epsilon})\|\, d\sigma \\
&\leqslant 2\epsilon\gamma + LK\int_0^{s+\epsilon\theta} \sup_{\sigma\in[t-r,s]}\|y(\sigma)\|\, d\sigma + (s-t)\|f_0(0)\| \\
&\leqslant 2\epsilon\gamma + LK(s+\epsilon\theta)\sup_{\sigma\in[t-r,s]}\|y(\sigma)\| + (s-t)\|f_0(0)\| \\
&\leqslant 2\epsilon\gamma + LK[s+\epsilon\theta-(t+\epsilon\theta)]\sup_{\sigma\in[t-r,s]}\|y(\sigma)\| \\
&\quad + (s-t)\|f_0(0)\| \\
&= 2\epsilon\gamma + LK(s-t)\sup_{\sigma\in[t-r,s]}\|y(\sigma)\| + (s-t)\|f_0(0)\|.
\end{aligned}$$

In either case, we have

$$\|y_{s,\epsilon}-y_{t,\epsilon}\|_\infty \leqslant 2\epsilon\gamma + LK(s-t)\sup_{\sigma\in[t-r,s]}\|y(\sigma)\| + (s-t)\|f_0(0)\|, \tag{3.72}$$

which implies $\|y_{s,\epsilon}-y_{t,\epsilon}\|_\infty \leqslant 2\gamma\epsilon$, as $s-t\to 0^+$. Hence, the mapping $[0,M]\ni t\mapsto y_{t,\epsilon}$ is continuous, where $y_{t,\epsilon}$ is a solution of the averaged FDE (3.68).

Now, we are in position to present a version of [82, Lemma 3.1] for Perron–Stieltjes integrals. This result can also be found in [84] for the case

$X = \mathbb{R}^n$. It is presented here for Banach space-valued functions and it is essential to the proofs of the nonperiodic averaging principles coming next.

Lemma 3.39: *Let $B \subset X$ be open. Suppose the function $f : G([-r,0],B) \times [0,\infty) \to X$ is Perron–Stieltjes integrable with respect to a nondecreasing function $h_1 : [0,\infty) \to \mathbb{R}$. Suppose, further, the limit*

$$f_0(\psi) = \lim_{T\to\infty} \frac{1}{T} \int_0^T f(\psi,s)dh_1(s), \quad \psi \in G([-r,0],B), \tag{3.73}$$

exists and is well defined. Then, for every $t, \alpha > 0$, the equality

$$\lim_{\epsilon\to 0^+} \frac{\epsilon}{\alpha} \int_{t/\epsilon}^{t/\epsilon+\alpha/\epsilon} f(\psi,s)dh_1(s) = f_0(\psi), \quad \psi \in G([-r,0],B),$$

holds, where the Perron–Stieltjes integral on the left-hand side exists and is well defined.

Proof. From (3.73), we derive that, for $t, \alpha > 0$ and $\psi \in G([-r,0],B)$, we have

$$\lim_{\epsilon\to 0^+} \frac{1}{t/\epsilon + \alpha/\epsilon} \int_0^{t/\epsilon+\alpha/\epsilon} f(\psi,s)dh_1(s) = f_0(\psi) \quad \text{and} \tag{3.74}$$

$$\lim_{\epsilon\to 0^+} \frac{\epsilon}{t} \int_0^{t/\epsilon} f(\psi,s)dh_1(s) = f_0(\psi). \tag{3.75}$$

Thus,

$$\begin{aligned}
&\lim_{\epsilon\to 0^+} \left[\frac{1}{t/\epsilon + \alpha/\epsilon} \int_0^{t/\epsilon+\alpha/\epsilon} f(\psi,s)dh_1(s) - \frac{1}{t/\epsilon}\int_0^{t/\epsilon} f(\psi,s)dh_1(s)\right] \\
&= \lim_{\epsilon\to 0^+} \left[\frac{1}{t/\epsilon + \alpha/\epsilon} \int_0^{t/\epsilon+\alpha/\epsilon} f(\psi,s)dh_1(s) - f_0(\psi)\right] \\
&\quad + \lim_{\epsilon\to 0^+} \left[f_0(\psi) - \frac{\epsilon}{t}\int_0^{t/\epsilon} f(\psi,s)dh_1(s)\right] = 0,
\end{aligned}$$

whence, for every $\epsilon > 0$, we obtain

$$\begin{aligned}
&\frac{\epsilon}{\alpha} \int_{t/\epsilon}^{t/\epsilon+\alpha/\epsilon} f(\psi,s)dh_1(s) = \frac{1}{\alpha/\epsilon} \int_0^{t/\epsilon+\alpha/\epsilon} f(\psi,s)dh_1(s) \\
&\quad - \frac{1}{\alpha/\epsilon}\int_0^{t/\epsilon} f(\psi,s)dh_1(s) \\
&= \left(\frac{t/\epsilon + \alpha/\epsilon}{t/\epsilon + \alpha/\epsilon}\right) \frac{1}{\alpha/\epsilon} \int_0^{t/\epsilon+\alpha/\epsilon} f(\psi,s)dh_1(s) - \left(\frac{t/\epsilon}{t/\epsilon}\right) \frac{1}{\alpha/\epsilon}\int_0^{t/\epsilon} f(\psi,s)dh_1(s) \\
&= \frac{1}{t/\epsilon + \alpha/\epsilon}\left(\frac{t}{\alpha} + 1\right) \int_0^{t/\epsilon+\alpha/\epsilon} f(\psi,s)dh_1(s) - \frac{t}{\alpha} \cdot \frac{1}{t/\epsilon}\int_0^{t/\epsilon} f(\psi,s)dh_1(s)
\end{aligned}$$

$$= \frac{1}{t/\epsilon + \alpha/\epsilon} \int_0^{t/\epsilon+\alpha/\epsilon} f(\psi, s)dh_1(s)$$
$$+ \frac{t}{\alpha} \left[\frac{1}{t/\epsilon + \alpha/\epsilon} \int_0^{t/\epsilon+\alpha/\epsilon} f(\psi, s)dh_1(s) - \frac{1}{t/\epsilon} \int_0^{t/\epsilon} f(\psi, s)dh_1(s) \right].$$

Then, combining (3.74) and (3.75), we get

$$\lim_{\epsilon \to 0^+} \left[\frac{\epsilon}{\alpha} \int_{t/\epsilon}^{t/\epsilon+\alpha/\epsilon} f(\psi, s)dh_1(s) - f_0(\psi) \right]$$
$$= \lim_{\epsilon \to 0^+} \left[\frac{1}{t/\epsilon + \alpha/\epsilon} \int_0^{t/\epsilon+\alpha/\epsilon} f(\psi, s)dh_1(s) - f_0(\psi) \right]$$
$$+ \lim_{\epsilon \to 0^+} \frac{t}{\alpha} \left[\frac{1}{t/\epsilon + \alpha/\epsilon} \int_0^{t/\epsilon+\alpha/\epsilon} f(\psi, s)dh_1(s) - \frac{1}{t/\epsilon} \int_0^{t/\epsilon} f(\psi, s)dh_1(s) \right] = 0$$

and hence,

$$\lim_{\epsilon \to 0^+} \frac{\epsilon}{\alpha} \int_{t/\epsilon}^{t/\epsilon+\alpha/\epsilon} f(\psi, s)dh_1(s) = f_0(\psi)$$

concluding the proof. □

The next corollary follows easily from Lemma 3.39. See [82, 84].

Corollary 3.40: *Consider open sets $B \subset X$ and $P \subset G([-r, 0], B)$, and assume that the function $f : P \times [0, \infty) \to X$ is a Perron–Stieltjes integrable with respect to a nondecreasing function $h_1 : [0, \infty) \to \mathbb{R}$. Suppose, in addition, the limit*

$$f_0(\varphi) = \lim_{T \to \infty} \frac{1}{T} \int_0^T f(\varphi, s)dh_1(s) \quad \text{for every} \quad \varphi \in P$$

exists and is well defined. Then, for every $t, \alpha > 0$ and $y_t \in P$, we obtain

$$\lim_{\epsilon \to 0^+} \frac{\epsilon}{\alpha} \int_{t/\epsilon}^{t/\epsilon+\alpha/\epsilon} f(y_t, s)dh_1(s) = f_0(y_t),$$

where the Perron–Stieltjes integral on the left-hand side exists and is well defined.

The next corollary follows easily by the steps of the proof of Lemma 3.39.

Corollary 3.41: *Let $B \subset X$ be open and $\widetilde{P}_\epsilon \subset G([-\frac{r}{\epsilon}, 0], B)$ be open, and assume that $f : \widetilde{P}_\epsilon \times [0, \infty) \to X$ is a Perron–Stieltjes integrable function with respect to a nondecreasing function $h_1 : [0, \infty) \to \mathbb{R}$. Suppose, further, the limit*

$$f_0(\varphi) = \lim_{T \to \infty} \frac{1}{T} \int_0^T f(\varphi, s)dh_1(s) \quad \text{for every} \quad \varphi \in \widetilde{P}_\epsilon$$

exists and is well defined. Then, for every $t, \alpha > 0$ and $y_{t,\epsilon} \in \widetilde{P}_{\epsilon}$, we obtain

$$\lim_{\eta \to 0^+} \frac{\eta}{\alpha} \int_{t/\eta}^{t/\eta+\alpha/\eta} f(y_{t,\epsilon}, s) dh_1(s) = f_0(y_{t,\epsilon}),$$

where the Perron–Stieltjes integral on the left-hand side exists and is well defined.

The next lemma can be found in [84, Lemma 3.2] for the finite dimensional case. We adapt it here for Banach space-valued functions.

Lemma 3.42: *Let $B \subset X$ and $\widetilde{P}_{\epsilon} \subset G([-\frac{r}{\epsilon}, 0], B)$ be open sets. Assume that $f : \widetilde{P}_{\epsilon} \times [0, \infty) \to X$ satisfies conditions (J1) and (J2) and $h_1 : [0, \infty) \to \mathbb{R}$ satisfies condition (J5). Suppose, for each $\varphi \in \widetilde{P}_{\epsilon}$, the limit*

$$f_0(\varphi) = \lim_{T \to \infty} \frac{1}{T} \int_0^T f(\varphi, s) dh_1(s)$$

exists. Assume that $0 < M < \infty$ and $y : [-r, M] \to B$ is a maximal solution of the autonomous FDE

$$\begin{cases} \dot{y} = f_0(y_{t,\epsilon}), \\ y_0 = \epsilon\phi, \end{cases} \tag{3.76}$$

with maximal interval of existence being $[-r, M]$. Then, given $\epsilon > 0$, there exists $\xi(\epsilon) > 0$ such that

$$\left\| \epsilon \int_0^t f\left(y_{s,\epsilon}, \frac{s}{\epsilon}\right) dh_1\left(\frac{s}{\epsilon}\right) - \int_0^t f_0(y_{s,\epsilon}) ds \right\| < \xi(\epsilon), \quad t \in [0, M],$$

and $\xi(\epsilon)$ tends to zero, as $\epsilon \to 0^+$.

Proof. Given $\epsilon > 0$ and $t \in [0, \infty)$, let δ be a gauge of $[0, t]$ which corresponds to $\epsilon > 0$ in the definition of the Perron–Stieltjes integral $\int_0^t f\left(y_{\sigma,\epsilon}, \frac{\sigma}{\epsilon}\right) dh_1\left(\frac{\sigma}{\epsilon}\right)$. Take a δ-fine tagged division $d = (\tau_i, [s_{i-1}, s_i])$, $i = 1, 2, \ldots, |d|$, of $[0, t]$. Thus,

$$\begin{aligned} &\left\| \epsilon \int_0^t f\left(y_{s,\epsilon}, \frac{s}{\epsilon}\right) dh_1\left(\frac{s}{\epsilon}\right) - \int_0^t f_0(y_{s,\epsilon}) ds \right\| \\ &\leqslant \sum_{i=1}^{|d|} \left\| \epsilon \int_{s_{i-1}}^{s_i} \left[f\left(y_{s,\epsilon}, \frac{s}{\epsilon}\right) - f\left(y_{s_{i-1},\epsilon}, \frac{s}{\epsilon}\right) \right] dh_1\left(\frac{s}{\epsilon}\right) \right\| \\ &+ \sum_{i=1}^{|d|} \left\| \int_{s_{i-1}}^{s_i} [f_0(y_{s,\epsilon}) - f_0(y_{s_{i-1},\epsilon})] \, ds \right\| \\ &+ \sum_{i=1}^{|d|} \left\| \int_{s_{i-1}}^{s_i} f\left(y_{s_{i-1},\epsilon}, \frac{s}{\epsilon}\right) dh_1\left(\frac{s}{\epsilon}\right) - \int_{s_{i-1}}^{s_i} f_0(y_{s_{i-1},\epsilon}) ds \right\|. \end{aligned} \tag{3.77}$$

Assume, without loss of generality, that the gauge δ satisfies $\delta(\tau_i) < \frac{\epsilon}{2}$, for every $\tau_i \in [s_{i-1}, s_i]$ and $i = 1, 2, \ldots, |d|$. By (3.72), we obtain

$$\begin{aligned}\|y_{s,\epsilon} - y_{s_{i-1},\epsilon}\|_\infty &\leqslant LK(s - s_{i-1}) \sup_{\sigma\in[s_{i-1}-r,s]} \|y(\sigma)\| + (s - s_{i-1})\|f_0(0)\| + 2\epsilon\gamma \\ &< LK2\delta(\tau_i)\left(\sup_{\sigma\in[s_{i-1}-r,s_i]} \|y(\sigma)\|\right) + 2\delta(\tau_i)\|f_0(0)\| + 2\epsilon\gamma \\ &< LK\epsilon\left(\sup_{\sigma\in[s_{i-1}-r,s_i]} \|y(\sigma)\|\right) + \epsilon\|f_0(0)\| + 2\epsilon\gamma,\end{aligned}$$

for $i = 1, 2, \ldots, |d|$ and $s \in [s_{i-1}, s_i]$. Then, taking

$$D = LK\left(\sup_{\sigma\subset[-r,M]} \|y(\sigma)\|\right) + \|f_0(0)\| + 2\gamma,$$

we get $\sup_{s\in[s_{i-1},s_i]} \|y_{s,\epsilon} - y_{s_{i-1},\epsilon}\|_\infty \leqslant \epsilon D$, for $i = 1, 2, \ldots, |d|$, which together with conditions (J2) and (J5) imply

$$\begin{aligned}&\sum_{i=1}^{|d|} \left\| \int_{s_{i-1}}^{s_i} \epsilon\left[f\left(y_{s,\epsilon}, \frac{s}{\epsilon}\right) - f\left(y_{s_{i-1},\epsilon}, \frac{s}{\epsilon}\right)\right] dh_1\left(\frac{s}{\epsilon}\right)\right\| \\ &\leqslant \epsilon L \sum_{i=1}^{|d|} \sup_{\sigma\in[s_{i-1},s_i]} \|y_{\sigma,\epsilon} - y_{s_{i-1},\epsilon}\|_\infty \int_{s_{i-1}}^{s_i} dh_1\left(\frac{s}{\epsilon}\right) \\ &\leqslant \epsilon^2 DL \sum_{i=1}^{|d|} \left[h_1\left(\frac{s_i}{\epsilon}\right) - h_1\left(\frac{s_{i-1}}{\epsilon}\right)\right] \\ &= \epsilon^2 DL\left[h_1\left(\frac{t}{\epsilon}\right) - h_1(0)\right] = \epsilon DLt\left[\frac{h_1\left(\frac{t}{\epsilon}\right) - h_1(0)}{\frac{t}{\epsilon}}\right].\end{aligned}$$

By hypothesis (J5), we can choose $\epsilon > 0$ sufficiently small such that

$$\frac{h_1\left(\frac{t}{\epsilon}\right) - h_1(0)}{\frac{t}{\epsilon}} \leqslant K, \quad \text{for each } t \in [0, M].$$

Then,

$$\sum_{i=1}^{|d|} \left\| \int_{s_{i-1}}^{s_i} \epsilon\left[f\left(y_{s,\epsilon}, \frac{s}{\epsilon}\right) - f\left(y_{s_i,\epsilon}, \frac{s}{\epsilon}\right)\right] dh_1\left(\frac{s}{\epsilon}\right)\right\| \leqslant \epsilon DLtK \leqslant \epsilon DLMK.$$

On the other hand, for $i = 1, 2, \ldots, |d|$ and $s \in [s_{i-1}, s_i]$, by (3.70), we have

$$\begin{aligned}&\sum_{i=1}^{|d|} \left\| \int_{s_{i-1}}^{s_i} [f_0(y_{s,\epsilon}) - f_0(y_{s_{i-1},\epsilon})]\, ds\right\| \\ &\leqslant LK \sum_{i=1}^{|d|} \sup_{\sigma\in[s_{i-1},s_i]} \|y_{\sigma,\epsilon} - y_{s_{i-1},\epsilon}\|_\infty (s_i - s_{i-1}) < \epsilon DLKM.\end{aligned}$$

Claim. The sum

$$\sum_{i=1}^{|d|} \left\| \epsilon \int_{s_{i-1}}^{s_i} f\left(y_{s_{i-1},\epsilon}, \frac{s}{\epsilon}\right) dh_1\left(\frac{s}{\epsilon}\right) - \int_{s_{i-1}}^{s_i} f_0(y_{s_{i-1},\epsilon}) ds \right\|$$

can be made arbitrarily small by Corollary 3.40.

Indeed, for each $i = 1, 2, \dots, |d|$ and $\alpha_i = s_i - s_{i-1}$, we have

$$\begin{aligned}
&\sum_{i=1}^{|d|} \left\| \epsilon \int_{s_{i-1}}^{s_i} f\left(y_{s_{i-1},\epsilon}, \frac{s}{\epsilon}\right) dh_1\left(\frac{s}{\epsilon}\right) - \int_{s_{i-1}}^{s_i} f_0(y_{s_{i-1},\epsilon}) ds \right\| \\
&\quad = \sum_{i=1}^{|d|} \left\| \epsilon \int_{s_{i-1}}^{s_{i-1}+\alpha_i} f\left(y_{s_{i-1},\epsilon}, \frac{s}{\epsilon}\right) dh_1\left(\frac{s}{\epsilon}\right) - \int_{s_{i-1}}^{s_{i-1}+\alpha_i} f_0(y_{s_{i-1},\epsilon}) ds \right\| \\
&\quad = \sum_{i=1}^{|d|} \alpha_i \left\| \frac{\epsilon}{\alpha_i} \int_{s_{i-1}/\epsilon}^{s_{i-1}/\epsilon+\alpha_i/\epsilon} f\left(y_{s_{i-1},\epsilon}, s\right) dh_1(s) - f_0(y_{s_{i-1},\epsilon}) \right\|.
\end{aligned}$$

Define, for each $i = 1, 2, \dots, |d|$,

$$\beta_i(\epsilon) = \frac{\epsilon}{\alpha_i} \int_{s_{i-1}/\epsilon}^{s_{i-1}/\epsilon+\alpha_i/\epsilon} f(y_{s_{i-1},\epsilon}, s) dh_1(s) - f_0(y_{s_{i-1},\epsilon})$$

and set $\beta(\epsilon) = \max\{\|\beta_i(\epsilon)\| \,:\, i = 1, 2, \dots, |d|\}$. Then,

$$\sum_{i=1}^{|d|} \alpha_i \left\|\beta_i(\epsilon)\right\| \leqslant \beta(\epsilon) \sum_{i=1}^{|d|} (s_i - s_{i-1}) = \beta(\epsilon) t < \beta(\epsilon) M$$

proving the *Claim*.

Now, by Corollary 3.40, it is clear that $\beta(\epsilon) \to 0$ as $\epsilon \to 0^+$. Therefore,

$$\left\| \epsilon \int_0^t f\left(y_{s,\epsilon}, \frac{s}{\epsilon}\right) dh_1\left(\frac{s}{\epsilon}\right) - \int_0^t f_0(y_{s,\epsilon}) ds \right\| < 2\epsilon DLKM + \beta(\epsilon)M.$$

Then, setting

$$\xi(\epsilon) = 2\epsilon DLKM + \beta(\epsilon)M,$$

yields $\xi(\epsilon)$ tends to zero as $\epsilon \to 0^+$ and the inequality

$$\left\| \epsilon \int_0^t f\left(y_{s,\epsilon}, \frac{s}{\epsilon}\right) dh_1\left(\frac{s}{\epsilon}\right) - \int_0^t f_0(y_{s,\epsilon}) ds \right\| < \xi(\epsilon)$$

holds completing the proof. □

The main result of this section follows next. It is a nonperiodic averaging principle for measure FDEs, and it slightly differs from the version found in [84, Theorem 3.1] with respect to initial conditions and codomains.

Theorem 3.43: *Consider open sets $B \subset X$ and $\widetilde{P}_\epsilon \subset G([-\frac{r}{\epsilon}, 0], B)$, and assume that $f\colon \widetilde{P}_\epsilon \times [0,\infty) \to X$ satisfies conditions (J1) and (J2) and $g\colon \widetilde{P}_\epsilon \times [0,\infty) \times (0,\epsilon_0] \to X$*

satisfies the conditions (J3) and (J4). Suppose conditions (J5)–(J7) are fulfilled, where h_1 and h_2 are nondecreasing functions and $\phi \in P$. Suppose, further, for each $\varphi \in \widetilde{P}_{\epsilon}$, the limit

$$f_0(\varphi) = \lim_{T\to\infty} \frac{1}{T} \int_0^T f(\varphi, s)dh_1(s)$$

exists and is well defined. Let $M > 0$ and consider the nonautonomous measure FDE

$$\begin{cases} x(t)=\phi(0)+\int_0^t \epsilon f\left(x_{s,\epsilon}, \frac{s}{\epsilon}\right) dh_1\left(\frac{s}{\epsilon}\right)+\int_0^t \epsilon^2 g\left(x_{s,\epsilon}, \frac{s}{\epsilon}, \epsilon\right) dh_2\left(\frac{s}{\epsilon}\right), & t \in [0, M], \\ x_0 = \epsilon\phi, \end{cases} \tag{3.78}$$

where $x_{t,\epsilon}(\theta) = x(t + \epsilon\theta)$, for $\theta \in [-\frac{r}{\epsilon}, 0]$, and the averaged autonomous FDE

$$\begin{cases} \dot{y} = f_0\left(y_{t,\epsilon}\right), \\ y_0 = \epsilon\phi, \end{cases} \tag{3.79}$$

where $t \in [0, M]$. Suppose $[0, \overline{b})$ is the maximal interval of existence of the measure FDE (3.78) and $[0, b)$ is the maximal interval of existence of the FDE (3.79). Likewise, assume that $x_\epsilon : [-r, M] \to B$ is a maximal solution of the measure FDE (3.78) and $y : [-r, M] \to B$ is a maximal solution of averaged FDE (3.79), where $M > 0$ is such that $M < \min(\overline{b}, b)$. Then, for every $\eta > 0$, there exists $\epsilon_0 > 0$ such that for $\epsilon \in (0, \epsilon_0]$,

$$\|x_\epsilon(t) - y(t)\| < \eta, \quad \text{for every } t \in [0, M].$$

Proof. Notice that if $t = 0$, then $\|x_\epsilon(0) - y(0)\| = 0$. Moreover, given $t > 0$, conditions (J2), (J4), (J5), and Lemma 3.42 yield

$$\begin{aligned}
&\|x_\epsilon(t) - y(t)\| \\
&= \left\| \epsilon \int_0^t f\left((x_\epsilon)_{s,\epsilon}, \frac{s}{\epsilon}\right) dh_1\left(\frac{s}{\epsilon}\right) + \epsilon^2 \int_0^t g\left((x_\epsilon)_{s,\epsilon}, \frac{s}{\epsilon}, \epsilon\right) dh_2\left(\frac{s}{\epsilon}\right) - \int_0^t f_0(y_{s,\epsilon}) ds \right\| \\
&\leqslant \left\| \epsilon \int_0^t f\left((x_\epsilon)_{s,\epsilon}, \frac{s}{\epsilon}\right) - f\left(y_{s,\epsilon}, \frac{s}{\epsilon}\right) dh_1\left(\frac{s}{\epsilon}\right) \right\| \\
&+ \left\| \epsilon \int_0^t f\left(y_{s,\epsilon}, \frac{s}{\epsilon}\right) dh_1\left(\frac{s}{\epsilon}\right) - \int_0^t f_0(y_{s,\epsilon}) ds \right\| + \left\| \epsilon^2 \int_0^t g\left((x_\epsilon)_{s,\epsilon}, \frac{s}{\epsilon}, \epsilon\right) dh_2\left(\frac{s}{\epsilon}\right) \right\| \\
&\leqslant L\epsilon \int_0^t \|(x_\epsilon)_{s,\epsilon} - y_{s,\epsilon}\|_\infty dh_1\left(\frac{s}{\epsilon}\right) + \left\| \epsilon \int_0^t f\left(y_{s,\epsilon}, \frac{s}{\epsilon}\right) dh_1\left(\frac{s}{\epsilon}\right) - \int_0^t f_0(y_{s,\epsilon}) ds \right\| \\
&+ \epsilon^2 C\left(h_2\left(\frac{t}{\epsilon}\right) - h_2(0)\right) \\
&< L\epsilon \int_0^t \|(x_\epsilon)_{s,\epsilon} - y_{s,\epsilon}\|_\infty dh_1\left(\frac{s}{\epsilon}\right) + \xi(\epsilon) + \epsilon C t \frac{h_2\left(\frac{t}{\epsilon}\right) - h_2(0)}{\frac{t}{\epsilon}},
\end{aligned}$$

where $\xi(\epsilon)$ is given by Lemma 3.42.

By condition (J6), we can choose $\epsilon > 0$ sufficiently small such that

$$\frac{h_2\left(\frac{t}{\epsilon}\right) - h_2(0)}{\frac{t}{\epsilon}} \leqslant N, \quad \text{for every } t > 0.$$

Therefore, for $0 < t \leqslant M$, we obtain

$$\|x_\epsilon(t) - y(t)\| \leqslant L\epsilon \int_0^t \|(x_\epsilon)_{s,\epsilon} - y_{s,\epsilon}\|_\infty \, dh_1\left(\frac{s}{\epsilon}\right) + \xi(\epsilon) + \epsilon CMN.$$

From the fact that $(x_\epsilon)_0 = \epsilon\phi = y_0$, we obtain $\sup_{\sigma\in[-r,0]} \|x_\epsilon(\sigma) - y(\sigma)\| = 0$, whence, for $s \in [0, t]$, we have

$$\begin{aligned}
\|(x_\epsilon)_{s,\epsilon} - y_{s,\epsilon}\|_\infty &= \sup_{\theta\in[-\frac{r}{\epsilon},0]} \|x_\epsilon(s+\epsilon\theta) - y(s+\epsilon\theta)\| \\
&= \sup_{\sigma\in[s-r,s]} \|x_\epsilon(\sigma) - y(\sigma)\| \leqslant \sup_{\sigma\in[-r,s]} \|x_\epsilon(\sigma) - y(\sigma)\| \\
&= \sup_{\sigma\in[0,s]} \|x_\epsilon(\sigma) - y(\sigma)\|.
\end{aligned}$$

Therefore, we obtain

$$\|x_\epsilon(t) - y(t)\| \leqslant L\epsilon \int_0^t \sup_{\sigma\in[0,s]} \|x_\epsilon(\sigma) - y(\sigma)\| \, dh_1\left(\frac{s}{\epsilon}\right) + \xi(\epsilon) + \epsilon CMN$$

and hence, the Grönwall inequality for the Perron–Stieltjes integral (Theorem 1.52) implies that

$$\|x_\epsilon(t) - y(t)\| \leqslant \sup_{\tau\in[0,t]} \|x_\epsilon(\tau) - y(\tau)\| \leqslant e^{\epsilon L(h_1(t/\epsilon)-h_1(0))} \left[\xi(\epsilon) + \epsilon CMN\right].$$

Finally, by hypothesis (J5), it is possible to choose $\epsilon > 0$ sufficiently small such that

$$\epsilon L\left[h_1(t/\epsilon) - h_1(0)\right] = tL\frac{\left[h_1(t/\epsilon) - h_1(0)\right]}{t/\epsilon} \leqslant tLK \leqslant MLK.$$

Then, taking $\eta = e^{KLM}\left[\xi(\epsilon) + \epsilon CMN\right]$, we get

$$\|x_\epsilon(t) - y(t)\| \leqslant \eta, \quad \text{for all } t \in [0, M],$$

proving the result. □

In view of Remark 3.38, another averaging method follows next as an immediate consequence of Theorem 3.43. Compare with the initial conditions and codomains used in [84, Corollary 3.2].

Corollary 3.44: *Let $B \subset X$ and $P \subset G([-r, 0], B)$ be open sets, and assume that $f : P \times [0, \infty) \to X$ satisfies conditions (J1) and (J2) with P instead of $\widetilde{P}_\epsilon$ and $g : P \times [0, \infty) \times (0, \epsilon_0] \to X$ satisfies the conditions (J3) and (J4), with P instead* *of $\widetilde{P}_\epsilon$. Suppose conditions (J5)–(J7) are fulfilled, where $h_1, h_2 : [0, \infty) \to \mathbb{R}$ are nondecreasing functions and $\phi \in P$. Consider the nonautonomous measure FDE*

$$\begin{cases} y(t) = y(0) + \displaystyle\int_0^t \epsilon f(y_s, s)dh_1(s) + \epsilon^2 \int_0^t g(y_s, s, \epsilon)dh_2(s), \\ y_0 = \epsilon\phi, \end{cases} \tag{3.80}$$

and the averaged autonomous FDE

$$\begin{cases} \dot{x} = \epsilon f_0(x_t), \\ x_0 = \epsilon\phi, \end{cases} \tag{3.81}$$

where f_0 is given by

$$f_0(\psi) = \lim_{T\to\infty} \frac{1}{T} \int_0^T f(\psi, s)dh_1(s), \quad \textit{for every } \psi \in P.$$

Let $M > 0$ and $x_\epsilon, y : \left[-r, \frac{M}{\epsilon}\right] \to B$ be solutions of (3.80) and (3.81), respectively. Then, for every $\eta > 0$, there exists $\epsilon_0 > 0$ such that

$$\|x_\epsilon(t) - y(t)\| < \eta, \quad \textit{for every } t \in \left[-r, \frac{M}{\epsilon}\right] \textit{ and } \epsilon \in (0, \epsilon_0].$$

Now, we are able to present a nonperiodic averaging principle for impulsive measure FDEs, using the relations, provided by Theorem 3.6, between the solutions of these equations and the solutions of measure FDEs. A slightly different version can be found in [84, Theorem 4.2].

Theorem 3.45: *Consider $P = G([-r, 0], X)$ and the nonautonomous impulsive measure FDE*

$$\begin{cases} x_\epsilon(t) = x_\epsilon(0) + \epsilon \displaystyle\int_0^t f((x_\epsilon)_s, s)dh(s) + \epsilon^2 \int_0^t g((x_\epsilon)_s, s, \epsilon)dh(s) + \epsilon \sum_{\substack{k\in\mathbb{N} \\ t_k<t}} I_k(x_\epsilon(t_k)), \\ (x_\epsilon)_0 = \epsilon\phi, \end{cases} \tag{3.82}$$

where $\epsilon \in (0, \epsilon_0]$, $\phi \in P$ and the function $f : P \times [0, \infty) \to X$ satisfies conditions (J1) and (J2). Assume, further, that the function $h : [0, \infty) \to \mathbb{R}$ fulfills condition (J5) and the function $g : P \times [0, \infty) \times (0, \epsilon_0] \to X$ is bounded and satisfies condition (J3). Moreover, suppose the impulse operators $I_k : X \to X$ are bounded and Lipschitz-continuous, with the same Lipschitz constant for all $k \in \mathbb{N}$, and the limit

$$f_0(\psi) = \lim_{T\to\infty} \frac{1}{T} \int_0^T f(\psi, s)dh(s)$$

exists and is well defined for every $\psi \in P$. Denote

$$I_0(z) = \lim_{T\to\infty} \frac{1}{T} \sum_{\substack{k\in\mathbb{N} \\ 0\leqslant t_k<T}} I_k(z), \quad z \in X,$$

and consider the averaged autonomous problem

$$\begin{cases} y_\epsilon(t) = y_\epsilon(0) + \displaystyle\int_0^t \epsilon[f_0((y_\epsilon)_s) + I_0(y_\epsilon(s))]\, ds, \\ (y_\epsilon)_0 = \epsilon\phi, \end{cases} \tag{3.83}$$

where $\epsilon \in (0, \epsilon_0]$. Let $M > 0$ and $x_\epsilon, y_\epsilon : \left[-r, \frac{M}{\epsilon}\right] \to X$ be solutions of (3.82) and (3.83), respectively. Then, for every η, there exists $\epsilon_0 > 0$ such that

$$\|x_\epsilon(t) - y_\epsilon(t)\| < \eta, \quad \text{for every } t \in \left[-r, \frac{M}{\epsilon}\right] \text{ and } \epsilon \in (0, \epsilon_0].$$

Proof. Define a function $\widetilde{h} : [0, \infty) \to \mathbb{R}$ by

$$\widetilde{h}(t) = \begin{cases} h(t), & t \in [0, t_1], \\ h(t) + c_k, & t \in (t_k, t_{k+1}],\ k \in \mathbb{N}, \end{cases}$$

in such way that, for all $k \in \mathbb{N}$, the sequence $\{c_k\}_{k=1}^\infty$ satisfies $0 \leqslant c_k \leqslant c_{k+1}$ and $\Delta^+\widetilde{h}(t_k) = 1$. By definition, the function $\widetilde{h}$ is nondecreasing and left-continuous.

On the other hand, as a consequence of (J5) and the definition of $\widetilde{h}$, there exists a constant $C > 0$ such that, for all $\beta \geqslant 0$, we have

$$\limsup_{T\to\infty} \frac{\widetilde{h}(T+\beta) - \widetilde{h}(\beta)}{T} \leqslant C.$$

Then, from the hypotheses, we obtain

$$x_\epsilon(t) = x_\epsilon(0) + \int_0^t \left(\epsilon f((x_\epsilon)_s, s) + \epsilon^2 g((x_\epsilon)_s, s, \epsilon)\right) dh(s) + \sum_{\substack{k\in\mathbb{N} \\ 0\leqslant t_k<t}} \epsilon I_k(x_\epsilon(t_k)),$$

for all $\epsilon \in (0, \epsilon_0]$ and all $t \in \left[0, \frac{M}{\epsilon}\right]$.

Define a function

$$F_\epsilon(z, t) = \begin{cases} \epsilon f(z, t) + \epsilon^2 g(z, t, \epsilon), & t \notin \{t_1, t_2, \dots\}, \\ \epsilon I_k(z(0)), & t = t_k,\ k \in \mathbb{N}, \end{cases}$$

for all $z \in P$ and every $t \geqslant 0$. As a consequence of Theorem 3.6, we obtain

$$x_\epsilon(t) = x_\epsilon(0) + \int_0^t F_\epsilon((x_\epsilon)_s, s) d\widetilde{h}(s) \tag{3.84}$$

for all $\epsilon \in (0, \epsilon_0]$ and all $t \in \left[0, \frac{M}{\epsilon}\right]$. Therefore, for each $z \in P$ and $t \geqslant 0$, we can rewrite $F_\epsilon(z, t)$ as

$$F_\epsilon(z, t) = \epsilon\widetilde{f}(z, t) + \epsilon^2\widetilde{g}(z, t, \epsilon), \tag{3.85}$$

where

$$\widetilde{f}(z, t) = \begin{cases} f(z, t), & t \notin \{t_1, t_2, \dots\}, \\ I_k(z(0)), & t = t_k, \ k \in \mathbb{N}, \end{cases}$$

$$\widetilde{g}(z, t, \epsilon) = \begin{cases} g(z, t, \epsilon), & t \notin \{t_1, t_2, \dots\}, \\ 0, & t = t_k, \ k \in \mathbb{N}. \end{cases}$$

Thus, equalities (3.84) and (3.85) imply that for every $\epsilon \in (0, \epsilon_0]$, the function $x_\epsilon : \left[-r, \frac{M}{\epsilon}\right] \to X$ satisfies the initial value problem

$$\begin{cases} x_\epsilon(t) = x_\epsilon(0) + \epsilon \displaystyle\int_0^t \widetilde{f}((x_\epsilon)_s, s)d\widetilde{h}(s) + \epsilon^2 \int_0^t \widetilde{g}((x_\epsilon)_s, s, \epsilon)d\widetilde{h}(s), \\ (x_\epsilon)_0 = \epsilon\phi. \end{cases} \tag{3.86}$$

The definition of $\widetilde{f}$, together with the hypotheses and Lemma 3.9 imply condition (J2) is fulfilled. Then, by Lemma 3.5, we obtain

$$\begin{aligned} \int_0^T \widetilde{f}(z, s)d\widetilde{h}(s) &= \int_0^T f(z, s)dh(s) + \sum_{\substack{k\in\mathbb{N} \\ 0\leqslant t_k<T}} \widetilde{f}(z, t_k)\Delta^+\widetilde{h}(t_k) \\ &= \int_0^T f(z, s)dh(s) + \sum_{\substack{k\in\mathbb{N} \\ 0\leqslant t_k<T}} I_k(z(0)), \end{aligned}$$

for all $z \in P$. Therefore, the function $\widetilde{f}_0 : P \to X$, defined by

$$\widetilde{f}_0(y) = \lim_{T\to\infty} \frac{1}{T} \int_0^T \widetilde{f}(y, s)d\widetilde{h}(s), \quad y \in P,$$

satisfies

$$\widetilde{f}_0(y) = \lim_{T\to\infty} \frac{1}{T} \int_0^T f(y, s)dh(s) + \lim_{T\to\infty} \frac{1}{T} \sum_{\substack{k\in\mathbb{N} \\ 0\leqslant t_k<T}} I_k(y(0)) = f_0(y) + I_0(y(0)), \quad y \in P.$$

Last but not least, Corollary 3.44 yields that for every $\eta > 0$, there exists $\epsilon_0 > 0$ such that

$$\|x_\epsilon(t) - y_\epsilon(t)\| < \eta, \quad \text{for all } \epsilon \in (0, \epsilon_0] \text{ and } t \in \left[-r, \frac{M}{\epsilon}\right], \tag{3.87}$$

and the proof is complete. □

The next result, which we state without proof, is important to establish a non-periodic averaging principle for functional dynamic equations on time scales. A proof of it follows as in [218, Corollary 5.6] with obvious adaptations.

Lemma 3.46: *If* $\sup \mathbb{T} = \infty$, $\lim_{t\to\infty} \frac{\mu(t)}{t} = 0$ *and* $m : [t_0, \infty)_{\mathbb{T}} \to X$ *is* Δ*-integrable over every compact subinterval of* $[t_0, \infty)_{\mathbb{T}}$, *then*

$$\lim_{T\to\infty} \frac{1}{T} \int_{t_0}^{t_0+T} m^*(s) d\eta(s) = \lim_{\substack{T\to\infty \\ t_0+T\in\mathbb{T}}} \frac{1}{T} \int_{t_0}^{t_0+T} m(s)\Delta s,$$

where $\eta(t) = t^*$, *provided the limit on the right-hand side exists.*

A version of the next result for the finite dimensional case and different initial conditions can be found in [84], which, in turn, is inspired in [218, Theorem 5.3].

Theorem 3.47: *Let* $B \subset X$ *be open,* $P = G([-r, 0], B)$, $\mathbb{T}$ *be a time scale with* $\sup \mathbb{T} = \infty$, *and* $[t_0 - r, \infty)_{\mathbb{T}}$ *be a time scale interval, with* $t_0 \in \mathbb{T}$, $\epsilon_0, L > 0$. *Suppose* $\lim_{t\to\infty} \frac{\mu(t)}{t} = 0$, *where* μ *is the graininess function. Consider a function* $f : P \times [t_0, \infty)_{\mathbb{T}} \to X$ *and a bounded function* $g : P \times [t_0, \infty)_{\mathbb{T}} \times (0, \epsilon_0] \to X$ *for which the next conditions hold*

(NT1) *for every* $b > t_0$ *and* $y \in G([t_0 - r, b], B)$, *the functions* $t \mapsto f(y_t, t)$ *and* $t \mapsto g(y_t, t, \epsilon)$ *are regulated on* $[t_0, b]_{\mathbb{T}}$;

(NT2) *there exists a constant* $C > 0$ *such that for* $x, y \in P$ *and* $u_1, u_2 \in [t_0, \infty)_{\mathbb{T}}$ *such that* $u_1 \leqslant u_2$,

$$\left\| \int_{u_1}^{u_2} [f(x, s) - f(y, s)] \, \Delta s \right\| \leqslant C \int_{u_1}^{u_2} \|x - y\|_\infty \, \Delta s;$$

(NT3) *if* $z : [-r, 0] \to B$ *is a regulated function, then*

$$f_0(z) = \lim_{\substack{T\to\infty \\ t_0+T\in\mathbb{T}}} \frac{1}{T} \int_{t_0}^{t_0+T} f(z, s)\Delta s$$

exists and is well defined.

Let $\phi \in G([t_0 - r, t_0]_{\mathbb{T}}, B)$ *be bounded. Suppose for every* $\epsilon \in (0, \epsilon_0]$, *the functional dynamic equation on time scales*

$$\begin{cases} x(t) = x(t_0) + \epsilon \displaystyle\int_{t_0}^{t} f(x_s^*, s)\Delta s + \epsilon^2 \int_{t_0}^{t} g(x_s^*, s, \epsilon)\Delta s, \\ x(t) = \epsilon\phi(t), \quad t \in [t_0 - r, t_0]_{\mathbb{T}}, \end{cases} \tag{3.88}$$

has a solution $x_\epsilon : \left[t_0 - r, t_0 + \frac{M}{\epsilon}\right]_{\mathbb{T}} \to B$*, and the averaged FDE*

$$\begin{cases} \dot{y} = \epsilon f_0(y_t^*), \\ y(t) = \epsilon\phi(t), \quad t \in [t_0 - r, t_0]_{\mathbb{T}}, \end{cases} \tag{3.89}$$

has a solution $y_\epsilon : \left[t_0 - r, t_0 + \frac{M}{\epsilon}\right]_{\mathbb{T}} \to B$*. Then, for every* $\eta > 0$*, there exists* $\epsilon_0 > 0$ *such that for* $\epsilon \in (0, \epsilon_0]$*,*

$$\|x_\epsilon(t) - y_\epsilon(t)\| < \eta, \quad \textit{for every;} \quad t \in \left[t_0 - r, t_0 + \frac{M}{\epsilon}\right]_{\mathbb{T}}.$$

Proof. Note that we can consider $t_0 = 0$ without loss of generality. Otherwise, we can deal with the shifted problem with the time scale $\widetilde{\mathbb{T}} = \{t - t_0 : t \subset \mathbb{T}\}$ and the functions

$$\widetilde{f}(x,t) = f(x, t + t_0) \qquad \text{and} \qquad \widetilde{g}(x,t,\epsilon) = g(x, t + t_0, \epsilon),$$

where $\widetilde{f} : P \times [0,\infty)_{\widetilde{\mathbb{T}}} \to X$ and $\widetilde{g} : P \times [0,\infty)_{\widetilde{\mathbb{T}}} \times (0, \epsilon_0] \to X$. For $t \in [0, \infty)$, $z \in P$ and $\epsilon \in (0, \epsilon_0]$, define the extensions

$$f^*(z,t) = f(z,t^*) \qquad \text{and} \qquad g^*(z,t,\epsilon) = g(z,t^*,\epsilon).$$

Since $\lim_{t\to\infty} \frac{\mu(t)}{t} = 0$, where μ is the graininess function, there are numbers $D > 0$ and $\tau \in \mathbb{T}$ such that $\frac{\mu(t)}{t} \leqslant D$, for every $t \in [\tau, \infty)_{\mathbb{T}}$. If $t \in \mathbb{R}$ is such that $\rho(t^*) \leqslant t$, then

$$t^* - t \leqslant \sigma(\rho(t^*)) - \rho(t^*),$$

once the backward and forward jump operators (see Definition 3.11) satisfy $t^* \leqslant \sigma(\rho(t^*))$ and $t \geqslant \rho(t^*)$ by the definition of t^*. Therefore,

$$t^* \leqslant t + \mu(\rho(t^*)) \leqslant t + D\rho(t^*) \leqslant t + Dt = t(D+1).$$

For all $t \in [0, \infty)$, set $h(t) = t^*$. Then, for sufficiently large T and for all $a \geqslant 0$, we have

$$\frac{h(a+T) - h(a)}{T} = \frac{(a+T)^* - a^*}{T} \leqslant \frac{(a+T)(D+1) - a^*}{T},$$

whence

$$\limsup_{T\to\infty} \frac{h(a+T) - h(a)}{T} \leqslant \limsup_{T\to\infty} \frac{(a+T)(D+1) - a^*}{T} = D + 1,$$

which, in turn, shows that conditions (J5) and (J6) are fulfilled.

As a consequence of Theorems 3.25 and 3.46, we conclude that

$$\begin{aligned} f_0(z) &= \lim_{T\to\infty} \frac{1}{T} \int_0^T f(z,s)\Delta s = \lim_{T\to\infty} \frac{1}{T} \int_0^T f(z,s^*) dh(s) \\ &= \lim_{T\to\infty} \frac{1}{T} \int_0^T f^*(z,s) dh(s), \end{aligned}$$

for every $z \in P$. Then, using the fact that $x_\epsilon : \left[-r, \frac{M}{\epsilon}\right]_{\mathbb{T}} \to B$ is a solution of the nonautonomous functionals dynamic equation on time scale (3.88), Theorem 3.34 yields $x_\epsilon^* : [-r, \frac{M}{\epsilon}] \cap \mathbb{T}^* \to B$ is a solution on $\left[-r, \frac{M}{\epsilon}\right] \cap \mathbb{T}^*$ of the measure FDE

$$\begin{cases} x^*(t) = x^*(0) + \int_0^t \epsilon f^*(x_s^*, s)dh(s) + \epsilon^2 \int_0^t g^*(x_s^*, s, \epsilon)dh(s), \\ x_0^* = \epsilon\phi^*. \end{cases} \tag{3.90}$$

Finally, Lemma 3.32 implies that conditions (J1), (J2), (J3), and (J4) are also fulfilled. Therefore, Theorem 3.44 implies that for every $\eta > 0$, there exists $\epsilon_0 > 0$ such that for all $\epsilon \in (0, \epsilon_0]$ and $t \in \left[0, \frac{M}{\epsilon}\right]$, the inequality

$$\|x_\epsilon^*(t) - y_\epsilon(t)\| < \eta,$$

holds, where y_ϵ is a solution of the averaged autonomous functional dynamic equation on time scales (3.89). Noticing that $x_\epsilon^*(t) = x_\epsilon(t)$ for $t \in \left[0, \frac{M}{\epsilon}\right]_{\mathbb{T}}$ and using the initial condition, the statement follows. □

3.5 Continuous Dependence on Time Scales

This section is devoted to results on continuous dependence on time scales of solutions of dynamic equations on time scales. This type of result has been investigated by several researchers [1, 62, 157] because it plays an important role in applications.

Let X be a Banach space with norm $\|\cdot\|$. The idea behind this type of result is to prove that the solution of the initial value problem

$$\begin{cases} x^\Delta(t) = f(x(t), t), & t \in \mathbb{T}_n, \\ x(t_0) = x_0, & t_0 \in \mathbb{T}_n, \end{cases}$$

where $f : X \times \mathbb{T}_n \to X$ and $x_0 \in X$, converges uniformly, as $n \to \infty$, to the solution of the problem

$$\begin{cases} x^\Delta(t) = f(x(t), t), & t \in \mathbb{T}, \\ x(t_0) = x_0, & t_0 \in \mathbb{T}, \end{cases}$$

where $f : X \times \mathbb{T} \to X$ and $x_0 \in X$, whenever $d_H(\mathbb{T}_n, \mathbb{T}) \to 0$, as $n \to \infty$, with d_H denoting the Hausdorff metric (or $\mathbb{T}_n \to \mathbb{T}$, as $n \to \infty$ using the induced metric from the Fell topology – see [190], for details). We consider the Hausdorff topology and Hausdorff metric in which the distance between two sets is defined by

$$d_H(A, B) = \max\left\{ \sup_{a\in A} \inf_{b\in B} \{\|a - b\|\}, \sup_{b\in B} \inf_{a\in A} \{\|a - b\|\} \right\}.$$

Our goal, here, is to generalize these results. More precisely, assuming that $B \subset X$ is open, $O_n \subset G([t_0 - r, t_0 + \sigma]_{\mathbb{T}_n}, B)$ is open, $P_n = \{y_t : y \in O_n, t \in [t_0, t_0 + \sigma]\}$,

$f : P_n \times [t_0, t_0 + \sigma]_{\mathbb{T}_n} \to X$ is a function. As commented in Section 3.2, we shall consider that $I + I_k : B \to B$ for every $k \in \{1, \dots, m\}$, where $I : B \to B$ is the identity operator and $I_k : B \to X$ denotes the impulse operator, our aim is to provide sufficient conditions to ensure that the sequence of solutions of the impulsive functional dynamic on time scales

$$\begin{cases} x(t) = x(t_0) + \displaystyle\int_{t_0}^{t} f\left(x_s^{*_n}, s\right) \Delta s + \sum_{\substack{k \in \{1, \dots, m\} \\ t_k < t}} I_k(x(t_k)), & t \in [t_0, t_0 + \sigma]_{\mathbb{T}_n}, \\ x(t) = \phi(t), \quad t \in [t_0 - r, t_0]_{\mathbb{T}_n} \end{cases} \tag{3.91}$$

converges uniformly to the solution of the impulsive functional dynamic equation on time scales

$$\begin{cases} x(t) = x(t_0) + \displaystyle\int_{t_0}^{t} f\left(x_s^{*}, s\right) \Delta s + \sum_{\substack{k \in \{1, \dots, m\} \\ t_k < t}} I_k(x(t_k)), & t \in [t_0, t_0 + \sigma]_{\mathbb{T}}, \\ x(t) = \phi(t), \quad t \in [t_0 - r, t_0]_{\mathbb{T}}, \end{cases} \tag{3.92}$$

where $f : P \times [t_0, t_0 + \sigma]_{\mathbb{T}} \to X$ is a function, $O \subset G([t_0 - r, t_0 + \sigma]_{\mathbb{T}}, B)$ is open, $P = \{y_t : y \in O, t \in [t_0, t_0 + \sigma]\}$, and $d_H(\mathbb{T}_n, \mathbb{T}) \to 0$ as $n \to \infty$, with d_H denoting the Hausdorff metric.

The results of this section are based on [21] with slight differences in the proofs. Here, also, the conditions are more relaxed.

Before we continue, let us clarify the notations $*_n$ and $*$ which appear in Eqs. (3.91) and (3.92). These two operators $*_n : \mathbb{T}_n^* \to \mathbb{T}_n$ and $* : \mathbb{T}^* \to \mathbb{T}$ are defined, respectively, by

$$t^{*_n} = \{s \in \mathbb{T}_n : s \geqslant t\} \quad \text{and} \quad t^* = \{s \in \mathbb{T} : s \geqslant t\}.$$

In order to prove a continuous dependence result for impulsive functional dynamic equations on time scales, the strategy is to employ the relations between the solutions of these equations and the solutions of impulsive measure FDEs, and not only those relations but also the relations between the solutions of impulsive measure FDEs and the solutions of measure FDEs (see Theorems 3.6 and 3.34).

For each $n \in \mathbb{N}$, let $\mathbb{T}_n$ be a time scale such that σ_n is its corresponding forward jump operator. Consider, for each $n \in \mathbb{N}$, a function $g_n : [t_0, t_0 + \sigma] \to \mathbb{R}$ given by $g_n(t) = \inf\{s \in \mathbb{T}_n : s \geqslant t\}$ related to $\mathbb{T}_n$.

We start by proving that, under the conditions below, the sequence of solutions $\{x_n^{*_n}\}_{n \in \mathbb{N}}$ of the initial value problems (for $n \in \mathbb{N}$)

$$\begin{cases} x_n^{*_n}(t) = x_n^{*_n}(t_0) + \displaystyle\int_{t_0}^{t} f((x_n^{*_n})_s, s) dg_n(s), & t \in [t_0, t_0 + \sigma] \cap \mathbb{T}_n^*, \\ (x_n^{*_n})_{t_0} = \phi^{*_n} \end{cases} \tag{3.93}$$

converges uniformly to the solution of the initial value problem

$$\begin{cases} x^*(t) = x^*(t_0) + \int_{t_0}^{t} f(x_s^*, s)dg(s), & t \in [t_0, t_0 + \sigma] \cap \mathbb{T}^*, \\ x_{t_0}^* = \phi^*. \end{cases} \tag{3.94}$$

Assume that $O \subset G([t_0 - r, t_0 + \sigma], B)$ is open, $P = \{y_t : y \in O, t \in [t_0, t_0 + \sigma]\}$, $f : P \times [t_0, t_0 + \sigma] \to X$ is a function, and $g : [t_0, t_0 + \sigma] \to \mathbb{R}$ is a nondecreasing and left-continuous function. Moreover, assume that $f : P \times [t_0, t_0 + \sigma] \to X$ fulfills the following conditions:

(CD1) the function $s \mapsto f(z_s, s)$ is regulated on $[t_0, t_0 + \sigma]$ for every $z \in O$;
(CD2) there exists a regulated function $M : [t_0, t_0 + \sigma] \to \mathbb{R}$ such that, for every $u_1, u_2 \in [t_0, t_0 + \sigma]$, with $u_1 \leqslant u_2$, and every $z \in O$, we have

$$\left\| \int_{u_1}^{u_2} f(z_s, s)dg(s) \right\| \leqslant \int_{u_1}^{u_2} M(s)dg(s);$$

(CD3) there exists a regulated function $L : [t_0, t_0 + \sigma] \to \mathbb{R}$ such that, for every $u_1, u_2 \in [t_0, t_0 + \sigma]$, with $u_1 \leqslant u_2$, and every $z, y \in O$, we have

$$\left\| \int_{u_1}^{u_2} [f(z_s, s) - f(y_s, s)]\, dg(s) \right\| \leqslant \int_{u_1}^{u_2} L(s)\|z_s - y_s\|_\infty\, dg(s).$$

The next continuous dependence result for measure FDEs is based on its analogue from [21] with minor changes. Here, we consider weaker conditions on the right-hand sides of the equations. The proof we present follows some ideas of [221].

Theorem 3.48: *Let $B \subset X$ be open, $t_0, t_0 + \sigma \in \mathbb{T} \cap \mathbb{T}_n$ for every $n \in \mathbb{N}$, with $\sigma > 0$, $O \subset G([t_0 - r, t_0 + \sigma], B)$, $P = \{y_t : y \in O, t \in [t_0, t_0 + \sigma]\}$ and suppose the function $f : P \times [t_0, t_0 + \sigma] \to X$ satisfies conditions* (CD1), (CD2), *and* (CD3), *and the functions $g, g_n : [t_0, t_0 + \sigma] \to \mathbb{R}$ are nondecreasing and left-continuous. Assume, in addition, that $x_n^{*n} : [t_0 - r, t_0 + \sigma] \to B$ is a solution of the initial value problem*

$$\begin{cases} x_n^{*n}(t) = x_n^{*n}(t_0) + \int_{t_0}^{t} f((x_n^{*n})_s, s)dg_n(s), & t \in [t_0, t_0 + \sigma], \\ (x_n^{*n})_{t_0} = \phi_{t_0}^{*n}, \end{cases} \tag{3.95}$$

and $x^* : [t_0 - r, t_0 + \sigma] \to B$ is a solution of the measure FDE given by

$$\begin{cases} x^*(t) = x^*(t_0) + \int_{t_0}^{t} f(x_s^*, s)dg(s), & t \in [t_0, t_0 + \sigma], \\ x_{t_0}^* = \phi_{t_0}^*, \end{cases} \tag{3.96}$$

where $\phi_{t_0}^{*n}, \phi_{t_0}^* \in G([-r, 0], B)$. Moreover, suppose the sequence of functions $\{g_n\}_{n\in\mathbb{N}}$ converges uniformly to g, as $n \to \infty$, and the sequence of initial conditions $\{\phi_{t_0}^{*n}\}_{n\in\mathbb{N}}$

converges uniformly to $\phi^*_{t_0}$, as $n \to \infty$. Then, for every $\epsilon > 0$, there exists $N > 0$ sufficiently large such that for every $n > N$, we have

$$\|x_n^{*_n}(t) - x^*(t)\| < \epsilon, \quad \text{for all} \quad t \in [t_0, t_0 + \sigma].$$

Proof. Because the sequence of functions $\{g_n\}_{n\in\mathbb{N}}$ converges uniformly to g, given $\epsilon > 0$, there exists $N_1 > 0$ sufficiently large such that for every $n > N_1$,

$$|g_n(t) - g(t)| < \epsilon, \quad \text{for all } t \in [t_0, t_0 + \sigma].$$

Then, $\|g_n - g\|_\infty < \epsilon$ for sufficiently large n.

On the other hand, since the sequence of functions ϕ^{*_n} converges uniformly to ϕ^* as $n \to \infty$, there exists $N_2 > 0$ sufficiently large such that for every $n > N_2$,

$$\|\phi^{*_n}(t) - \phi^*(t)\| < \epsilon, \quad \text{for all } t \in [-r, 0].$$

Now, due to the fact that $\lim_{n\to\infty} g_n(t_0) = g(t_0)$ and $\lim_{n\to\infty} g_n(t_0 + \sigma) = g(t_0 + \sigma)$, the sequences $\{g_n(t_0)\}_{n\in\mathbb{N}}$ and $\{g_n(t_0 + \sigma)\}_{n\in\mathbb{N}}$ are necessarily bounded. Thus, there exists a constant $M_1 > 0$ such that

$$\text{var}_{t_0}^{t_0+\sigma}(g_n(t)) = g_n(t_0 + \sigma) - g_n(t_0) \leqslant M_1, \quad n \in \mathbb{N}. \tag{3.97}$$

Hence, by condition (CD1) and the facts that $f_n - g$ is a function of bounded variation and $s \mapsto f(x_s, s)$ is a regulated function on $[t_0, t_0 + \sigma]$ for every $x \in O$, it follows that the Perron–Stieltjes integral $\int_{t_0}^{t_0+\sigma} f(x_s^*, s)d(g_n - g)(s)$ exists.

Since $g_n - g$ converges uniformly to 0 as $n \to \infty$, Theorem 1.88 yields

$$\lim_{n\to\infty} \int_{t_0}^{t} f(x_s, s)d(g_n - g)(s) = 0$$

uniformly with respect to $t \in [t_0, t_0 + \sigma]$. Therefore, for the given $\epsilon > 0$, there exists $N_3 \in \mathbb{N}$ such that

$$\left\| \int_{t_0}^{t} f(x_s, s)d(g_n - g)(s) \right\| \leqslant \epsilon, \quad \text{for all } n \geqslant N_3 \text{ and } t \in [t_0, t_0 + \sigma]. \tag{3.98}$$

For $t \in [t_0, t_0 + \sigma]$ and $n > \max\{N_1, N_2, N_3\}$, we also have

$$\begin{aligned}
&\left\| x_n^{*_n}(t) - x^*(t) \right\| \\
&= \left\| x_n^{*_n}(t_0) - x^*(t_0) + \int_{t_0}^{t} f((x_n^{*_n})_s, s)dg_n(s) - \int_{t_0}^{t} f((x^*)_s, s)dg(s) \right\| \\
&\leqslant \|x_n^{*_n}(t_0) - x^*(t_0)\| + \left\| \int_{t_0}^{t} f((x_n^{*_n})_s, s)dg_n(s) - \int_{t_0}^{t} f((x^*)_s, s)dg(s) \right\| \\
&\leqslant \|\phi^{*_n}(t_0) - \phi^*(t_0)\| + \left\| \int_{t_0}^{t} f((x_n^{*_n})_s, s)dg_n(s) - \int_{t_0}^{t} f((x^*)_s, s)dg(s) \right\|
\end{aligned}$$

$$\leqslant \left\| \int_{t_0}^{t} f((x_n^{*n})_s, s) dg_n(s) - \int_{t_0}^{t} f((x^*)_s, s) dg_n(s) \right\| + \left\| \int_{t_0}^{t} f((x^*)_s, s) dg_n(s) - \int_{t_0}^{t} f((x^*)_s, s) dg(s) \right\| \leqslant \epsilon + \int_{t_0}^{t} L(s) \|(x_n^{*n})_s - (x^*)_s\|_\infty \, dg_n(s),$$

where we used condition (CD3) and (3.98) to obtain the last inequality. Thus,

$$\|x_n^{*n}(t) - x^*(t)\| \leqslant \epsilon + \int_{t_0}^{t} L(s) \|(x_n^{*n})_s - (x^*)_s\|_\infty \, dg_n(s).$$

Using the facts that $(x_n^{*n})_{t_0} = \phi_{t_0}^{*n}$ and $(x^*)_{t_0} = \phi_{t_0}^{*}$, and by uniform convergence $\phi_{t_0}^{*n} \to \phi_{t_0}^{*}$, it follows that for $n > n_0$,

$$\|(x_n^{*n})_s - x_s^*\|_\infty = \sup_{\theta \in [-r,0]} \|x_n^{*n}(s+\theta) - x^*(s+\theta)\| \leqslant \epsilon + \sup_{\eta \in [0,s]} \|x_n^{*n}(\eta) - x^*(\eta)\|,$$

whence

$$\|x_n^{*n}(t) - x^*(t)\| \leqslant \epsilon + \int_{t_0}^{t} L(s) \left[\epsilon + \sup_{\eta \in [0,s]} \|x_n^{*n}(\eta) - x^*(\eta)\| \right] dg_n(s).$$

Since L is a regulated function, there exists $\gamma > 0$ such that $\gamma = \sup_{s \in [t_0, t_0+\sigma]} L(s)$. Hence, for every $t \in [t_0, t_0 + \sigma]$, we have

$$\begin{aligned}
&\|x_n^{*n}(t) - x^*(t)\| \\
&\leqslant \epsilon + \epsilon \int_{t_0}^{t_0+\sigma} \sup_{s \in [t_0, t_0+\sigma]} L(s) dg_n(s) + \int_{t_0}^{t} L(s) \sup_{\eta \in [0,s]} \|x_n^{*n}(\eta) - x^*(\eta)\| \, dg_n(s) \\
&= \epsilon + \epsilon \int_{t_0}^{t_0+\sigma} \gamma \, dg_n(s) + \int_{t_0}^{t} L(s) \sup_{\eta \in [0,s]} \|x_n^{*n}(\eta) - x^*(\eta)\| \, dg_n(s) \\
&\leqslant \epsilon + \epsilon \int_{t_0}^{t_0+\sigma} \gamma \, dg_n(s) + \int_{t_0}^{t} \gamma \sup_{\eta \in [0,s]} \|x_n^{*n}(\eta) - x^*(\eta)\| \, dg_n(s).
\end{aligned}$$

Finally, the Grönwall inequality (Theorem 1.52) yields

$$\begin{aligned}
\|x_n^{*n}(t) - x^*(t)\| &\leqslant \epsilon \left[1 + \int_{t_0}^{t_0+\sigma} \gamma \, dg_n(s) \right] e^{\int_{t_0}^{t} \gamma \, dg_n(s)} \\
&\leqslant \epsilon \left[1 + \int_{t_0}^{t_0+\sigma} \gamma \, dg_n(s) \right] e^{\int_{t_0}^{t_0+\sigma} \gamma \, dg_n(s)} \\
&\leqslant \epsilon (1 + \gamma M_1) e^{\gamma M_1}
\end{aligned}$$

where in the last inequality we used (3.97). Since $\epsilon > 0$ is arbitrarily small, the statement follows. □

It is worth highlighting that one cannot suppress, from Theorem 3.48, the hypothesis on the uniform convergence of the sequence of functions $\{g_n\}_{n\in\mathbb{N}}$ to g, as $n \to \infty$, because this cannot be ensured only by using $d(\mathbb{T}_n, \mathbb{T}) \to 0$, as $n \to \infty$. Below, we present an example, borrowed from [21], which illustrates this situation.

Example 3.49: Define the following time scales

$$\mathbb{T} = [0, a] \cup [a+1, b] \quad \text{and}$$
$$\mathbb{T}_n = [0, a + 1/n] \cup [a+1, b], \quad \text{for every } n \in \mathbb{N}.$$

Then, $d_H(\mathbb{T}, \mathbb{T}_n) = 1/n \to 0$, as $n \to \infty$. However, $g_n(a + 1/n) = a + 1/n$, for every $n \in \mathbb{N}$, while $g(a + 1/n) = a + 1$. Thus, for every $n \geqslant 2$, there exists $t \in [0, b]$ such that $g(t) - g_n(t) \geqslant 1/2$, implying that the sequence $\{g_n\}_{n\in\mathbb{N}}$ does not converge uniformly to g.

Even if we consider the Fell topology instead of the Hausdorff topology, the assumption on the uniform convergence of the sequence of functions $\{g_n\}_{n\in\mathbb{N}}$ is needed. The next example, also borrowed from [21], illustrates this fact.

Example 3.50: Consider $\mathbb{R}$ with the usual metric. Let $\mathrm{CL}(\mathbb{R})$ denote the set of all closed and nonempty subsets of $\mathbb{R}$ endowed with the Fell topology. Then, $\mathbb{T}_n = \{z + 1/n : z \in \mathbb{Z}\}$ converges to $\mathbb{Z}$, as $n \to \mathbb{N}$ (see [62, Lemma 4]). Besides, for each $n \in \mathbb{N}$, $g_n(z + 1/n) = z + 1/n$, while $g(z + 1/n) = z + 1$. Therefore, the sequence $\{g_n\}_{n\in\mathbb{N}}$ does not converge uniformly to g.

The next theorem, which appears in [21], is the main result of this section and it concerns continuous dependence on time scales of solutions of impulsive functional dynamic equations on time scales.

Theorem 3.51: *Let $t_0, t_0 + \sigma \in \mathbb{T} \cap \mathbb{T}_n$ for each $n \in \mathbb{N}$, with $\sigma > 0$, and consider an open set $B \subset X$. Suppose $x_n : [t_0 - r, t_0 + \sigma]_{\mathbb{T}_n} \to B$ is a solution of the impulsive functional dynamic equation on time scales*

$$\begin{cases} x_n(t) = x_n(t_0) + \displaystyle\int_{t_0}^{t} f_n((x_n^{*_n})_s, s)\Delta s + \sum_{\substack{k\in\{1,\ldots,m\} \\ t_k<t}} I_k(x_n(t_k)), & t \in [t_0, t_0 + \sigma]_{\mathbb{T}_n}, \\ x_n(t) = \phi(t), \quad t \in [t_0 - r, t_0]_{\mathbb{T}_n}, \end{cases} \tag{3.99}$$

where $P \subset G([-r, 0], B)$ *is open,* $O = G([t_0 - r, t_0 + \sigma], B)$, $\phi \in G([t_0 - r, t_0]_{\mathbb{T}_n \cap \mathbb{T}}, B)$, *and for each* $n \in \mathbb{N}$, *the function* $f_n : P \times [t_0, t_0 + \sigma]_{\mathbb{T}_n} \to X$, *satisfies the conditions:*

(CDT1) *the function* $s \mapsto f_n(x_s, s)$ *is regulated on* $[t_0, t_0 + \sigma]_{\mathbb{T}_n}$ *for every* $x \in O$;

(CDT2) *there exists a regulated function* $M : [t_0, t_0 + \sigma]_{\mathbb{T}_n} \to \mathbb{R}$ *such that, for all* $z \in O$ *and* $u_1, u_2 \in [t_0, t_0 + \sigma]_{\mathbb{T}_n}$, *with* $u_1 \leqslant u_2$, *we have*

$$\left\| \int_{u_1}^{u_2} f_n(z_s, s)\Delta s \right\| \leqslant \int_{u_1}^{u_2} M(s)\Delta s;$$

(CDT3) *there exists a regulated function* $L : [t_0, t_0 + \sigma]_{\mathbb{T}_n} \to \mathbb{R}$ *such that for all* $z, y \in O$ *and* $u_1, u_2 \in [t_0, t_0 + \sigma]_{\mathbb{T}_n}$, *with* $u_1 \leqslant u_2$, *we have*

$$\left\| \int_{u_1}^{u_2} [f_n(z_s, s) - f_n(y_s, s)]\, \Delta s \right\| \leqslant \int_{u_1}^{u_2} L(s)\|z_s - y_s\|_\infty\, \Delta s.$$

Suppose, in addition, $x : [t_0 - r, t_0 + \sigma]_{\mathbb{T}} \to B$ *is a solution of the impulsive functional dynamic equation on time scales*

$$\begin{cases} x(t) = x(t_0) + \displaystyle\int_{t_0}^{t} f(x_s^*, s)\Delta s + \sum_{\substack{k \in \{1,\dots,m\} \\ t_k < t}} I_k(x(t_k)), & t \in [t_0, t_0 + \sigma]_{\mathbb{T}}, \\ x(t) = \phi(t), \quad t \in [t_0 - r, t_0]_{\mathbb{T}}, \end{cases} \tag{3.100}$$

where $f : P \times [t_0, t_0 + \sigma]_{\mathbb{T}} \to X$ *satisfies conditions* (CDT1), (CDT2), *and* (CDT3), $\phi \in P$, *and for each* $k \in \{1, 2, \dots, m\}$, *impulse operator* $I_k : B \to X$ *is bounded and Lipschitz-continuous such that* $I + I_k : B \to B$, *where* $I : B \to B$ *is the identity operator. Assume that* $d_H(\mathbb{T}_n, \mathbb{T}) \to 0$ *as* $n \to \infty$ *and the sequence of functions* $\{g_n\}_{n \in \mathbb{N}}$ *converges uniformly to* g *as* $n \to \infty$. *Assume, further, that the sequence* $\{\phi^{*_n}\}_{n \in \mathbb{N}}$ *converges uniformly to* ϕ^* *as* $n \to \infty$. *Then, for every* $\epsilon > 0$, *there exists* $N > 0$ *sufficiently large such that for* $n > N$, *we have*

$$\|x_n(t) - x(t)\| < \epsilon \quad \textit{for} \quad t \in [t_0 - r, t_0 + \sigma]_{\mathbb{T}_n \cap \mathbb{T}}.$$

Proof. Since for each $n \in \mathbb{N}$, the function $f_n : P \times [t_0, t_0 + \sigma]_{\mathbb{T}} \to X$ satisfies conditions (CDT1), (CDT2), and (CDT3), Lemma 3.32 yields that the corresponding conditions (CD1), (CD2), and (CD3) are fulfilled for the extension of f_n. Therefore, all hypotheses of Theorem 3.48 are satisfied, and hence, the desired result follows immediately by applying the relations between the solutions of impulsive measure FDEs and the solutions of impulsive functional dynamic equations on time scales as described in Theorem 3.34. □

Criteria ensuring continuous dependence of solutions of dynamic equations on variable time scales have several applications in numerical analysis. It is a well-known fact that many differential equations cannot be solved analytically, however a numerical approximation to a certain solution is usually good enough to solve a problem described by models in engineering and sciences. In order to do this, it is possible to build up algorithms to compute such an approximation. Therefore, results as the ones presented in this chapter are very useful to the study of solutions of ODEs or dynamic equations on time scales, depending on the chosen time scale, without the necessity to solve the equations analytically.

In what follows, we present some examples which illustrate this fact and show the effectiveness of our results. Such examples can be found in [21, 59, 157, 190].

Example 3.52: Consider a simple autonomous linear dynamic equation given by

$$\begin{cases} x^{\Delta}(t) = ax(t), \\ x(0) = x_0. \end{cases} \tag{3.101}$$

If we solve Eq. (3.101) for the time scale $\mathbb{T} = \mathbb{R}$, we clearly obtain $x(t) = x_0 e^{at}$. On the other hand, solving Eq. (3.101) for the time scale $\mathbb{T}_n = \frac{1}{n}\mathbb{Z}$, where $n \in \mathbb{N}$, we obtain

$$y_n(t) = x_0\left(1 + \frac{a}{n}\right)^{nt} \quad \text{for every } t \in \mathbb{T}_n.$$

Therefore, $d_H(\mathbb{T}_n, \mathbb{R}) \to 0$, as $n \to \infty$, and moreover g_n converges to g uniformly as $n \to \infty$, where $g_n(t) = \inf\{s \in \mathbb{T}_n : s \geqslant t\}$ and $g(t) = \inf\{s \in \mathbb{R} : s \geqslant t\}$. Indeed, notice that if $t \in \mathbb{T}_n$, then $g_n(t) = g(t)$, while if $t \notin \mathbb{T}_n$, then $|g_n(t) - g(t)| \leqslant \frac{1}{n}$, for each $n \in \mathbb{N}$. Hence, for every $t \in \mathbb{R}$ and every $n \in \mathbb{N}$, we have $|g_n(t) - g(t)| < \frac{1}{n}$. It is also true that $\lim_{n\to\infty} y_n(t) = x(t)$.

Example 3.53: Consider a particular (logistic) initial value problem given by

$$\begin{cases} x^{\Delta}(t) = 4x\left(\frac{3}{4} - x\right), \\ x(0) = x_0. \end{cases} \tag{3.102}$$

Suppose $\mathbb{T}_n = \frac{1}{n}\mathbb{Z}_+$, for $n \in \mathbb{N}$. Then, evaluating the solution, one obtains

$$x^{\Delta}(t) = \frac{x(\sigma(t)) - x(t)}{\mu(t)} = \frac{x\left(t + \frac{1}{n}\right) - x(t)}{\frac{1}{n}} = 4x(t)\left[\frac{3}{4} - x(t)\right],$$

which implies that, for each $n \in \mathbb{N}$,

$$x\left(t + \frac{1}{n}\right) = \frac{4}{n}x(t)\left[\frac{3}{4} - x(t)\right] + x(t) = \frac{4}{n}x(t)\left[\frac{3+n}{4} - x(t)\right],$$

which is obtained by iterating the following equation (see [190], for details)

$$x_n(t) = \frac{4}{n}x(t)\left[\frac{3+n}{4} - x(t)\right]$$

for each $n \in \mathbb{N}$. Finally, taking $n \to \infty$, the solutions tend to the solution of the logistic differential equation on $\mathbb{R}_+$ and $d_H(\frac{1}{n}\mathbb{Z}_+, \mathbb{R}_+) \to 0$ as $n \to \infty$ (see [190]). Similarly, one can show that $g_n \to g$ uniformly, as $n \to \infty$, where $g_n(t) = \inf\{s \in \mathbb{T}_n : s \geqslant t\}$ and $g(t) = \inf\{s \in \mathbb{R} : s \geqslant t\}$.

4

Generalized Ordinary Differential Equations

Everaldo M. Bonotto[1], *Márcia Federson*[2], *and Jaqueline G. Mesquita*[3]

[1]*Departamento de Matemática Aplicada e Estatística, Instituto de Ciências Matemáticas e de Computação (ICMC), Universidade de São Paulo, São Carlos, SP, Brazil*
[2]*Departamento de Matemática, Instituto de Ciências Matemáticas e de Computação (ICMC), Universidade de São Paulo, São Carlos, SP, Brazil*
[3]*Departamento de Matemática, Instituto de Ciências Exatas, Universidade de Brasília, Brasília, DF, Brazil*

The goal of this chapter is to introduce a new class of integral equations known as *generalized ordinary differential equations* for functions taking values in a Banach space. Throughout the book, we use the short form "*generalized ODEs*" to refer to these equations.

It is well-known that the theory of generalized ordinary differential equations (ODEs) goes back to 1957 with the Czech mathematician Jaroslav Kurzweil. See his papers [147–149] from 1957 and 1958. Kurzweil's main initial idea was to generalize some results on continuous dependence, with respect to initial conditions, of solutions of classic ODEs in order to obtain averaging principles. Therefore, in [147], Kurzweil introduced the concept of generalized ODEs for vector-valued and Banach space-valued functions.

It is known that generalized ODEs are heavily based on the non-absolute integration theory of J. Kurzweil and R. Henstock, as presented in Chapters 1 and 2. Recall that the main feature of the Kurzweil–Henstock integral is to cope with highly oscillating functions and not only functions with many jumps.

One of the main advantages in working with generalized ODEs lies on the fact that such equations contain "*limiting equations*" of ODEs and other equations, as pointed out by Z. Artstein in [9], for instance. In such paper, Artstein proved that when the right-hand side of an ODE satisfies some Carathéodory- and Lipschitz-type conditions, the "*limiting equation*" may no longer be viewed as "an ODE." This inconvenient situation does not occur when we work in the framework of generalized ODEs. The reader may want to check both papers [4, 9] for more details. As a matter of fact, it has been observed, over the years,

Generalized Ordinary Differential Equations in Abstract Spaces and Applications, First Edition.
Edited by Everaldo M. Bonotto, Márcia Federson, and Jaqueline G. Mesquita.

that different types of differential equations can be identified with generalized ODEs, that is, they can be treated via the theory of these equations. Such is the case, for example, of classic ODEs, impulsive differential equations (IDEs) and measure differential equations (MDEs), as mentioned in the book [209] by Štefan Schwabik. Not only can these equations be seen as generalized ODEs, but also functional differential equations (FDEs) and, more generally, measure FDEs on neutral type, for instance.

The aim of the present chapter is to detail the relation between generalized ODEs and measure FDEs. Then, by means of the results of Chapter 3, relations between generalized ODEs and dynamic equations on time scales can be derived, as well as relations between generalized ODEs and impulsive FDEs. The relations between generalized ODEs and measure FDEs of neutral type are left to Chapter 15. It is also possible to generalize this correspondence, in order to establish a relation between generalized ODEs and other types of measure FDEs such as measure FDEs with infinite delays, measure FDEs with time-dependent delays, and measure FDEs with state-dependent delays (see [100, 116, 220]). Through these relations, we are able to obtain the analogues of the results for generalized ODEs for all these types of special equations.

We begin this chapter with a section where we recall the basic concepts and results of the theory of generalized ODEs. Such results are borrowed from [209] and adapted to the Banach space-valued case. They are essential to the proofs of the correspondences between the solutions of generalized ODEs and the solutions of measure FDEs.

4.1 Fundamental Properties

Let X be a Banach space endowed with the norm $\|\cdot\|$, $O \subset X$ be an open set, $I \subset \mathbb{R}$ be an interval, and $F : O \times I \to X$ be a function.

Definition 4.1: We say that $x : I \to X$ is a *solution of the generalized ODE*

$$\frac{dx}{d\tau} = DF(x, t) \tag{4.1}$$

on the interval I, whenever $(x(t), t) \in O \times I$ and

$$x(b) - x(a) = \int_a^b DF(x(\tau), t), \quad \text{for all } a, b \in I, \tag{4.2}$$

where the integral is in the sense of Definition 2.1, with $U(\tau, t) = F(x(\tau), t)$.

It is important to notice that any generalized ODE is a type of integral equation and the notation in the integrand of (4.1) does not mean that x is differentiable with respect to τ. It is only symbolical and follows the notation of the pioneer papers by Kurzweil (see [147–150], for instance).

The first result of this chapter is, therefore, an immediate consequence of the previous definition (Definition 4.1). An interested reader can check its proof in [209, Proposition 3.5] for the case where $X = \mathbb{R}^n$. We omit its proof for the Banach space-valued case here, since it follows similarly as in [209, Proposition 3.5] with obvious adaptations.

Theorem 4.2: *If $x : I \to X$ is a solution of the generalized ODE*

$$\frac{dx}{d\tau} = DF(x,t) \tag{4.3}$$

on I, then for every fixed $\gamma \in I$, we have

$$x(s) = x(\gamma) + \int_{\gamma}^{s} DF(x(\tau),t), \quad s \in I. \tag{4.4}$$

Reciprocally, if $x : I \to X$ satisfies the integral equation (4.4) *for some $\gamma \in I$ and every $s \in I$, and $(x(t),t) \in O \times I$, then x is a solution of generalized ODE* (4.3).

Throughout this chapter, we assume that $\Omega = O \times I$. Next, we present a class of right-hand sides F, which plays an important role within generalized ODEs.

Definition 4.3: Assume that $h : I \to \mathbb{R}$ is a nondecreasing function and $\omega : [0,\infty) \to [0,\infty)$ is a continuous and increasing function such that $\omega(0) = 0$. We say that a function $F : \Omega \to X$ *belongs to the class $\mathcal{F}(\Omega,h,\omega)$*, whenever, for all $x,y \in O$ and all $s_1, s_2 \in I$, we have

$$\|F(x,s_2) - F(x,s_1)\| \leqslant |h(s_2) - h(s_1)|, \tag{4.5}$$

$$\|F(x,s_2) - F(x,s_1) - F(y,s_2) + F(y,s_1)\| \leqslant \omega(\|x-y\|)|h(s_2) - h(s_1)|. \tag{4.6}$$

When $\omega : [0,\infty) \to [0,\infty)$ is the identity function, we write simply $\mathcal{F}(\Omega,h)$ instead of $\mathcal{F}(\Omega,h,\omega)$.

The next result, adapted from [209, Proposition 3.6] to Banach space-valued functions, describes how the solutions of (4.1) inherit the properties of the right-hand sides $F : \Omega \to X$. In particular, if F is continuous with respect to the second variable, then any solution $x : I \to O$ of the generalized ODE (4.1) is a continuous function.

Theorem 4.4: *Let $O \subset X$ be open and $x : [a,b] \to X$ be a solution of the generalized ODE* (4.1) *on $[a,b] \subset I$. Then, for every $\sigma \in [a,b]$,*

$$\lim_{s\to\sigma}[x(s) - F(x(\sigma),s) + F(x(\sigma),\sigma)] = x(\sigma).$$

Proof. Since $x : [a,b] \to X$ is a solution of the generalized ODE (4.1), we have $x(t) \in O$ for every $t \in [a,b]$. Let $\sigma \in [a,b]$ be fixed. According to Theorem 4.2,

$$x(s) - \int_{\sigma}^{s} DF(x(\tau),t) = x(\sigma),$$

which implies

$$x(s) - F(x(\sigma), s) + F(x(\sigma), \sigma) - x(\sigma)$$
$$- \int_{\sigma}^{s} DF(x(\tau), t) + F(x(\sigma), s) - F(x(\sigma), \sigma) = 0,$$

for every $s \in [a, b]$.

On the other hand, by Theorem 2.12, we obtain

$$\lim_{s\to\sigma} \left[\int_{\sigma}^{s} DF(x(\tau), t) - F(x(\sigma), s) + F(x(\sigma), \sigma) \right] = 0$$

ensuring the existence of the limit

$$\lim_{s\to\sigma} [x(s) - F(x(\sigma), s) + F(x(\sigma), \sigma) - x(\sigma)],$$

which, in turn, yields

$$\lim_{s\to\sigma} [x(s) - F(x(\sigma), s) + F(x(\sigma), \sigma) - x(\sigma)] = 0$$

concluding the proof. □

A proof of the next result can be carried out by straightforward adaptation of [209, Lemma 3.9] to the Banach space-valued case.

Lemma 4.5: *Assume that $F : \Omega \to X$ satisfies condition* (4.5) *and let* $[a, b] \subset I$. *If $x : [a, b] \to X$ is such that $x(t) \in O$ for every $t \in [a, b]$ and if the integral $\int_a^b DF(x(\tau), t)$ exists, then for every $s_1, s_2 \in [a, b]$, we have*

$$\left\| \int_{s_1}^{s_2} DF(x(\tau), t) \right\| \leqslant |h(s_1) - h(s_2)|.$$

Proof. By condition (4.5), we obtain

$$|t - \tau| \|F(x(\tau), t) - F(x(\tau), \tau)\| \leqslant |t - \tau| |h(t) - h(\tau)|$$

for every $\tau, t \in [a, b]$. In addition, the integral $\int_a^b dh(s)$ exists and

$$\int_{s_1}^{s_2} dh(t) = h(s_2) - h(s_1), \quad s_1, s_2 \in [a, b].$$

Then, the statement follows from Theorem 2.15. □

The next lemma gives us another useful estimate. Its proof for the Banach space-valued case follows the same ideas of [178, Lemma 5].

Lemma 4.6: *Assume that $F : \Omega \to X$ belongs to the class $\mathcal{F}(\Omega, h, \omega)$. If the functions $x, y : [a, b] \subset I \to O$ are regulated, then*

$$\left\| \int_a^b D[F(x(\tau), s) - F(y(\tau), s)] \right\| \leqslant \int_a^b \omega(\|x(s) - y(s)\|) dh(s).$$

Proof. We start by pointing out that the function $\omega(\|x(\cdot)-y(\cdot)\|)$ is regulated, since it is a composition of a continuous function and a regulated function. Therefore, from this fact and using that h is a nondecreasing function, the Perron–Stieltjes integral on the right-hand side exists. Then, choosing an arbitrary tagged division $d=(\tau_i,[s_{i-1},s_i])$ of $[a,b]$, we get the following estimates

$$\begin{aligned}
&\left\|\sum_{i=1}^{|d|}[F(x(\tau_i),s_i)-F(x(\tau_i),s_{i-1})-F(y(\tau_i),s_i)+F(y(\tau_i),s_{i-1})]\right\|\\
&\quad\leqslant\sum_{i=1}^{|d|}\left\|F(x(\tau_i),s_i)-F(x(\tau_i),s_{i-1})-F(y(\tau_i),s_i)+F(y(\tau_i),s_{i-1})\right\|\\
&\quad\leqslant\sum_{i=1}^{|d|}\omega(\|x(\tau_i)-y(\tau_i)\|)(h(s_i)-h(s_{i-1})),
\end{aligned}$$

where we also used the fact that $F\in\mathcal{F}(\Omega,h,\omega)$.

On the other hand, given an $\epsilon>0$, there is a tagged division $d=(\tau_i,[s_{i-1},s_i])\in TD_{[a,b]}$ such that

$$\begin{aligned}
&\left\|\int_a^b D[F(x(\tau),s)-F(y(\tau),s)]\right.\\
&\quad\left.-\sum_{i=1}^{|d|}[F(x(\tau_i),s_i)-F(x(\tau_i),s_{i-1})-F(y(\tau_i),s_i)+F(y(\tau_i),s_{i-1})]<\epsilon\right\|
\end{aligned}\tag{4.7}$$

and

$$\left|\int_a^b\omega(\|x(s)-y(s)\|)dh(s)-\sum_{i=1}^{|d|}\omega(\|x(\tau_i)-y(\tau_i)\|)(h(s_i)-h(s_{i-1}))\right|<\epsilon.\tag{4.8}$$

As a consequence of (4.7) and (4.8), we obtain

$$\begin{aligned}
&\left\|\int_a^b D[F(x(\tau),s)-F(y(\tau),s)]\right\|\\
&\quad\leqslant\left\|\sum_{i=1}^{|d|}[F(x(\tau_i),s_i)-F(x(\tau_i),s_{i-1})-F(y(\tau_i),s_i)+F(y(\tau_i),s_{i-1})]\right\|\\
&\quad\quad+\left\|\int_a^b D[F(x(\tau),s)-F(y(\tau),s)]\right.\\
&\quad\quad\left.-\sum_{i=1}^{|d|}[F(x(\tau_i),s_i)-F(x(\tau_i),s_{i-1})-F(y(\tau_i),s_i)+F(y(\tau_i),s_{i-1})]\right\|\\
&\quad<\sum_{i=1}^{|d|}\omega(\|x(\tau_i)-y(\tau_i)\|)(h(s_i)-h(s_{i-1}))+\epsilon
\end{aligned}$$

$$\leqslant \left| \sum_{i=1}^{|d|} \omega(\|x(\tau_i) - y(\tau_i)\|)(h(s_i) - h(s_{i-1})) - \int_a^b \omega(\|x(s) - y(s)\|)dh(s) \right|$$
$$+ \int_a^b \omega(\|x(s) - y(s)\|)dh(s) + \epsilon < 2\epsilon + \int_a^b \omega(\|x(s) - y(s)\|)dh(s).$$

By the arbitrariness of ϵ, we obtain the desired result. □

Now, we present a result, borrowed from [4], which guarantees the existence of the integral involved in the definition of a solution of the generalized ODE (4.1) (see Definition 4.1).

Theorem 4.7: *Let $F \in \mathcal{F}(\Omega, h)$ and $[a, b] \subset I$. If $x : [a, b] \to X$ is the uniform limit of a sequence $\{x_k\}_{k\in\mathbb{N}}$ of step functions $x_k : [a, b] \to X$ such that $(x(s), s) \in \Omega$ and $(x_k(s), s) \in \Omega$, for every $k \in \mathbb{N}$ and for every $s \in [a, b]$, then the integral $\int_a^b DF(x(\tau), s)$ exists and*

$$\int_a^b DF(x(\tau), s) = \lim_{k\to\infty} \int_a^b DF(x_k(\tau), s).$$

Proof. At first, notice that the limits $F(x, s^+)$ and $F(x, s^-)$ exist for every $x \in O$, since F satisfies condition (4.5) and h admits one-sided limits at every point of $[a, b] \subset I$.

Now, let us show that the Kurzweil integral $\int_a^b DF(\varphi(\tau), s)$ exists, for every finite step function $\varphi : [a, b] \to O$. If φ is a finite step function, then there is a tagged division $d = (\tau_i, [s_{i-1}, s_i])$ of $[a, b]$ such that $\varphi(s) = c_i \in O$ for $s \in (s_{i-1}, s_i)$, $i = 1, \dots, |d|$, where c_i is a constant for each $i = 1, \dots, |d|$. Suppose $(\varphi(s), s) \in \Omega$, for every $s \in [a, b]$. By the definition of the Kurzweil integral, if $s_{i-1} < \sigma_1 < \sigma_2 < s_i$, then clearly the integral $\int_{\sigma_1}^{\sigma_2} DF(\varphi(\tau), s)$ exists and

$$\int_{\sigma_1}^{\sigma_2} DF(\varphi(\tau), s) = F(c_i, \sigma_2) - F(c_i, \sigma_1).$$

Let $\sigma_0 \in (s_{i-1}, s_i)$ be given. Therefore,

$$\lim_{t\to s_{i-1}^+} \left[\int_t^{\sigma_0} DF(\varphi(\tau), s) + F(\varphi(s_{i-1}), t) - F(\varphi(s_{i-1}), s_{i-1}) \right]$$
$$= \lim_{t\to s_{i-1}^+} \left[F(c_i, \sigma_0) - F(c_i, t) + F(\varphi(s_{i-1}), t) - F(\varphi(s_{i-1}), s_{i-1}) \right]$$
$$= F(c_i, \sigma_0) - F(c_i, s_{i-1}^+) + F(\varphi(s_{i-1}), s_{i-1}^+) - F(\varphi(s_{i-1}), s_{i-1}).$$

Thus, by Theorem 2.9, the Kurzweil integral $\int_{s_{i-1}}^{\sigma_0} DF(\varphi(\tau), s)$ exists and it equals the previous computed limit.

Analogously, one can prove that the Kurzweil integral $\int_{\sigma_0}^{s_i} DF(\varphi(\tau), s)$ exists and

$$\int_{\sigma_0}^{s_i} DF(\varphi(\tau), s) = F(c_i, s_i^-) - F(c_i, \sigma_0) - F(\varphi(s_i), s_i^-) + F(\varphi(s_i), s_i).$$

By Theorem 2.6, the Kurzweil integral $\int_{s_{i-1}}^{s_i} DF(\varphi(\tau), s)$ exists and

$$\begin{aligned}\int_{s_{i-1}}^{s_i} DF(\varphi(\tau), s) = {} & F(c_i, s_i^-) - F(c_i, \sigma_0) - F(\varphi(s_i), s_i^-) + F(\varphi(s_i), s_i) \\ & + F(c_i, \sigma_0) - F(c_i, s_{i-1}^+) + F(\varphi(s_{i-1}), s_{i-1}^+) - F(\varphi(s_{i-1}), s_{i-1}) \\ = {} & F(c_i, s_i^-) - F(c_i, s_{i-1}^+) + F(\varphi(s_{i-1}), s_{i-1}^+) - F(\varphi(s_{i-1}), s_{i-1}) \\ & - F(\varphi(s_i), s_i^-) + F(\varphi(s_i), s_i).\end{aligned}$$

Proceeding as before, for every interval $[s_{i-1}, s_i]$, with $i = 1, \dots, |d|$, the Kurzweil integral $\int_a^b DF(\varphi(\tau), s)$ exists, by Theorem 2.6, and the equality

$$\begin{aligned}\int_a^b DF(\varphi(\tau), s) = {} & \sum_{i=1}^{|d|} [F(c_i, s_i^-) - F(c_i, s_{i-1}^+)] \\ & + \sum_{i=1}^{|d|} [F(\varphi(s_{i-1}), s_{i-1}^+) - F(\varphi(s_{i-1}), s_{i-1}) - F(\varphi(s_i), s_i^-) + F(\varphi(s_i), s_i)]\end{aligned}$$

holds, proving that the Kurzweil integral $\int_a^b DF(\varphi(\tau), s)$ exists for every finite step function $\varphi : [a, b] \to O$. From this fact, the Kurzweil integral $\int_a^b DF(x_k(\tau), s)$ exists for each $k \in \mathbb{N}$. Since $\{x_k\}_{k\in\mathbb{N}}$ tends uniformly to x, as $k \to \infty$, given $\epsilon > 0$, let $k_0 \in \mathbb{N}$ be such that for $k \geq k_0$, we have

$$\|x_k(s) - x(s)\| < \frac{\epsilon}{2[h(b) - h(a)] + 1}, \qquad \text{for every } s \in [a, b].$$

By the existence of $\int_a^b DF(x_k(\tau), s)$, let δ be a gauge on $[a, b]$ such that, for $k \geq k_0$, we get

$$\left\| \sum_{i=1}^{|d'|} [F(x_k(\tau_i), t_i) - F(x_k(\tau_i), t_{i-1})] - \int_a^b DF(x_k(\tau), s) \right\| < \frac{\epsilon}{2}$$

for every δ-fine tagged division $d' = (\tau_i, [t_{i-1}, t_i]) \in TD_{[a,b]}$, $i = 1, 2, \dots, |d'|$. Then, for every $k \geq k_0$, we obtain

$$\begin{aligned}& \left\| \sum_{i=1}^{|d'|} [F(x(\tau_i), t_i) - F(x(\tau_i), t_{i-1})] - \int_a^b DF(x_k(\tau), s) \right\| \\ & \leqslant \sum_{i=1}^{|d'|} \| F(x(\tau_i), t_i) - F(x(\tau_i), t_{i-1}) - F(x_k(\tau_i), t_i) + F(x_k(\tau_i), t_{i-1}) \| \\ & \quad + \left\| \sum_{i=1}^{|d'|} [F(x_k(\tau_i), t_i) - F(x_k(\tau_i), t_{i-1})] - \int_a^b DF(x_k(\tau), s) \right\|\end{aligned}$$

$$\leqslant \sum_{i=1}^{|d'|}[h(t_i) - h(t_{i-1})] \max_{1\leqslant i\leqslant |d'|} \|x(\tau_i) - x_k(\tau_i)\| + \frac{\epsilon}{2}$$

$$= [h(b) - h(a)] \max_{1\leqslant i\leqslant |d'|} \|x(\tau_i) - x_k(\tau_i)\| + \frac{\epsilon}{2} < \epsilon$$

and the proof is complete. □

Since regulated functions are the uniform limit of step functions, we have the following immediate consequence from Theorem 4.7 (see Theorem 1.4). Its version for $X = \mathbb{R}^n$ can be found in [209, Corollary 3.16] and the proof for the Banach space-valued case follows analogously. Thus, we omit its proof here. But we point out that the function $s \mapsto \int_a^s DF(x(\tau), t) \in X$ is of bounded variation in $[a, b]$ is a consequence of Lemma 4.5.

Corollary 4.8: *Let $F \in \mathcal{F}(\Omega, h)$, $x : [a, b] \to X$ be a regulated function on $[a, b] \subset I$ (in particular, a function of bounded variation in $[a, b]$) and $(x(s), s) \in \Omega$ for every $s \in [a, b]$. Then, the integral $\int_a^b DF(x(\tau), t)$ exists and the function $s \mapsto \int_a^s DF(x(\tau), t) \in X$ is of bounded variation in $[a, b]$ and, hence, also regulated.*

The next lemma ensures us that the solutions of the generalized ODE (4.3) inherit the regularity of the function h from condition (4.5). Further, by condition (4.5), the next result also ensures us that the solutions of the generalized ODE (4.3) are of bounded variation. These results can be found in [209, Lemma 3.10 and Corollary 3.11] for the case $X = \mathbb{R}^n$.

Lemma 4.9: *Suppose $F : \Omega \to X$ satisfies condition (4.5). If $[a, b] \subset I$ and $x : [a, b] \to X$ is a solution of the generalized ODE (4.1), then, for every $s_1, s_2 \in [a, b]$,*

$$\|x(s_1) - x(s_2)\| \leqslant |h(s_2) - h(s_1)|. \tag{4.9}$$

Moreover, x is of bounded variation in $[a, b]$ and $\mathrm{var}_a^b(x) \leqslant h(b) - h(a) < \infty$. If, in addition, h is continuous at $t \in [a, b]$, then t is a continuity point of the solution $x : [a, b] \to X$.

Proof. The first part of the statement follows by combining Lemma 4.5 and the definition of solution of a generalized ODE (Definition 4.1). For the second part, let $d = (s_i)$ be an arbitrary division of the interval $[a, b]$. By (4.9), we have

$$\sum_{i=1}^{|d|} \|x(s_i) - x(s_{i-1})\| \leqslant \sum_{i=1}^{|d|} |h(s_i) - h(s_{i-1})| = h(b) - h(a).$$

Considering the supremum over all divisions of $[a, b]$, the desired result follows. The third part of the lemma follows as an immediate consequence of (4.9). □

The next result, whose proof can be found in [209, Lemma 3.12] for the case where $X = \mathbb{R}^n$, describes the relation between the discontinuities of a solution of the generalized ODE (4.1) and the discontinuities of the function F on the right-hand side of (4.1). It also shows that the solutions of generalized ODEs may not be continuous in general.

Lemma 4.10: *Let $[a, b] \subset I$ and $x : [a, b] \to X$ be a solution of the generalized ODE* (4.1). *Assume that $F : \Omega \to X$ satisfies condition* (4.5). *Then,*

$$x(s^+) - x(s) = \lim_{\sigma \to s^+} x(\sigma) - x(s) = F(x(s), s^+) - F(x(s), s), \quad s \in [a, b),$$

$$x(s) - x(s^-) = x(s) - \lim_{\sigma \to s^-} x(\sigma) = F(x(s), s) - F(x(s), s^-), \quad s \in (a, b],$$

where $F(x, s^+) = \lim_{\sigma \to s^+} F(x, \sigma)$ for $s \in [a, b)$ and $F(x, s^-) = \lim_{\sigma \to s^-} F(x, \sigma)$ for $s \in (a, b]$.

Proof. Notice that the limits $F(x, s^+)$ and $F(x, s^-)$ exist for every $x \in O$, since F satisfies (4.5) and h has one-sided limits at every point of $[a, b] \subset I$. Since x is a solution of the generalized ODE (4.1) on $[a, b]$, we have $(x(t), t) \in O \times [a, b]$ and

$$x(\sigma) - x(s) = \int_s^\sigma DF(x(\tau), t)$$

for $a \leqslant s < \sigma \leqslant b$. Hence,

$$\lim_{\sigma \to s^+} x(\sigma) - x(s) = \lim_{\sigma \to s^+} \int_s^\sigma DF(x(\tau), t) = \lim_{\sigma \to s^+} [F(x(s), \sigma) - F(x(\sigma), s)],$$

proving the first equality of the statement. Proceeding analogously, one can show that the second equality also holds. □

4.2 Relations with Measure Differential Equations

Our aim in this section is to show that measure differential equations (we write simply MDEs) are special cases of generalized ODEs. The correspondence between these two classes of equations was first presented in [209] considering the relation between the Lebesgue–Stieltjes integral and the Kurzweil integral. Later, the authors of [78] extended the correspondence for Perron–Stieltjes integrals, but for n-finite dimensional space-valued functions. Here, we present such relations for the infinite dimensional case.

As in previous chapters, we consider X a Banach space endowed with norm $\| \cdot \|$ and an open set $O \subset X$.

Let $t_0 \in \mathbb{R}$ and $\tau_0 \geq t_0$. Then, consider an MDE whose integral form is given by

$$x(t) = x(\tau_0) + \int_{\tau_0}^t f(x(s), s) dg(s), \quad t \geq \tau_0, \tag{4.10}$$

where $f : O \times [t_0, t_0 + \sigma] \to X$ and $g : [t_0, t_0 + \sigma] \to \mathbb{R}$ are functions, and the integral on the right-hand side is considered in the sense of Perron–Stieltjes.

Given functions $f : O \times [t_0, t_0 + \sigma] \to X$ and $g : [t_0, t_0 + \sigma] \to \mathbb{R}$, we assume the following conditions:

(A1) The function $g : [t_0, t_0 + \sigma] \to \mathbb{R}$ is nondecreasing and left-continuous;

(A2) The Perron–Stieltjes integral $\int_{s_1}^{s_2} f(x(s), s)dg(s)$ exists, for all $x \in G([t_0, t_0 + \sigma], O)$ and all $s_1, s_2 \in [t_0, t_0 + \sigma]$;

(A3) There exists a Perron–Stieltjes integrable function $M : [t_0, t_0 + \sigma] \to \mathbb{R}$ with respect to g such that

$$\left\| \int_{s_1}^{s_2} f(x(s), s)dg(s) \right\| \leqslant \int_{s_1}^{s_2} M(s)dg(s),$$

for all $x \in G([t_0, t_0 + \sigma], O)$ and all $s_1, s_2 \in [t_0, t_0 + \sigma]$, with $s_1 \leqslant s_2$;

(A4) There exists a Perron–Stieltjes integrable function $L : [t_0, t_0 + \sigma] \to \mathbb{R}$ with respect to g such that

$$\left\| \int_{s_1}^{s_2} [f(x(s), s) - f(z(s), s)]\, dg(s) \right\| \leqslant \|x - z\|_\infty \int_{s_1}^{s_2} L(s)dg(s),$$

for all $x, z \in G_0([t_0, t_0 + \sigma], O)$ and all $s_1, s_2 \in [t_0, t_0 + \sigma]$, with $s_1 \leqslant s_2$.

The next result ensures that if the function $f : O \times [t_0, t_0 + \sigma] \to X$ satisfies conditions (A2)–(A4), and $g : [t_0, t_0 + \sigma] \to \mathbb{R}$ satisfies condition (A1), then the function F given by $F(x, t) = \int_{\tau_0}^{t} f(x, s)dg(s)$, for $(x, t) \in O \times [t_0, t_0 + \sigma]$, belongs to the class $\mathcal{F}(O \times [t_0, t_0 + \sigma], h)$, where $h(t) = \int_{\tau_0}^{t} (M(s) + L(s))dg(s)$, $t \in [t_0, t_0 + \sigma]$.

Theorem 4.11: *Suppose that $f : O \times [t_0, t_0 + \sigma] \to X$ satisfies the conditions* (A2)–(A4), *and $g : [t_0, t_0 + \sigma] \to \mathbb{R}$ satisfies the condition* (A1). *Choose an arbitrary $\tau_0 \in [t_0, t_0 + \sigma]$ and define $F : O \times [t_0, t_0 + \sigma] \to X$ by*

$$F(x, t) = \int_{\tau_0}^{t} f(x, s)dg(s), \quad (x, t) \in O \times [t_0, t_0 + \sigma]. \tag{4.11}$$

Then, $F \in \mathcal{F}(O \times [t_0, t_0 + \sigma], h)$, where $h : [t_0, t_0 + \sigma] \to \mathbb{R}$ given by

$$h(t) = \int_{\tau_0}^{t} (M(s) + L(s))dg(s), \quad t \in [t_0, t_0 + \sigma] \tag{4.12}$$

is a nondecreasing function.

Proof. Notice that by conditions (A3) and (A4), the function h is nondecreasing. Indeed, let $s_1, s_2 \in [t_0, t_0 + \sigma]$ be such that $s_1 \leqslant s_2$ and $x \in G([t_0, t_0 + \sigma], O)$. Then

$$0 \leqslant \left\| \int_{s_1}^{s_2} f(x(s), s)dg(s) \right\| \leqslant \int_{s_1}^{s_2} M(s)dg(s)$$
$$= \int_{\tau_0}^{s_2} M(s)dg(s) - \int_{\tau_0}^{s_1} M(s)dg(s),$$

which implies

$$\int_{\tau_0}^{s_2} M(s)dg(s) \geq \int_{\tau_0}^{s_1} M(s)dg(s),$$

showing that $t \mapsto \int_{\tau_0}^{t} M(s)dg(s)$ is a nondecreasing function.

Similarly, one can show that $t \mapsto \int_{\tau_0}^{t} L(s)dg(s)$ is a nondecreasing function. Therefore, h given by (4.12) is a nondecreasing function.

Once M and L are Perron–Stieltjes integrable functions, h is well-defined, and because g is left-continuous, so is h.

Now, let us verify that F is well-defined. In fact, given $t \in [t_0, \infty)$ and $x \in O$, define the auxiliary function

$$\begin{aligned} \alpha_x &: [t_0, \infty) \to O \\ &s \mapsto \alpha_x(s) = x. \end{aligned} \tag{4.13}$$

At first, note that $\alpha_x \in G_0([t_0, \infty), O)$ and, hence, $\alpha_x \in G([t_0, \infty), O)$. Using (A2), we have the existence of the integral $\int_{\tau_0}^{t} f(x, s)dg(s)$ and

$$\int_{\tau_0}^{t} f(\alpha_x(s), s)dg(s) = \int_{\tau_0}^{t} f(x, s)dg(s),$$

which implies that F is well-defined.

Given an arbitrary $x \in O$, by condition (A3), we have

$$\begin{aligned} \|F(x, s_2) - F(x, s_1)\| &= \left\| \int_{\tau_0}^{s_2} f(x, s)dg(s) - \int_{\tau_0}^{s_1} f(x, s)dg(s) \right\| \\ &= \left\| \int_{s_1}^{s_2} f(x, s)dg(s) \right\| \\ &= \left\| \int_{s_1}^{s_2} f(\alpha_x(s), s)dg(s) \right\| \\ &\leqslant \int_{s_1}^{s_2} M(s)dg(s) \\ &\leqslant \int_{s_1}^{s_2} [M(s) + L(s)]\, dg(s) = h(s_2) - h(s_1), \end{aligned}$$

for $t_0 \leqslant s_1 \leqslant s_2 \leqslant t_0 + \sigma$. On the other hand, by condition (A4), if x, $y \in O$ and $t_0 \leqslant s_1 \leqslant s_2 \leqslant t_0 + \sigma$, then

$$\begin{aligned} &\|F(x, s_2) - F(x, s_1) - F(y, s_2) + F(y, s_1)\| \\ &= \left\| \int_{s_2}^{s_1} [f(x, s) - f(y, s)]\, dg(s) \right\| \\ &= \left\| \int_{s_2}^{s_1} [f(\alpha_x(s), s) - f(\alpha_y(s), s)]\, dg(s) \right\| \end{aligned}$$

$$\leqslant \left\|\alpha_x - \alpha_y\right\|_\infty \int_{s_1}^{s_2} L(s)dg(s)$$
$$\leqslant \|x - y\| \, (h(s_2) - h(s_1))$$

and the statement follows. □

Remark 4.12: The function g in Theorem 4.11 is left-continuous by hypothesis. Then, by relation (4.12), h is also left-continuous.

The next result describes the relation between the Perron–Stieltjes integral and the Kurzweil integral. A similar version of this result was proved in [209, Proposition 5.12] when f satisfies the usual conditions of Carathéodory and in that result, the relation between the Lebesgue–Stieltjes integral and the Kurzweil integral is provided. Later, in [78], a version of this correspondence for Perron–Stieltjes integral and Kurzweil integral was given. The result that we present here is borrowed from [78], however this version is for functions taking values in a Banach space.

Theorem 4.13: *Assume that $f : O \times [t_0, t_0 + \sigma] \to X$ satisfies conditions* (A2)–(A4), *and $g : [t_0, t_0 + \sigma] \to \mathbb{R}$ satisfies condition* (A1). *Let $F : O \times [t_0, t_0 + \sigma] \to X$ be defined by* (4.11). *If $x : [t_0, t_0 + \sigma] \to O$ is a regulated function (in particular, a function of bounded variation) on $[t_0, t_0 + \sigma]$, then both the Kurzweil integral $\int_{t_0}^{t_0+\sigma} DF(x(\tau), t)$ and the Perron–Stieltjes integral $\int_{t_0}^{t_0+\sigma} f(x(s), s)dg(s)$ exist and they have the same value.*

Proof. Let $F : O \times [t_0, t_0 + \sigma] \to X$ be defined by (4.11). Then, by Theorem 4.11, $F \in \mathcal{F}(O \times [t_0, t_0 + \sigma], h)$, where h is given by (4.12).

Suppose $x : [t_0, t_0 + \sigma] \to O$ is a regulated function. Then, Corollary 4.8 implies that the Kurzweil integral $\int_{t_0}^{t_0+\sigma} DF(x(\tau), t)$ exists. We want to show that

$$\int_{t_0}^{t_0+\sigma} DF(x(\tau), t) = \int_{t_0}^{t_0+\sigma} f(x(s), s)dg(s).$$

Claim 1. For any finite step function $\varphi : [t_0, t_0 + \sigma] \to O$, we have

$$\int_{t_0}^{t_0+\sigma} f(\varphi(s), s)dg(s) = \int_{t_0}^{t_0+\sigma} DF(\varphi(\tau), t).$$

Let $\varphi : [t_0, t_0 + \sigma] \to O$ be a finite step function. Then, φ is a regulated function on $[t_0, t_0 + \sigma]$ and, hence, the Kurzweil integral $\int_{t_0}^{t_0+\sigma} DF(\varphi(\tau), t)$ exists and the Perron–Stieltjes integral $\int_{t_0}^{t_0+\sigma} f(\varphi(s), s)dg(s)$ also exists. Still because φ is a step function, there exists a division

$$t_0 = s_0 < s_1 < \cdots < s_n = t_0 + \sigma$$

and $c_1, c_2, \dots, c_n \in O$ such that

$$\varphi(s) = c_i \quad \text{for every } s \in (s_{i-1}, s_i).$$

If $s_{k-1} < \sigma_1 < \sigma_2 < s_k$, then

$$\int_{\sigma_1}^{\sigma_2} DF(\varphi(\tau), t) = \int_{\sigma_1}^{\sigma_2} DF(c_k, t) = F(c_k, \sigma_2) - F(c_k, \sigma_1).$$

Once

$$\begin{aligned} F(c_k, \sigma_2) - F(c_k, \sigma_1) &= \int_{t_0}^{\sigma_2} f(c_k, s)dg(s) - \int_{t_0}^{\sigma_1} f(c_k, s)dg(s) \\ &= \int_{\sigma_1}^{\sigma_2} f(c_k, s)dg(s) = \int_{\sigma_1}^{\sigma_2} f(\varphi(s), s)dg(s), \end{aligned}$$

we obtain

$$\int_{\sigma_1}^{\sigma_2} f(\varphi(s), s)dg(s) = F(c_k, \sigma_2) - F(c_k, \sigma_1) = \int_{\sigma_1}^{\sigma_2} DF(\varphi(\tau), t).$$

Let $k \in \{1, 2, \dots, n\}$ and $\sigma \in (s_{k-1}, s_k)$ be given. By Corollary 2.14, we obtain

$$\begin{aligned} \int_{s_{k-1}}^{\sigma} f(\varphi(s), s)dg(s) - f(\varphi(s_{k-1}), s_{k-1})\Delta^+ g(s_{k-1}) &= \lim_{\xi \to s_{k-1}^+} \int_{\xi}^{\sigma} f(\varphi(s), s)d(s) \\ &= \lim_{\xi \to s_{k-1}^+} F(c_k, \sigma) - F(c_k, \xi) \\ &= F(c_k, \sigma) - F(c_k, s_{k-1}^+), \end{aligned}$$

that is,

$$\int_{s_{k-1}}^{\sigma} f(\varphi(s), s)dg(s) = F(c_k, \sigma) - F(c_k, s_{k-1}^+) + f(\varphi(s_{k-1}), s_{k-1})\Delta^+ g(s_{k-1}). \tag{4.14}$$

On the other hand, Theorem 2.9 yields

$$\begin{aligned} \int_{s_{k-1}}^{\sigma} DF(\varphi(\tau), t) &= \lim_{\xi \to s_{k-1}^+} \left[\int_{\xi}^{\sigma} DF(\varphi(\tau), t) + F(\varphi(s_{k-1}), \xi) - F(\varphi(s_{k-1}), s_{k-1}) \right] \\ &= \lim_{\xi \to s_{k-1}^+} \left[\int_{\xi}^{\sigma} DF(c_k, t) + \int_{t_0}^{\xi} f(\varphi(s_{k-1}), s)dg(s) \right. \\ &\qquad \left. - \int_{t_0}^{s_{k-1}} f(\varphi(s_{k-1}), s)dg(s) \right] \\ &= \lim_{\xi \to s_{k-1}^+} \left[F(c_k, \sigma) - F(c_k, \xi) + \int_{s_{k-1}}^{\xi} f(\varphi(s_{k-1}), s)dg(s) \right] \\ &= F(c_k, \sigma) - F(c_k, s_{k-1}^+) + f(\varphi(s_{k-1}), s_{k-1})\Delta^+ g(s_{k-1}), \end{aligned} \tag{4.15}$$

where the last equality follows from Corollary 2.14. Therefore, (4.14) and (4.15) yield

$$\int_{s_{k-1}}^{\sigma} f(\varphi(s), s)dg(s) = \int_{s_{k-1}}^{\sigma} DF(\varphi(\tau), t),$$

for every $k \in \{1, 2, \ldots, n\}$ and for all $\sigma \in (s_{k-1}, s_k)$.

In a similar way, one can show that

$$\int_{\sigma}^{s_k} f(\varphi(s), s)dg(s) = \int_{\sigma}^{s_k} DF(\varphi(\tau), t),$$

for each $k \in \{1, 2, \ldots, n\}$ and for all $\sigma \in (s_{k-1}, s_k)$. Thus, we obtain

$$\begin{aligned}\int_{a}^{b} f(\varphi(s), s)dg(s) &= \sum_{k=1}^{n} \int_{s_{k-1}}^{s_k} f(\varphi(s), s)dg(s)\\ &= \sum_{k=1}^{n} \int_{s_{k-1}}^{s_k} DF(\varphi(\tau), t)\\ &= \int_{a}^{b} DF(\varphi(\tau), t),\end{aligned}$$

getting *Claim 1*.

Since $x : [t_0, t_0 + \sigma] \to O$ is a regulated function, there exists a sequence of finite step functions $\varphi_k : [t_0, t_0 + \sigma] \to O$, $k \in \mathbb{N}$ (see the comments after Theorem 1.4), such that

$$\|\varphi_k - x\|_\infty = \sup_{s\in[t_0,t_0+\sigma]} |\varphi_k(s) - x(s)| \quad \text{and} \quad \lim_{k\to\infty} \|\varphi_k - x\|_\infty = 0.$$

By *Claim 1*,

$$\int_{t_0}^{t_0+\sigma} f(\varphi_k(s), s)dg(s) = \int_{t_0}^{t_0+\sigma} DF(\varphi_k(\tau), t) \tag{4.16}$$

and, by Theorem 4.7, we have

$$\lim_{k\to\infty} \int_{t_0}^{t_0+\sigma} DF(\varphi_k(\tau), t) = \int_{t_0}^{t_0+\sigma} DF(x(\tau), t). \tag{4.17}$$

Then, the equalities (4.16) and (4.17) yield

$$\begin{aligned}\lim_{k\to\infty} \int_{t_0}^{t_0+\sigma} f(\varphi_k(s), s)dg(s) &= \lim_{k\to\infty} \int_{t_0}^{t_0+\sigma} DF(\varphi_k(\tau), t)\\ &= \int_{t_0}^{t_0+\sigma} DF(x(\tau), t).\end{aligned} \tag{4.18}$$

On the other hand, condition (A4) implies

$$\left\| \int_{t_0}^{t_0+\sigma} [f(\varphi_k(s), s) - f(x(s), s)]\, dg(s) \right\| \leqslant \|\varphi_k - x\|_\infty \int_{t_0}^{t_0+\sigma} L(s)dg(s),$$

whence

$$\lim_{k\to\infty}\int_{t_0}^{t_0+\sigma} f(\varphi_k(s), s)dg(s) = \int_{t_0}^{t_0+\sigma} f(x(s), s)dg(s). \tag{4.19}$$

Finally, equalities (4.18) and (4.19) yield

$$\int_{t_0}^{t_0+\sigma} DF(x(\tau), t) = \int_{t_0}^{t_0+\sigma} f(x(s), s)dg(s)$$

and the proof is finished. □

The next result gives us a correspondence between the solutions of the integral form of an MDE of type

$$x(t) = x(\tau_0) + \int_{\tau_0}^{t} f(x(s), s)dg(s), \quad t \geq \tau_0, \tag{4.20}$$

and the solutions of the generalized ODE

$$\frac{dx}{d\tau} = DF(x, t),$$

where $F : O \times [t_0, t_0 + \sigma] \to X$ is given by $F(x, t) = \int_{\tau_0}^{t} f(x, s)dg(s)$ for all $(x, t) \in O \times [t_0, t_0 + \sigma]$ and $g : [t_0, t_0 + \sigma] \to \mathbb{R}$, is nondecreasing and left-continuous. This result is inspired by [209, Theorem 5.17], but here the version is for a general Banach space and the integral that we are considering in (4.20) is in the sense of Perron–Stieltjes instead of Lebesgue–Stieltjes.

Theorem 4.14: *Assume that $f : O \times [t_0, t_0 + \sigma] \to X$ fulfills conditions* (A2)–(A4), *and $g : [t_0, t_0 + \sigma] \to \mathbb{R}$ fulfills condition* (A1). *Then, the function $x : I \to X$ is a solution of the MDE* (4.20) *on $I \subset [t_0, t_0 + \sigma]$ if and only if it is a solution of the generalized ODE*

$$\frac{dx}{d\tau} = DF(x, t) \tag{4.21}$$

on I, where the function F is given by (4.11).

Proof. Suppose $x : I \to X$ is a solution of the MDE (4.20). Thus $(x(t), t) \in O \times I$, see Definition 3.2. By Theorem 4.13, the Kurzweil integral $\int_{s_1}^{s_2} DF(x(\tau), t)$ exists, for every $s_1, s_2 \in I$, and

$$x(s_2) - x(s_1) = \int_{s_1}^{s_2} f(x(s), s)dg(s) = \int_{s_1}^{s_2} DF(x(\tau), t),$$

whence x is a solution of the generalized ODE (4.21).

Reciprocally, if x is a solution of the generalized ODE (4.21), then Lemma 4.9 implies that $x : I \to O$ is a regulated function and, by Theorem 4.13, x satisfies the MDE (4.20) proving the statement. □

4.3 Relations with Measure FDEs

Our goal in this section is to establish a correspondence between the solutions of a certain measure FDE and the solutions of a generalized ODE.

Let $t_0 \in \mathbb{R}$, $r, \sigma > 0$ and consider sets

$$O \subset G([t_0 - r, t_0 + \sigma], X) \quad \text{and}$$
$$P = \{y_t : y \in O,\ t \in [t_0, t_0 + \sigma]\} \subset G([-r, 0], X).$$

We shall consider in this section that O is an open subset of $G([t_0 - r, t_0 + \sigma], X)$. Assume that $g : [t_0, t_0 + \sigma] \to \mathbb{R}$ is a nondecreasing function, $f : P \times [t_0, t_0 + \sigma] \to X$ is a function, and consider the integral form of a measure FDE given by

$$y(t) = y(t_0) + \int_{t_0}^{t} f(y_s, s)dg(s), \quad t \in [t_0, t_0 + \sigma]. \tag{4.22}$$

We want to show that, under certain conditions, Eq. (4.22) can be related with a generalized ODE of the form

$$\frac{dx}{d\tau} = DF(x, t), \tag{4.23}$$

where $F : O \times [t_0, t_0 + \sigma] \to G([t_0 - r, t_0 + \sigma], X)$ is defined by

$$F(x, t)(\vartheta) = \begin{cases} 0, & t_0 - r \leqslant \vartheta \leqslant t_0, \\ \displaystyle\int_{t_0}^{\vartheta} f(x_s, s)dg(s), & t_0 \leqslant \vartheta \leqslant t \leqslant t_0 + \sigma, \\ \displaystyle\int_{t_0}^{t} f(x_s, s)dg(s), & t \leqslant \vartheta \leqslant t_0 + \sigma, \end{cases} \tag{4.24}$$

for every $x \in O$ and $t \in [t_0, t_0 + \sigma]$.

In the sequel, we explain the prolongation property of a subset of O of the space of regulated functions $G([t_0 - r, t_0 + \sigma], X)$. Such property is crucial to guarantee that the relation between the solutions of the equations is well-defined. It also ensures that if $y \in O$, then for every $t \in [t_0, t_0 + \sigma]$, the function

$$x(t)(\vartheta) = \begin{cases} y(\vartheta), & \vartheta \in [t_0 - r, t], \\ y(t), & \vartheta \in [t, t_0 + \sigma], \end{cases}$$

is such that $x(t) \in O$.

Definition 4.15: Let $O \subset G([t_0 - r, t_0 + \sigma], X)$ be open. The set O is said to have *the prolongation property* if, for every $y \in O$ and every $\bar{t} \in [t_0 - r, t_0 + \sigma]$, the function $\bar{y}$ given by

$$\bar{y}(t) = \begin{cases} y(t), & t_0 - r \leqslant t \leqslant \bar{t}, \\ y(\bar{t}), & \bar{t} < t \leqslant t_0 + \sigma \end{cases}$$

also belongs to O.

It is worth mentioning that there are many interesting sets satisfying this property. For instance, if $B \subset X$, then it is possible to show that the set $G([t_0 - r, t_0 + \sigma], B)$ has the prolongation property. The same applies to $C([t_0 - r, t_0 + \sigma], B)$, the space of continuous functions from $[t_0 - r, t_0 + \sigma]$ to B. In particular, any open ball in $G([t_0 - r, t_0 + \sigma], X)$ has this property. Therefore, this property fits very well our goals, since we deal with regulated functions (for instance, the solutions are regulated functions as well as the initial conditions).

From now on, we consider some conditions on the function $f : P \times [t_0, t_0 + \sigma] \to X$. We assume that, for every $y, z \in O$ and $s_1, s_2 \in [t_0, t_0 + \sigma]$, with $s_1 \leqslant s_2$, we have

(A) the Perron–Stieltjes integral $\int_{s_1}^{s_2} f(y_s, s)dg(s)$ exists;
(B) there exists a Perron–Stieltjes integrable function $M : [t_0, t_0 + \sigma] \to \mathbb{R}$ such that

$$\left\| \int_{s_1}^{s_2} f(y_s, s)dg(s) \right\| \leqslant \int_{s_1}^{s_2} M(s)dg(s);$$

(C) there exists a Perron–Stieltjes integrable function $L : [t_0, t_0 + \sigma] \to \mathbb{R}$ such that

$$\left\| \int_{s_1}^{s_2} [f(y_s, s) - f(z_s, s)]\, dg(s) \right\| \leqslant \int_{s_1}^{s_2} L(s) \|y_s - z_s\|_\infty \, dg(s).$$

At this point, it is important to notice that condition (A) requires the existence of the Perron–Stieltjes integral $\int_{t_0}^{t_0+\sigma} f(y_s, s)dg(s)$, for every $y \in O$, and we highlight that the class of functions, which are integrable in the sense of Perron–Stieltjes, is quite large, including functions that do not "behave well," not only having many discontinuities, but also being highly oscillating (see Chapter 1, for more details). Notice, as well, that Theorem 3.3 guarantees that the integral on the right-hand side of condition (C) is well-defined and exists.

The next result describes a relation between the regularity of the function F that appears on the right-hand side of the generalized ODE (4.23) and the regularity of the function that appears on the right-hand side of the measure FDE (4.22). The version presented here is more general than the one from [85].

Lemma 4.16: *Assume that $O \subset G([t_0 - r, t_0 + \sigma], X)$ is an open set and consider $P = \{y_t : y \in O, \ t \in [t_0, t_0 + \sigma]\}$. Suppose $g : [t_0, t_0 + \sigma] \to \mathbb{R}$ is a nondecreasing function and $f : P \times [t_0, t_0 + \sigma] \to X$ satisfies conditions (A)–(C). Then, the function $F : O \times [t_0, t_0 + \sigma] \to G([t_0 - r, t_0 + \sigma], X)$, given by (4.24) belongs to the class $\mathcal{F}(O \times [t_0, t_0 + \sigma], h)$, where*

$$h(t) = \int_{t_0}^{t} [M(s) + L(s)]\, dg(s), \quad t \in [t_0, t_0 + \sigma].$$

Proof. By condition (A), F is well-defined and the integrals that appear in its definition exist. For $y \in O$ and $t_0 \leqslant s_1 < s_2 \leqslant t_0 + \sigma$, by the definition of F, we have

$$F(y, s_2)(\vartheta) - F(y, s_1)(\vartheta) = \begin{cases} 0, & t_0 - r \leqslant \vartheta \leqslant s_1, \\ \int_{s_1}^{\vartheta} f(y_s, s)dg(s), & s_1 \leqslant \vartheta \leqslant s_2, \\ \int_{s_1}^{s_2} f(y_s, s)dg(s), & s_2 \leqslant \vartheta \leqslant t_0 + \sigma. \end{cases}$$

Then, by conditions (B) and (C), for all $y, z \in O$ and $t_0 \leqslant s_1 < s_2 \leqslant t_0 + \sigma$, we obtain

$$\begin{aligned} \|F(y, s_2) - F(y, s_1)\|_\infty &= \sup_{t_0 - r \leqslant \vartheta \leqslant t_0 + \sigma} \|F(y, s_2)(\vartheta) - F(y, s_1)(\vartheta)\| \\ &= \sup_{s_1 \leqslant \vartheta \leqslant s_2} \left\| \int_{s_1}^{\vartheta} f(y_s, s)dg(s) \right\| \\ &\leqslant \int_{s_1}^{s_2} M(s)dg(s) \leqslant h(s_2) - h(s_1) \end{aligned}$$

and, also,

$$\begin{aligned} &\|F(y, s_2) - F(y, s_1) - F(z, s_2) + F(z, s_1)\|_\infty \\ &= \sup_{s_1 \leqslant \vartheta \leqslant s_2} \left\| \int_{s_1}^{\vartheta} [f(y_s, s) - f(z_s, s)]\, dg(s) \right\| \\ &\leqslant \sup_{s_1 \leqslant \vartheta \leqslant s_2} \int_{s_1}^{\vartheta} L(s)\|y_s - z_s\|_\infty\, dg(s) \\ &\leqslant \|y - z\|_\infty \int_{s_1}^{s_2} L(s)dg(s) \leqslant \|y - z\|_\infty (h(s_2) - h(s_1)), \end{aligned}$$

proving the statement. □

The next lemma brings up an important property concerning the solution of the generalized ODE (4.23), where F is given by (4.24). This result, which can be found in [85] for the case n-dimensional, is important to prove the correspondence between the equations.

Lemma 4.17: *Suppose the open subset $O \subset G([t_0 - r, t_0 + \sigma], X)$ has the prolongation property and consider $P = \{y_t : y \in O,\ t \in [t_0, t_0 + \sigma]\}$. Assume that $\phi \in P$, $g : [t_0, t_0 + \sigma] \to \mathbb{R}$ is a nondecreasing function and $f : P \times [t_0, t_0 + \sigma] \to X$ satisfies condition (A)–(C). Consider $F : O \times [t_0, t_0 + \sigma] \to G([t_0 - r, t_0 + \sigma], X)$ given by (4.24) and assume that $x : [t_0, t_0 + \sigma] \to X$ is a solution of*

$$\frac{dx}{d\tau} = DF(x, t) \tag{4.25}$$

with initial condition

$$x(t_0)(\vartheta) = \begin{cases} \phi(\vartheta), & \vartheta \in [t_0 - r, t_0], \\ x(t_0)(t_0), & \vartheta \in [t_0, t_0 + \sigma]. \end{cases}$$

If $v, \vartheta \in [t_0, t_0 + \sigma]$, *then*

$$x(v)(\vartheta) = \begin{cases} x(v)(v), & \vartheta \geq v, \\ x(\vartheta)(\vartheta), & v \geq \vartheta. \end{cases} \tag{4.26}$$

Proof. Let $\vartheta \geq v$. Assuming that x is a solution of the generalized ODE (4.25) (hence, $x(t) \in O$ for every $t \in [t_0, t_0 + \sigma]$, see Definition 4.1), we obtain

$$x(v)(v) = x(t_0)(v) + \int_{t_0}^{v} DF(x(\tau), t)(v) \quad \text{and}$$

$$x(v)(\vartheta) = x(t_0)(\vartheta) + \int_{t_0}^{v} DF(x(\tau), t)(\vartheta).$$

Using the fact that $x(t_0)(\vartheta) = x(t_0)(v)$, the generalized ODE (4.25) yields the following equality

$$x(v)(\vartheta) - x(v)(v) = \int_{t_0}^{v} DF(x(\tau), t)(\vartheta) - \int_{t_0}^{v} DF(x(\tau), t)(v). \tag{4.28}$$

Because the Kurzweil integral $\int_{t_0}^{v} DF(x(\tau), t)$ exists, given $\epsilon > 0$, there is a gauge δ on $[t_0, v]$ such that, if $d = (\tau_i, [s_{i-1}, s_i])$ is a δ-fine tagged division of $[t_0, v]$, then

$$\left\| \sum_{i=1}^{|d|} [F(x(\tau_i), s_i) - F(x(\tau_i), s_{i-1})] - \int_{t_0}^{v} DF(x(\tau), t) \right\|_{\infty} < \epsilon. \tag{4.29}$$

Combining (4.28) and (4.29), we obtain

$$\begin{aligned} &\|x(v)(\vartheta) - x(v)(v)\| \\ &< 2\epsilon + \left\| \sum_{i=1}^{|d|} [F(x(\tau_i), s_i) - F(x(\tau_i), s_{i-1})](\vartheta) \right. \\ &\qquad \left. - \sum_{i=1}^{|d|} [F(x(\tau_i), s_i) - F(x(\tau_i), s_{i-1})](v) \right\|. \end{aligned}$$

On the other hand, using the definition of F, we conclude that

$$F(x(\tau_i), s_i)(\vartheta) - F(x(\tau_i), s_{i-1})(\vartheta) = F(x(\tau_i), s_i)(v) - F(x(\tau_i), s_{i-1})(v),$$

for every $i \in \{1, \dots, |d|\}$. Therefore, $\|x(v)(\vartheta) - x(v)(v)\| < 2\epsilon$ and, hence, by the arbitrariness of $\epsilon > 0$, we get the first equality of the statement.

Now, suppose $\vartheta \leqslant v$. Similarly as before, we obtain

$$x(v)(\vartheta) = x(t_0)(\vartheta) + \int_{t_0}^{v} DF(x(\tau), t)(\vartheta) \quad \text{and}$$

$$x(\vartheta)(\vartheta) = x(t_0)(\vartheta) + \int_{t_0}^{\vartheta} DF(x(\tau), t)(\vartheta),$$

whence

$$x(v)(\vartheta) - x(\vartheta)(\vartheta) = \int_{\vartheta}^{v} DF(x(\tau), t)(\vartheta).$$

Suppose $d = (\tau_i, [s_{i-1}, s_i])$ is an arbitrary tagged division of $[\vartheta, v]$. By the definition of F, for each $i = 1, 2, \ldots, |d|$,

$$F(x(\tau_i), s_i)(\vartheta) - F(x(\tau_i), s_{i-1})(\vartheta) = 0,$$

which yields $\int_{\vartheta}^{v} DF(x(\tau), t)(\vartheta) = 0$. Thus, since x is a solution of the generalized ODE (4.25), $x(v)(\vartheta) = x(\vartheta)(\vartheta)$ concluding the proof. □

The next two theorems are the main results of this chapter and they concern the correspondence between the solutions of certain measure FDEs and the solutions of nonautonomous generalized ODEs. The versions presented here are more general than those from [85], since conditions (B) and (C) on the function f involve functions M and L instead of constants. Besides, we consider Banach space-valued functions here.

Theorem 4.18: *Assume that O is an open subset of $G([t_0 - r, t_0 + \sigma], X)$ with the prolongation property, $P = \{y_t : y \in O,\ t \in [t_0, t_0 + \sigma]\}$, $\phi \in P$, $g : [t_0, t_0 + \sigma] \to \mathbb{R}$ is a nondecreasing function, $f : P \times [t_0, t_0 + \sigma] \to X$ satisfies conditions* (A)–(C). *Let $F : O \times [t_0, t_0 + \sigma] \to G([t_0 - r, t_0 + \sigma], X)$ be given by* (4.24). *Let $y \in O$ be a solution of the measure FDE*

$$\begin{cases} y(t) = y(t_0) + \displaystyle\int_{t_0}^{t} f(y_s, s)dg(s), & t \in [t_0, t_0 + \sigma], \\ y_{t_0} = \phi. \end{cases} \tag{4.30}$$

For every $t \in [t_0, t_0 + \sigma]$, let

$$x(t)(\vartheta) = \begin{cases} y(\vartheta), & \vartheta \in [t_0 - r, t], \\ y(t), & \vartheta \in [t, t_0 + \sigma]. \end{cases}$$

Then, the function $x : [t_0, t_0 + \sigma] \to O$ is a solution of the generalized ODE

$$\frac{dx}{d\tau} = DF(x, t).$$

Proof. Let $\epsilon > 0$ and define a function $h : [t_0, t_0 + \sigma] \to \mathbb{R}$ by

$$h(t) = \int_{t_0}^{t} M(s)dg(s), \quad \text{for } t \in [t_0, t_0 + \sigma].$$

Since g is nondecreasing, h is also a nondecreasing function. Then, given $v \in [t_0, t_0 + \sigma]$, the function h admits only a finite number of points $t \in [t_0, v]$ such that $\Delta^+ h(t) \geq \epsilon$, see Theorem 1.4.

Let us denote these points by $t_1, \dots, t_m$. Consider a gauge δ on $[t_0, v]$ such that

$$\delta(\tau) < \min\left\{\frac{t_k - t_{k-1}}{2} : k = 2, \dots, m\right\}, \quad \text{for } \tau \in [t_0, v], \quad \text{and}$$

$$\delta(\tau) < \min\left\{|\tau - t_k| : k = 1, \dots, m\right\}, \quad \text{for } \tau \in [t_0, v] \setminus \{t_k\}_{k=1}^m.$$

If a point-interval pair $(\tau, [c, d])$ is δ-fine, then $[c, d]$ contains at most one of the points $t_1, \dots, t_m$, and if $t_k \in [c, d]$, then $\tau = t_k$.

Using the equality $y_{t_k} = x(t_k)_{t_k}$, for $k \in \{1, \dots, m\}$, by Corollary 2.14, we obtain

$$\lim_{t \to t_k+} \int_{t_k}^{t} L(s)\|y_s - x(t_k)_s\|_\infty \, dg(s) = L(t_k)\|y_{t_k} - x(t_k)_{t_k}\|_\infty \Delta^+ g(t_k) = 0$$

for every $k \in \{1, \dots, m\}$. Therefore, we can pick up a gauge δ on $[t_0, v]$ such that

$$\int_{t_k}^{t_k+\delta(t_k)} L(s)\|y_s - x(t_k)_s\|_\infty \, dg(s) < \frac{\epsilon}{2m+1}, \quad k \in \{1, \dots, m\}. \tag{4.31}$$

By condition (B), we have, for $\tau \in [t_0, v] \setminus \{t_k\}_{k=1}^m$,

$$\|y(\tau^+) - y(\tau)\| = \left\| \lim_{t \to \tau^+} \int_{\tau}^{t} f(y_s, s)dg(s) \right\| \leqslant \lim_{t \to \tau^+} \int_{\tau}^{t} M(s)dg(s) \leqslant h(\tau^+) - h(\tau),$$

whence

$$\|y(\tau^+) - y(\tau)\| \leqslant \Delta^+ h(\tau) < \epsilon,$$

for every $\tau \in [t_0, v] \setminus \{t_k\}_{k=1}^m$. Therefore, we choose a gauge δ satisfying

$$\|y(\rho) - y(\tau)\| \leqslant \epsilon \tag{4.32}$$

for every $\tau \in [t_0, v] \setminus \{t_k\}_{k=1}^m$ and $\rho \in [\tau, \tau + \delta(\tau))$.

Assume, now, that $d = (\tau_i, [s_{i-1}, s_i])$, $i = 1, \dots, |d|$ is a δ-fine tagged division of $[t_0, v]$. From the fact that y is a solution of the measure FDE (4.30) and by the definition of x, we obtain

$$[x(s_i) - x(s_{i-1})](\vartheta) = \begin{cases} 0, & \vartheta \in [t_0 - r, s_{i-1}], \\ \int_{s_{i-1}}^{\vartheta} f(y_s, s)dg(s), & \vartheta \in [s_{i-1}, s_i], \\ \int_{s_{i-1}}^{s_i} f(y_s, s)dg(s), & \vartheta \in [s_i, v]. \end{cases}$$

By the definition of F, we have, for $i = 1, 2, \dots, |d|$,

$$[F(x(\tau_i), s_i) - F(x(\tau_i), s_{i-1})](\vartheta) = \begin{cases} 0, & \vartheta \in [t_0 - r, s_{i-1}], \\ \int_{s_{i-1}}^{\vartheta} f(x(\tau_i)_s, s)dg(s), & \vartheta \in [s_{i-1}, s_i], \\ \int_{s_{i-1}}^{s_i} f(x(\tau_i)_s, s)dg(s), & \vartheta \in [s_i, v]. \end{cases}$$

Thus, we get

$$\begin{aligned} &[x(s_i) - x(s_{i-1})](\vartheta) - [F(x(\tau_i), s_i) - F(x(\tau_i), s_{i-1})](\vartheta) \\ &= \begin{cases} 0, & \vartheta \in [t_0 - r, s_{i-1}], \\ \int_{s_{i-1}}^{\vartheta} [f(y_s, s) - f(x(\tau_i)_s, s)]\, dg(s), & \vartheta \in [s_{i-1}, s_i], \\ \int_{s_{i-1}}^{s_i} [f(y_s, s) - f(x(\tau_i)_s, s)]\, dg(s), & \vartheta \in [s_i, v], \end{cases} \end{aligned}$$

whence

$$\begin{aligned} &\left\| x(s_i) - x(s_{i-1}) - [F(x(\tau_i), s_i) - F(x(\tau_i), s_{i-1})] \right\|_\infty \\ &= \sup_{\vartheta \in [t_0 - r, v]} \left\| [x(s_i) - x(s_{i-1})](\vartheta) - [F(x(\tau_i), s_i) - F(x(\tau_i), s_{i-1})](\vartheta) \right\| \\ &= \sup_{\vartheta \in [s_{i-1}, s_i]} \left\| \int_{s_{i-1}}^{\vartheta} [f(y_s, s) - f(x(\tau_i)_s, s)]\, dg(s) \right\|. \end{aligned}$$

By the definition of x, it follows that $x(\tau_i)_s = y_s$ whenever $s \leqslant \tau_i$. Hence,

$$\begin{aligned} &\int_{s_{i-1}}^{\vartheta} [f(y_s, s) - f(x(\tau_i)_s, s)]\, dg(s) \\ &= \begin{cases} 0, & \vartheta \in [s_{i-1}, \tau_i], \\ \int_{\tau_i}^{\vartheta} [f(y_s, s) - f(x(\tau_i)_s, s)]\, dg(s), & \vartheta \in [\tau_i, s_i]. \end{cases} \end{aligned}$$

Then, by condition (C), we obtain

$$\begin{aligned} \left\| \int_{\tau_i}^{\vartheta} [f(y_s, s) - f(x(\tau_i)_s, s)]\, dg(s) \right\| &\leqslant \int_{\tau_i}^{\vartheta} L(s) \|y_s - x(\tau_i)_s\|_\infty\, dg(s) \\ &\leqslant \int_{\tau_i}^{s_i} L(s) \|y_s - x(\tau_i)_s\|_\infty\, dg(s). \end{aligned}$$

Using the definition of the gauge δ, we have the following possibilities:

(i) $[s_{i-1}, s_i] \cap \{t_1, \dots, t_m\} = \{t_k\}$, for some $k = \{1, \dots, m\}$, where $t_k = \tau_i$;
(ii) $[s_{i-1}, s_i] \cap \{t_1, \dots, t_m\} = \emptyset$.

In case (i), we have, from (4.31),

$$\int_{\tau_i}^{s_i} L(s) \|y_s - x(\tau_i)_s\|_\infty\, dg(s) \leqslant \frac{\epsilon}{2m + 1},$$

whence

$$\|x(s_i) - x(s_{i-1}) - [F(x(\tau_i), s_i) - F(x(\tau_i), s_{i-1})]\|_\infty \leqslant \frac{\epsilon}{2m+1}. \tag{4.33}$$

In case (ii), by (4.32), we obtain, for $s \in [\tau_i, s_i]$,

$$\|y_s - x(\tau_i)_s\|_\infty = \sup_{\rho\in[\tau_i,s]} \|y(\rho) - y(\tau_i)\| \leqslant \epsilon.$$

Thus, as a consequence, we have

$$\|x(s_i) - x(s_{i-1}) - [F(x(\tau_i), s_i) - F(x(\tau_i), s_{i-1})]\|_\infty \leqslant \epsilon \int_{\tau_i}^{s_i} L(s)dg(s). \tag{4.34}$$

Then, from inequalities (4.33) and (4.34) and the fact that case (i) occurs at most $2m$ times, we conclude that

$$\left\| x(v) - x(t_0) - \sum_{i=1}^{|d|} [F(x(\tau_i), s_i) - F(x(\tau_i), s_{i-1})] \right\|_\infty$$

$$\leqslant \epsilon \int_{t_0}^{t_0+\sigma} L(s)dg(s) + \frac{2m\epsilon}{2m+1}.$$

At last, by the arbitrariness of ϵ,

$$x(v) - x(t_0) = \int_{t_0}^{v} DF(x(\tau), t)$$

and we complete the proof. □

Theorem 4.19: *Assume that O is an open subset of $G([t_0 - r, t_0 + \sigma], X)$ with the prolongation property, $P = \{y_t : y \in O,\ t \in [t_0, t_0 + \sigma]\}$, $\phi \in P$, $g : [t_0, t_0 + \sigma] \to \mathbb{R}$ is a nondecreasing function and $f : P \times [t_0, t_0 + \sigma] \to X$ satisfies conditions (A)–(C). Let $F : O \times [t_0, t_0 + \sigma] \to G([t_0 - r, t_0 + \sigma], X)$ be given by (4.24). Let $x : [t_0, t_0 + \sigma] \to X$ be a solution of the generalized ODE*

$$\frac{dx}{d\tau} = DF(x, t),$$

with initial condition

$$x(t_0)(\vartheta) = \begin{cases} \phi(\vartheta - t_0), & t_0 - r \leqslant \vartheta \leqslant t_0, \\ \phi(0), & t_0 \leqslant \vartheta \leqslant t_0 + \sigma. \end{cases} \tag{4.35}$$

Then, the function $y \in O$, defined by

$$y(\vartheta) = \begin{cases} x(t_0)(\vartheta), & t_0 - r \leqslant \vartheta \leqslant t_0, \\ x(\vartheta)(\vartheta), & t_0 \leqslant \vartheta \leqslant t_0 + \sigma, \end{cases} \tag{4.36}$$

is a solution of the measure FDE

$$\begin{cases} y(t) = y(t_0) + \displaystyle\int_{t_0}^{t} f(y_s, s)dg(s), & t \in [t_0, t_0 + \sigma], \\ y_{t_0} = \phi. \end{cases} \tag{4.37}$$

Proof. We prove that y defined by (4.36) is a solution of the measure FDE (4.37). At first, note that $x(t) \in O$ for all $t \in [t_0, t_0 + \sigma]$ as x is a solution of $\frac{dx}{d\tau} = DF(x, t)$. By the definitions of y and $x(t_0)$, it is clear that $y_{t_0} = \phi$ is satisfied. By Lemma 4.17 and the definition of y, we have

$$\begin{aligned} y(v) - y(t_0) - \int_{t_0}^{v} f(y_s, s)dg(s) &= x(v)(v) - x(t_0)(t_0) - \int_{t_0}^{v} f(y_s, s)dg(s) \\ &= x(v)(v) - x(t_0)(v) - \int_{t_0}^{v} f(y_s, s)dg(s) \\ &= \left[\int_{t_0}^{v} DF(x(\tau), t)\right](v) - \int_{t_0}^{v} f(y_s, s)dg(s). \end{aligned} \tag{4.38}$$

Define the function $h : [t_0, t_0 + \sigma] \to \mathbb{R}$ by

$$h(t) = \int_{t_0}^{t} [M(s) + L(s)]\, dg(s), \quad t \in [t_0, t_0 + \sigma].$$

Since h is nondecreasing, given $v \in [t_0, t_0 + \sigma]$ and $\epsilon > 0$, there is only a finite number of points $t \in [t_0, v]$ for which $\Delta^+ h(t) \geq \epsilon$. As in the previous theorem, we denote such points by $t_1, \dots, t_m$, and we consider a gauge $\delta : [t_0, v] \to (0, \infty)$ on $[t_0, v]$ satisfying

(i) $\delta(\tau) < \min \left\{ \frac{t_k - t_{k-1}}{2} : k = 2, \dots, m \right\}, \tau \in [t_0, v]$;
(ii) $\delta(\tau) < \min \left\{ |\tau - t_k| : k = 1, \dots, m \right\}, \tau \in [t_0, v] \setminus \{t_1, \dots, t_m\}$;
(iii) $\int_{t_k}^{t_k + \delta(t_k)} L(s) \| y_s - x(t_k)_s \|_\infty \, dg(s) < \frac{\epsilon}{2m+1}, k \in \{1, \dots, m\}$.

Because conditions (A)–(C) are fulfilled, Lemma 4.16 implies that the function F given by (4.24) belongs to the class $\mathcal{F}(O \times [t_0, t_0 + \sigma], h)$, where h is defined as earlier.

Arguing as in the proof of the previous theorem, we can assume that the gauge δ satisfies

$$\|h(\rho) - h(\tau)\| \leqslant \epsilon,$$

for every $\rho \in [\tau, \tau + \delta(\tau))$ and $\tau \in [t_0, v] \setminus \{t_k\}_{k=1}^{m}$.

By the existence of the Kurzweil integral $\int_{t_0}^{v} DF(x(\tau), t)$, the gauge δ can be chosen such that

$$\left\| \int_{t_0}^{v} DF(x(\tau), t) - \sum_{i=1}^{|d|} \left[F(x(\tau_i), s_i) - F(x(\tau_i), s_{i-1}) \right] \right\|_\infty < \epsilon \tag{4.39}$$

for every δ-fine tagged division $d = (\tau_i, [s_{i-1}, s_i])$ of $[t_0, v]$, with $i = 1, 2, \dots, |d|$. Then, (4.38) and (4.39) yield

$$\left\| y(v) - y(t_0) - \int_{t_0}^{v} f(y_s, s)dg(s) \right\|$$

$$= \left\| \left[\int_{t_0}^{v} DF(x(\tau), t) \right] (v) - \int_{t_0}^{v} f(y_s, s) dg(s) \right\|$$

$$< \epsilon + \left\| \sum_{i=1}^{|d|} \left[F(x(\tau_i), s_i) - F(x(\tau_i), s_{i-1}) \right] (v) - \int_{t_0}^{v} f(y_s, s) dg(s) \right\|$$

$$\leqslant \epsilon + \sum_{i=1}^{|d|} \left\| \left[F(x(\tau_i), s_i) - F(x(\tau_i), s_{i-1}) \right] (v) - \int_{s_{i-1}}^{s_i} f(y_s, s) dg(s) \right\| .$$

On the other hand, from the definition of F, we obtain

$$\left[F(x(\tau_i), s_i) - F(x(\tau_i), s_{i-1}) \right] (v) = \int_{s_{i-1}}^{s_i} f(x(\tau_i)_s, s) dg(s).$$

This leads us to the following equalities

$$\left\| \left[F(x(\tau_i), s_i) - F(x(\tau_i), s_{i-1}) \right] (v) - \int_{s_{i-1}}^{s_i} f(y_s, s) dg(s) \right\|$$

$$= \left\| \int_{s_{i-1}}^{s_i} f(x(\tau_i)_s, s) dg(s) - \int_{s_{i-1}}^{s_i} f(y_s, s) dg(s) \right\|$$

$$= \left\| \int_{s_{i-1}}^{s_i} [f(x(\tau_i)_s, s) - f(y_s, s)] \, dg(s) \right\| .$$

According to Lemma 4.17 and by the definition of y, for every $i \in \{1, \ldots, |d|\}$, $x(\tau_i)_s = x(s)_s = y_s$ for $s \in [s_{i-1}, \tau_i]$ and $y_s = x(s)_s = x(s_i)_s$ for $s \in [\tau_i, s_i]$. Then,

$$\left\| \int_{s_{i-1}}^{s_i} [f(x(\tau_i)_s, s) - f(y_s, s)] \, dg(s) \right\| = \left\| \int_{\tau_i}^{s_i} [f(x(\tau_i)_s, s) - f(y_s, s)] \, dg(s) \right\|$$

$$= \left\| \int_{\tau_i}^{s_i} [f(x(\tau_i)_s, s) - f(x(s_i)_s, s)] \, dg(s) \right\|$$

$$\leqslant \int_{\tau_i}^{s_i} L(s) \| x(\tau_i)_s - x(s_i)_s \|_\infty \, dg(s),$$

where we use condition (C) to get the last inequality.

As in the proof of the previous result, let us consider two cases:

(1) $[s_{i-1}, s_i] \cap \{t_1, \ldots, t_m\} = \{t_k\}$ for some $k = \{1, \ldots, m\}$, and $t_k = \tau_i$;
(2) $[s_{i-1}, s_i] \cap \{t_1, \ldots, t_m\} = \emptyset$.

In case (1), we obtain the following inequality from the definition of the gauge δ

$$\int_{\tau_i}^{s_i} L(s) \| y_s - x(\tau_i)_s \|_\infty \, dg(s) \leqslant \frac{\epsilon}{2m+1},$$

which implies

$$\left\| \left(F(x(\tau_i), s_i) - F(x(\tau_i), s_{i-1}) \right) (v) - \int_{s_{i-1}}^{s_i} f(y_s, s) dg(s) \right\| \leqslant \frac{\epsilon}{2m+1}.$$

In case (2), Lemma 4.9 implies

$$\|x(s_i)_s - x(\tau_i)_s\|_\infty \leqslant \|x(s_i) - x(\tau_i)\|_\infty \leqslant h(s_i) - h(\tau_i) \leqslant \epsilon,$$

for every $s \in [\tau_i, s_i]$. Then,

$$\left\| \left(F(x(\tau_i), s_i) - F(x(\tau_i), s_{i-1})\right)(v) - \int_{s_{i-1}}^{s_i} f(y_s, s)dg(s) \right\| \leqslant \epsilon \int_{\tau_i}^{s_i} L(s)dg(s).$$

Finally, from cases (1) and (2) and using the fact that case (1) occurs at most $2m$ times, we conclude that

$$\sum_{i=1}^{|d|} \left\| \left(F(x(\tau_i), s_i) - F(x(\tau_i), s_{i-1})\right)(v) - \int_{s_{i-1}}^{s_i} f(y_s, s)dg(s) \right\|$$
$$\leqslant \epsilon \int_{t_0}^{t_0+\sigma} L(s)dg(s) + \frac{2m\epsilon}{2m+1} < \epsilon \left(1 + \int_{t_0}^{t_0+\sigma} L(s)dg(s)\right).$$

Thus,

$$\left\| y(v) - y(t_0) - \int_{t_0}^{v} f(y_s, s)dg(s) \right\| < \epsilon \left(2 + \int_{t_0}^{t_0+\sigma} L(s)dg(s)\right)$$

and, since $\epsilon > 0$ is arbitrary, the statement follows. □

Remark 4.20: In Theorem 4.19, a careful examination of the definition of y given by

$$y(\vartheta) = \begin{cases} x(t_0)(\vartheta), & t_0 - r \leqslant \vartheta \leqslant t_0, \\ x(\vartheta)(\vartheta), & t_0 \leqslant \vartheta \leqslant t_0 + \sigma, \end{cases}$$

shows us that it can be replaced by

$$y(\vartheta) = x(t_0 + \sigma)(\vartheta), \quad \text{for } t_0 - r \leqslant \vartheta \leqslant t_0 + \sigma,$$

due to the properties of x in Lemma 4.17. See [85] for details.

Remark 4.21: Let $x_0 \in O$ be arbitrary. Define $\phi : [-r, 0] \to X$ by $\phi(\theta) = x_0(t_0 - \theta)$ and $\widetilde{x} : [t_0 - r, t_0 + \sigma] \to X$ by

$$\widetilde{x}(\theta) = \begin{cases} \phi(\theta - t_0), & t_0 - r \leqslant \theta \leqslant t_0, \\ x_0(t_0), & t_0 \leqslant \theta \leqslant t_0 + \sigma. \end{cases} \tag{4.40}$$

Let $y \in O$ be a solution of the measure FDE (4.30) with initial condition $y_{t_0} = \phi$. By Theorem 4.18, the function $\widetilde{x} : [t_0, t_0 + \sigma] \to O$ defined by

$$\widetilde{x}(t)(\theta) = \begin{cases} y(\theta), & t_0 - t \leqslant \theta \leqslant t, \\ y(t), & t \leqslant \theta \leqslant t_0 + \sigma, \end{cases}$$

is a solution of the generalized ODE

$$\frac{dx}{d\tau} = DF(x, t),$$

with initial condition $\widetilde{x}(t_0) = \widetilde{x}_0$, where F is given by (4.24).

The correspondence between generalized ODEs and measure FDEs presented here takes into account finite delays. However, it is also possible to extend this correspondence to measure FDEs with infinite delays (see, e.g. [220]) and also to measure FDEs with time-dependent delays, as well as to measure FDEs with state-dependent delays (see, e.g. [100, 116]).

5

Basic Properties of Solutions

Everaldo M. Bonotto[1], Márcia Federson[2], Luciene P. Gimenes (in memorian)[3], Rogelio Grau[4], Jaqueline G. Mesquita[5], and Eduard Toon[6]

[1]*Departamento de Matemática Aplicada e Estatística, Instituto de Ciências Matemáticas e de Computação (ICMC), Universidade de São Paulo, São Carlos, SP, Brazil*
[2]*Departamento de Matemática, Instituto de Ciências Matemáticas e de Computação (ICMC), Universidade de São Paulo, São Carlos, SP, Brazil*
[3]*Departamento de Matemática, Centro de Ciências Exatas, Universidade Estadual de Maringá, Maringá, PR, Brazil*
[4]*Departamento de Matemáticas y Estadística, División de Ciencias Básicas, Universidad del Norte, Barranquilla, Colombia*
[5]*Departamento de Matemática, Instituto de Ciências Exatas, Universidade de Brasília, Brasília, DF, Brazil*
[6]*Departamento de Matemática, Instituto de Ciências Exatas, Universidade Federal de Juiz de Fora, Juiz de Fora, MG, Brazil*

In the Chapter 4, we saw the definition of nonautonomous generalized ordinary differential equations (ODEs) for Banach space-valued functions and an important relation with a special class of differential equations, namely, measure functional differential equations, which we refer to simply as measure FDEs. The present chapter is, therefore, devoted to basic properties of nonautonomous generalized ODEs and to applying some of the results to measure FDEs and functional dynamic equations on time scales. We recall the fact, presented in Chapter 3, that functional dynamic equations on time scales can be regarded as measure FDEs which, in turn, by virtue of the results in Chapter 4, can be regarded as generalized ODEs.

In Section 5.1, we deal with local existence and uniqueness of solutions of generalized ODEs and we also bring up analogous results to measure FDEs and functional dynamic equations on time scales. In Section 5.2, we cover results on prolongation of solutions and existence of a maximal solution. Still in Section 5.2, we present the analogues for measure differential equations (MDEs) and dynamic equations on time scales.

Generalized Ordinary Differential Equations in Abstract Spaces and Applications, First Edition.
Edited by Everaldo M. Bonotto, Márcia Federson, and Jaqueline G. Mesquita.

Throughout this chapter, X denotes a Banach space endowed with norm $\|\cdot\|$ and, as in previous chapters, we denote by $\|\cdot\|_\infty$ the supremum norm in the space of regulated functions, whose main properties can be found in Chapter 1.

5.1 Local Existence and Uniqueness of Solutions

In this section, we present a version of an existence and uniqueness result for solutions of a generalized ODE whose functions take values in a Banach space X. We also include results on the local existence and uniqueness of solutions of measure FDEs and functional dynamic equations on time scales.

As in Chapter 4, we consider a nonautonomous generalized ODE of the form

$$\frac{dx}{d\tau} = DF(x, t), \tag{5.1}$$

where $\Omega = O \times [a, b]$, with $O \subset X$ an open set, and $F : \Omega \to X$ being an X-valued function.

Up to this moment, we do not have any information about the existence of a solution of Eq. (5.1). The following result gives us an answer. For a proof of it, see [88, Theorem 2.15] for the infinite dimensional case, where the solution belongs to the space of functions of bounded variation. The reader may also consult [209] for the finite dimensional case, where the solution also belongs to the space of functions of bounded variation. Here, unlike what was done before, we present a proof for the existence and uniqueness of a solution that belongs to the space of regulated functions in the infinite dimensional case.

Theorem 5.1: *Let $F : \Omega \to X$ belong to the class $\mathcal{F}(\Omega, h)$, where the function $h : [a, b] \to \mathbb{R}$ is left-continuous (that is, $h(t^-) = h(t)$ for all $t \in (a, b]$). If $(\widetilde{x}, t_0) \in \Omega$ is such that*

$$\widetilde{x}_+ = \widetilde{x} + F(\widetilde{x}, t_0^+) - F(\widetilde{x}, t_0) \in O,$$

then there exists a $\Delta > 0$ such that, on the interval $[t_0, t_0 + \Delta]$, there exists a unique solution $x : [t_0, t_0 + \Delta] \to X$ of the generalized ODE (5.1) for which $x(t_0) = \widetilde{x}$.

Proof. At first, we consider t_0 as a point of continuity of h, that is, $h(t_0^+) = h(t_0)$.

Let $\Delta > 0$ be such that $[t_0, t_0 + \Delta] \subset (a, b)$, $h(t_0 + \Delta) - h(t_0) < \frac{1}{2}$ and also $x \in O$ whenever $\| x - \widetilde{x} \| \leqslant h(t_0 + \Delta) - h(t_0)$.

Our goal is to apply the Banach Fixed Point Theorem to show the existence and uniqueness of a solution to (5.1). In order to do that, consider Q as the set of all functions $z : [t_0, t_0 + \Delta] \to O$ such that $z \in G([t_0, t_0 + \Delta], O)$ and

$$\| z(t) - \widetilde{x} \| \leqslant h(t) - h(t_0), \quad \text{for every } t \in [t_0, t_0 + \Delta].$$

One can easily check that $Q \subset G([t_0, t_0 + \Delta], O)$ is closed (recall that, $G([t_0, t_0 + \Delta], O) = \{y \in G([t_0, t_0 + \Delta], X) : y(t) \in O \text{ for all } t \in [t_0, t_0 + \Delta]\}$, see Remark 1.2).

For all $s \in [t_0, t_0 + \Delta]$ and $z \in Q$, define an operator $T : Q \to G([t_0, t_0 + \Delta], X)$ by

$$Tz(s) = \widetilde{x} + \int_{t_0}^{s} DF(z(\tau), t),$$

where the integral on the right-hand side exists due to Corollary 4.8. By Lemma 4.5, for $s \in [t_0, t_0 + \Delta]$, we have

$$\| Tz(s) - \widetilde{x} \| = \left\| \int_{t_0}^{s} DF(z(\tau), t) \right\| \leqslant h(s) - h(t_0).$$

Therefore, T maps Q into itself.

Consider $t_0 \leqslant s_1 \leqslant t_0 + \Delta$ and $z_1, z_2 \in Q$. By Theorem 2.15 or Lemma 4.6,

$$\begin{aligned}
\| Tz_2(s_1) - Tz_1(s_1) \| &= \left\| \int_{t_0}^{s_1} D[F(z_2(\tau), t) - F(z_1(\tau), t)] \right\| \\
&\leqslant \left\| \int_{t_0}^{s_1} D[\| z_2(\tau) - z_1(\tau) \| \, h(t)] \right\| \\
&\leqslant \sup_{\tau \in [t_0, t_0 + \Delta]} \| z_2(\tau) - z_1(\tau) \| \cdot \left[h(s_1) - h(t_0) \right] \\
&= \| z_2 - z_1 \|_\infty \cdot \left[h(s_1) - h(t_0) \right].
\end{aligned}$$

Thus,

$$\| Tz_2 - Tz_1 \|_\infty \leqslant \| z_2 - z_1 \|_\infty \left[h(t_0 + \Delta) - h(t_0) \right] < \frac{1}{2} \| z_2 - z_1 \|_\infty,$$

and, hence, T is a contraction. Then, the Banach Fixed Point Theorem yields the result.

In the sequel, we need to analyze the case where t_0 is not a point of continuity of the function h. We define an auxiliary function $\widetilde{h} : [a, b] \to \mathbb{R}$ by

$$\widetilde{h}(t) = \begin{cases} h(t), & t \leqslant t_0 \\ h(t) - h(t_0^+) + h(t_0), & t > t_0. \end{cases}$$

Notice that $\widetilde{h}$ is nondecreasing and left-continuous and continuous at t_0.

For all $x \in O$, define

$$\widetilde{F}(x, t) = \begin{cases} F(x, t), & t \leqslant t_0, \\ F(x, t) - [F(\widetilde{x}, t_0^+) - F(\widetilde{x}, t_0)], & t > t_0, \end{cases}$$

It is not difficult to prove that $\widetilde{F} \in \mathcal{F}(\Omega, \widetilde{h})$. Thus, a solution z of the generalized ODE

$$\begin{cases} \frac{dz}{d\tau} = D\widetilde{F}(z,t), \\ z(t_0) = \widetilde{x}_+, \end{cases}$$

exists by the first part of the proof. Then, defining $x(t_0) = \widetilde{x}$ and $x(t) = z(t)$ for $t > t_0$, we obtain a solution of the generalized ODE (5.1) for which $x(t_0) = \widetilde{x}$. □

At this point, it is worth noticing that one can describe precisely the type of discontinuity of a solution of the generalized ODE (5.1). Indeed, the assumption on the left-continuity of the function h in Theorem 5.1 implies that the solution of (5.1) is left-continuous as well (see Lemma 4.9). This means that, given a solution x of (5.1), the left-limit $x(\sigma^-)$ exists for every σ in the domain of x. In addition, the number $\Delta > 0$ depends on the function h.

The second fact that we can infer from Theorem 5.1 is that the condition

$$\widetilde{x} + F(\widetilde{x}, t_0^+) - F(\widetilde{x}, t_0) \in O$$

is sufficient, but not necessary. The next example, borrowed from [78, Example 2.18], shows us this fact.

Example 5.2: Consider $X = \mathbb{R}$ with norm $|\cdot|$ (absolute value) and $\Omega = (-1,1) \times [0, 1]$. Let $\varphi : [0,1] \to \mathbb{R}$ be a function defined by $\varphi(t) = t - 1$, for $0 < t \leqslant 1$, and $\varphi(0) = 0$, and let $F : \Omega \to \mathbb{R}$ be defined by $F(x,t) = \varphi(t)$ for all $(x,t) \in \Omega$. Note that F is constant with respect to the first variable. Consider a function $h : [0,1] \to \mathbb{R}$ given by

$$h(t) = \begin{cases} 2t + 1, & 0 < t \leqslant 1, \\ h(0) = 0. \end{cases}$$

By definition, the function h is left-continuous on $(0,1]$ and increasing on $[0,1]$. Consider the generalized ODE given by

$$\begin{cases} \frac{dx}{d\tau} = DF(x,t) = D[\varphi(t)], \\ x(0) = 0. \end{cases} \tag{5.2}$$

We claim that $F \in \mathcal{F}(\Omega, h)$ whose proof we divide into two parts: (i) and (ii).

(i) We assert that, for all $(x, t_2), (x, t_1) \in \Omega$,

$$|F(x,t_2) - F(x,t_1)| \leqslant |h(t_2) - h(t_1)|.$$

Indeed. Let $0 < t_1 \leqslant t_2 \leqslant 1$ and $x \in (-1, 1)$. Then,

$$\begin{aligned} |F(x,t_2) - F(x,t_1)| &= |\varphi(t_2) - \varphi(t_1)| = |t_2 - 1 - (t_1 - 1)| = |t_2 - t_1| \\ &\leqslant |2(t_2 - t_1)| = |2t_2 + 1 - (2t_1 + 1)| \\ &= |h(t_2) - h(t_1)|. \end{aligned}$$

Now, let $t_1 = 0$, $0 < t_2 \leqslant 1$ and $x \in (-1, 1)$. In this case, we obtain

$$\begin{aligned} |F(x,t_2) - F(x,t_1)| &= |\varphi(t_2) - \varphi(t_1)| = |t_2 - 1| \\ &= 1 - t_2 < 1 < 2t_2 + 1 \\ &= h(t_2) - h(0) = |h(t_2) - h(t_1)|. \end{aligned}$$

Notice that the case $t_1 = t_2 = 0$ is trivial. Hence, the assertion follows.

(ii) We assert that, for all $(x, t_2), (x, t_1), (y, t_2), (y, t_1) \in \Omega$,

$$|F(x,t_2) - F(x,t_1) - F(y,t_2) + F(y,t_1)| \leqslant |h(t_2) - h(t_1)|\,|x - y|,$$

which comes directly from the fact that

$$\begin{aligned} |F(x,t_2) - F(x,t_1) - F(y,t_2) + F(y,t_1)| &= |\varphi(t_2) - \varphi(t_1) - \varphi(t_2) + \varphi(t_1)| \\ &= 0 \leqslant |h(t_2) - h(t_1)|\,|x - y|. \end{aligned}$$

This concludes the proof of the claim that $F \in \mathcal{F}(\Omega, h)$.

Now, we prove that $0 + F(0, 0^+) - F(0, 0) \notin (-1, 1)$. In fact, by the definition of F,

$$F(0,0^+) - F(0,0) = \lim_{t \to 0^+} F(0,t) - F(0,0) = \lim_{t \to 0^+} \varphi(t) = -1.$$

Hence,

$$0 + F(0,0^+) - F(0,0) = -1 \notin (-1, 1).$$

We proceed so as to show that the function φ, defined in the beginning of this example, is the unique solution of the generalized ODE (5.2) on $[0, 1]$.

At first, let us prove that φ is a solution of the generalized ODE (5.2) on $[0, 1]$. If $s \in [0, 1]$, then

$$\varphi(s) - \varphi(0) = \int_0^s D[\varphi(t)] = \int_0^s DF(\varphi(\tau), t),$$

which proves that φ is a solution of (5.2) on $[0, 1]$.

Suppose $x : [0, 1] \to \mathbb{R}$ is also a solution of (5.2) on $[0, 1]$. Then, given $s \in [0, 1]$, we obtain

$$x(s) = 0 + \int_0^s \underbrace{DF(x(\tau), t)}_{\varphi(t)} = \int_0^s D[\varphi(t)] = \varphi(s) - \underbrace{\varphi(0)}_{0} = \varphi(s),$$

that is, $x(s) = \varphi(s)$ for all $s \in [0, 1]$. By the definition of the Kurzweil integral, we have $\varphi(s) - \varphi(0) = \int_0^s D[\varphi(t)]$. This completes our example.

5.1.1 Applications to Other Equations

In this section, we apply Theorems 4.18, 4.19, and 5.1 in order to prove an existence and uniqueness result for measure FDEs. Then, as a direct consequence of Theorems 3.30 and 5.3, we prove an analogue for functional dynamic equations on time scales.

The following result is a generalization of [85, Theorem 5.3], since here we consider conditions (A)–(C) from Chapter 4, which are more general than the conditions considered in [85]. Moreover, we prove this result for functions taking values in an infinite dimensional Banach space.

Recall that the prolongation property is given in Definition 4.15.

Theorem 5.3: *Let $r, \sigma > 0$ and $t_0 \in \mathbb{R}$. Assume that $O \subset G([t_0 - r, t_0 + \sigma], X)$ is an open set with the prolongation property and consider the set $P = \{x_t : x \in O,\ t \in [t_0, t_0 + \sigma]\}$. Assume, further, that the function $g : [t_0, t_0 + \sigma] \to \mathbb{R}$ is nondecreasing and left-continuous, $f : P \times [t_0, t_0 + \sigma] \to X$ satisfies conditions (A)–(C) from Chapter 4. Let $F : O \times [t_0, t_0 + \sigma] \to G([t_0 - r, t_0 + \sigma], X)$ be given by (4.24). If $\phi \in P$ is such that the function*

$$z(t) = \begin{cases} \phi(t - t_0), & t \in [t_0 - r, t_0], \\ \phi(0) + f(\phi, t_0)\Delta^+ g(t_0), & t \in (t_0, t_0 + \sigma] \end{cases}$$

belongs to O, then there exist a $\delta > 0$ and a function $y : [t_0 - r, t_0 + \delta] \to X$, which is the unique solution of the measure FDE

$$\begin{cases} y(t) = y(t_0) + \displaystyle\int_{t_0}^{t} f(y_s, s)dg(s), \\ y_{t_0} = \phi. \end{cases}$$

Proof. By Lemma 4.16, the function F belongs to the class $\mathcal{F}(O \times [t_0, t_0 + \sigma], h)$, with

$$h(t) = \int_{t_0}^{t} (M(s) + L(s))dg(s), \quad t \in [t_0, t_0 + \sigma].$$

Define

$$x_0(\vartheta) = \begin{cases} \phi(\vartheta - t_0), & \vartheta \in [t_0 - r, t_0], \\ \phi(0), & \vartheta \in [t_0, t_0 + \sigma]. \end{cases}$$

Then, it is not difficult to see that $x_0 \in O$.

We proceed to prove that $x_0 + F(x_0, t_0^+) - F(x_0, t_0) \in O$. Clearly $F(x_0, t_0) = 0$. Note that the limit $F(x_0, t_0^+)$ is taken with respect to the supremum norm and we know that such limit exists because $F \in \mathcal{F}(O \times [t_0, t_0 + \sigma], h)$. Therefore, F is regulated with respect to the second variable and, hence, we can calculate the pointwise limit $F(x_0, t_0^+)(\vartheta)$ for every $\vartheta \in [t_0 - r, t_0 + \sigma]$. Then, Corollary 2.14 yields

$$F(x_0, t_0^+)(\vartheta) = \begin{cases} 0, & t \in [t_0 - r, t_0], \\ f(\phi, t_0)\Delta^+ g(t_0), & t \in (t_0, t_0 + \sigma]. \end{cases}$$

This shows that $x_0 + F(x_0, t_0^+) - F(x_0, t_0) = z \in O$. Thus, all hypotheses of Theorem 5.1 are satisfied. Hence, there exist $\delta > 0$ and a unique solution $x : [t_0, t_0 + \delta] \to X$ satisfying the generalized ODE

$$\begin{cases} \dfrac{dx}{d\tau} = DF(x, t), \\ x(t_0) = x_0. \end{cases} \tag{5.3}$$

Recall by the Definition 4.1 that $(x(t), t) \in O \times [t_0, t_0 + \delta]$. Finally, Theorem 4.19 implies that the function $y : [t_0 - r, t_0 + \delta] \to X$ given by

$$y(\vartheta) = \begin{cases} x(t_0)(\vartheta), & t_0 - r \leqslant \vartheta \leqslant t_0, \\ x(\vartheta)(\vartheta), & t_0 \leqslant \vartheta \leqslant t_0 + \delta \end{cases}$$

is a solution of the measure FDE

$$\begin{cases} y(t) = y(t_0) + \displaystyle\int_{t_0}^{t} f(y_s, s)dg(s), \\ y_{t_0} = \phi. \end{cases}$$

Note that this solution is unique. Otherwise, by Theorem 4.18, x would not the only solution of the generalized ODE (5.3), which would be a contradiction. □

The next existence and uniqueness result is specialized for functional dynamic equations on time scales. The version presented here is more general than the one presented in [85, Theorem 5.5], because we deal with arbitrary functions M and L instead of constants (see hypotheses (ii) and (iii) below). Nevertheless, the proof follows similarly as in [85, Theorem 5.5].

Theorem 5.4: *Let $[t_0 - r, t_0 + \sigma]_{\mathbb{T}}$ be a time scale interval, with $t_0, t_0 + \sigma \in \mathbb{T}$ and $\sigma > 0$, $B \subset X$ be open, $O = G([t_0 - r, t_0 + \sigma], B)$ and $P = G([-r, 0], B)$. Consider a function $f : P \times [t_0, t_0 + \sigma]_{\mathbb{T}} \to X$ satisfying the following conditions:*

(i) *The Perron Δ-integral $\int_{t_0}^{t_0+\sigma} f(y_s, s)\Delta s$ exists for every $y \in O$;*

(ii) *There exists a Perron Δ-integrable function $M : [t_0, t_0 + \sigma]_{\mathbb{T}} \to \mathbb{R}$ such that for every $y \in O$ and $u_1, u_2 \in [t_0, t_0 + \sigma]_{\mathbb{T}}$, with $u_1 \leqslant u_2$, we have*

$$\left\| \int_{u_1}^{u_2} f(y_s, s)\Delta s \right\| \leqslant \int_{u_1}^{u_2} M(s)\Delta s;$$

(iii) *There exists a Perron Δ-integrable function $L : [t_0, t_0 + \sigma]_{\mathbb{T}} \to \mathbb{R}$ such that for every $y, z \in O$ and $u_1, u_2 \in [t_0, t_0 + \sigma]_{\mathbb{T}}$, with $u_1 \leqslant u_2$, we have*

$$\left\| \int_{u_1}^{u_2} [f(y_s, s) - f(z_s, s)]\, \Delta s \right\| \leqslant \int_{u_1}^{u_2} L(s) \parallel y_s - z_s \|_\infty \Delta s.$$

If $\phi : [t_0 - r, t_0]_{\mathbb{T}} \to B$ is a regulated function for which $\phi(t_0) + f(\phi^*_{t_0}, t_0)\mu(t_0) \in B$ (where, as usual, $\mu : \mathbb{T} \to [0, \infty)$ is the graininess function), then there exist $\delta > 0$ satisfying $\delta \geqslant \mu(t_0)$ and $t_0 + \delta \in \mathbb{T}$, and a function $y : [t_0 - r, t_0 + \delta]_{\mathbb{T}} \to B$, which is the unique solution of the functional dynamic equation on time scales

$$\begin{cases} y(t) = y(t_0) + \displaystyle\int_{t_0}^{t} f(y^*_s, s)\Delta s, & t \in [t_0, t_0 + \delta], \\ y(t) = \phi(t), & t \in [t_0 - r, t_0]_{\mathbb{T}}. \end{cases}$$

Proof. Let $g(t) = t^*$ and $f^*(y, t) = f(y, t^*)$, for every $t \in [t_0, t_0 + \sigma]$ and $y \in P$. By the definition of g, $\Delta^+ g(t_0) = \mu(t_0)$. Then, conditions (i), (ii) and (iii) and Lemma 3.32 yield that the function f^* fulfills conditions (A)–(C), $g : [t_0, t_0 + \sigma] \to \mathbb{R}$ is non-decreasing and left-continuous. By Lemma 3.20, since $\phi : [t_0 - r, t_0]_{\mathbb{T}} \to B$ is a regulated function, $\phi^* : [t_0 - r, t_0] \to B$ is also a regulated function.

Define a function $z : [t_0, t_0 + \sigma] \to X$ by

$$z(t) = \begin{cases} \phi^*(t), & t \in [t_0 - r, t_0], \\ \phi^*(t_0) + f(\phi^*_{t_0}, t_0)\Delta^+ g(t_0), & t \in (t_0, t_0 + \sigma]. \end{cases} \tag{5.4}$$

If $t \in [t_0 - r, t_0]$, then by hypothesis,

$$\phi^*(t) = \phi(t^*) \in B \quad \text{for every } t \in [t_0 - r, t_0].$$

If $t \in (t_0, t_0 + \sigma]$, then

$$\phi^*(t_0) + f(\phi^*_{t_0}, t_0)\Delta^+ g(t_0) = \phi^*(t_0) + f(\phi^*_{t_0}, t_0)\mu(t_0) \in B.$$

Therefore, z defined by (5.4) belongs to B. In this way, all the assumptions of Theorem 5.3 are satisfied. Thus, there exist $\delta_1 > 0$ and a function $y : [t_0 - r, t_0 + \delta_1] \to B$, which is a unique solution of the measure FDE

$$\begin{cases} y(t) = y(t_0) + \displaystyle\int_{t_0}^{t} f^*(y_s, s)dg(s), & t \in [t_0, t_0 + \delta_1], \\ y_{t_0} = \phi^*_{t_0}. \end{cases} \tag{5.5}$$

If t_0 is right-dense, then there exists $\tau \in \mathbb{T}$ such that $t_0 < \tau < t_0 + \delta_1$. Take $\delta = \tau - t_0$. Notice that $\delta > 0$ and $t_0 + \delta = \tau \in \mathbb{T}$. Since $[t_0 - r, t_0 + \delta] \subset [t_0 - r, t_0 + \delta_1]$, $y|_{[t_0-r,t_0+\delta]}$ is also a solution of (5.5) on $[t_0 - r, t_0 + \delta]$. Then, Theorem 3.30 yields $y|_{[t_0-r,t_0+\delta]} = x^*$, where $x : [t_0 - r, t_0 + \delta]_{\mathbb{T}} \to B$ is a solution of the functional dynamic equation on time scales

$$\begin{cases} x(t) = x(t_0) + \displaystyle\int_{t_0}^{t} f(x_s^*, s)\Delta s, & t \in [t_0, t_0 + \delta]_{\mathbb{T}}, \\ x(t) = \phi(t), & t \in [t_0 - r, t_0]_{\mathbb{T}}. \end{cases} \tag{5.6}$$

Again by Theorem 3.30, we obtain the uniqueness of the solution x.

Now, assume that t_0 is right-scattered. Without loss of generality, one can take $\delta \geqslant \mu(t_0)$. Otherwise, set

$$y(\sigma(t_0)) = \phi(t_0) + f(\phi_{t_0}^*, t_0)\mu(t_0)$$

to get a solution defined in $[t_0 - r, t_0 + \mu(t_0)]_{\mathbb{T}}$. Arguing as before, we get $y|_{[t_0-r,t_0+\delta]} = x^*$, where $x : [t_0 - r, t_0 + \delta]_{\mathbb{T}} \to B$ is a solution of Eq. (5.6). Using Theorem 3.30 once again, we obtain the uniqueness of the solution x and the statement holds. □

5.2 Prolongation and Maximal Solutions

In this section, we bring up results concerning the prolongation of solutions of generalized ODEs for Banach space-valued functions. These results are fundamental in the study of asymptotic properties of solutions of generalized ODEs, such as stability, boundedness, and controllability of solutions as one can check in Chapters 8, 11, and 12. We also collect results on prolongation of solutions specialized for MDEs and for dynamic equations on time scales.

Let $O \subset X$ be an open set, $[t_0, \infty) \subset \mathbb{R}$ and $\Omega = O \times [t_0, \infty)$. Consider the generalized ODE

$$\frac{dx}{d\tau} = DF(x, t), \tag{5.7}$$

where $F \in \mathcal{F}(\Omega, h)$ and $h : [t_0, \infty) \to \mathbb{R}$ is nondecreasing and left-continuous. Throughout this section, for $t_0 < \beta < \vartheta < \infty$, define $\Gamma_{\beta,\vartheta}^{\infty} = \{[\beta, \vartheta], [\beta, \vartheta), [\beta, \infty)\}$. The next result presents conditions that guarantee the prolongation of a solution of the nonautonomous generalized ODE (5.7). Its proof is inspired in [209, Proposition 4.15]. Such result is crucial in extending solutions to unbounded intervals. We point out that the version we present for Banach space-valued functions was firstly proved in [78, Theorem 3.1]. We also notice that a version of for bounded intervals and finite dimensional space-valued functions can be found in [209, Lemma 4.4].

Theorem 5.5: *Let $F \in \mathcal{F}(\Omega, h)$, where the function $h : [t_0, \infty) \to \mathbb{R}$ is nondecreasing and left-continuous. If $x : [\gamma, \beta) \to X$ and $y : I \to X$, $I \in \Gamma^{\infty}_{\beta,\vartheta}$, are solutions of the generalized ODE (5.7) on $[\gamma, \beta)$ and I, respectively, where $t_0 \leqslant \gamma < \beta < \vartheta < \infty$ and if the limit $\lim_{t\to\beta^-} x(t)$ exists and $\lim_{t\to\beta^-} x(t) = y(\beta)$, then the function $z : [\gamma, \beta) \cup I \to X$, defined by $z(t) = x(t)$, for $t \in [\gamma, \beta)$, and $z(t) = y(t)$, for $t \in I$, is a solution of the generalized ODE (5.7) on $[\gamma, \beta) \cup I$, where $[\gamma, \beta) \cup I = [\gamma, \vartheta]$, for $I = [\beta, \vartheta]$, $[\gamma, \beta) \cup I = [\gamma, \vartheta)$, for $I = [\beta, \vartheta)$, and $[\gamma, \beta) \cup I = [\gamma, \infty)$, for $I = [\beta, \infty)$.*

Proof. Assume that the limit $\lim_{t\to\beta^-} x(t)$ exists and $\lim_{t\to\beta^-} x(t) = y(\beta)$. Define a function $z : [\gamma, \beta) \cup I \to X$ by $z(t) = x(t)$, for $t \in [\gamma, \beta)$, and $z(t) = y(t)$, for $t \in I$. Then, z is well-defined and it is a regulated function, since x and y are regulated functions by Lemma 4.9. Now, we show that z is, in fact, a solution of the generalized ODE (5.7) on $[\gamma, \beta) \cup I$. Note that $z(t) \in O$ for all $t \in [\gamma, \beta) \cup I$ since x and y are solutions of the generalized ODE (5.7), see Definition 4.1.

Because β is an accumulation point of the set $[\gamma, \beta)$, there is a sequence $\{t_n\}_{n\in\mathbb{N}} \subset [\gamma, \beta)$ such that $t_n \to \beta$ as $n \to \infty$. Thus, once $\lim_{t\to\beta^-} x(t) = y(\beta)$, we obtain

$$\lim_{n\to\infty} x(t_n) = y(\beta). \tag{5.8}$$

Let $s_1, s_2 \in [\gamma, \beta) \cup I$. The first possibility we consider is $s_1 \in [\gamma, \beta)$ and $s_2 \in I$. Then, for $n \in \mathbb{N}$ sufficiently large, we conclude that $t_n \in (s_1, \beta)$ and

$$\begin{aligned}
\int_{s_1}^{s_2} DF(z(\tau), t) &= \int_{s_1}^{\beta} DF(z(\tau), t) + \int_{\beta}^{s_2} DF(z(\tau), t) \\
&= \int_{s_1}^{t_n} DF(x(\tau), t) + \int_{t_n}^{\beta} DF(z(\tau), t) + \int_{\beta}^{s_2} DF(y(\tau), t) \\
&= x(t_n) - x(s_1) + \int_{t_n}^{\beta} DF(z(\tau), t) + y(s_2) - y(\beta).
\end{aligned}$$

Using the definition of the function z, we obtain

$$\int_{s_1}^{s_2} DF(z(\tau), t) = x(t_n) - y(\beta) + z(s_2) - z(s_1) + \int_{t_n}^{\beta} DF(z(\tau), t). \tag{5.9}$$

Then, by Lemma 4.5,

$$\left\| \int_{t_n}^{\beta} DF(z(\tau), t) \right\| \leqslant |h(\beta) - h(t_n)|,$$

for sufficiently large n. Furthermore, since h is left-continuous and $t_n \to \beta$, as $n \to \infty$, $t_n < \beta$, for all $n \in \mathbb{N}$, we obtain $\lim_{n\to\infty} |h(\beta) - h(t_n)| = 0$. Thus,

$$\lim_{n\to\infty} \int_{t_n}^{\beta} DF(z(\tau), t) = 0. \tag{5.10}$$

Finally, taking the limit as $n \to \infty$ in (5.9) and using equalities (5.8) and (5.10), we conclude that

$$\int_{s_1}^{s_2} DF(z(\tau), t) = z(s_2) - z(s_1), \tag{5.11}$$

for every $s_1, s_2 \in [\gamma, \beta) \cup I$ such that $s_1 \in [\gamma, \beta)$ and $s_2 \in I$.

In the other cases for which $s_1, s_2 \in [\gamma, \beta)$ and $s_1, s_2 \in I$, Eq. (5.11) is easily verified. Hence, z is a solution of the generalized ODE (5.7) on $[\gamma, \beta) \cup I$. □

At this point, we highlight the fact that [209, Lemma 4.4] can be derived as a consequence of our Theorem 5.5. This fact is shown in Corollary 5.6 below which is a version of [209, Lemma 4.4] for $I = [\beta, \vartheta]$. The proof presented here and borrowed from [78, Corollary 3.2] is slightly different though.

Corollary 5.6: *Let $F \in \mathcal{F}(\Omega, h)$, where the function $h : [t_0, \infty) \to \mathbb{R}$ is nondecreasing and left-continuous. If $x : [\gamma, \beta] \to X$ and $y : I \to X$, $I \in \Gamma^{\infty}_{\beta,\vartheta}$, are solutions of the generalized ODE (5.7) on $[\gamma, \beta]$ and I respectively, where $t_0 \leqslant \gamma < \beta < \vartheta < \infty$ and $x(\beta) = y(\beta)$, then the function $z : [\gamma, \beta] \cup I \to X$, given by $z(t) = x(t)$, for $t \in [\gamma, \beta]$, and $z(t) = y(t)$, for $t \in I$, is a solution of the generalized ODE (5.7) on $[\gamma, \beta] \cup I$.*

Proof. Suppose $x(\beta) = y(\beta)$. Due to the fact that $F \in \mathcal{F}(\Omega, h)$ and h is a left-continuous function on $[t_0, \infty)$, Lemma 4.9 implies that x is left-continuous on $(\gamma, \beta]$. Thus, $\lim_{t\to\beta^-} x(t) = x(\beta)$. Hence, $\lim_{t\to\beta^-} x(t) = y(\beta)$, once $x(\beta) = y(\beta)$. Moreover, because $x : [\gamma, \beta] \to X$ is a solution of the generalized ODE (5.7) on $[\gamma, \beta]$, $x|_{[\gamma,\beta)}$ is a solution of the generalized ODE (5.7) on $[\gamma, \beta)$. Then, all hypotheses of Theorem 5.5 are fulfilled and, therefore, the function z, defined at the beginning of its proof, is a solution of the generalized ODE (5.7) on $[\gamma, \beta] \cup I$. □

In the sequel, we intend to prove that, under certain conditions, the unique solution $x : [\tau_0, \tau_0 + \Delta] \to X$ of the generalized ODE (5.7) satisfying $x(\tau_0) = x_0$, which is ensured by Theorem 5.1, can be extended intervals containing $[\tau_0, \tau_0 + \Delta]$ up to a maximal interval of existence, at least while the graph of the solution does not reach the boundary of Ω.

Consider a pair $(x_0, \tau_0) \in \Omega$ satisfying

$$x_0 + F(x_0, \tau_0^+) - F(x_0, \tau_0) \in O. \tag{5.12}$$

For a fixed pair $(x_0, \tau_0) \in \Omega$, define the set S_{τ_0,x_0} all functions $x : I_x \subset [t_0, \infty) \to X$, with I_x being an interval and $\tau_0 = \min I_x$ such that x is a solution on I_x of the generalized ODE (5.7), with initial condition $x(\tau_0) = x_0$. Given two elements $x : I_x \to X$ and $z : I_z \to X$ of S_{τ_0,x_0}, we say that x is *smaller or equal* to z ($x \preccurlyeq z$), if and only if $I_x \subset I_z$ and $z|_{I_x} = x$. The fact that the relation $\preccurlyeq$ defines a partial order relation in S_{τ_0,x_0} follows straightforwardly. See, for instance, [78, Proposition 3.4 and Remark 3.5]. As a matter of fact, the relation $\preccurlyeq$ defines a total order in S_{τ_0,x_0}.

In order to investigate the forward maximal solution, one needs to require condition (5.12). Otherwise, $x(t) \notin O$ for $t > \tau_0$, contradicting the definition of a solution of a generalized ODE. Therefore, from now on, we assume that (5.12) is fulfilled for every $x \in O$ and every $t \in [t_0, \infty)$, which we summarize in the following definition,

$$\Omega = \Omega_F = \left\{(x,t) \in \Omega : x + F(x,t^+) - F(x,t) \in O\right\}, \tag{5.13}$$

meaning that there are no points in Ω at which the solution of the generalized ODE (5.7) escapes from the set O.

Now we recall the definitions, introduced in [78], of a proper prolongation or prolongation to the right of a solution and, then, of a maximal forward solution of a nonautonomous generalized ODE.

Definition 5.7: Let $\tau_0 \geqslant t_0$, $I \subset [t_0, \infty)$ and $x : I \to X$ be a solution of (5.7) on the interval I, with $\tau_0 = \min I$. A solution $y : J \to X, J \subset [t_0, \infty)$, with $\tau_0 = \min J$, of the generalized ODE (5.7) is called a *prolongation to the right* of x, if $I \subset J$ and $x(t) = y(t)$ for all $t \in I$. If $I \subsetneq J$, then y is called a *proper prolongation of x to the right* or *proper right prolongation of x*. Similarly, we define a proper prolongation of a solution to the left.

Definition 5.8: Let $(x_0, \tau_0) \in \Omega$. We say that $x : J \to X$ is a *maximal forward solution* or simply *maximal solution* of the generalized ODE

$$\begin{cases} \dfrac{dx}{d\tau} = DF(x,t), \\ x(\tau_0) = x_0, \end{cases} \tag{5.14}$$

if $x \in S_{\tau_0,x_0}$ and, for every $z : I \to X$ in S_{τ_0,x_0} such that $x \preccurlyeq z$, we have $x = z$. This means that x is a maximal solution of generalized ODE (5.14), whenever $x \in S_{\tau_0,x_0}$ and there is no proper right prolongation of x.

Remark 5.9: When $x : J \to X$ is a maximal solution of the generalized ODE 5.14 and $\sup J = \infty$, x is called a *global forward solution* on J.

The next auxiliary result is important to the proof of the existence and uniqueness of a maximal solution of the generalized ODE (5.14), for the case where X is a Banach space and $\Omega = O \times [t_0, \infty)$, with O being an open subset of X. We call the reader's attention to the fact that, in the classical theory of ODEs, the Grönwall inequality is a powerful tool to prove this type of result. Here, since up to now there is no Grönwall inequality available for Kurzweil integrals of the form $\int DU(\tau,t)$, the proof of the following result, borrowed from [78], is arduous.

Lemma 5.10: *Let $F \in \mathcal{F}(\Omega, h)$, where the function $h : [t_0, \infty) \to \mathbb{R}$ is nondecreasing and left-continuous and $\Omega = \Omega_F$, where Ω_F is given by (5.13). Let $(x_0, \tau_0) \in \Omega$ and consider the generalized ODE (5.14). If $y : J_y \to X$ and $z : J_z \to X$ are solutions of the generalized ODE (5.14), where J_y and J_z are intervals such that $\tau_0 = \min J_y = \min J_z$, then $y(t) = z(t)$, for all $t \in J_y \cap J_z$.*

Proof. Fix $(x_0, \tau_0) \in \Omega$ and assume that $y : J_y \to X$ and $z : J_z \to X$ are solutions of the generalized ODE (5.14), where J_y and J_z are intervals such that $\tau_0 = \min J_y = \min J_z$. Therefore,

$$y(\tau_0) = z(\tau_0) = x_0. \tag{5.15}$$

We want to show that $y(t) = z(t)$, for all $t \in J_y \cap J_z$. Notice that $J_y \cap J_z$ is an interval of the form $[\tau_0, d]$, or $[\tau_0, d)$, $d \leqslant \infty$. Thus, we need to consider two situations.

The first situation occurs when $J_y \cap J_z = [\tau_0, d]$. Let Λ be the set of all $t \in [\tau_0, d]$ such that $y(s) = z(s)$, for every $s \in [\tau_0, t]$. Since $\tau_0 \in \Lambda$, we have $\Lambda \neq \emptyset$. Then, taking $\lambda = \sup \Lambda$, we have $\lambda \leqslant d$ and $[\tau_0, \lambda) \subset \Lambda$. Therefore,

$$y(t) = z(t), \quad t \in [\tau_0, \lambda). \tag{5.16}$$

Owing to the facts that F belongs to the class $\mathcal{F}(\Omega, h)$, with h being a left-continuous function, and the functions y and z are solutions of the generalized ODE (5.14) on $J_y \cap J_z$, Lemma 4.9 implies that y and z are left-continuous on $(\tau_0, \lambda]$. Thus,

$$y(\lambda) = \lim_{t \to \lambda^-} y(t) = \lim_{t \to \lambda^-} z(t) = z(\lambda),$$

whence

$$y(\lambda) = z(\lambda). \tag{5.17}$$

In view of (5.16) and (5.17), we have

$$\lambda \in \Lambda \quad \text{and} \quad [\tau_0, \lambda] \subset \Lambda. \tag{5.18}$$

Now, we want to prove that $\lambda = d$. In order to do so, we suppose, by contradiction, that $\lambda < d$. Because $(y(\lambda), \lambda) \in \Omega_F = \Omega$, Theorem 5.1 yields that there are $\delta > 0$ (we can consider $\lambda + \delta < d$, for instance) and a unique solution $x : [\lambda, \lambda + \delta] \to X$ of the generalized ODE

$$\begin{cases} \dfrac{dx}{d\tau} = DF(x, t), \\ x(\lambda) = y(\lambda) = z(\lambda). \end{cases} \tag{5.19}$$

On the other hand, $y|_{[\lambda,\lambda+\delta]}$ and $z|_{[\lambda,\lambda+\delta]}$ are solutions of (5.19). Thus, by the uniqueness of a solution,

$$x(t) = z(t) = y(t), \quad t \in [\lambda, \lambda + \delta], \tag{5.20}$$

and, hence, (5.18) and (5.20) imply $\lambda + \delta \in \Lambda$, which contradicts the definition of λ. As a matter of fact, the contradiction came from the assumption that $\lambda < d$. This means $d = \lambda$ and $y(t) = z(t)$, for all $t \in J_y \cap J_z$.

Let us now consider the second situation, where $J_y \cap J_z = [\tau_0, d)$, with $d \leqslant \infty$. At first, we prove that, for every $\tau \in (\tau_0, d)$, we have $y(s) = z(s)$, for every $s \in [\tau_0, \tau)$.

Define a set Λ of all $t \in (\tau_0, d)$ such that $y(s) = z(s)$, for all $s \in [\tau_0, t)$. We proceed so as to show that Λ is not empty. Notice that $(x_0, \tau_0) \in \Omega = \Omega_F$ and, therefore, we can apply Theorem 5.1 to obtain $\eta > 0$ (we can take $\tau_0 + \eta < d$, for instance) and a unique solution $x : [\tau_0, \tau_0 + \eta] \to X$ of the generalized ODE

$$\begin{cases} \dfrac{dx}{d\tau} = DF(x, t), \\ x(\tau_0) = x_0 = y(\tau_0) = z(\tau_0). \end{cases} \tag{5.21}$$

However, once $y|_{[\tau_0,\tau_0+\eta]}$ and $z|_{[\tau_0,\tau_0+\eta]}$ are solutions of (5.21), we conclude that

$$x(t) = y(t) = z(t), \quad t \in [\tau_0, \tau_0 + \eta]. \tag{5.22}$$

Furthermore, $y(t) = z(t)$ for all $t \in [\tau_0, \tau_0 + \eta)$, which means that $\tau_0 + \eta \in \Lambda$. Therefore, Λ is not empty.

Consider $\lambda = \sup \Lambda$. In such a case, $\lambda \leqslant d$ and $(\tau_0, \lambda) \subset \Lambda$. Let us prove that $\lambda = d$. Once again, we suppose to the contrary that $\lambda < d$. We want to show that $y(t) = z(t)$, for $t \in [\tau_0, \lambda)$, and $y(\lambda) = z(\lambda)$. Let $t \in [\tau_0, \lambda)$. If $t = \tau_0$, the conclusion comes straightforwardly as a consequence of (5.15). However, if $t \in (\tau_0, \lambda)$, then $t \in \Lambda$ (since $(\tau_0, \lambda) \subset \Lambda$). Thus, $y(s) = z(s)$, for all $s \in [\tau_0, t)$.

Knowing that $y|_{(\tau_0,t]}$ and $z|_{(\tau_0,t]}$ are left-continuous functions, one gets

$$y(t) = \lim_{s \to t^-} y(s) = \lim_{s \to t^-} z(s) = z(t).$$

Therefore, $y(t) = z(t)$, for all $t \in [\tau_0, \lambda)$ and, hence, by the left-continuity of the functions $y|_{(\tau_0,\lambda]}$ and $z|_{(\tau_0,\lambda]}$, we have $y(\lambda) = z(\lambda)$. Then,

$$y(t) = z(t), \quad t \in [\tau_0, \lambda]. \tag{5.23}$$

Since $(y(\lambda), \lambda) \in \Omega_F = \Omega$, Theorem 5.1 yields that there are $\delta > 0$ (we can take $\lambda + \delta < d$, for instance) and a unique solution $x : [\lambda, \lambda + \delta] \to X$ of the generalized ODE

$$\begin{cases} \dfrac{dx}{d\tau} = DF(x, t), \\ x(\lambda) = y(\lambda) = z(\lambda). \end{cases} \tag{5.24}$$

In addition, because $z|_{[\lambda,\lambda+\delta]}$ and $y|_{[\lambda,\lambda+\delta]}$ are solutions of (5.24), the uniqueness of solution implies

$$x(t) = z(t) = y(t), \quad t \in [\lambda, \lambda + \delta]. \tag{5.25}$$

Then, (5.23) and (5.25) imply $\lambda + \delta \in \Lambda$, which contradicts the definition of λ. This contradiction implies that $d = \lambda$ and $\Lambda = (\tau_0, d)$.

Thus, $y(t) = z(t)$, for all $t \in [\tau_0, d) = J_y \cap J_z$, and we finished the proof. □

In view of the previous lemma, our job turned out to be easier. The next result brings up sufficient conditions, which guarantee existence and uniqueness of a maximal solution of the generalized ODE (5.7) with initial condition $x(\tau_0) = x_0$. A version of such result for the finite dimensional case, that is, $X = \mathbb{R}^n$ and, moreover, $\Omega = O \times (a, b)$, with $O = B_c = \{x \in \mathbb{R}^n : \| x \| < c\}$, and $(a, b) \subset \mathbb{R}$ can be found in [209, Proposition 4.13]. Here, we present a version for the infinite dimensional case, that is, X is any Banach space and also $\Omega = O \times [t_0, \infty)$, where $O \subset X$ is open. Such a version can be found in [78, Theorem 3.9].

Theorem 5.11: *Let $F \in \mathcal{F}(\Omega, h)$, where the function $h : [t_0, \infty) \to \mathbb{R}$ is nondecreasing and left-continuous. If $\Omega = \Omega_F$, then, for every $(x_0, \tau_0) \in \Omega$, there exists a unique maximal solution $x : J \to X$ of the generalized ODE (5.7), where $x(\tau_0) = x_0$ and J is an interval such that $\tau_0 = \min J$.*

Proof. Consider $\Omega = \Omega_F$ and let $(x_0, \tau_0) \in \Omega$ be fixed. At first, we prove the existence of a maximal solution. In order to do that, consider a set S of all functions $x : J_x \subset [t_0, \infty) \to X$ such that J_x is an interval with $\tau_0 = \min J_x$ and x is a solution of the generalized ODE (5.7), with $x(\tau_0) = x_0$. It is clear that S is nonempty by the local existence and uniqueness of solution guaranteed by Theorem 5.1.

Let $J = \bigcup_{y \in S} J_y$ and define a function $x : J \to X$ by the relation $x(t) = y(t)$, where $y \in S$ and $t \in J_y$. Note that if $y, z \in S$, then $y(s) = z(s)$, for all $s \in J_y \cap J_z$, by Lemma 5.10. Therefore, x is well-defined. Note, in addition, that $(x(t), t) \in O \times J$ and that J is an interval with $\tau_0 = \min J$. This follows from the fact that J is a union connected to a common point and x is a maximal solution of the generalized ODE (5.7), proving the existence of a maximal solution.

Now, we prove the uniqueness of such maximal solution. Suppose $x_1 : J_1 \to X$ and $x_2 : J_2 \to X$ are two maximal solutions of the generalized ODE (5.7) with $x_1(\tau_0) = x_2(\tau_0) = x_0$ and J_1, J_2 are intervals satisfying $\tau_0 = \min J_1 = \min J_2$. By Lemma 5.10, we need to prove that $x_1(t) = x_2(t)$, for all $t \in J_1 \cap J_2$. In order to do this, define a function $x_3 : J_3 \to X$, $J_3 = J_1 \cup J_2$ by $x_3(t) = x_1(t)$, for $t \in J_1$, and

$x_3(t) = x_2(t)$, for $t \in J_2$. Clearly, x_3 is a solution of the generalized ODE (5.7) with initial condition $x_3(\tau_0) = x_0$, $J_1, J_2 \subset J_3$ and $x_3|_{J_1} = x_1$, $x_3|_{J_2} = x_2$. Finally, because the solutions x_1 and x_2 are assumed to be maximal, we have $J_3 = J_2 = J_1$ and $x_3(t) = x_2(t) = x_1(t)$, for all $t \in J_3$. Therefore, $x_1 = x_2$. □

Next, we present a result which says that the maximal interval J of the existence and uniqueness of a maximal solution is half-open. For a finite dimensional version of this fact, see [209, Proposition 4.14], where the author considered $X = \mathbb{R}^n$ and $\Omega = O \times (a, b)$, where $O = B_c \subset \mathbb{R}^n$ and $(a, b) \subset \mathbb{R}$. The version presented here and borrowed from [78, Theorem 3.10] holds for the infinite dimensional case.

Theorem 5.12: *Consider $F \in \mathcal{F}(\Omega, h)$, where the function $h : [t_0, \infty) \to \mathbb{R}$ is nondecreasing and left-continuous and $\Omega = \Omega_F$. Suppose $(x_0, \tau_0) \in \Omega$ and $x : J \to X$ is the maximal solution of the generalized ODE (5.7), with $x(\tau_0) = x_0$, and J is an interval for which $\tau_0 = \min J$. Then, $J = [\tau_0, \omega)$ with $\omega \leqslant \infty$.*

Proof. It is clear that $J \subset [t_0, \infty)$. Let $\omega = \sup J$. Then, $\omega \leqslant \infty$. If $\omega = \infty$, then the result follows. Consider $\omega < \infty$. We want to prove that $\omega \notin J$. Suppose, to the contrary, that $\omega \in J$. Therefore, $J = [\tau_0, \omega]$. By the definition of a solution, $(x(\omega), \omega) \in \Omega = \Omega_F$. Thus, Theorem 5.1 yields there exists $\eta > 0$ such that there exists a unique solution $z : [\omega, \omega + \eta] \to X$ on $[\omega, \omega + \eta]$ of the generalized ODE (5.7) such that $z(\omega) = x(\omega)$. Therefore, Corollary 5.6 implies

$$y(t) = \begin{cases} x(t), & t \in [\tau_0, \omega], \\ z(t), & t \in [\omega, \omega + \eta], \end{cases}$$

is a solution of the generalized ODE (5.7) satisfying $y(\tau_0) = x_0$. Notice that y is a proper right prolongation of x. Such prolongation is assumed to be maximal and, hence, we have a contradiction. Therefore, $\omega \notin J$ and $J = [\tau_0, \omega)$. □

A version of [209, Proposition 4.15] for functions F taking values in an infinite dimensional Banach space X is presented next. This result can also be found in [78].

Theorem 5.13: *Let $F \in \mathcal{F}(\Omega, h)$, where $h : [t_0, \infty) \to \mathbb{R}$ is a nondecreasing and left-continuous function and $\Omega = \Omega_F$. Assume that $(x_0, \tau_0) \in \Omega$ and $x : [\tau_0, \omega) \to X$ is the maximal solution of the generalized ODE (5.7) with $x(\tau_0) = x_0$. Then, for every compact set $K \subset \Omega$, there exists $t_K \in [\tau_0, \omega)$ such that $(x(t), t) \notin K$, for all $t \in (t_K, \omega)$.*

Proof. We proceed with the proof by contradiction. Assume that there exist a compact set $K \subset \Omega$ and a sequence $\{t_n\}_{n \in \mathbb{N}} \subset [\tau_0, \omega)$ such as $\lim_{n \to \infty} t_n = \omega$ and

$(x(t_n), t_n) \in K$, for all $n \in \mathbb{N}$. We divided the proof into two cases, namely, $\omega = \infty$ and $\omega < \infty$.

Case 1: $\omega = \infty$.

Since K is compact, the sequence $\{(x(t_n), t_n)\}_{n\in\mathbb{N}}$ admits a convergent subsequence, which we denote, again, by $\{(x(t_n), t_n)\}_{n\in\mathbb{N}}$. Thus, there exists $(y, \tau) \in K$ such that

$$\lim_{n\to\infty} (x(t_n), t_n) = (y, \tau).$$

In particular,

$$\lim_{n\to\infty} t_n = \tau,$$

which contradicts the fact that $\lim_{n\to\infty} t_n = \omega = \infty$.

Case 2: $\omega < \infty$.

By the compactness of K, the sequence $\{(x(t_n), t_n)\}_{n\in\mathbb{N}}$ has a convergent subsequence that we still denote by $\{(x(t_n), t_n)\}_{n\in\mathbb{N}}$. Therefore, there exists $(y, \tau) \in K \subset \Omega$ such that $\lim_{n\to\infty}(x(t_n), t_n) = (y, \tau)$ and, hence,

$$\lim_{n\to\infty} t_n = \tau.$$

By the uniqueness of the limit, $\tau = \omega$. Once $(y, \omega) = (y, \tau) \in \Omega = \Omega_F$, we obtain

$$y + F(y, \omega^+) - F(y, \omega) \in O.$$

Then, Theorem 5.1 implies there exist $\eta > 0$ and a unique solution $z : [\omega, \omega + \eta] \to X$ of the generalized ODE (5.7) for which $z(\omega) = y$.

Consider a function $u : [\tau_0, \omega + \eta] \to X$ given by $u(t) = x(t)$, for $t \in [\tau_0, \omega)$, and $u(t) = z(t)$, for $t \in [\omega, \omega + \eta]$. Then, Theorem 5.5 yields that $u : [\tau_0, \omega + \eta] \to X$ is a solution of the generalized ODE (5.7), with initial condition $u(\tau_0) = x_0$. This leads us to a contradiction, because u is a proper right prolongation of the solution x, which in turn, is assumed to be maximal. The statement follows then. □

The previous theorem enables us to present two consequences. The first one follows immediately and, hence, we omit its proof. The reader may want to consult [78] though.

Corollary 5.14: *Consider $F \in \mathcal{F}(\Omega, h)$, where $h : [t_0, \infty) \to \mathbb{R}$ is a nondecreasing and left-continuous function and $\Omega = \Omega_F$. Take $(x_0, \tau_0) \in \Omega$ and let $x : [\tau_0, \omega) \to X$ be the maximal solution of the generalized ODE (5.7) with $x(\tau_0) = x_0$. If $x(t)$ belongs to a compact $N \subset O$, for all $t \in [\tau_0, \omega)$, then $\omega = \infty$.*

Corollary 5.15: *Assume that $F \in \mathcal{F}(\Omega, h)$, where the function $h : [t_0, \infty) \to \mathbb{R}$ is nondecreasing and left-continuous and $\Omega = \Omega_F$. Take $(x_0, \tau_0) \in \Omega$ and let $x : [\tau_0, \omega) \to X$ be the maximal solution of the generalized ODE (5.7) with initial condition $x(\tau_0) = x_0$. If $\omega < \infty$. Then, the following statements hold*

(i) *the limit* $\lim_{t\to\omega^-} x(t)$ *exists;*
(ii) $(y,\omega) \in \partial\Omega$ *and* $\lim_{t\to\omega^-}(x(t),t) = (y,\omega)$, *where* $y = \lim_{t\to\omega^-} x(t)$.

Proof. Consider $\omega < \infty$ and let $\epsilon > 0$ be given. Since $\omega \in (\tau_0,\infty)$ and h is left-continuous on (τ_0,∞), the limit $\lim_{t\to\omega^-} h(t)$ exists. Thus, by the Cauchy condition, there exists $\delta > 0$ (we can take $\tau_0 < \omega - \delta$) such that $|h(t) - h(s)| < \epsilon$, for all $t,s \in (\omega-\delta,\omega)$. Thus, by Lemma 4.9,

$$\|x(t) - x(s)\| \leqslant |h(t) - h(s)| < \epsilon, \quad \text{for every } t,s \in (\omega-\delta,\omega).$$

Again, by the Cauchy condition, $\lim_{t\to\omega^-} x(t)$ exists. Therefore, there exists $y \in X$ such that

$$y = \lim_{t\to\omega^-} x(t) \tag{5.26}$$

which proves item (i).

Now, let us prove (ii). Notice that $\lim_{t\to\omega^-}(x(t),t) = (y,\omega)$. Since ω is an accumulation point of the set $[\tau_0,\omega)$, there is a sequence $\{t_n\}_{n\in\mathbb{N}} \subset [\tau_0,\omega)$ such that $\lim_{n\to\infty} t_n = \omega$. This fact together with (5.26) imply $\lim_{n\to\infty} x(t_n) = y$. Then, once $\{(x(t_n),t_n)\}_{n\in\mathbb{N}} \subset \Omega$ and $\lim_{n\to\infty}(x(t_n),t_n) = (y,\omega)$, it follows that

$$(y,\omega) \in \overline{\Omega}. \tag{5.27}$$

We want to show that $(y,\omega) \notin \Omega$. Suppose, to the contrary, that $(y,\omega) \in \Omega$. By Theorem 5.1, there exist $\Delta > 0$ and a unique solution $z : [\omega,\omega+\Delta] \to X$ of the generalized ODE (5.7) satisfying $z(\omega) = y$. Define a function $u : [t_0,\omega+\Delta] \to X$ by

$$u(t) = \begin{cases} x(t), & t \in [\tau_0,\omega), \\ z(t), & t \in [\omega,\omega+\Delta]. \end{cases}$$

Then, Theorem 5.5 implies that u is a solution of the generalized ODE (5.7) with initial condition $u(\tau_0) = x(\tau_0) = x_0$. But this leads us to a contradiction, since u is a proper right prolongation of the solution x, which, in turn, is assumed to be maximal. Therefore, $(y,\omega) \notin \Omega$, that is, $(y,\omega) \in \Omega^c$ and this means that

$$(y,\omega) \in \overline{\Omega^c}. \tag{5.28}$$

Finally, (5.27) and (5.28) yield $(y,\omega) \in \partial\Omega$, which concludes the proof. □

Another consequence of Theorem 5.13, says that, even when $O = X$, it is possible to ensure that the maximal solution is defined on the whole interval $[\tau_0,\infty)$, whenever $x(\tau_0) = x_0$, that is, we have a global forward solution. This result is of extreme importance for the study of stability, boundedness, and control theory in the setting of generalized ODEs. Additionally, if $\Omega = N \times [t_0,\infty)$ and $F \in \mathcal{F}(\Omega,h)$, where N is a compact subset of X, then, by the definition of a solution, $(x(t), t) \in \Omega$, for every $t \in [\tau_0, \omega)$. In particular, $x(t) \in N$, for every $t \in [\tau_0, \omega)$ and, hence, $\omega = \infty$. It is enough to follow the ideas of Corollary 5.14. See also [78].

Corollary 5.16: *If $\Omega = X \times [t_0, \infty)$ and $F \in \mathcal{F}(\Omega, h)$, where $h : [t_0, \infty) \to \mathbb{R}$ is a nondecreasing and left-continuous function, then for every $(x_0, \tau_0) \in \Omega$, there exists a unique global forward solution on $[\tau_0, \infty)$ of the generalized ODE (5.7) with initial condition $x(\tau_0) = x_0$.*

Proof. We start by claiming that $\Omega = \Omega_F$. In order to prove our claim, it is enough to show that $\Omega \subset \Omega_F$. Take an arbitrary pair $(z_0, s_0) \in \Omega$. Since h is nondecreasing, the limit $\lim_{s\to s_0^+} h(s)$ exists. Then, the Cauchy condition implies that, for every $\epsilon > 0$, there exists $\delta > 0$ such that if $s, t \in (s_0, s_0 + \delta)$, then $|h(t) - h(s)| < \epsilon$. Since $F \in \mathcal{F}(\Omega, h)$, we obtain

$$\|F(z_0, t) - F(z_0, s)\| \leqslant |h(t) - h(s)| < \epsilon, \quad \text{for every } t, s \in (s_0, s_0 + \delta).$$

Therefore, the limit $\lim_{s\to s_0^+} F(z_0, s)$ exists. Denote such limit by $F(z_0, s_0^+)$. Then, $z_0 + F(z_0, s_0^+) - F(z_0, s_0) \in X$ and, hence, $(z_0, s_0) \in \Omega_F$. The claim is, therefore, true.

Now, take a pair $(x_0, \tau_0) \in \Omega$ and assume that $x : [\tau_0, \omega) \to X$ is the maximal solution of the generalized ODE (5.7) satisfying $x(\tau_0) = x_0$. The existence of such a solution is guaranteed by Theorems 5.11 and 5.12. We want to show that $\omega = \infty$.

Assume, to the contrary, that $\omega < \infty$. Then, Corollary 5.15 implies the limit $\lim_{t\to\omega^-} x(t)$ exists. Set

$$y = \lim_{t\to\omega^-} x(t) \in X. \tag{5.29}$$

Once ω is an accumulation point of the set $[\tau_0, \omega)$, there is a sequence $\{t_n\}_{n\in\mathbb{N}} \subset [\tau_0, \omega)$ such that $\lim_{n\to\infty} t_n = \omega$. Then, (5.29) yields $\lim_{n\to\infty} x(t_n) = y$.

Due to the fact that $N = \{x(t_n)\}_{n\in\mathbb{N}} \cup \{y\}$ is a compact subset of X, the set $N \times [\tau_0, \omega]$ is also a compact set in Ω. Thus, by Theorem 5.13, there exists $\hat{t} \in [\tau_0, \omega)$ such that

$$(x(t), t) \notin N \times [\tau_0, \omega], \quad \text{for all } t \in (\hat{t}, \omega),$$

which contradicts the fact that $x(t_n) \in N$, for all $n \in \mathbb{N}$ and $\lim_{n\to\infty} t_n = \omega$. Hence, $\omega = \infty$ necessarily. □

5.2.1 Applications to MDEs

In this section, we apply the results on prolongation and maximal solutions of generalized ODEs for MDEs. This is possible because of the relations between the solutions of one type of equation and the solution of another type of equation. These relations are set out in detail in Chapter 4.

The results presented here are based on [78], where the authors considered the case where $X = \mathbb{R}^n$. Here, we present them for the case of an arbitrary Banach space X. Thus, as in the previous chapters, X denotes a Banach space equipped with norm $\|\cdot\|$ and $O \subset X$ denotes an arbitrary open set. We also consider the integral form of a measure differential equation (we write MDE, for short) as follows:

$$x(t) = x(\tau_0) + \int_{\tau_0}^{t} f(x(s), s)dg(s), \quad t \geqslant \tau_0, \tag{5.30}$$

where $\tau_0 \geqslant t_0, f : O \times [t_0, \infty) \to X$ and $g : [t_0, \infty) \to \mathbb{R}$ are functions and the integral on the right-hand side is understood in the Perron–Stieltjes' sense as presented in Chapter 1.

We recall the reader that $G([t_0, \infty), X)$ denotes the vector space of all functions $x : [t_0, \infty) \to X$ such that $x|_{[\alpha,\beta]}$ belongs to the space $G([\alpha, \beta], X)$, for all $[\alpha, \beta] \subset [t_0, \infty)$. We use the symbol $G_0([t_0, \infty), X)$ to denote the vector space of all functions $x \in G([t_0, \infty), X)$ such that $\sup\limits_{s\in[t_0,\infty)} e^{-(s-t_0)} \| x(s) \| < \infty$. When this space is endowed with the norm

$$\|x\|_{[t_0,\infty)} = \sup_{s\in[t_0,\infty)} e^{-(s-t_0)} \| x(s) \|, \quad x \in G_0([t_0, \infty), X),$$

it becomes a Banach space. See the comments in the end of Subsection 1.1.1.

Given $x,\ z \in G_0([t_0, \infty), O)$ and $s_1,\ s_2 \in [t_0, \infty)$, with $s_1 \leqslant s_2$, we shall consider that the functions $f : O \times [t_0, \infty) \to X$ and $g : [t_0, \infty) \to \mathbb{R}$ satisfy the following conditions:

(A1) $g : [t_0, \infty) \to \mathbb{R}$ is nondecreasing and left-continuous on (t_0, ∞);
(A2) The Perron–Stieltjes integral $\int_{s_1}^{s_2} f(x(s), s)dg(s)$ exists;
(A3) There exists a locally Perron–Stieltjes integrable function $M : [t_0, \infty) \to \mathbb{R}$ with respect to g such that

$$\left\| \int_{s_1}^{s_2} f(x(s), s)dg(s) \right\| \leqslant \int_{s_1}^{s_2} M(s)dg(s);$$

(A4) There exists a locally Perron–Stieltjes integrable function $L : [t_0, \infty) \to \mathbb{R}$ with respect to g such that

$$\left\| \int_{s_1}^{s_2} [f(x(s), s) - f(z(s), s)]\, dg(s) \right\| \leqslant \|x - z\|_{[t_0,\infty)} \int_{s_1}^{s_2} L(s)dg(s).$$

The first result we present tells us about some properties of a function, which later will be used as the right-hand side of the generalized ODE, which corresponds to the MDE (5.30).

Lemma 5.17: *Let $f : O \times [t_0, \infty) \to X$ satisfy conditions (A2)–(A4) and g satisfy condition (A1). Then, the function $F : O \times [t_0, \infty) \to X$ defined by*

$$F(x,t) = \int_{t_0}^{t} f(x,s)dg(s),$$

for $(x,t) \in O \times [t_0, \infty)$, is an element of the class $\mathcal{F}(\Omega, h)$, where $\Omega = O \times [t_0, \infty)$ and $h(t) = \int_{t_0}^{t} [M(s) + L(s)]\, dg(s)$, $t \in [t_0, \infty)$, is a nondecreasing and left-continuous function. In addition, $\Omega = \Omega_F$.

Proof. The proof that $F \in \mathcal{F}(\Omega, h)$ is analogous to the proof of [78, Theorem 4.2]. On the other hand, Corollary 2.14 yields

$$z_0 + F(z_0, s_0^+) - F(z_0, s_0) = z_0 + f(z_0, s_0)\Delta^+ g(s_0) \in O,$$

for all $(z_0, s_0) \in \Omega$. Then, $z_0 + F(z_0, s_0^+) - F(z_0, s_0) \in O$, for all $(z_0, s_0) \in \Omega$, whence $\Omega = \Omega_F$. □

The next result is similar to Theorem 4.14 with obvious adaptations. Recall the concepts of solutions of MDEs and generalized ODEs in Definitions 3.2 and 4.1.

Theorem 5.18: *Assume that $f : O \times [t_0, \infty) \to X$ satisfies conditions (A2)–(A4), and $g : [t_0, \infty) \to \mathbb{R}$ satisfies condition (A1). Then, $x : I \to X$ is a solution of the MDE (5.30) on $I \subset [t_0, \infty)$ if and only if x is a solution on I of the generalized ODE*

$$\frac{dx}{d\tau} = DF(x,t),$$

where F is given as in Lemma 5.17.

The next two definitions bring up the concepts of prolongation to the right and maximal solution in the context of MDEs.

Definition 5.19: Let $\tau_0 \geqslant t_0$ and $x : J \to X$ be a solution of the MDE (5.30) on the interval $J \subset [t_0, \infty)$, with $\tau_0 = \min J$. A solution $y : \hat{J} \to X$, with $\hat{J} \subset [t_0, \infty)$ and $\tau_0 = \min \hat{J}$, of the MDE (5.30) is called a *prolongation to the right* of x, whenever $J \subset \hat{J}$ and $x(t) = y(t)$, for all $t \in J$. In particular, when $J \subsetneq \hat{J}$, y is called a *proper prolongation* of x *to the right* or *proper right prolongation.*

Analogously, one can define a proper prolongation of a solution to the left.

Definition 5.20: Let $(x_0, \tau_0) \in O \times [t_0, \infty)$ and $I \subset [t_0, \infty)$ be an interval for which $\tau_0 = \min I$. A solution $y : I \to X$ of the MDE

$$y(t) = x_0 + \int_{\tau_0}^{t} f(y(s), s)dg(s)$$

is called a *maximal solution*, if there exists no proper prolongation of y to the right. If, moreover, sup $I = \infty$, then y is called a global forward solution on I.

The next result ensures the existence and uniqueness of the maximal solution of the MDE (5.30).

Theorem 5.21: *Suppose $f : O \times [t_0, \infty) \to X$ satisfies conditions* (A2)–(A4), *and $g : [t_0, \infty) \to \mathbb{R}$ satisfies condition* (A1). *Assume, in addition, that for all $(z_0, s_0) \in O \times [t_0, \infty)$, we have $z_0 + f(z_0, s_0)\Delta^+ g(s_0) \in O$. Then, for every $(x_0, \tau_0) \in O \times [t_0, \infty)$, there exists a unique maximal solution $x : J \to X$ of the MDE* (5.30), *with initial condition $x(\tau_0) = x_0$, and J an interval with $\tau_0 = \min J$.*

Proof. Consider the generalized ODE

$$\begin{cases} \dfrac{dx}{d\tau} = DF(x, t), \\ x(\tau_0) = x_0, \end{cases} \tag{5.31}$$

where F is given as in Lemma 5.17 and, by the same lemma, $F \in \mathcal{F}(\Omega, h)$ and $\Omega = \Omega_F$, where $\Omega = O \times [t_0, \infty)$ and h is a nondecreasing and left-continuous function. Since all the hypotheses of Theorem 5.11 are satisfied, there exists a unique maximal solution $x : J \to X$ of the generalized ODE (5.31), where J is an interval such that $\tau_0 = \min J$. Then, Theorem 5.18 implies $x : J \to X$ is also a solution of the MDE (5.30) with $x(s_0) = x_0$.

We proceed so as to prove that the solution x of the MDE (5.30) does not have a proper prolongation to the right. Suppose, to the contrary, that there exists a solution $\hat{x} : \hat{J} \to X$ of the MDE (5.30) satisfying $\hat{x}(\tau_0) = x_0$, where $\hat{J}$ is an interval such that $\tau_0 = \min \hat{J}$ and $J \subsetneq \hat{J}$, that is, $\hat{x}$ extends x properly. Then, Theorem 4.18 yields $\hat{x}$ is a solution of the generalized ODE (5.31), with $\hat{x}(\tau_0) = x_0$, contradicting the maximality of the solution x. Hence, x does not admit proper extension to the right, that is, x is a maximal solution of the MDE (5.30) satisfying $x(\tau_0) = x_0$. Finally, Theorems 5.11 and 5.18 yield the uniqueness of the maximal solution x of the MDE (5.30). □

Remark 5.22: Notice that by the proof of Theorem 5.21, $x : J \to X$ is the maximal solution of the MDE (5.30) with $x(\tau_0) = x_0$, if and only if $x : J \to X$ is the maximal solution of the generalized ODE (5.31), where F is given by Lemma 5.17.

Similarly as we did for generalized ODEs, the next result says that the maximal interval J of the existence and uniqueness of a maximal solution of the MDE (5.30), with right-hand side given as in Lemma 5.17, is half-open.

Theorem 5.23: *Suppose $f : O \times [t_0, \infty) \to X$ satisfies conditions* (A2)–(A4), *and $g : [t_0, \infty) \to \mathbb{R}$ satisfies condition* (A1). *Assume, further, that for every pair $(z_0, s_0) \in O \times [t_0, \infty)$, $z_0 + f(z_0, s_0)\Delta^+ g(s_0) \in O$. Moreover, suppose $(x_0, \tau_0) \in O \times [t_0, \infty)$ and $x : J \to X$ is the maximal solution of the MDE* (5.30), *with initial condition $x(\tau_0) = x_0$, and J is an interval for which $\tau_0 = \min J$. Then, $J = [\tau_0, \omega)$, with $\omega \leqslant \infty$.*

Proof. Consider $(x_0, \tau_0) \in O \times [t_0, \infty)$ and suppose $x : J \to X$ is the maximal solution of the MDE (5.30), with initial condition $x(\tau_0) = x_0$, where J is an interval with $\tau_0 = \min J$. By Remark 5.22, $x : J \to X$ is the maximal solution of the generalized ODE (5.31), where F is given by Lemma 5.17, $\Omega = \Omega_F$ and $F \in \mathcal{F}(\Omega, h)$, with $\Omega = O \times [t_0, \infty)$. Moreover, the function $h : [t_0, \infty) \to \mathbb{R}$ is nondecreasing and left-continuous. Thus, all hypotheses of Theorem 5.12 are fulfilled and, hence, $J = [\tau_0, \omega)$, where $\omega \leqslant \infty$. □

The next result guarantees that the graph of the maximal solution escapes from compact sets at finite time.

Theorem 5.24: *Suppose $f : O \times [t_0, \infty) \to X$ satisfies conditions* (A2)–(A4), *and $g : [t_0, \infty) \to \mathbb{R}$ satisfies condition* (A1). *Assume, further, that for all $(z_0, s_0) \in O \times [t_0, \infty)$, $z_0 + f(z_0, s_0)\Delta^+ g(s_0) \in O$. Let $(x_0, \tau_0) \in O \times [t_0, \infty)$ and $x : [\tau_0, \omega) \to X$ be the maximal solution of the MDE* (5.30) *with initial condition $x(\tau_0) = x_0$. Then, for every compact set $K \subset O \times [t_0, \infty)$, there exists $t_K \in [\tau_0, \omega)$ such that $(x(t), t) \notin K$, for all $t \in (t_K, \omega)$.*

Proof. Let K be a compact subset of $O \times [t_0, \infty)$. Consider the function $F : O \times [t_0, \infty) \to X$ defined in Lemma 5.17, which implies $\Omega = \Omega_F$ and $F \in \mathcal{F}(\Omega, h)$, where $\Omega = O \times [t_0, \infty)$ and $h : [t_0, \infty) \to \mathbb{R}$ is a nondecreasing and left-continuous function. By Remark 5.22, $x : [\tau_0, \omega) \to X$ is the maximal solution of the generalized ODE (5.31). Therefore, once all hypotheses of Theorem 5.13 are fulfilled, there exists $t_K \in [\tau_0, \omega)$ such that $(x(t), t) \notin K$, for all $t \in (t_K, \omega)$. □

As a consequence of Theorem 5.24, we have the following result, which gives conditions for the existence of a global forward solution of the MDE (5.30) with right-hand side given as in Lemma 5.17 and initial condition $x(\tau_0) = x_0$.

Corollary 5.25: *Suppose $f : O \times [t_0, \infty) \to X$ satisfies conditions* (A2)–(A4), *and $g : [t_0, \infty) \to \mathbb{R}$ satisfies condition* (A1). *Assume that, for all $(z_0, s_0) \in O \times [t_0, \infty)$, we have $z_0 + f(z_0, s_0)\Delta^+ g(s_0) \in O$. Suppose $(x_0, \tau_0) \in O \times [t_0, \infty)$ and $x : [\tau_0, \omega) \to X$ is the maximal solution of the MDE* (5.30) *with $x(\tau_0) = x_0$. If $x(t) \in N$ for all $t \in [\tau_0, \omega)$, where N is a compact subset of O, then $\omega = \infty$ and we have a global forward solution of the MDE* (5.30) *with initial condition $x(\tau_0) = x_0$.*

Proof. Let $x(t) \in N$, for all $t \in [\tau_0, \omega)$, where N is a compact subset of O. Consider the function $F : O \times [t_0, \infty) \to X$ defined in Lemma 5.17 which satisfies $\Omega = \Omega_F$ and $F \in \mathcal{F}(\Omega, h)$, where $\Omega = O \times [t_0, \infty)$ and $h : [t_0, \infty) \to \mathbb{R}$ is a nondecreasing and left-continuous function. By Remark 5.22, $x : [\tau_0, \omega) \to X$ is the maximal solution of generalized ODE (5.31). Thus, by Corollary 5.14, $\omega = \infty$. □

The next corollary shows that if ω is finite, then the limit $y = \lim_{t \to \omega^-} x(t)$ exists and $(y, \omega) \in \partial\Omega$.

Corollary 5.26: *Suppose $f : O \times [t_0, \infty) \to X$ satisfies conditions (A2)–(A4), and $g : [t_0, \infty) \to \mathbb{R}$ satisfies condition (A1). Suppose, in addition, for all $(z_0, s_0) \in O \times [t_0, \infty)$, $z_0 + f(z_0, s_0)\Delta^+ g(s_0) \in O$. Take $(x_0, \tau_0) \in O \times [t_0, \infty)$ and let $x : [\tau_0, \omega) \to X$ be the maximal solution of the MDE (5.30), with $x(\tau_0) = x_0$. If $\omega < \infty$. Then the following conditions hold.*

(i) *The limit $\lim_{t \to \omega^-} x(t)$ exists;*
(ii) *$(y, \omega) \in \partial\Omega$ and $\lim_{t \to \omega^-} (x(t), t) = (y, \omega)$, where $y = \lim_{t \to \omega^-} x(t)$.*

Proof. Take $\omega < \infty$ and a pair $(x_0, \tau_0) \in O \times [t_0, \infty)$. Let $x : [\tau_0, \omega) \to X$ be the maximal solution of (5.30) with $x(\tau_0) = x_0$. Remark 5.22 implies that $x : [\tau_0, \omega) \to X$ is the maximal solution of the generalized ODE (5.31), where $F : O \times [t_0, \infty) \to X$, defined in Lemma 5.17, satisfies $\Omega = \Omega_F$ and $F \in \mathcal{F}(\Omega, h)$, where $\Omega = O \times [t_0, \infty)$ and $h : [t_0, \infty) \to \mathbb{R}$ is a nondecreasing and left-continuous function. Therefore, all the hypotheses of Corollary 5.15 are satisfied and the proof is complete. □

Corollary 5.27: *Suppose $f : X \times [t_0, \infty) \to X$ satisfies conditions (A2)–(A4) (with $O = X$), and $g : [t_0, \infty) \to \mathbb{R}$ satisfies condition (A1). Then, for every $(x_0, \tau_0) \in X \times [t_0, \infty)$, the maximal solution of the MDE (5.30), with initial condition $x(\tau_0) = x_0$, is defined on $[\tau_0, \infty)$, that is, the maximal solution is, in fact, a global forward solution.*

Proof. Consider a pair $(x_0, \tau_0) \in X \times [t_0, \infty)$ and let $F : O \times [t_0, \infty) \to X$ be defined as in Lemma 5.17 which implies $\Omega = \Omega_F$ and $F \in \mathcal{F}(\Omega, h)$, with $\Omega = X \times [t_0, \infty)$ and $h : [t_0, \infty) \to \mathbb{R}$ being a nondecreasing and left-continuous function.

Hence, Corollary 5.16 implies there exists a unique global forward solution $x : [\tau_0, \infty) \to X$ of the generalized ODE (5.31). Then, by Remark 5.22, $x : [\tau_0, \infty) \to X$ is also a global forward solution of the MDE (5.30), with initial condition $x(\tau_0) = x_0$. □

5.2.2 Applications to Dynamic Equations on Time Scales

In this section, we make use of the relation, detailed in Chapter 3, between the solutions of MDEs and the solutions of dynamic equations on time scales and we present results on prolongation and maximal solutions in the framework of dynamic equations on time scales. The theory presented here is based on [78], but we deal with infinite dimensional space-valued functions.

Let $\mathbb{T}$ be a time scale such that $t_0 \in \mathbb{T}$ and consider the dynamic equation on time scales given by

$$x(t) = x(t_0) + \int_{t_0}^{t} f(x^*(s), s)\Delta s, \quad t \in [t_0, \infty)_{\mathbb{T}}, \tag{5.32}$$

where $f : O \times [t_0, \infty)_{\mathbb{T}} \to X$ is a function, $O \subset X$ is an open subset and, as usual, $x^* : \mathbb{T}^* \to X$ is defined by $x^*(t) = x(t^*)$.

For all $y,\ w \in G_0([t_0, \infty)_{\mathbb{T}}, O)$ and all $s_1,\ s_2 \in [t_0, \infty)_{\mathbb{T}}$, with $s_1 \leqslant s_2$, we assume that the function $f : O \times [t_0, \infty)_{\mathbb{T}} \to X$ satisfies the conditions:

(B1) The Perron Δ-integral $\int_{s_1}^{s_2} f(y(s), s)\Delta s$ exists;

(B2) There exists a locally Perron Δ-integrable function $M : [t_0, \infty)_{\mathbb{T}} \to \mathbb{R}$ such that

$$\left\| \int_{s_1}^{s_2} f(y(s), s)\Delta s \right\| \leqslant \int_{s_1}^{s_2} M(s)\Delta s;$$

(B3) There exists a locally Perron Δ-integrable function $L : [t_0, \infty)_{\mathbb{T}} \to \mathbb{R}$ such that

$$\left\| \int_{s_1}^{s_2} [f(y(s), s) - f(w(s), s)]\, \Delta s \right\| \leqslant \|y - w\|_{[t_0,\infty)} \int_{s_1}^{s_2} L(s)\Delta s.$$

In the following result, we present a correspondence between the solutions of a MDE and the solutions of a dynamic equation on time scales. The reader may consult [78] for a proof of this result in the setting of $\mathbb{R}^n$-valued functions. The proof for infinite dimensional case follows similarly.

Theorem 5.28: *Let $\mathbb{T}$ be a time scale such that* $\sup \mathbb{T} = \infty$ *and let $[t_0, \infty)_{\mathbb{T}}$ be a time scale interval. Let $O \subset X$ be an open subset and $f : O \times [t_0, \infty)_{\mathbb{T}} \to X$ be a function. Assume that, for every $x \in G([t_0, \infty)_{\mathbb{T}}, X)$, the function $t \mapsto f(x(t), t)$ is Perron Δ-integrable on $[s_1, s_2]_{\mathbb{T}}$, for every $s_1, s_2 \in [t_0, \infty)_{\mathbb{T}}$. Define $g : [t_0, \infty) \to \mathbb{R}$ by $g(s) = s^*$, for every $s \in [t_0, \infty)$. Let $J \subset [t_0, \infty)$ be a nondegenerate interval such that $J \cap \mathbb{T}$ is nonempty and, for each $t \in J$, $t^* \in J \cap \mathbb{T}$. If $x : J \cap \mathbb{T} \to X$ is a solution of the dynamic equation on time scales*

$$\begin{cases} x(t) = x(\tau_0) + \displaystyle\int_{\tau_0}^{t} f(x^*(s), s)\Delta s, & t \in J \cap \mathbb{T}, \\ x(\tau_0) = x_0, \end{cases} \tag{5.33}$$

where $x_0 \in O$ *and* $\tau_0 \in J \cap \mathbb{T}$, *then* $x^* : J \to X$ *is a solution of the MDE*

$$y(t) = x_0 + \int_{\tau_0}^{t} f^*(y(s), s)dg(s) = x_0 + \int_{\tau_0}^{t} f(y(s), s^*)dg(s). \tag{5.34}$$

Conversely, if $y : J \to X$ *satisfies the MDE* (5.34), *then it must have the form* $y = x^*$, *where* $x : J \cap \mathbb{T} \to X$ *is a solution of the dynamic equation on time scales* (5.33).

The concepts of prolongation to the right and maximal solution in the setting of dynamic equations on time scales are described in the next two definitions.

Definition 5.29: Let $\mathbb{T}$ be a time scale such that $\sup \mathbb{T} = \infty$. Consider a solution $x : I_{\mathbb{T}} \to X$, with $I_{\mathbb{T}} \subset [t_0, \infty)_{\mathbb{T}}$, of the dynamic equation on time scale (5.32) on the interval $I_{\mathbb{T}}$, with $\tau_0 = \min I_{\mathbb{T}}$. The solution $y : J_{\mathbb{T}} \to X$, with $J_{\mathbb{T}} \subset [t_0, \infty)_{\mathbb{T}}$ and $\tau_0 = \min J_{\mathbb{T}}$, of the dynamic equation on time scales (5.32) is called a *prolongation* of x *to the right*, if $I_{\mathbb{T}} \subset J_{\mathbb{T}}$ and $x(t) = y(t)$ for all $t \in I_{\mathbb{T}}$. If $I_{\mathbb{T}} \subsetneq J_{\mathbb{T}}$, then y is called a *proper prolongation* of x *to the right*. In an analogous way, one can define a proper prolongation of a solution to the left.

Definition 5.30: Let $\mathbb{T}$ be a time scale such that $\sup \mathbb{T} = \infty$. Consider a pair $(x_0, \tau_0) \in O \times [t_0, \infty)_{\mathbb{T}}$. The solution $y : I_{\mathbb{T}} \to X$, with $I_{\mathbb{T}} \subset [t_0, \infty)_{\mathbb{T}}$ and $\tau_0 = \min I_{\mathbb{T}}$, of the dynamic equation on time scales

$$\begin{cases} x(t) = x(\tau_0) + \int_{\tau_0}^{t} f(x^*(s), s)\Delta s, \\ x(\tau_0) = x_0, \end{cases}$$

is called *maximal solution*, if there is no proper prolongation of y to the right. If, in addition, $\sup I = \infty$, then y is called a global forward solution on I.

Before proving the first result on prolongation of solutions of dynamic equations on time scales, we present a lemma that relates conditions (A2)–(A4) from the Subsection 5.2.1 to conditions (B1)–(B3) from the beginning of this subsection. The lemma implies that, if conditions (B1)–(B3) are fulfilled for f, M, and L, then so do conditions (A2)–(A4) for f^*, M^*, and L^*. It also generalizes Lemma 3.32 to unbounded intervals and functions without delays. Moreover, it is essential in the remainder of this section. We omit its proof here in order not to be repetitive, because it is essentially the proof of Lemma 3.32. But the reader may find a proof in [78, Theorem 5.8].

Lemma 5.31: *Let* $\mathbb{T}$ *be a time scale such that* $\sup \mathbb{T} = \infty$ *and* $t_0 \in \mathbb{T}$, *and* $f : O \times [t_0, \infty)_{\mathbb{T}} \to X$ *be a function. Define* $g(t) = t^*$ *and* $f^*(y, t) = f(y, t^*)$ *for every* $y \in O$ *and* $t \in [t_0, \infty)$. *Then, the following assertions hold.*

(i) *If $f : O \times [t_0, \infty)_{\mathbb{T}} \to X$ satisfies condition (B1), then the Perron–Stieltjes integral $\int_{t_1}^{t_2} f^*(x(s), s)dg(s)$ exists, for all $x \in G([t_0, \infty), O)$ and for all $t_1, t_2 \in [t_0, \infty)$.*

(ii) *If $f : O \times [t_0, \infty)_{\mathbb{T}} \to X$ satisfies condition (B2), then $f^* : O \times [t_0, \infty) \to X$ satisfies*

$$\left\| \int_{t_1}^{t_2} f^*(x(s), s)dg(s) \right\| \leqslant \int_{t_1}^{t_2} M^*(s)dg(s),$$

for all $t_1, t_2 \in [t_0, \infty)$, $t_1 \leqslant t_2$, and for all $x \in G([t_0, \infty), O)$, where $g(t) = t^*$ and $M^* : [t_0, \infty) \to \mathbb{R}$ is given by $M^*(s) = M(s^*)$.

(iii) *If $f : O \times [t_0, \infty)_{\mathbb{T}} \to X$ satisfies condition (B3), then $f^* : O \times [t_0, \infty) \to X$ satisfies*

$$\left\| \int_{t_1}^{t_2} [f^*(x(s), s) - f^*(z(s), s)]\, dg(s) \right\| \leqslant \|x - z\|_{[t_0,\infty)} \int_{t_1}^{t_2} L^*(s)dg(s),$$

for all $t_1, t_2 \in [t_0, \infty)$, $t_1 \leqslant t_2$, and for all $x, z \in G_0([t_0, \infty), O)$, where $g(t) = t^*$ and $L^* : [t_0, \infty) \to \mathbb{R}$ is given by $L^*(s) = L(s^*)$.

Theorem 5.32: *Consider a time scale $\mathbb{T}$ with $\sup \mathbb{T} = \infty$ and $t_0 \in \mathbb{T}$. Suppose $f : O \times [t_0, \infty)_{\mathbb{T}} \to X$ satisfies conditions (B1)–(B3) and assume that, for all $(z_0, s_0) \in O \times [t_0, \infty)_{\mathbb{T}}$, $z_0 + f(z_0, s_0)\mu(s_0) \in O$ (where, as usual, $\mu : \mathbb{T} \to [0, \infty)$ is the graininess function). Then, for all $(x_0, \tau_0) \in O \times [t_0, \infty)_{\mathbb{T}}$, there exists a unique maximal solution $x : [\tau_0, \omega)_{\mathbb{T}} \to X$, $\omega \leqslant \infty$, of the dynamic equation on time scales given by*

$$\begin{cases} x(t) = x(\tau_0) + \displaystyle\int_{\tau_0}^{t} f(x^*(s), s)\Delta s, \\ x(\tau_0) = x_0. \end{cases} \tag{5.35}$$

When $\omega < \infty$, we have $\omega \in \mathbb{T}$ and ω is left-dense.

Proof. Fix a pair $(x_0, \tau_0) \in O \times [t_0, \infty)_{\mathbb{T}}$.

Existence. Define $f^* : O \times [t_0, \infty) \to X$ by $f^*(x, t) = f(x, t^*)$, for all $(x, t) \in O \times [t_0, \infty)$ and $g(t) = t^*$, for every $t \in [t_0, \infty)$. Because f satisfies conditions (B1)–(B3), Lemma 5.31 implies f^* satisfies conditions (A2)–(A4). By definition, g is a nondecreasing and left-continuous function on (t_0, ∞). Notice that, for every $(z_0, s_0) \in O \times [t_0, \infty)$, we have

$$z_0 + f^*(z_0, s_0)\Delta^+ g(s_0) = \begin{cases} z_0 + f(z_0, s_0)\mu(s_0) \in O, & s_0 \notin \mathbb{T}, \\ z_0 \in O, & s_0 \in \mathbb{T}. \end{cases} \tag{5.36}$$

Then, for each $(z_0, s_0) \in O \times [t_0, \infty)$, $z_0 + f^*(z_0, s_0)\Delta^+ g(s_0) \in O$. Therefore f^* and g satisfy all the hypotheses of Theorems 5.21 and 5.23 and, hence, there exists a unique maximal solution $y : [\tau_0, \omega) \to X$, $\omega \leqslant \infty$, of the MDE

$$y(t) = x_0 + \int_{\tau_0}^{t} f^*(y(s), s)dg(s). \tag{5.37}$$

Now, we need to analyze two cases with respect to ω.

Case 1: $\omega = \infty$. According to Theorem 5.28, $y : [\tau_0, \infty) \to X$ must have the form $y = x^*$, where $x : [\tau_0, \infty)_{\mathbb{T}} \to X$ is a solution of the dynamic equation on time scales (5.35). Note that, in fact, x is a global forward solution.

Case 2: $\omega < \infty$. In order to prove that $\omega \in \mathbb{T}$ and ω is left-dense, we need to show two claims are true.

Claim 1. $\omega \in \mathbb{T}$. Suppose to the contrary, that is, $\omega \notin \mathbb{T}$ and define a set $H = \{s \in \mathbb{T} : s < \omega\}$, which is nonempty, once $\tau_0 \in H$. Since $\omega \notin \mathbb{T}$, $H = \mathbb{T} \cap (-\infty, \omega]$ and, hence, H is a closed subset of $\mathbb{R}$. Let $\beta = \sup H$. Since H is closed, $\beta \in H$. In addition, $\beta \leqslant \omega$. However, $\omega \notin \mathbb{T}$. Therefore, $\beta < \omega$. Because g is constant on $(\beta, \omega]$, we have $\int_\tau^t f(y(s), s)dg(s) = 0$, for all $\tau, t \in (\beta, \omega]$. Let $\eta \in (\beta, \omega)$ be fixed and define a function $u : [\tau_0, \omega] \to X$ by

$$u(t) = \begin{cases} y(t), & t \in [\tau_0, \omega), \\ y(\eta), & t = \omega. \end{cases}$$

Note that u is well-defined and $u|_{[\tau_0,\omega)} = y$. We want to prove that u is a solution of the MDE (5.37) on $[\tau_0, \omega]$. Let $s_1, s_2 \in [\tau_0, \omega]$ be such that $s_1 \in [\tau_0, \omega)$ and $s_2 = \omega$. Thus,

$$\begin{aligned} u(s_2) - u(s_1) &= \int_{s_1}^{\eta} f^*(y(s), s)dg(s) + \underbrace{\int_{\eta}^{\omega} f^*(y(s), s)dg(s)}_{=0} \\ &= \int_{s_1}^{\omega} f^*(y(s), s)dg(s) = \int_{s_1}^{s_2} f^*(u(s), s)dg(s). \end{aligned}$$

The case where $s_1, s_2 \in [\tau_0, \omega)$ follows from the definition of u. Thus, u is a solution of the MDE (5.37) on $[\tau_0, \omega]$. Note that u is a proper right proper prolongation of $y : [\tau_0, \omega) \to X$, contradicting the fact that y is the maximal solution of the MDE (5.37). Thus, $\omega \in \mathbb{T}$ and *Claim 1* holds.

Claim 2. ω is left-dense. We need to prove that $\rho(\omega) = \omega$. Assume that $\rho(\omega) < \omega$. Then, g is constant on $(\rho(\omega), \omega]$ and, hence, as in *Claim 1* (using $\beta = \rho(\omega)$), one can conclude that there exists a proper right prolongation of $y : [\tau_0, \omega) \to X$, contradicting the fact that y is the maximal solution of the MDE (5.37). Therefore, ω is left-dense and *Claim 2* holds.

By Theorem 5.28, $y : [\tau_0, \omega) \to X$ is of the form $y = x^*$, where $x : [\tau_0, \omega)_{\mathbb{T}} \to X$ is a solution of the dynamic equation on time scales (5.35). We point out that $x : [\tau_0, \omega)_{\mathbb{T}} \to X$ is a maximal solution of the dynamic equation on time scales (5.35). Indeed, assume that $z : J_{\mathbb{T}} \to X$ is a proper right prolongation of $x : [\tau_0, \omega)_{\mathbb{T}} \to X$.

Without loss of generality we can take $J_{\mathbb{T}} = [\tau_0, \omega]_{\mathbb{T}}$. Because $z : [\tau_0, \omega]_{\mathbb{T}} \to X$ is a solution of the dynamic equation on time scales (5.35), Theorem 5.28 implies that $z^* : [\tau_0, \omega] \to X$ is solution of MDE (5.37).

Notice that $z^*|_{[\tau_0,\omega)} = y$. Therefore, $z^* : [\tau_0, \omega] \to X$ is a proper right prolongation of $y : [\tau_0, \omega) \to X$, in contradiction to the fact that y is the maximal solution of the MDE (5.37). Therefore, $x : [\tau_0, \omega)_{\mathbb{T}} \to X$ is a maximal solution of the dynamic equation on time scales (5.35).

Uniqueness. Assume that $\psi : [\tau_0, \widetilde{\omega})_{\mathbb{T}} \to X$ is also a maximal solution of the dynamic equation on time scales (5.35). We need to prove that

$$x(t) = \psi(t), \quad \text{for all } t \in [\tau_0, \omega)_{\mathbb{T}} \cap [\tau_0, \widetilde{\omega})_{\mathbb{T}}. \tag{5.38}$$

By Theorem 5.28, $\psi^* : [\tau_0, \widetilde{\omega}) \to X$ is a solution of the MDE (5.37). On the other hand, $y : [\tau_0, \omega) \to X$ is the unique maximal solution of (5.37). Therefore, $y(t) = \psi^*(t)$, for every $t \in [\tau_0, \omega) \cap [\tau_0, \widetilde{\omega})$ and hence, $y(t) = \psi^*(t)$, for all $t \in [\tau_0, \omega)_{\mathbb{T}} \cap [\tau_0, \widetilde{\omega})_{\mathbb{T}}$. Then, for $t \in [\tau_0, \omega)_{\mathbb{T}} \cap [\tau_0, \widetilde{\omega})_{\mathbb{T}}$, $x(t) = x^*(t) = y(t) = \psi^*(t) = \psi(t)$ and (5.38) holds.

Now, define $\gamma : I_{\mathbb{T}} \to X$, with $I = [\tau_0, \omega) \cup [\tau_0, \widetilde{\omega})$, by

$$\gamma(t) = \begin{cases} x(t), & t \in [\tau_0, \omega)_{\mathbb{T}}, \\ \psi(t), & t \in [\tau_0, \widetilde{\omega})_{\mathbb{T}}. \end{cases}$$

According to (5.38), γ is well-defined. Notice that γ is a solution of (5.35) and $\gamma|_{[\tau_0,\omega)_{\mathbb{T}}} = x$, $\gamma|_{[\tau_0,\widetilde{\omega})_{\mathbb{T}}} = \psi$. Because x and ψ are maximal solutions of (5.35), we have $I_{\mathbb{T}} = [\tau_0, \omega)_{\mathbb{T}} = [\tau_0, \widetilde{\omega})_{\mathbb{T}}$ and, hence, $\gamma(t) = x(t) = \psi(t)$, for all $t \in I_{\mathbb{T}}$, that is, $x = \psi$, proving the uniqueness of x. □

The next corollary follows straightforwardly from Theorem 5.32.

Corollary 5.33: *Consider a time scale $\mathbb{T}$ for which* $\sup \mathbb{T} = \infty$ *and $t_0 \in \mathbb{T}$. Assume that $f : O \times [t_0, \infty)_{\mathbb{T}} \to X$ satisfies conditions* (B1)–(B3). *Assume, further, that for all $(z_0, s_0) \in O \times [t_0, \infty)_{\mathbb{T}}$, we have $z_0 + f(z_0, s_0)\mu(s_0) \in O$. Let $(x_0, \tau_0) \in O \times [t_0, \infty)_{\mathbb{T}}$ and $x : [\tau_0, \omega)_{\mathbb{T}} \to X$ be the maximal solution of the dynamic equation on time scales* (5.35) *(which is ensured by Theorem 5.32). If each point of $\mathbb{T}$ is left-scattered, then $\omega = \infty$ and we have a global forward solution of Eq.* (5.35).

Remark 5.34: In the proof of Theorem 5.32, $y = x^* : [\tau_0, \omega) \to X$ is the maximal solution of the MDE (5.37) if and only if $x : [\tau_0, \omega)_{\mathbb{T}} \to X$ is the maximal solution of the dynamic equation on time scales (5.35), where $(x_0, \tau_0) \in O \times [t_0, \infty)_{\mathbb{T}}$.

The theorem below shows that the graph of the maximal solution of the dynamic equation on time scales (5.35) fulfills the following property: for every compact $K \subset O \times [t_0, \infty)_{\mathbb{T}}$, there exists t_K such that if $t \in (t_K, \omega)_{\mathbb{T}}$, then $(x(t), t) \notin K$.

Theorem 5.35: *Consider a time scale $\mathbb{T}$ for which* $\sup \mathbb{T} = \infty$ *and* $t_0 \in \mathbb{T}$. *Assume that* $f : O \times [t_0, \infty)_{\mathbb{T}} \to X$ *fulfills conditions* (B1)–(B3) *and, for all* $(z_0, s_0) \in O \times [t_0, \infty)_{\mathbb{T}}$, *we have* $z_0 + f(z_0, s_0)\mu(s_0) \in O$. *Take* $(x_0, \tau_0) \in O \times [t_0, \infty)_{\mathbb{T}}$ *and let* $x : [\tau_0, \omega)_{\mathbb{T}} \to X$ *be the maximal solution of the dynamic equation on time scales* (5.35). *Then, for every compact set* $K \subset O \times [t_0, \infty)_{\mathbb{T}}$, *there exists* $t_K \in [\tau_0, \omega)$ *such that* $(x(t), t) \notin K$, *for* $t \in (t_K, \omega) \cap \mathbb{T}$.

Proof. Suppose K is a compact subset of $O \times [t_0, \infty)_{\mathbb{T}}$. In particular, $K \subset O \times [t_0, \infty)$. Let $(x_0, \tau_0) \in O \times [t_0, \infty)_{\mathbb{T}}$ and $x : [\tau_0, \omega)_{\mathbb{T}} \to X$ be the maximal solution of the dynamic equation on time scales (5.35). By Remark 5.34, $x^* : [\tau_0, \omega) \to X$ is the maximal solution of the MDE

$$y(t) = x_0 + \int_{\tau_0}^{t} f^*(y(s), s)dg(s), \quad t \geqslant \tau_0,$$

where $f^* : O \times [t_0, \infty) \to X$ is given by $f^*(x, t) = f(x, t^*)$, for all $(x, t) \in O \times [t_0, \infty)$ and $g(t) = t^*$, for all $t \in [t_0, \infty)$.

Since f satisfies conditions (B1)–(B3), by Lemma 5.31, f^* satisfies conditions (A2)–(A4). By definition, the function g is nondecreasing and left-continuous on (t_0, ∞). Using the same arguments as in the proof of Theorem 5.32, one can show that $z_0 + f^*(z_0, s_0)\Delta^+ g(s_0) \in O$, for all $(z_0, s_0) \in O \times [t_0, \infty)$. Thus, f^*, g, and x^* satisfy the hypotheses of Theorem 5.24 and, hence, there exists $t_K \in [\tau_0, \omega)$ such that $(x^*(t), t) \notin K$, for all $t \in (t_K, \omega)$. In particular, $(x(t), t) = (x(t^*), t) = (x^*(t), t) \notin K$, for all $t \in (t_K, \omega) \cap \mathbb{T}$ and this completes the proof. □

Notice that the set $(t_K, \omega) \cap \mathbb{T}$ from the proof of Theorem 5.35 is nonempty. In fact, if $\omega < \infty$, then Theorem 5.32 implies that $\omega \in \mathbb{T}$ and ω is left-dense. Therefore, $(t_K, \omega) \cap \mathbb{T} \neq \emptyset$. On the other hand, if $\omega = \infty$, then $(t_K, \infty) \cap \mathbb{T} \neq \emptyset$, since $\sup \mathbb{T} = \infty$.

Theorem 5.36: *Let* $\mathbb{T}$ *be a time scale such that* $\sup \mathbb{T} = \infty$ *and* $t_0 \in \mathbb{T}$. *Suppose* $f : O \times [t_0, \infty)_{\mathbb{T}} \to X$ *satisfies conditions* (B1)–(B3). *Moreover, assume that, for all* $(z_0, s_0) \in O \times [t_0, \infty)_{\mathbb{T}}$, $z_0 + f(z_0, s_0)\mu(s_0) \in O$. *Take* $(x_0, \tau_0) \in O \times [t_0, \infty)_{\mathbb{T}}$ *and suppose* $x : [\tau_0, \omega)_{\mathbb{T}} \to X$ *is the unique maximal solution of the dynamic equation on time scales* (5.35) *with initial condition* $x(\tau_0) = x_0$. *If* $x(t) \in N$, *for all* $t \in [\tau_0, \omega)_{\mathbb{T}}$, *where N is a compact subset of O, then* $\omega = \infty$ *and we have a global forward solution.*

Proof. Let $(x_0, \tau_0) \in O \times [t_0, \infty)_{\mathbb{T}}$ and $x : [\tau_0, \omega)_{\mathbb{T}} \to X$ be the unique maximal solution of (5.35) satisfying $x(\tau_0) = x_0$. Assume, in addition, that $x(t) \in N$, for any $t \in [\tau_0, \omega)_{\mathbb{T}}$, where N is a compact subset of O. By Remark 5.34, $x^* : [\tau_0, \omega) \to X$ is the unique maximal solution of

$$y(t) = x_0 + \int_{\tau_0}^{t} f^*(y(s), s)dg(s),$$

where $f^* : O \times [t_0, \infty)_{\mathbb{T}} \to X$ is given by $f^*(x, t) = f(x, t^*)$ for all $(x, t) \in O \times [t_0, \infty)$ and $g(t) = t^*$ for all $t \in [t_0, \infty)$.

Note that $x^*(s) \in N$, for all $s \in [\tau_0, \omega)$, since $x(t) \in N$, for all $t \in [\tau_0, \infty)_{\mathbb{T}}$. On the other hand, one can prove, similarly as in the proof of Theorem 5.32, that f^* satisfies conditions (A2)–(A4), g is a nondecreasing and left-continuous function on (τ_0, ∞) and, moreover, $z_0 + f^*(z_0, \tau_0)\Delta^+ g(\tau_0) \in O$, for all $(z_0, \tau_0) \in O \times [t_0, \infty)$. Thus, the functions $f^* : O \times [t_0, \infty) \to X$, $g : [t_0, \infty) \to \mathbb{R}$ and $x^* : [\tau_0, \omega) \to X$ satisfy all hypotheses of Corollary 5.25. Then $\omega = \infty$ and the desired result follows. □

Next, we present a result that guarantees that the graph of the maximal solution of the dynamic equation on time scales (5.35) converges to a point of the boundary of $\Omega_{\mathbb{T}}$, as $t \to \omega^-$.

Theorem 5.37: *Consider a time scale $\mathbb{T}$ for which $\sup \mathbb{T} = \infty$ and $t_0 \in \mathbb{T}$. Suppose $f : O \times [t_0, \infty)_{\mathbb{T}} \to X$ satisfies conditions (B1)–(B3) and, for all $(z_0, s_0) \in O \times [t_0, \infty)_{\mathbb{T}}$, we have $z_0 + f(z_0, s_0)\mu(s_0) \in O$. Take $(x_0, \tau_0) \in O \times [t_0, \infty)_{\mathbb{T}}$ and let $x : [\tau_0, \omega)_{\mathbb{T}} \to X$ be the maximal solution of the dynamic equation on time scales (5.35). If $\omega < \infty$. Then, the following statements hold:*

(i) *the limit $\lim_{t\to\omega^-} x(t)$ exists;*
(ii) *$(y, \omega) \in \partial\Omega_{\mathbb{T}}$ and $\lim_{t\to\omega^-}(x(t), t) = (y, \omega)$, where $y = \lim_{t\to\omega^-} x(t)$ and $\Omega_{\mathbb{T}} = O \times [t_0, \infty)_{\mathbb{T}}$.*

Proof. At first, let $\Omega_{\mathbb{T}} = O \times [t_0, \infty)_{\mathbb{T}}$ and $x : [\tau_0, \omega)_{\mathbb{T}} \to X$ be the maximal solution of the dynamic equation on time scales (5.35). Suppose, further, $\omega < \infty$.

Let $f^* : O \times [t_0, \infty) \to X$ be defined by $f^*(x, t) = f(x, t^*)$, for all $(x, t) \in O \times [t_0, \infty)$ and $g : [t_0, \infty) \to \mathbb{R}$ be given by $g(t) = t^*$, for all $t \in [t_0, \infty)$. By the same arguments as those used in the proof of Theorem 5.32, one can show that f^* fulfills conditions (A2)–(A4) and g fulfills condition (A1). In addition,

$$z_0 + f^*(z_0, s_0)\Delta^+ g(s_0) \in O, \quad \text{for all } (z_0, s_0) \in O \times [t_0, \infty).$$

On the other hand, Remark 5.34 yields $x^* : [\tau_0, \omega) \to X$ is the maximal solution of

$$y(t) = x_0 + \int_{\tau_0}^{t} f^*(y(s), s)dg(s), \quad t \geqslant \tau_0.$$

Once all hypotheses of Corollary 5.26 are fulfilled, the limit $\lim_{t\to\omega^-} x^*(t)$ exists and, moreover,

$$(y, \omega) \in \partial\Omega \quad \text{and} \quad \lim_{t\to\omega^-}(x^*(t), t) = (y, \omega), \tag{5.39}$$

where $y = \lim_{t\to\omega^-} x^*(t)$ and $\Omega = O \times [t_0, \infty)$. By Theorem 5.32, $\omega \in \mathbb{T}$ and ω is left-dense.

Now, we prove item (i). Let $\{t_k\}_{k\in\mathbb{N}}$ be a sequence in $[\tau_0, \omega)_{\mathbb{T}}$ such that $\lim_{n\to\infty} t_k = \omega$. Then,

$$x(t_k) = x(t_k^*) = \lim_{k\to\infty} x^*(t_k) = y,$$

that is, the limit $\lim_{t\to\omega^-} x(t)$ exists and $\lim_{t\to\omega^-} x(t) = y$.

Now, let us prove item (ii). Since ω is left-dense, there exists a sequence $\{s_k\}_{k\in\mathbb{N}}$ in $[\tau_0, \omega)_{\mathbb{T}}$ satisfying $\lim_{k\to\infty} s_k = \omega$. Because $\{(x(s_k), s_k)\}_{k\in\mathbb{N}} \in \Omega_{\mathbb{T}}$ and $\lim_{k\to\infty}(x(s_k), s_k) = (y, \omega)$, we obtain

$$(y, \omega) \in \overline{\Omega_{\mathbb{T}}}. \tag{5.40}$$

On the other hand, by (5.39), $(y, \omega) \in \overline{\Omega^c}$. But $\overline{\Omega^c} \subset \overline{\Omega^c_{\mathbb{T}}}$, because $\Omega_{\mathbb{T}} \subset \Omega$. Thus,

$$(y, \omega) \in \overline{\Omega^c_{\mathbb{T}}}. \tag{5.41}$$

At last, item (ii) follows from (5.40) and (5.41). □

When $O = X$, it is possible ensure that the domain of the maximal solution of the dynamic equation on time scales (5.35) equals $[\tau_0, \infty)_{\mathbb{T}}$. This is the content of the next theorem.

Theorem 5.38: *Let $\mathbb{T}$ be a time scale such that* $\sup \mathbb{T} = \infty$ *and $t_0 \in \mathbb{T}$. Suppose $f : X \times [t_0, \infty)_{\mathbb{T}} \to X$ satisfies conditions* (B1)–(B3) *(with $O = X$). Then, for every $(x_0, \tau_0) \in X \times [t_0, \infty)_{\mathbb{T}}$, there exists a unique maximal solution on $[\tau_0, \infty)_{\mathbb{T}}$ of the dynamic equation on time scales* (5.35) *fulfilling $x(\tau_0) = x_0$.*

Proof. Consider a given fixed pair $(x_0, \tau_0) \in X \times [t_0, \infty)_{\mathbb{T}}$. Thus, $(x_0, \tau_0) \in X \times [t_0, \infty)$. Define $f^* : X \times [t_0, \infty) \to X$ by $f^*(x, t) = f(x, t^*)$, for all $(x, t) \in X \times [t_0, \infty)$. Similarly as in the proof of Theorem 5.32, one can show that f^* satisfies conditions (A2)–(A4) and g satisfies condition (A1). Furthermore, $z_0 + f^*(z_0, s_0)\Delta^+ g(s_0) \in O$, for all $(z_0, s_0) \in O \times [t_0, \infty)$. Then, once all hypotheses of Corollary 5.27 are satisfied, there exists a unique maximal solution $y : [\tau_0, \infty) \to X$ of

$$y(t) = x_0 + \int_{\tau_0}^{t} f^*(y(s), s)dg(s).$$

According to Theorem 5.28, y must have the form $y = x^* : [\tau_0, \infty) \to X$, where $x : [\tau_0, \infty)_{\mathbb{T}} \to X$ is a solution of (5.35), with initial condition $x(\tau_0) = x_0$. Then, Remark 5.34 yields the result. □

6

Linear Generalized Ordinary Differential Equations

Everaldo M. Bonotto[1], Rodolfo Collegari[2], Márcia Federson[3], and Miguel V. S. Frasson[1]

[1]*Departamento de Matemática Aplicada e Estatística, Instituto de Ciências Matemáticas e de Computação (ICMC), Universidade de São Paulo, São Carlos, SP, Brazil*
[2]*Faculdade de Matemática, Universidade Federal de Uberlândia, Uberlândia, MG, Brazil*
[3]*Departamento de Matemática, Instituto de Ciências Matemáticas e de Computação (ICMC), Universidade de São Paulo, São Carlos, SP, Brazil*

The investigation of linear equations in the framework of generalized ordinary differential equations (ODEs), which we refer to as linear generalized ODEs, is very important as they also are in the setting of classical ODEs. Linear equations are easier to deal with and one usually obtain a wider knowledge about them as, for instance, every linear ODE with constant coefficients can be solved and every maximal solution of such equation is defined in the entire real line. The same applies to linear generalized ODEs.

This chapter, whose main references are [45, 209–211], deals with linear generalized ODEs presenting an appropriate environment where any initial value problem (IVP) for linear generalized ODEs admits a unique global solution. We also recall the notion of fundamental operator associated to a linear generalized ODE and we establish a variation-of-constants formula for linear perturbed generalized ODEs, extending the result from [45]. Lastly, we apply the results to measure functional differential equations (as before, we write measure FDEs, for short).

Let X be a Banach space and $L(X)$ be the Banach space of bounded linear operators from X into itself and equip these spaces with the usual operator norm. As described in Chapter 1, $BV([a,b],X)$ denotes the Banach space of functions $x : [a,b] \to X$ of bounded variation endowed with the variation norm.

Generalized Ordinary Differential Equations in Abstract Spaces and Applications, First Edition.
Edited by Everaldo M. Bonotto, Márcia Federson, and Jaqueline G. Mesquita.

Given a compact interval $[a, b] \subset \mathbb{R}$ and a function $F : X \times [a, b] \to X$, a generalized ODE

$$\frac{dx}{d\tau} = DF(x, t),$$

is called a *linear generalized ODE*, if there exist functions $A : [a, b] \to L(X)$ and $B : [a, b] \to X$ such that $F(x, t) = A(t)x + b(t)$. When $B \equiv 0$, the generalized ODE

$$\frac{dx}{d\tau} = D[A(t)x] \tag{6.1}$$

is called *linear homogeneous generalized ODE* and, in case $B \not\equiv 0$, the generalized ODE

$$\frac{dx}{d\tau} = D[A(t)x + B(t)] \tag{6.2}$$

is called *linear nonhomogeneous generalized ODE*.

Recall that, according to Definition 4.1, a function $x : [a, b] \to X$ is a *solution* of (6.2) on $[a, b]$, whenever

$$x(s_2) - x(s_1) = \int_{s_1}^{s_2} D[A(t)x(\tau)] + B(s_2) - B(s_1), \quad s_1, s_2 \in [a, b], \tag{6.3}$$

where the integral is in the sense of Kurzweil as in Definition 2.1.

Note that the integral term on the right-hand side of the last equality coincides with the abstract Perron–Stieltjes integral, since the Riemann-type sums for $\int_a^b D[A(t)x(\tau)]$ have the form $\sum [A(s_i) - A(s_{i-1})]x(\tau_i)$. Then we shall use a more conventional notation of the integral, namely, $\int_a^b d[A(t)]x(t)$, and, hence, (6.3) becomes

$$x(s_2) - x(s_1) = \int_{s_1}^{s_2} d[A(s)]x(s) + B(s_2) - B(s_1), \quad s_1, s_2 \in [a, b].$$

Proposition 6.1: *Suppose* $A \in BV([a, b], L(X))$ *and* $B \in BV([a, b], X)$. *If* $x : [a, b] \to X$ *is a solution of* (6.2) *on* $[a, b]$, *then* $x \in BV([a, b], X)$.

A proof of Proposition 6.1 follows the steps of [209, Lemma 6.1] with obvious adaptations for the Banach space.

Next we present a result on the existence and uniqueness of a global solution for an initial value problem (we write IVP, for short) for the linear generalized ODE (6.1). In order to do that, we introduce some conditions through the next definition.

Definition 6.2: Let $J \subset \mathbb{R}$ be an interval, $A : J \to L(X)$ be a function, $\Delta^+ A(t) = \lim_{s \to t^+} A(s) - A(t)$, $\Delta^- A(t) = A(t) - \lim_{s \to t^-} A(s)$ and $I \in L(X)$ be the identity operator. We say that

(i) A satisfies condition (Δ^+) on J, if $[I + \Delta^+ A(t)]^{-1} \in L(X)$, for all $t \in J \setminus \{\sup J\}$.

(ii) A satisfies condition (Δ^-) on J, if $[I - \Delta^- A(t)]^{-1} \in L(X)$, for all $t \in J \setminus \{\inf J\}$.
(iii) A satisfies condition (Δ), whenever A satisfies both conditions (Δ^+) and (Δ^-).

The next theorem concerns the backward and forward solutions of a linear non-homogeneous generalized ODE.

Theorem 6.3: *Let $(t_0, x_0) \in [a,b] \times X$, $B \in BV([a,b], X)$ and $A \in BV([a,b], L(X))$. Consider the IVP*

$$\begin{cases} \dfrac{dx}{d\tau} = D[A(t)x + B(t)], \\ x(t_0) = x_0. \end{cases} \tag{6.4}$$

(i) *If A satisfies (Δ^+) on $[a, t_0]$, then the IVP (6.4) has a unique solution on $[a, t_0]$.*
(ii) *If A satisfies (Δ^-) on $[t_0, b]$, then the IVP (6.4) has a unique solution on $[t_0, b]$.*
(iii) *If A satisfies (Δ) on $[a,b]$, then the IVP (6.4) has a unique solution on $[a,b]$.*

For a proof, the reader may consult [211, Propositions 2.8, 2.9, and Theorem 2.10].

Remark 6.4: Let $J \subset \mathbb{R}$ be an interval and denote by $BV_{loc}(J, X)$ the set of all functions $F : J \to X$ of locally bounded variation, that is, F is of bounded variation in each compact interval $[a,b] \subset J$. If $A \in BV_{loc}(\mathbb{R}, L(X))$ satisfies condition (Δ) on $\mathbb{R}$ and $B \in BV_{loc}(\mathbb{R}, X)$, then Theorem 6.3 guarantees existence and uniqueness of global forward solution of the IVP (6.4), with $(t_0, x_0) \in \mathbb{R} \times X$.

Remark 6.5: If $A \in BV([a,b], L(X))$, then, by Theorem 1.4, the sets

$$\{t \in [a,b) : \|\Delta^+ A(t)\| \geqslant 1\} \quad \text{and} \quad \{t \in (a,b] : \|\Delta^- A(t)\| \geqslant 1\}$$

are finite. Therefore, there exists a finite set $F \subset [a,b]$ such that the operators $[I + \Delta^+ A(t)]$ and $[I - \Delta^- A(t)]$ are invertible for every $t \in [a,b] \setminus F$. This means that when $A \in BV([a,b], L(X))$, condition (Δ) is satisfied except on a finite subset of $[a,b]$.

6.1 The Fundamental Operator

As we look at Eq. (6.1), we can also consider the equation

$$\frac{dz}{d\tau} = D[A(t)z], \tag{6.5}$$

with $z \in L(X)$. A solution of (6.5) on the interval $[a, b]$ is an operator-valued function $z : [a, b] \to L(X)$ such that

$$z(s_2) - z(s_1) = \int_{s_1}^{s_2} d[A(s)]z(s), \quad s_1, s_2 \in [a, b].$$

Thus, if we fix $(t_0, z_0) \in [a, b] \times L(X)$ and we assume that $A \in BV([a, b], L(X))$ satisfies condition (Δ), then Theorem 6.3 implies that the IVP

$$\begin{cases} \dfrac{dz}{d\tau} = D[A(t)z], \\ z(t_0) = z_0 \end{cases} \tag{6.6}$$

has a unique solution $\Phi : [a, b] \to L(X)$. Furthermore, $\Phi : [a, b] \to L(X)$ is a solution of the IVP (6.6), if and only if it satisfies the integral equation

$$\Phi(t) = z_0 + \int_{t_0}^{t} d[A(s)]\Phi(s), \quad t \in [a, b]. \tag{6.7}$$

This leads us to the following result. See [45, Theorem 4.3].

Theorem 6.6: *Assume that $A \in BV([a, b], L(X))$ satisfies condition (Δ). Then there exists a uniquely determined operator $U : [a, b] \times [a, b] \to L(X)$, called fundamental operator of the linear generalized ODE (6.1), such that*

$$U(t, s) = I + \int_{s}^{t} d[A(r)]U(r, s), \quad t, s \in [a, b]. \tag{6.8}$$

Note that, if $U : [a, b] \times [a, b] \to L(X)$ is the fundamental operator given by (6.8), then, for each $s \in [a, b]$, the operator $U(\cdot, s) : [a, b] \to L(X)$ is the unique solution of the IVP

$$\begin{cases} \dfrac{dz}{d\tau} = D[A(t)z], \\ z(s) = I. \end{cases}$$

Moreover, $U(\cdot, s) : [a, b] \to L(X)$ is of bounded variation in $[a, b]$, for each $s \in [a, b]$.

As a consequence, we have the following straightforward result.

Proposition 6.7: *Let $\tilde{x} \in X$, $s \in [a, b]$ and suppose $A \in BV([a, b], L(X))$ satisfies condition (Δ). Then the unique solution $x : [a, b] \to X$ of the IVP*

$$\begin{cases} \dfrac{dx}{d\tau} = D[A(t)x], \\ x(s) = \tilde{x} \end{cases}$$

is given by $x(t) = U(t,s)\widetilde{x}$, *for all* $t \in [a,b]$, *where* $U : [a,b] \times [a,b] \to L(X)$ *is given by* (6.8).

The next proposition gives us some basic properties of the fundamental operator. The proof follows the steps of [209, Theorem 6.15] with obvious adaptations for the Banach space-valued case.

Proposition 6.8: *Suppose* $A \in BV([a,b], L(X))$ *satisfies condition* (Δ). *Then the operator* $U : [a,b] \times [a,b] \to L(X)$ *given by* (6.8) *satisfies the following properties:*

(i) $U(t,t) = I$, *for all* $t \in [a,b]$;
(ii) *There exists a constant* $M > 0$ *such that, for all* $t, s \in [a,b]$, *we have*

$$\|U(t,s)\| \leqslant M, \quad \mathrm{var}_a^b U(t,\cdot) \leqslant M, \quad \text{and} \quad \mathrm{var}_a^b U(\cdot,s) \leqslant M;$$

(iii) $U(t,s) = U(t,r)U(r,s)$, *for all* $t, r, s \in [a,b]$;
(iv) *There exists* $[U(t,s)]^{-1} \in L(X)$ *and* $[U(t,s)]^{-1} = U(s,t)$, *for all* $t, s \in [a,b]$;
(v) *We have, for all* $t, s \in [a,b]$,

$$U(t^+,s) = [I + \Delta^+ A(t)]U(t,s) \text{ and } U(t^-,s) = [I - \Delta^- A(t)]U(t,s),$$
$$U(t,s^+) = U(t,s)[I + \Delta^+ A(s)]^{-1} \text{ and } U(t,s^-) = U(t,s)[I - \Delta^- A(s)]^{-1}.$$

6.2 A Variation-of-Constants Formula

In this section, we present a variation-of-constants formula for a *linear perturbed generalized ODE* of type

$$\frac{dx}{d\tau} = D[A(t)x + F(x,t)],$$

and in order to do that, we need some auxiliary results, which we present in the sequel.

The next three lemmas guarantee that the integrals involved in the variation-of-constants formula are well defined. The first lemma can be found in [45, Lemma 4.5].

Lemma 6.9: *Suppose* $A \in BV([a,b], L(X))$ *satisfies condition* (Δ) *and let* $U : [a,b] \times [a,b] \to L(X)$ *be given by* (6.8). *Then, for* $\varphi \in G([a,b], X)$, *the function* $\widehat{\varphi} : [a,b] \to X$ *given by* $\widehat{\varphi}(r) = \int_a^r d_s[U(r,s)]\varphi(s)$ *is regulated.*

Proof. Fix $\varphi \in G([a,b], X)$, $r \in [a,b)$ and let $\{r_n\}_{n\in\mathbb{N}} \subset [r,b]$ be a decreasing sequence that converges to r. Note that the integral $\int_a^r d_s[U(r^+,s)]\varphi(s)$ makes sense since, by Proposition 6.8 and the Theorem of Helly (Theorem 1.34), $U(r^+,\cdot) \in BV([a,b], L(X))$.

Defining $\widetilde{U} : [a,b] \times [a,b] \to L(X)$ by

$$\widetilde{U}(\sigma,s) = \begin{cases} U(\sigma,s), & a \leqslant s \leqslant \sigma \leqslant b \\ I, & a \leqslant \sigma < s \leqslant b \end{cases}$$

we obtain

$$\begin{aligned}\widehat{\varphi}(r) &= \int_a^r d_s[U(r,s)]\,\varphi(s) = \int_a^r d_s[U(r,s)]\,\varphi(s) + \int_r^b d_s[I]\,\varphi(s) \\ &= \int_a^b d_s[\widetilde{U}(r,s)]\,\varphi(s)\end{aligned}$$

and, in order to prove that $\widehat{\varphi}$ is regulated, it is enough to show that $\widetilde{U}(r_n, s) \to \widetilde{U}(r^+, s)$, uniformly on $[a,b]$, as $n \to \infty$.

Let $M > 0$ be such that $\|U(r,s)\| \leqslant M$, for all $s \in [a,b]$. By Proposition 6.8, we have

$$\begin{aligned}\sup_{s\in[a,b]} \left\|\widetilde{U}(r_n,s) - \widetilde{U}(r^+,s)\right\| &= \sup_{s\in[a,r]} \|U(r_n,s) - U(r^+,s)\| \\ &= \sup_{s\in[a,r]} \|U(r_n,r)U(r,s) - [I + \Delta^+A(r)]U(r,s)\| \\ &= \sup_{s\in[a,r]} \|U(r_n,r) - I - \Delta^+A(r)\|\|U(r,s)\| \\ &\leqslant M\left\|\int_r^{r_n} d[A(\tau)]U(\tau,r) - \Delta^+A(r)\right\| \to 0,\end{aligned}$$

as $n \to \infty$. Using the Uniform Convergence Theorem (see Theorem 1.51), we obtain

$$\lim_{n\to\infty} \int_a^b d_s\left[\widetilde{U}(r_n,s) - \widetilde{U}(r^+,s)\right]\,\varphi(s) = 0$$

which concludes the existence of the right-hand side limit $\lim_{s\to r^+}\widehat{\varphi}(s)$ for all $r \in [a,b)$. Analogously, we prove that the left-hand side limit $\lim_{s\to r^-}\widehat{\varphi}(s)$ exists for all $r \in (a,b]$. □

As a consequence of Theorem 2.9, we have the following useful result.

Lemma 6.10: *Let $K \in BV([a,b], L(X))$ and $x_0 \in X$. Then*

$$\int_a^b d[K(r)]\chi_{\{a\}}(r)x_0 = [K(a^+) - K(a)]x_0 \quad and$$

$$\int_a^b d[K(r)]\chi_{\{b\}}(r)x_0 = [K(b) - K(b^-)]x_0.$$

Moreover, for every $c \in (a,b)$ and every $(\alpha,\beta) \subset [a,b]$, we have

$$\int_a^b d[K(r)]\chi_{\{c\}}(r)x_0 = [K(c^+) - K(c^-)]x_0 \quad and$$

$$\int_a^b d[K(r)]\chi_{(\alpha,\beta)}(r)x_0 = [K(\beta^-) - K(\alpha^+)]x_0.$$

Next result can be found in [45, Lemma 4.6].

Lemma 6.11: *Let $A \in BV([a,b], L(X))$ satisfy condition (Δ), $K \in BV([a,b], L(X))$ and $U : [a,b] \times [a,b] \to L(X)$ be given by (6.8). Then the function $\hat{U} : [a,b] \to L(X)$ defined by*

$$\hat{U}(s) = \int_s^b d[K(r)]U(r,s)$$

is of bounded variation in $[a,b]$. Moreover, for every $c \in (a,b]$, we have

$$\lim_{s \to c^-} \hat{U}(s) = [K(c) - K(c^-)]U(c,c^-) + \int_c^b d[K(r)]U(r,c^-), \tag{6.9}$$

and, for $c \in [a,b)$, we have

$$\lim_{s \to c^+} \hat{U}(s) = \int_c^b d[K(r)]U(r,c^+) - [K(c^+) - K(c)]U(c,c^+). \tag{6.10}$$

Proof. At first, note that $\hat{U}$ is well defined, since $K, U(\cdot,s) \in BV([a,b], L(X))$ for all $s \in [a,b]$. Similarly as in the proof of Lemma 6.9, define $\tilde{U} : [a,b] \times [a,b] \to L(X)$ by

$$\tilde{U}(\sigma,s) = \begin{cases} U(\sigma,s), & a \leqslant s \leqslant \sigma \leqslant b, \\ I, & a \leqslant \sigma < s \leqslant b. \end{cases}$$

Note that, for each $\sigma \in [a,b]$, $\mathrm{var}_a^b(\tilde{U}(\sigma,\cdot)) = \mathrm{var}_a^\sigma(U(\sigma,\cdot))$. Also, for every $s \in [a,b]$,

$$\int_a^b d[K(r)]\,\tilde{U}(r,s) = \int_a^s d[K(r)]\,I + \int_s^b d[K(r)]\,U(r,s) = K(s) - K(a) + \hat{U}(s). \tag{6.11}$$

Then, if $d = (\alpha_j) \in D_{[a,b]}$, we obtain

$$\sum_{j=1}^{|d|} \left\| \hat{U}(\alpha_j) - \hat{U}(\alpha_{j-1}) \right\| \leqslant \sum_{j=1}^{|d|} \left\| \int_a^b d[K(r)](\tilde{U}(r,\alpha_j) - \tilde{U}(r,\alpha_{j-1})) \right\| + \mathrm{var}_a^b(K)$$

$$\leqslant \mathrm{var}_a^b(K) \sup_{r \in [a,b]} \sum_{j=1}^{|d|} \left\| \tilde{U}(r,\alpha_j) - \tilde{U}(r,\alpha_{j-1}) \right\| + \mathrm{var}_a^b(K)$$

$$\leqslant \operatorname{var}_a^b(K) \sup_{r\in[a,b]} \operatorname{var}_a^b\left(\widetilde{U}(r,\cdot)\right) + \operatorname{var}_a^b(K)$$
$$= \operatorname{var}_a^b(K) \sup_{r\in[a,b]} \operatorname{var}_a^b(U(r,\cdot)) + \operatorname{var}_a^b(K)$$
$$= \operatorname{var}_a^b(K) \left(\sup_{r\in[a,b]} \operatorname{var}_a^b(U(r,\cdot)) + 1 \right),$$

which implies that $\hat{U} \in BV([a,b], L(X))$.

Now, let $c \in (a,b]$ be fixed. By (6.11) and Theorem 1.51, we obtain

$$\lim_{s\to c^-} \hat{U}(s) = K(a) - K(c^-) + \lim_{s\to c^-} \int_a^b d[K(r)]\, \widetilde{U}(r,s)$$
$$= K(a) - K(c^-) + \int_a^b d[K(r)]\, \widetilde{U}(r,c^-), \tag{6.12}$$

where $\widetilde{U}(\cdot,c^-) : [a,b] \to L(X)$ is given by

$$\widetilde{U}(r,c^-) = \begin{cases} U(r,c^-), & a < c \leqslant r \leqslant b, \\ I, & a \leqslant r < c \leqslant b. \end{cases}$$

By Theorem 2.9 we have

$$\int_a^b d[K(r)]\, \widetilde{U}(r,c^-)$$
$$= \int_a^c d[K(r)]\, \widetilde{U}(r,c^-) + \int_c^b d[K(r)]\, U(r,c^-)$$
$$= \lim_{s\to c^-} \left[\int_a^s d[K(r)]I + [K(c) - K(s)]U(c,c^-) \right] + \int_c^b d[K(r)]U(r,c^-)$$
$$= K(c^-) - K(a) + [K(c) - K(c^-)]U(c,c^-) + \int_c^b d[K(r)]U(r,c^-). \tag{6.13}$$

Thus, equalities (6.12) and (6.13) yield (6.9), for every $c \in (a,b]$. In an analogous way, one can prove that (6.10) is verified for every $c \in [a,b)$. □

The following two results are very useful to obtain the variation-of-constants formula for a linear perturbed generalized ODE. They can be found [45, Lemma 4.8 and Corollary 4.9].

Proposition 6.12: *Let $A \in BV([a,b], L(X))$ satisfy condition (Δ) and consider functions $K \in BV([a,b], L(X))$ and $U : [a,b] \times [a,b] \to L(X)$ given by (6.8). Then, for $\varphi \in G([a,b], X)$, we have*

$$\int_a^b d[K(r)] \left(\int_a^r d_s[U(r,s)]\varphi(s) \right)$$
$$= \int_a^b d[K(s)]\varphi(s) + \int_a^b d_s \left[\int_s^b d_r[K(r)]U(r,s) \right] \varphi(s). \tag{6.14}$$

Proof. At first, note that Lemmas 6.9 and 6.11 guarantee that all the integrals involved in (6.14) are well defined. Let $(\alpha, \beta) \subset [a, b]$ and $x_0 \in X$ be fixed. According to Theorem 2.9 and Lemma 6.10, we obtain

$$\int_a^r d_s[U(r,s)]\chi_{(\alpha,\beta)}(s)x_0 = \begin{cases} 0, & a \leqslant r \leqslant \alpha, \\ x_0 - U(r,\alpha^+)x_0, & \alpha < r < \beta, \\ U(r,\beta^-)x_0 - U(r,\alpha^+)x_0, & \beta \leqslant r \leqslant b, \end{cases}$$

that is,

$$\int_a^r d_s[U(r,s)]\chi_{(\alpha,\beta)}(s)x_0 = \chi_{(\alpha,\beta)}(r)x_0 + \chi_{[\beta,b]}(r)U(r,\beta^-)x_0 - \chi_{(\alpha,b]}(r)U(r,\alpha^+)x_0.$$

Then, by Lemma 6.10, we have

$$\begin{aligned} &\int_a^b d[K(r)]\left(\int_a^r d_s[U(r,s)]\chi_{(\alpha,\beta)}(s)x_0\right) \\ &= \int_a^b d[K(r)]\left(\chi_{(\alpha,\beta)}(r)x_0 + \chi_{[\beta,b]}(r)U(r,\beta^-)x_0 - \chi_{(\alpha,b]}(r)U(r,\alpha^+)x_0\right) \\ &= [K(\beta^-) - K(\alpha^+)]x_0 + \int_a^b d[K(r)]\chi_{[\beta,b]}(r)U(r,\beta^-)x_0 \\ &\quad - \int_a^b d[K(r)]\chi_{(\alpha,b]}(r)U(r,\alpha^+)x_0. \end{aligned} \tag{6.15}$$

Let us calculate the integrals in (6.15). Using Lemma 6.10, we have

$$\begin{aligned} &\int_a^b d[K(r)]\chi_{[\beta,b]}(r)U(r,\beta^-)x_0 \\ &= \int_a^\beta d[K(r)]\chi_{\{\beta\}}(r)U(r,\beta^-)x_0 + \int_\beta^b d[K(r)]U(r,\beta^-)x_0 \\ &= [K(\beta) - K(\beta^-)]U(\beta,\beta^-)x_0 + \int_\beta^b d[K(r)]U(r,\beta^-)x_0 \end{aligned}$$

and also

$$\begin{aligned} &\int_a^b d[K(r)]\chi_{(\alpha,b]}(r)U(r,\alpha^+)x_0 \\ &= \int_\alpha^b d[K(r)]\chi_{(\alpha,b]}(r)U(r,\alpha^+)x_0 \\ &= \int_\alpha^b d[K(r)]U(r,\alpha^+)x_0 - \int_\alpha^b d[K(r)]\chi_{\{\alpha\}}(r)U(r,\alpha^+)x_0 \\ &= \int_\alpha^b d[K(r)]U(r,\alpha^+)x_0 - [K(\alpha^+) - K(\alpha)]U(\alpha,\alpha^+)x_0. \end{aligned}$$

Thus,

$$\begin{aligned}
&\int_a^b d[K(r)]\left(\int_a^r d_s[U(r,s)]\chi_{(\alpha,\beta)}(s)x_0\right)\\
&= [K(\beta^-)-K(\alpha^+)]x_0 + [K(\beta)-K(\beta^-)]U(\beta,\beta^-)x_0\\
&\quad + \int_\beta^b d[K(r)]U(r,\beta^-)x_0\\
&\quad - \left(\int_\alpha^b d[K(r)]U(r,\alpha^+)x_0 - [K(\alpha^+)-K(\alpha)]U(\alpha,\alpha^+)x_0\right).
\end{aligned} \tag{6.16}$$

On the other hand, by Lemma 6.10, we obtain

$$\int_a^b d[K(s)]\chi_{(\alpha,\beta)}(s)x_0 = [K(\beta^-)-K(\alpha^+)]x_0 \tag{6.17}$$

and, by Lemma 6.11, we obtain

$$\begin{aligned}
&\int_a^b d_s\left[\int_s^b d[K(r)]U(r,s)\right]\chi_{(\alpha,\beta)}(s)x_0\\
&= \lim_{s\to\beta^-}\left[\int_s^b d[K(r)]U(r,s)\right]x_0 - \lim_{s\to\alpha^+}\left[\int_s^b d[K(r)]U(r,s)\right]x_0\\
&= [K(\beta)-K(\beta^-)]U(\beta,\beta^-)x_0 + \int_\beta^b d[K(r)]\left(\lim_{s\to\beta^-}U(r,s)x_0\right)\\
&\quad - \left(\int_\alpha^b d[K(r)]U(r,\alpha^+)x_0 - [K(\alpha^+)-K(\alpha)]U(\alpha,\alpha^+)x_0\right).
\end{aligned} \tag{6.18}$$

Thus, equalities (6.16)–(6.18) yield (6.14), for every $\varphi(s) = \chi_{(\alpha,\beta)}(s)x_0$. Following analogous steps, one can prove that equality (6.14) holds, for every $\varphi(s) = \chi_{\{\gamma\}}(s)x_0$ for each $\gamma \in [a,b]$.

Finally, given $\varphi \in G([a,b],X)$, let $\{\varphi_n\}_{n\in\mathbb{N}}$ be a sequence of step functions which is uniformly convergent in $[a,b]$ to φ. Since each φ_n is a step function, equality (6.14) is satisfied for all $n \in \mathbb{N}$ and by the Uniform Convergence Theorem (Theorem 1.51), (6.14) is also fulfilled for $\varphi \in G([a,b],X)$. □

Corollary 6.13: *Let $A \in BV([a,b],L(X))$ satisfy condition (Δ) and $U : [a,b]\times[a,b] \to L(X)$ be given by (6.8). Then, for $\varphi \in G([a,b],X)$, we have*

$$\int_a^b d[A(r)]\left(\int_a^r d_s[U(r,s)]\varphi(s)\right) = \int_a^b d[A(s)]\varphi(s) + \int_a^b d_s[U(b,s)]\varphi(s).$$

Proof. The proof follows considering $K = A$ in Proposition 6.12. □

A variation-of-constants formula for linear perturbed generalized ODEs follows next.

Theorem 6.14: *Let $A \in BV([a,b], L(X))$ satisfy condition (Δ) and consider a function $F : X \times [a,b] \to X$ fulfilling $\|F(x,t_2) - F(x,t_1)\| \leqslant |h(t_2) - h(t_1)|$, for all $(x,t_2), (x,t_1) \in X \times [a,b]$ and for some nondecreasing function $h : [a,b] \to \mathbb{R}$. Let $U : [a,b] \times [a,b] \to L(X)$ be given by (6.8). Then, $x \in G([a,b], X)$ is a solution of the linear perturbed generalized ODE*

$$\begin{cases} \dfrac{dx}{d\tau} = D[A(t)x + F(x,t)], \\ x(t_0) = \widetilde{x}, \end{cases} \tag{6.19}$$

if and only if it is a solution of the integral equation

$$x(t) = U(t,t_0)\widetilde{x} + \int_{t_0}^{t} DF(x(\tau),s) - \int_{t_0}^{t} d_\sigma[U(t,\sigma)]\left(\int_{t_0}^{\sigma} DF(x(\tau),s)\right), \tag{6.20}$$

for all $t \in [a,b]$.

Proof. Let $x \in G([a,b], X)$ be a solution of the IVP (6.19) and define $y : [a,b] \to X$ by

$$y(t) = U(t,t_0)\widetilde{x} + \int_{t_0}^{t} DF(x(\tau),s) - \int_{t_0}^{t} d_\sigma[U(t,\sigma)]\left(\int_{t_0}^{\sigma} DF(x(\tau),s)\right), \quad t \in [a,b].$$

Using (6.8) and Corollary (6.13), for all $t \in [a,b]$, we have

$$\begin{aligned}
\int_{t_0}^{t} d[A(r)]y(r) &= \int_{t_0}^{t} d[A(r)]U(r,t_0)\widetilde{x} + \int_{t_0}^{t} d[A(r)] \int_{t_0}^{r} DF(x(\tau),s) \\
&\quad - \int_{t_0}^{t} d[A(r)] \int_{t_0}^{r} d_\sigma[U(r,\sigma)]\left(\int_{t_0}^{\sigma} DF(x(\tau),s)\right) \\
&= \int_{t_0}^{t} d[A(r)]U(r,t_0)\widetilde{x} + \int_{t_0}^{t} d[A(r)] \int_{t_0}^{r} DF(x(\tau),s) \\
&\quad - \int_{t_0}^{t} d[A(r)] \int_{t_0}^{r} DF(x(\tau),s) \\
&\quad - \int_{t_0}^{t} d_\sigma[U(t,\sigma)]\left(\int_{t_0}^{\sigma} DF(x(\tau),s)\right) \\
&= U(t,t_0)\widetilde{x} - \widetilde{x} - \int_{t_0}^{t} d_\sigma[U(t,\sigma)]\left(\int_{t_0}^{\sigma} DF(x(\tau),s)\right) \\
&= y(t) - \widetilde{x} - \int_{t_0}^{t} DF(x(\tau),s),
\end{aligned}$$

and, hence,

$$x(t) - y(t) = \int_{t_0}^{t} d[A(r)](x(r) - y(r)), \quad t \in [a, b].$$

Since the unique solution of the IVP

$$\begin{cases} \dfrac{dz}{d\tau} = D[A(t)z], \\ z(t_0) = 0 \end{cases}$$

is the trivial solution, $x(t) = y(t)$ for all $t \in [a, b]$, which implies that x is a solution of (6.20). On the other hand, let $x \in G([a, b], X)$ be a solution of (6.20) and define $\varphi(\sigma) = \int_{t_0}^{\sigma} DF(x(\tau), s)$ for $\sigma \in [a, b]$. Note that, by Corollary 1.48, $\varphi \in BV([a, b], X)$. Then, using Corollary 6.13, we obtain

$$\int_{t_0}^{t} d[A(r)]x(r) = U(t, t_0)\widetilde{x} - \widetilde{x} - \int_{t_0}^{t} d_\sigma[U(t, \sigma)]\varphi(\sigma) = x(t) - \widetilde{x} - \int_{t_0}^{t} DF(x(\tau), s),$$

for all $t \in [a, b]$, which means x is a solution of (6.19). □

As a consequence we have a variation-of-constants formula for a linear nonhomogeneous generalized ODE.

Corollary 6.15: *Let $A \in BV([a, b], L(X))$ satisfy condition (Δ) and consider a function $b \in BV([a, b], X)$. Let $U : [a, b] \times [a, b] \to L(X)$ be given by (6.8). Then, the unique solution $x \in BV([a, b], X)$ of the IVP*

$$\begin{cases} \dfrac{dx}{d\tau} = D[A(t)x + b(t)], \\ x(t_0) = \widetilde{x} \end{cases} \tag{6.21}$$

is given by

$$x(t) = U(t, t_0)\widetilde{x} + b(t) - b(t_0) - \int_{t_0}^{t} d_\sigma[U(t, \sigma)] \left(b(\sigma) - b(t_0)\right), \quad t \in [a, b]. \tag{6.22}$$

6.3 Linear Measure FDEs

In Chapter 4, there is a whole description of the correspondence between generalized ODEs and measure FDEs. Here, we are going to describe the correspondence between linear generalized ODEs and linear measure FDEs. The main reference for this section is [45, Section 5], for the case where $X = \mathbb{R}^n$.

Let $r, \sigma > 0$, $t_0 \in \mathbb{R}$ and X be a Banach space. Recall that, given a function $y : [t_0 - r, t_0 + \sigma] \to X$, the memory of y at a point $t \in [t_0, t_0 + \sigma]$ can be described by a function $y_t : [-r, 0] \to X$ defined by $y_t(\theta) = y(t + \theta)$, for all $\theta \in [-r, 0]$, and denote by $\mathfrak{L}(G([-r, 0], X), X)$ the space of all linear functions from $G([-r, 0], X)$ to X.

Consider the *linear measure FDE*

$$y(t) = y(t_0) + \int_{t_0}^{t} \ell(s) y_s \, dg(s), \tag{6.23}$$

and the IVP

$$\begin{cases} y(t) = \phi(0) + \int_{t_0}^{t} \ell(s) y_s \, dg(s), \\ y_{t_0} = \phi, \end{cases} \tag{6.24}$$

where $g : [t_0, t_0 + \sigma] \to \mathbb{R}$ is nondecreasing, $\ell : [t_0, t_0 + \sigma] \to \mathfrak{L}(G([-r, 0], X), X)$, $\phi \in G([-r, 0], X)$ and the integral in the right-hand side of (6.23), when it exists, is considered in the sense of Perron–Stieltjes.

Recall that, according to Definition 3.1, a function $y : [t_0 - r, t_0 + \sigma] \to X$ is a *solution* of the IVP (6.24), if $y(t) = \phi(t - t_0)$ for every $t \in [t_0 - r, t_0]$, the Perron–Stieltjes integral $\int_{t_0}^{t_0+\sigma} \ell(s) y_s \, dg(s)$ exists, and the equality (6.23) holds for every $t \in [t_0, t_0 + \sigma]$.

Next we introduce the following conditions on the function ℓ.

We say that $\ell : [t_0, t_0 + \sigma] \to \mathfrak{L}(G([-r, 0], X), X)$ satisfies condition

(L1) If the Perron–Stieltjes integral $\int_{t_0}^{t_0+\sigma} \ell(s) y_s \, dg(s)$ exists for every $y \in G([t_0 - r, t_0 + \sigma], X)$;

(L2) If there exists a Perron–Stieltjes integrable function $M : [t_0, t_0 + \sigma] \to \mathbb{R}$ such that

$$\left\| \int_{s_1}^{s_2} \ell(s) y_s \, dg(s) \right\| \leqslant \int_{s_1}^{s_2} M(s) \|y_s\|_\infty \, dg(s)$$

for all $s_1, s_2 \in [t_0, t_0 + \sigma]$ with $s_1 \leqslant s_2$ and $y \in G([t_0 - r, t_0 + \sigma], X)$.

Similarly as defined in Section 4.3, for all $y \in G([t_0 - r, t_0 + \sigma], X)$ and $t \in [t_0, t_0 + \sigma]$, define $F(y, t) : [t_0 - r, t_0 + \sigma] \to X$ by

$$F(y, t)(\vartheta) = \begin{cases} 0, & t_0 - r \leqslant \vartheta \leqslant t_0, \\ \displaystyle\int_{t_0}^{\vartheta} \ell(s) y_s \, dg(s), & t_0 \leqslant \vartheta \leqslant t \leqslant t_0 + \sigma, \\ \displaystyle\int_{t_0}^{t} \ell(s) y_s \, dg(s), & t_0 \leqslant t \leqslant \vartheta \leqslant t_0 + \sigma. \end{cases} \tag{6.25}$$

Proposition 6.16: *Suppose $\ell : [t_0, t_0 + \sigma] \to \mathfrak{L}(G([-r, 0], X), X)$ satisfies conditions (L1) and (L2) and let $F(y, t) : [t_0 - r, t_0 + \sigma] \to X$, for all $y \in G([t_0 - r, t_0 + \sigma], X)$ and all $t \in [t_0, t_0 + \sigma]$, be given by (6.25). Then, for each $t \in [t_0, t_0 + \sigma]$, the function $A(t) : G([t_0 - r, t_0 + \sigma], X) \to G([t_0 - r, t_0 + \sigma], X)$ given* *by* $[A(t)y](\vartheta) = F(y,t)(\vartheta)$ *is a bounded linear operator on* $G([t_0 - r, t_0 + \sigma], X)$. *Moreover,* $A : [t_0, t_0 + \sigma] \to L(G([t_0 - r, t_0 + \sigma], X))$ *is of bounded variation.*

Proof. For each $t \in [t_0, t_0 + \sigma]$, the linearity of both the integral and the operator $\ell(t)$ involved in (6.25) implies the linearity of $A(t)$. For $t \in [t_0, t_0 + \sigma]$ and $y \in G([t_0 - r, t_0 + \sigma], X)$, we have

$$\begin{aligned}\|A(t)y\|_\infty &= \sup_{\vartheta \in [t_0 - r, t_0 + \sigma]} \|[A(t)y](\vartheta)\| \leqslant \left\| \int_{t_0}^{t} \ell(s) y_s \, dg(s) \right\| \\ &\leqslant \left(\int_{t_0}^{t} M(s) dg(s) \right) \|y\|_\infty\end{aligned}$$

that is, $A(t)$ is a bounded linear operator on $G([t_0 - r, t_0 + \sigma], X)$. Besides, for all $y \in G([t_0 - r, t_0 + \sigma], X)$ and $t_0 \leqslant s_1 \leqslant s_2 \leqslant t_0 + \sigma$,

$$\begin{aligned}\|[A(s_2) - A(s_1)]y\|_\infty &= \sup_{\vartheta \in [t_0 - r, t_0 + \sigma]} \|[A(s_2)y](\vartheta) - [A(s_1)y](\vartheta)\| \\ &= \sup_{\vartheta \in [s_1, s_2]} \|[A(s_2)y](\vartheta) - [A(s_1)y](\vartheta)\| \\ &= \sup_{\vartheta \in [s_1, s_2]} \left\| \int_{s_1}^{\vartheta} \ell(s) y_s \, dg(s) \right\| \leqslant \left(\int_{s_1}^{s_2} M(s) dg(s) \right) \|y\|_\infty,\end{aligned}$$

which implies

$$\|A(s_2) - A(s_1)\| \leqslant \int_{s_1}^{s_2} M(s) dg(s). \tag{6.26}$$

Thus, $\mathrm{var}_{t_0}^{t_0+\sigma}(A) \leqslant \int_{t_0}^{t_0+\sigma} M(s) dg(s)$, which completes the proof. □

Consider, again, the linear generalized ODE

$$\begin{cases} \dfrac{dx}{d\tau} = D[A(t)x], \\ x(t_0) = \widetilde{x}, \end{cases} \tag{6.27}$$

where $A : [t_0, t_0 + \sigma] \to L(G([t_0 - r, t_0 + \sigma], X))$ is given by

$$[A(t)y](\vartheta) = \begin{cases} 0, & t_0 - r \leqslant \vartheta \leqslant t_0, \\ \displaystyle\int_{t_0}^{\vartheta} \ell(s) y_s \, dg(s), & t_0 \leqslant \vartheta \leqslant t \leqslant t_0 + \sigma, \\ \displaystyle\int_{t_0}^{t} \ell(s) y_s \, dg(s), & t_0 \leqslant t \leqslant \vartheta \leqslant t_0 + \sigma, \end{cases} \tag{6.28}$$

$\ell : [t_0, t_0 + \sigma] \to \mathfrak{L}(G([-r, 0], X), X)$ satisfies conditions (L1) and (L2) and $\widetilde{x} \in G([t_0 - r, t_0 + \sigma], X)$ is given by

$$\widetilde{x}(\vartheta) = \begin{cases} \phi(\vartheta - t_0), & t_0 - r \leqslant \vartheta \leqslant t_0, \\ \phi(0), & t_0 \leqslant \vartheta \leqslant t_0 + \sigma. \end{cases}$$

Note that a solution of the IVP (6.27) is a function $x : [t_0, t_0 + \sigma] \to G([t_0 - r, t_0 + \sigma], X)$ which satisfies the integral equation

$$x(t) = \widetilde{x} + \int_{t_0}^{t} d[A(s)]x(s), \quad t \in [t_0, t_0 + \sigma].$$

We refer to the IVP (6.27) as the linear generalized ODE associated to the linear measure FDE (6.24). The next auxiliary lemma follows as in Lemma 4.17.

Lemma 6.17: *If $x : [t_0, t_0 + \sigma] \to G([t_0 - r, t_0 + \sigma], X)$ is a solution of the linear generalized ODE (6.27) then, for $v \in [t_0, t_0 + \sigma]$, we have*

$$x(v)(\vartheta) = \begin{cases} x(v)(v), & v \leqslant \vartheta, \\ x(\vartheta)(\vartheta), & \vartheta \leqslant v. \end{cases}$$

The following result establishes a correspondence between a solution of the linear measure FDE (6.24) and a solution of the linear generalized ODE (6.27). The proof follows the same steps of Theorems 4.18 and 4.19. A similar result for the case of infinite delays can be found in [179, Theorems 4.4 and 4.5] when $X = \mathbb{R}^n$.

Theorem 6.18: *Suppose $\ell : [t_0, t_0 + \sigma] \to \mathfrak{L}(G([-r, 0], X), X)$ satisfies conditions (L1) and (L2).*

(i) *If $y : [t_0 - r, t_0 + \sigma] \to X$ is a solution of the linear measure FDE (6.24), then the function $x : [t_0, t_0 + \sigma] \to G([t_0 - r, t_0 + \sigma], X)$, given by*

$$x(t)(\vartheta) = \begin{cases} y(\vartheta), & \vartheta \in [t_0 - r, t], \\ y(t), & \vartheta \in [t, t_0 + \sigma], \end{cases}$$

is a solution of the linear generalized ODE (6.27).

(ii) *If $x : [t_0, t_0 + \sigma] \to G([t_0 - r, t_0 + \sigma], X)$ is a solution of the linear generalized ODE (6.27), then the function $y : [t_0 - r, t_0 + \sigma] \to X$, given by*

$$y(\vartheta) = \begin{cases} x(t_0)(\vartheta), & t_0 - r \leqslant \vartheta \leqslant t_0, \\ x(\vartheta)(\vartheta), & t_0 \leqslant \vartheta \leqslant t_0 + \sigma, \end{cases}$$

is a solution of the linear measure FDE (6.24).

We finish this section by presenting a result on the existence and uniqueness of solution of the linear generalized ODE (6.27). In order to obtain that, we require a stronger condition than (L2), since (L2) does not imply condition (Δ).

We say that $\ell : [t_0, t_0 + \sigma] \to \mathfrak{L}(G([-r, 0], X), X)$ satisfies condition

(L2*) if there exists a Perron integrable function $M : [t_0, t_0 + \sigma] \to \mathbb{R}$ such that

$$\left\| \int_{s_1}^{s_2} \ell(s) y_s \, dg(s) \right\| \leqslant \int_{s_1}^{s_2} M(s) \|y_s\|_\infty \, dg(s)$$

for all $s_1, s_2 \in [t_0, t_0 + \sigma]$, with $s_1 \leqslant s_2$ and all $y \in G([t_0 - r, t_0 + \sigma], X)$ and, for all $t \in [t_0, t_0 + \sigma]$, there exists $\xi = \xi(t) > 0$ such that

$$\int_{t_0}^{t_0+\xi} M(s) dg(s) < 1, \qquad \int_{t_0+\sigma-\xi}^{t_0+\sigma} M(s) dg(s) < 1$$

and

$$\int_{t-\xi}^{t+\xi} M(s) dg(s) < 1, \quad \text{for} \quad t \in (t_0, t_0 + \sigma).$$

Theorem 6.19: *Assume that $\ell : [t_0, t_0 + \sigma] \to \mathfrak{L}(G([-r, 0], X), X)$ satisfies conditions (L2) and (L2*). Then the linear generalized ODE (6.27) has a unique solution on $[t_0, t_0 + \sigma]$.*

Proof. Consider $t \in [t_0, t_0 + \sigma]$ and $\xi > 0$ according to (L2*). Using (6.26), we get

$$\|A(t + \xi) - A(t - \xi)\| \leqslant \int_{t-\xi}^{t+\xi} M(s) dg(s) < 1.$$

Thus, $\|\Delta^+ A(t)\|, \|\Delta^- A(t)\| < 1$ for all $t \in [t_0, t_0 + \sigma]$, which implies that (Δ) is satisfied and the result follows straightforwardly from Proposition 6.16 and Theorem 6.3. □

6.4 A Nonlinear Variation-of-Constants Formula for Measure FDEs

Let $r, \sigma > 0$, $t_0 \in \mathbb{R}$ and X be a Banach space. Consider the linear measure FDE

$$y(t) = y(t_0) + \int_{t_0}^{t} \ell(s) y_s \, dg(s), \tag{6.29}$$

the associated IVP

$$\begin{cases} y(t) = \phi(0) + \int_{t_0}^{t} \ell(s) y_s \, dg(s), \\ y_{t_0} = \phi \end{cases} \tag{6.30}$$

and the *linear perturbed measure FDE*

$$\begin{cases} y(t) = \phi(0) + \int_{t_0}^{t} \ell(s) y_s \, dg(s) + \int_{t_0}^{t} f(y_s, s) dg(s), \quad t \in [t_0, t_0 + \sigma], \\ y_{t_0} = \phi, \end{cases} \tag{6.31}$$

where $\phi \in G([-r,0],X)$, $\ell : [t_0, t_0+\sigma] \to \mathfrak{L}(G([-r,0],X),X)$, $f : G([-r,0],X) \times [t_0, t_0+\sigma] \to X$, $g : [t_0, t_0+\sigma] \to \mathbb{R}$ is nondecreasing and we consider Perron–Stieltjes integration.

Moreover, define, for $y \in G([t_0 - r, t_0+\sigma],X)$ and $t \in [t_0, t_0+\sigma]$,

$$F(y,t)(\vartheta) = \begin{cases} 0, & t_0 - r \leqslant \vartheta \leqslant t_0, \\ \displaystyle\int_{t_0}^{\vartheta} f\left(y_s, s\right) dg(s), & t_0 \leqslant \vartheta \leqslant t, \\ \displaystyle\int_{t_0}^{t} f\left(y_s, s\right) dg(s), & t \leqslant \vartheta \leqslant t_0 + \sigma. \end{cases}$$

Throughout this section we are going to assume that conditions (L2) and (L2*) are satisfied for $\ell : [t_0, t_0+\sigma] \to \mathfrak{L}(G([-r,0],X),X)$ and conditions (A)–(C) from Section 4.3 are satisfied for $f : G([-r,0],X) \times [t_0, t_0+\sigma] \to X$.

By Theorem 6.18, there is a correspondence between the linear measure FDE (6.30) and the linear generalized ODE (6.27). On the other hand, by Theorems 4.18 and 4.19, there is a correspondence between Eq. (6.31) and the following linear perturbed generalized ODE

$$\begin{cases} \dfrac{dx}{d\tau} = D[A(t)x + F(x,t)], \\ x(t_0) = \widetilde{x}, \end{cases} \tag{6.32}$$

where $A : [t_0, t_0+\sigma] \to L(G([t_0 - r, t_0+\sigma],X))$ is given by (6.28) and $\widetilde{x}$ is given by

$$\widetilde{x}(\vartheta) = \begin{cases} \phi(\vartheta - t_0), & t_0 - r \leqslant \vartheta \leqslant t_0, \\ \phi(0), & t_0 \leqslant \vartheta \leqslant t_0 + \sigma. \end{cases} \tag{6.33}$$

Next, we define the fundamental operator for the linear measure FDE (6.29). In order to do that we will also assume, throughout this section, the following condition:

(E) For all $t_0 \in \mathbb{R}$, $\phi \in G([-r,0],X)$ and $\ell : [t_0, t_0+\sigma] \to \mathfrak{L}(G([-r,0],X),X)$, the IVP

$$\begin{cases} y(t) = \phi(0) + \int_{t_0}^{t} \ell(s) y_s \, dg(s), \\ y_{t_0} = \phi, \end{cases}$$

has a unique solution on $[t_0 - r, t_0 + \sigma]$.

Definition 6.20: For each $s \in \mathbb{R}$ and each $t \in [s, s+\sigma]$, let $T(t,s) : G([-r,0],X) \to G([-r,0],X)$ be defined by $T(t,s)\phi = y_t$, where $y : [s-r, s+\sigma] \to X$ is the solution of the linear measure FDE

$$\begin{cases} y(t) = \phi(0) + \int_{s}^{t} \ell(\xi) y_\xi \, dg(\xi), \\ y_s = \phi. \end{cases}$$

The family of operators $\{T(t,s) : t \geqslant s\}$ is called *fundamental operator* of the linear measure FDE (6.29).

The next proposition gives us some basic properties of the fundamental operator introduced in Definition 6.20. The result follows straightforward from the definition of the fundamental operator and condition (E).

Proposition 6.21: *The fundamental operator* $\{T(t,s) : t \geqslant s\}$ *of the linear measure FDE* (6.29) *satisfies the following properties:*

(i) $T(t,t) = I$, *for all* $t \in \mathbb{R}$;
(ii) $T(t,s) = T(t,r)T(r,s)$, *for all* $t \geqslant r \geqslant s$;
(iii) *If* $y : [t_0 - r, t_0 + \sigma] \to X$ *is the solution of* (6.30) *then* $T(t,s)y_s = y_t$ *and* $y(t) = (T(t,s)\phi)(0)$, *for all* $t, s \in [t_0, t_0 + \sigma]$, *with* $t \geqslant s$.

In the sequel we present some auxiliary results that will be useful to obtain a variation-of-constants formula for a linear perturbed measure FDE.

Lemma 6.22: *Let* $y : [t_0 - r, t_0 + \sigma] \to X$ *and* $x : [t_0, t_0 + \sigma] \to G([t_0 - r, t_0 + \sigma], X)$ *be corresponding solutions of* (6.31) *and* (6.32), *respectively. Then* $F(y,t)(\vartheta) = \int_{t_0}^{t} DF(x(\tau), s)(\vartheta)$, *for all* $t_0 \leqslant t \leqslant t_0 + \sigma$.

A proof of Lemma 6.22 can be carried out as in [45, Lemma 6.1].

Note that, for all $y \in G([t_0 - r, t_0 + \sigma], X)$ and $t \in [t_0, t_0 + \sigma]$, $F(y,t)$ is a function from $[t_0 - r, t_0 + \sigma]$ to X and, for each $s \in [t_0, t_0 + \sigma]$, we denote $F(y,t)_s : [-r, 0] \to X$ by $F(y,t)_s(\theta) = F(y,t)(s + \theta)$.

The next two lemmas can be found in [45, Lemmas 6.4 and 6.5], respectively, for the case $X = \mathbb{R}^n$.

Lemma 6.23: *Let* $y : [t_0 - r, t_0 + \sigma] \to X$ *and* $x : [t_0, t_0 + \sigma] \to G([t_0 - r, t_0 + \sigma], X)$ *be corresponding solutions of* (6.31) *and* (6.32), *respectively. Then, for* $t_0 \leqslant s \leqslant t \leqslant t_0 + \sigma$ *and* $t_0 \leqslant w \leqslant t_0 + \sigma$, *we have*

$$U(t,s)F(y,w)(t) = \widetilde{T}(t,s)F(y,w)(0),$$

where $\widetilde{T}(t,s)F(y,w) = T(t,s)F(y,w)_s$, *for all* $t, s, w \in [t_0, t_0 + \sigma]$, *with* $t \geqslant s$, $\{T(t,s) : t \geqslant s\}$ *is the fundamental operator of the linear measure FDE* (6.23) *and* $U(t,s)$ *is the fundamental operator of the corresponding linear generalized ODE* (6.27).

Proof. Note that, by Proposition 6.21, $T(t,s)(F(y,w)_s)(0)$ describes the solution of the linear measure FDE

$$\begin{cases} y(t) = \phi(0) + \int_s^t \ell(\xi) y_\xi \, dg(\xi), \\ y_s = F(y,w)_s. \end{cases}$$

In addition, using Proposition 6.7, we have $U(t,s)(F(y,w))$, which describes the solution of the corresponding linear generalized ODE

$$\begin{cases} \dfrac{dx}{d\tau} = D[A(t)x], \\ x(s) = F(y,w). \end{cases}$$

By Theorem 6.18, $U(t,s)F(y,w)(t) = T(t,s)(F(y,w)_s)(0)$ for all $t_0 \leqslant s \leqslant t \leqslant t_0 + \sigma$ and $t_0 \leqslant w \leqslant t_0 + \sigma$ and the proof is finished. □

Lemma 6.24: *Let $y : [t_0 - r, t_0 + \sigma] \to X$ and $x : [t_0, t_0 + \sigma] \to G([t_0 - r, t_0 + \sigma], X)$ be corresponding solutions of (6.31) and (6.32), respectively. Then, for $t_0 \leqslant t \leqslant t_0 + \sigma$,*

$$\int_{t_0}^{t} d_s[U(t,s)]F(y,s)(t) = \int_{t_0}^{t} d_s[\widetilde{T}(t,s)]F(y,s)(0),$$

where $\widetilde{T}(t,s)F(y,w) = T(t,s)F(y,w)_s$, for all $t, s, w \in [t_0, t_0 + \sigma]$, with $t \geqslant s$, $\{T(t,s) : t \geqslant s\}$ is the fundamental operator of the linear measure FDE (6.23) and $U(t,s)$ is the fundamental operator of the corresponding linear generalized ODE (6.27).

Proof. Let $\epsilon > 0$ be given. Then, by the definition of the Kurzweil integral, there exists a gauge δ on $[t_0, t]$ such that, for every δ-fine tagged division $d = (\tau_i, [s_{i-1}, s_i])$ of $[t_0, t]$, we have

$$\left\| \sum_{i=1}^{|d|} [U(t,s_i) - U(t,s_{i-1})]F(y,\tau_i) - \int_{t_0}^{t} d_s[U(t,s)]F(y,s) \right\|_\infty < \epsilon.$$

By Lemma 6.23,

$$\begin{aligned} &\sum_{i=1}^{|d|} [U(t,s_i) - U(t,s_{i-1})]F(y,\tau_i)(t) \\ &\quad = \sum_{i=1}^{|d|} [T(t,s_i)(F(y,\tau_i)_{s_i}) - T(t,s_{i-1})(F(y,\tau_i)_{s_{i-1}})](0) \\ &\quad = \sum_{i=1}^{|d|} \left[\widetilde{T}(t,s_i) - \widetilde{T}(t,s_{i-1})\right] F(y,\tau_i)(0), \end{aligned}$$

which implies

$$\int_{t_0}^{t} d_s[U(t,s)]F(y,s)(t) = \int_{t_0}^{t} d_s\left[\widetilde{T}(t,s)\right] F(y,s)(0)$$

and completes the proof. □

The next result, which gives us a variation-of-constants formula for the linear perturbed measure FDE (6.31), follows from Theorem 6.14 and the correspondence of equations described in Theorem 6.18.

Theorem 6.25: *Let $y : [t_0 - r, t_0 + \sigma] \to X$ be a solution of the linear perturbed measure FDE (6.31). Then, for $t_0 \leqslant t \leqslant t_0 + \sigma$, we have*

$$y(t) = T(t, t_0)\phi(0) + \int_{t_0}^{t} f(y_s, s)dg(s) - \int_{t_0}^{t} d_s[\widetilde{T}(t, s)]F(y, s)(0),$$

where $\widetilde{T}(t, s)F(y, w) = T(t, s)F(y, w)_s$, for all $t, s, w \in [t_0, t_0 + \sigma]$, with $t \geqslant s$, $\{T(t, s) : t \geqslant s\}$ is the fundamental operator of the linear measure FDE (6.29).

Proof. Given $t \in [t_0, t_0 + \sigma]$, Theorem 6.14 implies

$$x(t)(t) = U(t, t_0)\widetilde{x}(t) + \int_{t_0}^{t} DF(x(\tau), s)(t) - \int_{t_0}^{t} d_s[U(t, s)] \left(\int_{t_0}^{s} DF(x(\tau), u) \right)(t),$$

where $U(t, s)$ is the fundamental operator of the linear generalized ODE (6.27), which corresponds to the linear measure FDE (6.29).

By Proposition 6.7, Theorem 6.18 and Proposition 6.21, we get

$$U(t, t_0)\widetilde{x}(t) = T(t, t_0)\phi(0).$$

Also, Lemmas 6.22 and 6.24 imply

$$\int_{t_0}^{t} DF(x(\tau), s)(t) = \int_{t_0}^{t} f(y_s, s)dg(s) \quad \text{and}$$

$$\int_{t_0}^{t} d_s[U(t, s)] \left(\int_{t_0}^{s} DF(x(\tau), u) \right)(t) = \int_{t_0}^{t} d_s[\widetilde{T}(t, s)]F(y, s)(0).$$

Finally, by Theorem 4.19, $y(t) = x(t)(t)$ for all $t \in [t_0, t_0 + \sigma]$ and thus

$$y(t) = T(t, t_0)\phi(0) + \int_{t_0}^{t} f(y_s, s)dg(s) - \int_{t_0}^{t} d_s[\widetilde{T}(t, s)]F(y, s)(0)$$

for all $t \in [t_0, t_0 + \sigma]$, which finishes the proof. □

7

Continuous Dependence on Parameters

Suzete M. Afonso[1], *Everaldo M. Bonotto*[2], *Márcia Federson*[3], *and Jaqueline G. Mesquita*[4]

[1] *Departamento de Matemática, Instituto de Geociências e Ciências Exatas, Universidade Estadual Paulista "Júlio de Mesquita Filho" (UNESP), Rio Claro, SP, Brazil*
[2] *Departamento de Matemática Aplicada e Estatística, Instituto de Ciências Matemáticas e de Computação (ICMC), Universidade de São Paulo, São Carlos, SP, Brazil*
[3] *Departamento de Matemática, Instituto de Ciências Matemáticas e de Computação (ICMC), Universidade de São Paulo, São Carlos, SP, Brazil*
[4] *Departamento de Matemática, Instituto de Ciências Exatas, Universidade de Brasília, Brasília, DF, Brazil*

In this chapter, our goal is to investigate results on continuous dependence on parameters for generalized ordinary differential equations (ODEs) taking values in a Banach space. Combining these results with the relations between generalized ODEs and measure functional differential equations (FDEs) described in Chapter 4, we are able to translate the obtained results to measure FDEs.

Most of the results presented in this chapter are generalizations of the those found in [209] and they were presented in [4, 85, 86, 95, 177]. However, we also include a new result, namely, Theorem 7.4, on the convergence of solutions of a nonautonomous generalized ODE. Such result ensures that, under certain conditions, if a sequence of solutions $x_k : [a, b] \to X$ of the nonautonomous generalized ODE

$$\frac{dx}{d\tau} = DF_k(x, t), \quad k \in \mathbb{N},$$

converges to a function x_0, where $[a, b]$ is a compact interval of the real line and X is an arbitrary Banach space, then this convergence is uniform and $x_0 : [a, b] \to X$ is a solution of the "*limiting equation*," where X is a Banach space. In order to obtain such a result, we assume some regularity on a sequence of right-hand sides F_k, with $k \in \mathbb{N}$.

Another interesting result we would like to highlight in this chapter is Theorem 7.7, which ensures that, under certain conditions, if $x_0 : [a, b] \to X$ is a

Generalized Ordinary Differential Equations in Abstract Spaces and Applications, First Edition.
Edited by Everaldo M. Bonotto, Márcia Federson, and Jaqueline G. Mesquita.

solution on $[a, b]$ of the nonautonomous generalized ODE given by

$$\frac{dx}{d\tau} = DF_0(x, t),$$

then there exists a sufficiently large $k \in \mathbb{N}$ such that the generalized ODE

$$\frac{dx}{d\tau} = DF_k(x, t)$$

possesses a solution $x_k : [a, b] \to X$, which converges uniformly to x_0 on $[a, b]$, as $k \to \infty$.

The previous results are stated not only for generalized ODEs, but also for measure FDEs. As we mentioned, this is done by means of the correspondence, presented in Chapter 4, between the solutions of these two types of equations. It is important to note that, whenever we translate a result for generalized ODEs to some other type of equation, we get a very general result involving nonabsolute integrable functions. This is due to the fact that, in the process of transferring the results from a nonautonomous generalized ODE to some other nonautonomous equation, the main properties of the Kurzweil integration theory are preserved and inherited by the other equation. For example, once the right-hand sides of generalized ODEs are Kurzweil integrable functions as presented in Chapter 2, the translation of results allows the right-hand sides of, say, measure FDEs to be Perron–Stieltjes integrable functions, meaning that they may be of unbounded variation and, hence, highly oscillating (see Remark 1.39).

7.1 Basic Theory for Generalized ODEs

In this section, our goal is to investigate continuous dependence results on parameters for generalized ODEs in a Banach space X, equipped with norm $\| \cdot \|$. Throughout this section, we assume $\Omega = O \times [a, b]$, where $O \subset X$ is an open set.

We start by presenting some auxiliary results.

Lemma 7.1: *Consider a sequence of functions $F_k : \Omega \to X$ such that $F_k \in \mathcal{F}(\Omega, h_k)$, where $\{h_k\}_{k\in\mathbb{N}}$ is an equiregulated sequence of nondecreasing and left-continuous functions $h_k : [a, b] \to \mathbb{R}$ such that $h_k(b) - h_k(a) \leqslant c$, for some $c > 0$, for all $k \in \mathbb{N}_0$. If $\{F_k\}_{k\in\mathbb{N}}$ converges pointwisely to F_0 on Ω, then*

$$\lim_{s\to t^+} F_k(x, s) = F_k(x, t^+)$$

uniformly, for all $x \in O$ and $k \in \mathbb{N}_0$, and

$$\lim_{k\to\infty} F_k(x, t^+) = F_0(x, t^+),$$

for all $(x, t) \in \Omega$.

Proof. Fix $t \in [a, b)$. Once $F_k \in \mathcal{F}(\Omega, h_k)$, we have, for all $(x, s), (x, t) \in \Omega$ and $k \in \mathbb{N}_0$,

$$\|F_k(x, s) - F_k(x, t)\| \leqslant |h_k(s) - h_k(t)|.$$

The fact that the sequence $\{h_k\}_{k\in\mathbb{N}}$ is equiregulated implies that for all $\epsilon > 0$, there exists $\delta > 0$ such that if $t < s < t + \delta < b$, then

$$\|F_k(x, s) - F_k(x, t)\| \leqslant |h_k(s) - h_k(t)| < \epsilon,$$

for all $x \in O$ and $k \in \mathbb{N}_0$. This means that the convergence $\lim_{s\to t^+} F_k(x, s) = F_k(x, t^+)$ is uniform for every $x \in O$ and $k \in \mathbb{N}_0$. Thus, by Moore–Osgood theorem (see [19], for instance), we get

$$\begin{aligned}\lim_{k\to\infty} F_k(x, t^+) &= \lim_{k\to\infty}\lim_{s\to t^+} F_k(x, s) = \lim_{s\to t^+}\lim_{k\to\infty} F_k(x, s)\\ &= \lim_{s\to t^+} F_0(x, s) = F_0(x, t^+),\end{aligned}$$

obtaining the desired result. □

The next result is a modified version of [2, Lemma A1] and it can be found in [177].

Theorem 7.2: *Consider a sequence of functions $F_k : \Omega \to X$ such that $F_k \in \mathcal{F}(\Omega, h_k)$ for every $k \in \mathbb{N}_0$, $\{F_k\}_{k\in\mathbb{N}}$ converges pointwisely to F_0 on Ω and*

$$\lim_{k\to\infty} F_k(x, t^+) = F_0(x, t^+), \quad t \in [a, b),$$

holds for all $x \in O$. Moreover, assume that $h_k : [a, b] \to \mathbb{R}$ is a nondecreasing and left-continuous function such that $h_k(b) - h_k(a) \leqslant c$, for some $c > 0$ and every $k \in \mathbb{N}_0$. Let $\psi_k \in G([a, b], X)$, $k \in \mathbb{N}$, be such that $\lim_{k\to\infty}\|\psi_k - \psi_0\|_\infty = 0$. Assume that $(\psi_k(s), s) \in C \times [a, b]$ for all $k \in \mathbb{N}$, where C is a closed subset in X such that $C \subset O$. Then, for every $s_1, s_2 \in [a, b]$, we have

$$\lim_{k\to\infty}\left\|\int_{s_1}^{s_2} DF_k(\psi_k(\tau), s) - \int_{s_1}^{s_2} DF_0(\psi_0(\tau), s)\right\| = 0. \tag{7.1}$$

Proof. We will prove that the estimate (7.1) holds for $s_1 = a$ and $s_2 = b$. The case $s_1, s_2 \in (a, b)$ follows similarly and, thus, we omit its proof here.

From the fact that ψ_0 is the uniform limit of regulated functions on $[a, b]$, it follows that $\psi_0 \in G([a, b], X)$, since $G([a, b], X)$ is complete. In addition, the existence of the Kurzweil integral $\int_a^b DF_k(\psi_k(\tau), s)$, for every $k \in \mathbb{N}_0$, is ensured by Corollary 4.8, because $(\psi_k(s), s) \in C \times [a, b] \subset O \times [a, b]$ for all $k \in \mathbb{N}$. Note, also, that $(\psi_0(s), s) \in C \times [a, b]$.

Let $\epsilon > 0$ be given. Since C is complete, $\psi_0 \in G([a,b], C)$. Theorem 1.4 yields there is a step function $y : [a,b] \to C$ such that

$$\|y - \psi_0\|_\infty = \sup_{a \leqslant t \leqslant b} \|y(t) - \psi_0(t)\| < \frac{\epsilon}{3c}.$$

Moreover, since $\lim_{k\to\infty} \|\psi_k - \psi_0\|_\infty = 0$, there exists $N_0 \in \mathbb{N}$ such that for $k > N_0$,

$$\|\psi_k - \psi_0\|_\infty < \frac{\epsilon}{3c},$$

whence

$$\begin{aligned}
&\left\| \int_a^b DF_k(\psi_k(\tau), s) - \int_a^b DF_0(\psi_0(\tau), s) \right\| \\
&\leqslant \left\| \int_a^b D[F_k(\psi_k(\tau), s) - F_k(\psi_0(\tau), s)] \right\| + \left\| \int_a^b D[F_k(\psi_0(\tau), s) - F_k(y(\tau), s)] \right\| \\
&+ \left\| \int_a^b D[F_k(y(\tau), s) - F_0(y(\tau), s)] \right\| + \left\| \int_a^b D[F_0(y(\tau), s) - F_0(\psi_0(\tau), s)] \right\|. \qquad (7.2)
\end{aligned}$$

Now, let us consider the first summand on the right-hand side of the inequality in (7.2). By Lemma 4.6, we have

$$\begin{aligned}
\left\| \int_a^b D[F_k(\psi_k(\tau), s) - F_k(\psi_0(\tau), s)] \right\| &\leqslant \int_a^b \|\psi_k(\tau) - \psi_0(\tau)\| dh_k(\tau) \\
&\leqslant \|\psi_k - \psi_0\|_\infty (h_k(b) - h_k(a)) < \frac{\epsilon}{3}.
\end{aligned}$$

Similarly, for the second and fourth summands on the right-hand side of (7.2), we have the following estimates

$$\left\| \int_a^b D[F_k(\psi_0(\tau), s) - F_k(y(\tau), s)] \right\| < \frac{\epsilon}{3c}(h_k(b) - h_k(a)) \leqslant \frac{\epsilon}{3}$$

$$\left\| \int_a^b D[F_0(y(\tau), s) - F_0(\psi_0(\tau), s)] \right\| < \frac{\epsilon}{3c}(h_0(b) - h_0(a)) \leqslant \frac{\epsilon}{3}.$$

In consequence, we obtain

$$\left\| \int_a^b DF_k(\psi_k(\tau), s) - \int_a^b DF_0(\psi_0(\tau), s) \right\| < \epsilon + \left\| \int_a^b D[F_k(y(\tau), s) - F_0(y(\tau), s)] \right\|.$$

Now, let us consider the Kurzweil integral $\int_a^b D[F_k(y(\tau), s) - F_0(y(\tau), s)]$. From the fact that $y : [a,b] \to C$ is a step function, there exists a finite number of points

$$a = r_0 < r_1 < r_2 < \cdots < r_{p-1} < r_p = b$$

such that for $\tau \in (r_{j-1}, r_j)$, $j \in \{1, 2, \dots, p\}$, the equality $y(\tau) = c_j \in C$ holds. Thus,

$$\int_a^b DF_k(y(\tau), s) = \sum_{j=1}^{p} \int_{r_{j-1}}^{r_j} DF_k(c_j, s).$$

By Theorem 2.12, we get

$$\int_{r_{j-1}}^{r_j} DF_k(y(\tau),t) = F_k\left(c_j, r_j^-\right) - F_k\left(c_j, r_{j-1}^+\right) + F_k\left(y(r_{j-1}), r_{j-1}^+\right)$$
$$- F_k\left(y\left(r_{j-1}\right), r_{j-1}\right) - F_k\left(y(r_j), r_j^-\right) + F_k\left(y\left(r_j\right), r_j\right),$$

which implies, by hypotheses, that

$$\lim_{k\to\infty} \int_{r_{j-1}}^{r_j} D[F_k(y(\tau),t) - F_0(y(\tau),t)] = 0.$$

Combining the previous estimates, the statement follows by the arbitrariness of $\epsilon > 0$. □

The following result is an immediate consequence of Theorem 7.2. It can be found in [4].

Corollary 7.3: *Assume that for each $k \in \mathbb{N}_0$, $F_k : \Omega \to X$ belongs to the class $\mathcal{F}(\Omega, h)$, where $h : [a,b] \to \mathbb{R}$ is a nondecreasing and left-continuous function. Assume, further, that $\{F_k\}_{k\in\mathbb{N}}$ converges uniformly to F_0 on Ω. Let C be a closed subset in X such that $C \subset O$. Then, for every $s_1, s_2 \in [a,b]$, we have*

$$\lim_{k\to\infty} \left\| \int_{s_1}^{s_2} DF_k(\psi_k(\tau), s) - \int_{s_1}^{s_2} DF_0(\psi_0(\tau), s) \right\| = 0, \tag{7.3}$$

provided one of the following conditions hold:

(i) $\psi_k \in G([a,b],X)$ *for all* $k \in \mathbb{N}$, $(\psi_k(s),s) \in C \times [a,b]$ *for all* $k \in \mathbb{N}$, *and* $\lim_{k\to\infty} \|\psi_k - \psi_0\|_\infty = 0$;
(ii) $\psi_k \in BV([a,b],X)$ *for all* $k \in \mathbb{N}$, $(\psi_k(s),s) \in C \times [a,b]$ *for all* $k \in \mathbb{N}$, *and* $\lim_{k\to\infty} \|\psi_k - \psi_0\|_{BV} = 0$.

Proof. If item (i) holds, then the result follows from Lemma 7.1 and Theorem 7.2 for the case where $h_k = h$, for all $k \in \mathbb{N}_0$.

Now, suppose item (ii) holds. We firstly recall that $BV([a,b],X) \subset G([a,b],X)$. Thus, for every $t \in [a,b]$, we have

$$\begin{aligned}
\|\psi_k(t) - \psi_0(t)\| &\leqslant \|\psi_k(a) - \psi_0(a)\| + \|\psi_k(t) - \psi_0(t) - (\psi_k(a) - \psi_0(a))\| \\
&\leqslant \|\psi_k(a) - \psi_0(a)\| + var_a^t(\psi_k - \psi_0) \\
&\leqslant \|\psi_k(a) - \psi_0(a)\| + var_a^b(\psi_k - \psi_0) \\
&= \|\psi_k - \psi_0\|_{BV},
\end{aligned}$$

which yields $\lim_{k\to\infty} \|\psi_k - \psi_0\|_\infty = 0$. Hence, condition (i) is valid and the result follows. □

Now, we present a result on continuous dependence of solutions on parameters for generalized ODEs. Such result, borrowed from [177], is a generalization of [95, Theorem 2.4]. Although the proof follows the ideas presented in [95], it is somehow different because it uses Theorem 7.2.

Theorem 7.4: *Let $\{h_k\}_{k\in\mathbb{N}}$ be a sequence of nondecreasing and left-continuous functions $h_k : [a,b] \to \mathbb{R}$ such that $h_k(b) - h_k(a) \leqslant c$, for some $c > 0$ and every $k \in \mathbb{N}_0$. Assume that, for every $k \in \mathbb{N}$, $F_k : \Omega \to X$ belongs to the class $\mathcal{F}(\Omega, h_k)$ and $\{F_k\}_{k\in\mathbb{N}}$ converges pointwisely to F_0 on Ω. Moreover, suppose for all $x \in O$, we have*

$$\lim_{k\to\infty} F_k(x, t^+) = F_0(x, t^+), \quad t \in [a, b).$$

For every $k \in \mathbb{N}$, let $x_k : [a,b] \to X$ be a solution of the generalized ODE

$$\frac{dx}{d\tau} = DF_k(x, t). \tag{7.4}$$

Assume that there exists a closed subset C in X, $C \subset O$, such that $x_k(t) \in C$ for all $t \in [a,b]$ and all $k \in \mathbb{N}$. If there exists a function $x_0 : [a,b] \to X$ such that $\{x_k\}_{k\in\mathbb{N}}$ converges uniformly to x_0 on $[a,b]$, then x_0 is a solution on $[a,b]$ of the generalized ODE

$$\frac{dx}{d\tau} = DF_0(x, t). \tag{7.5}$$

Proof. Since the functions $x_k : [a,b] \to X$ are regulated and $\{x_k\}_{k\in\mathbb{N}}$ converges uniformly to x_0 on $[a,b]$, $x_0 : [a,b] \to X$ is a regulated function on $[a,b]$ and $(x_0(s), s) \in C \times [a,b]$. Hence, Corollary 4.8 implies the Kurzweil integral $\int_{s_1}^{s_2} DF_0(x_0(\tau), t)$ exists for every $s_1, s_2 \in [a,b]$.

Using the definition of a solution of the generalized ODE (7.4), for each $k \in \mathbb{N}$ and for every $s_1, s_2 \in [a,b]$,

$$x_k(s_2) - x_k(s_1) = \int_{s_1}^{s_2} DF_k(x_k(\tau), t).$$

Thus, by Theorem 7.2, we obtain

$$\lim_{k\to\infty} \left\| \int_{s_1}^{s_2} DF_k(x_k(\tau), t) - \int_{s_1}^{s_2} DF_0(x_0(\tau), t) \right\| = 0,$$

for all $s_1, s_2 \in [a,b]$. Therefore,

$$x_0(s_2) - x_0(s_1) = \int_{s_1}^{s_2} DF_0(x_0(\tau), t),$$

for all $s_1, s_2 \in [a,b]$ and, hence, x_0 is a solution of the generalized ODE (7.5) on $[a,b]$. □

The next result is an immediate consequence of Theorem 7.4, which concerns continuous dependence on parameters for solutions of nonautonomous generalized ODEs. Such result, borrowed from [177], ensures that, under certain conditions, if a sequence of solutions $x_k : [a,b] \to X$ of the generalized ODE

$$\frac{dx}{d\tau} = DF_k(x,t)$$

converges to a function x_0, then this convergence is uniform and $x_0 : [a,b] \to X$ is a solution of the "*limiting equation.*"

Corollary 7.5: *Consider a sequence of functions $F_k : \Omega \to X$ such that $F_k \in \mathcal{F}(\Omega, h_k)$, $k \in \mathbb{N}$. Suppose the following conditions are fulfilled:*

(a) *The initial sequence of functions $h_k : [a,b] \to \mathbb{R}$, $k \in \mathbb{N}$, is equiregulated;*
(b) *The sequence $\{h_k(a)\}_{k\in\mathbb{N}}$ is bounded;*
(c) *For every $k \in \mathbb{N}$, the function h_k is nondecreasing and left-continuous satisfying $h_k(b) - h_k(a) \leqslant c$ for some $c > 0$;*
(d) *$\{F_k\}_{k\in\mathbb{N}}$ converges pointwisely to F_0 on Ω;*
(e) *For each $k \in \mathbb{N}$, $x_k : [a,b] \to X$ is a solution of the generalized ODE*

$$\frac{dx}{d\tau} = DF_k(x,t) \tag{7.6}$$

on the interval $[a,b]$, there exists a closed subset C in X, $C \subset O$, such that $x_k(t) \in C$, for all $t \in [a,b]$ and all $k \in \mathbb{N}$, and $\{x_k\}_{k\in\mathbb{N}}$ converges pointwisely to $x_0 : [a,b] \to X$.

Then, the following assertions hold:

(i) *The sequence $\{h_k\}_{k\in\mathbb{N}}$ has a subsequence that converges uniformly to a nondecreasing function h_0;*
(ii) *For every $s_1, s_2 \in [a,b]$,*

$$\|x_0(s_2) - x_0(s_1)\| \leqslant |h_0(s_2) - h_0(s_1)|,$$

where h_0 is the function described in (i);
(iii) *$\lim_{k\to\infty} x_k(t) = x_0(t)$ uniformly on $[a,b]$;*
(iv) *x_0 is a solution on $[a,b]$ of the generalized ODE*

$$\frac{dx}{d\tau} = DF_0(x,t).$$

Proof. Conditions (b) and (c) imply that for every $k \in \mathbb{N}$, h_k is a nondecreasing function and the sequence $\{h_k(a)\}_{k\in\mathbb{N}}$ is bounded. Hence, the set $\{h_k(t) : t \in [a,b]\}$ is also bounded for every $k \in \mathbb{N}$ and $t \in [a,b]$.

Since the sequence $\{h_k\}_{k\in\mathbb{N}}$ is equiregulated, by Corollary 1.19, there exists a uniformly convergent subsequence $\{h_{n_k}\}_{k\in\mathbb{N}}$. For the sake of simplicity, we denote this subsequence by $\{h_k\}_{k\in\mathbb{N}}$. Let $h_0 : [a,b] \to \mathbb{R}$ be a function for which the sequence $\{h_k\}_{k\in\mathbb{N}}$ converges uniformly. Since $\{h_k\}_{k\in\mathbb{N}}$ is a sequence of nondecreasing functions, h_0 is also a nondecreasing function and, hence, item (i) follows.

Suppose $a \leqslant s_1 < s_2 \leqslant b$. For every $k \in \mathbb{N}$, Lemma 4.9 yields the estimate

$$\|x_k(s_2) - x_k(s_1)\| \leqslant |h_k(s_2) - h_k(s_1)|.$$

Applying the limit as $k \to \infty$ in the previous inequality, we get by (i), the following estimate

$$\|x_0(s_2) - x_0(s_1)\| \leqslant |h_0(s_2) - h_0(s_1)|,$$

for $s_1, s_2 \in [a,b]$ and, therefore, (ii) follows.

In sequel, our goal is to prove (iii). Using Lemma 4.9, we get

$$\|x_k(s_2) - x_k(s_1)\| \leqslant |h_k(s_2) - h_k(s_1)|,$$

for $s_1, s_2 \in [a,b]$. Since the sequence of functions $\{h_k\}_{k\in\mathbb{N}}$ is equiregulated, the previous inequality, together with Corollary 1.1 and the fact that $\lim_{k\to\infty} x_k(t) = x_0(t)$ pointwisely for every $t \in [a,b]$, imply that $\{x_k\}_{k\in\mathbb{N}}$ converges uniformly to x_0 on $[a,b]$.

In order to prove (iv), note that, by the definition of a solution of the generalized ODE (7.6), we have, for each $k \in \mathbb{N}$, $(x_k(t), t) \in C \times [a,b] \subset O \times [a,b]$ and

$$x_k(s_2) - x_k(s_1) = \int_{s_1}^{s_2} DF_k(x_k(\tau), s), \tag{7.7}$$

for every $s_1, s_2 \in [a,b]$. Because $F_k \in \mathcal{F}(\Omega, h_k)$, for each $k \in \mathbb{N}$, the sequence $\{h_k\}_{k\in\mathbb{N}}$ is equiregulated and $\{F_k\}_{k\in\mathbb{N}}$ converges pointwisely to F_0, Corollary 1.1 implies that the sequence of the right-hand side, $\{F_k\}_{k\in\mathbb{N}}$, converges uniformly to F_0. Thus, Lemma 7.1 yields $\lim_{k\to\infty} F_k(x, t^+) = F_0(x, t^+)$. Therefore, all hypotheses of Theorem 7.2 are fulfilled and, hence, for all $s_1, s_2 \in [a,b]$, we have

$$\lim_{k\to\infty} \left\| \int_{s_1}^{s_2} DF_k(x_k(\tau), s) - \int_{s_1}^{s_2} DF_0(x_0(\tau), s) \right\| = 0. \tag{7.8}$$

Finally, from (7.7), we obtain

$$\begin{aligned}\left\| x_0(s_2) - x_0(s_1) - \int_{s_1}^{s_2} DF_0(x_0(\tau), s) \right\| &\leqslant \|x_k(s_2) - x_0(s_2)\| + \|x_k(s_1) - x_0(s_1)\| \\ &+ \left\| \int_{s_1}^{s_2} DF_k(x_k(\tau), s) - \int_{s_1}^{s_2} DF_0(x_0(\tau), s) \right\|.\end{aligned}$$

Then, taking $k \to \infty$, it follows from (iii) and (7.8) that

$$x_0(s_2) - x_0(s_1) = \int_{s_1}^{s_2} DF_0(x_0(\tau), s),$$

for all $s_1, s_2 \in [a,b]$, proving the result. □

The next result is a consequence of Corollary 7.5 for the case where $h_k = h$, for all $k \in \mathbb{N}_0$. In such case, the equiregulatedness of $\{h_k\}_{k\in\mathbb{N}}$ follows from Corollary 1.19. This result can be found in [4].

Corollary 7.6: *Assume that, for each $k \in \mathbb{N}_0$, $F_k : \Omega \to X$ belongs to the class $\mathcal{F}(\Omega, h)$, where $h : [a,b] \to \mathbb{R}$ is a nondecreasing and left-continuous function, and $\{F_k\}_{k\in\mathbb{N}}$ converges pointwisely to F_0 on Ω. Let $x_k : [a, b] \to X$, $k \in \mathbb{N}$, be solutions of the generalized ODE*

$$\frac{dx}{d\tau} = DF_k(x, t)$$

on $[a, b]$ such that $\{x_k\}_{k\in\mathbb{N}}$ converges pointwisely to $x_0 : [a, b] \to X$. Assume that there exists a closed subset C in X, with $C \subset O$, such that $x_k(t) \in C$ for all $t \in [a, b]$ and all $k \in \mathbb{N}$. Then, $x_0 : [a, b] \to X$ satisfies:

(i) *$\|x_0(s_2) - x_0(s_1)\| \leqslant h(s_2) - h(s_1)$, for $s_1 \leqslant s_2$, with $s_1, s_2 \in [a, b]$;*
(ii) *$\lim_{k\to\infty} x_k(s) = x_0(s)$ uniformly on $[a, b]$;*
(iii) *x_0 is a solution on $[a, b]$ of the generalized ODE*

$$\frac{dx}{d\tau} = DF_0(x, t).$$

The next result is a generalization of [209, Theorem 8.6] for the case of Banach space-valued functions. Here, different arguments from those of [209] are used.

Theorem 7.7: *Assume that, for each $k \in \mathbb{N}$, the function $F_k : \Omega \to X$ belongs to the class $\mathcal{F}(\Omega, h)$, where $h : [a, b] \to \mathbb{R}$ is a nondecreasing and left-continuous function, and $\{F_k\}_{k\in\mathbb{N}}$ converges pointwisely to F_0 on Ω. Let $x_0 : [a, b] \to X$ be a solution of the generalized ODE*

$$\frac{dx}{d\tau} = DF_0(x, t) \tag{7.9}$$

on $[a, b]$, satisfying the following uniqueness property:

(U) *if $z : [a, \gamma] \to X$, $[a, \gamma] \subset [a, b]$, is a solution of (7.9) such that $z(a) = x_0(a)$, then $z(t) = x_0(t)$ for every $t \in [a, \gamma]$.*

Assume, further, that there exists $\rho > 0$ such that if $s \in [a, b]$ and $\|y - x_0(s)\| < \rho$, then $(y, s) \in \Omega$. Let $\{y_k\}_{k\in\mathbb{N}} \subset O$ satisfy $\lim_{k\to\infty} y_k = x_0(a)$. Then, there exists a positive integer k_0 such that, for all $k > k_0$, there exists a solution $x_k : [a, b] \to X$ of the generalized ODE

$$\frac{dx}{d\tau} = DF_k(x, t), \tag{7.10}$$

with $x_k(a) = y_k$, such that $\{x_k\}_{k\in\mathbb{N}}$ converges uniformly to x_0 on $[a, b]$.

Proof. Let $y \in O$ be such that $\|y - x_0(a)\| < \frac{\rho}{2}$ (or $\|y - x_0(a^+)\| < \frac{\rho}{2}$, where $x_0(a^+) = x_0(a) + F_0(x_0(a), a^+) - F_0(x_0(a), a)$). By hypothesis, $(y, a) \in \Omega$. Since $\lim_{k\to\infty} y_k = x_0(a)$, the pointwise convergence of $\{F_k\}_{k\in\mathbb{N}}$ yields

$$\lim_{k\to\infty}[y_k + F_k(y_k, a^+) - F_k(y_k, a)] = x_0(a) + F_0(x_0(a), a^+) - F_0(x_0(a), a). \tag{7.11}$$

Indeed, it is enough to observe that, applying (4.6) to F_k and using the pointwise convergence of F_k to F_0, for each $\epsilon > 0$,

$$\begin{aligned}
&\|[y_k + F_k(y_k, a^+) - F_k(y_k, a)] - [x_0(a) + F_0(x_0(a), a^+) - F_0(x_0(a), a)]\| \\
&\quad \leqslant \|y_k - x_0(a)\| + \|F_k(y_k, a^+) - F_k(y_k, a) - F_0(x_0(a), a^+) + F_0(x_0(a), a)\| \\
&\quad \leqslant \|y_k - x_0(a))\| + \|y_k - x_0(a)\|(h(a^+) - h(a)) + \epsilon,
\end{aligned}$$

for sufficiently large k.

The convergence in (7.11) implies the existence of $k_1 \in \mathbb{N}$ such that for $k > k_1$, we have $(y_k, a) \in \Omega$ and $(y_k + F_k(y_k, a^+) - F_k(y_k, a), a) \in \Omega$. Since O is open, one can obtain $d > a$ such that, if $t \in [a, d]$ and

$$\left\|z - \left[y_k + F_k(y_k, a^+) - F_k(y_k, a)\right]\right\| \leqslant h(t) - h(a^+),$$

then $(z, t) \in \Omega$ for all $k > k_1$. By Theorem 5.1, for each $k > k_1$, there is a unique solution $x_k : [a, d] \to X$ of the generalized ODE (7.10) on $[a, d]$ for which $x_k(a) = y_k$. As mentioned in [209, Theorem 8.6], the solutions x_k of the generalized ODE (7.10) exist on the same interval $[a, d]$, for all $k > k_1$, since the choice of d depends only on the function h.

Now, we claim that $\lim_{k\to\infty} x_k(t) = x_0(t)$ for every $t \in [a, d]$. Indeed, fix $t \in [a, d]$. By Theorem 7.2,

$$\eta_k = \left\|\int_a^t DF_k(x_0(\tau), s) - \int_a^t DF_0(x_0(\tau), s)\right\| \to 0, \quad \text{as} \quad k \to \infty. \tag{7.12}$$

Thus, for $k > k_1$, we have

$$\begin{aligned}
\|x_k(t) - x_0(t)\| &= \left\|x_k(a) + \int_a^t DF_k(x_k(\tau), s) - x_0(a) - \int_a^t DF_0(x_0(\tau), s)\right\| \\
&\leqslant \|x_k(a) - x_0(a)\| + \left\|\int_a^t DF_k(x_k(\tau), s) - \int_a^t DF_k(x_0(\tau), s)\right\| \\
&\quad + \left\|\int_a^t DF_k(x_0(\tau), s) - \int_a^t DF_0(x_0(\tau), s)\right\| \\
&\leqslant \|y_k - x_0(a)\| + \int_a^t \|x_k(s) - x_0(s)\| \, dh(s) + \eta_k,
\end{aligned}$$

where we used Lemma 4.6 and (7.12) to obtain the last inequality. Then, for $k > k_1$, by the Grönwall inequality (see Theorem 1.52), we obtain

$$\|x_k(t) - x_0(t)\| \leqslant \left(\|y_k - x_0(a)\| + \eta_k\right) e^{h(t)-h(a)} \to 0 \quad \text{as} \quad k \to \infty, \tag{7.13}$$

where we used (7.12) and the fact that $\lim_{k\to\infty} y_k = x_0(a)$ to obtain (7.13). Therefore, $\lim_{k\to\infty} x_k(t) = x_0(t)$, for each $t \in [a, d]$.

Let us verify that $\lim_{k\to\infty} x_k(t) = x_0(t)$ on $[a, b]$. Suppose to the contrary, that is, there exists $d^* \in (a, b)$ satisfying the following property: for every $d < d^*$, there is a solution x_k of the generalized ODE (7.10) with $x_k(a) = y_k$, defined on $[a, d]$, for $k \in \mathbb{N}$ sufficiently large, such that $\lim_{k\to\infty} x_k(t) = x_0(t)$ for $t \in [a, d]$, but this convergence is not true on $[a, d']$ for $d' > d^*$. According to Lemma 4.9, the inequality $\|x_k(s_2) - x_k(s_1)\| \leqslant |h(s_2) - h(s_1)|$ holds for $s_2, s_1 \in [a, d^*)$ and $k \in \mathbb{N}$ sufficiently large. Therefore, for $k \in \mathbb{N}$ sufficiently large, the limit $x_k(d^{*-})$ exists. Using the fact that the solution x_0 is left-continuous, we get

$$\lim_{k\to\infty} x_k(d^{*-}) = x_0(d^{*-}) = x_0(d^*).$$

Set $x_k(d^*) = x_k(d^{*-})$. Then, $\lim_{k\to\infty} x_k(d^*) = x_0(d^*)$, which implies that the statement holds on $[a, d^*]$. Since $d^* < b$, one can use the same argument, with initial condition $\lim_{k\to\infty} x_k(d^*) = x_0(d^*)$ and the local uniqueness given by condition (U), to conclude that the statement is also valid on $[d^*, d^* + \delta]$, for some $\delta > 0$, which contradicts the definition of d^*. Hence, there is $k_0 \in \mathbb{N}$ such that x_k is defined on $[a, b]$, for all $k > k_0$, and $\lim_{k\to\infty} x_k(t) = x_0(t)$ on $[a, b]$.

Finally, set $C = \{z \in X : \|z - x_0(s)\| \leqslant \frac{\rho}{2}, s \in [a, b]\} \subset O$. By (7.13), we have $x_k(t) \in C$ for every $t \in [a, b]$ and k sufficiently large. Then, Corollary 7.6 yields this convergence is uniform on $[a, b]$. □

The next result is a generalization of [209, Theorem 8.8] for Banach spaces whose proof follows the same steps as the finite dimensional case. Thus, we provide only few comments here.

Theorem 7.8: *Assume that for each $k \in \mathbb{N}_0$, the function $F_k : \Omega \to X$ belongs to the class $\mathcal{F}(\Omega, h_k)$, where $h_k : [a, b] \to \mathbb{R}$ is nondecreasing and left-continuous. Suppose $h_0 : [a, b] \to \mathbb{R}$ is nondecreasing and continuous on $[a, b]$ and, for every $a \leqslant t_1 \leqslant t_2 \leqslant b$, we have*

$$\limsup_{k\to\infty} (h_k(t_2) - h_k(t_1)) \leqslant h_0(t_2) - h_0(t_1)$$

and $\{F_k\}_{k\in\mathbb{N}}$ converges pointwisely to F_0 on Ω. Let $x_0 : [a, b] \to X$ be a solution on $[a, b]$ of the generalized ODE

$$\frac{dx}{d\tau} = DF_0(x, t), \tag{7.14}$$

which has the uniqueness property (U) described in Theorem 7.7. Assume, further, that there is $\rho > 0$ such that if $s \in [a, b]$ and $\|y - x(s)\| < \rho$, then $(y, s) \in \Omega$. Let

$\{y_k\}_{k\in\mathbb{N}} \in O$ satisfy $\lim_{k\to\infty} y_k = x_0(a)$. Then, for every $\mu > 0$, there exists $k_* \in \mathbb{N}$ such that for $k > k_*$, there exists a solution x_k on $[a,b]$ of the generalized ODE

$$\frac{dx}{d\tau} = DF_k(x,t), \tag{7.15}$$

for which $x_k(a) = y_k$ and

$$\|x_k(s) - x_0(s)\| < \mu, \quad \text{for all } s \in [a,b].$$

Proof. The existence of the solutions x_k of the generalized ODE (7.15) for sufficiently large $k \in \mathbb{N}$ and the pointwise convergence $\lim_{k\to\infty} x_k(s) = x_0(s)$, for $s \in [a,b]$, follows similarly as in the proof of Theorem 7.7 with minor changes. The remaining part follows exactly as in the proof of [209, Theorem 8.8]. □

7.2 Applications to Measure FDEs

In this section, our goal is to prove some results on continuous dependence on parameters for measure FDEs, using what we obtained in Section 7.1 and the correspondence between generalized ODEs and measure FDEs presented in Chapter 4.

Let $r, \sigma > 0$. Similarly as in Chapter 4, we assume that $O \subset G([t_0 - r, t_0 + \sigma], \mathbb{R}^n)$ has the prolongation property and we borrow the same conditions (A)–(C) to the next results of this section.

The first result we present is borrowed from [85].

Theorem 7.9: *Assume that $O = G([t_0 - r, t_0 + \sigma], \mathbb{R}^n)$, $P = \{y_t : y \in O, t \in [t_0, t_0 + \sigma]\}$, $g : [t_0, t_0 + \sigma] \to \mathbb{R}$ is a nondecreasing left-continuous function, and $f_k : P \times [t_0, t_0 + \sigma] \to \mathbb{R}^n$, $k \in \mathbb{N}_0$, are functions that satisfy conditions (A)–(C) for the same functions M and L. Suppose, in addition, that for every $y \in O$,*

$$\lim_{k\to\infty} \int_{t_0}^{t} f_k(y_s, s)dg(s) = \int_{t_0}^{t} f_0(y_s, s)dg(s)$$

uniformly with respect to $t \in [t_0, t_0 + \sigma]$. Let $y_k \in O$, $k \in \mathbb{N}$, be solutions of the measure FDEs

$$\begin{cases} y_k(t) = y_k(t_0) + \displaystyle\int_{t_0}^{t} f_k((y_k)_s, s)dg(s), & t \in [t_0, t_0 + \sigma], \\ (y_k)_{t_0} = \phi_k. \end{cases}$$

Consider a sequence of functions $\phi_k \in P$, $k \in \mathbb{N}$, such that $\{\phi_k\}_{k\in\mathbb{N}}$ converges to ϕ_0 uniformly on $[-r, 0]$. If there exists a function y_0 such that $\{y_k\}_{k\in\mathbb{N}}$ converges pointwisely to y_0 on $[t_0, t_0+\sigma]$, then y_0 is a solution of the measure FDE

$$\begin{cases} y_0(t) = y_0(t_0) + \displaystyle\int_{t_0}^{t} f_0((y_0)_s, s)dg(s), & t \in [t_0, t_0+\sigma], \\ (y_0)_{t_0} = \phi_0. \end{cases}$$

Proof. For each $k \in \mathbb{N}$, define an auxiliary function $F_k : O \times [t_0, t_0+\sigma] \to G([t_0 - r, t_0+\sigma], \mathbb{R}^n)$ by

$$F_k(x,t)(\vartheta) = \begin{cases} 0, & t_0 - r \leqslant \vartheta \leqslant t_0, \\ \displaystyle\int_{t_0}^{\vartheta} f_k(x_s, s)dg(s), & t_0 \leqslant \vartheta \leqslant t \leqslant t_0+\sigma, \\ \displaystyle\int_{t_0}^{t} f_k(x_s, s)dg(s), & t \leqslant \vartheta \leqslant t_0+\sigma. \end{cases} \tag{7.16}$$

By hypotheses, $\{F_k\}_{k\in\mathbb{N}}$ converges uniformly to F_0, for every $(x,t) \in O \times [t_0, t_0+\sigma]$, where

$$F_0(x,t)(\vartheta) = \begin{cases} 0, & t_0 - r \leqslant \vartheta \leqslant t_0, \\ \displaystyle\int_{t_0}^{\vartheta} f_0(x_s, s)dg(s), & t_0 \leqslant \vartheta \leqslant t \leqslant t_0+\sigma, \\ \displaystyle\int_{t_0}^{t} f_0(x_s, s)dg(s), & t \leqslant \vartheta \leqslant t_0+\sigma, \end{cases} \tag{7.17}$$

whence $F_0 \in \mathcal{F}(O \times [t_0, t_0+\sigma], h)$. Then, the Moore–Osgood theorem (see [19]) yields

$$\lim_{k\to\infty} F_k(x, t^+) = F_0(x, t^+),$$

for every $(x,t) \in O \times [t_0, t_0+\sigma)$.

For every $k \in \mathbb{N}_0$ and $t \in [t_0, t_0+\sigma]$, define

$$x_k(t)(\vartheta) = \begin{cases} y_k(\vartheta), & \vartheta \in [t_0 - r, t], \\ y_k(t), & \vartheta \in [t, t_0+\sigma]. \end{cases} \tag{7.18}$$

Applying Theorem 4.18, one concludes that for each $k \in \mathbb{N}$, $x_k : [t_0, t_0+\sigma] \to O$ is a solution of the generalized ODE

$$\frac{dx}{d\tau} = DF_k(x,t).$$

On the other hand, for each $k \in \mathbb{N}$ and $t_0 \leqslant t_1 \leqslant t_2 \leqslant t_0+\sigma$,

$$\|y_k(t_2) - y_k(t_1)\| = \left\| \int_{t_1}^{t_2} f_k((y_k)_s, s)dg(s) \right\| \leqslant \int_{t_1}^{t_2} M(s)dg(s) \leqslant \eta(K(t_2) - K(t_1)),$$

where $\eta(t) = t$, for all $t \in [0, \infty)$, and $K(t) = \int_{t_0}^{t} M(s)dg(s) + t$, for all $t \in [t_0, t_0 + \sigma]$. One can easily check that K and η are increasing functions fulfilling $\eta(0) = 0$. Moreover, the sequence $\{y_k(t_0)\}_{k\in\mathbb{N}}$ is bounded, since $y_k(t_0) = \phi_k(0)$ and $\phi_k \to \phi_0$ uniformly on $[-r, 0]$. Hence, Corollary 1.19 implies that $\{y_k\}_{k\in\mathbb{N}}$ admits a subsequence that is uniformly convergent on $[t_0, t_0 + \sigma]$. For simplicity, let us denote this subsequence again by $\{y_k\}_{k\in\mathbb{N}}$. By the property of the initial condition, this uniform convergence can be extended to the entire interval $[t_0 - r, t_0 + \sigma]$. From this fact and by (7.18), we conclude that $\{x_k\}_{k\in\mathbb{N}}$ converges to a certain x_0 uniformly on $[t_0, t_0 + \sigma]$, where $x_0 : [t_0, t_0 + \sigma] \to O$ is given by

$$x_0(t)(\vartheta) = \begin{cases} y_0(\vartheta), & \vartheta \in [t_0 - r, t], \\ y_0(t), & \vartheta \in [t, t_0 + \sigma]. \end{cases} \tag{7.19}$$

Finally, by Theorem 7.4, x_0 is a solution of

$$\frac{dx}{d\tau} = DF_0(x, t)$$

on $[t_0, t_0 + \sigma]$, whence, by Theorem 4.19, $y_0 : [t_0 - r, t_0 + \sigma] \to \mathbb{R}^n$ satisfies

$$\begin{cases} y_0(t) = y_0(t_0) + \displaystyle\int_{t_0}^{t} f_0((y_0)_s, s)dg(s), & t \in [t_0, t_0 + \sigma], \\ (y_0)_{t_0} = \phi_0, \end{cases}$$

and the statement follows. □

In the next result, we present the analogue version of Theorem 7.7 for measure FDEs. This formulation is presented for the first time here. Its proof follows directly from the correspondence between generalized ODEs and measure FDEs described by Theorems 4.18 and 4.19, and from Theorem 7.7.

Theorem 7.10: *Suppose* $O = G([t_0 - r, t_0 + \sigma], \mathbb{R}^n)$, $P = \{y_t : y \in O, t \in [t_0, t_0 + \sigma]\}$, $g : [t_0, t_0 + \sigma] \to \mathbb{R}$ *is a nondecreasing and left-continuous function, and for each* $k \in \mathbb{N}$, $f_k : P \times [t_0, t_0 + \sigma] \to \mathbb{R}^n$ *satisfies conditions (A)–(C) for the same functions M and L. Suppose, further, for every* $y \in O$,

$$\lim_{k\to\infty} \int_{t_0}^{t} [f_k(y_s, s) - f_0(y_s, s)]\, dg(s) = 0, \quad t \in [t_0, t_0 + \sigma]. \tag{7.20}$$

Let $y_0 : [t_0 - r, t_0 + \sigma] \to \mathbb{R}^n$ *be the unique solution of*

$$\begin{cases} y_0(t) = y_0(t_0) + \displaystyle\int_{t_0}^{t} f_0((y_0)_s, s)dg(s), & t \in [t_0, t_0 + \sigma], \\ (y_0)_{t_0} = \phi_0, \end{cases} \tag{7.21}$$

where $\phi_0 \in P$. *Consider* $\{\phi_k\}_{k\in\mathbb{N}}$ *is a sequence of regulated and left-continuous functions in P such that converges uniformly to* ϕ_0 *on* $[-r, 0]$. *Let* $\{z_k\}_{k\in\mathbb{N}} \subset O$ *satisfy*

$\lim_{k\to\infty} z_k(t_0+\theta) = \phi_0(\theta)$ *on* $[-r,0]$. *Then, for sufficiently large* $k \in \mathbb{N}$, *there exists a solution* y_k *of*

$$\begin{cases} y_k(t) = y_k(t_0) + \int_{t_0}^{t} f_k((y_k)_s, s)dg(s), & t \in [t_0, t_0+\sigma], \\ (y_k)_{t_0} = \phi_k. \end{cases} \tag{7.22}$$

Moreover, the sequence $\{y_k\}_{k\in\mathbb{N}}$ *converges uniformly to* y_0 *on* $[t_0-r, t_0+\sigma]$.

Proof. The main idea behind this proof is to apply the correspondence between measure FDEs and generalized ODEs, as well as to apply Theorem 7.7 .

Let us consider for each $k \in \mathbb{N}$, and each function f_k, the function $F_k : O \times [t_0, t_0+\sigma] \to G([t_0-r, t_0+\sigma], \mathbb{R}^n)$ given by (7.16). Clearly, conditions (A)–(C), together with Lemma 4.16, imply that $F_k \in \mathcal{F}(O \times [t_0, t_0+\sigma], h)$, for every $k \in \mathbb{N}$, with

$$h(t) = \int_{t_0}^{t} (L(s) + M(s))dg(s), \quad \text{for } t \in [t_0, t_0+\sigma].$$

By the hypotheses and by definition, $\{F_k\}_{k\in\mathbb{N}}$ converges uniformly to F_0, given by (7.17) for every $(x,t) \in O \times [t_0, t_0+\sigma]$. Thus, $F_0 \in \mathcal{F}(O \times [t_0, t_0+\sigma], h)$. Applying the Moore–Osgood theorem (see [19]), we obtain

$$\lim_{k\to\infty} F_k(x, t^+) = F_0(x, t^+), \quad \text{for every } (x,t) \in O \times [t_0, t_0+\sigma).$$

Let y_0 be the unique solution of the measure FDE (7.21). Define $x_0 : [t_0, t_0+\sigma] \to O$ as follows

$$x_0(t)(\vartheta) = \begin{cases} y_0(\vartheta), & \vartheta \in [t_0-r, t], \\ y_0(t), & \vartheta \in [t, t_0+\sigma]. \end{cases}$$

Then, Theorem 4.18 yields x_0 is the unique solution of the generalized ODE (7.9) on $[t_0, t_0+\sigma]$.

Consider the following generalized ODE

$$\frac{dx}{d\tau} = DF_k(x, t), \tag{7.23}$$

on $[t_0, t_0+\sigma]$, with initial condition

$$z_k(t_0)(\vartheta) = \begin{cases} \phi_k(\vartheta - t_0), & \vartheta \in [t_0-r, t_0], \\ \phi_k(0), & \vartheta \in [t_0, t_0+\sigma]. \end{cases}$$

Notice that for $\vartheta \in [t_0-r, t_0]$, we have

$$\lim_{k\to\infty} z_k(t_0)(\vartheta) = \lim_{k\to\infty} \phi_k(\vartheta - t_0) = \phi_0(\vartheta - t_0) = x_0(t_0)(\vartheta)$$

and for $\vartheta \in [t_0, t_0+\sigma]$, we obtain

$$\lim_{k\to\infty} z_k(t_0)(\vartheta) = \lim_{k\to\infty} \phi_k(0) = \phi_0(0) = x_0(t_0)(\vartheta).$$

Hence, for $\vartheta \in [t_0 - r, t_0 + \sigma]$, we have

$$\lim_{k\to\infty} z_k(t_0)(\vartheta) = x_0(t_0)(\vartheta),$$

where x_0 is the solution of (7.23), by the uniqueness. Therefore, applying Theorem 7.7, there exists a positive integer k_0 such that for all $k > k_0$, there exists a solution x_k of the generalized ODE (7.23) satisfying $\lim_{k\to\infty} x_k(s) = x_0(s)$ for every $s \in [t_0, t_0 + \sigma]$.

Define, for each $k > k_0$, the function

$$y_k(\vartheta) = \begin{cases} x_k(t_0)(\vartheta), & \vartheta \in [t_0 - r, t_0], \\ x_k(\vartheta)(\vartheta), & \vartheta \in [t_0, t_0 + \sigma]. \end{cases} \tag{7.24}$$

Using Theorem 4.19 again, we conclude that y_k is a solution of the measure FDE (7.22) on $[t_0 - r, t_0 + \sigma]$. Then, by (7.24) and using the assumptions on the initial condition, we obtain

$$\lim_{k\to\infty} (y_k)_{t_0}(\theta) = \lim_{k\to\infty} \phi_k(\theta) = \phi_0(\theta) = (y_0)_{t_0}(\theta),$$

for $\theta \in [-r, 0]$. Hence,

$$\lim_{k\to\infty} y_k(s) = y_0(s), \quad \text{for } s \in [t_0 - r, t_0].$$

Finally, combining this with the definition of y_k and the fact that $\lim_{k\to\infty} x_k(s) = x_0(s)$, for every $s \in [t_0, t_0 + \sigma]$, we conclude that

$$\lim_{k\to\infty} y_k(t) = y_0(t), \quad \text{for every } t \in [t_0 - r, t_0 + \sigma],$$

and the proof is finished. □

8

Stability Theory

Suzete M. Afonso[1], *Fernanda Andrade da Silva*[2], *Everaldo M. Bonotto*[3], *Márcia Federson*[2], *Luciene P. Gimenes (in memorian)*[4], *Rogelio Grau*[5], *Jaqueline G. Mesquita*[6], *and Eduard Toon*[7]

[1]*Departamento de Matemática, Instituto de Geociências e Ciências Exatas, Universidade Estadual Paulista "Júlio de Mesquita Filho" (UNESP), Rio Claro, SP, Brazil*
[2]*Departamento de Matemática, Instituto de Ciências Matemáticas e de Computação (ICMC), Universidade de São Paulo, São Carlos, SP, Brazil*
[3]*Departamento de Matemática Aplicada e Estatística, Instituto de Ciências Matemáticas e de Computação (ICMC), Universidade de São Paulo, São Carlos, SP, Brazil*
[4]*Departamento de Matemática, Centro de Ciências Exatas, Universidade Estadual de Maringá, Maringá, PR, Brazil*
[5]*Departamento de Matemáticas y Estadística, División de Ciencias Básicas, Universidad del Norte, Barranquilla, Colombia*
[6]*Departamento de Matemática, Instituto de Ciências Exatas, Universidade de Brasília, Brasília, DF, Brazil*
[7]*Departamento de Matemática, Instituto de Ciências Exatas, Universidade Federal de Juiz de Fora, Juiz de Fora, MG, Brazil*

This chapter is dedicated to the study of the stability theory for generalized ODEs. Here, we consider X a Banach space, equipped with norm $\|\cdot\|$, and we assume that $\Omega = O \times [t_0, \infty)$, where $t_0 \geqslant 0$, and O is an open subset of X such that $0 \in O$ (neutral element of X). We also assume that $F : \Omega \to X$ belongs to the class $\mathcal{F}(\Omega, h)$ introduced in Definition 4.3, where $h : [t_0, \infty) \to \mathbb{R}$ is a left-continuous and nondecreasing function. Throughout this chapter, we deal with stability issues concerning nonautonomous generalized ODEs of type

$$\frac{dx}{d\tau} = DF(x, t), \tag{8.1}$$

in the aforementioned setting. We focus our attention on the study of the stability of the trivial solution of (8.1) for which we consider that the following condition is fulfilled:

(ZS) $F(0, t_2) - F(0, t_1) = 0$ for all $t_1, t_2 \in [t_0, \infty)$.

Generalized Ordinary Differential Equations in Abstract Spaces and Applications, First Edition.
Edited by Everaldo M. Bonotto, Márcia Federson, and Jaqueline G. Mesquita.

Condition (ZS) guarantees that indeed $x \equiv 0$ is a solution of (8.1). In order to check this fact, it is enough to take $U(\tau, t) = F(x(\tau), t)$ in the definition of the Kurzweil integral $\int_{t_1}^{t_2} DU(x, t)$ (see Definition 2.1), with $t_1, t_2 \in [t_0, \infty)$, $t_1 < t_2$, and to verify that the Riemann-type sum, which approximates the integral is only formed by differences of the form

$$U(\tau_i, s_i) - U(\tau_i, s_{i-1}) = F(x(\tau_i), s_i) - F(x(\tau_i), s_{i-1}).$$

Therefore, assumption (ZS) implies that

$$\int_{t_1}^{t_2} DF(0, t) = F(0, t_2) - F(0, t_1) = 0, \quad t_1, t_2 \in [t_0, \infty),$$

which, in turn, implies that $x \equiv 0$ satisfies the generalized ODE (8.1) on $[t_0, \infty)$.

Notice that if we add to $F(x(\tau), t)$ a function that depends only on the variable x, then the solutions of (8.1) do not change. In particular, if we subtract $F(x(\tau), t_0)$ from $F(x(\tau), t)$, then we obtain a normalized representation F_1 of F fulfilling $F_1(z, t_0) = 0$ for every $z \in O$. This fact shows us that, in order to get stability results for any solution of the generalized ODE (8.1), it is sufficient to obtain the same results for the trivial solution of (8.1).

The results on the stability of the trivial solution in the framework of the generalized ODE (8.1) are inspired in the theory, developed by Aleksandr M. Lyapunov on the stability of solutions for classic ODEs. See [166, 167, 170], for instance. In his famous PhD thesis [170], Lyapunov proposed two methods to establish stability of solutions of ODEs. The first method, also known as Indirect Method of Lyapunov, consists of studying the stability of equilibrium points of a nonlinear system by analyzing the stability of the trivial solution for the corresponding linearized system. The second method, which is now referred to as the Lyapunov Stability Criterion or Direct Method of Lyapunov, makes use of a Lyapunov functional to determine criteria for stability.

The results that guarantee that "*the existence of a Lyapunov functional implies stability*" are known as Lyapunov-type theorems. On the other hand, the results that show that "*stability implies the existence of a Lyapunov functional*" are known as converse Lyapunov theorems. Converse Lyapunov theorems confirm the effectiveness of the Direct Method of Lyapunov. If the stability of the trivial solution of the generalized ODE (8.1) guarantees the existence of a Lyapunov functional, we can assure that it is always possible to find such a functional, although this may be an arduous task.

The study of converse Lyapunov theorems for generalized ODEs started in 1984, when Štefan Schwabik introduced the concept of variational stability, which generalizes Lyapunov stability and integral stability, and he established converse Lyapunov theorems for generalized ODEs in the finite dimensional case (see [208]). Later, the authors of [89] established converse Lyapunov theorems for a class of functional differential equations (we write FDEs) with Perron integrable right-hand sides. Recently, motivated by the works [89, 208], the authors of [7] stated converse Lyapunov theorems on regular stability and integral stability for measure FDEs. This was done via generalized ODEs.

In what follows, we present a concept of a Lyapunov functional for generalized ODEs, which we use throughout this chapter.

Recall that O is an open subset of the Banach space X.

Definition 8.1: Let $B \subseteq O$ be any subset. We say that the function $V : [t_0, \infty) \times B \to \mathbb{R}_+$ is a *Lyapunov functional with respect to the generalized ODE* (8.1), if the following conditions are satisfied:

(LF1) $V(\cdot, x) : [t_0, \infty) \to \mathbb{R}_+$ is left-continuous on (t_0, ∞), for all $x \in B$;

(LF2) there exists a continuous strictly increasing function $b : \mathbb{R}_+ \to \mathbb{R}_+$ satisfying $b(0) = 0$ such that

$$V(t, x) \geqslant b(\| x \|),$$

for all $t \in [t_0, \infty)$ and all $x \in B$;

(LF3) for every solution $x : I \to B$ of the generalized ODE (8.1), defined on the interval $I \subset [t_0, \infty)$, the derivative

$$D^+V(t, x(t)) = \limsup_{\eta \to 0^+} \frac{V(t + \eta, x(t + \eta)) - V(t, x(t))}{\eta} \leqslant 0$$

holds for all $t \in I \setminus \sup I$, that is, the right derivative of V along every solution of the generalized ODE (8.1) is non-positive.

We point out that if $x : I \subset [t_0, \infty) \to B$ is a solution of the generalized ODE (8.1), condition (LF3) of Definition 8.1 holds and the function $I \ni t \mapsto V(t, x(t))$ is left-continuous, then

(R1) $V(t_2, x(t_2)) \leqslant V(t_1, x(t_1))$ for all $t_1, t_2 \in I$, with $t_1 \leqslant t_2$.

On the other hand, it is clear that if the solution $x : I \subset [t_0, \infty) \to B$ of the generalized ODE (8.1) satisfies condition (R1), then it satisfies condition (LF3) as well. Therefore, conditions (LF3) and (R1) are somehow related and this allows us to work with the most convenient condition for each type of stability.

This chapter deals with three notions of stability in the framework of generalized ODEs: variational stability, due to Schwabik, Lyapunov stability and also regular stability as introduced in [87]. With respect to Lyapunov stability (also known as uniform stability), we also present results for measure differential equations (as before, we write simply MDEs) and for dynamic equations on time scales. We organized this chapter so that its first section contains results on variational stability and variational asymptotic stability for generalized ODEs. Still in this section we make an account of Lyapunov-type theorems and converse Lyapunov theorems. In Section 8.2, we deal with uniform stability and uniform asymptotic stability of the trivial solution in the setting of the generalized ODE (8.1). Then, using the correspondence between the solutions of generalized ODEs and MDEs (respectively, dynamic equations on time scales) and the results from Section 8.2, we describe Lyapunov-type theorems for MDEs in Section 8.3 (respectively, for dynamic equations on time scales in Section 8.4). Lastly but not least, the results concerning regular stability for generalized ODEs, including a Lyapunov-type theorems and a converse Lyapunov theorems, are brought together in Section 8.5.

It is worth mentioning that the concepts of variational and regular stability concern, respectively, the local behavior of a function of bounded variation and a regulated function, which is initially close to the trivial solution $x \equiv 0$, while the concepts of Lyapunov stability give us the asymptotic behavior of solutions of generalized ODEs.

8.1 Variational Stability for Generalized ODEs

This section is devoted to presenting results on the variational stability of the trivial solution of the generalized ODE (8.1).

The concept of variational stability for ordinary differential equations was introduced by H. Okamura in [191] in 1943, who called it *strong stability*, see [204, 238]. Later, in the year 1959, I. Vrkoč considered, in [234], Carathéodory equations and proved that Okamura's strong stability is equivalent to his concept of integral stability.

For generalized ODEs, the concept of variational stability was introduced by Š. Schwabik in the work [208] from 1984. The starting point of Schwabik's approach was the paper [234] of I. Vrkoč on integral stability, besides the paper [39] by S.-N. Chow and J. A. Yorke from 1974, in which there is an improvement of Vrkoč's results.

As we mentioned in Lemma 4.9, every solution $x : [\alpha, \beta] \to O$ of the generalized ODE (8.1), with $t_0 \leqslant \alpha < \beta < \infty$, is of bounded variation. For this reason, it is natural to measure the distance between two solutions of (8.1) using the variation norm, as Schwabik pointed out in [208], and it seems to be very plausible to use the concept of variational stability (strong stability) introduced by Okamura handled by Vrkoč and others.

We recall that the set $BV([\alpha, \beta], X)$, with $-\infty < \alpha < \beta < \infty$, denotes the Banach space of all functions $x : [\alpha, \beta] \to X$ of bounded variation, endowed with the standard variation norm

$$\| x\|_{BV} = \| x(\alpha) \| + \mathrm{var}_{\alpha}^{\beta}(x), \text{ for all } x \in BV([\alpha, \beta], X),$$

where $\mathrm{var}_{\alpha}^{\beta}(x)$ denotes the variation of the function x on $[\alpha, \beta]$. See Chapter 1 for more details.

The next concepts of stability are borrowed from [208]. See, also, [209]. As previously mentioned, these concepts take into consideration the variation of solutions of the Eq. (8.1) around the trivial solution $x \equiv 0$.

Definition 8.2: The trivial solution $x \equiv 0$ of the generalized ODE (8.1) is

(i) *Variationally stable*, if for every $\epsilon > 0$, there exists $\delta = \delta(\epsilon) > 0$ such that if $\bar{x} : [\gamma, v] \to O$, $t_0 \leqslant \gamma < v < \infty$, is a function of bounded variation in $[\gamma, v]$ and

$$\| \bar{x}(\gamma) \| < \delta \quad \text{and} \quad \mathrm{var}_{\gamma}^{v} \left(\bar{x}(s) - \int_{\gamma}^{s} DF(\bar{x}(\tau), t) \right) < \delta,$$

then

$$\| \bar{x}(t) \| < \epsilon, \text{ for every } t \in [\gamma, v];$$

(ii) *Variationally attracting*, if there exists $\delta_0 > 0$ and for every $\epsilon > 0$, there exist $T = T(\epsilon) \geqslant 0$ and $\rho = \rho(\epsilon) > 0$ such that if $\bar{x} : [\gamma, v] \to O$, with $t_0 \leqslant \gamma < v < \infty$, is a function of bounded variation in $[\gamma, v]$ and

$$\| \bar{x}(\gamma) \| < \delta_0 \quad \text{and} \quad var_{\gamma}^{v} \left(\bar{x}(s) - \int_{\gamma}^{s} DF(\bar{x}(\tau), t) \right) < \rho,$$

then

$$\| \bar{x}(t) \| < \epsilon, \text{ for every } t \in [\gamma, v] \cap [\gamma + T, \infty);$$

(iii) *Variationally asymptotically stable*, if it is both variationally stable and variationally attracting.

Note that if $\bar{x} : [\gamma, v] \to O$, with $t_0 \leqslant \gamma < v < \infty$, is a solution of the generalized ODE (8.1), then

$$\bar{x}(s) - \bar{x}(\gamma) = \int_{\gamma}^{s} DF(\bar{x}(\tau), t), \text{ for } s \in [\gamma, v],$$

and, hence,

$$\mathrm{var}_{\gamma}^{v} \left(\bar{x}(s) - \int_{\gamma}^{s} DF(\bar{x}(\tau), t) \right) = \mathrm{var}_{\gamma}^{v}(\bar{x}(\gamma)) = 0.$$

8.1.1 Direct Method of Lyapunov

We turn our attention to Lyapunov-type theorems for the generalized ODE (8.1).

We start by presenting an auxiliary result, namely, Lemma 8.3, which describes some properties of a possible candidate to a Lyapunov functional. Such result can be found in [208, 209] for the case where the function V is defined in $[0,\infty)\times\mathbb{R}^n$. Here, we consider V defined in the set $[t_0,\infty)\times X$, where X is a Banach space. Moreover, we consider $\Omega = O\times[t_0,\infty)$, where O is an open subset of X such that $0\in O$ and $t_0\geqslant 0$, $F\in\mathcal{F}(\Omega,h)$, and F satisfies (ZS). We omit its proof, since it follows similarly as in [209, Lemma 10.12] (see also [208, Lemma 1]).

Lemma 8.3: *Let $F\in\mathcal{F}(\Omega,h)$. Assume that $V:[t_0,\infty)\times X\to\mathbb{R}$ is such that $V(\cdot,x):[t_0,\infty)\to\mathbb{R}$ is left-continuous on (t_0,∞), for all $x\in X$, and satisfies*

$$|V(t,z)-V(t,y)|\leqslant K\parallel z-y\parallel,\quad z,y\in X,\ t\in[t_0,\infty),\tag{8.2}$$

where $K>0$ is a constant. Suppose, in addition, there exists a function $\Phi:X\to\mathbb{R}$ such that for every solution $x:I\to O$, with $I\subset[t_0,\infty)$, of the generalized ODE (8.1), we have

$$D^+V(t,x(t)) = \limsup_{\eta\to 0^+}\frac{V(t+\eta,x(t+\eta))-V(t,x(t))}{\eta}\leqslant\Phi(x(t)),\quad t\in I\setminus\sup I.\tag{8.3}$$

If $\bar{x}:[\gamma,v]\to O$, $t_0\leqslant\gamma<v<\infty$, is left-continuous on $(\gamma,v]$ and of bounded variation in $[\gamma,v]$, then

$$V(v,\bar{x}(v))-V(\gamma,\bar{x}(\gamma))\leqslant K\,\mathrm{var}_\gamma^v\left(\bar{x}(s)-\int_\gamma^s DF(\bar{x}(\tau),t)\right)+M(v-\gamma),\tag{8.4}$$

where $M=\sup_{t\in[\gamma,v]}\Phi(\bar{x}(t))$.

Sufficient conditions for the trivial solution of the generalized ODE (8.1) to be variationally stable (see Definition 8.2-(i)) are provided in the next result. Š. Schwabik proved this criterion for the case where $X=\mathbb{R}^n$. Due to the similarity with that result, we omit its proof here and we refer the interested reader to [208, Theorem 1] or [209, Theorem 10.13] for more details.

Theorem 8.4: *Let $F\in\mathcal{F}(\Omega,h)$ satisfy condition (ZS) and consider $V:[t_0,\infty)\times\overline{B_\rho}\to\mathbb{R}$ a Lyapunov functional, where $\overline{B_\rho}=\{y\in X:\parallel y\parallel\leqslant\rho\}\subset O$. Assume that V satisfies the additional conditions:*

(i) *$V(t,0)=0$, $t\in[t_0,\infty)$;*

(ii) *There is a constant* $K > 0$ *such that, for every* $t \in [t_0, \infty)$ *and every* $z, y \in \overline{B_\rho}$,

$$|V(t,z) - V(t,y)| \leqslant K \parallel z - y \parallel .$$

Then, the trivial solution $x \equiv 0$ *of the generalized ODE* (8.1) *is variationally stable.*

Under certain additional conditions, the trivial solution of the generalized ODE (8.1) is variationally asymptotically stable. This result was also stated by Š. Schwabik in [208, Theorem 2] for the case where $X = \mathbb{R}^n$. The reader can come out with an analogous proof of [208, Theorem 2] to the next result (see also [209, Theorem 10.14]).

Theorem 8.5: *Assume that* $F \in \mathcal{F}(\Omega, h)$ *satisfies* (ZS) *and* $V : [t_0, \infty) \times \overline{B_\rho} \to \mathbb{R}$ *is a Lyapunov functional, where* $\overline{B_\rho} = \{y \in X : \parallel y \parallel \leqslant \rho\} \subset O$. *Assume, in addition, that* V *satisfies conditions* (i) *and* (ii) *from Theorem 8.4, and that there is a continuous function* $\Phi : X \to \mathbb{R}$, *with* $\Phi(0) = 0$ *and* $\Phi(x) > 0$ *for* $x \neq 0$, *such that for every solution* $x : I \to B_\rho$ *of the generalized ODE* (8.1), *with* $I \subset [t_0, \infty)$, *we have*

$$D^+V(t,x(t)) = \limsup_{\eta \to 0+} \frac{V(t+\eta, x(t+\eta)) - V(t,x(t))}{\eta} \leqslant -\Phi(x(t)), \quad t \in I \setminus \sup I. \tag{8.5}$$

Then, the trivial solution $x \equiv 0$ *of the generalized ODE* (8.1) *is variationally asymptotically stable.*

8.1.2 Converse Lyapunov Theorems

In this subsection, we prove that the variational stability and the variational asymptotic stability imply the existence of a Lyapunov functional with the properties described in Theorems 8.4 and 8.5, respectively. The results described in this subsection are borrowed from [208].

Throughout this subsection, we consider $\Omega = O \times [t_0, \infty)$ where O is an open subset of X such that $0 \in O$ and $t_0 \geqslant 0$, $F \in \mathcal{F}(\Omega, h)$ satisfies (ZS), and we assume the existence of a local solution of the generalized ODE (8.1) on $[s, s + \delta(s)]$ for all $s \geqslant t_0$, where $\delta(s) > 0$ depends on s. The reader may want to consult Theorem 5.1 for a sufficient condition for this existence.

First, we introduce a modified notion of the variation of a given function.

Definition 8.6: Let $-\infty < a < b < \infty$ and $f : [a, b] \to X$ be a given function. For a given division $d = (t_i) \in D_{[a,b]}$ and for every $\lambda \geqslant 0$, we define

$$\sum_{i=1}^{|d|} e^{-\lambda(t_i - t_{i-1})} \parallel f(t_i) - f(t_{i-1}) \parallel = v_\lambda(f, D) \quad \text{and}$$

$$\mathrm{e}_\lambda \mathrm{var}_a^b(f) = \sup_{d \in D_{[a,b]}} v_\lambda(f, D).$$

The number $\mathrm{e}_\lambda \mathrm{var}_a^b(f)$ is called the e_λ-*variation* of the function f over $[a, b]$.

The next result relates the concept of the e_λ-variation with the concept of variation of a given function. For a proof of this fact, we refer the interested reader to [209, Lemma 10.16] for the case where X is the n-dimensional Euclidean space. The proof for the case where X is an infinite dimensional Banach space is analogous and, therefore, it will be omitted here.

Lemma 8.7: *Let $f : [a, b] \to X$ be a given function with $-\infty < a < b < \infty$. Then,*

$$e^{-\lambda(b-a)} \mathrm{var}_a^b(f) \leqslant e_\lambda \mathrm{var}_a^b(f) \leqslant \mathrm{var}_a^b(f), \tag{8.6}$$

for all $\lambda \geqslant 0$. Moreover, if $a \leqslant c \leqslant b$, then the identity

$$e_\lambda \mathrm{var}_a^b(f) = e^{-\lambda(b-c)} e_\lambda \mathrm{var}_a^c(f) + e_\lambda \mathrm{var}_c^b(f) \tag{8.7}$$

holds for all $\lambda \geqslant 0$.

The following result gives us an important propriety of e_λ-variation and follows directly from Lemma 8.7.

Corollary 8.8: *If $a \leqslant c \leqslant b$ and $\lambda \geqslant 0$, then $e_\lambda \mathrm{var}_a^c(f) \leqslant e_\lambda \mathrm{var}_a^b(f)$.*

Now, let us consider a special set of functions of locally bounded variation. Let $a > 0$ be such that $B_a = \{x \in X : \| x \| < a\} \subset O$. For $t > t_0$ and $x \in B_a$, we consider the set

(SA) $A_a(t, x)$ of all functions $\varphi : [t_0, \infty) \to X$ of locally bounded variation in $[t_0, \infty)$, such that $\varphi(t_0) = 0$, $\varphi(t) = x$, $\sup_{s \in [t_0, t]} \| \varphi(s) \| < a$ and φ is left-continuous on (t_0, ∞).

Moreover, for all $\lambda \geqslant 0$, $s \geqslant t_0$, and $x \in B_a$, define $V_\lambda : [t_0, \infty) \times B_a \to \mathbb{R}$ by

$$V_\lambda(s, x) = \begin{cases} \inf\limits_{\varphi \in A_a(s,x)} \left\{ \mathrm{e}_\lambda \mathrm{var}_{t_0}^s \left(\varphi(\sigma) - \int_{t_0}^{\sigma} DF(\varphi(\tau), t) \right) \right\}, & s > t_0, \\ \| x \|, & s = t_0. \end{cases} \tag{8.8}$$

Since φ is of locally bounded variation, the Kurzweil integral $\int_{t_0}^{\sigma} DF(\varphi(\tau), t)$ exists for all $\sigma \in [t_0, s]$ and the function

$$[t_0, s] \ni \sigma \mapsto \varphi(\sigma) - \int_{t_0}^{\sigma} DF(\varphi(\tau), t)$$

is also of bounded variation in $[t_0, s]$ for all $s \in [t_0, \infty)$, by Corollary 4.8. Furthermore, the e_λ-variation of this function is bounded. Therefore, V_λ is well-defined for all $a > 0$ and $\lambda \geqslant 0$. Additionally, the zero function $\varphi \equiv 0$ belongs to the set $A_a(s, 0)$ and, hence,

$$V_\lambda(s, 0) = 0, \quad \text{for all } s \geqslant t_0 \text{ and } \lambda \geqslant 0. \tag{8.9}$$

On the other hand, for all $s \geqslant t_0$ and $x \in B_a$, we have $V_\lambda(s, x) \geqslant 0$, as

$$e_\lambda \text{var}_{t_0}^{s} \left(\varphi(\sigma) - \int_{t_0}^{\sigma} DF(\varphi(\tau), t) \right) \geqslant 0, \quad \text{for all } \varphi \in A_a(s, x).$$

Lemma 8.9: *Let $F \in \mathcal{F}(\Omega, h)$. For $x, y \in B_a$, $s \in [t_0, \infty)$ and $\lambda \geqslant 0$, the functional V_λ given by (8.8) satisfies the inequality*

$$|V_\lambda(s, x) - V_\lambda(s, y)| \leqslant \| x - y \| .$$

Proof. If $s = t_0$, then

$$V_\lambda(t_0, y) - V_\lambda(t_0, x) = \| y \| - \| x \| \leqslant \| y - x \|$$

and the proof is complete.

Now, assume that $s > t_0$. Let $\eta > 0$ be such that $0 < \eta < s - t_0$ and let $\varphi \in A_a(s, x)$ be an arbitrary function, where $A_a(s, x)$ is defined in (SA).

Define a function $\varphi_\eta : [t_0, \infty) \to X$ by

$$\varphi_\eta(\sigma) = \begin{cases} \varphi(\sigma), & \text{if } \sigma \in [t_0, s - \eta], \\ \varphi(s - \eta) + \dfrac{y - \varphi(s - \eta)}{\eta}(\sigma - s + \eta), & \text{if } \sigma \in [s - \eta, s], \\ y, & \text{if } \sigma \in (s, \infty). \end{cases}$$

Clearly, $\varphi_\eta \in A_a(s, y)$. Then, by (8.7), we have

$$\begin{aligned}
V_\lambda(s, y) &\leqslant e_\lambda \text{var}_{t_0}^{s} \left(\varphi_\eta(\sigma) - \int_{t_0}^{\sigma} DF(\varphi_\eta(\tau), t) \right) \\
&= e^{-\lambda\eta} e_\lambda \text{var}_{t_0}^{s-\eta} \left(\varphi_\eta(\sigma) - \int_{t_0}^{\sigma} DF(\varphi_\eta(\tau), t) \right) \\
&\quad + e_\lambda \text{var}_{s-\eta}^{s} \left(\varphi_\eta(\sigma) - \int_{t_0}^{\sigma} DF(\varphi_\eta(\tau), t) \right) \\
&= e^{-\lambda\eta} e_\lambda \text{var}_{t_0}^{s-\eta} \left(\varphi(\sigma) - \int_{t_0}^{\sigma} DF(\varphi(\tau), t) \right) \\
&\quad + e_\lambda \text{var}_{s-\eta}^{s} \left(\varphi_\eta(\sigma) - \int_{t_0}^{\sigma} DF(\varphi_\eta(\tau), t) \right) \\
&\leqslant e^{-\lambda\eta} e_\lambda \text{var}_{t_0}^{s-\eta} \left(\varphi(\sigma) - \int_{t_0}^{\sigma} DF(\varphi(\tau), t) \right)
\end{aligned}$$

$$+ \operatorname{var}_{s-\eta}^{s}(\varphi_\eta) + \operatorname{var}_{s-\eta}^{s}\left(\int_{t_0}^{\sigma} DF(\varphi_\eta(\tau), t)\right)$$

$$\leqslant e^{-\lambda\eta} \mathrm{e}_\lambda \operatorname{var}_{t_0}^{s-\eta}\left(\varphi(\sigma) - \int_{t_0}^{\sigma} DF(\varphi(\tau), t)\right)$$

$$+ \| y - \varphi(s-\eta) \| + h(s) - h(s-\eta),$$

where the last inequality follows from Lemma 4.5. By the fact that $\eta > 0$, we obtain

$$e^{-\lambda\eta} \mathrm{e}_\lambda \operatorname{var}_{t_0}^{s-\eta}\left(\varphi(\sigma) - \int_{t_0}^{\sigma} DF(\varphi(\tau), t)\right) \leqslant \mathrm{e}_\lambda \operatorname{var}_{t_0}^{s}\left(\varphi(\sigma) - \int_{t_0}^{\sigma} DF(\varphi(\tau), t)\right).$$

Therefore, we conclude that

$$V_\lambda(s, y) \leqslant \mathrm{e}_\lambda \operatorname{var}_{t_0}^{s}\left(\varphi(\sigma) - \int_{t_0}^{\sigma} DF(\varphi(\tau), t)\right) + \| y - \varphi(s-\eta) \| + h(s) - h(s-\eta).$$

Then, as $\eta \to 0^+$, we obtain

$$V_\lambda(s, y) \leqslant \mathrm{e}_\lambda \operatorname{var}_{t_0}^{s}\left(\varphi(\sigma) - \int_{t_0}^{\sigma} DF(\varphi(\tau), t)\right) + \| y - x \| .$$

Since the choice of $\varphi \in A_a(s, x)$ is arbitrary, we can take the infimum for all $\varphi \in A_a(s, x)$ on the right-hand side of the last inequality, obtaining

$$V_\lambda(s, y) \leqslant V_\lambda(s, x) + \| y - x \| .$$

Analogously, we can prove that $V_\lambda(s, x) \leqslant V_\lambda(s, y) + \| y - x \|$, whence it follows that $|V_\lambda(s, x) - V_\lambda(s, y)| \leqslant \| x - y \|$. □

The proof of the next result can be found in [209, Lemma 10.20] for the case where X is the n-dimensional Euclidean space. The proof for the case where X is an arbitrary infinite dimensional Banach space is similar. For this reason, we omit it here.

Lemma 8.10: *Let $F \in \mathcal{F}(\Omega, h)$, $y \in B_a$, $s, r \in [t_0, \infty)$, $\lambda \geqslant 0$ and V_λ be defined by* (8.8). *Then, the following inequality holds*

$$|V_\lambda(r, y) - V_\lambda(s, y)| \leqslant (1 - e^{-\lambda|r-s|})a + |h(r) - h(s)|.$$

Corollary 8.11 in the sequel is an immediate consequence of Lemmas 8.9 and 8.10.

Corollary 8.11: *Let $F \in \mathcal{F}(\Omega, h)$, $\lambda \geqslant 0$ and V_λ be defined by* (8.8). *Then, for all $x, y \in B_a$ and all $r, s \in [t_0, \infty)$, we have*

$$|V_\lambda(s, x) - V_\lambda(r, y)| \leqslant \| x - y \| + (1 - e^{-\lambda|r-s|})a + |h(r) - h(s)|.$$

The next result deals with the right derivative of V_λ.

Lemma 8.12: *Let $F \in \mathcal{F}(\Omega, h)$. If $x : [s, s+\delta(s)] \to O$ is a solution of the generalized ODE (8.1), with $s \geqslant t_0$ and $\delta(s) > 0$, then the function V_λ defined by (8.8), for all $\lambda \geqslant 0$, satisfies*

$$\limsup_{\eta \to 0^+} \frac{V_\lambda(s+\eta, x(s+\eta)) - V_\lambda(s, x(s))}{\eta} \leqslant -\lambda V_\lambda(s, x(s)).$$

Proof. Choose an arbitrary $x_0 \in O$ and let $a > 0$ be such that $a > \| x_0 \| + h(s+1) - h(s)$. Now, let $\varphi \in A_a(s, x_0)$ be arbitrary and $x : [s, s+\delta(s)] \to O$ be the solution of the generalized ODE (8.1) with $x(s) = x_0$.

Let $0 < \eta < \min\{\delta(s), 1\}$ and define

$$\varphi_\eta(\sigma) = \begin{cases} \varphi(\sigma), & \text{if } \sigma \in [t_0, s], \\ x(\sigma), & \text{if } \sigma \in [s, s+\eta], \\ x(s+\eta), & \text{if } \sigma \in (s+\eta, \infty). \end{cases}$$

Then, $\varphi(s) = x(s) = \varphi_\eta(s) = x_0$ and $\varphi_\eta \in A_a(s+\eta, x(s+\eta))$. Therefore, using Lemma 4.5 and the fact that x is solution of the generalized ODE (8.1), we obtain

$$\begin{aligned} \| x(\sigma) \| &= \left\| x_0 + \int_s^\sigma DF(x(\tau), t) \right\| \\ &\leqslant \| x_0 \| + h(\sigma) - h(s) \\ &\leqslant \| x_0 \| + h(s+1) - h(s) < a, \end{aligned}$$

for $\sigma \in [s, s+\eta]$, provided h is nondecreasing. Hence, by (8.7), we have

$$\begin{aligned} V_\lambda(s+\eta, x(s+\eta)) &\leqslant \mathrm{e}_\lambda \mathrm{var}_{t_0}^{s+\eta} \left(\varphi_\eta(\sigma) - \int_{t_0}^\sigma DF(\varphi_\eta(\tau), t) \right) \\ &= e^{-\lambda\eta} \mathrm{e}_\lambda \mathrm{var}_{t_0}^{s} \left(\varphi(\sigma) - \int_{t_0}^\sigma DF(\varphi(\tau), t) \right) \\ &\quad + \mathrm{e}_\lambda \mathrm{var}_{s}^{s+\eta} \left(x(\sigma) - \int_{t_0}^s DF(\varphi(\tau), t) - \int_s^\sigma DF(x(\tau), t) \right) \\ &= e^{-\lambda\eta} \mathrm{e}_\lambda \mathrm{var}_{t_0}^{s} \left(\varphi(\sigma) - \int_{t_0}^\sigma DF(\varphi(\tau), t) \right) \\ &\quad + \mathrm{e}_\lambda \mathrm{var}_{s}^{s+\eta} \left(x_0 - \int_{t_0}^s DF(\varphi(\tau), t) \right) \\ &= e^{-\lambda\eta} \mathrm{e}_\lambda \mathrm{var}_{t_0}^{s} \left(\varphi(\sigma) - \int_{t_0}^\sigma DF(\varphi(\tau), t) \right), \end{aligned}$$

since

$$\mathrm{var}_{s}^{s+\eta} \left(x_0 - \int_{t_0}^s DF(\varphi(\tau), t) \right) = 0.$$

Taking the infimum for every $\varphi \in A_a(s, x_0)$ on the right-hand side of the inequality obtained above, we get

$$V_\lambda(s+\eta, x(s+\eta)) \leqslant e^{-\lambda\eta} V_\lambda(s, x_0) = e^{-\lambda\eta} V_\lambda(s, x(s)),$$

consequently,

$$V_\lambda(s+\eta, x(s+\eta)) - V_\lambda(s, x(s)) \leqslant (e^{-\lambda\eta} - 1) V_\lambda(s, x(s))$$

and we finally obtain

$$\limsup_{\eta\to 0^+} \frac{V_\lambda(s+\eta, x(s+\eta)) - V_\lambda(s, x(s))}{\eta} \leqslant -\lambda V_\lambda(s, x(s)),$$

because $\lim_{\eta\to 0^+} \frac{e^{-\lambda\eta}-1}{\eta} = -\lambda$. □

In order to establish converse theorems, we need a different concept of variational stability and an auxiliary result. For this purpose, consider the following nonhomogeneous generalized ODE

$$\frac{dx}{d\tau} = D[F(x, t) + P(t)], \tag{8.10}$$

where $F \in \mathcal{F}(\Omega, h)$ satisfies condition (ZS) and $P : [t_0, \infty) \to X$ is a Kurzweil integrable function.

Definition 8.13: The trivial solution $x \equiv 0$ of the generalized ODE (8.1) is

(i) *Variationally stable with respect to perturbations*, if for every $\epsilon > 0$, there exists $\delta = \delta(\epsilon) > 0$ such that if $\| x_0 \| < \delta$ and $P \in G^-([\gamma, v], X)$, with $\mathrm{var}_\gamma^v(P) < \delta$, then

$$\| x(t, \gamma, x_0) \| = \| x(t) \| < \epsilon, \quad \text{for every } t \in [\gamma, v],$$

where $x(\cdot, \gamma, x_0)$ is the solution of the nonhomogeneous generalized ODE (8.10), with $x(\gamma, \gamma, x_0) = x_0$ and $[\gamma, v] \subset [t_0, \infty)$;

(ii) *Variationally attracting with respect to perturbations*, if there is a $\widetilde{\delta} > 0$ and for every $\epsilon > 0$, there exist $T = T(\epsilon) \geqslant 0$ and $\rho = \rho(\epsilon) > 0$ such that if $\| x_0 \| < \widetilde{\delta}$ and $P \in G^-([\gamma, v], X)$ with $\mathrm{var}_\gamma^v(P) < \rho$, then

$$\| x(t, \gamma, x_0) \| = \| x(t) \| < \epsilon, \quad \text{for every } t \geqslant \gamma + T, \ t \in [\gamma, v],$$

where $x(\cdot, \gamma, x_0)$ is the solution of the nonhomogeneous generalized ODE (8.10) with $x(\gamma, \gamma, x_0) = x_0$ and $[\gamma, v] \subset [t_0, \infty)$;

(iii) *Variationally asymptotically stable with respect to perturbations*, if it is both variationally stable and variationally attracting with respect to perturbations.

The proof of the next result can be found in [89, Proposition 3.1].

Proposition 8.14: *Let $F \in \mathcal{F}(\Omega, h)$ satisfy condition (ZS). The following statements hold.*

(i) *The trivial solution $x \equiv 0$ of the generalized ODE (8.1) is variationally stable if and only if it is variationally stable with respect to perturbations.*
(ii) *The trivial solution $x \equiv 0$ of the generalized ODE (8.1) is variationally attracting if and only if it is variationally attracting with respect to perturbations.*
(iii) *The trivial solution $x \equiv 0$ of the generalized ODE (8.1) is variationally asymptotically stable if and only if it is variationally asymptotically stable with respect to perturbations.*

Now, we are able to prove the converse theorems for variational stability and variational asymptotic stability of the trivial solution $x \equiv 0$ of the generalized ODE (8.1). The next results are borrowed from [89].

Theorem 8.15: *Let $F \in \mathcal{F}(\Omega, h)$ satisfy condition (ZS). If the trivial solution $x \equiv 0$ of the generalized ODE (8.1) is variationally stable, then, for every $a > 0$ such that $B_a = \{x \in X : \| x \| < a\} \subset O$, there exists a function $V : [t_0, \infty) \times B_a \to \mathbb{R}$, satisfying the following conditions:*

(i) *For every $x \in B_a$, the function $V(\cdot, x)$ is left-continuous on (t_0, ∞);*
(ii) *$V(t, 0) = 0$ and $|V(t, x) - V(t, y)| \leqslant \| x - y \|$ for all $x, y \in B_a$, $t \in [t_0, \infty)$;*
(iii) *For every solution $x : [s, s + \delta(s)] \to B_a$, with $\delta(s) > 0$ and $s \geqslant t_0$, of the generalized ODE (8.1), we have*

$$\limsup_{\eta \to 0^+} \frac{V(s + \eta, x(s + \eta)) - V(s, x(s))}{\eta} \leqslant 0,$$

that is, the right derivative of V along every solution $x(\cdot)$ of (8.1) is non-positive;
(iv) *There exists a continuous increasing function $b : \mathbb{R}_+ \to \mathbb{R}_+$ such that $b(0) = 0$ and $b(\| x \|) \leqslant V(t, x)$, for all $x \in B_a$ and $t \in [t_0, \infty)$.*

Proof. Let $V_0(s, x)$ be defined by (8.8) with $\lambda = 0$. Let us define

$$V(s, x) = V_0(s, x), \quad \text{for all } s \geqslant t_0 \text{ and } x \in B_a.$$

Item (i) is a consequence of Corollary 8.11. By Lemma 8.9 and Eq. (8.9), we obtain item (ii). By Lemma 8.12, item (iii) is also satisfied. Lastly, we need to verify that condition (iv) holds. In order to do that, we use the variational stability of the solution $x \equiv 0$ of the generalized ODE (8.1). We assume that item (iv) does not hold. Therefore, there exist $\epsilon > 0$, with $0 < \epsilon < a$, and a sequence $\{(t_k, x_k)\}_{k \in \mathbb{N}}$ such that $\epsilon < \| x_k \| < a$, $t_k \to \infty$, and $V(t_k, x_k) \to 0$ as $k \to \infty$.

By Proposition 8.14 (i), we can assume that the trivial solution $x \equiv 0$ of the generalized ODE (8.1) is variationally stable with respect to perturbations. Then, take $\delta = \delta(\epsilon) > 0$ as in Definition 8.13 (i). Let $k_0 \in \mathbb{N}$ be such that $V(t_k, x_k) < \delta$ for all $k > k_0$. Hence, there exists $\varphi_k \in A_a(t_k, x_k)$ satisfying

$$\text{var}_{t_0}^{t_k}\left(\varphi_k(\sigma) - \int_{t_0}^{\sigma} DF(\varphi_k(\tau), t)\right) < \delta.$$

Define a function $P : [t_0, \infty) \to X$ by

$$P(\sigma) = \begin{cases} \varphi_k(\sigma) - \displaystyle\int_{t_0}^{\sigma} DF(\varphi_k(\tau), t), & \text{for } \sigma \in [t_0, t_k], \\ x_k - \displaystyle\int_{t_0}^{t_k} DF(\varphi_k(\tau), t), & \text{for } \sigma \in [t_k, \infty). \end{cases}$$

Hence,

$$\text{var}_{t_0}^{\infty}(P) = \text{var}_{t_0}^{t_k}\left(\varphi_k(\sigma) - \int_{t_0}^{\sigma} DF(\varphi_k(\tau), t)\right) < \delta$$

and the function P is left-continuous. For $\sigma \in [t_0, t]$, we have

$$\begin{aligned}
\varphi_k(\sigma) &= \varphi_k(\sigma) + \int_{t_0}^{\sigma} DF(\varphi_k(\tau), t) - \int_{t_0}^{\sigma} DF(\varphi_k(\tau), t) \\
&= \varphi_k(t_0) + \int_{t_0}^{\sigma} DF(\varphi_k(\tau), t) + P(\sigma) \\
&= \varphi_k(t_0) + \int_{t_0}^{\sigma} DF(\varphi_k(\tau), t) + P(\sigma) - P(t_0) \\
&= \varphi_k(t_0) + \int_{t_0}^{\sigma} D[F(\varphi_k(\tau), t) + P(t)],
\end{aligned}$$

since $\varphi_k \in A_a(t_k, x_k)$ and $P(t_0) = \varphi_k(t_0) = 0$. Therefore, φ_k is a solution of the nonhomogeneous generalized ODE

$$\frac{dy}{d\tau} = D[F(y, t) + P(t)].$$

Then, once the trivial solution $x \equiv 0$ of the generalized ODE (8.1) is variationally stable with respect to perturbations, we conclude that $\| \varphi_k(s) \| < \epsilon$ for all $s \in [t_0, t_k]$. In particular, we have $\| \varphi_k(t_k) \| = \| x_k \| < \epsilon$, which contradicts our assumption. □

Theorem 8.16: *Let $F \in \mathcal{F}(\Omega, h)$ satisfy condition (ZS). If the trivial solution $x \equiv 0$ of the generalized ODE (8.1) is variationally asymptotically stable, then there exist*

$a > 0$ such that $B_a = \{x \in X : \| x \| < a\} \subset O$ and a function $V : [t_0, \infty) \times B_a \to \mathbb{R}$ satisfying the following conditions:

(i) *for every $x \in B_a$, the function $V(\cdot, x)$ is left-continuous on (t_0, ∞);*
(ii) *$V(t, 0) = 0$ and $|V(t,x) - V(t,y)| \leqslant \| x - y \|$ for all $x, y \in B_a$, $t \in [t_0, \infty)$;*
(iii) *for every solution $x : [s, s+\delta(s)] \to B_a$, with $\delta(s) > 0$ and $s \geqslant t_0$, of the generalized ODE (8.1) we have*

$$\limsup_{\eta \to 0^+} \frac{V(s+\eta, x(s+\eta)) - V(s, x(s))}{\eta} \leqslant -t_0 V(t, x(s));$$

(iv) *there exists a continuous increasing function $b : \mathbb{R}_+ \to \mathbb{R}_+$ such that $b(0) = 0$ and $b(\| x \|) \leqslant V(t,x)$ for every $x \in B_a$, $t \in [t_0, \infty)$.*

Proof. By Proposition 8.14 (ii), the trivial solution $x \equiv 0$ of the generalized ODE (8.1) is variationally attracting with respect to perturbations, since it is variationally asymptotically stable. Take $\widetilde{\delta}$ as in Definition 8.13 (i) and $0 < a < \widetilde{\delta}$. Let $V_{t_0}(s, x)$ be defined by (8.8) with $\lambda = t_0$. We define

$$V(s,x) = V_{t_0}(s,x), \text{ for all } s \geqslant t_0 \text{ and } x \in B_a. \tag{8.11}$$

Items (i), (ii), and (iii) follow in the same way as in the proof of Theorem 8.15. Assume that condition (iv) does not hold. Then, there exists ϵ, with $0 < \epsilon < a$, and a sequence $\{(t_k, x_k)\}_{k \in \mathbb{N}}$ such that $\epsilon \leqslant \| x_k \| < a$, $t_k \to \infty$ and $V(t_k, x_k) \to 0$ as $k \to \infty$.

Once the trivial solution $x \equiv 0$ of the generalized ODE (8.1) is variationally attracting with respect to perturbations, there are $T = T(\epsilon) \geqslant 0$ and $\rho(\epsilon) > 0$ such that if $\| y_0 \| < \widetilde{\delta}$ and $P \in G^-([\alpha, \beta], X)$, with $\mathrm{var}_\alpha^\beta(P) < \rho(\epsilon)$, then

$$\| y(t, \alpha, y_0) \| < \epsilon, \quad \text{for all } t \in [\alpha, \beta] \cap [\alpha + T, \infty) \text{ and } \alpha \geqslant t_0,$$

where $y(\cdot, \alpha, y_0)$ is a solution of the nonhomogeneous generalized ODE

$$\frac{dy}{d\tau} = D[F(y,t) + P(t)]$$

satisfying $y(\alpha, \alpha, y_0) = y_0$.

Choose $k_0 \in \mathbb{N}$ such that $t_k > T + t_0$ and $V(t_k, x_k) < \rho(\epsilon) e^{-t_0 T}$ for all $k \in \mathbb{N}$, $k \geqslant k_0$. Fix $k \geqslant k_0$. By the definition of V given by (8.11), we can select $\varphi \in A_a(t_k, x_k)$ such that

$$\mathrm{e}_{t_0} \mathrm{var}_{t_0}^{t_k} \left(\varphi(\sigma) - \int_{t_0}^{\sigma} DF(\varphi(\tau), t) \right) < \rho(\epsilon) e^{-t_0 T}.$$

Defining $\alpha = t_k - T$, we have $\alpha > t_0$, since $t_k > T + t_0$. Hence,

$$\mathrm{e}_{t_0} \mathrm{var}_{\alpha}^{t_k} \left(\varphi(\sigma) - \int_{t_0}^{\sigma} DF(\varphi(\tau), t) \right) < \rho(\epsilon) e^{-t_0 T}.$$

Moreover, by (8.6), we also have

$$\begin{aligned}
e^{-t_0 T}\operatorname{var}_{\alpha}^{t_k}\left(\varphi(\sigma)-\int_{t_0}^{\sigma} DF(\varphi(\tau),t)\right) &= e^{-t_0(t_k-\alpha)}\operatorname{var}_{\alpha}^{t_k}\left(\varphi(\sigma)-\int_{t_0}^{\sigma} DF(\varphi(\tau),t)\right)\\
&\overset{(8.6)}{\leqslant} e_{t_0}\operatorname{var}_{\alpha}^{t_k}\left(\varphi(\sigma)-\int_{t_0}^{\sigma} DF(\varphi(\tau),t)\right)\\
&\leqslant \rho(\epsilon)e^{-t_0 T}.
\end{aligned}$$

Therefore,

$$\operatorname{var}_{\alpha}^{t_k}\left(\varphi(\sigma)-\int_{t_0}^{\sigma} DF(\varphi(\tau),t)\right) < \rho(\epsilon). \tag{8.12}$$

For $\sigma \in [\alpha, t_k]$, define

$$P(\sigma) = \varphi(\sigma) - \int_{t_0}^{\sigma} DF(\varphi(\tau), t).$$

Then, $P \in G^{-}([\alpha, t_k], X)$ and, by (8.12), $\operatorname{var}_{\alpha}^{t_k}(P) < \rho(\epsilon)$. Furthermore, notice that $\varphi : [\alpha, t_k] \to X$ is a solution of the nonhomogeneous generalized ODE

$$\frac{dy}{d\tau} = D[F(y,t) + P(t)].$$

Indeed, we have

$$\begin{aligned}
\varphi(\sigma) &= \varphi(\sigma) + \int_{t_0}^{\sigma} DF(\varphi(\tau),t) - \int_{t_0}^{\sigma} DF(\varphi(\tau),t)\\
&= \int_{t_0}^{\sigma} DF(\varphi(\tau),t) + P(\sigma)\\
&= P(\alpha) + \int_{t_0}^{\alpha} DF(\varphi(\tau),t) + \int_{\alpha}^{\sigma} DF(\varphi(\tau),t) + P(\sigma) - P(\alpha)\\
&= \varphi(\alpha) + \int_{\alpha}^{\sigma} D[F(\varphi(\tau),t) + P(t)].
\end{aligned}$$

Since $\varphi \in A_a(t_k, x_k)$ and $a < \widetilde{\delta}$, we get $\| \varphi(\alpha) \| < \widetilde{\delta}$ and, by definition of the variational attractivity of the trivial solution, the inequality $\| \varphi(t) \| < \epsilon$ is satisfied, for every $t \geqslant \alpha + T$. This is also valid for $t = t_k = \alpha + T$, that is, $\| \varphi(t_k) \| = \| x_k \| < \epsilon$ and this contradicts the fact that $\| x_k \| \geqslant \epsilon$. □

8.2 Lyapunov Stability for Generalized ODEs

This section is devoted to the study of Lyapunov (sometimes known as uniform) stability theory in the context of generalized ODEs and it is divided into two subsections.

Recall that we are considering that the function $F : \Omega \to X$ belongs to the class $\mathcal{F}(\Omega, h)$, where $\Omega = O \times [t_0, \infty)$, O is an open subset of X such that $0 \in O$, $h : [t_0, \infty) \to \mathbb{R}$ is a left-continuous and nondecreasing function, and F satisfies (ZS). Denote by $x(\cdot) = x(\cdot, s_0, x_0)$ the maximal solution $x : [s_0, \omega(s_0, x_0)) \to O$ of the generalized ODE (8.1), with initial condition $x(s_0) = x_0$, where $t_0 \leqslant s_0 < \omega(s_0, x_0) \leqslant \infty$. The existence of such a solution is assumed throughout this section and can be guaranteed by Theorem 5.11, if we consider $\Omega = \Omega_F$, where Ω_F is given by (5.13).

In the following lines, we present the classical definitions of stability, uniform stability and uniform asymptotic stability for the trivial solution of the generalized ODE (8.1), which were first introduced in [80].

Definition 8.17: The trivial solution of the generalized ODE (8.1) is

(i) *Stable*, if for any $s_0 \geqslant t_0$ and $\epsilon > 0$, there exists $\delta = \delta(\epsilon, s_0) > 0$ such that if $x_0 \in O$ with $\| x_0 \| < \delta$, then

$$\| x(t, s_0, x_0) \| = \| x(t) \| < \epsilon, \quad \text{for all } t \in [s_0, \omega(s_0, x_0)),$$

where $x(\cdot, s_0, x_0)$ is the solution of the generalized ODE (8.1) with $x(s_0, s_0, x_0) = x_0$;

(ii) *Uniformly stable (Lyapunov stable)*, if it is stable with δ independent of s_0;

(iii) *Uniformly asymptotically stable*, if there exists $\delta_0 > 0$ such that for every $\epsilon > 0$, there exists $T = T(\epsilon) \geqslant 0$ such that if $s_0 \geqslant t_0$ and $x_0 \in O$, with $\| x_0 \| < \delta_0$, then

$$\| x(t, s_0, x_0) \| < \epsilon, \quad \text{for all } t \in [s_0, \omega(s_0, x_0)) \cap [s_0 + T(\epsilon), \infty),$$

where $x(\cdot, s_0, x_0)$ is the solution of the generalized ODE (8.1) with $x(s_0, s_0, x_0) = x_0$.

8.2.1 Direct Method of Lyapunov

By using the Lyapunov direct method, we present criteria of uniform stability and uniform asymptotic stability for generalized ODEs. The results described here do not require any Lipschitz-type condition concerning the Lyapunov functional. See [80] for more details.

Theorem 8.18: *Let $F \in \mathcal{F}(\Omega, h)$ satisfy condition (ZS) and consider $V : [t_0, \infty) \times \overline{B}_\rho \to \mathbb{R}$ a Lyapunov functional with respect to the generalized ODE (8.1), where $\rho > 0$, $\overline{B}_\rho = \{x \in X : \| x \| \leqslant \rho\}$, and $\overline{B}_\rho \subset O$. Assume that*

(i) *There exists a continuous increasing function $a : \mathbb{R}_+ \to \mathbb{R}_+$ such that $a(0) = 0$ and $V(t, x) \leqslant a(\| x \|)$, for all $t \in [t_0, \infty)$ and $x \in \overline{B}_\rho$;*

(ii) *The function $I \ni t \mapsto V(t, x(t))$, $I \subset [t_0, \infty)$, is nonincreasing along every solution $x : I \to \overline{B}_\rho$ of the generalized ODE (8.1).*

Then, the trivial solution $x \equiv 0$ of the generalized ODE (8.1) is uniformly stable.

Proof. By the definition of a Lyapunov functional with respect to the generalized ODE (8.1) (see Definition 8.1, condition (LF2)), there exists a continuous strictly increasing function $b : \mathbb{R}_+ \to \mathbb{R}_+$ such that

$$V(t,x) \geqslant b(\| x \|) \quad \text{for all } (t,x) \in [t_0,\infty) \times \overline{B}_\rho. \tag{8.13}$$

Since $a(0) = 0$ and a is continuous, for all $\epsilon > 0$, $a|_{[0,\epsilon]}$ is uniformly continuous, $b(\epsilon) > 0$ and, therefore, there exists δ depending on ϵ such that $a(\delta) < b(\epsilon)$.

Let $x_0 \in \overline{B}_\rho$ and $x(t) = x(t, s_0, x_0)$, with $t \in [s_0, \omega(s_0, x_0))$, be the maximal solution of the generalized ODE (8.1). Assume that $\| x(s_0) \| = \| x_0 \| < \delta$. Condition (ii) and (8.13) imply that

$$b(\| x(t) \|) \leqslant V(t, x(t)) \leqslant V(s_0, x_0) \leqslant a(\| x_0 \|) < a(\delta) < b(\epsilon),$$

for all $t \in [s_0, \omega(s_0, x_0))$. Then, because b is an increasing function, we conclude that $\| x(t) \| < \epsilon$, for all $t \in [s_0, \omega(s_0, x_0))$, whence it follows that the trivial solution of the generalized ODE (8.1) is uniformly stable. □

In the sequel, we present an example of Lyapunov functional, which does not satisfy Lipschitz-type condition. See [80, Example 2.6].

Example 8.19: Consider $\Omega = (-1,1) \times [0,\infty)$ and let $F : \Omega \to \mathbb{R}$ be given by

$$F(x,t) = -ktx,$$

where $k > 0$ is fixed. Define $h : [0,\infty) \to \mathbb{R}$ by

$$h(t) = kt, \quad \text{for all } t \in [0,\infty).$$

Note that, by definition, the function h is left-continuous on $(0,\infty)$ and increasing on $[0,\infty)$. Also, it is easy to prove that $F \in \mathcal{F}(\Omega, h)$. Consider the generalized ODE

$$\frac{dx}{d\tau} = DF(x,t) = D[-ktx]. \tag{8.14}$$

The generalized ODE (8.14) is associated with the following autonomous ODE

$$\frac{dx}{dt} = -kx. \tag{8.15}$$

Notice that x is a solution of the generalized ODE (8.14) if and only if x is a solution of the autonomous ODE (8.15) (see [80, Remark 2.7]). Therefore, $x(t) = x_0 e^{-k(t-s_0)}$ is the solution of the generalized ODE (8.14), with initial condition $x(s_0) = x_0 \in (-1,1)$.

Define $V : [0,\infty) \times \mathbb{R} \to \mathbb{R}$ by

$$V(t,x) = x^2.$$

Since V does not depend on the first variable, the function $V(\cdot, x) : [0,\infty) \to \mathbb{R}$ is left-continuous on $(0,\infty)$, for all $x \in \mathbb{R}$.

In addition, if $x : [\alpha,\beta] \to (-1,1)$, $[\alpha,\beta] \subset \mathbb{R}_+$, is a solution of the generalized ODE (8.14), then

$$\begin{aligned} D^+V(t,x(t)) =& \limsup_{\eta\to 0^+} \frac{x_0^2 e^{-2k(t+\eta-s_0)} - x_0^2 e^{-2k(t-s_0)}}{\eta} \\ =& -2kx_0^2 e^{-2k(t-s_0)} \leqslant 0, \end{aligned}$$

for all $t \in [\alpha,\beta)$, that is, the right derivative of V is non-positive along every solution of the generalized ODE (8.14). Hence, condition (LF3) of Definition 8.1 is satisfied. Moreover, the function

$$[\alpha,\beta] \ni t \mapsto V(t,x(t)) = x^2(t) \in (0,\infty)$$

is left-continuous on $[\alpha,\beta)$. Then, V satisfies condition (R1) and, therefore, the function $[\alpha,\beta] \ni t \mapsto V(t,x(t)) \in \mathbb{R}$ is nonincreasing, getting the condition (ii) of Theorem 8.18.

Now, consider the continuous increasing function $b : \mathbb{R}_+ \to \mathbb{R}_+$ given by

$$b(s) = \frac{1}{2}s^2, \quad \text{for all } s \in \mathbb{R}_+.$$

Note that condition (LF2) is satisfied, since

$$V(t,x) = x^2 = |x^2| \geqslant \frac{1}{2}|x^2| = b(|x|),$$

for all $(t,x) \in \mathbb{R}_+ \times \mathbb{R}$. Therefore, $V(t,x) = x^2$ is a Lyapunov functional with respect to the generalized ODE (8.14).

Let $a : \mathbb{R}_+ \to \mathbb{R}_+$ be given by

$$a(s) = \frac{3}{2}s^2, \quad \text{for all } s \in \mathbb{R}.$$

Then, a is a continuous increasing function, $a(0) = 0$, and for all $x \in \mathbb{R}$, we have

$$V(t,x) = x^2 = |x|^2 \leqslant \frac{3}{2}|x|^2 = a(|x|),$$

whence it follows that the condition (i) of Theorem 8.18 holds.

In contrast, $V(t,x) = x^2$ is not a Lipschitz function. Indeed, assume that there is $L > 0$ such that $|V(t,x) - V(t,y)| \leqslant L|x-y|$, for all $(t,x) \in [0,\infty) \times \mathbb{R}$. Taking $x = L+1$ and $y = 0$, we conclude that $L+1 \leqslant L$, which obviously does not occur, provided $L > 0$.

In the sequel, we present a Lyapunov-type theorem on uniform asymptotic stability of the trivial solution of the generalized ODE (8.1). Such a result can be found on [80, Theorem 3.6].

Theorem 8.20: *Let F satisfy condition (ZS) and consider $V : [t_0,\infty) \times \overline{B}_\rho \to \mathbb{R}$ a Lyapunov functional with respect to the generalized ODE (8.1), where $\rho > 0$, $\overline{B}_\rho = \{x \in X : \| x \| \leqslant \rho\}$, and $\overline{B}_\rho \subset O$. Assume the following conditions hold:*

(i) *There exists a continuous increasing function* $a : \mathbb{R}_+ \to \mathbb{R}_+$ *such that* $a(0) = 0$ *and, for all solution* $x : I \to \overline{B}_\rho$ *of the generalized ODE* (8.1), *with* $I \subset [t_0, \infty)$, *we have*

$$V(t, x(t)) \leqslant a(\| x(t) \|), \quad \text{for all } t \in I;$$

(ii) *There exists a continuous function* $\Phi : X \to \mathbb{R}$ *satisfying*

$$\Phi(0) = 0 \quad \text{and} \quad \Phi(x) > 0, \text{ for } x \neq 0,$$

such that, for all solution $x : I \to \overline{B}_\rho$ *of the generalized ODE* (8.1), *with* $I \subset [t_0, \infty)$, *we have*

$$V(s, x(s)) - V(t, x(t)) \leqslant (s - t)\,(-\Phi(x(t))), \quad \text{for all } t, s \in I, \; t \leqslant s.$$

Then, the trivial solution $x \equiv 0$ *of the generalized ODE* (8.1) *is uniformly asymptotically stable.*

Proof. Let $\delta_0 = \frac{\rho}{2}$ and $\epsilon > 0$. Once all hypotheses from Theorem 8.18 are satisfied, the trivial solution $x \equiv 0$ of the generalized ODE (8.1) is uniformly stable and, therefore, there exists $\delta(\epsilon) = \delta \in (0, \rho)$ such that if $\tau_0 \geqslant t_0$ and $y_0 \in \overline{B}_\rho$ with $\| y_0 \| < \delta$, then

$$\| x(t, \tau_0, y_0) \| < \epsilon \quad \text{for all } \; t \in [\tau_0, \omega(\tau_0, y_0)). \tag{8.16}$$

Define $N = \sup\{-\Phi(y) : \; \delta(\epsilon) \leqslant \| y \| < \rho\} < 0$ and $T(\epsilon) = -\frac{a(\delta_0)}{N} > 0$.

Let $s_0 \geqslant t_0$, $x_0 \in \overline{B}_\rho$ and $x : [s_0, \omega(s_0, x_0)) \to O$ be the maximal solution of the generalized ODE (8.1) with initial condition $x(s_0) = x_0$ such that $\| x_0 \| < \delta_0$. We need to show that $\| x(t) \| < \epsilon$, for all $t \in [s_0, \omega(s_0, x_0)) \cap [s_0 + T(\epsilon), \infty)$.

At first, notice that if $T(\epsilon) \geqslant \omega(s_0, x_0) - s_0$, then $[s_0, \omega(s_0, x_0)) \cap [s_0 + T(\epsilon), \infty) = \emptyset$. Therefore, we may assume that $T(\epsilon) < \omega(s_0, x_0) - s_0$.

Claim. There exists $\bar{t} \in [s_0, s_0 + T(\epsilon)]$ such that $\| x(\bar{t}) \| < \delta(\epsilon)$.

Indeed, assume that $\| x(s) \| \geqslant \delta(\epsilon)$, for all $s \in [s_0, s_0 + T(\epsilon)]$. By hypotheses (i) and (ii), we have

$$\begin{aligned} V(s_0 + T(\epsilon), x(s_0 + T(\epsilon))) &\leqslant V(s_0, x(s_0)) + T(\epsilon)(-\Phi(x(s_0))) \\ &\leqslant a(\| x(s_0) \|) + T(\epsilon)N \\ &\leqslant a(\delta_0) + \left(-\frac{a(\delta_0)}{N}\right) N = 0. \end{aligned} \tag{8.17}$$

On the other hand, condition (LF2) implies that

$$V(s_0 + T(\epsilon), x(s_0 + T(\epsilon))) \geqslant b(\| x(s_0 + T(\epsilon)) \|) \geqslant b(\delta(\epsilon)) > 0,$$

which contradicts (8.17) and ensures the veracity of the *Claim*.

Finally, set $y_0 = x(\bar{t})$. By (8.16), we have $\| x(t, \bar{t}, y_0) \| < \epsilon$, for all $t \in [\bar{t}, \omega(\bar{t}, y_0))$. Consequently, $\| x(t) \| < \epsilon$ for all $t \in [s_0 + T(\epsilon), \omega(s_0, x_0))$, as $\bar{t} \in [s_0, s_0 + T(\epsilon)]$ and $T(\epsilon) < \omega(s_0, x_0) - s_0$. This shows that the trivial solution $x \equiv 0$ of the generalized ODE (8.1) is uniformly asymptotically stable. □

8.3 Lyapunov Stability for MDEs

In this section, we use the results established in the previous sections to obtain Lyapunov stability results for MDEs. The results described in this section are borrowed from [80] and adapted to Banach space-valued case.

As before, we consider a Banach space X with norm $\| \cdot \|$ and $O \subset X$ as being an open set containing the origin. Now, let us consider the following integral form of an MDE of type

$$x(t) = x(\tau_0) + \int_{\tau_0}^{t} f(x(s), s)\, dg(s), \qquad t \geqslant \tau_0, \tag{8.18}$$

where $\tau_0 \geqslant t_0 \geqslant 0, f : O \times [t_0, \infty) \to X$, and $g : [t_0, \infty) \to \mathbb{R}$.

We remind the reader that $G([t_0, \infty), X)$ denotes the vector space of all functions $x : [t_0, \infty) \to X$ such that $x \in G([\alpha, \beta], X)$, for all $[\alpha, \beta] \subset [t_0, \infty)$. Moreover, the set $G_0([t_0, \infty), X)$ denotes the vector space of all functions $x \in G([t_0, \infty), X)$ such that

$$\sup_{s \in [t_0, \infty)} e^{-(s-t_0)} \| x(s) \| < \infty.$$

This space, equipped with the norm

$$\| x \|_{[t_0, \infty)} = \sup_{s \in [t_0, \infty)} e^{-(s-t_0)} \| x(s) \|,$$

for $x \in G_0([t_0, \infty), X)$, becomes a Banach space (see Proposition 1.9). We also assume the following conditions:

(A1) The function $g : [t_0, \infty) \to \mathbb{R}$ is nondecreasing and left-continuous on (t_0, ∞);

(A2) The Perron–Stieltjes integral $\int_{s_1}^{s_2} f(x(s), s)\, dg(s)$ exists, for all $x \in G([t_0, \infty), O)$ and all $s_1, s_2 \in [t_0, \infty)$;

(A3) There exists a locally Perron–Stieltjes integrable function $M : [t_0, \infty) \to \mathbb{R}$ with respect to g such that

$$\left\| \int_{s_1}^{s_2} f(x(s), s)\, dg(s) \right\| \leqslant \int_{s_1}^{s_2} M(s)\, dg(s),$$

for all $x \in G([t_0, \infty), O)$ and all $s_1, s_2 \in [t_0, \infty)$, with $s_1 \leqslant s_2$;

(A4) There exists a locally Perron–Stieltjes integrable function $L : [t_0, \infty) \to \mathbb{R}$ with respect to g such that

$$\left\| \int_{s_1}^{s_2} [f(x(s), s) - f(z(s), s)] \, dg(s) \right\| \leqslant \|x - z\|_{[t_0,\infty)} \int_{s_1}^{s_2} L(s) \, dg(s),$$

for all $x, z \in G_0([t_0, \infty), O)$ and all $s_1, s_2 \in [t_0, \infty)$, with $s_1 \leqslant s_2$.

Under the aforementioned conditions, it is known that $x : I \to O$ is a solution of the MDE (8.18) on $I \subset [t_0, \infty)$ if and only if x is a solution of the generalized ODE

$$\frac{dx}{d\tau} = DF(x, t) \tag{8.19}$$

on I, where the function $F : O \times [t_0, \infty) \to X$ is given by

$$F(x, t) = \int_{\tau_0}^{t} f(x, s) \, dg(s) \tag{8.20}$$

for all $(x, t) \in O \times [t_0, \infty)$, see Theorem 5.18.

Throughout this section, we assume that $f(0, t) = 0$, for all $t \geqslant t_0$. This implies that the function $x \equiv 0$ is a solution of the MDE (8.18). Moreover, we denote by $x(\cdot) = x(\cdot, s_0, x_0)$ the unique maximal solution $x : [s_0, \omega(s_0, x_0)) \to O$ of the MDE (8.18) with initial condition $x(s_0) = x_0$. A sufficient condition for the existence and uniqueness of this maximal solution can be found in Theorem 5.21.

For simplicity of notation, when it is clear, we write only ω instead of $\omega(s_0, x_0)$.

Next, we present the definitions of Lyapunov stability, uniform stability, and uniform asymptotic stability of the trivial solution of the MDE (8.18).

Definition 8.21: The trivial solution of the MDE (8.18) is

(i) *Stable*, if for every $s_0 \geqslant t_0$ and $\epsilon > 0$, there exists $\delta = \delta(\epsilon, s_0) > 0$ such that if $x_0 \in O$ with $\| x_0 \| < \delta$, then

$$\| x(t, s_0, x_0) \| = \| x(t) \| < \epsilon, \quad \text{for all } t \in [s_0, \omega(s_0, x_0)),$$

where $x(\cdot, s_0, x_0)$ is the solution of the MDE (8.18) with $x(s_0, s_0, x_0) = x_0$;

(ii) *Uniformly stable (Lyapunov stable)*, if it is stable and δ is independent of s_0;

(iii) *Uniformly asymptotically stable*, if there exists $\delta_0 > 0$ and for every $\epsilon > 0$, there exists $T = T(\epsilon) \geqslant 0$ such that if $s_0 \geqslant t_0$ and $x_0 \in O$, with $\| x_0 \| < \delta_0$, then

$$\| x(t, s_0, x_0) \| = \| x(t) \| < \epsilon, \quad \text{for all } t \in [s_0, \omega(s_0, x_0)) \cap [s_0 + T, \infty),$$

where $x(\cdot, s_0, x_0)$ is the solution of the MDE (8.18) such that $x(s_0, s_0, x_0) = x_0$.

In the sequel, we present a concept of Lyapunov functional with respect to the MDE (8.18).

Definition 8.22: We say that $U : [t_0, \infty) \times \overline{B}_\rho \to \mathbb{R}$, where $\rho > 0$ and $\overline{B}_\rho = \{x \in X : \| x \| \leqslant \rho\} \subset O$, is a *Lyapunov functional with respect to the MDE* (8.18), if the following conditions are satisfied:

(LFM1) For all $x \in \overline{B}_\rho$, the function $U(\cdot, x) : [t_0, \infty) \to \mathbb{R}$ is left-continuous on (t_0, ∞);

(LFM2) There exists a continuous increasing function $b : \mathbb{R}_+ \to \mathbb{R}_+$ satisfying $b(0) = 0$ such that

$$U(t,x) \geqslant b(\| x \|), \quad \text{for every } (t,x) \in [t_0, \infty) \times \overline{B}_\rho;$$

(LFM3) For every solution $x : I \subset [t_0, \infty) \to \overline{B}_\rho$ of the MDE (8.18),

$$D^+U(t,x(t)) = \limsup_{\eta \to 0^+} \frac{U(t+\eta, x(t+\eta)) - U(t,x(t))}{\eta} \leqslant 0$$

holds for all $t \in I \setminus \sup I$, that is, the right derivative of U along every solution of the MDE (8.18) is non-positive.

8.3.1 Direct Method of Lyapunov

In this subsection, we present direct methods of Lyapunov for uniform stability and for uniform asymptotic stability of the trivial solution of the MDE (8.18). These results are borrowed from [80].

Theorem 8.23: *Assume that $f : O \times [t_0, \infty) \to X$ satisfies conditions* (A2), (A3), *and* (A4), *$g : [t_0, \infty) \to \mathbb{R}$ satisfies condition* (A1), *and $f(0,t) = 0$ for every $t \in [t_0, \infty)$. Furthermore, let $U : [t_0, \infty) \times \overline{B}_\rho \to \mathbb{R}$ be a Lyapunov functional with respect to the MDE* (8.18), *where $\rho > 0$ and $\overline{B}_\rho = \{x \in X : \| x \| \leqslant \rho\} \subset O$. Assume, in addition, that U satisfies the conditions:*

(i) *There exists a continuous increasing function $a : \mathbb{R}_+ \to \mathbb{R}_+$ such that $a(0) = 0$ and*

$$U(t,x) \leqslant a(\| x \|), \quad \textit{for every } (t,x) \in [t_0, \infty) \times \overline{B}_\rho;$$

(ii) *The function $I \ni t \mapsto U(t, x(t))$, where $I \subset [t_0, \infty)$ is an interval, is nonincreasing along every solution $x : I \to \overline{B}_\rho$ of the MDE* (8.18).

Then, the trivial solution $x \equiv 0$ of the MDE (8.18) *is uniformly stable.*

Proof. Let $F : O \times [t_0, \infty) \to X$ be defined by (8.20). Since the function $f : O \times [t_0, \infty) \to X$ satisfies conditions (A2), (A3), (A4), and the function $g : [t_0, \infty) \to$

$\mathbb{R}$ satisfies condition (A1), it follows from Lemma 5.17 that $F \in \mathcal{F}(\Omega, h)$, where $\Omega = O \times [t_0, \infty)$ and the function $h : [t_0, \infty) \to \mathbb{R}$ is given by

$$h(t) = \int_{t_0}^{t} (M(s) + L(s))\, dg(s).$$

Note that h is clearly left-continuous on (t_0, ∞), once g is left-continuous on (t_0, ∞). In addition, as $f(0, t) = 0$ for all $t \in [t_0, \infty)$, we have $F(0, t_2) - F(0, t_1) = 0$, for all $t_2, t_1 \geqslant t_0$.

Using the correspondence between the solutions of the generalized ODE (8.19) and the solutions of the MDE (8.18) (see Theorem 5.18), we infer that $U : [t_0, \infty) \times \overline{B}_\rho \to \mathbb{R}$ is a Lyapunov functional with respect to the generalized ODE (8.19), where F is given by (8.20). Moreover, conditions (i) and (ii) allow us to conclude that U also fulfills all conditions from Theorem 8.18. In this way, once all hypotheses from Theorem 8.18 are satisfied, the trivial solution $x \equiv 0$ of the generalized ODE (8.19) is uniformly stable.

Finally, using again the correspondence between the solutions of the generalized ODE (8.19) and the solutions of the MDE (8.18) (see Theorem 5.18), we conclude that the trivial solution $x \equiv 0$ of the MDE (8.18) is also uniformly stable. □

The following result ensures that the trivial solution $x \equiv 0$ of the MDE (8.18) is uniformly asymptotically stable under certain conditions.

Theorem 8.24: *Suppose $f : O \times [t_0, \infty) \to X$ satisfies conditions* (A2), (A3), *and* (A4), *$g : [t_0, \infty) \to \mathbb{R}$ satisfies condition* (A1), *and $f(0, t) = 0$ for every $t \in [t_0, \infty)$. Suppose $U : [t_0, \infty) \times \overline{B}_\rho \to \mathbb{R}$, $\rho > 0$ and $\overline{B}_\rho = \{x \in X : \| x \| \leqslant \rho\} \subset O$, satisfies conditions* (LFM1) *and* (LFM2) *from Definition 8.22 and condition* (i) *from Theorem 8.23. Moreover, suppose there exists a continuous function $\Phi : X \to \mathbb{R}$ satisfying $\Phi(0) = 0$ and $\Phi(z) > 0$ for $z \neq 0$, such that for every maximal solution $x(\cdot) = x(\cdot, s_0, x_0)$, $(x_0, s_0) \in \overline{B}_\rho \times [t_0, \infty)$, of the MDE* (8.18), *we have*

$$U(s, x(s)) - U(t, x(t)) \leqslant (s - t)\,(-\Phi(x(t)))\,, \tag{8.21}$$

for all $t, s \in [s_0, \omega)$ with $t \leqslant s$. Then, the trivial solution $x \equiv 0$ of the MDE (8.18) *is uniformly asymptotically stable.*

Proof. Let $F : O \times [t_0, \infty) \to X$ be the function defined by (8.20). Using the same arguments as in the proof of Theorem 8.23, we have $F \in \mathcal{F}(\Omega, h)$, where $\Omega = O \times [t_0, \infty)$ and the function $h : [t_0, \infty) \to \mathbb{R}$ is given by

$$h(t) = \int_{t_0}^{t} (M(s) + L(s))\, dg(s).$$

By the correspondence between the solutions of the generalized ODE (8.19) and the solutions of the MDE (8.18) (Theorem 5.18), we can verify that $U : [t_0, \infty) \times$

$\overline{B}_\rho \to \mathbb{R}$ is a Lyapunov functional with respect to the generalized ODE (8.19) and it satisfies all hypotheses of Theorem 8.20, whence it follows that the trivial solution $x \equiv 0$ of the generalized ODE (8.19) is uniformly asymptotically stable. Again, the correspondence between the solutions of the generalized ODE (8.19) and the solutions of the MDE (8.18) (Theorem 5.18) helps us to conclude that the trivial solution $x \equiv 0$ of the MDE (8.18) is uniformly asymptotically stable. □

8.4 Lyapunov Stability for Dynamic Equations on Time Scales

In this section, we recall the concepts of Lyapunov stability in the framework of dynamic equations on time scales. These concepts were first dealt with in such an environment in [80].

Let $\mathbb{T}$ be a time scale and X be a Banach space. We recall that if $f : \mathbb{T} \to X$ is a function, then $f^* : \mathbb{T}^* \to X$ is defined by

$$f^*(t) = f(t^*), \quad \text{for all } t \in \mathbb{T}^*,$$

where $\mathbb{T}^*$ is given by

$$\mathbb{T}^* = \begin{cases} (-\infty, \sup \mathbb{T}), & \text{if } \sup \mathbb{T} < \infty, \\ (-\infty, \infty), & \text{otherwise,} \end{cases}$$

and $t^* = \inf \{s \in \mathbb{T} : s \geqslant t\}$, for all $t \in \mathbb{R}$ such that $t \leqslant \sup \mathbb{T}$.

Throughout this section, we consider a time scale $\mathbb{T}$ such that $\sup \mathbb{T} = \infty$, $t_0 \in \mathbb{T}$, $t_0 \geqslant 0$ and a dynamic equation on time scales of type

$$x(t) = x(t_0) + \int_{t_0}^{t} f(x(s), s)\Delta s, \quad t \in [t_0, \infty)_{\mathbb{T}}, \tag{8.22}$$

where the integral on the right-hand side is in the sense of Perron Δ-integral, $f : B_c \times [t_0, \infty)_{\mathbb{T}} \to X$, $B_c = \{x \in X : \| x \| < c\}$, and $\| \cdot \|$ is a norm in X.

We recall that $G([t_0, \infty)_{\mathbb{T}}, B_c)$ denotes the space of all functions $x : [t_0, \infty)_{\mathbb{T}} \to B_c$, which are regulated on $[t_0, \infty)_{\mathbb{T}}$ (the concept of regulated function defined on a time scale can be found in Definition 3.14). Moreover, the set $G_0([t_0, \infty)_{\mathbb{T}}, X)$ denotes the vector space of all functions $x \in G([t_0, \infty)_{\mathbb{T}}, X)$ such that

$$\sup_{s \in [t_0, \infty)_{\mathbb{T}}} e^{-(s-t_0)} \| x(s) \| < \infty.$$

From now on, we consider that the function $f : B_c \times [t_0, \infty)_{\mathbb{T}} \to X$ fulfills the following conditions:

(C1) The Perron Δ-integral $\int_{s_1}^{s_2} f(y(s), s)\Delta s$ exists, for all $y \in G([t_0, \infty)_{\mathbb{T}}, B_c)$ and all $s_1, s_2 \in [t_0, \infty)_{\mathbb{T}}$;

(C2) There is a locally Perron Δ-integrable function $M : [t_0, \infty)_{\mathbb{T}} \to \mathbb{R}$ such that

$$\left\| \int_{s_1}^{s_2} f(y(s), s)\Delta s \right\| \leqslant \int_{s_1}^{s_2} M(s)\Delta s,$$

for all $y \in G([t_0, \infty)_{\mathbb{T}}, B_c)$ and all $s_1, s_2 \in [t_0, \infty)_{\mathbb{T}}, s_1 \leqslant s_2$;

(C3) There is a locally Perron Δ-integrable function $L : [t_0, \infty)_{\mathbb{T}} \to \mathbb{R}$ such that

$$\left\| \int_{s_1}^{s_2} [f(y(s), s) - f(w(s), s)]\Delta s \right\| \leqslant \|y - w\|_{[t_0,\infty)} \int_{s_1}^{s_2} L(s)\Delta s,$$

for all $y, w \in G_0([t_0, \infty)_{\mathbb{T}}, B_c)$ and all $s_1, s_2 \in [t_0, \infty)_{\mathbb{T}}, s_1 \leqslant s_2$.

Under the previous conditions, we recall a correspondence between dynamic equations on time scales and MDEs (see Theorem 3.30 for more details).

Let $I \subset [t_0, \infty)$ be a nondegenerate interval such that $I \cap \mathbb{T}$ is nonempty and for each $t \in I$, we have $t^* \in I \cap \mathbb{T}$. If $x : I \cap \mathbb{T} \to X$ is a solution of the dynamic equation on time scales (8.22), then $x^* : I \to X$ is a solution of the MDE

$$y(t_2) - y(t_1) = \int_{t_1}^{t_2} f^*(y(s), s)\, dg(s), \qquad t_1, t_2 \in I. \tag{8.23}$$

Conversely, if $y : I \to X$ satisfies the MDE (8.23), then it must have the form $y = x^*$, where $x : I \cap \mathbb{T} \to X$ is a solution of the dynamic equation on time scales (8.22). Here, $f^* : B_c \times [t_0, \infty) \to X$ and $g : [t_0, \infty) \to \mathbb{R}$ are given by $f^*(x, s) = f(x, s^*)$ and $g(s) = s^*$, respectively.

In what follows, let us assume that $f(0, t) = 0$ for every $t \in [t_0, \infty)_{\mathbb{T}}$. This condition implies that $x \equiv 0$ is a solution of the dynamic equation on time scales (8.22). Moreover, we denote by $x(\cdot) = x(\cdot, s_0, x_0)$ the unique maximal solution $x : [s_0, \omega(s_0, x_0))_{\mathbb{T}} \to X$ of the dynamic equation on time scales (8.22) with $x(s_0) = x_0$.

Lyapunov stability concepts for the trivial solution $x \equiv 0$ of the dynamic equation on time scales (8.22), introduced in [80], are presented in the sequel.

Definition 8.25: Let $\mathbb{T}$ be a time scale such that $\sup \mathbb{T} = \infty$. The trivial solution of the dynamic equation on time scales (8.22) is

(i) *Stable*, if for every $s_0 \in \mathbb{T}$ with $s_0 \geqslant t_0$ and $\epsilon > 0$, then there exists $\delta = \delta(\epsilon, s_0) > 0$ such that for $x_0 \in B_c$ with $\| x_0 \| < \delta$, we have

$$\| x(t, s_0, x_0) \| = \| x(t) \| < \epsilon, \quad \text{for all } t \in [s_0, \omega(s_0, x_0))_{\mathbb{T}},$$

where $x(\cdot, s_0, x_0)$ is the solution of the dynamic equation on time scales (8.22) with $x(s_0, s_0, x_0) = x_0$;

(ii) *Uniformly stable (Lyapunov stable)*, if it is stable with δ independent of s_0;

(iii) *Uniformly asymptotically stable*, if there exists $\delta_0 > 0$ and for every $\epsilon > 0$, there exists $T = T(\epsilon) \geqslant 0$ such that if $s_0 \in \mathbb{T}$, $s_0 \geqslant t_0$ and $x_0 \in B_c$, with $\| x_0 \| < \delta_0$, then

$$\| x(t, s_0, x_0) \| = \| x(t) \| < \epsilon, \quad \text{for all } t \in [s_0, \omega(s_0, x_0)) \cap [s_0 + T, \infty) \cap \mathbb{T},$$

where $x(\cdot, s_0, x_0)$ is the solution of the dynamic equation on time scales (8.22) with $x(s_0, s_0, x_0) = x_0$.

Definition 8.26: We say that $U : [t_0, \infty)_{\mathbb{T}} \times \overline{B}_\rho \to \mathbb{R}$, where $0 < \rho < c$ and $\overline{B}_\rho = \{x \in X : \| x \| \leqslant \rho\}$, is a *Lyapunov functional with respect to the dynamic equation on time scales* (8.22), if the following conditions are satisfied:

(LFD1) The function $t \mapsto U(t, x)$ is left-continuous on $(t_0, \infty)_{\mathbb{T}}$, for all $x \in \overline{B}_\rho$;
(LFD2) There exists a continuous increasing function $b : \mathbb{R}_+ \to \mathbb{R}_+$ satisfying $b(0) = 0$ such that

$$U(t, x) \geqslant b(\| x \|), \quad \text{for every } (t, x) \in [t_0, \infty)_{\mathbb{T}} \times \overline{B}_\rho;$$

(LFD3) For every solution $z : [s_0, \infty)_{\mathbb{T}} \to \overline{B}_\rho$, $s_0 \geqslant t_0$, of the dynamic equation on time scales (8.22), we have

$$U(s, z(s)) - U(t, z(t)) \leqslant 0, \quad \text{for every } s, t \in [s_0, \infty)_{\mathbb{T}}, \text{ with } t \leqslant s.$$

8.4.1 Direct Method of Lyapunov

In this subsection, we estimate the region of uniform stability using Lyapunov's direct method. The results described in this subsection are presented in [80].

The following result is concerned with the uniform stability of the trivial solution of the dynamic equation on time scales (8.22).

Theorem 8.27: *Let* $\mathbb{T}$ *be a time scale such that* $\sup \mathbb{T} = \infty$ *and* $[t_0, \infty)_{\mathbb{T}}$ *be a time scale interval. Assume that* $f : B_c \times [t_0, \infty)_{\mathbb{T}} \to X$ *satisfies conditions* (C1), (C2), (C3), *and* $f(0, t) = 0$ *for every* $t \in [t_0, \infty)_{\mathbb{T}}$. *Let* $U : [t_0, \infty)_{\mathbb{T}} \times \overline{B}_\rho \to X$, $0 < \rho < c$, *be a Lyapunov functional with respect to the dynamic equation on time scales* (8.22), *where* $\overline{B}_\rho = \{x \in X : \| x \| \leqslant \rho\}$. *Moreover, assume that* U *satisfies the following conditions:*

(i) *There exists a continuous increasing function* $a : \mathbb{R}_+ \to \mathbb{R}_+$ *such that* $a(0) = 0$ *and for all* $(t, x) \in [t_0, \infty)_{\mathbb{T}} \times \overline{B}_\rho$, *we have*

$$U(t, x) \leqslant a(\| x \|); \tag{8.24}$$

(ii) *The function* $[s_0, \omega)_{\mathbb{T}} \ni t \mapsto U(t, x(t))$ *is nonincreasing, for every solution* $x : [s_0, \omega)_{\mathbb{T}} \to X$ *of the dynamic equation on time scales (8.22) with* $s_0 \in [t_0, \infty)_{\mathbb{T}}$.

Then, the trivial solution $x \equiv 0$ *of the dynamic equation on time scales (8.22) is uniformly stable.*

Proof. Let $f^* : B_c \times [t_0, \infty) \to X$ and $g : [t_0, \infty) \to [t_0, \infty)_{\mathbb{T}}$ be the functions defined by $f^*(x, t) = f(x, t^*)$, for all $(x, t) \in B_c \times [t_0, \infty)$, and $g(t) = t^*$, for all $t \in [t_0, \infty)$. Since f satisfies conditions (C1), (C2), and (C3), it follows that f^* satisfies conditions (A2), (A3), and (A4) described in Section 8.3. In addition, by the definition of g, it follows that g satisfies (A1). Since $f(0, t) = 0$, for all $t \in [t_0, \infty)_{\mathbb{T}}$, we get $f^*(0, t) = 0$, for all $t \in [t_0, \infty)$.

Define

$$U^*(t, x) = U(t^*, x), \quad \text{for every } (t, x) \in [t_0, \infty) \times \overline{B}_\rho.$$

We will show that $U^* : [t_0, \infty) \times \overline{B}_\rho \to \mathbb{R}$ is a Lyapunov functional with respect to the MDE

$$y(t) = y(\tau_0) + \int_{\tau_0}^{t} f^*(y(s), s)\, dg(s), \quad \text{for all } t,\ \tau_0 \in [t_0, \infty), \tag{8.25}$$

and U^* satisfies conditions (i) and (ii) of Theorem 8.23. Indeed, since $U^*(t, x) = U(g(t), x)$ and the functions $U(\cdot, x)$, and $g(t) = t^*$ are left-continuous on (t_0, ∞), then the function $U^*(\cdot, x) : [t_0, \infty) \to \mathbb{R}$ is also left-continuous on (t_0, ∞), for all $x \in \overline{B}_\rho$.

Moreover, there exists a continuous and increasing function $b : \mathbb{R}_+ \to \mathbb{R}_+$ with $b(0) = 0$ such that

$$U(s, x) \geqslant b(\| x \|), \quad \text{for every } (s, x) \in [t_0, \infty)_{\mathbb{T}} \times \overline{B}_\rho.$$

Thus, for every $(t, x) \in [t_0, \infty) \times \overline{B}_\rho$, we conclude

$$U^*(t, x) = U(t^*, x) \geqslant b(\| x \|).$$

Let $(x_0, s_0) \in \overline{B}_\rho \times [t_0, \infty)$ and $y = y(\cdot, s_0, x_0) : [s_0, \omega) \to \overline{B}_\rho$ be the solution of the MDE (8.25). By the fact that $y(t) \in \overline{B}_\rho \subset B_c$, for $t \in [s_0, \omega)$, we have $\omega = \infty$. Now, as $y : [s_0, \infty) \to \overline{B}_\rho$ is a global forward solution of (8.25), we have

$$y(\tau) = y(s_0) + \int_{s_0}^{\tau} f(y(s), s^*)\, dg(s), \quad \tau \in [s_0, \infty),$$

with $g(s) = s^*$.

Claim. $U^*(s, y(s)) - U^*(t, y(t)) \leqslant 0$ for all $s_0 \leqslant t < s < \infty$.

In order to prove the *Claim*, let us consider two cases.

At first, we assume that $s_0 \in [t_0, \infty) \setminus \mathbb{T}$. Thus, g is constant on $[s_0, s_0^*]$, whence it follows that $t^* = s_0^*$ for all $t \in [s_0, s_0^*]$ and, consequently, $y(\tau) = y(s_0^*)$ for all $\tau \in [s_0, s_0^*]$.

On the other hand, using the correspondence between solutions of the dynamic equation on time scales (8.22) and solutions of the MDE (8.25), $y|_{[s_0^*,\infty)}$ must possess the form

$$y|_{[s_0^*,\infty)} = x^*, \tag{8.26}$$

where $x : [s_0^*, \infty)_{\mathbb{T}} \to X$ is a solution of the dynamic equation on time scales (8.22).

Besides, x is also the global forward solution of the dynamic equation on time scales (8.22) through $(x(s_0^*), s_0^*)$, and $y(s_0^*) = x^*(s_0^*) = x(s_0^*)$.

When $s_0 \leqslant t < s \leqslant s_0^*$, we obtain

$$\begin{aligned} U^*(s, y(s)) - U^*(t, y(t)) =& U(s^*, y(s)) - U(t^*, y(t)) \\ =& U(s^*, x(s^*)) - U(t^*, x(t^*)) = 0. \end{aligned}$$

When $s_0 \leqslant t \leqslant s_0^* < s$, we have

$$\begin{aligned} U^*(s, y(s)) - U^*(t, y(t)) =& U(s^*, y(s)) - U(t^*, y(t)) \\ =& U(s^*, x^*(s)) - U(s_0^*, y(s_0^*)) \\ =& U(s^*, x(s^*)) - U(s_0^*, x(s_0^*)) \\ =& U(s^*, x(s^*)) - U(t^*, x(t^*)) \leqslant 0. \end{aligned}$$

When $s_0^* \leqslant t < s$, then, by (8.26), we derive

$$\begin{aligned} U^*(s, y(s)) - U^*(t, y(t)) =& U(s^*, y(s)) - U(t^*, y(t)) \\ =& U(s^*, x(s^*)) - U(t^*, x(t^*)) \leqslant 0, \end{aligned}$$

which proves the *Claim* for the case where $s_0 \in [t_0, \infty) \setminus \mathbb{T}$.

Now, suppose $s_0 \in \mathbb{T}$. Hence, $s_0 = s_0^* \in \mathbb{T}$ and, consequently, $[s_0, \infty)_{\mathbb{T}} = [s_0^*, \infty)_{\mathbb{T}}$. Also, $y : [s_0, \infty) \to X$ is such that $y = x^*$, where $x : [s_0, \infty)_{\mathbb{T}} \to X$ is a global forward solution of the dynamic equation on time scales (8.22) through $(y(s_0), s_0)$. Thus,

$$\begin{aligned} U^*(s, y(s)) - U^*(t, y(t)) &= U^*(s, x^*(s)) - U^*(t, x^*(t)) \\ &= U(s^*, x(s^*)) - U(t^*, x(t^*)) \leqslant 0, \end{aligned}$$

for every $s_0 \leqslant t < s < \infty$, whence it follows the truthfulness of the *Claim* for the case where $s_0 \in \mathbb{T}$. Hence, U^* satisfies condition (ii) of Theorem 8.23.

Analogously, one can prove that

$$\limsup_{\eta \to 0^+} \frac{U^*(t+\eta, y(t+\eta)) - U^*(t, y(t))}{\eta} \leqslant 0,$$

for every $s_0 \leqslant t < \infty$.

Thus, U^* is a Lyapunov functional with respect to the MDE (8.25).

Now, by hypothesis (i), there exists a continuous and increasing function $a : \mathbb{R}_+ \to \mathbb{R}_+$ for which $a(0) = 0$ and condition (8.24) is fulfilled. Then, for all solutions $y : I \to \overline{B}_\rho$, $I \subset [t_0, \infty)$, of the MDE (8.25), we have

$$U^*(t, y(t)) = U(t^*, y(t)) \leqslant a(\| y(t) \|), \quad \text{for all } t \in I.$$

Once all conditions of Theorem 8.23 are satisfied, the trivial solution $x^* \equiv 0$ of the MDE (8.25) is uniformly stable. Using the correspondence between solutions of the dynamic equation on time scales (8.22) and solutions of the MDE (8.25), the trivial solution $x \equiv 0$ of the dynamic equation on time scales (8.22) is uniformly stable. □

Theorem 8.28: *Let $\mathbb{T}$ be a time scale such that* $\sup \mathbb{T} = \infty$ *and $[t_0, \infty)_\mathbb{T}$ be a time scale interval. Assume that $f : B_c \times [t_0, \infty)_\mathbb{T} \to X$ satisfies conditions* (C1), (C2), (C3), *and $f(0, t) = 0$ for every $t \in [t_0, \infty)_\mathbb{T}$. Suppose $U : [t_0, \infty)_\mathbb{T} \times \overline{B}_\rho \to \mathbb{R}$, with $0 < \rho < c$ and $\overline{B}_\rho = \{x \in X : \| x \| \leqslant \rho\}$, satisfies conditions* (LFD1) *and* (LFD2) *from Definition 8.26, and condition* (i) *from Theorem 8.27. Moreover, suppose there exists a continuous function $\phi : X \to \mathbb{R}$ satisfying $\phi(0) = 0$ and $\phi(v) > 0$ for $v \neq 0$, such that for every maximal solution $x(\cdot) = x(\cdot, s_0, x_0)$, $(x_0, s_0) \in \overline{B}_\rho \times [t_0, \infty)_\mathbb{T}$, of the dynamic equation on time scales* (8.22), *we have*

$$U(s^*, x(s^*)) - U(t^*, x(t^*)) \leqslant (s - t)^*(-\phi(x(t^*))), \tag{8.27}$$

for all $t, s \in [s_0, \omega)_\mathbb{T}$, with $t \leqslant s$. Then, the trivial solution $x \equiv 0$ of the dynamic equation on time scales (8.22) *is uniformly asymptotically stable.*

Proof. Let $f^* : B_c \times [t_0, \infty) \to X$ and $g : [t_0, \infty) \to [t_0, \infty)_\mathbb{T}$ be the functions defined by $f^*(x, t) = f(x, t^*)$, for all $(x, t) \in B_c \times [t_0, \infty)$, and $g(t) = t^*$, for all $t \in [t_0, \infty)$.

Using the same arguments as in the proof of Theorem 8.27, we can prove that f^* satisfies conditions (A2)–(A4) and $f^*(0, t) = 0$ for every $t \in [t_0, \infty)$.

Let us define

$$U^*(t, x) = U(t^*, x), \quad \text{for every } (t, x) \in [t_0, \infty) \times \overline{B}_\rho.$$

Then, we can show that U^* satisfies conditions (LFM1) and (LFM2) from Definition 8.22, using similar arguments as those employed in the proof of Theorem 8.27. According to hypothesis (i) from Theorem 8.27, we can also prove that U^* satisfies hypothesis (i) from Theorem 8.23, as in the proof of Theorem 8.27.

Now, we will show that U^* satisfies condition (8.21). Consider $(x_0, s_0) \in \overline{B}_\rho \times [t_0, \infty)$ and $y = y(\cdot, s_0, x_0) : [s_0, \omega) \to \overline{B}_\rho$ be the maximal solution of the MDE (8.25). Since $y(t) \in \overline{B}_\rho \subset B_c$, we have $\omega = \infty$. Let us consider two cases.

At first, we assume that $s_0 \in [t_0, \infty) \setminus \mathbb{T}$. By similar arguments used in the proof of Theorem 8.27, we can prove that $y|_{[s_0^*, \infty)}$ is such that $y|_{[s_0^*, \infty)} = x^*$, where

$x : [s_0^*, \infty)_{\mathbb{T}} \to X$ is the global forward solution of the dynamic equation on time scales (8.22) through $(x(s_0^*), s_0^*)$.

If $s_0 \leqslant t < s \leqslant s_0^*$, we get

$$
\begin{aligned}
U^*(s, y(s)) - U^*(t, y(t)) &= U(s^*, y(s)) - U(t^*, y(t)) \\
&= U(s^*, x(s^*)) - U(t^*, x(t^*)) = 0.
\end{aligned}
$$

Using this fact together with (8.27), we obtain $-\phi(x(t^*)) = 0$ and, hence, $x(t^*) = 0$, whence $y(t) = 0$, since $x(t^*) = x(s_0^*) = y(s_0^*) = y(t)$. Therefore,

$$
U^*(s, y(s)) - U^*(t, y(t)) = (s - t)(-\phi(y(t))) = 0.
$$

If $s_0 \leqslant t \leqslant s_0^* < s$, we have

$$
\begin{aligned}
U^*(s, y(s)) - U^*(t, y(t)) &= U(s^*, x(s^*)) - U(t^*, x(t^*)) \\
&\leqslant (s - t)^*(-\phi(x(t^*))) \\
&= (s - t)^*(-\phi(y(t))) \\
&\leqslant (s - t)(-\phi(y(t))).
\end{aligned}
$$

Besides, if $s_0^* \leqslant t < s$, we obtain

$$
\begin{aligned}
U^*(s, y(s)) - U^*(t, y(t)) &= U(s^*, x(s^*)) - U(t^*, x(t^*)) \\
&\leqslant (s - t)^*(-\phi(x(t^*))) \\
&= (s - t)^*(-\phi(y(t))) \\
&\leqslant (s - t)(-\phi(y(t))).
\end{aligned}
$$

Thus,

$$
U^*(s, y(s)) - U^*(t, y(t)) \leqslant (s - t)(-\phi(y(t))),
$$

for all $s_0 \leqslant t < s < \infty$.

Let us suppose $s_0 \in \mathbb{T}$. Thus, $[s_0, \infty)_{\mathbb{T}} = [s_0^*, \infty)_{\mathbb{T}}$. Repeating arguments used in the proof of Theorem 8.27, we can show that $y : [s_0, \infty) \to X$ is such that $y = x^*$, where $x : [s_0, \infty)_{\mathbb{T}} \to X$ is the global forward solution of dynamic equation on time scales (8.22) through $(x(s_0), s_0)$. Hence, we obtain

$$
\begin{aligned}
U^*(s, y(s)) - U^*(t, y(t)) &= U^*(s, x^*(s)) - U^*(t, x^*(t)) \\
&= U(s^*, x(s^*)) - U(t^*, x(t^*)) \\
&\leqslant (s - t)^*(-\phi(y(t))) \\
&\leqslant (s - t)(-\phi(y(t))).
\end{aligned}
$$

Therefore,

$$
U^*(s, y(s)) - U^*(t, y(t)) \leqslant (s - t)(-\phi(y(t))),
$$

for all $s_0 \leqslant t < s < \infty$.

Since all hypotheses of Theorem 8.24 are satisfied, the trivial solution $x^* \equiv 0$ of the MDE (8.25) is uniformly asymptotically stable. Using the correspondence between solutions of the dynamic equation on time scales (8.22) and solutions of the MDE (8.25) (see Theorem 3.30), the trivial solution $x \equiv 0$ of the dynamic equation on time scales (8.22) is uniformly asymptotically stable. □

8.5 Regular Stability for Generalized ODEs

This section concerns some results on regular stability for the trivial solution of generalized ODEs. The concepts of regular stability for generalized ODEs were introduced by the authors of [87].

Throughout this section, we assume that the function $F : \Omega \to X$ belongs to the class $\mathcal{F}(\Omega, h)$, where $\Omega = O \times [t_0, \infty)$, O is an open subset of X such that $0 \in O$, and $h : [t_0, \infty) \to \mathbb{R}$ is a left-continuous and nondecreasing function. Moreover, we suppose F satisfies condition (ZS), that is,

$$F(0, t_2) - F(0, t_1) = 0, \quad \text{for all } t_2, t_1 \in [t_0, \infty).$$

Consider the generalized ODE (8.1) and the nonhomogeneous generalized ODE

$$\frac{dx}{d\tau} = D[F(x, t) + P(t)], \tag{8.28}$$

where $P : [t_0, \infty) \to X$ is a Kurzweil integrable function. We also assume that, for all $\alpha \geqslant t_0$ and for all $x_0 \in X$, there exist local solutions, x of the generalized ODE (8.1), and $\bar{x}$ of the nonhomogeneous generalized ODE (8.28), on $[\alpha, \beta] \subset [t_0, \infty)$, with $x(\alpha) = x_0 = \bar{x}(\alpha)$. Sufficient conditions for the existence and uniqueness of these solutions are described in Theorem 5.1.

Now, we recall concepts concerning regular stability of the trivial solution of the generalized ODE (8.1).

Definition 8.29: The trivial solution $x \equiv 0$ of the generalized ODE (8.1) is

(i) *Regularly stable*, if for every $\epsilon > 0$, there exists $\delta = \delta(\epsilon) \in (0, \epsilon)$ such that if the function $\bar{x} : [\alpha, \beta] \subset [t_0, \infty) \to X$ is regulated, left-continuous on $(\alpha, \beta]$ and satisfies

$$\| \bar{x}(\alpha) \| < \delta \quad \text{and} \quad \sup_{s \in [\alpha, \beta]} \left\| \bar{x}(s) - \bar{x}(\alpha) - \int_\alpha^s DF(\bar{x}(\tau), t) \right\| < \delta,$$

then

$$\| \bar{x}(t) \| < \epsilon, \quad \text{for all } t \in [\alpha, \beta];$$

(ii) *Regularly attracting*, if there exists $\delta_0 > 0$ and for every $\epsilon > 0$, there exist $T = T(\epsilon) \geqslant 0$ and $\rho = \rho(\epsilon) > 0$ such that if $\overline{x} : [\alpha, \beta] \subset [t_0, \infty) \to X$ is a regulated function, left-continuous on $(\alpha, \beta]$ and satisfies

$$\| \overline{x}(\alpha) \| < \delta_0 \quad \text{and} \quad \sup_{s \in [\alpha,\beta]} \left\| \overline{x}(s) - \overline{x}(\alpha) - \int_\alpha^s DF(\overline{x}(\tau), t) \right\| < \rho,$$

then

$$\| \overline{x}(t) \| < \epsilon, \quad \text{for all } t \in [\alpha, \beta] \cap [\alpha + T, \infty);$$

(iii) *Regularly asymptotically stable*, if it is regularly stable and regularly attracting.

For regular stability by perturbations, we have the following concepts.

Definition 8.30: The trivial solution $x \equiv 0$ of the generalized ODE (8.1) is

(i) *Regularly stable with respect to perturbations*, if for every $\epsilon > 0$, there exists $\delta = \delta(\epsilon) > 0$ such that, if $\| x_0 \| < \delta$ and $P \in G^-([\alpha, \beta], X)$, with $\sup_{s \in [\alpha,\beta]} \| P(s) - P(\alpha) \| < \delta$, then

$$\| \overline{x}(t, \alpha, x_0) \| < \epsilon, \quad \text{for all } t \in [\alpha, \beta],$$

where $\overline{x}(\cdot, \alpha, x_0)$ is the solution of the nonhomogeneous generalized ODE (8.28), with initial condition $\overline{x}(\alpha, \alpha, x_0) = x_0$ and $[\alpha, \beta] \subset [t_0, \infty)$;

(ii) *Regularly attracting with respect to perturbations*, if there exists $\widetilde{\delta} > 0$ and for every $\epsilon > 0$, there exist $T = T(\epsilon) \geqslant 0$ and $\rho = \rho(\epsilon) > 0$ such that, if $\| x_0 \| < \widetilde{\delta}$ and $P \in G^-([\alpha, \beta], X)$, with $\sup_{s \in [\alpha,\beta]} \| P(s) - P(\alpha) \| < \rho$, then

$$\| \overline{x}(t, \alpha, x_0) \| < \epsilon, \quad \text{for all } t \in [\alpha, \beta] \cap [\alpha + T, \infty),$$

where $\overline{x}(\cdot, \alpha, x_0)$ is the solution of the nonhomogeneous generalized ODE (8.28), with initial condition $\overline{x}(\alpha, \alpha, x_0) = x_0$ and $[\alpha, \beta] \subset [t_0, \infty)$;

(iii) *Regularly asymptotically stable with respect to perturbations*, if it is regularly stable with respect to perturbations and regularly attracting with respect to perturbations.

Note that the concept of regular stability is more general than the concept of variational stability defined in Section 8.1. Similar statements hold for regular attractivity and regular asymptotic stability.

The following result shows us that there is an equivalence between the concepts of stability given in Definitions 8.29 and 8.30. This result can be found in [87, Theorem 4.7].

Theorem 8.31: *Let $F \in \mathcal{F}(\Omega, h)$ satisfy condition* (ZS). *The following statements are true.*

(i) *The trivial solution $x \equiv 0$ of the generalized ODE (8.1) is regularly stable if and only if it is regularly stable with respect to perturbations.*
(ii) *The trivial solution $x \equiv 0$ of the generalized ODE (8.1) is regularly attracting if and only if it is regularly attracting with respect to perturbations.*
(iii) *The trivial solution $x \equiv 0$ of the generalized ODE (8.1) is regularly asymptotically stable if and only if it is regularly asymptotically stable with respect to perturbations.*

Proof. We start by proving (i). At first, suppose that the trivial solution $x \equiv 0$ of the generalized ODE (8.1) is regularly stable. Let $\epsilon > 0$ and $\delta = \delta(\epsilon) > 0$ be as in Definition 8.29(i) and $\bar{x} : [\alpha, \beta] \to X$ be a solution of the nonhomogeneous generalized ODE (8.28), with $P \in G^-([\alpha, \beta], X)$. Thus, by the definition of solution of (8.28), we have

$$\bar{x}(s) - \bar{x}(\alpha) = \int_\alpha^s DF(\bar{x}(\tau), t) + P(s) - P(\alpha),$$

for all $s \in [\alpha, \beta]$. Notice that, for all $s_1, s_2 \in [\alpha, \beta]$, we have

$$\begin{aligned} \| \bar{x}(s_2) - \bar{x}(s_1) \| &\leqslant \left\| \int_{s_1}^{s_2} DF(\bar{x}(\tau), t) \right\| + \| P(s_2) - P(s_1) \| \\ &\leqslant \| h(s_2) - h(s_1) \| + \| P(s_2) - P(s_1) \|, \end{aligned}$$

where the last inequality follows from Lemma 4.5. Thus, every point in $[\alpha, \beta]$ at which h and P are continuous is a point of continuity of the solution $\bar{x}$. Hence, once h and P are left-continuous on $(\alpha, \beta]$, $\bar{x}$ is also left-continuous on $(\alpha, \beta]$.

Assume that $\| \bar{x}(\alpha) \| < \delta$ and $\sup_{s \in [\alpha, \beta]} \| P(s) - P(\alpha) \| < \delta$. Then,

$$\sup_{s \in [\alpha, \beta]} \left\| \bar{x}(s) - \bar{x}(\alpha) - \int_\alpha^s DF(\bar{x}(\tau), t) \right\| = \sup_{s \in [\alpha, \beta]} \| P(s) - P(\alpha) \| < \delta,$$

and, since the trivial solution of the generalized ODE (8.1) is regularly stable, we conclude that $\| \bar{x}(t) \| < \epsilon$ for every $t \in [\alpha, \beta]$, which implies that trivial solution $x \equiv 0$ of (8.1) is regularly stable with respect to perturbations.

Reciprocally, assume that the trivial solution $x \equiv 0$ is regularly stable with respect to perturbations. Let $\epsilon > 0$ and $\delta = \delta(\epsilon) > 0$ be as in Definition 8.30(i) and let $\bar{x} : [\alpha, \beta] \subset [t_0, \infty) \to X$ be a regulated function, which is left-continuous on $(\alpha, \beta]$ such that

$$\| \bar{x}(\alpha) \| < \delta \quad \text{and} \quad \sup_{s \in [\alpha, \beta]} \left\| \bar{x}(s) - \bar{x}(\alpha) - \int_\alpha^s DF(\bar{x}(\tau), t) \right\| < \delta.$$

Let $P : [\alpha, \beta] \to X$ be defined by

$$P(s) = P(\alpha) + \bar{x}(s) - \bar{x}(\alpha) - \int_\alpha^s DF(\bar{x}(\tau), t), \quad \text{for all } s \in [\alpha, \beta].$$

Once $F \in \mathcal{F}(\Omega, h)$ and $\overline{x}, h \in G^{-}([\alpha, \beta], X)$, we have $P \in G^{-}([\alpha, \beta], X)$. Moreover, for all $s \in [\alpha, \beta]$,

$$\overline{x}(s) = \overline{x}(\alpha) + \int_{\alpha}^{s} DF(\overline{x}(\tau), t) + P(s) - P(\alpha),$$

and, therefore, $\overline{x}$ is a solution of the nonhomogeneous generalized ODE (8.28). On the other hand,

$$\sup_{s \in [\alpha, \beta]} \| P(s) - P(\alpha) \| = \sup_{s \in [\alpha, \beta]} \left\| \overline{x}(s) - \overline{x}(\alpha) - \int_{\alpha}^{s} DF(\overline{x}(\tau), t) \right\| < \delta.$$

Since the trivial solution $x \equiv 0$ of the generalized ODE (8.1) is regularly stable with respect to perturbations, we have $\| \overline{x}(t) \| < \epsilon$, for all $t \in [\alpha, \beta]$, which allows us to conclude that the trivial solution $x \equiv 0$ of the generalized ODE (8.1) is regularly stable.

The proof of item (ii) is analogous to the proof of item (i) and we leave it aside. Finally, item (iii) follows from items (i) and (ii). □

8.5.1 Direct Method of Lyapunov

This subsection is devoted to presenting Lyapunov-type theorems for the generalized ODE (8.1), using the concepts introduced in Section 8.5. These results can be found in [7] and [87].

Recall that we are considering $F \in \mathcal{F}(\Omega, h)$ satisfying condition (ZS), where $\Omega = O \times [t_0, \infty)$, $O \subset X$ is an open subset containing the origin, and $h : [t_0, \infty) \to \mathbb{R}$ is a nondecreasing and left-continuous function.

In what follows, we present a result that will be essential to our purposes and it can be found in [7, Lemma 3.7].

Lemma 8.32: *Let $F \in \mathcal{F}(\Omega, h)$, where $h : [t_0, \infty) \to \mathbb{R}$ is a nondecreasing and left-continuous function, and let $V : [t_0, \infty) \times X \to \mathbb{R}$ be a functional. Assume that:*

(i) *for each left-continuous function $z : [\alpha, \beta] \subset [t_0, \infty) \to X$ on $(\alpha, \beta]$, the function $[\alpha, \beta] \ni t \mapsto V(t, z(t))$ is left-continuous on $(\alpha, \beta]$;*

(ii) *for all regulated functions $x, y : [\alpha, \beta] \subset [t_0, \infty) \to X$, the inequality*

$$|V(t, x(t)) - V(t, y(t)) - V(s, x(s)) + V(s, y(s))| \leqslant \sup_{\xi \in [s,t]} \| x(\xi) - y(\xi) \| \tag{8.29}$$

holds for all $\alpha \leqslant s < t \leqslant \beta$;

(iii) *given $(s_0, x_0) \in [t_0, \infty) \times X$, the function $[s_0, \omega(s_0, x_0)) \ni t \mapsto V(t, x(t))$ is nonincreasing along every solution $x(\cdot, s_0, x_0) : [s_0, \omega(s_0, x_0)) \to X$ of the generalized ODE (8.1) with $x(s_0) = x_0$.*

If, moreover, $\overline{x} : [\alpha, \beta] \subset [t_0, \infty) \to X$ *is regulated and left-continuous on* $(\alpha, \beta]$*, then*

$$V(v, \overline{x}(v)) - V(\gamma, \overline{x}(\gamma)) \leqslant 2 \sup_{\xi \in [\alpha,\beta]} \left\| \overline{x}(\xi) - \overline{x}(\alpha) - \int_\alpha^\xi DF(\overline{x}(\tau), t) \right\|$$

for all $v, \gamma \in [\alpha, \beta]$ *with* $\gamma \leqslant v$*.*

Proof. Let $\overline{x} : [a, b] \to X$ be a regulated and left-continuous function on $[a, b)$. By Corollary 4.8, the integral $\int_\alpha^\beta DF(\overline{x}(\tau), t)$ exists. Set $\sigma \in [\alpha, \beta]$. By Theorem 5.1, there exists a local solution $x : [\sigma, \sigma + \eta_1(\sigma)] \to X$ of the generalized ODE (8.1), with initial condition $x(\sigma) = \overline{x}(\sigma)$.

Consider $\eta > 0$ such that $\eta \leqslant \eta_1(\sigma)$ and $\sigma + \eta \leqslant \beta$. Hence, the Kurzweil integral $\int_\sigma^{\sigma+\eta} D[F(\overline{x}(\tau), t) - F(x(\tau), t)]$ exists by Theorem 2.5.

By assumption (iii), we obtain

$$V(\sigma + \eta, x(\sigma + \eta)) - V(\sigma, x(\sigma)) \leqslant 0. \tag{8.30}$$

Besides, by Corollary 2.8, we have

$$\left\| F(\overline{x}(\sigma), s) - F(\overline{x}(\sigma), \sigma) - \int_\sigma^s DF(\overline{x}(\tau), t) \right\| < \frac{\eta\epsilon}{2} \quad \text{and} \tag{8.31}$$

$$\left\| F(x(\sigma), s) - F(x(\sigma), \sigma) - \int_\sigma^s DF(x(\tau), t) \right\| < \frac{\eta\epsilon}{2}, \tag{8.32}$$

for all $s \in [\sigma, \sigma + \eta]$.

Furthermore, by (8.31) and (8.32), we obtain

$$\begin{aligned}
&\sup_{s\in[\sigma,\sigma+\eta]} \left\| \int_\sigma^s D[F(\overline{x}(\tau), t) - F(x(\tau), t)] \right\| \\
&\quad - \sup_{s\in[\sigma,\sigma+\eta]} \| F(\overline{x}(\sigma), s) - F(\overline{x}(\sigma), \sigma) - F(x(\sigma), s) + F(x(\sigma), \sigma) \| \\
&\leqslant \sup_{s\in[\sigma,\sigma+\eta]} \left(\left\| \int_\sigma^s D[F(\overline{x}(\tau), t) - F(x(\tau), t)] \right. \right. \\
&\quad \left. \left. -(F(\overline{x}(\sigma), s) - F(\overline{x}(\sigma), \sigma) - F(x(\sigma), s) + F(x(\sigma), \sigma)) \right\| \right) \\
&\leqslant \sup_{s\in[\sigma,\sigma+\eta]} \left\| F(\overline{x}(\sigma), s) - F(\overline{x}(\sigma), \sigma) - \int_\sigma^s DF(\overline{x}(\tau), t) \right\| \\
&\quad + \sup_{s\in[\sigma,\sigma+\eta]} \left\| F(x(\sigma), s) - F(x(\sigma), \sigma) - \int_\sigma^s DF(x(\tau), t) \right\| < \eta\epsilon.
\end{aligned} \tag{8.33}$$

Moreover, as $\overline{x}(\sigma) = x(\sigma)$ and $F \in \mathcal{F}(\Omega, h)$, we get

$$\begin{aligned}
&\sup_{s\in[\sigma,\sigma+\eta]} \| F(\overline{x}(\sigma), s) - F(\overline{x}(\sigma), \sigma) - F(x(\sigma), s) + F(x(\sigma), \sigma) \| \\
&\leqslant \| \overline{x}(\sigma) - x(\sigma) \| \sup_{s\in[\sigma,\sigma+\eta]} |h(s) - h(\sigma)| = 0.
\end{aligned} \tag{8.34}$$

Replacing (8.34) in (8.33), we obtain

$$\sup_{s\in[\sigma,\sigma+\eta]}\left\|\int_{\sigma}^{s} D[F(\overline{x}(\tau),t)-F(x(\tau),t)]\right\| < \epsilon\eta. \tag{8.35}$$

Since $x(\sigma)=\overline{x}(\sigma)$, using (8.29), we obtain

$$\begin{aligned}
&V(\sigma+\eta,\overline{x}(\sigma+\eta))-V(\sigma+\eta,x(\sigma+\eta))\\
&=V(\sigma+\eta,\overline{x}(\sigma+\eta))-V(\sigma+\eta,x(\sigma+\eta))-V(\sigma,\overline{x}(\sigma))+V(\sigma,x(\sigma))\\
&\leqslant \left|V(\sigma+\eta,\overline{x}(\sigma+\eta))-V(\sigma+\eta,x(\sigma+\eta))-V(\sigma,\overline{x}(\sigma))+V(\sigma,x(\sigma))\right|\\
&\leqslant \sup_{\xi\in[\sigma,\sigma+\eta]} \|\, \overline{x}(\xi)-x(\xi)\,\| .
\end{aligned} \tag{8.36}$$

Now, by Eqs. (8.30), (8.35), (8.36) and noticing that x is a solution of the generalized ODE (8.1), we have

$$\begin{aligned}
&V(\sigma+\eta,\overline{x}(\sigma+\eta))-V(\sigma,\overline{x}(\sigma))\\
&=V(\sigma+\eta,\overline{x}(\sigma+\eta))-V(\sigma+\eta,x(\sigma+\eta))+V(\sigma+\eta,x(\sigma+\eta))-V(\sigma,x(\sigma))\\
&\leqslant \sup_{\xi\in[\sigma,\sigma+\eta]}\|\,\overline{x}(\xi)-x(\xi)\,\|=\sup_{\xi\in[\sigma,\sigma+\eta]}\|\,\overline{x}(\xi)-\overline{x}(\sigma)+x(\sigma)-x(\xi)\,\|\\
&=\sup_{\xi\in[\sigma,\sigma+\eta]}\left\|\overline{x}(\xi)-\overline{x}(\sigma)-\int_{\sigma}^{\xi}DF(x(\tau),t)\right\|\\
&=\sup_{\xi\in[\sigma,\sigma+\eta]}\left\|\overline{x}(\xi)-\overline{x}(\sigma)-\int_{\sigma}^{\xi}DF(\overline{x}(\tau),t)+\int_{\sigma}^{\xi}D[F(\overline{x}(\tau),t)-F(x(\tau),t)]\right\|\\
&\leqslant \sup_{\xi\in[\sigma,\sigma+\eta]}\left\|\overline{x}(\xi)-\overline{x}(\sigma)-\int_{\sigma}^{\xi}DF(\overline{x}(\tau),t)\right\|\\
&+\sup_{\xi\in[\sigma,\sigma+\eta]}\left\|\int_{\sigma}^{\xi}D[F(\overline{x}(\tau),t)-F(x(\tau),t)]\right\|\\
&\leqslant \sup_{\xi\in[\sigma,\sigma+\eta]}\left\|\overline{x}(\xi)-\overline{x}(\sigma)-\int_{\sigma}^{\xi}DF(\overline{x}(\tau),t)\right\|+\epsilon\eta.
\end{aligned}$$

Making $\epsilon\to 0$, we derive that

$$V(\sigma+\eta,\overline{x}(\sigma+\eta))-V(\sigma,\overline{x}(\sigma))\leqslant \sup_{\xi\in[\sigma,\sigma+\eta]}\left\|\overline{x}(\xi)-\overline{x}(\sigma)-\int_{\sigma}^{\xi}DF(\overline{x}(\tau),t)\right\|. \tag{8.37}$$

Let $P:[\alpha,\beta]\to X$ be defined by

$$P(s)=\overline{x}(s)-\overline{x}(\alpha)-\int_{\alpha}^{s}DF(\overline{x}(\tau),t),\quad \text{for all } s\in[\alpha,\beta].$$

Note that

$$\begin{aligned}
\|\,P(s_2)-P(s_1)\,\| &\leqslant \|\,\overline{x}(s_2)-\overline{x}(s_1)\,\|+\left\|\int_{s_1}^{s_2}DF(\overline{x}(\tau),t)\right\|\\
&\leqslant \|\,\overline{x}(s_2)-\overline{x}(s_1)\,\|+\|\,h(s_2)-h(s_1)\,\|
\end{aligned}$$

for all $s_1, s_2 \in [\alpha, \beta]$, where the last inequality follows from Lemma 4.5. Thus, every point in $[\alpha, \beta]$ at which h and $\overline{x}$ are continuous is a point of continuity of the function P and, because h and $\overline{x}$ are left-continuous on $(\alpha, \beta]$, P is also left-continuous on $(\alpha, \beta]$. In addition, once $\overline{x}$ is regulated, the function $s \mapsto \int_\alpha^s DF(\overline{x}(\tau), t)$ is also regulated and, hence, P is regulated. Note, as well, that

$$\begin{aligned}
&\| P(\xi) - P(\sigma) \| \\
&= \left\| \overline{x}(\xi) - \overline{x}(\alpha) - \int_\alpha^\xi DF(\overline{x}(\tau), t) - \left(\overline{x}(\sigma) - \overline{x}(\alpha) - \int_\alpha^\sigma DF(\overline{x}(\tau), t) \right) \right\| \\
&\leqslant \left\| \overline{x}(\xi) - \overline{x}(\alpha) - \int_\alpha^\xi DF(\overline{x}(\tau), t) \right\| + \left\| \overline{x}(\sigma) - \overline{x}(\alpha) - \int_\alpha^\sigma DF(\overline{x}(\tau), t) \right\|.
\end{aligned} \tag{8.38}$$

For $\sigma \in [\alpha, \beta]$ fixed, define $f : [\alpha, \beta] \to \mathbb{R}$ by

$$f(t) = \begin{cases} \sup_{\xi \in [t,\sigma]} \| P(\xi) - P(\sigma) \|, & t \in [\alpha, \sigma], \\ \sup_{\xi \in [\sigma,t]} \| P(\xi) - P(\sigma) \|, & t \in [\sigma, \beta]. \end{cases}$$

In this way, f is well-defined and it is left-continuous on $(\alpha, \beta]$, provided P is left-continuous on $(\alpha, \beta]$. Furthermore,

$$f(\sigma + \eta) - f(\sigma) = \sup_{\xi \in [\sigma, \sigma+\eta]} \left\| \overline{x}(\xi) - \overline{x}(\sigma) - \int_\sigma^\xi DF(\overline{x}(\tau), t) \right\|.$$

By the last equality and (8.37), we obtain

$$V(\sigma + \eta, \overline{x}(\sigma + \eta)) - V(\sigma, \overline{x}(\sigma)) \leqslant f(\sigma + \eta) - f(\sigma). \tag{8.39}$$

Using Proposition 1.8, we conclude that

$$V(v, \overline{x}(v)) - V(\gamma, \overline{x}(\gamma)) \leqslant f(v) - f(\gamma),$$

for all $\gamma, v \in [\alpha, \beta]$, with $\gamma \leqslant v$.

To finalize the proof, we need to show that

$$f(v) - f(\gamma) \leqslant 2 \sup_{\xi \in [\alpha,\beta]} \left\| \overline{x}(\xi) - \overline{x}(\alpha) - \int_\alpha^\xi DF(\overline{x}(\tau), t) \right\|,$$

for all $\gamma, v \in [\alpha, \beta]$, with $\gamma \leqslant v$. Let us consider three cases.

Case 1: If $\gamma < v \leqslant \sigma$, then

$$\begin{aligned}
f(v) - f(\gamma) &= \sup_{\xi \in [v,\sigma]} \| P(\xi) - P(\sigma) \| - \sup_{\xi \in [\gamma,\sigma]} \| P(\xi) - P(\sigma) \| \\
&\leqslant \sup_{\xi \in [\gamma,\sigma]} \| P(\xi) - P(\sigma) \| - \sup_{\xi \in [\gamma,\sigma]} \| P(\xi) - P(\sigma) \| \\
&= 0 \leqslant 2 \sup_{\xi \in [\alpha,\beta]} \left\| \overline{x}(\xi) - \overline{x}(\alpha) - \int_\alpha^\xi DF(\overline{x}(\tau), t) \right\|.
\end{aligned}$$

Case 2: If $\gamma < \sigma \leqslant \upsilon$, then

$$\begin{aligned}
f(\upsilon) - f(\gamma) &= \sup_{\xi \in [\sigma,\upsilon]} \| P(\xi) - P(\sigma) \| - \sup_{\xi \in [\gamma,\sigma]} \| P(\xi) - P(\sigma) \| \\
&\leqslant \sup_{\xi \in [\sigma,\upsilon]} \| P(\xi) - P(\sigma) \| \\
&\overset{(8.38)}{\leqslant} \sup_{\xi \in [\sigma,\upsilon]} \left(\left\| \overline{x}(\xi) - \overline{x}(\alpha) - \int_\alpha^\xi DF(\overline{x}(\tau), t) \right\| \right. \\
&\quad + \left. \left\| \overline{x}(\sigma) - \overline{x}(\alpha) - \int_\alpha^\sigma DF(\overline{x}(\tau), t) \right\| \right) \\
&\leqslant 2 \sup_{\xi \in [\alpha,\beta]} \left\| \overline{x}(\xi) - \overline{x}(\alpha) - \int_\alpha^\xi DF(\overline{x}(\tau), t) \right\| .
\end{aligned}$$

Case 3: If $\sigma \leqslant \gamma < \upsilon$, then

$$\begin{aligned}
f(\upsilon) - f(\gamma) &= \sup_{\xi \in [\sigma,\upsilon]} \| P(\xi) - P(\sigma) \| - \sup_{\xi \in [\sigma,\gamma]} \| P(\xi) - P(\sigma) \| \\
&\leqslant \sup_{\xi \in [\sigma,\gamma]} \| P(\xi) - P(\sigma) \| + \sup_{\xi \in [\gamma,\upsilon]} \| P(\xi) - P(\sigma) \| \\
&\quad - \sup_{\xi \in [\sigma,\gamma]} \| P(\xi) - P(\sigma) \| \\
&= \sup_{\xi \in [\gamma,\upsilon]} \| P(\xi) - P(\sigma) \| \\
&\overset{(8.38)}{\leqslant} \sup_{\xi \in [\gamma,\upsilon]} \left(\left\| \overline{x}(\xi) - \overline{x}(\alpha) - \int_\alpha^\xi DF(\overline{x}(\tau), t) \right\| \right. \\
&\quad + \left. \left\| \overline{x}(\sigma) - \overline{x}(\alpha) - \int_\alpha^\sigma DF(\overline{x}(\tau), t \right\| \right) \\
&\leqslant 2 \sup_{\xi \in [\alpha,\beta]} \left\| \overline{x}(\xi) - \overline{x}(\alpha) - \int_\alpha^\xi DF(\overline{x}(\tau), t) \right\|
\end{aligned}$$

and the proof is complete. □

The next result can be found in [7, Theorem 3.8] and it is known as Lyapunov-type theorem for regular stability. This shows that the existence of a Lyapunov functional with respect to the generalized ODE (8.1) satisfying some properties implies that the trivial solution $x \equiv 0$ of the generalized ODE (8.1) is regularly stable.

Theorem 8.33: *Let $F \in \mathcal{F}(\Omega, h)$ satisfy condition* (ZS) *and let $V : [t_0, \infty) \times X \to \mathbb{R}$ be a Lyapunov functional with respect to the generalized ODE* (8.1) *satisfying the conditions of Lemma 8.32. Moreover, assume that:*

(i) $V(t, 0) = 0$, *for all* $t \in [t_0, \infty)$;

(ii) *there exists a continuous increasing function* $a : \mathbb{R}_+ \to \mathbb{R}_+$ *satisfying* $a(0) = 0$, *such that*

$$V(t, z) \leqslant a(\| z \|), \quad \text{for all } z \in X \text{ and } t \in [t_0, \infty).$$

Then, the trivial solution $x \equiv 0$ *of the generalized ODE* (8.1) *is regularly stable.*

Proof. By condition (LF2) of Definition 8.1, there exists an increasing function $b : \mathbb{R}_+ \to \mathbb{R}_+$ satisfying

$$b(\| x \|) \leqslant V(t,x), \quad \text{for every } (t,x) \in [t_0, \infty) \times X.$$

Let $s_0 \geqslant t_0$ and $\epsilon > 0$. By the fact that $a(0) = 0$, a is continuous and increasing and $b(\epsilon) > 0$, there exists $\delta = \delta(\epsilon)$ such that $0 < \delta < \frac{b(\epsilon)}{3}$ and $a(\delta) < \frac{b(\epsilon)}{3}$.

We want to prove that

$$\| \bar{x}(t) \| < \epsilon \quad \text{for all } t \in [\alpha, \beta],$$

where $t_0 \leqslant \alpha < \beta < \infty$ and $\bar{x} : [\alpha, \beta] \to X$ is a regulated function and left-continuous on $(\alpha, \beta]$ satisfying

$$\| \bar{x}(\alpha) \| < \delta \quad \text{and} \quad \sup_{s \in [\alpha,\beta]} \left\| \bar{x}(s) - \bar{x}(\alpha) - \int_\alpha^s DF(\bar{x}(\tau), t) \right\| < \delta.$$

By Lemma 8.32, condition (ii), and the fact that $a(\delta) < \frac{b(\epsilon)}{3}$, we have

$$\begin{aligned} V(t, \bar{x}(t)) &\leqslant V(\alpha, \bar{x}(\alpha)) + 2 \sup_{s \in [\alpha,\beta]} \left\| \bar{x}(s) - \bar{x}(\alpha) - \int_\alpha^s DF(\bar{x}(\tau), t) \right\| \\ &\leqslant a(\| \bar{x}(\alpha) \|) + 2\delta \leqslant a(\delta) + 2\delta < 3\frac{b(\epsilon)}{3} = b(\epsilon), \end{aligned}$$

for all $t \in [\alpha, \beta]$. On the other hand,

$$b(\| \bar{x}(t) \|) \leqslant V(t, \bar{x}(t)) \leqslant b(\epsilon), \quad \text{for all } t \in [\alpha, \beta].$$

Since b is increasing, it follows that $\| \bar{x}(t) \| < \epsilon$ for all $t \in [\alpha, \beta]$. Therefore, the trivial solution $x \equiv 0$ of the generalized ODE (8.1) is regularly stable. □

The next result can be found in [87, Theorem 6.4] and it shows that the existence of a Lyapunov functional with an additional condition implies regular asymptotic stability for the generalized ODE (8.1).

Theorem 8.34: *Let* $F \in \mathcal{F}(\Omega, h)$ *satisfy condition* (ZS) *and consider* $V : [t_0, \infty) \times \overline{B}_\rho \to \mathbb{R}$ *a Lyapunov functional with respect the generalized ODE* (8.1) *satisfying the conditions of Theorem 8.33, where* $\overline{B}_\rho = \{x \in X : \| x \| \leqslant \rho\} \subset O$. *Moreover, assume that:*

(i) *there exists a continuous function* $\Phi : X \to \mathbb{R}_+$ *satisfying* $\Phi(0) = 0$ *and* $\Phi(x) > 0$ *for* $x \neq 0$, *such that for every solution* $x : [\alpha, \beta] \subset [t_0, \infty) \to X$ *of the generalized ODE* (8.1), *we have, for every* $t \in [\alpha, \beta)$,

$$D^+V(t,x(t)) = \limsup_{\eta \to 0^+} \frac{V(t+\eta, x(t+\eta)) - V(t, x(t))}{\eta} \leqslant -\Phi(x(t)). \quad (8.40)$$

Then, the trivial solution $x \equiv 0$ *of the generalized ODE* (8.1) *is regularly asymptotically stable.*

Proof. By Theorem 8.33, it remains to prove that the solution $x \equiv 0$ of the generalized ODE is regularly attracting.

At first, notice that, using the ideas of Lemma 8.32 and Eq. (8.40), we can prove that, for every regulated function $y : [\alpha, \beta] \subset [t_0, \infty) \to X$ left-continuous on $(\alpha, \beta]$, we have

$$V(v, y(v)) - V(\gamma, y(\gamma)) \leqslant \sup_{\xi \in [\alpha,\beta]} \left\| y(\xi) - y(\alpha) - \int_\alpha^\xi DF(y(\tau), t) \right\| + M(v - \gamma), \quad (8.41)$$

for all $v, \gamma \in [\alpha, \beta]$, with $\gamma \leqslant v$ and $M = \sup_{\xi \in [\alpha,\beta]} \Phi(y(\xi))$.

From the regular stability of the trivial solution $x \equiv 0$ of the generalized ODE (8.1), it follows that, for every $a > 0$, with $a < \rho$, there exists $\delta_0 \in (0, a)$ such that if $\overline{x} \in G^-([s_0, s_1], X)$ is such that

$$\| \overline{x}(s_0) \| < \delta_0 \quad \text{and} \quad \sup_{s \in [s_0,s_1]} \left\| \overline{x}(s) - \overline{x}(s_0) - \int_{s_0}^s DF(\overline{x}(\tau), t) \right\| < \delta_0,$$

then

$$\| \overline{x}(t) \| < a, \quad \text{for all } t \in [s_0, s_1],$$

that is, $x(t) \in B_a = \{x \in X : \| x \| < a\} \subset B_\rho = \{x \in X : \| x \| < \rho\} \subset \overline{B}_\rho \subset O$, for every $t \in [s_0, s_1]$.

Let $\epsilon > 0$ be arbitrary. Using again the regular stability of the trivial solution $x \equiv 0$ of the generalized ODE (8.1), there is $\delta = \delta(\epsilon) > 0$ such that, for every function $\overline{x} \in G^-([s_2, s_3], X)$ with

$$\| \overline{x}(s_2) \| < \delta(\epsilon) \quad \text{and} \quad \sup_{s \in [s_2,s_3]} \left\| \overline{x}(s) - \overline{x}(s_2) - \int_{s_2}^s DF(\overline{x}(\tau), t) \right\| < \delta(\epsilon), \quad (8.42)$$

then

$$\| \overline{x}(t) \| < \epsilon, \quad (8.43)$$

for every $t \in [s_2, s_3]$. Define $\rho(\epsilon) = \min\{\delta_0, \delta(\epsilon)\} > 0$ and $T(\epsilon) = -\left(\frac{\delta_0 + \rho(\epsilon)}{N}\right)$, where $N = \sup\{-\Phi(x) : \rho(\epsilon) \leqslant \| x \| < \epsilon\} = -\inf\{\Phi(x) : \rho(\epsilon) \leqslant \| x \| < \epsilon\} < 0$.

Assume that $\bar{x} : [s_0, s_1] \to X$ is a regulated function on $[s_0, s_1]$ and left-continuous on $(s_0, s_1]$ such that $\| \bar{x}(s_0) \| < \delta_0$ and

$$\sup_{s\in[s_2,s_3]} \left\| \bar{x}(s) - \bar{x}(s_2) - \int_{s_2}^{s} DF(\bar{x}(\tau), t) \right\| < \rho(\epsilon). \tag{8.44}$$

Suppose that $T(\epsilon) < s_1 - s_0$. We assure that there is $\bar{t} \in [s_0, s_0 + T(\epsilon)]$ such that $\| \bar{x}(\bar{t}) \| < \rho(\epsilon)$. Assume the contrary, that is, $\| \bar{x}(s) \| \geqslant \rho(\epsilon)$ for all $s \in [s_0, s_0 + T(\epsilon)]$. By (8.41), we have

$$\begin{aligned}
&V(s_0 + T(\epsilon), \bar{x}(s_0 + T(\epsilon))) \\
&\leqslant V(s_0, \bar{x}(s_0)) + \sup_{s\in[s_0,s_0+T(\epsilon)]} \left\| \bar{x}(s) - \bar{x}(s_0) - \int_{s_0}^{s} DF(\bar{x}(\tau), t) \right\| + NT(\epsilon) \\
&\leqslant \| \bar{x}(s_0) \| + \rho(\epsilon) + N\left(\frac{-(\delta_0 + \rho(\epsilon))}{N} \right) \\
&\leqslant \delta_0 + \rho(\epsilon) + N\left(\frac{-(\delta_0 + \rho(\epsilon))}{N} \right) = 0,
\end{aligned}$$

which contradicts the inequality

$$V(s_0 + T(\epsilon), \bar{x}(s_0 + T(\epsilon))) \geqslant b(\| \bar{x}(s_0 + T(\epsilon)) \|) \geqslant b(\rho(\epsilon)) > 0.$$

Therefore, there exists $\bar{t} \in [s_0, s_0 + T(\epsilon)]$ such that $\| \bar{x}(\bar{t}) \| < \rho(\epsilon)$. Once (8.42) holds in view of the choice of $\rho(\epsilon)$ and, moreover, (8.43) holds for the case where $s_2 = \bar{t}$ and $s_3 = s_1$, by (8.44), we obtain

$$\| \bar{x}(t) \| < \epsilon, \quad \text{for } t \in [\bar{t}, s_1].$$

Consequently,

$$\| \bar{x}(t) \| < \epsilon, \quad \text{for } t > s_0 + T(\epsilon),$$

since $\bar{t} \in [s_0, s_0 + T(\epsilon)]$, and this completes the proof. □

8.5.2 Converse Lyapunov Theorem

Motivated by Š. Schwabik's work, [208], where he proved converse Lyapunov theorems concerning variational stability, the authors of [7] established converse Lyapunov theorems concerning regular stability. In this subsection, we describe such results.

Throughout this subsection, we assume that $F \in \mathcal{F}(\Omega, h)$ satisfies condition (ZS), where $\Omega = X \times [t_0, \infty)$. By Corollary 5.16, for every $(x_0, s_0) \in \Omega$, there exists a unique global forward solution $x : [s_0, \infty) \to X$ of the generalized ODE (8.1).

Given $s \geqslant t_0$ and $x \in X$, define

$$A(s, x) = \{ \varphi \in G([t_0, \infty), X) : \varphi(t_0) = 0, \ \varphi(s) = x, \ \varphi \text{ is left-continuous on } (t_0, \infty) \} \tag{8.45}$$

and $V : [t_0, \infty) \times X \to \mathbb{R}$ by

$$V(s,x) = \begin{cases} \inf\limits_{\varphi \in A(s,x)} \left\{ \sup\limits_{\sigma \in [t_0,s]} \left\| \varphi(\sigma) - \int_{t_0}^{\sigma} DF(\varphi(\tau), t) \right\| \right\}, & \text{if } s > t_0, \\ \| x \|, & \text{if } s = t_0. \end{cases} \tag{8.46}$$

Remark 8.35: Since $\varphi \in A(s,x)$ is a regulated function, by Corollary 4.8, the Kurzweil integral $\int_{t_0}^{\sigma} DF(\varphi(\tau), t)$ exists, for every $\sigma \in [t_0, \infty)$, and the mapping $s \ni [t_0, \infty) \mapsto \int_{t_0}^{s} DF(\varphi(\tau), t)$ is regulated, as well as the mapping $s \ni [t_0, \infty) \mapsto \varphi(s) - \int_{t_0}^{s} DF(\varphi(\tau), t)$. On the other hand, since $G([a,b], X) \subset B([a,b], X)$, where $B([a,b], X)$ denotes the space of all bounded function $f : [a,b] \to X$, we have

$$\sup_{\sigma \in [a,b]} \left\| \varphi(\sigma) - \int_{t_0}^{\sigma} DF(\varphi(\tau), t) \right\| < \infty, \quad \text{for all } [a,b] \subset [t_0, \infty).$$

In particular,

$$\sup_{\sigma \in [t_0,s]} \left\| \varphi(\sigma) - \int_{t_0}^{\sigma} DF(\varphi(\tau), t) \right\| < \infty, \quad \text{for all } s \geqslant t_0.$$

Therefore, the function V, given by (8.46), is well-defined for all $(s,x) \in [t_0, \infty) \times X$.

Our next goal is to show that if the trivial solution of the generalized ODE (8.1) is regularly stable, then V, given by (8.46), is a Lyapunov functional with respect to the generalized ODE (8.1) and it satisfies all conditions of Theorem 8.33.

The next result shows that V, given by (8.46), satisfies condition (i) of Theorem 8.33.

Lemma 8.36: *The function V defined in* (8.46) *satisfies:*

(i) $V(s,0) = 0$, *for all* $s \geqslant t_0$;
(ii) $V(s,x) \geqslant 0$, *for all* $x \in X$ *and all* $s \geqslant t_0$.

Proof. Since $\varphi \equiv 0 \in A(s,0)$, we have $V(s,0) = 0$, for all $s \geqslant t_0$. Now, we notice that item (ii) follows from the definition of V. □

The next result gives an estimate for the function V defined by (8.46).

Lemma 8.37: *Let $F \in \mathcal{F}(\Omega, h)$ and $V : [t_0, \infty) \times X \to \mathbb{R}$ be defined by* (8.46). *Then, for all $x, y \in X$ and all $s \in [t_0, \infty)$, we have*

$$V(s,y) - V(s,x) \leqslant \| y - x \| .$$

Proof. If $s = t_0$, then

$$V(t_0, y) - V(t_0, x) = \| y \| - \| x \| \leqslant \| y - x \|$$

and the proof is complete.

Now, assume that $s > t_0$. Let $\eta > 0$ be such that $0 < \eta < s - t_0$ and $\varphi \in A(s, x)$ be an arbitrary function. Define $\varphi_\eta : [t_0, \infty) \to X$ by

$$\varphi_\eta(\sigma) = \begin{cases} \varphi(\sigma), & \text{if } \sigma \in [t_0, s-\eta], \\ \varphi(\sigma) + \dfrac{\sigma - s + \eta}{\eta}(y - x), & \text{if } \sigma \in [s-\eta, s], \\ y, & \text{if } \sigma \in (s, \infty). \end{cases} \tag{8.47}$$

It is not difficult to see that $\varphi_\eta \in A(s, y)$. By the definition of V in (8.46),

$$V(s, y) \leqslant \sup_{\sigma \in [t_0, s]} \left\| \varphi_\eta(\sigma) - \int_{t_0}^{\sigma} DF(\varphi_\eta(\tau), t) \right\|. \tag{8.48}$$

Moreover,

$$\begin{aligned} &\sup_{\sigma \in [t_0, s]} \left\| \varphi_\eta(\sigma) - \int_{t_0}^{\sigma} DF(\varphi_\eta(\tau), t) \right\| - \sup_{\sigma \in [t_0, s]} \left\| \varphi(\sigma) - \int_{t_0}^{\sigma} DF(\varphi(\tau), t) \right\| \\ &\leqslant \sup_{\sigma \in [t_0, s]} \left\| \varphi_\eta(\sigma) - \int_{t_0}^{\sigma} DF(\varphi_\eta(\tau), t) - \varphi(\sigma) + \int_{t_0}^{\sigma} DF(\varphi(\tau), t) \right\| \\ &\leqslant \sup_{\sigma \in [t_0, s-\eta]} \left\| \varphi_\eta(\sigma) - \int_{t_0}^{\sigma} DF(\varphi_\eta(\tau), t) - \varphi(\sigma) + \int_{t_0}^{\sigma} DF(\varphi(\tau), t) \right\| \\ &\quad + \sup_{\sigma \in [s-\eta, s]} \left\| \varphi_\eta(\sigma) - \int_{t_0}^{\sigma} DF(\varphi_\eta(\tau), t) - \varphi(\sigma) + \int_{t_0}^{\sigma} DF(\varphi(\tau), t) \right\|. \end{aligned} \tag{8.49}$$

By (8.47), (8.48), and (8.49), we obtain

$$\begin{aligned} V(s, y) &\leqslant \sup_{\sigma \in [t_0, s-\eta]} \left\| \varphi(\sigma) - \int_{t_0}^{\sigma} DF(\varphi(\tau), t) - \varphi(\sigma) + \int_{t_0}^{\sigma} DF(\varphi(\tau), t) \right\| \\ &\quad + \sup_{\sigma \in [s-\eta, s]} \left\| \varphi(\sigma) + \frac{\sigma - s + \eta}{\eta}(y - x) - \int_{t_0}^{\sigma} DF(\varphi_\eta(\tau), t) \right. \\ &\quad \left. - \varphi(\sigma) + \int_{t_0}^{\sigma} DF(\varphi(\tau), t) \right\| + \sup_{\sigma \in [t_0, s]} \left\| \varphi(\sigma) - \int_{t_0}^{\sigma} DF(\varphi(\tau), t) \right\| \\ &= \sup_{\sigma \in [s-\eta, s]} \left\| \frac{\sigma - s + \eta}{\eta}(y - x) - \int_{t_0}^{s-\eta} DF(\varphi(\tau), t) - \int_{s-\eta}^{\sigma} DF(\varphi_\eta(\tau), t) \right. \\ &\quad \left. + \int_{t_0}^{s-\eta} DF(\varphi(\tau), t) + \int_{s-\eta}^{\sigma} DF(\varphi(\tau), t) \right\| \\ &\quad + \sup_{\sigma \in [t_0, s]} \left\| \varphi(\sigma) - \int_{t_0}^{\sigma} DF(\varphi(\tau), t) \right\| \end{aligned}$$

$$= \sup_{\sigma\in[s-\eta,s]} \left\| \frac{\sigma - s + \eta}{\eta}(y-x) - \int_{s-\eta}^{\sigma} DF(\varphi_\eta(\tau), t) + \int_{s-\eta}^{\sigma} DF(\varphi(\tau), t) \right\|$$

$$+ \sup_{\sigma\in[t_0,s]} \left\| \varphi(\sigma) - \int_{t_0}^{\sigma} DF(\varphi(\tau), t) \right\|$$

$$\leqslant \sup_{\sigma\in[s-\eta,s]} \left\| \frac{\sigma - s + \eta}{\eta}(y-x) \right\| + \sup_{\sigma\in[s-\eta,s]} \left\| \int_{s-\eta}^{\sigma} DF(\varphi_\eta(\tau), t) \right\|$$

$$+ \sup_{\sigma\in[s-\eta,s]} \left\| \int_{s-\eta}^{\sigma} DF(\varphi(\tau), t) \right\| + \sup_{\sigma\in[t_0,s]} \left\| \varphi(\sigma) - \int_{t_0}^{\sigma} DF(\varphi(\tau), t) \right\|$$

$$\leqslant \| y - x \| + 2 \sup_{\sigma\in[s-\eta,s]} |h(\sigma) - h(s-\eta)| + \sup_{\sigma\in[t_0,s]} \left\| \varphi(\sigma) - \int_{t_0}^{\sigma} DF(\varphi(\tau), t) \right\|$$

$$= \| y - x \| + 2|h(s) - h(s-\eta)| + \sup_{\sigma\in[t_0,s]} \left\| \varphi(\sigma) - \int_{t_0}^{\sigma} DF(\varphi(\tau), t) \right\|,$$

where the inequality

$$\left\| \int_{s-\eta}^{\sigma} DF(\varphi(\tau), t) \right\| \leqslant |h(\sigma) - h(s-\eta)|$$

follows from Lemma 4.5.

Making $\eta \to 0^+$, we obtain

$$V(s, y) \leqslant \| y - x \| + \sup_{\sigma\in[t_0,s]} \left\| \varphi(\sigma) - \int_{t_0}^{\sigma} DF(\varphi(\tau), t) \right\|.$$

Then, taking the infimum for all $\varphi \in A(s, x)$ on the right-hand side of the last inequality, we conclude that

$$V(s, y) \leqslant \| y - x \| + V(s, x)$$

and the proof is complete. □

The proof of the next result follows from Lemma 8.37 and from the fact that V, defined by (8.46), satisfies $V(s, 0) = 0$ for all $s \geqslant t_0$.

Corollary 8.38: *If $F \in \mathcal{F}(\Omega, h)$, then $V : [t_0, \infty) \times X \to \mathbb{R}$ defined by (8.46) satisfies*

$$V(s, x) \leqslant \| x \|, \quad \textit{for all } x \in X \textit{ and } s \in [t_0, \infty).$$

The next result will be of interest for the forthcoming considerations.

Lemma 8.39: *Let $F \in \mathcal{F}(\Omega, h)$ and $V : [t_0, \infty) \times X \to \mathbb{R}$ be defined by (8.46). Then, the function $t \mapsto V(t, x(t))$ is nonincreasing along every global forward* *solution $x : [t_0, \infty) \to X$, which is denoted by $x(\cdot, s_0, x_0)$, for $(s_0, x_0) \in [t_0, \infty) \times X$, of the generalized ODE (8.1), with initial condition $x(s_0) = x_0$.*

Proof. Consider $x : [s_0, \infty) \to X$ being the global forward solution of the generalized ODE (8.1) with initial condition $x(s_0) = x_0$. Let $t_1, t_2 \in [s_0, \infty)$ be such that $t_2 > t_1$ and consider $\varphi \in A(t_1, x(t_1))$.

Let $\phi : [t_0, \infty) \to X$ be defined by

$$\phi(\sigma) = \begin{cases} \varphi(\sigma), & \text{if } \sigma \in [t_0, t_1], \\ x(\sigma), & \text{if } \sigma \in [t_1, t_2], \\ x(t_2), & \text{if } \sigma \in (t_2, \infty). \end{cases}$$

By Lemma 4.9, x is left-continuous on $(t_1, t_2]$ and, by definition of the set $A(t_1, x(t_1))$, φ is left-continuous on (t_0, ∞) (see (8.45)). Moreover, since $\phi(t_2) = \phi(t) = x(t_2)$ for all $t \geqslant t_2$, we conclude that ϕ is left-continuous on (t_0, ∞) and, therefore, $\phi \in A(t_2, x(t_2))$. Thus, V given by (8.46) satisfies

$$V(t_2, x(t_2)) \leqslant \sup_{\sigma \in [t_0, t_2]} \left\| \phi(\sigma) - \int_{t_0}^{\sigma} DF(\phi(\tau), s) \right\|. \tag{8.50}$$

Now, define $f : [t_0, t_2] \to X$ by

$$f(\sigma) = \phi(\sigma) - \int_{t_0}^{\sigma} DF(\phi(\tau), t) \quad \text{for all } \sigma \in [t_0, t_2].$$

Notice that, for all $t, s \in [t_0, t_2]$, with $s < t$, we have

$$\begin{aligned} \| f(t) - f(s) \| &= \left\| \phi(t) - \int_{t_0}^{t} DF(\phi(\tau), v) - \left(\phi(s) - \int_{t_0}^{s} DF(\phi(\tau), v) \right) \right\| \\ &= \left\| \phi(t) - \phi(s) - \int_{s}^{t} DF(\phi(\tau), v) \right\| \\ &\leqslant \| \phi(t) - \phi(s) \| + \left\| \int_{s}^{t} DF(\phi(\tau), v) \right\| \\ &\leqslant \| \phi(t) - \phi(s) \| + |h(t) - h(s)|, \end{aligned} \tag{8.51}$$

where the last inequality follows from Lemma 4.5. Therefore, every point in $[t_0, t_2]$ at which h and φ are continuous is a point of continuity of the function f and, because h and φ are left-continuous on $(t_0, t_2]$, the function f is also left-continuous on $(t_0, t_2]$. Besides, f is regulated, since ϕ is regulated. Therefore, by Proposition 1.6, we can consider two cases with respect to

$$\sup_{\sigma \in [t_0, t_2]} \left\| \phi(\sigma) - \int_{t_0}^{\sigma} DF(\phi(\tau), s) \right\|.$$

Case 1: Suppose, for some $v \in [t_0, t_2]$, we have

$$\sup_{\sigma\in[t_0,t_2]} \left\| \phi(\sigma) - \int_{t_0}^{\sigma} DF(\phi(\tau), s) \right\| = \left\| \phi(v) - \int_{t_0}^{v} DF(\phi(\tau), s) \right\|.$$

Then, either $v \in [t_0, t_1]$ or $v \in [t_1, t_2]$.
Firstly, consider $v \in [t_0, t_1]$, then

$$\sup_{\sigma\in[t_0,t_2]} \left\| \phi(\sigma) - \int_{t_0}^{\sigma} DF(\phi(\tau), s) \right\| = \sup_{\sigma\in[t_0,t_1]} \left\| \phi(\sigma) - \int_{t_0}^{\sigma} DF(\phi(\tau), s) \right\|.$$

By the fact that $\phi(t) = \varphi(t)$ for all $t \in [t_0, t_1]$, we get

$$\sup_{\sigma\in[t_0,t_2]} \left\| \phi(\sigma) - \int_{t_0}^{\sigma} DF(\phi(\tau), s) \right\| = \sup_{\sigma\in[t_0,t_1]} \left\| \varphi(\sigma) - \int_{t_0}^{\sigma} DF(\varphi(\tau), s) \right\|.$$

From this and (8.50), we derive that

$$V(t_2, x(t_2)) \leqslant \sup_{\sigma\in[t_0,t_1]} \left\| \varphi(\sigma) - \int_{t_0}^{\sigma} DF(\varphi(\tau), s) \right\|.$$

Then, taking the infimum for all $\varphi \in A(t_1, x(t_1))$ on the right-hand side of the last inequality, we have $V(t_2, x(t_2)) \leqslant V(t_1, x(t_1))$.

On the other hand, since $\phi(t) = x(t)$ for all $t \in [t_1, t_2]$, then, if $v \in [t_1, t_2]$, it follows that

$$\begin{aligned} \left\| \phi(v) - \int_{t_0}^{v} DF(\phi(\tau), s) \right\| &= \left\| \phi(v) - \int_{t_0}^{t_1} DF(\phi(\tau), s) - \int_{t_1}^{v} DF(\phi(\tau), s) \right\| \\ &= \left\| x(v) - \int_{t_0}^{t_1} DF(\varphi(\tau), s) - \int_{t_1}^{v} DF(x(\tau), s) \right\|. \end{aligned} \tag{8.52}$$

By definition of solution of the generalized ODE (8.1), we have

$$x(v) - \int_{t_1}^{v} DF(x(\tau), s) = x(t_1) = \varphi(t_1). \tag{8.53}$$

Replacing (8.53) in (8.52), we get

$$\begin{aligned} \left\| \phi(v) - \int_{t_0}^{v} DF(\phi(\tau), s) \right\| &= \left\| \varphi(t_1) - \int_{t_0}^{t_1} DF(\varphi(\tau), s) \right\| \\ &\leqslant \sup_{\sigma\in[t_0,t_1]} \left\| \varphi(\sigma) - \int_{t_0}^{\sigma} DF(\varphi(\tau), s) \right\|. \end{aligned} \tag{8.54}$$

Now, (8.54) implies the following inequality

$$V(t_2, x(t_2)) \leqslant \sup_{\sigma\in[t_0,t_1]} \left\| \varphi(\sigma) - \int_{t_0}^{\sigma} DF(\varphi(\tau), s) \right\|.$$

Taking the infimum for all $\varphi \in A(t_1, x(t_1))$ on the right-hand side of the inequality above, we obtain $V(t_2, x(t_2)) \leqslant V(t_1, x(t_1))$.

Case 2: Suppose, for some $v \in [t_0, t_2)$, we have

$$\sup_{\sigma\in[t_0,t_2]} \left\| \phi(\sigma) - \int_{t_0}^{\sigma} DF(\phi(\tau), s) \right\| = \left\| \phi(v^+) - \lim_{\sigma\to v^+} \int_{t_0}^{\sigma} DF(\phi(\tau), s) \right\|.$$

Hence, either $v \in [t_0, t_1)$ or $v \in [t_1, t_2)$. If $v \in [t_0, t_1)$, then the proof follows as in case 1.

By Theorem 4.4 and Lemma 4.10, if $x : [t_0, \infty) \to X$ is the solution of the generalized ODE (8.1), then, for all $v \in [t_0, \infty)$, we have

$$x(v^+) - \lim_{\sigma\to v^+} \int_{t_1}^{\sigma} DF(x(\tau), s) = x(v) - \int_{t_1}^{v} DF(x(\tau), s). \tag{8.55}$$

Now, since $v \in [t_1, t_2)$ and $\phi|_{[t_1,t_2)} = x$, then

$$\begin{aligned}
\sup_{\sigma\in[t_0,t_2]} \left\| \phi(\sigma) - \int_{t_0}^{\sigma} DF(\phi(\tau), s) \right\| &= \left\| \phi(v^+) - \lim_{\sigma\to v^+} \int_{t_0}^{\sigma} DF(\phi(\tau), s) \right\| \\
&= \left\| x(v^+) - \lim_{\sigma\to v^+} \int_{t_0}^{\sigma} DF(\phi(\tau), s) \right\| \\
&= \left\| x(v^+) - \int_{t_0}^{t_1} DF(\varphi(\tau), s) - \lim_{\sigma\to v^+} \int_{t_1}^{\sigma} DF(x(\tau), s) \right\| \\
&\overset{(8.55)}{=} \left\| x(v) - \int_{t_0}^{t_1} DF(\varphi(\tau), s) - \int_{t_1}^{v} DF(x(\tau), s) \right\| \\
&\overset{(8.53)}{=} \left\| x(t_1) - \int_{t_0}^{t_1} DF(\varphi(\tau), s) \right\| \\
&= \left\| \varphi(t_1) - \int_{t_0}^{t_1} DF(\varphi(\tau), s) \right\| \\
&\leqslant \sup_{\sigma\in[t_0,t_1]} \left\| \varphi(\sigma) - \int_{t_0}^{\sigma} DF(\varphi(\tau), s) \right\|.
\end{aligned}$$

By (8.50),

$$V(t_2, x(t_2)) \leqslant \sup_{\sigma\in[t_0,t_1]} \left\| \varphi(\sigma) - \int_{t_0}^{\sigma} DF(\varphi(\tau), s) \right\|.$$

Thus, if we take the infimum for all $\varphi \in A(t_1, x(t_1))$ on the right-hand side of the last inequality, we conclude that $V(t_2, x(t_2)) \leqslant V(t_1, x(t_1))$. □

In what follows, we present a lemma which showed to be essential to prove Lemmas 8.41 and 8.42.

Lemma 8.40: *For all $s \geqslant t_0$ and all $x \in X$, the set $A(s, x)$ defined by (8.45) is closed.*

Proof. Consider a sequence $\{\varphi_n\}_{n\in\mathbb{N}}$ in $A(s,x)$ such that $\{\varphi_n\}_{n\in\mathbb{N}}$ converges to the function φ in $G^-([t_0,\infty),X)$ with the topology of locally uniform convergence. We want to show that $\varphi \in A(s,x)$. Notice that φ satisfies

$$\varphi(t_0) = \lim_{n\to\infty} \varphi_n(t_0) = \lim_{n\to\infty} 0 = 0 \quad \text{and}$$
$$\varphi(s) = \lim_{n\to\infty} \varphi_n(s) = \lim_{n\to\infty} x = x.$$

We want to prove that φ is left-continuous on (t_0,∞). In order to do that, we will show that φ is left-continuous on $(\alpha,\beta]$, for each $(\alpha,\beta] \subset (t_0,\infty)$. Indeed, given $t \in (\alpha,\beta] \subset (t_0,\infty)$, we apply the Moore–Osgood theorem (see, e.g., [19]) to obtain

$$\begin{aligned}\lim_{\xi\to t^-} \varphi(\xi) &= \lim_{\xi\to t^-}\lim_{n\to\infty} \varphi_n(\xi) = \lim_{n\to\infty}\lim_{\xi\to t^-} \varphi_n(\xi)\\ &= \lim_{n\to\infty} \varphi_n(t^-) = \lim_{n\to\infty} \varphi_n(t) = \varphi(t),\end{aligned}$$

for all $t \in (\alpha,\beta]$. Therefore, $\varphi(t_0) = 0$, $\varphi(s) = x$ and φ is left-continuous on (t_0,∞) which implies that $\varphi \in A(s,x)$ and $A(s,x)$ is closed. □

The next result shows that V given by (8.46) satisfies condition (LF1) of Definition 8.1.

Lemma 8.41: *Let $F \in \mathcal{F}(\Omega,h)$ and $V : [t_0,\infty)\times X \to \mathbb{R}$ be defined by (8.46). Then, the function $V(\cdot,y) : [t_0,\infty) \to \mathbb{R}$ is left-continuous on (t_0,∞), for all $y \in X$.*

Proof. Let $y \in X$ and $\sigma_0 \in (t_0,\infty)$ be given. By Lemma 8.40, there exists $\psi \in A(\sigma_0,y)$ satisfying

$$V(\sigma_0,y) = \sup_{\sigma\in[t_0,\sigma_0]} \left\| \psi(\sigma) - \int_{t_0}^{\sigma} DF(\psi(\tau),t) \right\|.$$

Since h and ψ are left-continuous on (t_0,∞), for all $\epsilon > 0$, there exists δ_0 such that if $t \in [\sigma_0-\delta_0,\sigma_0)$, then

$$|h(t) - h(\sigma_0)| < \epsilon \quad \text{and} \quad \| \psi(t) - \psi(\sigma_0) \| < \epsilon. \tag{8.56}$$

We need to prove that $|V(t,y) - V(\sigma_0,y)| < \epsilon$ for all $t \in [\sigma_0-\delta_0,\sigma_0)$. At first, we observe that

$$\begin{aligned}V(\sigma_0,y) &= \sup_{\sigma\in[t_0,\sigma_0]} \left\| \psi(\sigma) - \int_{t_0}^{\sigma} DF(\psi(\tau),s) \right\|\\ &\geqslant \sup_{\sigma\in[t_0,t]} \left\| \psi(\sigma) - \int_{t_0}^{\sigma} DF(\psi(\tau),s) \right\| \geqslant V(t,\psi(t)).\end{aligned}$$

On the other hand, by Lemma 8.37 and by Eq. (8.56), we have

$$\begin{aligned}V(t,y) - V(\sigma_0,y) &\leqslant V(t,y) - V(t,\psi(t)) \leqslant \| y - \psi(t) \|\\ &= \| \psi(\sigma_0) - \psi(t) \| < \epsilon\end{aligned}$$

for all $t \in [\sigma_0-\delta_0,\sigma_0)$.

It remains to show that $V(\sigma_0, y) - V(t, y) < \epsilon$ for all $t \in [\sigma_0 - \delta_0, \sigma_0)$.

Let $x : [t, \infty) \to X$ be the global forward solution of the generalized ODE (8.1) with initial condition $x(t) = y$. Then, by Lemma (8.39), $V(\sigma_0, x(\sigma_0)) - V(t, x(t)) \leqslant 0$, since $t < \sigma_0$.

Then,

$$V(\sigma_0, y) - V(t, y) = V(\sigma_0, y) - V(\sigma_0, x(\sigma_0)) + V(\sigma_0, x(\sigma_0)) - V(t, x(t))$$
$$\leqslant V(\sigma_0, y) - V(\sigma_0, x(\sigma_0)) \leqslant \| x(\sigma_0) - y \|,$$

where the last inequality follows from Lemma 8.37.

By the fact that x is the solution of the generalized ODE (8.1), with $x(t) = y$, it follows that

$$\| x(\sigma_0) - y \| = \left\| \int_t^{\sigma_0} DF(x(\tau), s) \right\| \leqslant |h(\sigma_0) - h(t)|,$$

where the last inequality follows from Lemma 4.5.

Therefore, by (8.56), we conclude that

$$V(\sigma_0, y) - V(t, y) \leqslant \| x(\sigma_0) - y \| \leqslant |h(\sigma_0) - h(t)| < \epsilon,$$

for all $t \in [\sigma_0 - \delta, \sigma_0)$, which completes the proof. □

In the sequel, we prove that if the trivial solution of the generalized ODE (8.1) is regularly stable, then V, defined by (8.46), satisfies condition (LF2) of Definition 8.1.

Lemma 8.42: *Let $F \in \mathcal{F}(\Omega, h)$ satisfy condition (ZS). If the trivial solution $x \equiv 0$ of the generalized ODE (8.1) is regularly stable, then V defined by (8.46) satisfies:*

(i) *There exists a continuous (strictly) increasing function $b : \mathbb{R}_+ \to \mathbb{R}_+$ satisfying $b(0) = 0$ such that*

$$V(t, x) \geqslant b(\| x \|), \quad \textit{for every } (t, x) \in [t_0, \infty) \times X.$$

Proof. Suppose that (i) does not hold. Then, there are $\epsilon > 0$ and a sequence of pairs $\{(t_k, x_k)\}_{k \in \mathbb{N}}$ in $[t_0, \infty) \times X$, such that

$$\epsilon \leqslant \| x_k \|, \tag{8.57}$$

$t_k \to \infty$ and $V(t_k, x_k) \to 0$, as $k \to \infty$.

On the other hand, since the trivial solution $x \equiv 0$ of the generalized ODE (8.1) is regularly stable, it follows from Theorem 8.31 that $x \equiv 0$ is regularly stable with respect to perturbation.

Take $\delta = \delta(\epsilon) > 0$ given in Definition 8.30(i). Since $V(t_k, x_k) \to 0$ as $t_k \to \infty$, there exists $k_0 \in \mathbb{N}$ such that $V(t_k, x_k) < \delta$, for all $k > k_0$. Moreover, since $A(t_k, x_k)$ is a closed set, for each $k > k_0$, there exists $\varphi_k \in A(t_k, x_k)$ such that

$$V(t_k, x_k) = \sup_{\sigma \in [t_0, t_k]} \left\| \varphi_k(\sigma) - \int_{t_0}^{\sigma} DF(\varphi_k(\tau), t) \right\| < \delta. \tag{8.58}$$

For each $k > k_0$, define $P_k : [t_0, t_k] \to X$ by

$$P_k(\sigma) = \varphi_k(\sigma) - \int_{t_0}^{\sigma} DF(\varphi_k(\tau), t), \quad \text{for } \sigma \in [t_0, t_k].$$

Then, $P_k(t_0) = \varphi_k(t_0) = 0$, since $\varphi_k \in A(t_k, x_k)$ and

$$\begin{aligned} \sup_{\sigma \in [t_0, t_k]} \| P_k(\sigma) - P_k(t_0) \| &= \sup_{\sigma \in [t_0, t_k]} \| P_k(\sigma) \| \\ &= \sup_{\sigma \in [t_0, t_k]} \left\| \varphi_k(\sigma) - \int_{t_0}^{\sigma} DF(\varphi_k(\tau), t) \right\| < \delta. \end{aligned}$$

Moreover, it is clear that $P \in G^-([t_0, t_k], X)$. Now, note, in addition, that for $\sigma \in [t_0, t_k]$, we have

$$\begin{aligned} \varphi_k(\sigma) &= \int_{t_0}^{\sigma} DF(\varphi_k(\tau), t) + \varphi_k(\sigma) - \int_{t_0}^{\sigma} DF(\varphi_k(\tau), t) + \varphi_k(t_0) \\ &= \varphi_k(t_0) + \int_{t_0}^{\sigma} DF(\varphi_k(\tau), t) + P_k(\sigma) - P_k(t_0) \\ &= \varphi_k(t_0) + \int_{t_0}^{\sigma} D[F(\varphi_k(\tau), t) + P_k(t)] \end{aligned}$$

which implies that $\varphi_k : [t_0, t_k] \to X$ is the solution of the nonhomogeneous generalized ODE

$$\frac{dx}{d\tau} = D[F(x, t) + P_k(t)],$$

with initial condition $\varphi_k(t_0) = 0$. Since the trivial solution of the generalized ODE is regularly stable with respect to perturbations, $P \in G^-([t_0, t_k], X)$ and $\| \varphi_k(t_0) \| = 0 < \delta$, we have $\| \varphi_k(t) \| < \epsilon$ for all $t \in [t_0, t_k]$. In particular, $\| \varphi_k(t_k) \| = \| x_k \| < \epsilon$ (since $\varphi_k \in A(t_k, x_k)$), which contradicts (8.57). Therefore, condition (i) is satisfied. □

The next theorem, established in [7], is known as converse Lyapunov theorem on regular stability for generalized ODEs because it shows that regular stability implies the existence of a Lyapunov functional with some properties described in Theorem 8.33.

Theorem 8.43: *Let $F \in \mathcal{F}(\Omega, h)$ satisfy condition (ZS). If the trivial solution $x \equiv 0$ of the generalized ODE (8.1) is regularly stable, then there exists a functional $V : [t_0, \infty) \times X \to \mathbb{R}$ satisfying the following conditions:*

(i) $V(\cdot,x) : [t_0,\infty) \to \mathbb{R}$ *is left-continuous on* (t_0,∞) *for all* $x \in X$;
(ii) *There exists a continuous (strictly) increasing function* $b : \mathbb{R}_+ \to \mathbb{R}_+$ *satisfying* $b(0) = 0$ *such that*

$$V(t,x) \geqslant b(\| x \|), \quad \text{for every } (t,x) \in [t_0,\infty) \times X;$$

(iii) *The function* $t \mapsto V(t,x(t))$ *is nonincreasing along every global forward solution* $x : [t_0,\infty) \to X$ *of the generalized ODE* (8.1) *with* $x(s_0) = x_0$;
(iv) $V(t,0) = 0$ *for all* $t \in [t_0,\infty)$;
(v) *There exists a continuous increasing function* $a : \mathbb{R}_+ \to \mathbb{R}_+$ *satisfying* $a(0) = 0$, *such that*

$$V(t,z) \leqslant a(\| z \|), \quad \text{for all } z \in X \text{ and } t \in [t_0,\infty);$$

(vi) *For every solution* $x : [\alpha,\beta] \subset [t_0,\infty) \to X$ *of the generalized ODE* (8.1)

$$D^+V(t,x(t)) = \limsup_{\eta\to 0^+} \frac{V(t+\eta,x(t+\eta)) - V(t,x(t))}{\eta} \leqslant 0, \quad t \in [\alpha,\beta)$$

holds, that is, the right derivative of V *along every solution of the generalized ODE* (8.1) *is non-positive.*

Proof. By Lemmas 8.39, 8.41, and 8.42, the function $V : [t_0,\infty) \times X \to \mathbb{R}$ defined by (8.46) satisfies items (i), (ii), and (iii). By Lemma 8.36 and Corollary 8.38, V satisfies items (iv) and (v). Finally, it is clear that if V satisfies (iii), then it satisfies (vi). □

The next statement is a Converse Lyapunov theorem on regular asymptotic stability of the trivial solution $x \equiv 0$ of the generalized ODE (8.1) and it was also proved in [7].

Theorem 8.44: *Let* $F \in \mathcal{F}(\Omega,h)$ *satisfy condition (ZS). If the trivial solution* $x \equiv 0$ *of the generalized ODE* (8.1) *is regularly asymptotically stable, then there exists a functional* $V : [t_0,\infty) \times X \to \mathbb{R}$ *satisfying the following conditions:*

(i) $V(t,0) = 0$ *for all* $t \in [t_0,\infty)$;
(ii) *There exists a continuous increasing function* $a : \mathbb{R}_+ \to \mathbb{R}_+$ *satisfying* $a(0) = 0$, *such that*

$$V(t,y) \leqslant a(\| y \|), \quad \text{for all } y \in X \text{ and } t \in [t_0,\infty);$$

(iii) *The function* $t \mapsto V(t,x(t))$ *is nonincreasing along every global forward solution* $x : [t_0,\infty) \to X$ *of the generalized ODE* (8.1) *with* $x(s_0) = x_0$;
(iv) $V(\cdot,x) : [t_0,\infty) \to \mathbb{R}$ *is left-continuous on* (t_0,∞) *for all* $x \in X$;

(v) *For every solution* $x : [\alpha, \beta] \subset [t_0, \infty) \to X$ *of* (8.1), *we have*

$$D^+V(t, x(t)) = \limsup_{\eta \to 0^+} \frac{V(t+\eta, x(t+\eta)) - V(t, x(t))}{\eta} \leqslant -V(t, x(t)), \quad t \in [\alpha, \beta);$$

(vi) *There exists a continuous (strictly) increasing function* $b : \mathbb{R}_+ \to \mathbb{R}_+$ *satisfying* $b(0) = 0$ *such that*

$$V(t, x) \geqslant b(\| x \|), \quad \textit{for every } (t, x) \in [t_0, \infty) \times X.$$

Proof. For $s \geqslant t_0$ and $x \in X$, define $V : [t_0, \infty) \times X \to \mathbb{R}$ by

$$V(s, x) = \begin{cases} \inf\limits_{\varphi \in A(s,x)} \left\{ \sup\limits_{\sigma \in [t_0, s]} \left\| \varphi(\sigma) - \int_{t_0}^{\sigma} DF(\varphi(\tau), t) \right\| e^{-s} \right\}, & \text{if } s > t_0, \\ \| x \|, & \text{if } s = t_0. \end{cases} \tag{8.59}$$

Notice that V is well-defined (see Remark 8.35). Similarly as in the proofs of Lemmas 8.36, 8.37, 8.39, and 8.41, one can prove that the function V, defined by (8.59), satisfies items (i), (ii), (iii), and (iv), respectively. Although the proof of item (v) is similar to the proof of Lemma 8.39, it shows the importance of the exponential function in the definition of the Lyapunov functional. For this reason, it is relevant to bring it up here.

In order to prove item (v), let $x : [\alpha, \beta] \subset [t_0, \infty) \to X$ be a solution of the generalized ODE (8.1) and $t \in [\alpha, \beta)$. Take $\varphi \in A(t, x(t))$ and consider $0 < \eta < t + \eta < \beta$. Define $\phi_\eta : [t_0, \infty) \to X$ by

$$\phi_\eta(\sigma) = \begin{cases} \varphi(\sigma), & \text{if } \sigma \in [t_0, t], \\ x(\sigma), & \text{if } \sigma \in [t, t+\eta], \\ x(t+\eta), & \text{if } \sigma \in (t+\eta, \infty). \end{cases}$$

Therefore, $\phi_\eta \in A(t+\eta, x(t+\eta))$ and

$$V(t+\eta, x(t+\eta)) \leqslant \sup_{\sigma \in [t_0, t+\eta]} \left\| \phi_\eta(\sigma) - \int_{t_0}^{\sigma} DF(\phi_\eta(\tau), s) \right\| e^{-(t+\eta)}.$$

As in the proof of Lemma 8.39, we consider two cases with respect to

$$\sup_{\sigma \in [t_0, t_2]} \left\| \phi_\eta(\sigma) - \int_{t_0}^{\sigma} DF(\phi_\eta(\tau), s) \right\| e^{-(t+\eta)}.$$

Case 1: Suppose, for some $v \in [t_0, t+\eta]$, we have

$$\sup_{\sigma \in [t_0, t+\eta]} \left\| \phi_\eta(\sigma) - \int_{t_0}^{\sigma} DF(\phi_\eta(\tau), s) \right\| e^{-(t+\eta)} = \left\| \phi_\eta(v) - \int_{t_0}^{v} DF(\phi_\eta(\tau), s) \right\| e^{-(t+\eta)}.$$

At first, assume that $v \in [t_0, t]$. Then, $\phi_\eta(s) = \varphi(s)$, for all $s \in [t_0, t]$, and

$$V(t+\eta, x(t+\eta)) \leqslant \left\| \varphi(v) - \int_{t_0}^{v} DF(\varphi(\tau), s) \right\| e^{-t} e^{-\eta}$$
$$\leqslant \left(\sup_{\sigma \in [t_0, t]} \left\| \varphi(\sigma) - \int_{t_0}^{\sigma} DF(\varphi(\tau), s) \right\| e^{-t} \right) e^{-\eta}.$$

Taking the infimum for all $\varphi \in A(t, x(t))$ on the right-hand side of the inequality above, we obtain

$$V(t+\eta, x(t+\eta)) \leqslant V(t, x(t))e^{-\eta}. \tag{8.60}$$

If $v \in [t, t+\eta]$, then the proof of (8.60) is analogous to the proof of Case 1 in Lemma 8.39, with the same adaptations that we did here. The same statement holds for the proof of Case 2.

Case 2: Suppose, for some $v \in [t_0, t+\eta]$, we have

$$\sup_{\sigma \in [t_0, t+\eta]} \left\| \phi_\eta(\sigma) - \int_{t_0}^{\sigma} DF(\phi_\eta(\tau), s) \right\| e^{-(t+\eta)}$$
$$= \left\| \phi_\eta(v^+) - \lim_{\sigma \to v^+} \int_{t_0}^{\sigma} DF(\phi_\eta(\tau), s) \right\| e^{-(t+\eta)}.$$

By (8.60), we obtain

$$V(t+\eta, x(t+\eta)) - V(t, x(t)) \leqslant V(t, x(t))(e^{-\eta} - 1)$$

and, moreover,

$$\limsup_{\eta \to 0^+} \frac{V(t+\eta, x(t+\eta)) - V(t, x(t))}{\eta} \leqslant \limsup_{\eta \to 0^+} \frac{V(t, x(t))(e^{-\eta} - 1)}{\eta}$$
$$= V(t, x(t)) \limsup_{\eta \to 0^+} \frac{(e^{-\eta} - 1)}{\eta}$$
$$= -V(t, x(t)),$$

which completes the proof of item (v).

The proof of item (vi) is quite similar to the proof of Lemma 8.42. We leave the details to the interested reader. However, we inform that item (vi) is proved in [7]. □

9

Periodicity

Marielle Ap. Silva[1], Everaldo M. Bonotto[2], Rodolfo Collegari[3], Márcia Federson[1], and Maria Carolina Mesquita[1]

[1]*Departamento de Matemática, Instituto de Ciências Matemáticas e de Computação (ICMC), Universidade de São Paulo, São Carlos, SP, Brazil*
[2]*Departamento de Matemática Aplicada e Estatística, Instituto de Ciências Matemáticas e de Computação (ICMC), Universidade de São Paulo, São Carlos, SP, Brazil*
[3]*Faculdade de Matemática, Universidade Federal de Uberlândia, Uberlândia, MG, Brazil*

The study of periodic solutions is an important and well-known branch of the theory of differential equations related, in a broad sense, to the study of periodic phenomena that arise in problems applied in technology, biology, and economics. For example, in order to investigate the impact of environmental factors, the requirement of periodic parameters is more realistic and important due to many periodic factors of the real world. We can mention, for instance seasonal effects of the climate, mating and harvesting habits, among others.

Let $T > 0$. We know that a function $f : \mathbb{R} \to \mathbb{R}$ is called T-periodic, if it satisfies

$$f(t+T) = f(t), \quad \text{for every } t \in \mathbb{R}.$$

For example, the trigonometric functions sine and cosine are 2π-periodic. But what can we say about the periodicity of a function which takes values in a finite dimensional space of higher dimension? The next example can motivate possible answers.

Example 9.1: Consider functions $g, h : \mathbb{R} \to \mathbb{R} \times \mathbb{R}$ given by

$$g(t) = (\cos(t), \sin(t)) \quad \text{and} \quad h(t) = (t^2, \sin(t))$$

whose trajectories are presented in Figures 9.1 and 9.2, respectively.

What can we say about the periodicity of such functions? Note that each coordinate of g is 2π-periodic, while the first coordinate of h is nonperiodic. If we extend

Generalized Ordinary Differential Equations in Abstract Spaces and Applications, First Edition.
Edited by Everaldo M. Bonotto, Márcia Federson, and Jaqueline G. Mesquita.

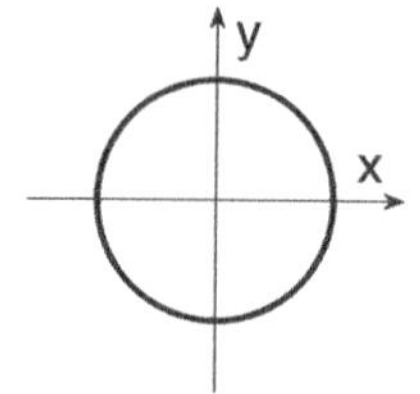

Figure 9.1 Trajectory of g.

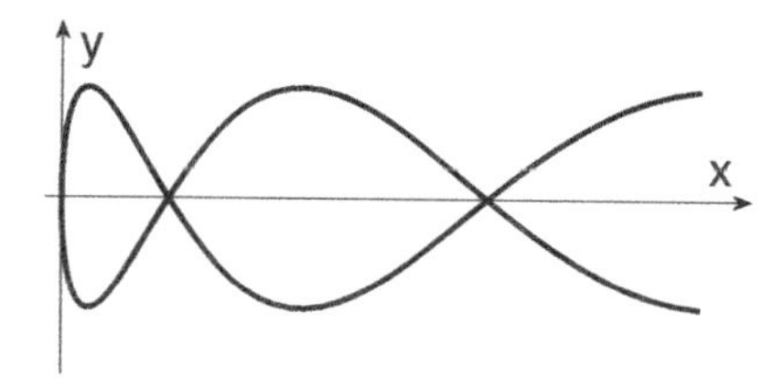

Figure 9.2 Trajectory of h.

the notion of periodicity to $\mathbb{R}^n$ requiring that $f : \mathbb{R} \to \mathbb{R}^n$ is T-periodic, whenever $f_i : \mathbb{R} \to \mathbb{R}$ is T-periodic, for every $i = 1, \dots, n$, then g is 2π-periodic, while h is nonperiodic.

There are many works concerning periodicity of solutions in the framework of classic ordinary differential equations (ODEs) and impulsive differential equations. We can mention, for instance, [99, 165, 188, 199, 228, 235, 236]. However, in these works, usual hypotheses on continuity or piecewise continuity on the right-hand sides of the equations are required. Here, on the contrary, it is enough to consider functions which are integrable in the sense of Perron or Perron–Stieltjes, allowing the right-hand sides to have many discontinuities and to be highly oscillating.

While, on the one hand, there are many works on periodic solutions of ODEs and impulsive differential equations, on the other hand, there are a few studies on the periodic behavior of solutions of generalized ODEs. In the framework of generalized ODEs, periodic boundary value problems in the finite dimensional case were investigated in [81] and [230], and the theory of affine-periodic solutions, which encompasses the notions of periodicity, quasiperiodicity, and antiperiodicity, was introduced in [77] for the space $\mathbb{C}^n$.

In this chapter, we start by investigating periodicity of solutions of linear generalized ODEs for functions taking values in $\mathbb{R}^n$. This is the content of Section 9.1, where we prove a result which relates the periodicity of solutions of linear nonhomogeneous generalized ODEs to the periodicity of solutions of linear homogeneous generalized ODEs. Not only that, but we also present a Floquet-type theorem which provides a characterization of the fundamental matrix of periodic linear generalized ODEs. This characterization, which we refer to as Floquet Theorem for generalized ODEs, is due to Š. Schwabik (see [207]). Here, we redeem the results of [207] and present them in English with a slightly different guise. We also include some applications for linear ODEs with impulses.

In the second part of this chapter, we focus our attention to the infinite dimensional case. Thus, in Section 9.2, we present a result on the existence of what we call a (θ, T)-periodic solution of a linear nonhomogeneous generalized ODE, with $T > 0$ and $\theta > 0$. Roughly speaking, the notion of (θ, T)-periodic solution means that a solution $x : [0, \infty) \to X$ of a linear nonhomogeneous generalized ODE is (θ, T)-periodic, whenever x is a solution and $x(t + T) = \theta x(t)$, for all $t \in [0, \infty)$. In particular, if we consider $\theta = 1$, we have the concept of periodic solutions, and for $\theta = -1$, we have the concept of antiperiodic solutions. Here, we get an existence result for the case where θ is a positive number. The main tool used here is the Banach Fixed Point Theorem.

An auxiliary result, namely Lemma 9.21, with some ideas coming from [77], relates a (θ, T)-periodic solution of a linear nonhomogeneous generalized ODE to a solution of a boundary value problem. Using such lemma and the Banach Fixed Point Theorem, we obtain conditions for the existence of a (θ, T)-periodic solution for linear generalized ODEs. We present an example to illustrate this result. Still in this section, we apply the results to measure differential equations (we write MDEs as in Chapter 3) which are known to be special cases of generalized ODEs (see Chapter 4). Recall that the reason for applying the results to MDEs is because they encompass many types of equations, such as classic ODEs, impulsive differential equations, and dynamic equations on time scales.

9.1 Periodic Solutions and Floquet's Theorem

The aim of this section is to present a study on periodic solutions for the following linear nonhomogeneous generalized ODE:

$$\frac{dx}{d\tau} = D[A(t)x + g(t)], \tag{9.1}$$

where $A : [0, \infty) \to L(\mathbb{R}^n)$ is a T-periodic operator and $g : [0, \infty) \to \mathbb{R}^n$ is a T-periodic function. In addition, we prove a Floquet-type theorem for the linear homogeneous generalized ODE

$$\frac{dx}{d\tau} = D[A(t)x], \tag{9.2}$$

which provides a characterization for its fundamental matrices. A first investigation of the Floquet Theorem for linear generalized ODEs was made by Š. Schwabik in [207].

Next, we present a concept of periodicity for functions defined on $[0, \infty)$ taking values in $\mathbb{R}^n$. This concept is similar to the concept already known for real-valued functions defined on $\mathbb{R}$.

Definition 9.2: A function $x : [0, \infty) \to \mathbb{R}^n$ is said to be *periodic*, if there exists $T > 0$ such that

$$x(t + T) = x(t),$$

for all $t \in [0, \infty)$. In this case, T is called a *period* of x, and x is called *T-periodic*. Moreover, we say that $A : [0, \infty) \to L(\mathbb{R}^n)$ is a *T-periodic operator*, if

$$A(t) = A(t + T),$$

for all $t \in [0, \infty)$.

Now, we provide necessary and sufficient conditions for a solution of linear nonhomogeneous generalized ODE (9.1) to be periodic. Theorem 9.3 is a well-known result for classical ODEs and a more general version of it for generalized ODEs can be found in [93].

Recall that $A \in BV_{loc}([0, \infty), L(\mathbb{R}^n))$, whenever $A : [0, \infty) \to L(\mathbb{R}^n)$ is of locally bounded variation in $[0, \infty)$, that is, for all compact interval $[a, b] \subset [0, \infty)$, $A \in BV([a, b], L(\mathbb{R}^n))$. Throughout this section, we assume that $A \in BV_{loc}([0, \infty), L(\mathbb{R}^n))$ satisfies condition (Δ) presented in Chapter 6. We draw the reader's attention to the fact that, under these conditions, the existence and uniqueness of a solution $x : [0, \infty) \to \mathbb{R}^n$ of the linear nonhomogeneous generalized ODE (9.1) is ensured by Remark 6.4.

Theorem 9.3: *Let $x : [0, \infty) \to \mathbb{R}^n$ be a solution of the linear nonhomogeneous generalized ODE* (9.1). *Assume that $A \in BV_{loc}([0, \infty), L(\mathbb{R}^n))$ is a T-periodic operator satisfying condition (Δ) on $[0, \infty)$ and $g : [0, \infty) \to \mathbb{R}^n$ is a T-periodic function. Then $x : [0, \infty) \to \mathbb{R}^n$ is a T-periodic solution of* (9.1) *if and only if $x(0) = x(T)$.*

Proof. Suppose $x : [0, \infty) \to \mathbb{R}^n$ is a T-periodic solution of (9.1). Taking $t = 0$, we have $x(0) = x(T)$. On the other hand, let us assume that $x : [0, \infty) \to \mathbb{R}^n$ is a solution of (9.1) satisfying $x(0) = x(T)$. If $y(t) = x(t + T)$, for all $t \in [0, \infty)$, then by the substitution theorem (Theorem 2.18), we have

$$y(s_2) - y(s_1) = x(s_2 + T) - x(s_1 + T) = \int_{s_1+T}^{s_2+T} D[A(s)x(s) + g(s)]$$
$$= \int_{s_1}^{s_2} D[A(s + T)x(s + T) + g(s + T)]$$
$$= \int_{s_1}^{s_2} D[A(s)y(s) + g(s)],$$

for every $s_1, s_2 \in [0, \infty)$. Therefore, $y : [0, \infty) \to \mathbb{R}^n$ is a solution of (9.1) and $y(0) = x(0 + T) = x(0)$. From the uniqueness of the solution of the initial value problem (see Remark 6.4), $x(t) = y(t)$, for all $t \in [0, \infty)$. Thus,

$$x(t) = y(t) = x(t + T), \quad t \in [0, \infty),$$

that is, $x : [0, \infty) \to \mathbb{R}^n$ is T-periodic. □

In what follows, we present some results and definitions on fundamental matrices. These notions are analogous to the classical theory of linear ODEs. The next result is presented in [209, Theorem 6.11].

Theorem 9.4: *Assume that $A \in BV_{loc}([0, \infty), L(\mathbb{R}^n))$ satisfies condition (Δ) on $[0, \infty)$. If $t_0 \in [0, \infty)$, then for every matrix $\widetilde{X} \in L(\mathbb{R}^n)$, there exists a uniquely determined matrix-valued function $X : [0, \infty) \to L(\mathbb{R}^n)$ such that*

$$X(t) = \widetilde{X} + \int_{t_0}^{t} d[A(s)]X(s),$$

for $t \in [0, \infty)$.

We say that a matrix-valued function $X : [0, \infty) \to L(\mathbb{R}^n)$ is a solution of the matrix equation

$$\frac{dX}{d\tau} = D[A(t)X] \tag{9.3}$$

if for every $s_1, s_2 \in [0, \infty)$, the identity

$$X(s_2) - X(s_1) = \int_{s_1}^{s_2} d[A(s)]X(s)$$

holds. Furthermore, a matrix-valued function $X : [0, \infty) \to L(\mathbb{R}^n)$ is said to be a *fundamental matrix* of the linear generalized ODE (9.2), if X is a solution of the matrix equation (9.3) and the matrix $X(t)$ is nonsingular for at least one value $t \in [0, \infty)$. A proof of the next result can be found in [209, Theorem 6.12].

Theorem 9.5: *Assume that $A \in BV_{loc}([0,\infty), L(\mathbb{R}^n))$ satisfies condition (Δ) on $[0,\infty)$. Then every fundamental matrix $X : [0,\infty) \to L(\mathbb{R}^n)$ of the linear generalized ODE (9.2) is nonsingular for all $t \in [0,\infty)$.*

Remark 9.6: Let $X, Y : [0,\infty) \to L(\mathbb{R}^n)$ be fundamental matrices of the linear generalized ODE (9.2). According to [209, Theorem 6.13] and [209, Lemma 6.18],

$$X(t)X^{-1}(s) = Y(t)Y^{-1}(s)$$

for all $t, s \in [0,\infty)$.

Theorem 9.7 gives us a variation-of-constants formula for a solution of the linear nonhomogeneous generalized ODE (9.1). For a proof of this fact, the reader may want to consult [209, Corollary 6.19].

Theorem 9.7: *Assume that $A \in BV_{loc}([0,\infty), L(\mathbb{R}^n))$ satisfies condition (Δ) on $[0,\infty)$, $t_0 \in [0,\infty)$, $\widetilde{X} \in \mathbb{R}^n$, and $g \in BV_{loc}([0,\infty), \mathbb{R}^n)$. If $X : [0,\infty) \to L(\mathbb{R}^n)$ is an arbitrary fundamental matrix of the linear generalized ODE (9.2), then the unique solution of the initial value problem*

$$\begin{cases} \dfrac{dx}{d\tau} = D[A(t)x + g(t)], \\ x(t_0) = \widetilde{X} \end{cases}$$

can be represented by

$$x(t) = g(t) - g(t_0) + X(t)\left(X^{-1}(t_0)\widetilde{X} - \int_{t_0}^{t} d[X^{-1}(s)](g(s) - g(t_0))\right),$$

for all $t \in [0,\infty)$.

The next result gives us necessary and sufficient conditions for the linear generalized ODE (9.2) to have a T-periodic nontrivial solution.

Proposition 9.8: *Let $A \in BV_{loc}([0,\infty), L(\mathbb{R}^n))$ be a T-periodic operator satisfying condition (Δ) on $[0,\infty)$ and $X : [0,\infty) \to L(\mathbb{R}^n)$ be a fundamental matrix of the linear generalized ODE (9.2). Then (9.2) has a T-periodic nontrivial solution if and only if*

$$\det[X(T) - X(0)] = 0.$$

Proof. Let $x : [0,\infty) \to \mathbb{R}^n$ be a T-periodic nontrivial solution of (9.2). Then, there exists $s \in [0,\infty)$ such that

$$x(s) = \widetilde{X} \neq 0.$$

Using Theorem 9.7 with $g \equiv 0$, we may write $x(t) = X(t)X^{-1}(s)\widetilde{X}$ for all $t \in [0, \infty)$. Set $K = X^{-1}(s)\widetilde{X}$. Then,

$$x(t) = X(t)X^{-1}(s)\widetilde{X} = X(t)K$$

for all $t \in [0, \infty)$. Using Theorem 9.3, we obtain

$$X(T)K = x(T) = x(0) = X(0)K,$$

that is, $[X(T) - X(0)]K = 0$. Thus, since $K \neq 0$, we conclude that $\det [X(T) - X(0)] = 0$.

Now, assume that $\det [X(T) - X(0)] = 0$. Then, one can obtain $z \in \mathbb{R}^n$, $z \neq 0$, such that $[X(T) - X(0)]z = 0$. Let $w = X(0)z$. Note that $x(t) = X(t)X^{-1}(0)w$ is a nontrivial solution of (9.2) defined on $[0, \infty)$. Besides, we have

$$x(T) = X(T)X^{-1}(0)w = X(T)z = X(0)z = w = x(0).$$

Then, Theorem 9.3 assures that x is T-periodic. □

Corollary 9.9: *Consider the linear generalized ODE* (9.2) *and suppose* $A(t) = A$, *for all* $t \in [0, \infty)$. *Then,* (9.2) *has a* T*-periodic nontrivial solution if and only if the matrix* $(I - e^{AT})$ *is singular.*

Proof. It is enough to note that $X(t) = e^{At}$ is a fundamental matrix of Eq. (9.2), where $A(t) = A$ is a constant matrix. □

Theorem 9.10 below exhibits sufficient conditions for Eq. (9.1) to admit a unique T-periodic solution.

Theorem 9.10: *Assume that* $A \in BV_{loc}([0, \infty), L(\mathbb{R}^n))$ *is a* T*-periodic operator satisfying condition* (Δ) *on* $[0, \infty)$ *and* $g \in BV_{loc}([0, \infty), \mathbb{R}^n)$ *is a* T*-periodic function. If* $x \equiv 0$ *is the unique* T*-periodic solution of the linear generalized ODE* (9.2), *then the linear nonhomogeneous generalized ODE* (9.1) *has a unique* T*-periodic solution.*

Proof. Let $X : [0, \infty) \to L(\mathbb{R}^n)$ be a fundamental matrix of the linear generalized ODE (9.2). By Proposition 9.8, we have $\det [X(T) - X(0)] \neq 0$. Define

$$z = X(0)[X(T) - X(0)]^{-1} \int_0^T d_s[X(T)X^{-1}(s)](g(s) - g(0)).$$

By the variation-of-constants formula in Theorem 9.7, the function

$$w(t) = X(t)X^{-1}(0)z + g(t) - g(0) - \int_0^t d_s[X(t)X^{-1}(s)](g(s) - g(0))$$

is a solution of (9.1) on $[0, \infty)$ such that $w(0) = z$. Besides,

$$
\begin{aligned}
w(T) &= X(T)X^{-1}(0)z + g(T) - g(0) - \int_0^T d_s[X(T)X^{-1}(s)](g(s) - g(0)) \\
&= X(T)X^{-1}(0)z - [X(T) - X(0)]X^{-1}(0)z \\
&= w(0).
\end{aligned}
$$

Thus, by Theorem 9.3, w is a T-periodic solution of (9.1).

In order to show the uniqueness, suppose w_1 and w_2 are T-periodic solutions of (9.1). Consequently, $w_1 - w_2$ is a T-periodic solution of (9.2). By hypothesis, we have $x = w_1 - w_2 \equiv 0$, that is, $w_1 \equiv w_2$. □

We are now in a position to present a characterization of fundamental matrices of linear homogeneous generalized ODE (9.2). But before this, we need an auxiliary result, whose proof can be found [38].

Lemma 9.11: *If C is a $n \times n$ matrix with* $\det C \neq 0$*, then there exists a matrix B such that $e^B = C$.*

Theorem 9.12 in the sequel, known as the Floquet Theorem, can also be found in [207].

Theorem 9.12: *Consider the linear generalized ODE (9.2), where the operator $A \in BV_{loc}([0,\infty), L(\mathbb{R}^n))$ is a T-periodic operator satisfying condition (Δ) on $[0, \infty)$. Let $X : [0, \infty) \to L(\mathbb{R}^n)$ be a fundamental matrix of (9.2). Then,*

(i) *$X(t + T)$ is also a fundamental matrix of (9.2);*
(ii) *there exist a T-periodic matrix-valued function $P : [0, \infty) \to L(\mathbb{R}^n)$ and a constant matrix B such that*

$$X(t) = P(t)e^{Bt},$$

for all $t \in [0, \infty)$.

Proof. Let us prove (i). Consider $X(t)$ a fundamental matrix of (9.2). Then, for all $s_1, s_2 \in [0, \infty)$, we have

$$X(s_2) - X(s_1) = \int_{s_1}^{s_2} d[A(s)]X(s).$$

According to Theorem 2.18, we obtain

$$
\begin{aligned}
X(s_2 + T) - X(s_1 + T) &= \int_{s_1+T}^{s_2+T} d[A(s)]X(s) = \int_{s_1}^{s_2} d[A(s + T)]X(s + T) \\
&= \int_{s_1}^{s_2} d[A(s)]X(s + T),
\end{aligned}
$$

since A is T-periodic from hypothesis. Thus, $X(t+T)$ is a fundamental matrix of (9.2).

Now, we prove (ii). By item (i) and Remark 9.6,

$$X(t)X^{-1}(0) = X(t+T)X^{-1}(T),$$

for all $t \in [0, \infty)$. Thus, the nonsingular matrix $C = X^{-1}(0)X(T)$ satisfies $X(t+T) = X(t)\,C$, for all $t \in [0, \infty)$. By Lemma 9.11, one can find a matrix B such that $e^{BT} = C$. Define $P(t) = X(t)\,e^{-Bt}$ for all $t \in [0, \infty)$. Thus,

$$P(t+T) = X(t+T)\,e^{-B(t+T)} = X(t)\,C\,C^{-1}\,e^{-Bt} = X(t)\,e^{-Bt} = P(t),$$

holds for all $t \in [0, \infty)$, which completes the proof. □

Remark 9.13: To prove the Floquet Theorem (Theorem 9.12) in the classical case, derivative properties, such as the chain rule, are strongly used (see [38] for instance). In the context of generalized ODEs, we only use properties of the Perron–Stieltjes integral, and this is the main difference in the proofs of the Floquet Theorem for ODEs and its version for generalized ODEs. Not only that, but the right-hand sides or coefficients of linear homogeneous generalized ODEs may be nonabsolute integrable.

9.1.1 Linear Differential Systems with Impulses

Let $F : [0, \infty) \to L(\mathbb{R}^n)$ be a locally Perron integrable $n \times n$ matrix-valued function. Consider a sequence $\{t_i\}_{i\in\mathbb{N}} \subset [0, \infty)$ such that

$$t_1 < t_2 < \cdots < t_i < \cdots$$

and $\lim_{i\to\infty} t_i = \infty$. For each $i \in \mathbb{N}$, let $B_i \in L(\mathbb{R}^n)$ be a matrix-valued function such that $I + B_i$ is nonsingular and I denotes the identity matrix.

We consider the following linear impulsive differential system:

$$\begin{cases} \dot{x}(t) = F(t)x, & t \neq t_i, \quad (9.4)\\ \Delta x(t_i) = x(t_i^+) - x(t_i) = B_i x(t_i), & i \in \mathbb{N}. \quad (9.5) \end{cases}$$

We shall assume that there exists $k \in \mathbb{N}$ such that the following conditions hold:

(C1) $0 < t_1 < \cdots < t_k < T$ and $t_{i+k} = t_i + T$ for all $i \in \mathbb{N}$;
(C2) $F(t+T) = F(t)$ for all $t \in [0, \infty)$ and $B_{i+k} = B_i$ for all $i \in \mathbb{N}$;
(C3) $\int_0^T F(s)ds + \sum_{i=1}^k B_i = 0$.

Now, define

$$A(t) = \int_0^t F(s)ds + \sum_{i=1}^{\infty} B_i H_{t_i}(t), \quad t \in [0, \infty), \tag{9.6}$$

where for each $i \in \mathbb{N}$, $H_{t_i}(t) = 0$ for $0 \leqslant t \leqslant t_i$ and $H_{t_i}(t) = 1$ for $t > t_i$.

Definition 9.14: Let $t_0 \in [0, \infty)$ and $x_0 \in \mathbb{R}^n$. A function $x : [0, \infty) \to \mathbb{R}^n$ is said to be a *solution of the initial value problem*

$$\begin{cases} \dot{x}(t) = F(t)x, & t \neq t_i, \\ \Delta x(t_i) = x(t_i^+) - x(t_i) = B_i x(t_i), & i \in \mathbb{N}, \\ x(t_0) = x_0, \end{cases} \tag{9.7}$$

if $\dot{x}(t) = F(t)x(t)$ for almost all $t \in [0, \infty) \setminus \{t_i : i \in \mathbb{N}\}$, $x(t_0) = x_0$ and

$$x(t_i^+) = \lim_{t \to t_i^+} x(t) = x(t_i) + B_i x(t_i), \quad \text{for } i \in \mathbb{N}.$$

According to [209, Example 6.20], we can state the following result.

Lemma 9.15: *The operator $A : [0, \infty) \to L(\mathbb{R}^n)$ given by (9.6) is of locally bounded variation and satisfies condition (Δ) on $[0, \infty)$. Moreover, $x : [0, \infty) \to \mathbb{R}^n$ is a solution of the impulsive system (9.4) and (9.5) if and only if x is a solution of the linear homogeneous generalized ODE*

$$\frac{dx}{d\tau} = D[A(t)x], \tag{9.8}$$

where A is given by (9.6).

The next result gives us sufficient conditions for the operator A to be periodic.

Proposition 9.16: *Assume that conditions (C1)–(C3) are satisfied. Then, the operator $A : [0, \infty) \to L(\mathbb{R}^n)$ given by (9.6) is T-periodic.*

Proof. Let $t \in [0, \infty)$. Then, either $t \in [0, t_1]$ or there exists a natural number $j \in \mathbb{N}$ such that $t_j < t \leqslant t_{j+1}$. Let us consider the second case. Thus,

$$t_{j+k} = t_j + T < t + T \leqslant t_{j+1} + T = t_{j+k+1},$$

by virtue of (C1). Consequently, using conditions (C2) and (C3), we have

$$\begin{aligned} A(t+T) &= \int_0^{t+T} F(s)ds + \sum_{i=1}^{\infty} B_i H_{t_i}(t+T) \\ &= \int_0^T F(s)ds + \sum_{i=1}^{k} B_i H_{t_i}(t+T) + \int_T^{t+T} F(s)ds + \sum_{i=k+1}^{\infty} B_i H_{t_i}(t+T) \\ &= \int_0^T F(s)ds + \sum_{i=1}^{k} B_i + \int_0^t F(s+T)ds + \sum_{i=k+1}^{j+k} B_i \end{aligned}$$

$$= \int_0^t F(s+T)ds + \sum_{i=1}^{j} B_{i+k} = \int_0^t F(s)ds + \sum_{i=1}^{j} B_i$$

$$= \int_0^t F(s)ds + \sum_{i=1}^{\infty} B_i H_{t_i}(t) = A(t),$$

which completes the proof. □

Next, we provide conditions for a solution of the impulsive system (9.4) and (9.5) to be periodic.

Theorem 9.17: *Let $x : [0, \infty) \to \mathbb{R}^n$ be a solution of the impulsive differential system (9.4) and (9.5). If conditions (C1)–(C3) are satisfied, then $x : [0, \infty) \to \mathbb{R}^n$ is T-periodic if and only if $x(0) = x(T)$.*

Proof. Let $x : [0, \infty) \to \mathbb{R}^n$ be a solution of the system (9.4) and (9.5). It is enough to prove the sufficient condition. Let us assume that $x(0) = x(T)$. Since $F : [0, \infty) \to L(\mathbb{R}^n)$ is a T-periodic function, using Lemma 9.15 and Proposition 9.16, we have $A \in BV_{loc}([0, \infty), L(\mathbb{R}^n))$ is a T-periodic operator which satisfies condition (Δ) on $[0, \infty)$. Thus, by Theorem 9.3, we conclude that $x : [0, \infty) \to \mathbb{R}^n$ is a T-periodic solution of (9.8) and, consequently, by Lemma 9.15, $x : [0, \infty) \to \mathbb{R}^n$ is a T-periodic solution of the system (9.4) and (9.5). □

In what follows, we describe a Floquet theory for the impulsive differential system (9.4) and (9.5).

Let $Z : [0, \infty) \to L(\mathbb{R}^n)$ be an arbitrary fundamental matrix of the impulsive system (9.4) and (9.5). Then,

$$Z(t) = Z(0) + \int_0^t F(r)Z(r)dr + \sum_{0<t_i<t} B_i Z(t_i),$$

for all $t \in [0, \infty)$ and $i \in \mathbb{N}$.

Now, let $X : [0, \infty) \to L(\mathbb{R}^n)$ be a fundamental matrix of (9.8). Therefore,

$$X(t) = X(0) + \int_0^t d[A(r)]X(r), \quad t \in [0, \infty).$$

We claim that

$$X(t)X^{-1}(0) = Z(t)Z^{-1}(0),$$

for all $t \in [0, \infty)$. Indeed, let $t \in [0, \infty)$. At first, by Corollary 2.14, note that

$$\int_{t_k}^{t_{k+1}} d\left[\sum_{i=1}^{\infty} B_i H_{t_i}(r)\right] X(r)$$

$$= \lim_{\gamma \to t_k^+} \int_{\gamma}^{t_{k+1}} d\left[\sum_{i=1}^{k} B_i\right] X(r) + \left[\sum_{i=1}^{\infty} B_i H_{t_i}(t_k^+) - \sum_{i=1}^{\infty} B_i H_{t_i}(t_k)\right] X(t_k)$$

$$= 0 + \left[\sum_{i=1}^{k} B_i - \sum_{i=1}^{k-1} B_i\right] X(t_k) = B_k X(t_k).$$

Thus, using Theorem 1.59, we obtain

$$\int_0^t d[A(r)]X(r) = \int_0^t d\left[\int_0^r F(s)ds\right] X(r) + \int_0^t d\left[\sum_{i=1}^{\infty} B_i H_{t_i}(r)\right] X(r)$$

$$= \int_0^t F(r)X(r)dr + \sum_{0<t_j<t} B_j X(t_j),$$

that is,

$$X(t) = X(0) + \int_0^t F(r)X(r)dr + \sum_{0<t_j<t} B_j X(t_j).$$

Finally, by Remark 9.6,

$$X(t)X^{-1}(0) = Z(t)Z^{-1}(0), \tag{9.9}$$

for all $t \in [0, \infty)$.

Theorem 9.18: *Assume that conditions* (C1)–(C3) *are satisfied. Let* $Z : [0, \infty) \to L(\mathbb{R}^n)$ *be a fundamental matrix of the impulsive differential system* (9.4) *and* (9.5). *Then, the system* (9.4) *and* (9.5) *has a* T*-periodic nontrivial solution if and only if*

$$\det[Z(T) - Z(0)] = 0.$$

Proof. Consider the linear generalized ODE (9.8), where $A \in BV_{loc}([0, \infty), L(\mathbb{R}^n))$ given by (9.6) is T-periodic and satisfies condition (Δ) on $[0, \infty)$. Let $X : [0, \infty) \to L(\mathbb{R}^n)$ be a fundamental matrix of (9.8). By Proposition 9.8, the generalized ODE (9.8) has a T-periodic nontrivial solution if and only if $\det[X(T) - X(0)] = 0$. Since $Z(t)Z^{-1}(0) = X(t)X^{-1}(0)$ for all $t \in [0, \infty)$ (see (9.9)), we have

$$\det[Z(T) - Z(0)] = \det[Z(T)Z^{-1}(0) - I] \det Z(0)$$
$$= \det[X(T)X^{-1}(0) - I] \det Z(0)$$
$$= \det[X(T) - X(0)] \det X^{-1}(0) \det Z(0).$$

Thus, by Lemma 9.15, the impulsive differential system (9.4) and (9.5) has a T-periodic nontrivial solution if and only if

$$\det[Z(T) - Z(0)] = 0,$$

and the proof is finished. □

Now, we are able to present a Floquet-type theorem for the impulsive differential system (9.4) and (9.5).

Theorem 9.19: *Consider the linear impulsive differential system* (9.4) *and* (9.5) *and assume that conditions* (C1)–(C3) *hold. If* $Z : [0,\infty) \to L(\mathbb{R}^n)$ *is a fundamental matrix of* (9.4)*, then there exist a T-periodic matrix-valued function* $P : [0,\infty) \to L(\mathbb{R}^n)$ *and a constant matrix B such that*

$$Z(t) = P(t)e^{Bt},$$

for all $t \in [0,\infty)$.

Proof. Consider the linear generalized ODE (9.8), where $A \in BV_{loc}([0,\infty), L(\mathbb{R}^n))$ given by (9.6) is T-periodic and satisfies condition (Δ) on $[0,\infty)$. Let $X : [0,\infty) \to L(\mathbb{R}^n)$ be a fundamental matrix of the Eq. (9.8). By (9.9),

$$X(t)X^{-1}(0) = Z(t)Z^{-1}(0), \quad t \in [0,\infty).$$

Note that

$$M(t) = X(t)X^{-1}(0)Z(0),$$

for $t \in [0,\infty)$, is also a fundamental matrix of the Eq. (9.8), since

$$\begin{aligned} M(t) - M(0) &= X(t)X^{-1}(0)Z(0) - Z(0) \\ &= \left[X(0) + \int_0^t d[A(r)]X(r)\right] X^{-1}(0)Z(0) - Z(0) \\ &= Z(0) + \int_0^t d[A(r)]X(r)X^{-1}(0)Z(0) - Z(0) \\ &= \int_0^t d[A(r)]M(r). \end{aligned}$$

Then, by the Floquet Theorem for linear generalized ODEs (Theorem 9.12), there exist a constant matrix B and a T-periodic function $P : [0,\infty) \to L(\mathbb{R}^n)$ such that

$$M(t) = P(t)e^{Bt},$$

for all $t \in [0,\infty)$. At least, since $Z(t) = X(t)X^{-1}(0)Z(0) = M(t)$ for all $t \in [0,\infty)$, the statement follows. □

9.2 (θ, T)-Periodic Solutions

In this section, X denotes a Banach space with norm $\|\cdot\|$. Let $T > 0$ and $\theta > 0$ be given. We want to investigate the existence of what we call (θ, T)-periodic solutions of the following linear nonhomogeneous generalized ODE:

$$\frac{dx}{d\tau} = D[A(t)x + f(t)], \tag{9.10}$$

where $A : [0, \infty) \to L(X)$ is an operator and $f : [0, \infty) \to X$ is a function. In order to do that, given $t_0 \in [0, \infty)$, we consider the boundary value problem

$$\begin{cases} \dfrac{dx}{d\tau} = D[A(t)x + f(t)] \\ x(t_0 + T) = \theta x(t_0), \end{cases}$$

and we show that the existence of a (θ, T)-periodic solution of the linear nonhomogeneous generalized ODE (9.10) is equivalent to the existence of a solution of the above boundary value problem. See Lemma 9.21 in the sequel.

In Theorem 9.22, by means of the Banach Fixed Point Theorem, we establish sufficient conditions which guarantee the existence of a (θ, T)-periodic solution of the linear nonhomogeneous generalized ODE (9.10), where θ is a positive number.

Using the correspondence between generalized ODEs and measure differential equations (we write MDEs for short), we apply the main results of this section to MDEs which are known to encompass not only ODEs and impulsive differential equations but also dynamic equations on time scales (see Chapter 3). Therefore, the theorems of this section can also be applied to these types of equations.

Let $a, b \in \mathbb{R}$, $a < b$. Recall that $BV([a, b], X)$ denotes the space of all functions from $[a, b]$ to X of bounded variation in $[a, b]$, which is a Banach space when equipped with the norm $\| \cdot \|_{BV}$ (see Chapter 1) and $BV_{loc}([a, \infty), X)$ is the set of functions of locally bounded variation in $[a, \infty)$, that is, they are of bounded variation on each compact interval $J \subset [a, \infty)$.

By considering the linear nonhomogeneous generalized ODE (9.10), we shall assume throughout this section the following general conditions:

(A1) $A(t + T) = A(t)$ for all $t \in [0, \infty)$.
(A2) $A \in BV_{loc}([0, \infty), L(X))$ and $f \in BV_{loc}([0, \infty), X)$.
(A3) A satisfies condition (Δ^+) on $[0, \infty)$ (see , p. 207).

According to Remark 6.4, for any $(x_0, t_0) \in X \times [0, \infty)$, the linear nonhomogeneous generalized ODE (9.10) admits a unique solution defined on $[t_0, \infty)$ through (x_0, t_0).

As presented in Chapter 6, a function $x : [t_0, \infty) \to X$, $t_0 \geqslant 0$, is a solution of the linear nonhomogeneous generalized ODE (9.10) on $[t_0, \infty)$, if it satisfies

$$x(t_2) = x(t_1) + \int_{t_1}^{t_2} d[A(s)]x(s) + f(t_2) - f(t_1), \quad \text{for all} \quad t_1, t_2 \in [t_0, \infty).$$

In what follows, we introduce the definition of a (θ, T)-periodic solution of the linear nonhomogeneous generalized ODE (9.10).

Definition 9.20: Let $t_0 \in [0, \infty)$, $\theta > 0$, and $T > 0$. A function $x : [t_0, \infty) \to X$ is said to be a *(θ, T)-periodic solution* of (9.10), if x is a solution of the linear nonhomogeneous generalized ODE (9.10) and, for all $t \in [t_0, \infty)$,

$$x(t + T) = \theta x(t).$$

The next result is presented in [77] in the framework of generalized ODEs for the case $X = \mathbb{C}^n$. Here, we present a version of it for Banach spaces in the framework of linear nonhomogeneous generalized ODEs. This result is useful to ensure the existence of a global (θ, T)-periodic solution of the linear nonhomogeneous generalized ODE (9.10).

Lemma 9.21: *Assume that conditions* (A1)–(A3) *are satisfied. Let* $t_0 \in [0, \infty)$, $\theta > 0$, *and* $T > 0$ *be such that*

$$f(t + T) = \theta f(t), \quad \text{for all } t \in [0, \infty).$$

Then, the existence of a (θ, T)*-periodic solution* $x : [t_0, \infty) \to X$ *of* (9.10) *is equivalent to the existence of a solution* $z : [t_0, t_0 + T] \to X$ *of the boundary value problem*

$$\begin{cases} \dfrac{dx}{d\tau} = D[A(t)x + f(t)] \\ x(t_0 + T) = \theta x(t_0). \end{cases} \tag{9.11}$$

Proof. Assume that $x : [t_0, \infty) \to X$ is a (θ, T)-periodic solution of (9.10). Then,

$$x(t_2) - x(t_1) = \int_{t_1}^{t_2} d[A(s)]x(s) + f(t_2) - f(t_1), \quad \text{for all } t_2, t_1 \in [t_0, \infty), \tag{9.12}$$

and

$$x(t + T) = \theta x(t), \quad \text{for all } t \in [t_0, \infty). \tag{9.13}$$

Now, we introduce an auxiliary function $z : [t_0, t_0 + T] \to X$ defined by

$$z(t) = \theta^{-1} x(t + T), \quad t \in [t_0, t_0 + T]. \tag{9.14}$$

We assert that z is a solution of the boundary value problem (9.11). Indeed, for every $t_1, t_2 \in [t_0, t_0 + T]$, we have

$$\begin{aligned} z(t_2) - z(t_1) &= \theta^{-1} x(t_2 + T) - \theta^{-1} x(t_1 + T) \\ &= \theta^{-1} \int_{t_1+T}^{t_2+T} d[A(s)]x(s) + \theta^{-1} f(t_2 + T) - \theta^{-1} f(t_1 + T). \end{aligned}$$

By Theorem 2.18, with $\psi(\xi) = \xi + T$, $\xi \in [t_0, t_0 + T]$, we obtain

$$\int_{t_1+T}^{t_2+T} d[A(s)]x(s) = \int_{\psi(t_1)}^{\psi(t_2)} d[A(s)]x(s) = \int_{t_1}^{t_2} d[A(s+T)]x(s+T), \qquad (9.15)$$

for all $t_1, t_2 \in [t_0, t_0 + T]$. Using condition (A1), (9.15) and the hypothesis on f, we conclude that

$$\begin{aligned} z(t_2) - z(t_1) &= \int_{t_1}^{t_2} d[A(s)]\theta^{-1}x(s+T) + \theta^{-1}\theta f(t_2) - \theta^{-1}\theta f(t_1) \\ &= \int_{t_1}^{t_2} d[A(s)]z(s) + f(t_2) - f(t_1), \end{aligned}$$

for all $t_1, t_2 \in [t_0, t_0 + T]$. Moreover, by (9.13) and (9.14), we have

$$z(t_0 + T) = \theta^{-1}x(t_0 + 2T) = \theta^{-1}\theta x(t_0 + T) = x(t_0 + T) = \theta z(t_0).$$

Thus, z is a solution of the boundary value problem (9.11).

On the other hand, suppose $z : [t_0, t_0 + T] \to X$ is a solution of the boundary value problem (9.11). Then, z is a solution of the linear nonhomogeneous generalized ODE (9.10) satisfying $z(t_0 + T) = \theta z(t_0)$. By Remark 6.4, there exists a unique global forward solution $x : [t_0, \infty) \to X$ of (9.10) with $x(t_0) = \theta^{-1}z(t_0 + T)$. Then, the uniqueness implies $x|_{[t_0,t_0+T]} = z$, that is, x is an extension of z.

We need to prove that $x(t + T) = \theta x(t)$ for all $t \in [t_0, \infty)$. Using the same steps as before, it is possible to prove that the function $\varphi(t) = \theta^{-1}x(t + T)$ is a solution of the linear nonhomogeneous generalized ODE (9.10) satisfying

$$\varphi(t_0) = \theta^{-1}x(t_0 + T) = \theta^{-1}z(t_0 + T) = x(t_0).$$

Notice that $\varphi(t)$ and $x(t)$ are solutions of the linear nonhomogeneous generalized ODE (9.10) with $\varphi(t_0) = x(t_0)$. Then, by the uniqueness of the initial value problem with Eq. (9.10) and initial condition $\varphi(t_0) = x(t_0)$, we can conclude $\varphi(t) = x(t)$, for all $t \in [t_0, \infty)$, that is,

$$x(t + T) = \theta x(t), \quad \text{for all } t \in [t_0, \infty),$$

and the proof is complete. □

It is important to mention that, due to Lemma 9.21, it is possible to change the problem of finding a (θ, T)-periodic solution of the linear nonhomogeneous generalized ODE (9.10) to finding a solution of the boundary value problem (9.11).

The next result ensures, under some conditions, the existence of a (θ, T)-periodic solution of the linear nonhomogeneous generalized ODE (9.10).

Theorem 9.22: *Suppose conditions* (A1)–(A3) *are satisfied. Let* $\theta > 0$ *be such that*

$$f(t + T) = \theta f(t), \quad \textit{for all } t \in [0, \infty).$$

Moreover, assume that for a given $t_0 \in [0,\infty)$, *we have* $\text{var}_{t_0}^{t_0+T}(A) < 1$ *and*

$$\int_{t_0}^{t_0+T} d[A(t)]z = 0, \quad \textit{for all } z \in X.$$

Then, the linear nonhomogeneous generalized ODE (9.10) *admits a* (θ, T)*-periodic solution defined on* $[t_0,\infty)$.

Proof. Define an operator $\Phi : BV([t_0, t_0 + T], X) \to BV([t_0, t_0 + T], X)$ by

$$\Phi(x)(t) = \int_{t_0}^{t} d[A(s)]x(s) + f(t), \quad t \in [t_0, t_0 + T].$$

Notice that the operator Φ is well-defined. Indeed, since (A2) holds, by Corollary 1.48, the integral $\int_{t_0}^{t} d[A(s)]x(s)$ exists for all $t \in [t_0, t_0 + T]$. Moreover, by Corollary 4.8, $\Phi(x) \in BV([t_0, t_0 + T], X)$.

By Corollary 1.48, we conclude

$$\left\| \int_{t_0}^{t} d[A(s)]x(s) \right\| \leqslant \text{var}_{t_0}^{t_0+T}(A)\, \|x\|_{\infty}, \quad \text{for every } t \in [t_0, t_0 + T].$$

Consequently,

$$\text{var}_{t_0}^{t_0+T}[\Phi(x) - \Phi(y)] \leqslant \text{var}_{t_0}^{t_0+T}(A)\, \|x - y\|_{\infty},$$

for all $x, y \in BV([t_0, t_0 + T], X)$. Thus, for any $x, y \in BV([t_0, t_0 + T], X)$, we have

$$\begin{aligned}\| \Phi(x) - \Phi(y)\|_{BV} &= \| \Phi(x)(t_0) - \Phi(y)(t_0) \| + \text{var}_{t_0}^{t_0+T}[\Phi(x) - \Phi(y)] \\ &\leqslant \text{var}_{t_0}^{t_0+T}(A)\, \|x - y\|_{\infty} \leqslant K\, \|x - y\|_{BV},\end{aligned}$$

where $K = \text{var}_{t_0}^{t_0+T}(A) < 1$ by hypothesis. Thus, Φ is a contraction on the Banach space $BV([t_0, t_0 + T], X)$. Therefore, by the Banach Fixed Point Theorem, Φ has a unique fixed point $\widetilde{X}$ in $BV([t_0, t_0 + T], X)$.

Claim. $\widetilde{X} : [t_0, t_0 + T] \to X$ is a solution of the boundary value problem (9.11).

In fact, since $\widetilde{X} \in BV([t_0, t_0 + T], X)$ is a fixed point of the operator Φ, we have

$$\widetilde{X}(t) = \int_{t_0}^{t} d[A(s)]\widetilde{X}(s) + f(t), \quad t \in [t_0, t_0 + T]. \tag{9.16}$$

Taking $t = t_0$, we obtain

$$\widetilde{X}(t_0) = f(t_0). \tag{9.17}$$

By (9.16) and (9.17), we have

$$\widetilde{X}(t) = \widetilde{X}(t_0) + \int_{t_0}^{t} d[A(s)]\widetilde{X}(s) + f(t) - f(t_0), \quad \text{for all } t \in [t_0, t_0 + T],$$

that is, $\widetilde{X} : [t_0, t_0 + T] \to X$ is a solution of the linear nonhomogeneous generalized ODE (9.10).

Now, taking $t = t_0 + T$ in (9.16) and using (9.17) and the hypotheses $f(t_0 + T) = \theta f(t_0)$ and $\int_{t_0}^{t} d[A(s)]\widetilde{X}(s) = 0$ for all $t \in [t_0, t_0 + T]$, we get

$$\widetilde{X}(t_0 + T) = f(t_0 + T) = \theta f(t_0) = \theta \widetilde{X}(t_0).$$

Thus, $\widetilde{X} : [t_0, t_0 + T] \to X$ is a solution of the boundary value problem (9.11). Then, by Lemma 9.21, there exists a (θ, T)-periodic solution of the linear nonhomogeneous generalized ODE (9.10) defined on $[t_0, \infty)$. □

In the sequel, we present an example, which illustrates Theorem 9.22.

Example 9.23: Consider the linear nonhomogeneous generalized ODE given by

$$\frac{dx}{d\tau} = D[\alpha \cos(t)x + e^t \cos(t)], \tag{9.18}$$

defined on $[0, \infty)$, where $0 < \alpha < \frac{1}{2\pi}$. For each $t \in [0, \infty)$, define an operator $A(t) : \mathbb{R} \to \mathbb{R}$ by $A(t)x = \alpha \cos(t)x$, $x \in \mathbb{R}$, and the function $f : [0, \infty) \to \mathbb{R}$ by $f(t) = e^t \cos(t)$. It is clear that, for all $t \in (0, \infty)$, we have

$$A(t + 2\pi) = A(t) \quad \text{and} \quad \left[I + \lim_{s \to t^+} A(s) - A(t)\right]^{-1} = I \in L(\mathbb{R}).$$

Note also that

$$\int_0^{2\pi} d[\alpha \cos(t)]z = \alpha[\cos(2\pi) - \cos(0)]z = 0, \quad \text{for all } z \in \mathbb{R}.$$

We assert that $A \in BV_{loc}([0, \infty), L(\mathbb{R}))$ and $f \in BV_{loc}([0, \infty), \mathbb{R})$. Indeed, let $a, b \in \mathbb{R}$, $a < b$, and $d = (t_i), j = 1, 2, \ldots, |d|$, be an arbitrary division of $[a, b]$. Then,

$$\begin{aligned} \| A(t_i) - A(t_{i-1}) \| &= \sup_{|x| \leqslant 1} |A(t_i)x - A(t_{i-1})x| \\ &= \sup_{|x| \leqslant 1} |\alpha \cos(t_i)x - \alpha \cos(t_{i-1})x| \\ &\leqslant \alpha |t_i - t_{i-1}| = \alpha(t_i - t_{i-1}), \end{aligned}$$

which implies that

$$\sum_{i=1}^{|d|} \| A(t_i) - A(t_{i-1}) \| \leqslant \sum_{i=1}^{|d|} \alpha(t_i - t_{i-1}) = \alpha(b - a),$$

that is, $A \in BV([a, b], L(\mathbb{R}))$. Hence, $A \in BV_{loc}([0, \infty), L(\mathbb{R}))$. On the other hand, since f is locally Lipschitz, it follows that $f \in BV_{loc}([0, \infty), \mathbb{R})$.

We conclude that conditions [A1]–[A3] are satisfied.

Now, set $\theta = e^{2\pi}$. Since $f(t + 2\pi) = e^{2\pi} f(t)$ for all $t \in [0, \infty)$ and

$$\text{var}_0^{2\pi}(A) \leqslant 2\pi\alpha < 1,$$

it follows from Theorem 9.22 that the linear nonhomogeneous generalized ODE (9.18) admits a $(\theta, 2\pi)$-periodic solution defined on $[0, \infty)$.

9.2.1 An Application to MDEs

In this subsection, we present an application of the results of Section 9.2 to MDEs. In order to do that, let us consider the following integral form of an MDE:

$$x(t) = x(t_0) + \int_{t_0}^{t} \mathcal{F}(s)x(s)ds + \int_{t_0}^{t} \mathcal{G}(s)x(s)du(s) + h(t) - h(t_0), \tag{9.19}$$

where $t_0 \geqslant 0$, $t \in [t_0, \infty)$, and $\mathcal{F} : [0, \infty) \to L(X)$, $\mathcal{G} : [0, \infty) \to L(X)$ and $u : [0, \infty) \to [0, \infty)$ are functions. We also assume the following conditions:

(E1) $\mathcal{F}(\cdot)$ is a T-periodic locally Perron integrable function over $[0, \infty)$;
(E2) u is of locally bounded variation in $[0, \infty)$, left-continuous on $(0, \infty)$ and, for all $t \in (0, \infty)$ and for some $\beta > 0$, $u(t + T) = u(t) + \beta$;
(E3) $\mathcal{G}(\cdot)$ is a T-periodic locally Perron–Stieltjes integrable function with respect to u over $[0, \infty)$;
(E4) there exists a locally Perron integrable function $M_1 : [0, \infty) \to \mathbb{R}$ such that for each $a, b \in [0, \infty)$, $a \leqslant b$, we have

$$\left\| \int_a^b \mathcal{F}(s)ds \right\| \leqslant \int_a^b M_1(s)ds;$$

(E5) E5 there exists a locally Perron–Stieltjes integrable function $M_2 : [0, \infty) \to \mathbb{R}$ such that for each $a, b \in [0, \infty)$, $a \leqslant b$, we have

$$\left\| \int_a^b \mathcal{G}(s)du(s) \right\| \leqslant \int_a^b M_2(s)du(s);$$

(E6) $h \in BV_{loc}([0, \infty), X)$ and there is $\theta > 0$, such that for all $t \in [0, \infty)$,

$$h(t + T) = \theta h(t);$$

(E7) for all $t \in [0, \infty)$, there exists $\xi = \xi(t) > 0$ such that

$$\int_t^{t+\xi} M_1(s)ds + \int_t^{t+\xi} M_2(s)du(s) < 1;$$

(E8) the following equalities hold

$$\int_{t_0}^{t_0+T} \mathcal{F}(s)ds = \int_{t_0}^{t_0+T} \mathcal{G}(s)du(s) = 0.$$

Let $t_0 \geqslant 0$. According to Definition 3.2, a function $x : [t_0, \infty) \to X$ is called a *solution of the initial value problem*

$$\begin{cases} x(t) = x(t_0) + \int_{t_0}^{t} \mathcal{F}(s)x(s)ds + \int_{t_0}^{t} \mathcal{G}(s)x(s)du(s) + h(t) - h(t_0), \\ x(t_0) = x_0, \end{cases} \tag{9.20}$$

on $[t_0, \infty)$, if for all $t \in [t_0, \infty)$, the Perron integral $\int_{t_0}^{t} \mathcal{F}(s)x(s)ds$ and the Perron–Stieltjes integral $\int_{t_0}^{t} \mathcal{G}(s)x(s)du(s)$ exist and x satisfies the following integral equation

$$x(t) = x_0 + \int_{t_0}^{t} \mathcal{F}(s)x(s)ds + \int_{t_0}^{t} \mathcal{G}(s)x(s)du(s) + h(t) - h(t_0), \quad t \in [t_0, \infty).$$

The next result exhibits a relation between the MDE (9.19) and its corresponding linear nonhomogeneous generalized ODE. See Theorem 4.14. Such result can also be found in [209] for the finite dimensional case, but the proof for functions taking values in a general Banach space X follows analogously.

Theorem 9.24: *Let $t_0 \in [0, \infty)$. The function $x : [t_0, \infty) \to X$ is a solution of the MDE (9.19), with initial condition $x(t_0) = x_0$ if and only if $x : [t_0, \infty) \to X$ is a solution of the linear nonhomogeneous generalized ODE*

$$\begin{cases} \frac{dx}{d\tau} = D[A(t)x + h(t)] \\ x(t_0) = x_0, \end{cases} \tag{9.21}$$

where, for all $t \in [0, \infty)$,

$$A(t) = \int_{t_0}^{t} \mathcal{F}(s)ds + \int_{t_0}^{t} \mathcal{G}(s)du(s), \tag{9.22}$$

Definition 9.25: Given $T > 0$, $\theta > 0$ and $t_0 \in [0, \infty)$, we say that a function $x : [t_0, \infty) \to X$ is a *(θ, T)-periodic solution of the MDE* (9.19), if x is a solution of (9.19) and for all $t \in [t_0, \infty)$,

$$x(t + T) = \theta x(t).$$

The next lemma exhibits some properties of the operator $A : [0, \infty) \to L(X)$ defined in (9.22).

Lemma 9.26: *Let $A : [0, \infty) \to L(X)$ be the operator given by (9.22). Then, A is T-periodic in $[0, \infty)$, $A \in BV_{loc}([0, \infty), X)$, and A satisfies the condition (Δ^+) on*

$[0,\infty)$ *(see p. 207). Moreover,*

$$\int_{t_0}^{t_0+T} d[A(s)]z = 0, \quad \text{for all } z \in X.$$

Proof. Let $[a,b] \subset [0,\infty)$ be given and consider a division $d = (t_i)$, $i = 1, 2, \ldots, |d|$, of $[a,b]$. By conditions [E4] and [E5], we have

$$\begin{aligned}
\sum_{i=1}^{|d|} \| A(t_i) - A(t_{i-1}) \| &\leqslant \sum_{i=1}^{|d|} \left\| \int_{t_{i-1}}^{t_i} \mathcal{F}(r)dr + \int_{t_{i-1}}^{t_i} \mathcal{G}(r)du(r) \right\| \\
&\leqslant \sum_{i=1}^{|d|} \left(\int_{t_{i-1}}^{t_i} M_1(r)dr + \int_{t_{i-1}}^{t_i} M_2(r)du(r) \right) \\
&\leqslant \int_a^b M_1(r)dr + \int_a^b M_2(r)du(r),
\end{aligned}$$

which is finite by hypothesis. Thus, $A \in BV_{loc}([0,\infty), L(X))$.

By condition [E7], for each $t \in [0,\infty)$, there exists $\xi = \xi(t) > 0$ such that $\| A(t+\xi) - A(t) \| < 1$. Consequently, $\| \Delta^+ A(t) \| < 1$ and, hence, A satisfies condition (Δ^+), presented in Chapter 6, on $[0,\infty)$.

Moreover, by conditions (E1)–(E5) and (E8), we have

$$\begin{aligned}
A(t+T) &= \int_{t_0}^{t+T} \mathcal{F}(s)ds + \int_{t_0}^{t+T} \mathcal{G}(s)du(s) \\
&= \int_{t_0}^{t_0+T} \mathcal{F}(s)ds + \int_{t_0+T}^{t+T} \mathcal{F}(s)ds + \int_{t_0}^{t_0+T} \mathcal{G}(s)du(s) + \int_{t_0+T}^{t+T} \mathcal{G}(s)du(s) \\
&= \int_{t_0}^{t} \mathcal{F}(s+T)ds + \int_{t_0}^{t} \mathcal{G}(s+T)du(s+T) \\
&= \int_{t_0}^{t} \mathcal{F}(s)ds + \int_{t_0}^{t} \mathcal{G}(s)du(s) = A(t),
\end{aligned}$$

where in the last inequality, we used the fact $u(s+T) = u(s) + \beta$, for all $s \in [0,\infty)$, and by definition of the integral, we have

$$\int_{t_0}^{t} \mathcal{G}(s)du_1(s) = 0,$$

with $u_1(s) = \beta$, for all $s \in [0,\infty)$. Thus, we conclude that A is T-periodic on $[0,\infty)$.

At last, by Proposition 1.67, we obtain

$$\int_{t_0}^{t_0+T} d[A(s)]z = \int_{t_0}^{t_0+T} \mathcal{F}(s)\,z\,ds + \int_{t_0}^{t_0+T} \mathcal{G}(s)\,z\,du(s)$$

for every $z \in X$. Thus, using condition [E8], we conclude that

$$\int_{t_0}^{t_0+T} d[A(s)]z = 0$$

for all $z \in X$, and the proof is complete. □

In view of the relations between the MDE (9.19) and the linear nonhomogeneous generalized ODE (9.21) and due to Lemma 9.21, we conclude that the existence of a (θ, T)-periodic solution $x : [t_0, \infty) \to X$ of the MDE (9.19) is equivalent to the existence of a solution of the boundary value problem

$$\begin{cases} x(t) = x(t_0) + \int_{t_0}^{t} \mathcal{F}(s)x(s)ds + \int_{t_0}^{t} \mathcal{G}(s)x(s)du(s) + h(t) - h(t_0), & t \in [t_0, t_0 + T], \\ x(t_0 + T) = \theta x(t_0). \end{cases} \tag{9.23}$$

In the next result, we consider the constant β given in condition [E2] and the functions M_1 and M_2 given by conditions (E4) and (E5), respectively.

Theorem 9.27: *Assume that conditions* [E1]–[E8] *are satisfied. If*

$$M_1(s) + M_2(s) \leqslant \frac{1}{\eta \max\{T, \beta\}}, \quad \textit{for all } s \in [0, \infty),$$

where $\eta > 2$, then the MDE (9.19) *admits a (θ, T)-periodic solution on* $[t_0, \infty)$.

Proof. We are going to use Theorem 9.22 to conclude the result. By Lemma 9.26, conditions [A1]–[A3] from the previous section are satisfied. In addition, condition [E6] says that there exists $\theta > 0$ such that $h(t + T) = \theta h(t)$ for all $t \in [0, \infty)$. Now, let $t_0 \in [0, \infty)$. Then,

$$\begin{aligned} \operatorname{var}_{t_0}^{t_0+T}(A) &\leqslant \int_{t_0}^{t_0+T} M_1(s)ds + \int_{t_0}^{t_0+T} M_2(s)du(s) \\ &\leqslant \frac{T}{\eta \max\{T, \beta\}} + \frac{[u(t_0 + T) - u(t_0)]}{\eta \max\{T, \beta\}} \leqslant \frac{2}{\eta} < 1, \end{aligned}$$

where we used the facts that $u(t_0 + T) - u(t_0) = \beta$ and $\eta > 2$. Thus, by Theorems 9.22 and 9.24, the MDE (9.19) admits a (θ, T)-periodic solution on $[t_0, \infty)$. □

10

Averaging Principles

Márcia Federson[1] *and Jaqueline G. Mesquita*[2]

[1]*Departamento de Matemática, Instituto de Ciências Matemáticas e de Computação (ICMC), Universidade de São Paulo, São Carlos, SP, Brazil*
[2]*Departamento de Matemática, Instituto de Ciências Exatas, Universidade de Brasília, Brasília, DF, Brazil*

Averaging methods have the purpose to simplify the analysis of nonautonomous differential systems through simpler autonomous differential systems obtained as an *"averaged"* equation of the original (or exact) equation. Roughly speaking, averaging methods enable one to replace a time-varying small perturbation, acting on a long time interval, by a time-invariant perturbation and, in this process, only a small error is introduced.

As we mentioned in Section 2.3, although the idea of averaging came first with Andrew Stephenson in his studies from the early 1900s (see [224, 225]), the term *"averaging"* goes back to the works of Pjotr Kapitza (see [139, 140]). The work by N. N. Krylov and N. N. Bogolyubov in [145], published in 1934, as well as the paper [20] from 1955 by N. N. Bogolyubov and A. Mitropols'kiĭ are also considered to be pioneers in averaging methods (also referred to as averaging principles).

When dealing with averaging methods, one is expected to describe the difference between the solutions of the exact system and the averaged system for a sufficiently small parameter on a finite interval. The following nonlinear system was considered in [20, 145], for instance,

$$\begin{cases} \dfrac{dx}{dt} = \dot{x} = \epsilon X(t,x), \\ x(0) = x_0, \end{cases} \tag{10.1}$$

where ϵ is a small parameter, x and X are n-dimensional vectors, and X is Lipschitzian on the second variable. The averaged equation for the ordinary differential equation (ODE) (10.1) is

$$\begin{cases} \dot{y} = \epsilon X_0(y), \\ y(0) = x_0, \end{cases} \tag{10.2}$$

Generalized Ordinary Differential Equations in Abstract Spaces and Applications, First Edition.
Edited by Everaldo M. Bonotto, Márcia Federson, and Jaqueline G. Mesquita.

with

$$X_0(x) = \lim_{T\to\infty} \frac{1}{T} \int_0^T X(t,x)dt,$$

that is, the right-hand side of the autonomous equation (10.2) is obtained by taking the average or mean of the right-hand side of the nonautonomous equation (10.1).

In the 1960s, V. I. Fodčuk [94], A. Halanay [113], J. K. Hale [114], G. N. Medvedev [175], and G. N. Medvedev, B. I. Morgunov, and V. M. Volosov [176] developed methods of averaging for certain functional differential equations (we write, for short, FDEs) with a small parameter. However, the averaged systems they considered were autonomous ODEs. Let us specify that situation with an example.

Consider the delay differential equation

$$\dot{x} = \epsilon f(t, x(t-r)), \tag{10.3}$$

where $r > 0$, and $\epsilon > 0$ is a small parameter. Consider the change of variables $t \mapsto t/\epsilon$ and $y(t) = x(t/\epsilon)$. Then,

$$x\left(\frac{t}{\epsilon} - r\right) = y(t - \epsilon r)$$

and, therefore, Eq. (10.3) can be rewritten as

$$\dot{y}(t) = \frac{1}{\epsilon}\dot{x}\left(\frac{t}{\epsilon}\right) = f\left(\frac{t}{\epsilon}, y(t-\epsilon r)\right).$$

Taking $\epsilon \to 0^+$, the delay becomes negligible and, hence, the averaged equation is an autonomous ODE given by

$$\dot{y} = f_0(y), \quad \text{with} \quad f_0(y) = \lim_{T\to\infty} \frac{1}{T} \int_0^T f(s,y)ds.$$

In the 1970s, the investigations about averaging methods for FDEs showed that the classic approximation by solutions of an autonomous ODE could be replaced by an approximation by solutions of an autonomous FDE, whenever one could treat separately the limiting process involving the delay and the averaging process. At this moment, it became clear that dealing with these two processes apart would enable one to get an infinite-dimensional phase space. Furthermore, a better approximation of solutions was obtained and such a fact could be verified by the order of the approximation and computational simulations. In this respect, the reader may want to consult the works of V. V. Strygin in [226] and of B. Lehman and S. P. Weibel in [162]. See also [160, 161, 163], for instance. In these papers, the authors considered a nonautonomous FDE of the form

$$\begin{cases} \dot{x} = \epsilon f\left(t, x_t\right), \\ x_0 = \phi, \end{cases}$$

where $\epsilon > 0$ is a small parameter and, as usual, $x_t(\theta) = x(t+\theta)$, for $\theta \in [-r, 0]$, with $r \geqslant 0$ and $t \geqslant 0$. The initial function ϕ is taken in the Banach space

$C = C([-r, 0], \mathbb{R}^n)$ of continuous functions from $[-r, 0]$ to $\mathbb{R}^n$, equipped with the supremum norm. The function $f: \mathbb{R} \times C \to \mathbb{R}^n$ is continuous on the first variable and Lipschitzian on the second variable. The averaged system is then given by

$$\begin{cases} \dot{y} = \epsilon f_0\left(y_t\right), \\ y_0 = \phi, \end{cases}$$

where, for every $\varphi \in C$, the limit

$$f_0(\varphi) = \lim_{T\to\infty} \frac{1}{T} \int_0^T f(s, \varphi) ds$$

exists.

In the late 1980s, D. D. Bainov and S. D. Milusheva (see [11]) considered an impulsive FDE of neutral type given by

$$\begin{cases} \dot{x}(t) = \epsilon X(t, x(t), x(\Delta(t, x(t)), \dot{x}(\Delta(t, x(t))))), & t > 0,\ t \neq \tau_i(x), \\ x(t) = \phi(t, \epsilon), & t \in [-r, 0], \\ \dot{x}(t) = \dot{\phi}(t, \epsilon), & t \in [-r, 0], \\ x_i^+ = x_i^- + \epsilon I_i(x_i^-), & i \in \mathbb{N}, \end{cases} \tag{10.4}$$

where $\epsilon > 0$ is a small parameter, $r > 0$, $t - r \leqslant \Delta(t, x(t)) \leqslant t$, $t \geqslant 0$, $\phi(t, \epsilon)$ is the initial function, the surfaces $\tau_i(x)$ are such that $\tau_i(x) < \tau_{i+1}(x)$, for $i \in \mathbb{N}$, and all $\tau_i(x)$ are in the half-space $t > 0$ for all x in some set $D \subset \mathbb{R}^n$ and $i \in \mathbb{N}$. The averaged system for (10.4) is given by

$$\begin{cases} \dot{y} = \epsilon X_0\left(y\right) + \epsilon I_0(y), \\ y\left(0\right) = x_0, \end{cases} \tag{10.5}$$

whenever the limits

$$\lim_{T\to\infty} \frac{1}{T} \int_t^{t+T} X(s, x, x, 0) ds = X_0(x) \quad \text{and} \quad \lim_{T\to\infty} \frac{1}{T} \sum_{t<t_i<t+T} I_i(x) = I_0(x)$$

exist. The authors also assumed that $X(t, x, y, z)$ and $\Delta(t, x)$ are continuous functions, $\phi(t, \epsilon)$ is continuously differentiable, the impulse operators $I_i(x)$, $i \in \mathbb{N}$, are continuous and the functions $\tau_i(x)$, $i \in \mathbb{N}$, are twice continuously differentiable. Then, they proved that, under certain conditions, for each $\eta > 0$ and each $L > 0$, there exists $\epsilon \in (0, \epsilon_0]$, $\epsilon_0 = \epsilon_0(\eta, L)$, such that if $0 < \epsilon \leqslant \epsilon_0$, then

$$\|x(t) - y(t)\| < \eta, \quad \text{for all } t \in \left[0, \frac{L}{\epsilon}\right],$$

where x and y are solutions of (10.4) and (10.5), respectively. This means that the solution of the averaged equation (10.5) is close to the solution of the original equation (10.4).

Our aim in the next sections is to present periodic and nonperiodic averaging principles for generalized ODEs. We also include one application for impulsive differential equations (IDEs, for short). We start by presenting periodic averaging principles and, then, we present nonperiodic averaging methods. All the results of this chapter can be found in [83, 178].

10.1 Periodic Averaging Principles

In this section, we present a periodic averaging principle for generalized ODEs described by

$$\frac{dx}{d\tau} = D\left[\epsilon F(x,t) + \epsilon^2 G(x,t,\epsilon)\right].$$

This result can be found in [178] for the case $X = \mathbb{R}^n$. Here, we present its version for Banach space-valued functions whose proof is essentially the same. We also add an application to IDEs.

Theorem 10.1: *Let $B \subset X$ be open, $\epsilon_0 > 0$ and $L > 0$ be given and consider $\Omega = B \times [0,\infty)$ and functions $F : \Omega \to X$ and $G : \Omega \times (0, \epsilon_0] \to X$ fulfilling the following conditions:*

(i) *there exist nondecreasing left-continuous functions $h_1, h_2 : [0,\infty) \to [0,\infty)$ such that F belongs to the class $\mathcal{F}(\Omega, h_1)$, and for every fixed $\epsilon \in (0, \epsilon_0]$, the function $(x,t) \mapsto G(x,t,\epsilon)$ belongs to the class $\mathcal{F}(\Omega, h_2)$;*

(ii) *$F(x,0) = 0$ and $G(x,0,\epsilon) = 0$ for every $x \in B$, $\epsilon \in (0, \epsilon_0]$;*

(iii) *there exist $T > 0$ and a bounded Lipschitz-continuous function $M : B \to X$ such that, for every $x \in B$ and every $t \in [0,\infty)$,*

$$F(x, t+T) - F(x,t) = M(x);$$

(iv) *there exists a constant $\alpha > 0$ such that, for every $j \in \mathbb{N}$,*

$$h_1(jT) - h_1((j-1)T) \leqslant \alpha;$$

(v) *there exists a constant $\beta > 0$ such that*

$$\left|\frac{h_2(t)}{t}\right| \leqslant \beta, \quad \text{for every} \quad t > 0.$$

Let

$$F_0(x) = \frac{F(x,T)}{T}, \quad \text{for all } x \in B.$$

Assume that for every $\epsilon \in (0, \epsilon_0]$, $x_\epsilon : [0, \frac{L}{\epsilon}] \to B$ is a solution of the generalized ODE

$$\begin{cases} \dfrac{dx}{d\tau} = D\left[\epsilon F(x,t) + \epsilon^2 G(x,t,\epsilon)\right], \\ x(0) = x_0(\epsilon), \end{cases} \tag{10.6}$$

and $y_\epsilon : [0, \frac{L}{\epsilon}] \to B$ *is a solution of the autonomous ODE*

$$\begin{cases} y'(t) = \epsilon F_0(y(t)), \\ y(0) = y_0(\epsilon). \end{cases} \tag{10.7}$$

Suppose, in addition, there exists a constant $J > 0$ *such that*

$$\|x_0(\epsilon) - y_0(\epsilon)\| \leqslant J\epsilon, \quad \text{for every } \epsilon \in (0, \epsilon_0].$$

Then, there exists a constant $K > 0$ *such that for every* $\epsilon \in (0, \epsilon_0]$ *and every* $t \in [0, \frac{L}{\epsilon}]$, *we have*

$$\|x_\epsilon(t) - y_\epsilon(t)\| \leqslant K\epsilon.$$

Proof. Since M is a bounded and Lipschitzian function, there are positive constants m and l for which

$$\|M(x)\| \leqslant m \quad \text{and} \quad \|M(x) - M(y)\| \leqslant l \, \|x - y\| \, ,$$

for all $x, y \in X$. Suppose $x \in B$. Then, from the definition of F_0 and by (ii), we have

$$\|F_0(x)\| = \left\| \frac{F(x,T)}{T} \right\| = \left\| \frac{F(x,T) - F(x,0)}{T} \right\| = \frac{\|M(x)\|}{T} \leqslant \frac{m}{T}. \tag{10.8}$$

Define a function $H : B \times [0, \infty) \to X$ by $H(x, t) = F_0(x)t$. Then, the definition of F_0 yields H can be written as

$$H(x,t) = \frac{F(x,T)}{T} t, \quad \text{for every } \ x \in B \times [0, \infty).$$

Thus,

$$\begin{aligned} \|H(x, s_2) - H(x, s_1)\| &= \frac{1}{T} \|F(x,T)s_2 - F(x,T)s_1\| \\ &= \frac{1}{T} \|F(x,T)\| \, (s_2 - s_1) \leqslant \frac{m}{T}(s_2 - s_1), \end{aligned}$$

and also

$$\begin{aligned} &\|H(x, s_2) - H(x, s_1) - H(y, s_2) + H(y, s_1)\| \\ &\quad = \frac{1}{T} \|F(x,T)s_2 - F(x,T)s_1 - F(y,T)s_2 + F(y,T)s_1\| \\ &\quad = \frac{1}{T} \|F(x,T) - F(y,T)\| \, (s_2 - s_1) \\ &\quad = \frac{1}{T} \|F(x,T) - F(x,0) + F(y,0) - F(y,T)\| \, (s_2 - s_1) \\ &\quad = \frac{1}{T} \|M(x) - M(y)\| \, (s_2 - s_1) \leqslant \frac{l}{T} \|x - y\| \, (s_2 - s_1), \end{aligned}$$

for all $x, y \in B$ and all $s_1, s_2 \in [0, \infty)$, with $s_1 \leqslant s_2$. Therefore, H is an element of the class $\mathcal{F}(\Omega, h_3)$, where $h_3(t) = (\frac{m+l}{T})t$ for $t \geqslant 0$.

Since x_ϵ is a solution of the nonautonomous generalized ODE (10.6) and y_ϵ is a solution of the ODE (10.7), we have, for every $t \in [0, \frac{L}{\epsilon}]$,

$$x_\epsilon(t) = x_0(\epsilon) + \epsilon \int_0^t DF(x_\epsilon(\tau), s) + \epsilon^2 \int_0^t DG(x_\epsilon(\tau), s, \epsilon), \quad \text{and}$$

$$y_\epsilon(t) = y_0(\epsilon) + \epsilon \int_0^t F_0(y_\epsilon(\tau))d\tau = y_0(\epsilon) + \epsilon \int_0^t D[F_0(y_\epsilon(\tau))s],$$

whence we obtain

$$\begin{aligned}
\|x_\epsilon(t) - y_\epsilon(t)\| &= \left\| x_0(\epsilon) - y_0(\epsilon) + \epsilon \int_0^t DF(x_\epsilon(\tau), s) \right. \\
&\quad \left. + \epsilon^2 \int_0^t DG(x_\epsilon(\tau), s, \epsilon) - \epsilon \int_0^t D[F_0(y_\epsilon(\tau))s] \right\| \\
&\leqslant J\epsilon + \epsilon \left\| \int_0^t D[F(x_\epsilon(\tau), s) - F(y_\epsilon(\tau), s)] \right\| \\
&\quad + \epsilon \left\| \int_0^t D[F(y_\epsilon(\tau), s) - F_0(y_\epsilon(\tau))s] \right\| \\
&\quad + \epsilon^2 \left\| \int_0^t DG(x_\epsilon(\tau), s, \epsilon) \right\|.
\end{aligned} \tag{10.9}$$

On the other hand, since G belongs to $\mathcal{F}(\Omega, h_2)$, Lemma 4.5 yields

$$\begin{aligned}
\epsilon^2 \left\| \int_0^t DG(x_\epsilon(\tau), s, \epsilon) \right\| &\leqslant \epsilon^2 (h_2(t) - h_2(0)) \\
&\leqslant \epsilon^2 h_2\left(\frac{L}{\epsilon}\right) = \epsilon L \frac{h_2(L/\epsilon)}{L/\epsilon} \leqslant \epsilon L \beta,
\end{aligned}$$

for all $t \in [0, \frac{L}{\epsilon}]$, where this last inequality follows from (v). Now, using the fact that F belongs to the class $\mathcal{F}(\Omega, h_1)$ and by Lemma 4.6, we obtain

$$\left\| \int_0^t D[F(x_\epsilon(\tau), s) - F(y_\epsilon(\tau), s)] \right\| \leqslant \int_0^t \|x_\epsilon(s) - y_\epsilon(s)\| \, dh_1(s). \tag{10.10}$$

Let us estimate the second term on the right-hand side of (10.9). In order to do that, we assume that p is the largest integer for which $pT \leqslant t$ (notice that p varies as t changes). Thus, for every $t \in [0, \frac{L}{\epsilon}]$, we have

$$\begin{aligned}
\int_0^t D[F(y_\epsilon(\tau), s) - F_0(y_\epsilon(\tau))s] &= \sum_{j=1}^{p} \int_{(j-1)T}^{jT} D[F(y_\epsilon(\tau), s) - F_0(y_\epsilon(\tau))s] \\
&\quad + \int_{pT}^{t} D[F(y_\epsilon(\tau), s) - F_0(y_\epsilon(\tau))s].
\end{aligned}$$

In addition, for every $j \in \{1, \dots, p\}$, we have

$$
\begin{aligned}
\int_{(j-1)T}^{jT} & D[F(y_\epsilon(\tau), s) - F_0(y_\epsilon(\tau))s] \\
&= \int_{(j-1)T}^{jT} D[F(y_\epsilon(\tau), s) - F(y_\epsilon(jT), s)] \\
&\quad - \int_{(j-1)T}^{jT} D[F_0(y_\epsilon(\tau))s - F_0(y_\epsilon(jT))s] \\
&\quad + \int_{(j-1)T}^{jT} D[F(y_\epsilon(jT), s) - F_0(y_\epsilon(jT))s].
\end{aligned} \tag{10.11}
$$

Applying Lemma 4.5 (see the estimate (10.10)) to the first integral on the right-hand side of (10.11), we get

$$
\left\| \int_{(j-1)T}^{jT} D[F(y_\epsilon(\tau), s) - F(y_\epsilon(jT), s)] \right\| \leqslant \int_{(j-1)T}^{jT} \|y_\epsilon(s) - y_\epsilon(jT)\| \, dh_1(s).
$$

Moreover, for every $x, y \in B$, we have

$$
\begin{aligned}
\|F_0(x) - F_0(y)\| &= \left\| \frac{F(x, T) - F(y, T)}{T} \right\| \\
&= \left\| \frac{F(x, T) - F(x, 0) - F(y, T) + F(y, 0)}{T} \right\| \\
&\leqslant \|x - y\| \, \frac{(h(T) - h(0))}{T}
\end{aligned} \tag{10.12}
$$

whence $F_0 : B \to X$ is continuous.

Using the fact that y_ϵ is a solution of the autonomous ODE (10.7), we obtain for every $s \in [(j-1)T, jT]$,

$$
\begin{aligned}
\|y_\epsilon(s) - y_\epsilon(jT)\| &\leqslant \left\| \int_{jT}^{s} \epsilon F_0(y_\epsilon(s)) ds \right\| \leqslant \epsilon \int_{jT}^{s} \|F_0(y_\epsilon(s))\| \, ds \\
&\leqslant \epsilon \int_{jT}^{s} \frac{m}{T} \, ds = \epsilon \frac{m}{T}(jT - s) \leqslant \epsilon m,
\end{aligned}
$$

where we used the estimate (10.8), and Corollary 1.48 together with the fact that $s \mapsto F_0(y_\epsilon(s))$ is continuous to achieve the second inequality. Therefore,

$$
\int_{(j-1)T}^{jT} \|y_\epsilon(s) - y_\epsilon(jT)\| \, dh_1(s) \leqslant \epsilon m \left[h_1(jT) - h_1((j-1)T)\right] \leqslant \epsilon m \alpha.
$$

A similar argument applied to the second integral on the right-hand side of (10.11) yields

$$
\begin{aligned}
\left\| \int_{(j-1)T}^{jT} D[F_0(y_\epsilon(\tau))s - F_0(y_\epsilon(jT))s] \right\| &\leqslant \epsilon m \left[h_3(jT) - h_3((j-1)T)\right] \\
&= \epsilon m(m + l).
\end{aligned}
$$

Now, we want to obtain an estimate for the third integral on the right-hand side of (10.11). Given an arbitrary $y \in B$, we have

$$\int_{(j-1)T}^{jT} D[F(y,s) - F_0(y)s] = F(y,jT) - F(y,(j-1)T) - F_0(y)T$$
$$= M(y) - F(y,T) = -F(y,0) = 0,$$

where we applied conditions (ii) and (iii). Taking p as the largest integer for which $pT \leqslant t \leqslant L/\epsilon$, we get

$$\left\| \sum_{i=1}^{p} \int_{(i-1)T}^{iT} D[F(y_\epsilon(\tau),s) - F_0(y_\epsilon(\tau))s] \right\| \leqslant p\epsilon m\alpha + p\epsilon m(m+l)$$
$$\leqslant \frac{Lm\alpha}{T} + \frac{m(m+l)L}{T},$$

combining the previous estimates. Then, using Lemma 4.5, we obtain

$$\left\| \int_{pT}^{t} D[F(y_\epsilon(\tau),s) - F_0(y_\epsilon(\tau))s] \right\| \leqslant \left\| \int_{pT}^{t} DF(y_\epsilon(\tau),s) \right\| + \left\| \int_{pT}^{t} D[F_0(y_\epsilon(\tau))s] \right\|$$
$$\leqslant h_1(t) - h_1(pT) + h_3(t) - h_3(pT)$$
$$\leqslant h_1(pT+T) - h_1(pT) + h_3(pT+T) - h_3(pT)$$
$$\leqslant \alpha + m + l,$$

since $F \in \mathcal{F}(\Omega, h_1)$ and $H(x,t) = F_0(x)t \in \mathcal{F}(\Omega, h_3)$, whence we conclude that

$$\left\| \int_{0}^{t} D[F(y_\epsilon(\tau),s) - F_0(y_\epsilon(\tau))s] \right\| \leqslant K_1,$$

for a certain positive constant K_1. Then,

$$\|x_\epsilon(t) - y_\epsilon(t)\| \leqslant \epsilon \int_0^t \|x_\epsilon(s) - y_\epsilon(s)\| \, dh_1(s) + \epsilon(J + K_1 + L\beta).$$

By Grönwall' inequality (Theorem 1.52), we have

$$\|x_\epsilon(t) - y_\epsilon(t)\| \leqslant e^{\epsilon(h_1(t)-h_1(0))}\epsilon(J + K_1 + L\beta).$$

Finally, note that

$$\epsilon[h_1(t) - h_1(0)] \leqslant \epsilon[h_1(L/\epsilon) - h_1(0)] \leqslant \epsilon[h_1(\lceil L/(\epsilon T)\rceil\, T) - h_1(0)]$$
$$\leqslant \epsilon \left\lceil \frac{L}{\epsilon T} \right\rceil \alpha \leqslant \epsilon \left(\frac{L}{\epsilon T} + 1 \right) \alpha \leqslant \left(\frac{L}{T} + \epsilon_0 \right) \alpha.$$

Then,

$$\|x_\epsilon(t) - y_\epsilon(t)\| \leqslant e^{\left(\frac{L}{T}+\epsilon_0\right)\alpha}\epsilon(J + K_1 + L\beta) = K\epsilon,$$

where $K = e^{\left(\frac{L}{T}+\epsilon_0\right)\alpha}(J + K_1 + L\beta)$ and the statement follows. □

10.1.1 An Application to IDEs

This section concerns an application of the previous result to IDEs. We start by presenting two auxiliary lemmas which relate T-periodic integrands to T-periodic Stieltjes-type integrals. These results are known for non-Stieltjes-type integrals. Next, they are presented within the framework of Perron–Stieltjes integrals.

Lemma 10.2: *Let $z: [0, \infty) \to X$ be a T-periodic function and $g: [0, \infty) \to \mathbb{R}$ be a nondecreasing function such that the Perron–Stieltjes integral $\int_0^t z(s)dg(s)$ exists for every $t \in [0, \infty)$. Assume, further, that there exists $\alpha > 0$ such that*

$$g(t+T) - g(t) = \alpha, \quad \text{for every } t \in [0, \infty).$$

Then, for each $n \in \mathbb{N}_0$, we have

$$\int_{nT}^{(n+1)T} z(s)dg(s) = \int_0^T z(s)dg(s).$$

Proof. Let $n \in \mathbb{N}_0$ be fixed. Since z is a T-periodic function, z is an nT-periodic function, that is,

$$z(s+nT) = z(s), \quad \text{for all } s \in [0, \infty).$$

On the other hand, let $\phi: [0, T] \to \mathbb{R}$ be given by $\phi(s) = s + nT$, for all $s \in [0, T]$. Then,

$$\begin{aligned}\int_{nT}^{(n+1)T} z(s)dg(s) &= \int_{nT}^{T+nT} z(s)dg(s) = \int_{\phi(0)}^{\phi(T)} z(\phi(s))\, dg(\phi(s)) \\ &= \int_0^T z(s+nT)dg(s) = \int_0^T z(s)dg(s),\end{aligned}$$

getting the desired result. □

Lemma 10.3: *Given an open set $B \subset X$, let $f: B \times [0, \infty) \to X$ be a T-periodic function with respect to the second variable and $g: [0, \infty) \to \mathbb{R}$ be a nondecreasing function such that the Perron–Stieltjes integral $\int_0^t f(x, s)dg(s)$ exists for every $t \in [0, \infty)$ and $x \in B$. Assume, in addition, that there exists $\alpha > 0$ such that*

$$g(t+T) - g(t) = \alpha, \quad \text{for every } t \in [0, \infty).$$

Then, for every $t \geqslant 0$ and $x \in B$, we have

$$\int_t^{t+T} f(x, s)dg(s) = \int_0^T f(x, s)dg(s).$$

Proof. Let $t \geqslant 0$ be given. Since $[0, \infty) = \cup_{n \in \mathbb{N}_0} [nT, (n+1)T]$, there exists $n \in \mathbb{N}_0$ such that $nT \leqslant t \leqslant (n+1)T$. In particular,

$$t \leqslant (n+1)T = nT + T \leqslant t + T,$$

due to the fact that $nT \leqslant t$, that is, $t \leqslant (n+1)T \leqslant t + T$. As a consequence, we get

$$\begin{aligned}\int_t^{t+T} f(x,s)dg(s) &= \int_t^{(n+1)T} f(x,s)dg(s) + \int_{(n+1)T}^{t+T} f(x,s)dg(s)\\ &= \int_t^{(n+1)T} f(x,s)dg(s) + \int_{nT+T}^{t+T} f(x,s)dg(s)\end{aligned}$$

Then, considering the following change of variable $\varphi(\xi) = \xi + T$, we obtain

$$\begin{aligned}\int_t^{t+T} f(x,s)dg(s) &= \int_t^{(n+1)T} f(x,s)dg(s) + \int_{\varphi(nT)}^{\varphi(t)} f(x,s)dg(s)\\ &= \int_t^{(n+1)T} f(x,s)dg(s) + \int_{nT}^{t} f(x,\varphi(s))dg(\varphi(s))\\ &= \int_t^{(n+1)T} f(x,s)dg(s) + \int_{nT}^{t} f(x,s+T)dg(s)\\ &= \int_t^{(n+1)T} f(x,s)dg(s) + \int_{nT}^{t} f(x,s)dg(s)\\ &= \int_{nT}^{(n+1)T} f(x,s)dg(s) = \int_0^{T} f(x,s)dg(s),\end{aligned}$$

where we used the T-periodicity of f and, to obtain the last equality, we employed Lemma 10.2.

□

Consider a function $f: X \times [0,\infty) \to X$, an increasing sequence of numbers $0 \leqslant t_1 < t_2 < \ldots$, and a sequence of mappings $I_i: X \to X$, with $i \in \mathbb{N}$, consider the IDE

$$\begin{cases} x'(t) = \epsilon f(x(t),t) + \epsilon^2 \tilde{g}(x(t),t,\epsilon), & t \neq t_i,\\ \Delta^+ x(t_i) = x(t_i^+) - x(t_i) = \epsilon I_i(x(t_i)), & i \in \mathbb{N}. \end{cases} \tag{10.13}$$

Throughout this subsection, we assume that f is T-periodic on the second variable and the impulse times are periodic in the sense that there exists $k \in \mathbb{N}$ such that, for every integer $i > k$, we have

$$t_i = t_{i-k} + T \quad \text{and} \quad I_i = I_{i-k}.$$

Let us assume, in addition, that the functions f and $\tilde{g}$ are continuous with respect to the second variable. Moreover, consider that the impulse operators I_i, $i \in \mathbb{N}$, are bounded and Lipschitz-continuous.

The first result we present is borrowed from [178], where it was proved for the n-dimensional case. Here, we extend it to the case of Banach space-valued function. The proof remains similar though.

Theorem 10.4: *Let $\Omega = X \times [0,\infty)$, $T > 0$, $\epsilon_0 > 0$, $L > 0$ and consider functions $f: \Omega \to X$ and $\tilde{g}: \Omega \times (0,\epsilon_0] \to X$ which are continuous on the second variable and satisfy the following conditions:*

(H1) *for every* $t \in [0, \infty)$, $x \in X$, *and* $\epsilon \in (0, \epsilon_0]$, *the Perron integrals*

$$\int_0^t f(x, s)ds \quad \text{and} \quad \int_0^t \tilde{g}(x, s, \epsilon)ds$$

exist;

(H2) *for every* $s_1, s_2 \in [0, \infty)$ *such that* $s_1 \leqslant s_2$ *and* $x, y \in X$, *there exists a constant* $C > 0$ *such that*

$$\left\| \int_{s_1}^{s_2} f(x, s)ds \right\| \leqslant \int_{s_1}^{s_2} C\, ds \quad \text{and}$$

$$\left\| \int_{s_1}^{s_2} [f(x, s) - f(y, s)]\, ds \right\| \leqslant \int_{s_1}^{s_2} C\, \|x - y\|\, ds;$$

(H3) *for every* $s_1, s_2 \in [0, \infty)$ *such that* $s_1 \leqslant s_2$, $x, y \in X$ *and* $\epsilon \in (0, \epsilon_0]$, *there exists a constant* $N > 0$ *such that*

$$\left\| \int_{s_1}^{s_2} \tilde{g}(x, s, \epsilon)ds \right\| \leqslant \int_{s_1}^{s_2} N\, ds \quad \text{and}$$

$$\left\| \int_{s_1}^{s_2} [\tilde{g}(x, s, \epsilon) - \tilde{g}(y, s, \epsilon)]\, ds \right\| \leqslant \int_{s_1}^{s_2} N\, \|x - y\|\, ds.$$

Assume, further, that f *is* T*-periodic with respect to the second variable. Take* $k \in \mathbb{N}$ *and consider* $0 \leqslant t_1 < t_2 < \cdots < t_k < T$ *and impulse operators* $I_i : X \to X$, $i \in \{1, 2, \ldots, k\}$ *which are bounded and Lipschitzian functions and* $\epsilon \in (0, \epsilon_0]$. *For every integer* $i > k$, *set*

$$t_i = t_{i-k} + T \quad \text{and} \quad I_i = I_{i-k}.$$

Define

$$f_0(x) = \frac{1}{T} \int_0^T f(x, s)ds \text{ and } I_0(x) = \frac{1}{T} \sum_{i=1}^{k} I_i(x)$$

for every $x \in X$. *Moreover, assume that, for every* $\epsilon \in (0, \epsilon_0]$, $x_\epsilon : [0, \frac{L}{\epsilon}] \to X$ *is a solution of the IDE*

$$\begin{cases} x'(t) = \epsilon f(x(t), t) + \epsilon^2 \tilde{g}(x(t), t, \epsilon) & t \neq t_i, \\ \Delta^+ x(t_i) = \epsilon I_i(x(t_i)), \quad i \in \mathbb{N}, \\ x(0) = x_0(\epsilon), \end{cases} \tag{10.14}$$

and $y_\epsilon : [0, \frac{L}{\epsilon}] \to X$ *is a solution of the ODE*

$$\begin{cases} y'(t) = \epsilon[f_0(y(t)) + I_0(y(t))], \\ y(0) = y_0(\epsilon). \end{cases} \tag{10.15}$$

If there is a constant $J > 0$ *such that*

$$\|x_0(\epsilon) - y_0(\epsilon)\| \leqslant J\epsilon \quad \text{for every } \epsilon \in (0, \epsilon_0],$$

then there exists a constant $K > 0$ *such that for every* $\epsilon \in (0, \epsilon_0]$ *and* $t \in [0, \frac{L}{\epsilon}]$,

$$\|x_\epsilon(t) - y_\epsilon(t)\| \leqslant K\epsilon.$$

Proof. Given $\epsilon \in (0, \epsilon_0]$, define two functions

$$F(x,t) = \int_0^t f(x,s)ds + \sum_{i=1}^{\infty} I_i(x) H_{t_i}(t) \quad \text{and}$$

$$G(x,t,\epsilon) = \int_0^t \tilde{g}(x,s,\epsilon)ds,$$

where $x \in X$, $t \in [0,\infty)$ and, for each $i \in \mathbb{N}$, H_{t_i} is the Heaviside function concentrated at t_i. Suppose x_ϵ is a solution of the generalized ODE given by

$$\frac{dx_\epsilon}{d\tau} = D[\epsilon F(x_\epsilon, t) + \epsilon^2 G(x_\epsilon, t, \epsilon)]$$

which is exactly the generalized ODE corresponding to the integral form of the IDE (10.14). The reader may want to check Chapters 3 and 4 for more details on the relations between several differential equations and generalized ODEs, and also [209] for this specific type of equation.

By the hypothesis, there is a positive constant D such that, for every $x, y \in X$, $t \in [0,\infty)$ and $i \in \mathbb{N}$, we have

$$\|I_i(x)\| \leqslant D \quad \text{and} \quad \|I_i(x) - I_i(y)\| \leqslant D \|x - y\|.$$

Define a function $h_1 : [0,\infty) \to \mathbb{R}$ by

$$h_1(t) = Ct + D\sum_{i=1}^{\infty} H_{t_i}(t).$$

Thus, h_1 is left-continuous and nondecreasing. Moreover, for all $0 \leqslant u \leqslant t$, we get

$$\begin{aligned}
\|F(x,t) - F(x,u)\| &= \left\| \int_u^t f(x,s)ds + \sum_{i=1}^{\infty} I_i(x)[H_{t_i}(t) - H_{t_i}(u)] \right\| \\
&\leqslant \left\| \int_u^t f(x,s)ds \right\| + \sum_{i=1}^{\infty} \|I_i(x)\| \, [H_{t_i}(t) - H_{t_i}(u)] \\
&\leqslant C(t-u) + D\sum_{i=1}^{\infty} \left[H_{t_i}(t) - H_{t_i}(u)\right] \\
&= h_1(t) - h_1(u)
\end{aligned}$$

and, also,

$$
\begin{aligned}
\|F(x,t) &- F(x,u) - F(y,t) + F(y,u)\| \\
&= \left\| \int_u^t [f(x,s) - f(y,s)]\, ds + \sum_{i=1}^{\infty} \left[I_i(x) - I_i(y)\right] \left[H_{t_i}(t) - H_{t_i}(u)\right] \right\| \\
&\leqslant \left\| \int_u^t f(x,s) - f(y,s) ds \right\| + \sum_{i=1}^{\infty} \|I_i(x) - I_i(y)\| \left[H_{t_i}(t) - H_{t_i}(u)\right] \\
&\leqslant \|x-y\| \left[C(t-u) + D \sum_{i=1}^{\infty} \left[H_{t_i}(t) - H_{t_i}(u)\right] \right] \\
&= \|x-y\| \, (h_1(t) - h_1(u)).
\end{aligned}
$$

Therefore, F belongs to the class $\mathcal{F}(\Omega, h_1)$.

Now, let us define $h_2 : [0, \infty) \to \mathbb{R}$ by $h_2(t) = Nt$. Then, for $0 \leqslant u \leqslant t$, we obtain

$$
\|G(x,t,\epsilon) - G(x,u,\epsilon)\| = \left\| \int_u^t \tilde{g}(x,s,\epsilon) ds \right\| \leqslant N(t-u) = h_2(t) - h_2(u).
$$

In addition, for $0 \leqslant u \leqslant t$ and $x, y \in X$, we have

$$
\begin{aligned}
\|G(x,t,\epsilon) &- G(x,u,\epsilon) - G(y,t,\epsilon) + G(y,u,\epsilon)\| \\
&= \left\| \int_u^t [\tilde{g}(x,s,\epsilon) - \tilde{g}(y,s,\epsilon)]\, ds \right\| \leqslant N \, \|x-y\| \, (t-u) \\
&= \|x-y\| \, [h_2(t) - h_2(u)].
\end{aligned}
$$

Hence, for every fixed $\epsilon \in (0, \epsilon_0]$, the function $(x,t) \mapsto G(x,t,\epsilon)$ belongs to the class $\mathcal{F}(\Omega, h_2)$.

Notice that $F(x,0) = 0$ and $G(x,0,\epsilon) = 0$ for all $x \in X$. Thus, for all $t \geqslant 0$ and $x \in X$, by Lemma 10.3, taking $g(s) = s$, we obtain

$$
\begin{aligned}
F(x,t+T) - F(x,t) &= \int_t^{t+T} f(x,s) ds + \sum_{i : t \leqslant t_i < t+T} I_i(x) \\
&= \int_0^T f(x,s) ds + \sum_{i : 0 \leqslant t_i < T} I_i(x).
\end{aligned}
$$

This shows that the difference $F(x,t+T) - F(x,t)$ does not depend on t, which allows us to set

$$
M(x) = F(x,t+T) - F(x,t), \quad \text{for all } x \in X.
$$

Then, for every $x, y \in X$, we have

$$
\|M(x)\| = \|F(x,T) - F(x,0)\| = \left\| \int_0^T f(x,s) ds + \sum_{i=1}^{k} I_i(x) \right\| \leqslant CT + kD
$$

and, also,

$$
\begin{aligned}
\|M(x) - M(y)\| &= \|F(x,T) - F(y,T) - F(x,0) + F(y,0)\| \\
&= \left\| \int_0^T (f(x,s) - f(y,s))ds + \sum_{i=1}^{k} (I_i(x) - I_i(y)) \right\| \\
&\leqslant \left\| \int_0^T f(x,s) - f(y,s)ds \right\| + \sum_{i=1}^{k} \|I_i(x) - I_i(y)\| \\
&\leqslant CT\,\|x-y\| + kD\,\|x-y\| \\
&= \|x-y\|\,(CT + kD).
\end{aligned}
$$

Therefore, M is bounded and Lipschitzian.

Take $j \in \mathbb{N}$. Then, we have

$$
\begin{aligned}
h_1(jT) - h_1((j-1)T) &= CjT + D\sum_{i=1}^{\infty} H_{t_i}(jT) - C(j-1)T - D\sum_{i=1}^{\infty} H_{t_i}((j-1)T) \\
&= CT + D \sum_{i:(j-1)T \leqslant t_i < jT} 1 = CT + Dk.
\end{aligned}
$$

Notice that $|h_2(t)/t| = N$, for all $t > 0$.

Since all the hypotheses of Theorem 10.1 are fulfilled, in order to finish the proof, it is enough to consider

$$
F_0(x) = \frac{F(x,T)}{T} = \frac{1}{T}\int_0^T f(x,s)ds + \frac{1}{T}\sum_{i=1}^{k} I_i(x) = f_0(x) + I_0(x)
$$

and, then, Theorem 10.1 yields the statement. □

10.2 Nonperiodic Averaging Principles

This section is dedicated to presenting two different forms of the same nonperiodic averaging principle for nonautonomous generalized ODEs of type

$$
\begin{cases}
\dfrac{dx}{d\tau} = D\left[\epsilon F\left(x, \dfrac{t}{\epsilon}\right)\right], \\
x(0) = x_0,
\end{cases}
$$

on $[0, \frac{L}{\epsilon}]$, where $\epsilon > 0$ is a small parameter and $F: \Omega \to X$, with $\Omega = B \times [0, \infty)$ and $B = \{x \in X : \|x\| < c\}$, $c > 0$. In the first result, we relate a solution of the above generalized ODE to a solution of the following averaged ODE:

$$
\dot{y} = F_0(y),
$$

where

$$F_0(x) = \lim_{r\to\infty} \frac{F(x,r)}{r}, \quad \text{for all } x \in B.$$

Such a result generalizes [[209], Theorem 8.12], and it can be found in [83]. The version presented here is more general, since our result holds for any generalized ODE, while in [83], the result is restricted to a specific generalized ODE. It is relevant to bring both results here (see Theorems 10.5 and 10.7), due to the different techniques used in each of the proofs.

Defining a function $G: \Omega \to X$ by

$$G(x,t)(\vartheta) = \epsilon F\left(x, \frac{t}{\epsilon}\right)\left(\frac{\vartheta}{\epsilon}\right), \quad \text{for every } (x,t) \in \Omega \text{ and } \vartheta \in [0,\infty),$$

we obtain, by a change of variable, that a solution of the above nonautonomous generalized ODE can be related to the solution of the autonomous ODE:

$$\dot{y} = G_0(y),$$

where

$$G_0(x)(\vartheta) = \lim_{\epsilon\to 0^+} \frac{G\left(x, \frac{t}{\epsilon}\right)\left(\frac{\vartheta}{\epsilon}\right)}{\frac{t}{\epsilon}}, \quad \text{for } x \in B \text{ and } \vartheta \in [0,\infty).$$

The next result is a generalization of the result found in [83] for functions taking values in a Banach space. The proof of this result is very similar to the one found in [83] with obvious adaptation, which, in turn, is inspired in the proof of a result found in [209].

Theorem 10.5: *Consider $\Omega = B \times [0,\infty)$ and $B = \{x \in X: \|x\| < c\}$, with $c > 0$, and assume that $F \in \mathcal{F}(\Omega, h)$, where $h: [0,\infty) \to \mathbb{R}$ is a left-continuous and nondecreasing function, and $F: B \times [0,\infty) \to X$ is such that $F(y,0) = 0$ for every $y \in B$. Assume, further, that there is a constant C such that, for every $\alpha \geqslant 0$,*

$$\limsup_{r\to\infty} \frac{h(r+\alpha) - h(\alpha)}{r} \leqslant C \tag{10.16}$$

and, for every $x \in B$,

$$\lim_{r\to\infty} \frac{F(x,r)}{r} = F_0(x).$$

Let $y: [0,\infty) \to X$ be the uniquely determined solution of the autonomous ODE

$$\begin{cases} \dot{y} = F_0(y), \\ y(0) = \tilde{y}, \end{cases} \tag{10.17}$$

which belongs to B together with its ρ-neighborhood such that $\rho > 0$, that is, there exists $\rho > 0$ such that

$$\{\tilde{x} \in X: \|\tilde{x} - y(t)\| < \rho, \quad \text{for every } t \in [0,\infty)\} \subset B.$$

Then, for every $\mu > 0$ and $L > 0$, there exists $\epsilon_0 > 0$ such that, for all $\epsilon \in (0, \epsilon_0)$,

$$\|x_\epsilon(t) - z_\epsilon(t)\| < \mu, \quad \text{for every } t \in \left[0, \frac{L}{\epsilon}\right],$$

where x_ϵ is a solution of the generalized ODE:

$$\begin{cases} \dfrac{dx}{d\tau} = D\left[\epsilon F(x,t)\right], \\ x_\epsilon(0) = y(0), \end{cases} \tag{10.18}$$

on $[0, \frac{L}{\epsilon}]$ and z_ϵ is a solution of the autonomous ODE

$$\begin{cases} \dot{z} = \epsilon F_0(z), \\ z_\epsilon(0) = y(0), \end{cases} \tag{10.19}$$

on $[0, \frac{L}{\epsilon}]$.

Proof. For $y \in B$, $t \in [0, \infty)$ and $\epsilon > 0$, consider the functions

$$G_\epsilon(y,t) = \epsilon F\left(y, \frac{t}{\epsilon}\right) \quad \text{and} \quad h_\epsilon(t) = \epsilon h\left(\frac{t}{\epsilon}\right).$$

Clearly, the function h_ϵ is nondecreasing and left-continuous on $[0, \infty)$.

Once $F \in \mathcal{F}(\Omega, h)$, we have, for every $x, y \in B$ and every $t_1, t_2 \in [0, \infty)$ such that $t_2 \geqslant t_1$,

$$\begin{aligned} \|G_\epsilon(y,t_2) - G_\epsilon(y,t_1)\| &= \epsilon \left\| F\left(y, \frac{t_2}{\epsilon}\right) - F\left(y, \frac{t_1}{\epsilon}\right) \right\| \\ &\leqslant \epsilon \left| h\left(\frac{t_2}{\epsilon}\right) - h\left(\frac{t_1}{\epsilon}\right) \right| = h_\epsilon(t_2) - h_\epsilon(t_1), \end{aligned}$$

and, also,

$$\begin{aligned} &\|G_\epsilon(y,t_2) - G_\epsilon(y,t_1) - G_\epsilon(x,t_2) + G_\epsilon(x,t_1)\| \\ &\quad = \epsilon \left\| F\left(y, \frac{t_2}{\epsilon}\right) - F\left(y, \frac{t_1}{\epsilon}\right) - F\left(x, \frac{t_2}{\epsilon}\right) + F\left(x, \frac{t_1}{\epsilon}\right) \right\| \\ &\quad \leqslant \|y - x\| \left[h_\epsilon(t_2) - h_\epsilon(t_1)\right]. \end{aligned}$$

Therefore, $G_\epsilon \in \mathcal{F}(\Omega, h_\epsilon)$, for all $\epsilon > 0$.

Now, consider $y \in B$. By the hypotheses, we have

$$\lim_{r \to \infty} \frac{F(y,r) - F(y,0)}{r} = \lim_{r \to \infty} \frac{F(y,r)}{r} = F_0(y).$$

Thus, for every $\eta > 0$, there exists $R > 0$ such that for $r > R$,

$$\begin{aligned} \|F_0(y)\| &\leqslant \left\| F_0(y) - \frac{F(y,r) - F(y,0)}{r} \right\| + \frac{\|F(y,r) - F(y,0)\|}{r} \\ &\leqslant \eta + \frac{h(r) - h(0)}{r} < 2\eta + C, \end{aligned}$$

where we used the estimates $\|F(y,r) - F(y,0)\| \leqslant h(r) - h(0)$ (because $F \in \mathcal{F}(\Omega, h)$) and (10.16). Then, since $\eta > 0$ can be taken arbitrarily small, we obtain

$$\|F_0(y)\| \leqslant C, \quad \text{for all } y \in B.$$

In an analogous way, for every $x, y \in B$ and every $\eta > 0$, there exists a positive constant R such that, whenever $r > R$, we have

$$\begin{aligned}\|F_0(x) - F_0(y)\| &< \eta + \frac{\|F(y,r) - F(y,0) - F(x,r) + F(x,0)\|}{r} \\ &\leqslant \eta + \|y - x\| \frac{h(r) - h(0)}{r} \leqslant \eta + C\|y - x\|.\end{aligned}$$

Then, the arbitrariness of $\eta > 0$ yields

$$\|F_0(x) - F_0(y)\| \leqslant C\|y - x\|, \quad \text{for all } x, y \in B.$$

Notice that, for $y \in B$ and $t > 0$, we have

$$\lim_{\epsilon \to 0^+} G_\epsilon(y,t) = \lim_{\epsilon \to 0^+} \epsilon F\left(y, \frac{t}{\epsilon}\right) = \lim_{\epsilon \to 0^+} t\frac{\epsilon}{t} F\left(y, \frac{t}{\epsilon}\right) = tF_0(y) \quad \text{and}$$
$$\lim_{\epsilon \to 0^+} G_\epsilon(y,0) = \lim_{\epsilon \to 0^+} \epsilon F(y,0) = 0.$$

Consider, for $y \in B$ and $t \geqslant 0$, a function given by $G_0(y,t) = tF_0(y)$. Then, it is clear that

$$\lim_{\epsilon \to 0^+} G_\epsilon(y,t) = G_0(y,t)$$

and, moreover, $G_0 \in \mathcal{F}(\Omega, h_0)$, with $h_0(t) = Ct$, for all $t \geqslant 0$.

For $0 \leqslant t_1 < t_2 < \infty$, we have

$$\begin{aligned}h_\epsilon(t_2) - h_\epsilon(t_1) &= \epsilon \left[h\left(\frac{t_2}{\epsilon}\right) - h\left(\frac{t_1}{\epsilon}\right)\right] \\ &= (t_2 - t_1)\frac{\epsilon}{(t_2 - t_1)}\left[h\left(\frac{t_2 - t_1}{\epsilon} + \frac{t_1}{\epsilon}\right) - h\left(\frac{t_1}{\epsilon}\right)\right].\end{aligned}$$

Then, the hypotheses yield

$$\limsup_{\epsilon \to 0^+} (h_\epsilon(t_2) - h_\epsilon(t_1)) \leqslant C(t_2 - t_1) = h_0(t_2) - h_0(t_1),$$

once

$$\lim_{\epsilon \to 0^+} \frac{t_2 - t_1}{\epsilon} = \infty.$$

From the fact that $y\colon [0, \infty) \to B$ is a solution of the autonomous ODE (10.17), we obtain, by the properties of the Kurzweil integral, the equalities

$$y(s_2) - y(s_1) = \int_{s_1}^{s_2} F_0(y(\tau))d\tau = \int_{s_1}^{s_2} D[F_0(y(\tau))t] = \int_{s_1}^{s_2} DG_0(y(\tau), t),$$

for all $s_1, s_2 \in [0, \infty)$. Thus, y is also a solution on $[0, \infty)$ of the generalized ODE:

$$\frac{dy}{d\tau} = DG_0(y, t)$$

and, by the hypotheses, such a solution is uniquely determined. Therefore, all conditions of Theorem 7.8 are satisfied for the parameter $\epsilon \to 0^+$. Then, Theorem 7.8 implies that, for every $\mu > 0$ and $L > 0$, there exists $\epsilon_0 > 0$ such that, for $\epsilon \in (0, \epsilon_0)$, there is a solution y_ϵ on the interval $[0, L]$ of the generalized ODE:

$$\frac{dy}{d\tau} = DG_\epsilon(y, t), \tag{10.20}$$

with initial condition $y_\epsilon(0) = y(0)$, and

$$\|y_\epsilon(s) - y(s)\| \leqslant \mu, \quad \text{for all } s \in [0, L]. \tag{10.21}$$

For the solution $y_\epsilon : [0, L] \to B$ of (10.20), we have

$$y_\epsilon(s_2) - y_\epsilon(s_1) = \int_{s_1}^{s_2} DG_\epsilon(y_\epsilon(\tau), t) = \epsilon \int_{s_1}^{s_2} DF\left(y_\epsilon(\tau), \frac{t}{\epsilon}\right)$$

whenever $s_1, s_2 \in [0, L]$. For $t \in [0, \frac{L}{\epsilon}]$, denote $x_\epsilon(t) = y_\epsilon(\epsilon t)$. Then, we get

$$\begin{aligned} x_\epsilon(t_2) - x_\epsilon(t_1) = y_\epsilon(\epsilon t_2) - y_\epsilon(\epsilon t_1) &= \epsilon \int_{\epsilon t_1}^{\epsilon t_2} DF\left(y_\epsilon(\sigma), \frac{s}{\epsilon}\right) \\ &= \epsilon \int_{\epsilon t_1}^{\epsilon t_2} DF\left(x_\epsilon\left(\frac{\sigma}{\epsilon}\right), \frac{s}{\epsilon}\right) \end{aligned}$$

for every $t_1, t_2 \in [0, \frac{L}{\epsilon}]$. Then, taking $\varphi(\sigma) = \frac{\sigma}{\epsilon}$ and applying Theorem 2.18, we obtain

$$\int_{\epsilon t_1}^{\epsilon t_2} DF\left(x_\epsilon\left(\frac{\sigma}{\epsilon}\right), \frac{s}{\epsilon}\right) = \int_{\varphi(\epsilon t_1)}^{\varphi(\epsilon t_2)} DF(x_\epsilon(\tau), t) = \int_{t_1}^{t_2} DF(x_\epsilon(\tau), t)$$

for any $t_1, t_2 \in [0, \frac{L}{\epsilon}]$ and $y_\epsilon(0) = x_\epsilon(0) = y(0)$. Therefore, the function $x_\epsilon : [0, \frac{L}{\epsilon}] \to B$ is a solution of the generalized ODE (10.18) on $[0, \frac{L}{\epsilon}]$.

Similarly, it can be shown that the function $z_\epsilon : [0, \frac{L}{\epsilon}] \to B$ given by $z_\epsilon(t) = y(\epsilon t)$ is a solution of the autonomous ODE (10.19) on $[0, \frac{L}{\epsilon}]$. Thus, (10.21) yields

$$\|x_\epsilon(t) - z_\epsilon(t)\| = \|y_\epsilon(\epsilon t) - y(\epsilon t)\| < \mu$$

for every $t \in [0, \frac{L}{\epsilon}]$, concluding the proof. □

As an immediate consequence of Theorem 10.5, we have the following result.

Corollary 10.6: *Consider $\Omega = B \times [0, \infty)$ and $B = \{x \in X : \|x\| < c\}$, with $c > 0$ and assume that $F \in \mathcal{F}(\Omega, h)$, where $h : [0, \infty) \to \mathbb{R}$ is a left-continuous and nondecreasing function, and $F : B \times [0, \infty) \to X$ is such that $F(y, 0) = 0$ for every $y \in B$.*

Assume, further, that there is a constant C such that, for every $\alpha \geqslant 0$,

$$\limsup_{r\to\infty} \frac{h(r+\alpha)-h(\alpha)}{r} \leqslant C \tag{10.22}$$

and, for every $x \in B$,

$$\lim_{r\to\infty} \frac{F(x,r)}{r} = F_0(x).$$

Let $y\colon [0,\infty) \to X$ be a uniquely determined solution of the autonomous ODE:

$$\begin{cases} \dot{y} = F_0(y), \\ y(0) = \tilde{y}, \end{cases} \tag{10.23}$$

which belongs to B together with its ρ-neighborhood, with $\rho > 0$, that is, there exists $\rho > 0$ such that

$$\{\tilde{x} \in X\colon \|\tilde{x} - y(t)\| < \rho, \quad \text{for every } t \in [0,\infty)\} \subset B.$$

Then, for every $\mu > 0$ and $L > 0$, there exists $\epsilon_0 > 0$ such that, for all $\epsilon \in (0, \epsilon_0)$,

$$\|x_\epsilon(t) - z_\epsilon(t)\| < \mu, \quad \text{for every } t \in [0, L],$$

where x_ϵ is a solution of the generalized ODE:

$$\begin{cases} \dfrac{dx}{d\tau} = D\left[\epsilon F\left(x, \dfrac{t}{\epsilon}\right)\right], \\ x_\epsilon(0) = y(0), \end{cases} \tag{10.24}$$

on $[0, L]$ and z_ϵ is a solution on $[0, L]$ of the autonomous ODE:

$$\begin{cases} \dot{z} = F_0(z), \\ z_\epsilon(0) = y(0). \end{cases} \tag{10.25}$$

The next result can be found in [83]. But the version presented here is more general, since it holds for a more general setting of generalized ODEs. This type of result can be useful when we want to apply the results from generalized ODEs to measure FDEs, using the correspondences presented in Chapter 4.

Theorem 10.7: *Consider a function $G\colon \Omega \to X$ such that $G \in \mathcal{F}(\Omega, h)$, where $\Omega = B \times [0,\infty)$, $B = \{x \in X\colon \|x\| < c\}$, with $c > 0$, and $h\colon [0,\infty) \to \mathbb{R}$ is a left-continuous and nondecreasing function. Suppose, for every $\alpha \geqslant 0$, we have*

$$\limsup_{\epsilon\to 0^+} \frac{h\left(\frac{t}{\epsilon}+\alpha\right) - h(\alpha)}{\frac{t}{\epsilon}} \leqslant C, \quad \text{where } C > 0 \text{ is a constant,} \tag{10.26}$$

and, for every $x \in B$, we have

$$\lim_{\epsilon\to 0^+} \frac{G\left(x, \frac{t}{\epsilon}\right)\left(\frac{\vartheta}{\epsilon}\right)}{\frac{t}{\epsilon}} = G_0(x)(\vartheta), \quad \text{for all } \vartheta \in [0,\infty). \tag{10.27}$$

Suppose, in addition, for each $y \in B$, the function $G(y,t)\colon [0,\infty) \to X$ given by $\vartheta \mapsto G(y,t)(\vartheta)$, is such that $G(y,0)(\vartheta) = 0$ for every $\vartheta \geqslant 0$ and $y \in B$. Let $y\colon [0,\infty) \to B$ be the unique solution of the autonomous ODE:

$$\begin{cases} \dot{y} = G_0(y), \\ y(0) = \tilde{y} \end{cases} \tag{10.28}$$

and assume that there exists $\rho > 0$ such that

$$\{\tilde{x} \in X\colon \|\tilde{x} - y(t)\| < \rho, \quad \text{for every } t \in [0,\infty)\} \subset B.$$

Then, for every $\mu > 0$ and every $L > 0$, there exists $\epsilon_0 > 0$ such that for $\epsilon \in (0, \epsilon_0)$,

$$\|x_\epsilon(t) - z_\epsilon(t)\| < \mu, \quad \text{for every } t \in \left[0, \frac{L}{\epsilon}\right],$$

where x_ϵ is a solution of the generalized ODE:

$$\frac{dx}{d\tau} = \epsilon D\left[G(x,t)\right] \tag{10.29}$$

on $\left[0, \frac{L}{\epsilon}\right]$ such that $x_\epsilon(0) = y(0)$ and z_ϵ is a solution on $[0, \frac{L}{\epsilon}]$ of the autonomous ODE:

$$\begin{cases} \dot{z} = \epsilon G_0(z), \\ z_\epsilon(0) = y(0). \end{cases} \tag{10.30}$$

Proof. For $y \in B$, $t \in [0,\infty)$, and $\epsilon > 0$, define the functions

$$h_\epsilon(t) = \epsilon h\left(\frac{t}{\epsilon}\right) \quad \text{and} \quad H_\epsilon(y,t)(\vartheta) = \begin{cases} 0, & \vartheta \in [-r, 0], \\ \epsilon G\left(y, \frac{t}{\epsilon}\right)\left(\frac{\vartheta}{\epsilon}\right), & \vartheta \in [0,\infty). \end{cases}$$

It is easy to see that h_ϵ is nondecreasing and left-continuous on $[0,\infty)$. Moreover, since $G \in \mathcal{F}(\Omega, h)$, we have, for every $x, y \in B$, $t_1, t_2 \in [0,\infty)$, and $\vartheta \in [0,\infty)$,

$$\begin{aligned} \|H_\epsilon(y,t_2)(\vartheta) - H_\epsilon(y,t_1)(\vartheta)\| &= \left\|\epsilon G\left(y, \frac{t_2}{\epsilon}\right)\left(\frac{\vartheta}{\epsilon}\right) - \epsilon G\left(y, \frac{t_1}{\epsilon}\right)\left(\frac{\vartheta}{\epsilon}\right)\right\| \\ &\leqslant \epsilon\left|h\left(\frac{t_2}{\epsilon}\right) - h\left(\frac{t_1}{\epsilon}\right)\right| = |h_\epsilon(t_2) - h_\epsilon(t_1)| \end{aligned}$$

and, also,

$$\begin{aligned} &\| H_\epsilon(x,t_2)(\vartheta) - H_\epsilon(x,t_1)(\vartheta) - H_\epsilon(y,t_2)(\vartheta) + H_\epsilon(y,t_1)(\vartheta)\| \\ &= \left\|\epsilon G\left(x, \frac{t_2}{\epsilon}\right)\left(\frac{\vartheta}{\epsilon}\right) - \epsilon G\left(x, \frac{t_1}{\epsilon}\right)\left(\frac{\vartheta}{\epsilon}\right) - \epsilon G\left(y, \frac{t_2}{\epsilon}\right)\left(\frac{\vartheta}{\epsilon}\right)\right. \\ &= \left. +\epsilon G\left(y, \frac{t_1}{\epsilon}\right)\left(\frac{\vartheta}{\epsilon}\right)\right\| \\ &\leqslant \|x - y\|_\infty \epsilon\left|h\left(\frac{t_2}{\epsilon}\right) - h\left(\frac{t_1}{\epsilon}\right)\right| = \|x - y\|\, |h_\epsilon(t_2) - h_\epsilon(t_1)|. \end{aligned}$$

Thus,

$$\|H_\epsilon(y,t_2) - H_\epsilon(y,t_1)\| \leqslant |h_\epsilon(t_2) - h_\epsilon(t_1)| \quad \text{and}$$
$$\|H_\epsilon(x,t_2) - H_\epsilon(x,t_1) - H_\epsilon(y,t_2) + H_\epsilon(y,t_1)\| \leqslant \|x-y\|\,|h_\epsilon(t_2) - h_\epsilon(t_1)|$$

and, therefore, $H_\epsilon \in \mathcal{F}(\Omega, h_\epsilon)$ for $\epsilon > 0$.

Let $y \in B$ and $t \in [0,\infty)$. Then, for every $\vartheta \in [0,\infty)$, we obtain

$$\lim_{\epsilon\to 0^+} \frac{G\left(y,\frac{t}{\epsilon}\right)\left(\frac{\vartheta}{\epsilon}\right) - G(y,0)\left(\frac{\vartheta}{\epsilon}\right)}{\frac{t}{\epsilon}} = \lim_{\epsilon\to 0^+} \frac{G\left(y,\frac{t}{\epsilon}\right)\left(\frac{\vartheta}{\epsilon}\right)}{\frac{t}{\epsilon}} = G_0(y)(\vartheta),$$

where we used (10.27) and the fact that $G(y,0)(\vartheta) = 0$ for every $\vartheta \in [0,\infty)$. Therefore, in virtue of (10.27), for every $\eta > 0$, there exists a sufficiently small $\epsilon > 0$ such that for $\vartheta \in [0,\infty)$, we obtain

$$\begin{aligned}\|G_0(y)(\vartheta)\| \leqslant{}& \left\|G_0(y)(\vartheta) - \frac{\epsilon}{t}\left[G\left(y,\frac{t}{\epsilon}\right)\left(\frac{\vartheta}{\epsilon}\right) - G(y,0)\left(\frac{\vartheta}{\epsilon}\right)\right]\right\| \\ &+ \frac{\epsilon}{t}\left\|G\left(y,\frac{t}{\epsilon}\right)\left(\frac{\vartheta}{\epsilon}\right) - G(y,0)\left(\frac{\vartheta}{\epsilon}\right)\right\| \\ \leqslant{}& \eta + \frac{\epsilon}{t}\left[h\left(\frac{t}{\epsilon}\right) - h(0)\right] < 2\eta + C,\end{aligned}$$

where we also used the fact that

$$\left\|G\left(y,\frac{t}{\epsilon}\right) - G(y,0)\right\| \leqslant h\left(\frac{t}{\epsilon}\right) - h(0),$$

once $G \in \mathcal{F}(\Omega, h)$. Thus, we have

$$\|G_0(y)\| \leqslant C, \quad \text{for every } y \in B,$$

because $\eta > 0$ can be taken arbitrarily small.

Similarly, if $x, y \in B$ and $t \in [0,\infty)$, then for every $\eta > 0$, there exists a sufficiently small $\epsilon > 0$ such that, for all $\vartheta \in [0,\infty)$,

$$\begin{aligned}&\|\, G_0(x)(\vartheta) - G_0(y)(\vartheta)\| \\ &\quad < \eta + \frac{\epsilon}{t}\left\|G\left(y,\frac{t}{\epsilon}\right)\left(\frac{\vartheta}{\epsilon}\right) - G(y,0)\left(\frac{\vartheta}{\epsilon}\right) - G\left(x,\frac{t}{\epsilon}\right)\left(\frac{\vartheta}{\epsilon}\right) + G(x,0)\left(\frac{\vartheta}{\epsilon}\right)\right\| \\ &\quad \leqslant \eta + \|x-y\|\,\frac{\epsilon}{t}\left[h\left(\frac{t}{\epsilon}\right) - h(0)\right] \\ &\quad \leqslant \eta + (\eta + C)\,\|y-x\| \leqslant \eta(1 + \|y-x\|) + C\,\|y-x\|,\end{aligned}$$

and, again, because $\eta > 0$ can be taken sufficiently small, we obtain

$$\|G_0(x) - G_0(y)\| \leqslant C\,\|y-x\|, \qquad x,y \in B. \tag{10.31}$$

Besides, for $y \in B$, $t \in (0, \infty)$, and $\vartheta \in [0, \infty)$, we have

$$\lim_{\epsilon \to 0^+} H_\epsilon(y, t)(\vartheta) = \lim_{\epsilon \to 0^+} \epsilon G\left(y, \frac{t}{\epsilon}\right)\left(\frac{\vartheta}{\epsilon}\right)$$
$$= \lim_{\epsilon \to 0^+} t\frac{\epsilon}{t} G\left(y, \frac{t}{\epsilon}\right)\left(\frac{\vartheta}{\epsilon}\right) = tG_0(y)(\vartheta)$$

and, for $t = 0$ and $\vartheta \in [0, \infty)$, we get

$$\lim_{\epsilon \to 0^+} H_\epsilon(y, 0)(\vartheta) = \lim_{\epsilon \to 0^+} \epsilon G(y, 0)\left(\frac{\vartheta}{\epsilon}\right) = 0,$$

where we recall that $G(y, 0)(\vartheta) = 0$ for every $\vartheta \geqslant 0$ and $y \in B$. Thus, defining $H_0(y, t) = tG_0(y)$, for $y \in B$ and $t \geqslant 0$, we obtain

$$\lim_{\epsilon \to 0^+} H_\epsilon(y, t) = H_0(y, t).$$

By (10.31), $H_0 \in \mathcal{F}(\Omega, h_0)$, where $h_0(t) = Ct$, $t \geqslant 0$. Furthermore, from the definition of h_ϵ, we have, for $0 \leqslant t_1 < t_2 < \infty$,

$$h_\epsilon(t_2) - h_\epsilon(t_1) = \epsilon\left[h\left(\frac{t_2}{\epsilon}\right) - h\left(\frac{t_1}{\epsilon}\right)\right]$$
$$= (t_2 - t_1)\frac{\epsilon}{t_2 - t_1}\left[h\left(\frac{t_2 - t_1}{\epsilon} + \frac{t_1}{\epsilon}\right) - h\left(\frac{t_1}{\epsilon}\right)\right]$$

and, by condition (10.26), we have

$$\limsup_{\epsilon \to 0^+}[h_\epsilon(t_2) - h_\epsilon(t_1)] \leqslant C(t_2 - t_1) = h_0(t_2) - h_0(t_1),$$

since

$$\lim_{\epsilon \to 0^+} \frac{t_2 - t_1}{\epsilon} = \infty.$$

Once $y \in \mathcal{O}$ is a solution of the autonomous ODE (10.28) and using the properties of the Kurzweil integral, we obtain

$$y(s_2) - y(s_1) = \int_{s_1}^{s_2} G_0(y(\tau))d\tau = \int_{s_1}^{s_2} D[G_0(y(\tau))t] = \int_{s_1}^{s_2} DH_0(y(\tau), t),$$

for every $s_1, s_2 \in [0, \infty)$. Thus, y is a solution of the generalized ODE:

$$\frac{dy}{d\tau} = DH_0(y, t)$$

such that $y(0) = \tilde{y}$ on $[0, \infty)$ and, by the hypotheses and Theorem 5.1, this solution is uniquely determined.

Thus, all hypotheses of Theorem 7.8 are fulfilled. Therefore, by Theorem 7.8, for every $\mu > 0$ and every $L > 0$, there is a $\epsilon_0 > 0$ such that for $\epsilon \in (0, \epsilon_0)$ and there exists a solution y_ϵ on the interval $[0, L]$ of the generalized ODE:

$$\frac{dx}{d\tau} = DH_\epsilon(x, t), \tag{10.32}$$

satisfying $y_\epsilon(0) = y(0)$ and

$$\|y_\epsilon(s) - y(s)\| \leqslant \mu, \quad \text{for every } s \in [0, L] \subset [0, \infty),$$

where y is solution of the averaged equation (10.28). Then, proceeding as in the previous theorem, we obtain the desired result. □

A direct consequence of Theorem 10.7 follows next.

Corollary 10.8: *Consider a function $G:\Omega \to X$ such that $G \in \mathcal{F}(\Omega, h)$, where $\Omega = B \times [0, \infty)$, $B = \{x \in X : \|x\| < c\}$, with $c > 0$, and $h:[0, \infty) \to \mathbb{R}$ is a left-continuous and nondecreasing function. Assume that, for every $\alpha \geqslant 0$,*

$$\limsup_{\epsilon \to 0^+} \frac{h\left(\frac{t}{\epsilon} + \alpha\right) - h(\alpha)}{\frac{t}{\epsilon}} \leqslant C, \quad \text{where } C > 0 \text{ is a constant,} \tag{10.33}$$

and, for every $x \in B$, we have

$$\lim_{\epsilon \to 0^+} \frac{G\left(x, \frac{t}{\epsilon}\right)\left(\frac{\vartheta}{\epsilon}\right)}{\frac{t}{\epsilon}} = G_0(x)(\vartheta), \quad \text{for all } \vartheta \in [0, \infty). \tag{10.34}$$

Suppose, in addition, for each $y \in B$, the function $G(y, 0):[0, \infty) \to X$ given by $\vartheta \mapsto G(y, t)(\vartheta)$, is such that $G(y, 0)(\vartheta) = 0$ for every $\vartheta \geqslant 0$ and $y \in B$. Let $y:[0, \infty) \to B$ be the unique solution of the autonomous ODE:

$$\begin{cases} \dot{y} = G_0(y)(\vartheta), \\ y(0) = \tilde{y} \end{cases} \tag{10.35}$$

and assume that there exists $\rho > 0$ such that

$$\{\tilde{x} \in X : \|\tilde{x} - y(t)\| < \rho \quad \text{for every} \quad t \in [0, \infty)\} \subset B.$$

Then, for every $\mu > 0$ and every $L > 0$, there exists $\epsilon_0 > 0$ such that for $\epsilon \in (0, \epsilon_0)$,

$$\|x_\epsilon(t) - z_\epsilon(t)\| < \mu, \quad \text{for every } t \in [0, L],$$

where $x_\epsilon:[0, L] \to X$ is a solution of the generalized ODE:

$$\begin{cases} \frac{dx}{d\tau} = \epsilon D\left[G\left(x, \frac{t}{\epsilon}\right)\left(\frac{\vartheta}{\epsilon}\right)\right], \\ x_\epsilon(0) = y(0), \end{cases} \tag{10.36}$$

on $[0, L]$ *such that* $x_\epsilon(0) = y(0)$ *and* $z_\epsilon : [0, L] \to X$ *is a solution of the autonomous ODE:*

$$\begin{cases} \dot{z} = G_0(z)(\vartheta), \\ z_\epsilon(0) = y(0) \end{cases} \tag{10.37}$$

on $[0, L]$.

We end this chapter by calling the reader's attention to the fact that, in Subsection 3.4.2, there are nonperiodic averaging principles specified for functional MDEs.

11

Boundedness of Solutions

Suzete M. Afonso[1], Fernanda Andrade da Silva[2], Everaldo M. Bonotto[3], Márcia Federson[4], Rogelio Grau[5], Jaqueline G. Mesquita[6], and Eduard Toon[7]

[1]Departamento de Matemática, Instituto de Geociências e Ciências Exatas, Universidade Estadual Paulista "Júlio de Mesquita Filho" (UNESP), Rio Claro, SP, Brazil
[2]Departamento de Matemática, Instituto de Ciências Matemáticas e de Computação (ICMC), Universidade de São Paulo, São Carlos, SP, Brazil
[3]Departamento de Matemática Aplicada e Estatística, Instituto de Ciências Matemáticas e de Computação (ICMC), Universidade de São Paulo, São Carlos, SP, Brazil
[4]Departamento de Matemática, Instituto de Ciências Matemáticas e de Computação (ICMC), Universidade de São Paulo, São Carlos, SP, Brazil
[5]Departamento de Matemáticas y Estadística, División de Ciencias Básicas, Universidad del Norte, Barranquilla, Colombia
[6]Departamento de Matemática, Instituto de Ciências Exatas, Universidade de Brasília, Brasília, DF, Brazil
[7]Departamento de Matemática, Instituto de Ciências Exatas, Universidade Federal de Juiz de Fora, Juiz de Fora, MG, Brazil

Concepts of boundedness of solutions in the setting of generalized ODEs were introduced in [2] and, since then, the theory has been explored occasionally. We can mention, for instance, [79]. The concepts we bring up here were inspired by the definitions of boundedness of solutions in the framework of impulsive functional differential equations (we write simply impulsive FDEs) explored by X. Fu and L. Zhang in [98], Z. Luo and J. Shen in [169], and I. Stamova in [223].

In the paper [223], I. Stamova proved several criteria, via Lyapunov's Direct Method, for the boundedness of solutions of a class of FDEs undergoing variable impulse perturbations. Boundedness results for nonautonomous generalized ODEs and the correspondence between the former and FDEs with variable impulses that allowed the authors of [2] to obtain similar criteria assuming weaker conditions.

In [79], the authors introduced new concepts of boundedness of solutions in the setting of measure differential equations (we write MDEs) and, motivated by the work [2], they provided results on boundedness of solutions for MDEs using the correspondence between the solutions of a class of these

Generalized Ordinary Differential Equations in Abstract Spaces and Applications, First Edition.
Edited by Everaldo M. Bonotto, Márcia Federson, and Jaqueline G. Mesquita.

equations and the solutions of a class of generalized ODEs. By virtue of this correspondence, the results obtained in [79] are more general than those found in the literature. The reader may want to consult [2, 33, 69] for instance. The authors of [79] also extended their results to dynamic equations on time scales, using the fact that the latter can be regarded as MDEs (see [219]).

Throughout this chapter, we consider that X is a Banach space with norm $\|\cdot\|$, $t_0 \geqslant 0$, and $\Omega = X \times [t_0, \infty)$. Let $F : \Omega \to X$ be a function defined for every $(x, t) \in \Omega$ and taking values in the Banach space X. We consider F as an element of the class $\mathcal{F}(\Omega, h)$ (see Definition 4.3), where $h : [t_0, \infty) \to \mathbb{R}$ is a left-continuous and nondecreasing function on (t_0, ∞), and the nonautonomous generalized ODE:

$$\frac{dx}{d\tau} = DF(x, t) \tag{11.1}$$

subject to the initial condition:

$$x(s_0) = x_0, \tag{11.2}$$

where $(x_0, s_0) \in \Omega$.

Anchored by Corollary 5.16, we may assume that, for every $(x_0, s_0) \in \Omega$, there exists a unique global forward solution $x : [s_0, \infty) \to X$ of the initial value problem (11.1)–(11.2). Then, for every $(x_0, s_0) \in \Omega$, we denote by $x(\cdot, s_0, x_0)$ the unique (global forward) solution of the generalized ODE (11.1) satisfying $x(s_0) = x_0$.

This chapter is organized as follows. In Section 11.1, we recall the concepts of uniform boundedness, quasiuniform boundedness, and uniform ultimate boundedness in the scenery of generalized ODEs. Moreover, criteria of uniform boundedness and uniform ultimate boundedness for the generalized ODE (11.1) are also included.

In Section 11.2, by using the results established in Section 11.1 and the correspondence between generalized ODEs and MDEs, we exhibit results on boundedness for MDEs. We conclude this chapter by showing the extension of one of these results to a certain impulsive differential equation (IDE) in Subsection 11.2.1.

11.1 Bounded Solutions and Lyapunov Functionals

In this section, we present some results concerning the boundedness of solutions for the generalized ODE (11.1) using Lyapunov functionals.

Next, we remind the concept of Lyapunov functional with respect to the generalized ODE (11.1) presented in Chapter 8, for the reader's convenience.

A function $V : [t_0, \infty) \times X \to \mathbb{R}$ is said to be a Lyapunov functional with respect to the generalized ODE (11.1), whenever the following conditions are fulfilled:

(LF1) $V(\cdot, x) : [t_0, \infty) \to \mathbb{R}$ is left-continuous on (t_0, ∞) for every $x \in X$;

(LF2) there exists a continuous strictly increasing function $b:\mathbb{R}_+ \to \mathbb{R}_+$ satisfying $b(0) = 0$ such that, for all $t \in [t_0, \infty)$ and $x \in X$, we have

$$V(t,x) \geqslant b(\|x\|);$$

(LF3) for every solution $x:[\gamma, v) \to X$ of the generalized ODE (11.1), with $[\gamma, v) \subseteq [t_0, \infty)$, we have

$$D^+V(t,x(t)) = \limsup_{\eta \to 0^+} \frac{V(t+\eta, x(t+\eta)) - V(t,x(t))}{\eta} \leqslant 0, \quad t \in [\gamma, v),$$

that is, the right derivative of V along every solution of the generalized ODE (11.1) is nonpositive.

The concepts of uniform boundedness for generalized ODEs introduced in [2] are presented in the sequel. Recall that by $x(\cdot, s_0, x_0)$, we mean the solution of the initial value problem (11.1)–(11.2).

Definition 11.1: The generalized ODE (11.1) is said to be

(i) *uniformly bounded*, if for every $\alpha > 0$, there exists $M = M(\alpha) > 0$ such that, for all $s_0 \in [t_0, \infty)$ and all $x_0 \in X$, with $\|x_0\| < \alpha$, we have

$$\|x(s, s_0, x_0)\| < M, \quad \text{for all } s \geqslant s_0;$$

(ii) *quasiuniformly ultimately bounded*, if there exists $B > 0$ such that for every $\alpha > 0$, there exists $T = T(\alpha) > 0$, such that for all $s_0 \in [t_0, \infty)$ and all $x_0 \in X$, with $\|x_0\| < \alpha$, we have

$$\|x(s, s_0, x_0)\| < B, \quad \text{for all } s \geqslant s_0 + T;$$

(iii) *uniformly ultimately bounded*, if it is uniformly bounded and quasiuniformly ultimately bounded.

The next auxiliary result was proved in [79, Lemma 3.4], and it is essential to derive the main results of this section.

Lemma 11.2: *Consider the generalized ODE (11.1), with $F \in \mathcal{F}(\Omega, h)$, and assume that the functional $V:[t_0, \infty) \times X \to \mathbb{R}$ satisfies the following conditions:*

(V1) *for each left-continuous function $z:[\gamma, v] \to X$ on $(\gamma, v]$, the function $[\gamma, v] \ni t \mapsto V(t, z(t)) \in \mathbb{R}_+$ is also left-continuous on $(\gamma, v]$;*

(V2) *for all functions $x, y:[\gamma, v] \to X$, $[\gamma, v] \subset [t_0, \infty)$, of bounded variation in $[\gamma, v]$, the following condition:*

$$\begin{aligned}&|V(t,x(t)) - V(t,y(t)) - V(s,x(s)) + V(s,y(s))| \\ &\leqslant \left(h_1(t) - h_1(s)\right) \sup_{\xi \in [\gamma, v]} \|x(\xi) - y(\xi)\|\end{aligned}$$

holds for all $\gamma \leqslant s < t \leqslant v$, where $h_1:[t_0, \infty) \to \mathbb{R}$ is a nondecreasing and left-continuous function;

(V3) *there exists a function $\Phi: X \to \mathbb{R}$ such that for all solution $z:[s_0, \infty) \to X$ of (11.1), with $s_0 \geqslant t_0$, and for all $s_0 \leqslant t < s < \infty$, we have*

$$V(s, z(s)) - V(t, z(t)) \leqslant (s - t)\Phi(z(t)).$$

If $\overline{x}: [\gamma, v] \to X$, $t_0 \leqslant \gamma < v < \infty$, is left-continuous on $(\gamma, v]$ and of bounded variation in $[\gamma, v]$, then

$$V(v, \overline{x}(v)) - V(\gamma, \overline{x}(\gamma))$$
$$\leqslant (h_1(v) - h_1(\gamma)) \sup_{s\in[\gamma,v]} \left\| \overline{x}(s) - \overline{x}(\gamma) - \int_\gamma^s DF(\overline{x}(\tau), t) \right\| + (v - \gamma)K, \quad (11.3)$$

where $K = \sup\left\{\Phi(\overline{x}(t)) : t \in [\gamma, v]\right\}$.

Proof. Let $[\gamma, v] \subset [t_0, \infty)$ and $\overline{x}:[\gamma, v] \to X$ be a function which is left-continuous on $(\gamma, v]$ and of bounded variation in $[\gamma, v] \subset [t_0, \infty)$. Corollary 4.8 guarantees the existence of the Kurzweil integral $\int_\gamma^v DF(\overline{x}(\tau), t)$.

Set

$$K = \sup\left\{\Phi(\overline{x}(t)) : t \in [\gamma, v]\right\}.$$

If $K = \infty$, then it is clear that inequality (11.3) is satisfied and the result follows. Therefore, we assume that $K < \infty$.

Given $\sigma \in [\gamma, v]$, the existence and uniqueness of a global forward solution $x:[\sigma, \infty) \to X$ of the generalized ODE (11.1) on $[\sigma, \infty)$ satisfying the initial condition $x(\sigma) = \overline{x}(\sigma)$ is ensured by Corollary 5.16, since $(\overline{x}(\sigma), \sigma) \in \Omega = X \times [t_0, \infty)$. For every $\eta_1 > 0$, $x|_{[\sigma,\sigma+\eta_1]}$ is also a solution of the generalized ODE (11.1). Thus, Corollary 4.8 and Theorem 2.5 imply that the Kurzweil integral $\int_\sigma^{\sigma+\eta_1} DF(x(\tau), t)$ exists.

Let $\eta_2 > 0$ be such that $\eta_2 \leqslant \eta_1$ and $\sigma + \eta_2 \leqslant v$. Then, the Kurzweil integrals

$$\int_\sigma^{\sigma+\eta_2} DF(\overline{x}(\tau), t) \quad \text{and} \quad \int_\sigma^{\sigma+\eta_2} DF(x(\tau), t)$$

also exist by the integrability on subintervals of the Kurzweil integral (see Theorem 2.5). Consequently, the integral

$$\int_\sigma^{\sigma+\eta_2} D[F(\overline{x}(\tau), t) - F(x(\tau), t)] \quad (11.4)$$

exists as well. Therefore, for $\epsilon > 0$, there is a gauge δ of the interval $[\sigma, \sigma + \eta_2]$ corresponding to ϵ in the definition of the integral (11.4). Without loss of generality, we can consider $\eta_2 < \delta(\sigma)$.

Let $\Phi : X \to \mathbb{R}$ be the function from assumption (V3). Then,

$$V(s, x(s)) - V(t, x(t)) \leqslant (s - t)\Phi(x(t)),$$

for every $\sigma \leqslant t < s < \infty$ and, consequently, for every $0 < \eta < \eta_2$, we obtain

$$V(\sigma + \eta, x(\sigma + \eta)) - V(\sigma, x(\sigma)) \leqslant \eta\Phi(x(\sigma)). \tag{11.5}$$

Now, observe that Corollary 2.8 provides the following relations:

$$\left\| F(\overline{x}(\sigma), s) - F(\overline{x}(\sigma), \sigma) - \int_{\sigma}^{s} DF(\overline{x}(\tau), t) \right\| < \frac{\eta\epsilon}{2[h_1(\sigma + \eta) - h_1(\sigma) + 1]} \tag{11.6}$$

and

$$\left\| F(x(\sigma), s) - F(x(\sigma), \sigma) - \int_{\sigma}^{s} DF(x(\tau), t) \right\| < \frac{\eta\epsilon}{2[h_1(\sigma + \eta) - h_1(\sigma) + 1]}, \tag{11.7}$$

for every $s \in [\sigma, \sigma + \eta]$.

Note also that

$$\begin{aligned}
&\sup_{s\in[\sigma,\sigma+\eta]} \left\| \int_{\sigma}^{s} D[F(\overline{x}(\tau), t) - F(x(\tau), t)] \right\| \\
&\quad - \sup_{s\in[\sigma,\sigma+\eta]} \| F(\overline{x}(\sigma), s) - F(\overline{x}(\sigma), \sigma) - F(x(\sigma), s) + F(x(\sigma), \sigma) \| \\
&\leqslant \sup_{s\in[\sigma,\sigma+\eta]} \left\| \int_{\sigma}^{s} D[F(\overline{x}(\tau), t) - F(x(\tau), t)] \right. \\
&\quad \left. - [F(\overline{x}(\sigma), s) - F(\overline{x}(\sigma), \sigma) - F(x(\sigma), s) + F(x(\sigma), \sigma)] \right\| \\
&\leqslant \sup_{s\in[\sigma,\sigma+\eta]} \left\| F(\overline{x}(\sigma), s) - F(\overline{x}(\sigma), \sigma) - \int_{\sigma}^{s} DF(\overline{x}(\tau), t) \right\| \\
&\quad + \sup_{s\in[\sigma,\sigma+\eta]} \left\| F(x(\sigma), s) - F(x(\sigma), \sigma) - \int_{\sigma}^{s} DF(x(\tau), t) \right\|.
\end{aligned} \tag{11.8}$$

In addition, since $F \in \mathcal{F}(\Omega, h)$ and $x(\sigma) = \overline{x}(\sigma)$, we have

$$\begin{aligned}
&\sup_{s\in[\sigma,\sigma+\eta]} \| F(\overline{x}(\sigma), s) - F(\overline{x}(\sigma), \sigma) - F(x(\sigma), s) + F(x(\sigma), \sigma) \| \\
&\quad \leqslant \| \overline{x}(\sigma) - x(\sigma) \| \sup_{s\in[\sigma,\sigma+\eta]} |h(s) - h(\sigma)| = 0,
\end{aligned} \tag{11.9}$$

where the last inequality follows from condition (4.6) of Definition 4.3. Therefore, (11.6), (11.7), (11.8), and (11.9) yield

$$\sup_{s\in[\sigma,\sigma+\eta]} \left\| \int_{\sigma}^{s} D[F(\overline{x}(\tau), t) - F(x(\tau), t)] \right\| \leqslant \frac{\eta\epsilon}{[h_1(\sigma + \eta) - h_1(\sigma) + 1]}. \tag{11.10}$$

Note that x is of bounded variation in $[\gamma, \upsilon]$. This fact is a consequence of Lemma 4.9, since $F \in \mathcal{F}(\Omega, h)$ and the function h is nondecreasing. Consequently, x is of bounded variation in $[\sigma, \sigma+\eta] \subset [\gamma, \upsilon]$. Thus, by assumption (V2) and by the relation $\overline{x}(\sigma) = x(\sigma)$, we obtain

$$\begin{aligned}
&V(\sigma+\eta, \overline{x}(\sigma+\eta)) - V(\sigma+\eta, x(\sigma+\eta))\\
&\quad = V(\sigma+\eta, \overline{x}(\sigma+\eta)) - V(\sigma+\eta, x(\sigma+\eta)) - V(\sigma, \overline{x}(\sigma)) + V(\sigma, x(\sigma))\\
&\quad \leqslant \left|V(\sigma+\eta, \overline{x}(\sigma+\eta)) - V(\sigma+\eta, x(\sigma+\eta)) - V(\sigma, \overline{x}(\sigma)) + V(\sigma, x(\sigma))\right|\\
&\quad \leqslant (h_1(\sigma+\eta) - h_1(\sigma)) \sup_{s\in[\sigma,\sigma+\eta]} \|\overline{x}(s) - x(s)\|\\
&\quad = (h_1(\sigma+\eta) - h_1(\sigma)) \sup_{s\in[\sigma,\sigma+\eta]} \|\overline{x}(s) - \overline{x}(\sigma) + x(\sigma) - x(s)\|\\
&\quad = (h_1(\sigma+\eta) - h_1(\sigma)) \sup_{s\in[\sigma,\sigma+\eta]} \left\|\overline{x}(s) - \overline{x}(\sigma) - \int_\sigma^s DF(x(\tau), t)\right\|,
\end{aligned}$$

which means that

$$\begin{aligned}
&V(\sigma+\eta, \overline{x}(\sigma+\eta)) - V(\sigma+\eta, x(\sigma+\eta))\\
&\quad \leqslant (h_1(\sigma+\eta) - h_1(\sigma)) \sup_{s\in[\sigma,\sigma+\eta]} \left\|\overline{x}(s) - \overline{x}(\sigma) - \int_\sigma^s DF(x(\tau), t)\right\|. \qquad (11.11)
\end{aligned}$$

Then, by (11.5), (11.10), and (11.11), we get

$$\begin{aligned}
&V(\sigma+\eta, \overline{x}(\sigma+\eta)) - V(\sigma, \overline{x}(\sigma))\\
&\quad = V(\sigma+\eta, \overline{x}(\sigma+\eta)) - V(\sigma+\eta, x(\sigma+\eta)) + V(\sigma+\eta, x(\sigma+\eta))\\
&\qquad - V(\sigma, x(\sigma))\\
&\quad \leqslant (h_1(\sigma+\eta) - h_1(\sigma)) \sup_{s\in[\sigma,\sigma+\eta]} \left\|\overline{x}(s) - \overline{x}(\sigma) - \int_\sigma^s DF(x(\tau), t)\right\| + \eta\Phi(x(\sigma))\\
&\quad \leqslant (h_1(\sigma+\eta) - h_1(\sigma)) \sup_{s\in[\sigma,\sigma+\eta]} \left\|\overline{x}(s) - \overline{x}(\sigma) - \int_\sigma^s DF(x(\tau), t)\right\| + \eta K\\
&\quad \leqslant (h_1(\sigma+\eta) - h_1(\sigma)) \sup_{s\in[\sigma,\sigma+\eta]} \left\|\overline{x}(s) - \overline{x}(\sigma) - \int_\sigma^s DF(\overline{x}(\tau), t)\right\|\\
&\qquad + (h_1(\sigma+\eta) - h_1(\sigma)) \sup_{s\in[\sigma,\sigma+\eta]} \left\|\int_\sigma^s D[F(\overline{x}(\tau), t) - F(x(\tau), t)]\right\| + \eta K\\
&\quad \leqslant (h_1(\sigma+\eta) - h_1(\sigma)) \sup_{s\in[\sigma,\sigma+\eta]} \left\|\overline{x}(s) - \overline{x}(\sigma) - \int_\sigma^s DF(\overline{x}(\tau), t)\right\| + \eta\epsilon + \eta K.
\end{aligned} \qquad (11.12)$$

Let us define $\Gamma : [\gamma, \upsilon] \to X$ by

$$\Gamma(s) = \overline{x}(s) - \int_\gamma^s DF(\overline{x}(\tau), t), \quad \text{for } s \in [\gamma, \upsilon].$$

The Kurzweil integral $\int_\gamma^\upsilon DF(\overline{x}(\tau), t)$ exists and the function $s \mapsto \int_\gamma^s DF(\overline{x}(\tau), t)$ is of bounded variation in $[\gamma, \upsilon]$ by Corollary 4.8, since $\overline{x}$ is a function of bounded variation in $[\gamma, v]$ and $(\overline{x}(s), s) \in \Omega$, for every $s \in [\gamma, v]$. Consequently, for each $s \in [\gamma, v]$, the existence of the Kurzweil integral $\int_\gamma^s DF(\overline{x}(\tau), t)$ is guaranteed by Theorem 2.5. Therefore, the function Γ is well defined and is of bounded variation in $[\gamma, v]$. Furthermore, Γ is left-continuous on $(\gamma, v]$, since $\overline{x}$ and h are left-continuous on $(\gamma, v]$ (see Lemma 4.5). Note also that, for $s \in [\gamma, v]$, we have

$$
\begin{aligned}
\Gamma(s) - \Gamma(\sigma) &= \overline{x}(s) - \overline{x}(\sigma) - \int_\gamma^s DF(\overline{x}(\tau), t) + \int_\gamma^\sigma DF(\overline{x}(\tau), t) \\
&= \overline{x}(s) - \overline{x}(\sigma) - \int_\sigma^s DF(\overline{x}(\tau), t).
\end{aligned} \tag{11.13}
$$

Now, if $f : [\gamma, v] \to \mathbb{R}$ is the function given by

$$
f(t) = \begin{cases} (h_1(t) - h_1(\sigma)) \sup\limits_{s \in [\gamma, t]} \|\Gamma(s) - \Gamma(\sigma)\| + \epsilon t + Kt, & t \in [\gamma, \sigma], \\ (h_1(t) - h_1(\sigma)) \sup\limits_{s \in [\sigma, t]} \|\Gamma(s) - \Gamma(\sigma)\| + \epsilon t + Kt, & t \in [\sigma, v], \end{cases}
$$

then f is well defined, and it is left-continuous on $(\gamma, v]$, since h_1 and Γ are left-continuous on $(\gamma, v]$. Moreover, the left-continuity of $\overline{x} : [\gamma, v] \to X$ together with assumption (V1) imply that the function $[\gamma, v] \ni t \mapsto V(t, \overline{x}(t)) \in \mathbb{R}_+$ is left-continuous on $(\gamma, v]$ as well.

By (11.12) and (11.13), we have

$$
\begin{aligned}
&V(\sigma + \eta, \overline{x}(\sigma + \eta)) - V(\sigma, \overline{x}(\sigma)) \\
&\leqslant (h_1(\sigma + \eta) - h_1(\sigma)) \sup_{s \in [\sigma, \sigma + \eta]} \|\Gamma(s) - \Gamma(\sigma)\| + \eta\epsilon + \eta K \\
&= f(\sigma + \eta) - f(\sigma).
\end{aligned}
$$

Since the functions $[\gamma, v] \ni t \mapsto V(t, \overline{x}(t))$ and $[\gamma, v] \ni t \mapsto f(t)$ fulfill all hypotheses of Proposition 1.8, we derive

$$
\begin{aligned}
&V(v, \overline{x}(v)) - V(\gamma, \overline{x}(\gamma)) \leqslant f(v) - f(\gamma) \\
&= (h_1(v) - h_1(\sigma)) \sup_{s \in [\sigma, v]} \|\Gamma(s) - \Gamma(\sigma)\| + \epsilon v + Kv \\
&\quad - (h_1(\gamma) - h_1(\sigma)) \sup_{s \in [\gamma, \gamma]} \|\Gamma(s) - \Gamma(\sigma)\| - \epsilon\gamma - K\gamma \\
&\leqslant (h_1(v) - h_1(\sigma)) \sup_{s \in [\gamma, v]} \|\Gamma(s) - \Gamma(\sigma)\| + \epsilon v + Kv \\
&\quad + (h_1(\sigma) - h_1(\gamma)) \sup_{s \in [\gamma, v]} \|\Gamma(s) - \Gamma(\sigma)\| - \epsilon\gamma - K\gamma \\
&= (h_1(v) - h_1(\gamma)) \sup_{s \in [\gamma, v]} \|\Gamma(s) - \Gamma(\sigma)\| + \epsilon(v - \gamma) + K(v - \gamma) \\
&= (h_1(v) - h_1(\gamma)) \sup_{s \in [\gamma, v]} \left\| \overline{x}(s) - \overline{x}(\sigma) - \int_\sigma^s DF(\overline{x}(\tau), t) \right\| \\
&\quad + K(v - \gamma) + \epsilon(v - \gamma),
\end{aligned}
$$

whence we conclude that inequality (11.3) holds, since $\epsilon > 0$ is arbitrary. □

The next theorem, established in [79, Theorem 3.5], provides us sufficient conditions which guarantee that the generalized ODE (11.1) is uniformly bounded. This result does not require the Lipschitz condition on the Lyapunov functional, which makes it more general than [2, Theorem 4.3].

Theorem 11.3: *Consider the generalized ODE (11.1), with $F \in \mathcal{F}(\Omega, h)$, and $V\colon \left[t_0, \infty\right) \times X \to \mathbb{R}$ a Lyapunov functional such that, for each left-continuous function $z\colon [\gamma, v] \subset [t_0, \infty) \to X$ on $(\gamma, v]$, the function $[\gamma, v] \ni t \mapsto V(t, z(t))$ is left-continuous on $(\gamma, v]$. Moreover, suppose V satisfies the following conditions:*

(UB1) *the function $b\colon \mathbb{R}_+ \to \mathbb{R}_+$ of condition (LF2) from the definition of Lyapunov functional is such that $b(s) \to \infty$ as $s \to \infty$;*

(UB2) *there exists a monotone increasing function $p\colon \mathbb{R}_+ \to \mathbb{R}_+$ such that $p(0) = 0$ and, for every pair $(t, z) \in [t_0, \infty) \times X$,*

$$V(t, z) \leqslant p(\|z\|); \tag{11.14}$$

(UB3) *for every global forward solution $z\colon [s_0, \infty) \to X$, $s_0 \geqslant t_0$, of the generalized ODE (11.1), we have, for every $s_0 \leqslant t < s < \infty$,*

$$V(s, z(s)) - V(t, z(t)) \leqslant 0.$$

Under these conditions, the generalized ODE (11.1) is uniformly bounded.

Proof. Let $\alpha > 0$. Since $p(\alpha) > 0$ and $b(s) \to \infty$ as $s \to \infty$, by assumption (UB1), there exists $M = M(\alpha) > 0$ such that $p(\alpha) < b(s)$, for every $s \geqslant M$. Therefore, for $s = M$, we have

$$p(\alpha) < b(M). \tag{11.15}$$

Given $s_0 \in [t_0, \infty)$ and $x_0 \in X$, denote by $x(\cdot) = x(\cdot, s_0, x_0)\colon [s_0, \infty) \to X$ the global forward solution of the generalized ODE (11.1) satisfying the initial condition $x(s_0) = x_0$, with $\|x_0\| < \alpha$.

Claim 1. For all $s \geqslant s_0$, we have

$$V(s, x(s)) < b(M). \tag{11.16}$$

In fact, using assumption (UB3), conditions (11.14) and (11.15), we obtain

$$V(s, x(s)) \leqslant V(s_0, x(s_0)) = V(s_0, x_0) \leqslant p(\|x_0\|) < p(\alpha) < b(M) \tag{11.17}$$

for every $s \geqslant s_0$.

Claim 2. For every $s \geqslant s_0$, we have

$$\|x(s)\| < M.$$

Indeed, otherwise, there would be $\bar{s} \in [s_0, \infty)$ such that $\|x(\bar{s})\| \geqslant M$. This and condition (LF2) from the definition of Lyapunov functional would imply

$$V(\bar{s}, x(\bar{s})) \geqslant b(\|x(\bar{s})\|) \geqslant b(M), \tag{11.18}$$

since b is an increasing function. Clearly, relation (11.18) contradicts (11.16) and this contradiction implies $\|x(s)\| < M$, for every $s \geqslant s_0$, which proves *Claim 2* and completes the proof. □

The next result, which was also proved in [79, Theorem 3.6], gives us sufficient conditions to ensure that the generalized ODE (11.1) is uniformly ultimately bounded.

Theorem 11.4: *Consider the generalized ODE* (11.1)*, with* $F \in \mathcal{F}(\Omega, h)$*, and* $V: [t_0, \infty) \times X \to \mathbb{R}$ *a Lyapunov functional. Assume that* V *satisfies assumptions* (V1) *and* (V2) *from Lemma 11.2, and also hypotheses* (UB1) *and* (UB2) *from Theorem 11.3. Moreover, assume that* V *satisfies the following condition:*

($\tilde{V}3$) *there exists a continuous function* $\Phi: X \to \mathbb{R}_+$*, with* $\Phi(0) = 0$ *and* $\Phi(x) > 0$ *whenever* $x \neq 0$*, such that, for every global forward solution* $z: [s_0, \infty) \to X$ *of* (11.1)*, with* $s_0 \geqslant t_0$*, we have*

$$V(s, z(s)) - V(t, z(t)) \leqslant (s - t)(-\Phi(z(t))), \quad s_0 \leqslant t < s < \infty.$$

Then, the generalized ODE (11.1) *is uniformly ultimately bounded.*

Proof. First of all, note that assumption ($\tilde{V}3$) implies

$$V(s, z(s)) - V(t, z(t)) \leqslant (s - t)(-\Phi(z(t))) \leqslant 0,$$

for every $s_0 \leqslant t < s < \infty$ and every global forward solution $z: [s_0, \infty) \to X$, with $s_0 \in [t_0, \infty)$, of the generalized ODE (11.1). Thus, since all hypotheses of Theorem 11.3 are satisfied, the generalized ODE (11.1) is uniformly bounded. In this way, it remains to prove that the generalized ODE (11.1) is quasiuniformly ultimately bounded.

Given $\alpha > 0$, $s_0 \in [t_0, \infty)$, and $x_0 \in X$, let $x(\cdot) = x(\cdot, s_0, x_0): [s_0, \infty) \to X$ be the solution of the generalized ODE (11.1) satisfying the initial condition $x(s_0) = x_0$, with $\|x_0\| < \alpha$. We can claim that there exists (by Definition 11.1 (i)) a positive number $M_1 = M_1(\alpha)$ such that

$$\|x(s, s_0, x_0)\| < M_1, \quad \text{for every } s \geqslant s_0,$$

since the generalized ODE (11.1) is uniformly bounded.

In addition, we can affirm that

($\tilde{A}1$) there exists $M_2 = M_2(\alpha) > 0$ such that $p(\alpha) < b(M_2)$,

using the same argument as in (11.15) from the proof of Theorem 11.3.

Let $[\bar{t}, \infty) \subset [t_0, \infty)$ and define $y: [\bar{t}, \infty) \to X$ by $y(s) = x(s, s_0, x_0)$ for every $s \in [\bar{t}, \infty)$. Since the generalized ODE (11.1) is uniformly bounded, if $\left\|y(\bar{t})\right\| < \mu$, where $\mu > 0$, then there exists $B = B(\mu) > 0$ such that

$$\|y(s)\| < B, \quad \text{for every } s \geqslant \bar{t}. \tag{11.19}$$

Without loss of generality, we can take $B \in (\mu, \infty)$ (otherwise, we could take $B > B'$ such that $B' \in (\mu, \infty)$).

Consider α and B as above and set

$$M = M(\alpha) = \max\left\{M_1(\alpha), M_2(\alpha)\right\} \quad \text{and} \quad \lambda = \min\{\alpha, \mu\}.$$

Since $(\tilde{A}1)$ holds and b is increasing, we have

$$\left\|x(s, s_0, x_0)\right\| < M, \quad \text{for every } s \geqslant s_0, \quad \text{and} \quad p(\alpha) < b(M). \tag{11.20}$$

Define

$$N = \sup\{-\Phi(z)\colon \lambda \leqslant \|z\| < M\} < 0 \quad \text{and} \quad T(\alpha) = -\frac{2b(M)}{N} > 0.$$

If we show that $\left\|x(s, s_0, x_0)\right\| < B$ for every $s \geqslant s_0 + T(\alpha)$, then we complete the proof. But, before that, we state that

$(\tilde{A}2)$ There exists $t' \in \left[s_0 + T(\alpha)/2, s_0 + T(\alpha)\right]$ such that $\|x(t')\| < \lambda$.

In fact, assume that assertion $(\tilde{A}2)$ is false, that is,

$$\|x(s)\| \geqslant \lambda, \quad \text{for every } s \in \left[s_0 + T(\alpha)/2, s_0 + T(\alpha)\right]. \tag{11.21}$$

Let $I_\alpha = \left[s_0 + T(\alpha)/2, s_0 + T(\alpha)\right]$. Since the function $x(\cdot) = x(\cdot, s_0, x_0)$ is solution of the initial value problem (11.1)–(11.2), $x(\cdot, s_0, x_0)|_{I_\alpha}$ is left-continuous on $(s_0 + T(\alpha)/2, s_0 + T(\alpha)]$ and of bounded variation in I_α. Therefore, applying Lemma 11.2, we obtain

$$\begin{aligned} V(s_0 + T(\alpha), x(s_0 + T(\alpha))) &\leqslant V\left(s_0 + \frac{T(\alpha)}{2}, x\left(s_0 + \frac{T(\alpha)}{2}\right)\right) \\ &+ \left(h_1(s_0 + T(\alpha)) - h_1\left(s_0 + \frac{T(\alpha)}{2}\right)\right) \\ &\times \underbrace{\sup_{s \in I_\alpha}\left\|x(s) - x\left(s_0 + \frac{T(\alpha)}{2}\right) - \int_{s_0 + \frac{T(\alpha)}{2}}^{s} DF(x(\tau, s_0, z_0), t)\right\|}_{0} \\ &+ \frac{T(\alpha)}{2} \sup\left\{-\Phi(x(s)) \colon s \in I_\alpha\right\}, \end{aligned}$$

and, consequently,

$$\begin{aligned} V(s_0 + T(\alpha), x(s_0 + T(\alpha))) &\leqslant V\left(s_0 + \frac{T(\alpha)}{2}, x\left(s_0 + \frac{T(\alpha)}{2}\right)\right) \\ &\quad + \frac{T(\alpha)}{2} \sup\left\{-\Phi(x(s))\colon s \in I_\alpha\right\} \\ &\leqslant V\left(s_0 + \frac{T(\alpha)}{2}, x\left(s_0 + \frac{T(\alpha)}{2}\right)\right) + \frac{T(\alpha)}{2} \sup\{-\Phi(z)\colon \lambda \leqslant \|z\| < M\}. \end{aligned} \tag{11.22}$$

The last inequality in (11.22) follows from the relations:

$$\lambda \overset{(11.21)}{\leqslant} \|x(s)\| = \|x(s, s_0, z_0)\| \overset{(11.20)}{<} M,$$

which hold for every $s \in I_\alpha$. Moreover, using the same argument as in (11.17) from the proof of Theorem 11.3, we get

$$V\left(s_0 + \frac{T(\alpha)}{2}, x\left(s_0 + \frac{T(\alpha)}{2}\right)\right) < b(M), \tag{11.23}$$

as (11.20) holds.

Thus, using (11.22) and (11.23), we have

$$\begin{aligned} V(s_0 + T(\alpha), x(s_0 + T(\alpha))) &< b(M) + \frac{T(\alpha)}{2} \sup\{-\Phi(z)\colon \lambda \leqslant \|z\| < M\} \\ &= b(M) + \frac{T(\alpha)}{2} N = b(M) - \frac{2b(M)}{2N} N = 0, \end{aligned}$$

which yields

$$V(s_0 + T(\alpha), x(s_0 + T(\alpha))) < 0. \tag{11.24}$$

On the other hand, condition (LF2) from the definition of Lyapunov functional and assumption (11.21) imply

$$V(s_0 + T(\alpha), x(t_0 + T(\alpha))) \geqslant b(\|x(s_0 + T(\alpha))\|) \geqslant b(\lambda) > 0,$$

which contradicts (11.24). From this, we infer that the assertion ($\tilde{\text{A}}2$) holds. Thus,

$$\|x(s, s_0, x_0)\| < B \quad \text{for } s \geqslant t',$$

since (11.19) holds for $\bar{t} = t'$. Furthermore,

$$\|x(s, s_0, x_0)\| < B \quad \text{for } s > t_0 + T(\alpha),$$

once $t' \in I_\alpha$. Therefore, the generalized ODE (11.1) is quasiuniformly ultimately bounded and the proof is complete. □

11.2 An Application to MDEs

Using the boundedness results for generalized ODEs proved in the previous section (namely Theorems 11.3 and 11.4), the authors of [79] obtained boundedness results for MDEs with functions taking values in $\mathbb{R}^n$. In this section, we exhibit the same results for MDEs with functions taking values in an arbitrary Banach space, whose proofs are similar to the ones presented in [79].

Let $(X, \|\cdot\|)$ be a Banach space. Given $t_0 \geqslant 0$ and a function $f : X \times [t_0, \infty) \to X$, we consider the integral form of a MDE of type

$$x(t) = x(s_0) + \int_{s_0}^{t} f(x(s), s)dg(s), \quad t \geqslant s_0, \tag{11.25}$$

where $s_0 \geqslant t_0$ and $g : [t_0, \infty) \to \mathbb{R}$. We assume that the following conditions are fulfilled:

(D1) the function $g : [t_0, \infty) \to \mathbb{R}$ is nondecreasing and left-continuous on (t_0, ∞);
(D2) the Perron–Stieltjes integral $\int_{s_1}^{s_2} f(x(s), s)dg(s)$ exists, for all $x \in G([t_0, \infty), X)$ and all $s_1, s_2 \in [t_0, \infty)$;
(D3) there exists a locally Perron–Stieltjes integrable function $M : [t_0, \infty) \to \mathbb{R}$ with respect to g such that

$$\left\| \int_{s_1}^{s_2} f(x(s), s)dg(s) \right\| \leqslant \int_{s_1}^{s_2} M(s)dg(s)$$

for all $x \in G([t_0, \infty), X)$ and all $s_1, s_2 \in [t_0, \infty)$, with $s_1 \leqslant s_2$;
(D4) there exists a locally Perron–Stieltjes integrable function $L : [t_0, \infty) \to \mathbb{R}$ with respect to g such that

$$\left\| \int_{s_1}^{s_2} [f(x(s), s) - f(z(s), s)]dg(s) \right\| \leqslant \|x - z\|_{[t_0,\infty)} \int_{s_1}^{s_2} L(s)dg(s)$$

for all $x, z \in G_0([t_0, \infty), X)$ and all $s_1, s_2 \in [t_0, \infty)$, with $s_1 \leqslant s_2$.

We remind the reader that the Banach space $(G_0([t_0, \infty), X), \|\cdot\|_{[t_0,\infty)})$ was described in Chapter 1, and so does the vector space $G([t_0, \infty), X)$.

Theorem 5.18 assures that $x : I \to X$ is a solution of the MDE (11.25) on $I \subset [t_0, \infty)$ if and only if x is a solution of the generalized ODE

$$\frac{dx}{d\tau} = DF(x, t) \tag{11.26}$$

on I, where the function $F : X \times [t_0, \infty) \to X$ is given by

$$F(x, t) = \int_{s_0}^{t} f(x, s)dg(s) \tag{11.27}$$

for all $(x, t) \in X \times [t_0, \infty)$.

Recall that $x(\cdot) = x(\cdot, s_0, x_0)$ denotes the unique global forward solution $x\colon [s_0, \infty) \to X$ of the MDE (11.25) with $x(s_0) = x_0$ which we are assuming to exist (see Corollary 5.27).

The concepts of uniform boundedness for MDEs were introduced in [79, Definition 3.5] and are presented in the sequel.

Definition 11.5: We say that the MDE (11.25) is

(i) *uniformly bounded*, if for every $\alpha > 0$,, there exists $M = M(\alpha) > 0$ such that, for every $s_0 \in [t_0, \infty)$ and for all $x_0 \in X$, with $\|x_0\| < \alpha$, we have

$$\|x(s, s_0, x_0)\| < M, \quad \text{for all } s \geqslant s_0;$$

(ii) *quasiuniformly ultimately bounded*, if there exists $B > 0$ such that for every $\alpha > 0$, there exists $T = T(\alpha) > 0$, such that for all $s_0 \in [t_0, \infty)$ and all $x_0 \in X$, with $\|x_0\| < \alpha$, we have

$$\|x(s, s_0, x_0)\| < B, \quad \text{for all } s \geqslant s_0 + T;$$

(iii) *uniformly ultimately bounded*, if it is uniformly bounded and quasiuniformly ultimately bounded.

The concept of a Lyapunov functional with respect to the MDE (11.25) was presented in Section 8.3. Nonetheless, we recall it here for the sake of convenience.

Definition 11.6: A functional $U\colon [t_0, \infty) \times X \to \mathbb{R}$ is said to be a Lyapunov functional with respect to the MDE (11.25), if the following conditions are satisfied:

(LFM1) for all $x \in X$, the function $U(\cdot, x)\colon [t_0, \infty) \to \mathbb{R}$ is left-continuous on (t_0, ∞);

(LFM2) there exists a continuous increasing function $b\colon \mathbb{R}_+ \to \mathbb{R}_+$ satisfying $b(0) = 0$ such that

$$U(t, x) \geqslant b(\|x\|), \quad \text{for every } (t, x) \in [t_0, \infty) \times X;$$

(LFM3) for every solution $x\colon [\gamma, v) \subset [t_0, \infty) \to X$ of the MDE (11.25),

$$D^+U(t, x(t)) = \limsup_{\eta \to 0^+} \frac{U(t + \eta, x(t + \eta)) - U(t, x(t))}{\eta} \leqslant 0$$

holds for all $t \in [\gamma, v)$, that is, the right derivative of U along every solution of the MDE (11.25) is nonpositive.

The next theorem ensures us that the MDE (11.25) is uniformly bounded under certain conditions. The version of this result for MDEs with functions taking values in $\mathbb{R}^n$ was proved in [79, Theorem 4.6]. The proof we present here was borrowed from [79].

Theorem 11.7: *Assume that the function $f: X \times [t_0, \infty) \to X$ satisfies conditions (D2), (D3), (D4), and the function $g: [t_0, \infty) \to \mathbb{R}$ satisfies condition (D1). Let $U: [t_0, \infty) \times X \to \mathbb{R}$ be a Lyapunov functional such that for each left-continuous function $z: [\gamma, v] \subset [t_0, \infty) \to X$ on $(\gamma, v]$, the function $[\gamma, v] \ni t \mapsto U(t, z(t))$ is left-continuous on $(\gamma, v]$. Moreover, suppose U satisfies the following conditions:*

(UBM1) *the function $b: \mathbb{R}_+ \to \mathbb{R}_+$ of condition* (LFM2) *from the definition of Lyapunov functional is such that $b(s) \to \infty$ as $s \to \infty$;*

(UBM2) *there exists a monotone increasing function $p: \mathbb{R}_+ \to \mathbb{R}_+$ such that $p(0) = 0$ and, for every pair $(t, z) \in [t_0, \infty) \times X$,*

$$U(t, z) \leqslant p(\|z\|); \tag{11.28}$$

(UBM3) *for every solution $z: [s_0, \infty) \to X$, $s_0 \geqslant t_0$, of the MDE (11.25), we have*

$$U(s, z(s)) - U(t, z(t)) \leqslant 0,$$

for every $s_0 \leqslant t < s < \infty$.

Then, the MDE (11.25) is uniformly bounded.

Proof. Let us consider $F: X \times [t_0, \infty) \to X$ the function defined by (11.27). Since the function $f: X \times [t_0, \infty) \to X$ satisfies conditions (D2), (D3), (D4), and the function $g: [t_0, \infty) \to \mathbb{R}$ satisfies condition (D1), it follows by Lemma 5.17 that $F \in \mathcal{F}(\Omega, h)$, where $\Omega = X \times [t_0, \infty)$ and the function $h: [t_0, \infty) \to \mathbb{R}$ is given by

$$h(t) = \int_{t_0}^{t} [M(s) + L(s)] dg(s).$$

Since g is left-continuous on (t_0, ∞), h is also left-continuous on (t_0, ∞).

Using the correspondence between the solutions of the generalized ODE (11.26) and the solutions of the MDE (11.25) guaranteed by Theorem 5.18, we can easily verify that U satisfies condition (UB3) from Theorem 11.3. Moreover, conditions (UBM1) and (UBM2) imply that U also satisfies conditions (UB1) and (UB2) from Theorem 11.3. Therefore, as all hypotheses from Theorem 11.3 are satisfied, the generalized ODE (11.25) is uniformly bounded.

Finally, the correspondence between the solutions of the generalized ODE (11.26) and the solutions of the MDE (11.25) allows us to conclude that the MDE (11.25) is also uniformly bounded, obtaining the desired result. □

The correspondence between the solutions of the generalized ODE (11.26) and the solutions of the MDE (11.25) together with Theorem 11.4 enables us to obtain the next criterion to guarantee the uniform ultimate boundedness of the MDE

(11.25). An analogous version of this result for MDEs with functions taking values in $\mathbb{R}^n$ was proved in [79, Theorem 4.7].

Theorem 11.8: *Assume that* $f: X \times [t_0, \infty) \to X$ *satisfies conditions* (D2), (D3), (D4), *and the function* $g: [t_0, \infty) \to \mathbb{R}$ *satisfies condition* (D1). *Let* $U: [t_0, \infty) \times X \to \mathbb{R}$ *be a Lyapunov functional such that for each left-continuous function* $z: [\gamma, v] \subset [t_0, \infty) \to X$ *on* $(\gamma, v]$, *the function* $[\gamma, v] \ni t \mapsto U(t, z(t))$ *is left-continuous on* $(\gamma, v]$. *Moreover, suppose* U *satisfies conditions* (UBM1) *and* (UBM2) *from Theorem 11.7, besides of the following conditions:*

(UBM1*) *for every* $x, y: [\gamma, v] \to X$ *of bounded variation in* $[\gamma, v]$, *with* $[\gamma, v] \subset [t_0, \infty)$, *we have*

$$|U(t, x(t)) - U(t, y(t)) - U(s, x(s)) + U(s, y(s))| \leqslant \left(\int_s^t P(\tau) du(\tau) \right) \sup_{\xi \in [\gamma, v]} \|x(\xi) - y(\xi)\|,$$

for every $\gamma \leqslant s < t \leqslant v$, *where* $u: [t_0, \infty) \to \mathbb{R}$ *is a nondecreasing and left-continuous function and* $P: [t_0, \infty) \to \mathbb{R}$ *is a locally Perron–Stieltjes integrable function with respect to* u;

(UBM2*) *there exists a continuous function* $\phi: X \to \mathbb{R}_+$, *with* $\phi(0) = 0$ *and* $\phi(x) > 0$, *whenever* $x \neq 0$, *such that for every solution* $z: [s_0, \infty) \to X$, $s_0 \geqslant t_0$, *of the MDE* (11.25), *we have*

$$U(s, z(s)) - U(t, z(t)) \leqslant (s - t)(-\phi(z(t))),$$

for every $s_0 \leqslant t < s < \infty$.

Then, the MDE (11.25) *is uniformly ultimately bounded.*

The proof of Theorem 11.8 basically follows the same procedure as the proof of Theorem 11.7. More specifically, making use of the correspondence between a solution of the MDE (11.25) and a solution of the generalized ODE (11.26), it is possible to show that U satisfies the conditions of Theorem 11.4 (which allows us to conclude that the generalized ODE (11.26) is uniformly ultimately bounded) and that the uniform ultimate boundedness of the generalized ODE (11.26) implies the uniform ultimate boundedness of the MDE (11.25).

Remark 11.9: Using the fact, observed in [85, 219], that MDEs encompass dynamic equations on time scales, the authors of [79] extended their results concerning boundedness of solutions of MDEs [79, Theorems 4.6 and 4.7] for dynamic equations on time scales.

11.2.1 An Example

Let us consider $\mathbb{R}$ with the absolute-value norm $|\cdot|$ and the following impulsive differential equation (we write IDE for short)

$$\begin{cases} \dot{x} = -b(t)\zeta(x), & t \neq t_k,\ t \geqslant 0, \\ \Delta(x(t_k)) = x(t_k^+) - x(t_k) = I_k(x(t_k)), & k \in \mathbb{N}, \end{cases} \tag{11.29}$$

under the following conditions:

(E1) $b:\mathbb{R}_+ \to \mathbb{R}_+$ is a nonnegative function, $\zeta:\mathbb{R} \to \mathbb{R}$ is a function fulfilling $\zeta(0) = 0$ and $x\zeta(x) > 0$, whenever $x \neq 0$, and the Perron integral $\int_{\tau_1}^{\tau_2} b(s)\zeta(x(s))ds$ exists, for all $x \in G(\mathbb{R}_+, \mathbb{R})$ and all $\tau_1, \tau_2 \in \mathbb{R}_+$, with $\tau_1 < \tau_2$;

(E2) there exists a locally Perron integrable function $m:\mathbb{R}_+ \to \mathbb{R}_+$ such that for all $\tau_1, \tau_2 \in \mathbb{R}_+$, with $\tau_1 < \tau_2$,

$$\left| \int_{\tau_1}^{\tau_2} b(s)\zeta(x(s))ds \right| \leqslant \int_{\tau_1}^{\tau_2} m(s)ds,$$

for all $x \in G(\mathbb{R}_+, \mathbb{R})$;

(E3) there exists a locally Perron integrable function $\ell:\mathbb{R}_+ \to \mathbb{R}_+$ such that for all $\tau_1, \tau_2 \in \mathbb{R}_+$, with $\tau_1 < \tau_2$,

$$\left| \int_{\tau_1}^{\tau_2} [b(s)\zeta(x(s)) - b(s)\zeta(z(s))]ds \right| \leqslant \|x - z\|_{[0,\infty)} \int_{\tau_1}^{\tau_2} \ell(s)ds,$$

for all $x, z \in G_0(\mathbb{R}_+, \mathbb{R})$;

(E4) $0 < t_1 < t_2 < \cdots < t_k < \ldots$ and $\lim_{k\to\infty} t_k = \infty$;

(E5) for all $k \in \mathbb{N}$, the impulse operators $I_k:\mathbb{R} \to \mathbb{R}$ satisfy $I_k(0) = 0$ and $xI_k(x) < 0$ for all $x \neq 0$. Moreover, there are constants $K_1 > 0$ and $0 < K_2 < 1$ such that, for all $k \in \mathbb{N}$ and all $x, y \in \mathbb{R}$, we have

$$|I_k(x)| \leqslant K_1 \quad \text{and} \quad |I_k(x) - I_k(y)| \leqslant K_2|x - y|.$$

Given $s_0 \geqslant 0$, we say that a function $x:[s_0, \infty) \to \mathbb{R}$ is a solution of (11.29), if x is differentiable almost everywhere on each interval $[0, t_1] \cap [s_0, \infty)$ and $(t_k, t_{k+1}] \cap [s_0, \infty)$ for $k \in \mathbb{N}$, $x'(t) = -b(t)\zeta(x(t))$, for almost all $t \in [s_0, \infty)$ and $x(t_k^+) = x(t_k) + I_k(x(t_k))$ if $t_k \in [s_0, \infty)$. It is worthwhile to mention that if $s_0 = 0$, then $t_k \in (s_0, \infty)$ by virtue of (E4).

If $x:[s_0, \infty) \to \mathbb{R}$ is solution of (11.29), with $s_0 \geqslant 0$, then x satisfies the following integral equation:

$$x(t) = x(s_0) - \int_{s_0}^{t} b(s)\zeta(x(s))ds + \sum_{k=1}^{\infty} I_k(x(t_k))H_{t_k}(t),$$

for all $t \geqslant s_0$, where H_{t_k} denotes the left-continuous Heaviside function concentrated at t_k, that is,

$$H_{t_k}(t) = \begin{cases} 0, & \text{for } 0 \leqslant t \leqslant t_k, \\ 1, & \text{for } t > t_k, \end{cases}$$

(see Eq. (3.12)).

Let $f: \mathbb{R} \times \mathbb{R}_+ \to \mathbb{R}$ and $g: \mathbb{R}_+ \to \mathbb{R}$ be functions defined by

$$f(x,t) = -b(t)\zeta(x) \quad \text{and} \quad g(t) = t.$$

Consider, also, the functions

$$\tilde{f}(x,t) = \begin{cases} f(x,t), & \text{if } t \in \mathbb{R}_+ \setminus \{t_1, t_2, \dots\}, \\ I_k(x(t_k)), & \text{if } t = t_k, \quad k \in \mathbb{N}, \end{cases}$$

and

$$\tilde{g}(t) = \begin{cases} t, & \text{if } t \in [0, t_1], \\ t + k, & \text{if } t \in (t_k, t_{k+1}], \ k \in \mathbb{N}. \end{cases}$$

Anchored by [86, Theorem 3.1], we can affirm that the IDE (11.29) can be transformed into the integral form of a measure functional differential equations without impulses as follows:

$$x(t) = x(s_0) + \int_{s_0}^{t} \tilde{f}(x(s), s) d\tilde{g}(s), \quad \text{for all } t \geqslant s_0. \tag{11.30}$$

Hence, x is solution of the IDE (11.29) if and only if x is solution of the MDE (11.30). Since conditions (E1)–(E5) hold, it is possible to prove that conditions (D1)–(D4) are satisfied with f and g replaced by $\tilde{f}$ and $\tilde{g}$, respectively, see [86, Lemma 3.3].

Our goal is to show that the IDE (11.29) is uniformly bounded, which means that for every $\alpha > 0$, there exists $M = M(\alpha) > 0$ such that, for all $s_0 \in [t_0, \infty)$ and all $x_0 \in \mathbb{R}$, with $|x_0| < \alpha$, we have

$$|x(s, s_0, x_0)| < M, \quad \text{for all } s \geqslant s_0,$$

where $x(\cdot) = x(\cdot, s_0, x_0)$ denotes the unique global forward solution $x: [s_0, \infty) \to \mathbb{R}$ of the IDE (11.29) such that $x(s_0) = x_0$.

Define $U: \mathbb{R}_+ \times \mathbb{R} \to \mathbb{R}_+$ by $U(t, x) = |x|$. Let us verify that the function U satisfies the conditions of Theorem 11.7.

It is clear that, for all $x \in \mathbb{R}$, the function $U(\cdot, x): [t_0, \infty) \to \mathbb{R}$ is left-continuous on (t_0, ∞), whence it follows that U satisfies condition (LFM1). Moreover, for each left-continuous function $z: [\gamma, v] \subset [t_0, \infty) \to \mathbb{R}$ on $(\gamma, v]$, the function $[\gamma, v] \ni t \mapsto U(t, z(t))$ is left-continuous on $(\gamma, v]$.

If we consider functions $b: \mathbb{R}_+ \to \mathbb{R}_+$ and $p: \mathbb{R}_+ \to \mathbb{R}_+$ defined by $b(s) = \frac{s}{2}$ and $p(s) = s$, we have

$$b(|x|) \leqslant U(t, x) \leqslant p(|x|),$$

with $b(0) = 0 = p(0)$ and $b(s) \to \infty$ as $s \to \infty$. This shows that U satisfies conditions (LFM2), (UBM1), and (UBM2).

Define an auxiliary function $V: \mathbb{R} \to \mathbb{R}_+$ by $V(x) = |x|, x \in \mathbb{R}$. Let $x: [s_0, \infty) \to \mathbb{R}$ be a solution of (11.30), with $s_0 \geqslant 0$. We claim that

$$D^+V(x(t)) = \limsup_{\eta \to 0^+} \frac{V(x(t+\eta)) - V(x(t))}{\eta} \leqslant 0, \quad \text{for all } t \geqslant s_0.$$

At first, we suppose $t \neq t_k$ for all $k \in \mathbb{N}$. It is known that $x \equiv 0$ is the unique solution of (11.29) such that $x(t) = 0$. Then, we may assume that $x(t) \neq 0$ or $x(t) = 0$.

(i) If $x(t) \neq 0$, then

$$D^+V(x(t)) = \operatorname{sgn}(x(t))x'(t) = -\operatorname{sgn}(x(t))b(t)\zeta(x(t)) \leqslant 0,$$

since b is nonnegative and $x\zeta(x) > 0$ for $x \neq 0$. As usual, the symbol $\operatorname{sgn}(z)$ denotes the sign of z.

(ii) If $x(t) = 0$, then

$$D^+V(x(t)) = 0.$$

Now, suppose $t = t_k$ for some $k \in \mathbb{N}$. We claim that $V(x(t_k^+)) \leqslant V(x(t_k))$. Indeed, if $x(t_k) = 0$, then $I_k(x(t_k)) = 0$ by condition (E5). Thus,

$$V(x(t_k^+)) = V(x(t_k) + I_k x(t_k)) = 0 = V(x(t_k)).$$

If $x(t_k) > 0$, then condition (E5) implies that $|I_k(x(t_k))| \leqslant K_2|x(t_k)| < |x(t_k)|$ and $I_k(x(t_k)) < 0$, that is,

$$-x(t_k) = -|x(t_k)| < I_k(x(t_k)) < 0,$$

whence it follows that

$$0 < x(t_k) + I_k(x(t_k)) < x(t_k).$$

Consequently,

$$\begin{aligned} V(x(t_k^+)) &= V(x(t_k) + I_k x(t_k)) \\ &= x(t_k) + I_k(x(t_k)) < x(t_k) = V(x(t_k)). \end{aligned}$$

Now, if $x(t_k) < 0$, then condition (E5) implies that $|I_k(x(t_k))| \leqslant K_2|x(t_k)| < |x(t_k)|$ and $I_k(x(t_k)) > 0$, that is,

$$0 < I_k(x(t_k)) \leqslant |I_k(x(t_k))| < |x(t_k)| = -x(t_k)$$

and, therefore,

$$x(t_k) < x(t_k) + I_k(x(t_k)) < 0.$$

This implies that

$$\begin{aligned} V(x(t_k^+)) &= V(x(t_k) + I_k x(t_k)) \\ &= |x(t_k) + I_k(x(t_k))| \leqslant |x(t_k)| = V(x(t_k)). \end{aligned}$$

Thus, once $V(x(t_k^+)) \leqslant V(x(t_k))$, we have

$$V(x(t_k + \eta)) \leqslant V(x(t_k)),$$

for sufficiently small $\eta > 0$. Consequently, for $t = t_k$, we get

$$D^+ V(x(t)) = \limsup_{\eta \to 0^+} \frac{V(x(t+\eta)) - V(x(t))}{\eta} \leqslant 0,$$

which completes the proof that $D^+V(x(t)) \leqslant 0$ for all $t \geqslant s_0$. Therefore,

$$V(x(t)) \leqslant V(x(s)), \quad \text{for all } t \geqslant s \geqslant s_0,$$

whence we obtain

$$U(t, x(t)) = V(x(t)) \leqslant V(x(s)) = U(s, x(s)),$$

for all $t \geqslant s \geqslant s_0$, showing that condition (UBM3) is satisfied.

Since all conditions of Theorem 11.7 are satisfied, we conclude that the MDE (11.30) is uniformly bounded and, consequently, the IDE (11.29) is uniformly bounded.

12

Control Theory

Fernanda Andrade da Silva[1], *Márcia Federson*[1], *and Eduard Toon*[2]

[1]*Departamento de Matemática, Instituto de Ciências Matemáticas e de Computação (ICMC), Universidade de São Paulo, São Carlos, SP, Brazil*
[2]*Departamento de Matemática, Instituto de Ciências Exatas, Universidade Federal de Juiz de Fora, Fora, MG, Brazil*

A control system is a time-evolving system over which one can act through an input or control function. The purpose of control theory is to analyze properties of such systems, with the intention of bringing a certain initial data to a certain final data. Observability in control theory is a measure of how well the states of a system can be inferred from some knowledge about its external outputs. Roughly speaking, observability means that, from system outputs, it is possible to determine the behavior of the entire system.

In 1991, Milan Tvrdý introduced a concept of complete controllability for generalized ODE defined in finite dimensional spaces (see [229]). Tvrdý considered the following integral equations:

$$\begin{cases} x(t) - x(0) - \displaystyle\int_0^t d[A(s)]x(s) + (\mathcal{B}u)(t) - (\mathcal{B}u)(0) = f(t) - f(0), \\ \qquad\qquad\qquad\qquad\qquad\qquad\qquad\qquad\qquad x \in \mathbb{G}_L^n,\ u \in \mathcal{U}, \\ Mx(0) + \displaystyle\int_0^1 K(s)d[x(s)] = r, \end{cases} \tag{12.1}$$

where $\mathcal{U} = L_2^n$, that is, U is the space of n-vector-valued functions which are square integrable over $[0,1]$ in the sense of Lebesgue, $\mathbb{G}_L^n$ is the linear space of n-vector-valued functions regulated on $[0,1]$ and left-continuous on $(0,1)$, M is a constant $m \times n$-matrix, $K(t)$ is an $m \times n$-matrix valued function of bounded variation in $[0,1]$, $r \in \mathbb{R}^n, f : [0,1] \to \mathbb{R}^n$ is regulated on $[0,1]$ and left-continuous on $(0,1)$, $A(t)$ is an $n \times n$-matrix-valued function of bounded variation in $[0,1]$, left-continuous on $(0,1)$, right-continuous at 0 such that $\det[I + \Delta^+ A(t)] \neq 0$ on

Generalized Ordinary Differential Equations in Abstract Spaces and Applications, First Edition.
Edited by Everaldo M. Bonotto, Márcia Federson, and Jaqueline G. Mesquita.

$[0, 1]$ and $\mathcal{B} \in L(\mathcal{U}, \mathbb{G}_L^n)$. System (12.1) is said to be completely controllable, if it possesses a solution in $\mathbb{G}_L^n \times \mathcal{U}$, for any $f \in \mathbb{G}_L^n$ and any $r \in \mathbb{R}^n$.

In this chapter, we introduce new concepts of controllability and observability for abstract generalized ODEs, and we investigate necessary and sufficient conditions for a system of nonhomogeneous generalized ODEs, with initial data, controls, and observations taking values in Banach spaces, to be exactly controllable, approximately controllable, or observable. These are the contents of Section 12.1. In Section 12.2, we include an application to ordinary differential equations (we write ODEs for short) with Perron integrable functions on the right-hand side. Moreover, Corollaries 12.7 and 12.9 show that the results described in Section 12.2 extend classical results on controllability and observability for ODEs presented in the literature (see [237], for instance).

12.1 Controllability and Observability

Our goal, in this section, is to establish necessary and sufficient conditions for a system of nonhomogeneous generalized ODEs to be controllable/observable (see Theorem 12.2 in the present section).

Let $\tilde{U}, X$, and Y be Banach spaces and $S \subset X$. We denote by $L(X, Y)$ the space of continuous linear mappings $T: X \to Y$. When $X = Y$, we write simply $L(X)$ instead of $L(X, X)$. Throughout this section, S denotes the space of initial data, $\tilde{U}$ is the control space, X is the evolution space, and Y is the observation space. Furthermore, we assume that $0 \in S$ (neutral element of X).

Recall that $BV_{loc}([t_0, \infty), X)$ denotes the vector space of all functions $f: [t_0, \infty) \to X$ such that x is of bounded variation in $[a, b]$, for all $[a, b] \subset [t_0, \infty)$.

Let $t_0 \geqslant 0$ and consider the nonhomogeneous generalized ODE:

$$\frac{dx}{d\tau} = D[A(t)x + B(t)u], \tag{12.2}$$

where $A: [t_0, \infty) \to L(X)$, $B: [t_0, \infty) \to L(\tilde{U}, X)$, and $u \in \tilde{U}$ are such that the Perron integral $\int_a^b B(s)uds$ exists for all $a, b \in [t_0, \infty)$ (see Remark 1.39 for the definition of $\int_a^b B(s)u\, ds$). Moreover, assume that $B \in BV_{loc}([t_0, \infty), L(\tilde{U}, X))$, $A \in BV_{loc}([t_0, \infty), L(X))$, and A satisfies the conditions of Definition 6.2 on $[t_0, \infty)$, that is,

(D) $[I + \Delta^+ A(t)]^{-1} \in L(X)$, for all $t \in [t_0, \infty)$ and $[I - \Delta^- A(t)]^{-1} \in L(X)$, for all $t \in (t_0, \infty)$, where $\Delta^+ A(t) = A(t^+) - A(t)$ and $\Delta^- A(t) = A(t) - A(t^-)$.

Denote by $x(\cdot) = x(\cdot, d, u)$ the solution x of the nonhomogeneous generalized ODE (12.2), with initial condition $x(t_0) = d$ and control u.

By Remark 6.4, we can assume that all solutions of the nonhomogeneous generalized ODE (12.2) are defined for all $t \in [t_0, \infty)$. By the variation-of-constants formula (see Theorem 6.14), the solution of the nonhomogeneous generalized ODE (12.2), with initial condition $x(t_0) = d$, is given by

$$x(t) = U(t,t_0)d + B(t)u - B(t_0)u - \int_{t_0}^{t} d_s[U(t,s)](B(t)u - B(t_0)u),$$

for all $t \in [t_0, \infty)$, where U is the fundamental operator of the linear generalized ODE

$$\frac{dx}{d\tau} = D[A(t)x]$$

and it is given by

$$U(t,s) = I + \int_{s}^{t} d[A(s)]U(r,s), \quad t,s \in [t_0,\infty), \tag{12.3}$$

as attested by Theorem 6.6.

The evolution map of the nonhomogeneous generalized ODE (12.2) at time $t \in [t_0, \infty)$ is given by $(d,u) \mapsto x(t,d,u) = F(t)d + G(t)u$, defined on $S \times \tilde{U}$ taking values in X, where

(i) for all $d \in S$, the mapping $[t_0,\infty) \ni t \mapsto F(t)d \in X$ is given by

$$F(t)d = U(t,t_0)d, \tag{12.4}$$

with $U(t,t_0)$ given by (12.3);

(ii) for all $t \in [t_0,\infty)$, the mapping $\tilde{U} \ni u \mapsto G(t)u \in X$ is given by

$$G(t)u = B(t)u - B(t_0)u - \int_{t_0}^{t} d_s[U(t,s)](B(t)u - B(t_0)u). \tag{12.5}$$

We also define an observer $C\colon [t_0,\infty) \to L(X,Y)$ and the observation at time $t \in [t_0,\infty)$ by

$$y(t,d,u) = C(t)x(t,d,u). \tag{12.6}$$

Now, we consider a system of nonhomogeneous generalized ODEs

$$\begin{aligned} \frac{dx}{d\tau} &= D[A(t)x + B(t)u], \\ y(t) &= C(t)x(t). \end{aligned} \tag{12.7}$$

We introduce, in the next lines, concepts of controllability and observability for system (12.7), where we assume that $A\colon [t_0,\infty) \to L(X)$, with $A \in BV_{loc}([t_0,\infty), L(X))$, satisfies condition (D) and C is given by (12.6). Consider $B\colon [t_0,\infty) \to L(\tilde{U},X)$, with $B \in BV_{loc}([t_0,\infty), L(\tilde{U},X))$, and assume that there exists the Perron integral $\int_a^b B(s)u\,ds$, for all $a,b \in [t_0,\infty)$ and all $u \in \tilde{U}$.

Definition 12.1: Let $t_0 < T < \infty$ be fixed. The data $d \in S$ is said to be

(i) *approximately controllable* at time T to a point $\tilde{x} \in X$, if there exists a sequence of controls $\{u_n\}_{n\in\mathbb{N}}$ in $\tilde{U}$ such that $x(T, d, u_n) \to \tilde{x}$ as $n \to \infty$;
(ii) *exactly controllable* at time T to $\tilde{x}$, if there exists a control $u \in \tilde{U}$ such that $x(T, d, u) = \tilde{x}$.

The system of generalized ODEs (12.7) is said to be *approximately controllable* (*exactly controllable*) at time T, if all points of S are approximately controllable (exactly controllable) at time T to all points of X. Accordingly, given $u \in \tilde{U}$, a state $d \in S$ is said to be *observable* at time T, if d can be uniquely determined by u and by the observation $y(T, d, u)$. The system (12.7) is said to be *observable* at time T, if all states in S are observable at time T.

For the next theorem, we define the mapping

$$\begin{aligned} \overline{F}(T): S &\to L([0,T],X) \\ d &\mapsto \overline{F}(T)d : [0,T] \to X \\ &\qquad t \mapsto (\overline{F}(T)d)(t) = F(t)d, \end{aligned} \tag{12.8}$$

where $F(t)d = U(t, t_0)d$ for all $d \in S$ and $t \in [t_0, \infty)$.

Theorem 12.2: *Consider the mappings G, C, and $\overline{F}$ given by* (12.5), (12.6), *and* (12.8), *respectively. The following assertions hold:*

(i) *The system of generalized ODEs* (12.7) *is approximately controllable at time T if and only if the range of $G(T)$ is everywhere dense in X.*
(ii) *The system of generalized ODEs* (12.7) *is exactly controllable at time T if and only if the mapping $G(T)$ is onto.*
(iii) *The system of generalized ODEs* (12.7) *is observable at time T if and only if the composite mapping $C(T)\overline{F}(T)$ is one-to-one.*

Proof. At first, we prove item (i). Let $\tilde{x} \in X$ be arbitrary. Since the system of generalized ODEs (12.7) is approximately controllable at time T, by Definition 12.1, $d = 0$ is approximately controllable at time T to $\tilde{x}$. Therefore, there exists a sequence $\{u_n\}_{n\in\mathbb{N}}$ in $\tilde{U}$ such that $x(T, 0, u_n) \to \tilde{x}$ as $n \to \infty$, that is, $G(T)u_n \to \tilde{x}$, as $n \to \infty$. Hence, the range of $G(T)$ is everywhere dense in X.

Conversely, for arbitrary $d \in S$ and $\tilde{x} \in X$, if the range of $G(T)$ is everywhere dense in X, then there exists a sequence $\{u_n\}_{n\in\mathbb{N}}$ in $\tilde{U}$ such that $G(T)u_n \to \tilde{x} - F(T)d$ as $n \to \infty$, that is, $x(T, d, u_n) \to \tilde{x}$, as $n \to \infty$.

As in item (i), in order to prove item (ii), we take $\tilde{x}$ arbitrarily and $d = 0$. Since the system of generalized ODEs (12.7) is exactly controllable at time T, there exists $u \in \tilde{U}$ such that $x(T, 0, u) = \tilde{x}$, that is, $G(T)u = \tilde{x}$. Therefore, $G(T)$ is onto.

Conversely, for arbitrary $d \in S$ and $\tilde{x} \in X$, let $z = \tilde{x} - F(T)d$. Since $G(T)$ is onto, there exists $u \in \tilde{U}$ such that $G(T)u = z = \tilde{x} - F(T)d$. Then, $x(T, d, u) = \tilde{x}$.

Finally, for (iii), let $d', d \in S$ be such that $d' \neq d$. Since the system of generalized ODEs (12.7) is observable at time T, by Definition 12.1, $y(T, d', 0) \neq y(T, d, 0)$, that is, $C(T)(\overline{F}(T)d) \neq C(T)(\overline{F}(T)d')$. Hence, $C(T)\overline{F}(T)$ is one-to-one.

Conversely, take $d \in S$ arbitrarily. Suppose the system of generalized ODEs (12.7) is nonobservable at time T. Then, there exists $d' \neq d \in S$ such that $y(T, d', 0) = y(T, d, 0)$, for $u = 0$. Thus, $C(T)(\overline{F}(T)(d' - d)) = 0$ in opposition to our initial hypothesis. □

12.2 Applications to ODEs

In this section, we show that when we consider a system of classic ODEs, our main results described in Section 12.1 generalize the results described in [237] (see [237, Theorems 1 and 2] and Corollaries 12.7 and 12.9 in this section).

Consider the system of ODEs:

$$\begin{aligned} \dot{\hat{x}} &= \hat{A}(t)\hat{x} + \hat{B}(t)u(t), \\ \hat{y}(t) &= C(t)\hat{x}, \end{aligned} \tag{12.9}$$

where $\dot{\hat{x}} = \frac{d}{dt}\hat{x}(t)$, $X = \mathbb{R}^n = S$, $Y = \mathbb{R}^r$, $\tilde{U} = \mathbb{R}^p$, $\mathcal{U}$ is the set of all control functions $u: [t_0, \infty) \to \tilde{U}$, $\hat{A}(\cdot)$, $\hat{B}(\cdot)u(\cdot)$, and $C(\cdot)$ are locally Perron integrable over $[t_0, \infty)$, and the system of generalized ODEs:

$$\begin{aligned} \frac{dx}{d\tau} &= D[A(t)x + B(t)u], \\ y(t) &= C(t)x(t), \end{aligned} \tag{12.10}$$

where

$$A(t) = \int_{t_0}^{t} \hat{A}(s)ds \quad \text{and} \quad B(t)u = \int_{t_0}^{t} \hat{B}(s)u(s)ds. \tag{12.11}$$

Notice that $A(\cdot)x$ and $B(\cdot)u$ are continuous on $[t_0, \infty)$ for all $x \in \mathbb{R}^n$ and all $u \in \mathcal{U}$ (see Corollary 2.14). Therefore, $B \in BV_{loc}([t_0, \infty), L(\mathcal{U}, \mathbb{R}^n))$, $A \in BV_{loc}([t_0, \infty), L(\mathbb{R}^n))$, and A satisfies condition (D) which implies that A and B satisfy all conditions of Section 12.1.

The next result gives a characterization for the solution of the nonhomogeneous generalized ODE (12.10).

Theorem 12.3: *Let $x: [t_0, \infty) \to X$ be the solution of the nonhomogeneous generalized ODE* (12.10) *with initial condition $x(t_0) = d \in S$. Then, x is given by*

$$x(t) = U(t, t_0)d + \int_{t_0}^{t} U(t, \tau)\hat{B}(\tau)u(\tau)d\tau,$$

for all $t \in [t_0, \infty)$, where U is given by (12.3), and the integral on the right-hand side is in Perron's sense.

Proof. By Theorem 6.14, the solution of the nonhomogeneous generalized ODE (12.10), with initial condition $x(t_0) = d$, satisfies the integral equation

$$x(t) = U(t, t_0)d + B(t)u - \int_{t_0}^{t} d_s[U(t, s)]B(s)u, \tag{12.12}$$

for all $t \in [t_0, \infty)$, see Eq. (6.22).

By the Integration by Parts Formula for the Perron-Stieltjes integral (see Corollary 1.56),

$$\begin{aligned}\int_{t_0}^{t} d_s[U(t, s)]B(s)u &= U(t, t)B(t)u - U(t, t_0)B(t_0)u - \int_{t_0}^{t} U(t, s)d[B(s)u] \\ &= B(t)u - \int_{t_0}^{t} U(t, s)\widehat{B}(s)u(s)ds. \end{aligned} \tag{12.13}$$

Then, replacing (12.13) in (12.12), we obtain

$$x(t) = U(t, t_0)d + \int_{t_0}^{t} U(t, s)\widehat{B}(s)u(s)ds, \tag{12.14}$$

where the integral in (12.14) is in Perron's sense. □

Remark 12.4: If Φ is the fundamental operator of the ODE (12.9), then

$$\Phi(t, s) = U(t, s), \quad \text{for all } t, s \in [t_0, \infty).$$

See [209, Example 6.2]. In this case, the functions G and F given by (12.5) and (12.4), respectively, can be rewritten as follows:

$$G(t)u = \int_{t_0}^{t} U(t, s)\widehat{B}(s)u(s)ds = \int_{t_0}^{t} \Phi(t, s)\widehat{B}(s)u(s)ds, \tag{12.15}$$

for all $t \in [t_0, \infty)$ and all $u \in \mathcal{U}$ and

$$F(t)d = \Phi(t, t_0)d, \quad \text{for all } t \in [t_0, \infty) \text{ and all } d \in S. \tag{12.16}$$

Notice that the definitions of exact controllability, approximate controllability, and of observability for ODEs are analogous to the same definitions for generalized ODEs. In the sequel, we describe necessary and sufficient conditions on controllability and observability for the system of ODEs (12.9). The proof is a consequence of Theorem 12.2 and Remark 12.4.

Corollary 12.5: *The following assertions hold.*

(i) *The system of ODEs (12.9) is approximately controllable at time T if and only if the range of $G(T)$, defined by (12.15), is everywhere dense in X.*
(ii) *The system of ODEs (12.9) is exactly controllable at time T if and only if the mapping $G(T)$, defined by (12.15), is onto.*
(iii) *The system of ODEs (12.9) is observable at time T if and only if the composite mapping $C(T)\overline{F}(T)$ is one-to-one, where $(\overline{F}(T)d)(t) = F(t)d$ for all $d \in S$ and $t \in [0, T]$ and F is given (12.16).*

In the sequel, we use the notation M' to denote the transpose of a given matrix M.

The next result relates conditions on exact controllability between systems of generalized ODEs (12.10) and ODEs (12.9).

Theorem 12.6: *Let $T \geqslant t_0$. The function $G(T)$, given by (12.15), is onto if and only if the rows of the matrix $U(t_0, T)\widehat{B}(T)$ are linearly independent, where U is given by (12.3).*

Proof. Assume that the rows of the matrix $U(t_0, T)\widehat{B}(T)$ are linearly independent. Then, the matrix

$$C(t_0, T) = \int_{t_0}^{T} U(T, \tau)\widehat{B}(T)\widehat{B}'(\tau)U'(t_0, \tau)d\tau$$

is positive definite. Let $x \in X$ be arbitrary and take

$$u(\tau) = \widehat{B}'(\tau)U'(t_0, \tau)C^{-1}(t_0, T)x, \quad t_0 \leqslant \tau \leqslant T.$$

By (12.15), we obtain

$$G(T)u = \int_{t_0}^{T} U(T, \tau)\widehat{B}(\tau)u(\tau)d\tau = x,$$

which shows that $G(T)$ is onto.

Conversely, if the rows of the matrix $U(t_0, T)\widehat{B}(T)$ are linearly dependent, then there exists $x \in X$, $x \neq 0$, such that

$$x'U(t_0, \tau)\widehat{B}(T) \equiv 0, \quad t_0 \leqslant \tau \leqslant T.$$

Since $G(T)$ is onto, there exists u such that $G(T)u = x$. Therefore,

$$\int_{t_0}^{T} U(T, \tau)\widehat{B}(T)u(\tau)d\tau = x.$$

Multiplying the above equation by $x'U(t_0, T)$, we obtain

$$\int_{t_0}^{T} x'U(t_0, \tau)\hat{B}(T)u(\tau)d\tau = x'U(t_0, T)x,$$

whence $0 = x'U(t_0, T)x$ which contradicts the fact that U is the fundamental operator of the generalized ODE

$$\frac{dx}{d\tau} = D[A(t)x],$$

where A is given by (12.11). □

As a consequence of Remark 12.4, Corollary 12.5, and Theorem 12.6, we obtain the next result.

Corollary 12.7: *The system of ODEs (12.9) is exactly controllable at time T if and only if the rows of the matrix $\Phi(t_0, T)\hat{B}(T)$ are linearly independent, where Φ is the fundamental operator of (12.9).*

The following result relates conditions about observability between systems of generalized ODEs (12.10) and ODEs (12.9).

Theorem 12.8: *Let $T \geqslant t_0$ and $u \equiv 0$ in (12.10). Then, $C(T)\overline{F}(T)$ is one-to-one if and only if the columns of the matrix $C(T)U(T, t_0)$ are linearly independent, where C is given by (12.11), $\overline{F}(T)$ by (12.16) and U is given by (12.3).*

Proof. Suppose $C(T)\overline{F}(T)$ is one-to-one and the columns of the matrix $C(T)U(T, t_0)$ are linearly dependent. Then, there exists $d \in S$, with $d \neq 0$, such that $C(T)U(T, t_0)d = 0$. On the other hand, $C(T)U(T, t_0)d = C(T)(\overline{F}(T)d)$. Thus, $C(T)(\overline{F}(T)d) = 0$ for $d \neq 0$, which contradicts the fact that $C(T)\overline{F}(T)$ is one-to-one.

Conversely, assume that the columns of the matrix $C(T)U(T, t_0)$ are linearly independent and $C(T)\overline{F}(T)$ is not one-to-one. Then, there exist $d', d \in S$ such that $d' \neq d$ and $C(T)(\overline{F}(T)d') = C(T)(\overline{F}(T)d)$. Hence, $C(T)(\overline{F}(T)(d' - d)) = 0$ implies $C(T)U(T, t_0)\overline{d} = 0$ for $\overline{d} = d' - d \neq 0$, which contradicts the fact that the columns of the matrix $C(T)U(T, t_0)$ are linearly independent. □

The proof of the next result follows directly from Remark 12.4, Corollary 12.5, and Theorem 12.8.

Corollary 12.9: *The system of ODEs (12.9) is observable at time T if and only if the columns of the matrix $C(T)\Phi(T, t_0)$ are linearly independent, where C is given by (12.11), $\overline{F}(T)$ by (12.16), and Φ is the fundamental operator of (12.9).*

13

Dichotomies

Everaldo M. Bonotto[1] *and Márcia Federson*[2]

[1] *Departamento de Matemática Aplicada e Estatística, Instituto de Ciências Matemáticas e de Computação (ICMC), Universidade de São Paulo, São Carlos, SP, Brazil*
[2] *Departamento de Matemática, Instituto de Ciências Matemáticas e de Computação (ICMC), Universidade de São Paulo, São Carlos, SP, Brazil*

The theory of exponential dichotomy is an important tool to the investigation of the behavior of nonautonomous differential equations. Exponential dichotomy is a kind of conditional stability which generalizes the notion of hyperbolicity of autonomous systems to nonautonomous systems. There are many important consequences for systems which admit a dichotomy. We can mention, for instance, the work [168] where the authors present a proof of Smale's Conjecture by means of exponential dichotomies.

The classical notion and basic properties of dichotomies in the context of ODEs can be found in [48] and [49], among other references. We also mention Kenneth J. Palmer who made a great contribution to the theory of exponential dichotomy in the framework of ODEs (see, for instance, [192, 193]). Later, Liviu H. Popescu improved results on exponential dichotomies to infinite dimensional Banach spaces, [197]. Good references containing results on dichotomies for nonautonomous systems are [117] and [200]. The reader may also want to consult the paper [10] by D. D. Bainov, S. I. Kostadinov, and N. Van Minh for the study of dichotomies in the setting of impulsive differential equations (IDEs), and the paper [43] by Charles V. Coffman and Juan Jorge Schäffer in the setting of difference equations.

In this chapter, following the ideas of [48], we present the theory of exponential dichotomy for a class of linear generalized ODEs. All the results presented in this chapter were borrowed from [29]. In Section 13.1, we recall the concept of exponential dichotomy, and we present sufficient conditions to obtain exponential

Generalized Ordinary Differential Equations in Abstract Spaces and Applications, First Edition.
Edited by Everaldo M. Bonotto, Márcia Federson, and Jaqueline G. Mesquita.

dichotomy. Section 13.2 deals with the relation between the exponential dichotomy of a linear generalized ODE and the existence of bounded solutions of its perturbation. Lastly, we apply the results to MDEs in Section 13.3 and to IDEs in Section 13.4.

13.1 Basic Theory for Generalized ODEs

This section concerns the theory of exponential dichotomy in the context of linear generalized ODEs. The reader may find all the definitions and results presented here in [29, Section 3].

Let $f: J \to \mathbb{M}_n(\mathbb{R})$ be a continuous function defined on an interval $J \subset \mathbb{R}$, where $\mathbb{M}_n(\mathbb{R})$ denotes the space of all $n \times n$ real matrices, and consider the following linear ordinary differential equation

$$x' = f(t)x. \tag{13.1}$$

Let $X(t)$ be a fundamental matrix of (13.1). In the classical theory of ordinary differential equations, the ODE (13.1) is said to admit an *exponential dichotomy* on J, if there exist positive constants K_1, K_2, α_1, and α_2, and a constant matrix P such that $P^2 = P$, satisfying the following conditions:

- $\|X(t)PX^{-1}(s)\| \leqslant K_1 e^{-\alpha_1(t-s)}$ for all $t, s \in J$, with $t \geqslant s$;
- $\|X(t)(I - P)X^{-1}(s)\| \leqslant K_2 e^{-\alpha_2(s-t)}$ for all $t, s \in J$, with $s \geqslant t$.

The matrix P is a projection onto the stable subspace of $\mathbb{R}^n$, and $I - P$ is a projection onto the unstable subspace of $\mathbb{R}^n$.

One of our interests is to handle the theory of dichotomies for ODEs of type (13.1) whose function f takes values in a Banach space and is integrable in a more general sense. Here we consider the nonabsolute integration in the senses of Perron and Perron–Stieltjes, following the notation and terminology of Chapter 1. In order to obtain results on dichotomies for a class of ODEs, IDEs, MDEs, MFDEs, etc., we translate the notion of dichotomies for a class of linear generalized ODEs following the notation and terminology of Chapters 2, 3, 4, and 6.

Given a Banach space X and an interval $J \subset \mathbb{R}$, we consider the following linear generalized ODE

$$\frac{dx}{d\tau} = D[A(t)x], \tag{13.2}$$

where $A: J \to L(X)$ is an operator satisfying the following general conditions:

(H_1) $A \in BV([a, b], L(X))$, for every interval $[a, b] \subset J$;
(H_2) A satisfies condition (Δ) presented in Definition 6.2.

Let $t_0 \in J$ and $x_0 \in X$. Under conditions (H_1) and (H_2), the linear generalized ODE (13.2) with initial condition $x(t_0) = x_0$ admits a unique global forward solution, see Remark 6.4.

According to Theorem 6.6, the linear generalized ODE (13.2) admits a fundamental operator $U: J \times J \to L(X)$. For a fixed arbitrary point $t_0 \in J$, we define the operators $\mathcal{U}, \mathcal{U}^{-1}: J \to L(X)$ by

$$\mathcal{U}(t) = U(t, t_0) \quad \text{and} \quad \mathcal{U}^{-1}(t) = U(t_0, t),$$

for all $t \in J$. Thus, if $x: J \to X$ is the solution of the linear generalized ODE (13.2) with initial condition $x(s) = x_0$, $x_0 \in X$ and $s \in J$, then

$$x(t) = \mathcal{U}(t)\mathcal{U}^{-1}(s)x_0 \quad \text{for all} \quad t \in J.$$

Next, we present the concept of exponential dichotomy.

Definition 13.1: The linear generalized ODE (13.2) admits an *exponential dichotomy* on J, if there exist positive constants K_1, K_2, α_1, and α_2 and a projection $P: X \to X$ $(P^2 = P)$ such that

(i) $\|\mathcal{U}(t)P\mathcal{U}^{-1}(s)\| \leqslant K_1 e^{-\alpha_1(t-s)}$, for all $t, s \in J$, with $t \geqslant s$;
(ii) $\|\mathcal{U}(t)(I-P)\mathcal{U}^{-1}(s)\| \leqslant K_2 e^{-\alpha_2(s-t)}$, for all $t, s \in J$, with $s \geqslant t$.

The proof of the next result is inspired in [48], see page 13 there.

Lemma 13.2: *Assume that the operator A, which satisfies conditions (H_1) and (H_2), is defined on $[0, \infty)$. If the linear generalized ODE (13.2) admits an exponential dichotomy on the interval $[a, \infty)$, $a \geqslant 0$, then it also admits an exponential dichotomy on $[0, \infty)$.*

Proof. By Definition 13.1, there are positive constants $K_1, K_2, \alpha_1, \alpha_2$, and a projection P such that

$$\|\mathcal{U}(t)P\mathcal{U}^{-1}(s)\| \leqslant K_1 e^{-\alpha_1(t-s)}, \qquad \text{for all } t, s \in [a, \infty), \text{ with } t \geqslant s;$$
$$\|\mathcal{U}(t)(I-P)\mathcal{U}^{-1}(s)\| \leqslant K_2 e^{-\alpha_2(s-t)}, \qquad \text{for all } t, s \in [a, \infty), \text{ with } s \geqslant t.$$

Now, by Proposition 6.8, item (ii), there exists $M \geqslant 1$ such that

$$\|\mathcal{U}(t)\| \leqslant M \quad \text{and} \quad \|\mathcal{U}^{-1}(t)\| \leqslant M, \text{ for all } t \in [0, a].$$

Let $t, s \in [0, \infty)$. Consider $M_1 = K_1 e^{\alpha_1 a} M^4$. We have some cases to consider:

- If $0 \leqslant a \leqslant s \leqslant t$, then

$$\begin{aligned}\|\mathcal{U}(t)P\mathcal{U}^{-1}(s)\| &\leqslant K_1 e^{-\alpha_1(t-s)} \\ &\leqslant M_1 e^{-\alpha_1(t-s)}.\end{aligned}$$

- If $0 \leqslant s \leqslant a \leqslant t$, then

$$\begin{aligned}\|\mathcal{U}(t)P\mathcal{U}^{-1}(s)\| &\leqslant \|\mathcal{U}(t)P\mathcal{U}^{-1}(a)\|\,\|\mathcal{U}(a)\mathcal{U}^{-1}(s)\|\\ &\leqslant K_1e^{-\alpha_1(t-a)}M^2\\ &\leqslant K_1e^{-\alpha_1(t-a)}M^2e^{\alpha_1 s}\\ &\leqslant M_1e^{-\alpha_1(t-s)}.\end{aligned}$$

- If $0 \leqslant s \leqslant t \leqslant a$, then

$$\begin{aligned}\|\mathcal{U}(t)P\mathcal{U}^{-1}(s)\| &\leqslant \|\mathcal{U}(t)\mathcal{U}^{-1}(a)\|\,\|\mathcal{U}(a)P\mathcal{U}^{-1}(a)\|\,\|\mathcal{U}(a)\mathcal{U}^{-1}(s)\|\\ &\leqslant M^2K_1M^2\\ &\leqslant M^4K_1e^{\alpha_1(a-t+s)}\\ &= M_1e^{-\alpha_1(t-s)}.\end{aligned}$$

Hence, $\|\mathcal{U}(t)P\mathcal{U}^{-1}(s)\| \leqslant M_1e^{-\alpha_1(t-s)}$, for all $t, s \in [0,\infty)$ with $t \geqslant s$. Analogously, taking $M_2 = K_2e^{\alpha_2 a}M^4$, we obtain $\|\mathcal{U}(t)(I-P)\mathcal{U}^{-1}(s)\| \leqslant M_2e^{-\alpha_2(s-t)}$ for all $t, s \in [0,\infty)$ with $t \leqslant s$. □

The first characterization of an exponential dichotomy for the linear generalized ODE (13.2) is given by the next result.

Proposition 13.3: *The linear generalized ODE (13.2) admits an exponential dichotomy on J if and only if there are positive constants L_1, L_2, M, β_1, and β_2, and a projection $P: X \to X$ such that, for all $\xi \in X$, the following estimates hold:*

(i) $\|\mathcal{U}(t)P\xi\| \leqslant L_1e^{-\beta_1(t-s)}\,\|\mathcal{U}(s)P\xi\|$, *for all $t, s \in J$, with $t \geqslant s$;*
(ii) $\|\mathcal{U}(t)(I-P)\xi\| \leqslant L_2e^{-\beta_2(s-t)}\,\|\mathcal{U}(s)(I-P)\xi\|$, *for all $t, s \in J$, with $t \leqslant s$;*
(iii) $\|\mathcal{U}(t)P\mathcal{U}^{-1}(t)\| \leqslant M$, *for all $t \in J$.*

Proof. Suppose the linear generalized ODE (13.2) has an exponential dichotomy on the interval J. By Definition 13.1, there are positive constants $K_1, K_2, \alpha_1, \alpha_2$, and a projection P such that

$$\|\mathcal{U}(t)P\mathcal{U}^{-1}(s)\| \leqslant K_1e^{-\alpha_1(t-s)}, \qquad \text{for all} \quad t, s \in J, \text{ with } t \geqslant s; \tag{13.3}$$

$$\|\mathcal{U}(t)(I-P)\mathcal{U}^{-1}(s)\| \leqslant K_2e^{-\alpha_2(s-t)}, \qquad \text{for all} \quad t, s \in J, \text{ with } s \geqslant t. \tag{13.4}$$

Set $L_1 = M = K_1$, $L_2 = K_2$, $\beta_1 = \alpha_1$, and $\beta_2 = \alpha_2$. Now, let $\xi \in X$. Using the estimates (13.3) and (13.4), we obtain

$$\begin{aligned}\|\mathcal{U}(t)P\xi\| &= \|\mathcal{U}(t)P\mathcal{U}^{-1}(s)\mathcal{U}(s)P\xi\| \leqslant K_1e^{-\alpha_1(t-s)}\,\|\mathcal{U}(s)P\xi\|\\ &= L_1e^{-\beta_1(t-s)}\,\|\mathcal{U}(s)P\xi\|,\end{aligned}$$

for all $t, s \in J$ with $t \geqslant s$, and

$$\begin{aligned}\|\mathcal{U}(t)(I-P)\xi\| &= \|\mathcal{U}(t)(I-P)\mathcal{U}^{-1}(s)\mathcal{U}(s)(I-P)\xi\| \\ &\leqslant K_2 e^{-\alpha_2(s-t)} \|\mathcal{U}(s)(I-P)\xi\| \\ &= L_2 e^{-\beta_2(s-t)} \|\mathcal{U}(s)(I-P)\xi\|,\end{aligned}$$

for all $t, s \in J$ with $t \leqslant s$. By (13.3), we have

$$\|\mathcal{U}(t)P\mathcal{U}^{-1}(t)\| \leqslant K_1 e^{-\alpha_1(t-t)} = K_1 = M,$$

for all $t \in J$.

For the sufficient condition, let us assume the existence of positive constants L_1, L_2, M, β_1, and β_2 satisfying conditions (i), (ii), and (iii). Set $K_1 = ML_1$ and $K_2 = (1+M)L_2$. If $t, s \in J$ with $t \geqslant s$, we obtain

$$\begin{aligned}\|\mathcal{U}(t)P\mathcal{U}^{-1}(s)\| &= \sup_{\|\xi\|\leqslant 1} \|\mathcal{U}(t)P\mathcal{U}^{-1}(s)\xi\| \\ &\leqslant \sup_{\|\xi\|\leqslant 1} L_1 e^{-\beta_1(t-s)} \|\mathcal{U}(s)P\mathcal{U}^{-1}(s)\xi\| \\ &\leqslant ML_1 e^{-\beta_1(t-s)} = K_1 e^{-\beta_1(t-s)}.\end{aligned}$$

On the other hand, if $t, s \in J$ with $t \leqslant s$, then

$$\begin{aligned}\|\mathcal{U}(t)(I-P)\mathcal{U}^{-1}(s)\| &= \sup_{\|\xi\|\leqslant 1} \|\mathcal{U}(t)(I-P)\mathcal{U}^{-1}(s)\xi\| \\ &\leqslant \sup_{\|\xi\|\leqslant 1} L_2 e^{-\beta_2(s-t)} \|\mathcal{U}(s)(I-P)\mathcal{U}^{-1}(s)\xi\| \\ &\leqslant (1+M)L_2 e^{-\beta_2(s-t)} = K_2 e^{-\beta_2(s-t)}.\end{aligned}$$

□

In what follows, we present the concept of bounded growth for linear generalized ODEs.

Definition 13.4: Let $J \subset \mathbb{R}$ be an interval such that $\sup J = \infty$. The linear generalized ODE (13.2) admits a bounded growth on J, if there are constants $h > 0$ and $C \geqslant 1$ such that for any solution $x: J \to X$ of (13.2), we have

$$\|x(t)\| \leqslant C \|x(s)\| \quad \text{for } s, t \in J, \text{ with } s \leqslant t \leqslant s+h.$$

Example 13.5: Consider the linear generalized ODE

$$\frac{dx}{d\tau} = D[A(t)x], \tag{13.5}$$

where $A: [0, \infty) \to \mathbb{R}$ is given by

$$A(t) = -\int_0^t a(s)ds, \text{ for all } t \geqslant 0.$$

Assume that $a\colon[0,\infty)\to\mathbb{R}$ is Perron integrable and $\|a\|_\infty = \sup_{s\in[0,\infty)}|a(s)| < \infty$. The function A satisfies conditions (H_1)–(H_2) and, moreover,

$$U(t,s) = \exp\left(-\int_s^t a(r)dr\right), \quad t,s\in[0,\infty),$$

is the fundamental operator of the linear generalized ODE (13.5). Note that, given $h>0$, if $0\leqslant t-s\leqslant h$, then

$$\|U(t,s)\| \leqslant e^{\|a\|_\infty h}.$$

Now, let $x\colon[0,\infty)\to\mathbb{R}$ be a solution of the linear generalized ODE (13.5), that is, $x(t)=U(t,t_0)x_0$ for all $t\in[0,\infty)$ and $x_0\in\mathbb{R}$. Then,

$$|x(t)| = |U(t,s)U(s,t_0)x_0| \leqslant e^{\|a\|_\infty h}|x(s)|,$$

whenever $0\leqslant s\leqslant t\leqslant s+h$. Hence, Eq. (13.5) admits a bounded growth on $[0,\infty)$.

Lemma 13.6: *Let $J\subset\mathbb{R}$ be an interval such that* $\sup J=\infty$. *The linear generalized ODE (13.2) admits a bounded growth on J, if and only if there exist constants $C\geqslant 1$ and $\mu\geqslant 0$ such that*

$$\|\mathcal{U}(t)\mathcal{U}^{-1}(s)\| \leqslant Ce^{\mu(t-s)} \quad \textit{for all}\ \ s,t\in J,\ \textit{with}\ t\geqslant s.$$

Proof. Assume that (13.2) admits a bounded growth on J. By Definition 13.4, there exist constants $C\geqslant 1$ and $h>0$ such that

$$\|\mathcal{U}(t)\mathcal{U}^{-1}(s)\xi\| \leqslant C\,\|\mathcal{U}(r)\mathcal{U}^{-1}(s)\xi\| \quad \text{for } \xi\in X,\ s,t,r\in J,\ \text{with}\ r\leqslant t\leqslant r+h.$$

Thus,

$$\|\mathcal{U}(t)\mathcal{U}^{-1}(s)\| \leqslant C\,\|\mathcal{U}(r)\mathcal{U}^{-1}(s)\| \quad \text{for } s,t,r\in J,\ \text{with}\ r\leqslant t\leqslant r+h. \tag{13.6}$$

Let $\mu'\geqslant 0$ be such that $e^{\mu'}=C$ and $\mu=\frac{\mu'}{h}$. Now, let $t,s\in J$ with $t\geqslant s$. There exists $n\in\mathbb{N}$ such that $t\in[s+nh,s+(n+1)h)$. Using (13.6), we have

$$\begin{aligned}\|\mathcal{U}(t)\mathcal{U}^{-1}(s)\| &\leqslant C\,\|\mathcal{U}(t-h)\mathcal{U}^{-1}(s)\| \leqslant\cdots\leqslant C^{n+1}\,\|\mathcal{U}(s)\mathcal{U}^{-1}(s)\| = C^{n+1}\\ &= C(e^{\mu'})^n = C(e^{\mu})^{hn} \leqslant Ce^{\mu(t-s)},\end{aligned}$$

which completes the necessary condition.

Now, let us assume that there are constants $C\geqslant 1$ and $\mu\geqslant 0$ such that

$$\|\mathcal{U}(t)\mathcal{U}^{-1}(s)\| \leqslant Ce^{\mu(t-s)} \quad \text{for all}\ \ s,t\in J,\ \text{with}\ t\geqslant s.$$

Let $x\colon J \to X$ be a solution of (13.2). Fix $r \in J$. Then $x(t) = \mathcal{U}(t)\mathcal{U}^{-1}(r)x(r)$ for all $t \in J$. Consider $h > 0$ and take $t, s \in J$ with $s \leqslant t \leqslant s + h$, then

$$\begin{aligned}\|x(t)\| &= \left\|\mathcal{U}(t)\mathcal{U}^{-1}(r)x(r)\right\| \leqslant \left\|\mathcal{U}(t)\mathcal{U}^{-1}(s)\right\| \left\|\mathcal{U}(s)\mathcal{U}^{-1}(r)x(r)\right\| \\ &\leqslant Ce^{\mu(t-s)}\, \|x(s)\| \leqslant Ce^{\mu h}\, \|x(s)\|\,.\end{aligned}$$

Since $Ce^{\mu h} \geqslant 1$, we conclude that the linear generalized ODE (13.2) admits a bounded growth on J. □

Next, we state an auxiliary result. This result can be found in [48], pages 11–14, and in [205].

Lemma 13.7: *Let $J = [a, \infty)$ and $a \geqslant 0$. Assume that there exist positive constants α_1, α_2, K_1, and K_2, and a projection $P\colon X \to X$ satisfying conditions:*

(i) $\|\mathcal{U}(t)P\xi\| \leqslant K_1 e^{-\alpha_1(t-s)}\, \|\mathcal{U}(s)P\xi\|$, *for all* $t \geqslant s \geqslant a$, $\xi \in X$;
(ii) $\|\mathcal{U}(t)(I-P)\xi\| \leqslant K_2 e^{-\alpha_2(s-t)}\, \|\mathcal{U}(s)(I-P)\xi\|$, *for all* $a \leqslant t \leqslant s$, $\xi \in X$.

Then the following conditions hold:

(a) *given $\theta \in (0, 1)$, there is $T > 0$ such that for any solution $x\colon J \to X$ of (13.2), we have*

$$\|x(s)\| \leqslant \theta \sup_{|u-s|\leqslant T} \|x(u)\|\,, \ \textit{for all } s \geqslant T;$$

(b) *if the linear generalized ODE (13.2) admits a bounded growth on $[a, \infty)$, $a \geqslant 0$, then there exists $M \geqslant 0$ such that $\left\|\mathcal{U}(t)P\mathcal{U}^{-1}(t)\right\| \leqslant M$, for all $t \in J = [a, \infty)$.*

Proof. Taking $K = \max\{K_1, K_2\}$ and $\alpha = \min\{\alpha_1, \alpha_2\}$, we may rewrite conditions (i) and (ii) as)

(i′) $\|\mathcal{U}(t)P\xi\| \leqslant Ke^{-\alpha(t-s)}\, \|\mathcal{U}(s)P\xi\|$, for all $t \geqslant s \geqslant a$, $\xi \in X$;
(ii′) $\|\mathcal{U}(t)(I-P)\xi\| \leqslant Ke^{-\alpha(s-t)}\, \|\mathcal{U}(s)(I-P)\xi\|$, for all $a \leqslant t \leqslant s$, $\xi \in X$.

(*a*) For a given $\theta \in (0, 1)$, let $T_1 \geqslant 0$ be such that

$$K^{-1}e^{\alpha T_1} - Ke^{-\alpha T_1} \geqslant 2\theta^{-1}.$$

Let $x\colon J \to X$ be a solution of the generalized ODE (13.2) and $s \geqslant a + T_1$. Define

$$x_1(t) = \mathcal{U}(t)P\mathcal{U}^{-1}(t)x(t) \quad \text{and} \quad x_2(t) = \mathcal{U}(t)(I-P)\mathcal{U}^{-1}(t)x(t),$$

for all $t \in J$. Note that $x(t) = x_1(t) + x_2(t)$ and

$$\begin{aligned}x(t) &= \mathcal{U}(t)\mathcal{U}^{-1}(s)x(s) = \mathcal{U}(t)\mathcal{U}^{-1}(s)(x_1(s) + x_2(s)) \\ &= \mathcal{U}(t)P\mathcal{U}^{-1}(s)x(s) + \mathcal{U}(t)(I-P)\mathcal{U}^{-1}(s)x(s).\end{aligned}$$

Case 1: $\|x_2(s)\| \geqslant \|x_1(s)\|$.

$$\begin{aligned}\|x(s+T_1)\| &= \|\mathcal{U}(s+T_1)P\mathcal{U}^{-1}(s)x(s) + \mathcal{U}(s+T_1)(I-P)\mathcal{U}^{-1}(s)x(s)\| \\ &\geqslant \|\mathcal{U}(s+T_1)(I-P)\mathcal{U}^{-1}(s)x(s)\| - \|\mathcal{U}(s+T_1)P\mathcal{U}^{-1}(s)x(s)\| \\ &\geqslant K^{-1}e^{\alpha T_1}\|\mathcal{U}(s)(I-P)\mathcal{U}^{-1}(s)x(s)\| \\ &\quad - Ke^{-\alpha T_1}\|\mathcal{U}(s)P\mathcal{U}^{-1}(s)x(s)\| \\ &= K^{-1}e^{\alpha T_1}\|x_2(s)\| - Ke^{-\alpha T_1}\|x_1(s)\| \\ &\geqslant \left(K^{-1}e^{\alpha T_1} - Ke^{-\alpha T_1}\right)\|x_2(s)\| \\ &\geqslant 2\theta^{-1}\|x_2(s)\| \\ &\geqslant \theta^{-1}\|x(s)\|.\end{aligned}$$

Hence, $\|x(s)\| \leqslant \theta\|x(s+T_1)\|$. Consequently,

$$s \geqslant a + T_1 \;\Rightarrow\; \|x(s)\| \leqslant \theta \sup_{|u-s|\leqslant T_1} \|x(u)\|.$$

Case 2: $\|x_2(s)\| \leqslant \|x_1(s)\|$.

$$\begin{aligned}\|x(s-T_1)\| &= \|\mathcal{U}(s-T_1)P\mathcal{U}^{-1}(s)x(s) + \mathcal{U}(s-T_1)(I-P)\mathcal{U}^{-1}(s)x(s)\| \\ &\geqslant \|\mathcal{U}(s-T_1)P\mathcal{U}^{-1}(s)x(s)\| - \|\mathcal{U}(s-T_1)(I-P)\mathcal{U}^{-1}(s)x(s)\| \\ &\geqslant K^{-1}e^{\alpha T_1}\|\mathcal{U}(s)P\mathcal{U}^{-1}(s)x(s)\| \\ &\quad - Ke^{-\alpha T_1}\|\mathcal{U}(s)(I-P)\mathcal{U}^{-1}(s)x(s)\| \\ &= K^{-1}e^{\alpha T_1}\|x_1(s)\| - Ke^{-\alpha T_1}\|x_2(s)\| \\ &\geqslant K^{-1}e^{\alpha T_1}\|x_1(s)\| - Ke^{-\alpha T_1}\|x_1(s)\| \\ &\geqslant 2\theta^{-1}\|x_1(s)\| \\ &\geqslant \theta^{-1}\|x(s)\|.\end{aligned}$$

Thus, $\|x(s)\| \leqslant \theta\|x(s-T_1)\|$. Hence,

$$s \geqslant a + T_1 \;\Rightarrow\; \|x(s)\| \leqslant \theta \sup_{|u-s|\leqslant T_1} \|x(u)\|.$$

Taking $T = a + T_1$, we conclude the proof of item (a).

(b) Since the linear generalized ODE (13.2) admits a bounded growth on $[a,\infty)$, it follows by Lemma 13.6 that there exist constants $C \geqslant 1$ and $\mu \geqslant 0$ such that

$$\|\mathcal{U}(t)\mathcal{U}^{-1}(s)\| \leqslant Ce^{\mu(t-s)} \quad \text{for all } t,s \in [a,\infty), \text{ with } t \geqslant s.$$

Let $t \in [a,\infty)$ and $h > 0$. Using conditions (i') and (ii'), we obtain

$$\|\mathcal{U}(t+h)P\mathcal{U}^{-1}(t)\| \leqslant Ke^{-\alpha h}\|\mathcal{U}(t)P\mathcal{U}^{-1}(t)\|$$

and

$$\|\mathcal{U}(t+h)(I-P)\mathcal{U}^{-1}(t)\| \geqslant K^{-1}e^{\alpha h}\|\mathcal{U}(t)(I-P)\mathcal{U}^{-1}(t)\|.$$

Define $\rho = \rho(t) = \left\|\mathcal{U}(t)(I-P)\mathcal{U}^{-1}(t)\right\|$, $\sigma = \sigma(t) = \left\|\mathcal{U}(t)P\mathcal{U}^{-1}(t)\right\|$ and consider h such that $\gamma = K^{-1}e^{\alpha h} - Ke^{-\alpha h} > 0$. Note that $|\rho - \sigma| \leqslant 1$ and

$$\begin{aligned}
&Ce^{\mu h}\left\|\rho^{-1}\mathcal{U}(t)(I-P)\mathcal{U}^{-1}(t) + \sigma^{-1}\mathcal{U}(t)P\mathcal{U}^{-1}(t)\right\| \\
&\quad \geqslant \left\|\mathcal{U}(t+h)\mathcal{U}^{-1}(t)[\rho^{-1}\mathcal{U}(t)(I-P)\mathcal{U}^{-1}(t) + \sigma^{-1}\mathcal{U}(t)P\mathcal{U}^{-1}(t)]\right\| \\
&\quad = \left\|\rho^{-1}\mathcal{U}(t+h)(I-P)\mathcal{U}^{-1}(t) + \sigma^{-1}\mathcal{U}(t+h)P\mathcal{U}^{-1}(t)\right\| \\
&\quad \geqslant \rho^{-1}\left\|\mathcal{U}(t+h)(I-P)\mathcal{U}^{-1}(t)\right\| - \sigma^{-1}\left\|\mathcal{U}(t+h)P\mathcal{U}^{-1}(t)\right\| \\
&\quad \geqslant K^{-1}e^{\alpha h} - Ke^{-\alpha h} = \gamma.
\end{aligned}$$

Consequently,

$$\begin{aligned}
\gamma C^{-1}e^{-\mu h} &\leqslant \left\|\rho^{-1}\mathcal{U}(t)(I-P)\mathcal{U}^{-1}(t) + \sigma^{-1}\mathcal{U}(t)P\mathcal{U}^{-1}(t)\right\| \\
&= \left\|\sigma^{-1}I + (\rho^{-1} - \sigma^{-1})\mathcal{U}(t)(I-P)\mathcal{U}^{-1}(t)\right\| \\
&\leqslant \sigma^{-1} + |\rho^{-1} - \sigma^{-1}|\rho = \sigma^{-1}(1 + \rho\sigma|\rho^{-1} - \sigma^{-1}|) \\
&= \sigma^{-1}(1 + |\rho\sigma(\rho^{-1} - \sigma^{-1})|) = \sigma^{-1}(1 + |\sigma - \rho|) \\
&\leqslant 2\sigma^{-1}.
\end{aligned}$$

Therefore, $\left\|\mathcal{U}(t)P\mathcal{U}^{-1}(t)\right\| = \sigma \leqslant 2\gamma^{-1}Ce^{\mu h}$ as desired. □

Our goal now is to present a result that gives us sufficient conditions to obtain exponential dichotomy for the linear generalized ODE (13.2) on $[0, \infty)$. Before that, we present an auxiliary result.

Lemma 13.8: *Let $J = [0, \infty)$. Assume that there are $T > 0$, $C > 1$, and $0 < \theta < 1$ such that any solution $x: [0, \infty) \to X$ of the linear generalized ODE* (13.2) *satisfies the conditions:*

(i) $\|x(t)\| \leqslant C\|x(s)\|$ *for* $0 \leqslant s \leqslant t \leqslant s + T$;
(ii) $\|x(t)\| \leqslant \theta \sup_{|u-t|\leqslant T} \|x(u)\|$ *for all* $t \geqslant T$.

If $x: [0, \infty) \to X$ is a bounded nontrivial solution of (13.2), *then*

$$\sup_{|u-t|\leqslant nT} \|x(u)\| \leqslant \theta \sup_{|u-t|\leqslant (n+1)T} \|x(u)\|, \quad n \in \mathbb{N},$$

provided $t \geqslant (n+1)T$.

Proof. Let x be a bounded nontrivial solution of the linear generalized ODE (13.2). Note that x is defined on $[0, \infty)$ as conditions H_1 and H_2 hold. Define the auxiliary function $\mu(s) = \sup_{u\geqslant s}\|x(u)\|$, for all $s \geqslant 0$. For $t \geqslant s + T$, it follows by condition (ii) that

$$\|x(t)\| \leqslant \theta \sup_{|u-t|<T} \|x(u)\| \leqslant \theta \sup_{u\geqslant s} \|x(u)\| = \theta\mu(s). \tag{13.7}$$

Consequently, $\sup_{t\geqslant s+T}\|x(t)\| \leqslant \theta\mu(s) < \mu(s)$ since $\theta < 1$, which implies

$$\mu(s) = \sup_{s\leqslant u\leqslant s+T} \|x(u)\| .$$

Also, by the definition of μ, we have $\|x(r)\| \leqslant \mu(s)$ for all $r \geqslant s$.

On the other hand, condition (i) yields

$$\|x(t)\| \leqslant \mu(s) = \sup_{s\leqslant u\leqslant s+T} \|x(u)\| \leqslant C\,\|x(s)\| , \quad 0 \leqslant s \leqslant t < \infty. \tag{13.8}$$

Now, let $t \geqslant (n+1)T$, $n \in \mathbb{N}$, and choose $\eta > 0$ such that $|\eta - t| \leqslant nT$. Note that $\eta \geqslant t - nT \geqslant T$. Using condition (ii),

$$\begin{aligned}\|x(\eta)\| &\leqslant \theta \sup_{\eta-T\leqslant u\leqslant \eta+T} \|x(u)\| \\ &\leqslant \theta \sup_{t-(n+1)T\leqslant u\leqslant t+(n+1)T} \|x(u)\| = \theta \sup_{|u-t|\leqslant (n+1)T} \|x(u)\| .\end{aligned}$$

Thus,

$$\sup_{|\eta-t|\leqslant nT} \|x(\eta)\| \leqslant \theta \sup_{|u-t|\leqslant (n+1)T} \|x(u)\|$$

and the result is proved. □

Theorem 13.9: *Let $J = [0,\infty)$ and assume that*

$$V_0 = \left\{x_0 \in X : \|x_0\| = 1 \text{ and } x(\cdot, x_0) \text{ is unbounded}\right\}$$

is a compact subset of X, where $x(t, x_0)$, $t \geqslant 0$, is the solution of (13.2) such that $x(0, x_0) = x_0$. Assume that there are $T > 0$, $C > 1$, and $0 < \theta < 1$ such that any solution $x : [0,\infty) \to X$ of the linear generalized ODE (13.2) satisfies the conditions:

(i) $\|x(t)\| \leqslant C\,\|x(s)\|$ *for* $0 \leqslant s \leqslant t \leqslant s + T$;
(ii) $\|x(t)\| \leqslant \theta \sup_{|u-t|\leqslant T} \|x(u)\|$ *for all* $t \geqslant T$.

Moreover, assume that for each $x_0 \in V_0$, there is an increasing sequence $\{t_n^{x_0}\}_{n\in\mathbb{N}} \subset \mathbb{R}_+$, with $t_{n+1}^{x_0} \leqslant t_n^{x_0} + T$, $n \in \mathbb{N}$, such that

$$\begin{aligned}&\left\|x(t, x_0)\right\| < \theta^{-n}C, \qquad \text{for all } t \in [0, t_n^{x_0}), \quad \text{and} \\ &\left\|x(t_n^{x_0}, x_0)\right\| \geqslant \theta^{-n}C, \quad n \in \mathbb{N}.\end{aligned}$$

Then the linear generalized ODE (13.2) admits an exponential dichotomy on $[0,\infty)$.

Proof. The proof follows the ideas of [48, Proposition 1, page 14]. At first, we prove the following auxiliary conditions:

(I) if $x : [0,\infty) \to X$ is a bounded nontrivial solution of (13.2), then there are $K > 1$ and $\alpha > 0$ such that $\|x(t)\| \leqslant Ke^{-\alpha(t-s)}\,\|x(s)\|$, for $0 \leqslant s \leqslant t < \infty$;

(II) if $x\colon [0,\infty) \to X$ is an unbounded solution of (13.2) such that $\|x(0)\| = 1$, then there are $K > 1$ and $\alpha > 0$ such that $\|x(t)\| \leqslant Ke^{-\alpha(s-t)}\|x(s)\|$ for $t_1^{x(0)} \leqslant t \leqslant s < \infty$.

Let us prove that (I) holds. For that, let $x\colon [0,\infty) \to X$ be a bounded nontrivial solution of the linear generalized ODE (13.2). Define $K = \theta^{-1}C > 1$ and $\alpha = -T^{-1}\ln\theta > 0$. Let $t \geqslant s$ and take $n \in \mathbb{N}_0$ such that

$$s + nT \leqslant t < s + (n+1)T.$$

In the case when $n = 0$, that is, $s \leqslant t < s + T$, we conclude that $\frac{(t-s)}{T} < 1$. Hence, taking into account that condition (i) holds, we obtain

$$\|x(t)\| \leqslant C\|x(s)\| = \theta^{-1}\theta C\|x(s)\| \leqslant \theta^{-1}C\theta^{\frac{(t-s)}{T}}\|x(s)\| = Ke^{-\alpha(t-s)}\|x(s)\|.$$

However, if $n \geqslant 1$, then the following two inequalities are valid:

$$t - nT \geqslant s \quad \text{and} \quad \frac{(t-s)}{T} < n + 1.$$

Using Lemma 13.8 and condition (ii), we get

$$\begin{aligned}\|x(t)\| &\leqslant \theta \sup_{|u-t|\leqslant T}\|x(u)\| \leqslant \theta^2 \sup_{|u-t|\leqslant 2T}\|x(u)\| \leqslant \cdots \leqslant \theta^n \sup_{|u-t|\leqslant nT}\|x(u)\| \\ &\leqslant \theta^n \sup_{u\geqslant s}\|x(u)\| \leqslant \theta^n C\|x(s)\| = \theta^{-1}C\theta^{n+1}\|x(s)\| \\ &\leqslant \theta^{-1}C\theta^{\frac{(t-s)}{T}}\|x(s)\| = Ke^{-\alpha(t-s)}\|x(s)\|.\end{aligned}$$

Thus, the statement (I) is proved.

Now, we verify that (II) holds. Set $K = \theta^{-1}C > 1$ and $\alpha = -T^{-1}\ln\theta > 0$. Let $x\colon [0,\infty) \to X$ be an unbounded solution of (13.2) satisfying the initial condition $\|x(0)\| = 1$. Condition (i) implies that

$$\|x(t)\| \leqslant C\|x(0)\| = C < \theta^{-1}C, \quad 0 \leqslant t \leqslant T. \tag{13.9}$$

Further, according to the hypothesis, we may obtain an increasing sequence $\{t_n^{x(0)}\}_{n\in\mathbb{N}} \subset \mathbb{R}_+$, which we denote simply by $\{t_n\}_{n\in\mathbb{N}}$, such that $t_{n+1} \leqslant t_n + T$ for all $n \in \mathbb{N}$, $\|x(t)\| < \theta^{-n}C$ for $t \in [0, t_n)$ and

$$\|x(t_n)\| \geqslant \theta^{-n}C, \quad n \in \mathbb{N}.$$

This last inequality together with (13.9) implies $t_1 > T$. Consequently, $T < t_1 < t_2 < \cdots < t_n < \cdots$. If $t_n \to \lambda$ as $n \to \infty$, for some λ, then $\lim_{s\to\lambda^-} x(s) = \infty$ which is a contradiction once the solution x is regulated. Hence, $t_n \to \infty$ as $n \to \infty$.

Let $t_1 \leqslant t \leqslant s$. We may assume, without loss of generality, that $t_m \leqslant t < t_{m+1}$ and $t_n \leqslant s < t_{n+1}$, for some $m, n \in \mathbb{N}$. In this way, we obtain $\frac{s-t}{T} < n - m + 1$ since

$$\begin{aligned}s - t &< t_{n+1} - t_m \leqslant t_n + T - t_m \leqslant t_{n-1} + 2T - t_m < \cdots \\ &< t_m + (n-m+1)T - t_m.\end{aligned}$$

Since condition (i) holds and taking into account the properties of the sequence $\{t_n\}_{n\in\mathbb{N}} \subset \mathbb{R}_+$, we conclude that

$$\begin{aligned} \|x(t)\| &< \theta^{-m-1}C = \theta^{n-m}C\theta^{-n-1} \leqslant \theta^{n-m}\left\|x(t_{n+1})\right\| \\ &\leqslant C\theta^{-1}\theta^{n-m+1}\|x(s)\| \leqslant C\theta^{-1}\theta^{\frac{(s-t)}{T}}\|x(s)\| \\ &= Ke^{-\alpha(s-t)}\|x(s)\|, \end{aligned}$$

and we conclude the proof of statement (II).

In order to show that the linear generalized ODE (13.2) admits an exponential dichotomy on $[0,\infty)$, let us consider the following subspace of X:

$$X_1 = \{\xi \in X : x(t,\xi) \text{ is bounded in } [0,\infty)\}.$$

Let X_2 be a subspace of X such that $X = X_1 \oplus X_2$. By hypothesis, for each $\xi \in X_2$ such that $\|\xi\| = 1$, there exists $t_1 = t_1(\xi)$ satisfying the following conditions:

$$x(t_1,\xi) \geqslant \theta^{-1}C \quad \text{and} \quad \|x(t,\xi)\| < \theta^{-1}C, \text{ for } 0 \leqslant t < t_1.$$

Claim. The set $\left\{t_1(\xi) : \xi \in X_2 \text{ and } \|\xi\| = 1\right\}$ is bounded.

Indeed, suppose to the contrary that there is a sequence $\{\xi_n\}_{n\in\mathbb{N}} \subset X_2$, $\left\|\xi_n\right\| = 1$, such that $t_1^{(n)} = t_1(\xi_n) \to \infty$ as $n \to \infty$. We may assume that $\xi_n \to \xi_0$ as $n \to \infty$, for some $\xi_0 \in X_2$, since V_0 is compact. Note that $\left\|\xi_0\right\| = 1$. Consequently, for every $t \geqslant 0$, we have

$$x(t,\xi_n) = U(t,0)\xi_n \to U(t,0)\xi_0 = x(t,\xi_0), \quad \text{as} \quad n \to \infty,$$

where $U: J \times J \to L(X)$ is the fundamental operator of (13.2). But this implies that

$$\left\|x(t,\xi_0)\right\| \leqslant \theta^{-1}C, \text{ for } 0 \leqslant t < \infty,$$

as $\left\|x(t,\xi_n)\right\| < \theta^{-1}C$ for all $0 \leqslant t < t_1^{(n)}$ and all $n \in \mathbb{N}$, which is a contradiction because $0 \neq \xi_0 \in X_2$. Hence,

$$\left\{t_1(\xi):\ \xi \in X_2 \text{and } \|\xi\| = 1\right\}$$

is bounded, that is, $T_1 = \sup_{\xi\in X_2, \|\xi\|=1} t_1(\xi) < \infty$.

Using condition (II), for each $a \in X_2$, $a \neq 0$, we obtain

$$\begin{aligned} \|x(t,a)\| = \|a\|\left\|x\left(t, \frac{a}{\|a\|}\right)\right\| &\leqslant \|a\|\, Ke^{-\alpha(s-t)}\left\|x\left(s, \frac{a}{\|a\|}\right)\right\| \\ &= Ke^{-\alpha(s-t)}\|x(s,a)\|, \end{aligned}$$

for $T_1 \leqslant t \leqslant s < \infty$.

Taking the projection $P \in L(X)$ from the decomposition $X = X_1 \oplus X_2$ on the subspace X_1, we conclude for each $\xi \in X$ that

$$\begin{aligned} &\|\mathcal{U}(t)P\xi\| \leqslant Ke^{-\alpha(t-s)}\|\mathcal{U}(s)P\xi\|, && t \geqslant s \geqslant T_1, \text{ and} \\ &\|\mathcal{U}(t)(I-P)\xi\| \leqslant Ke^{-\alpha(s-t)}\|\mathcal{U}(s)(I-P)\xi\|, && s \geqslant t \geqslant T_1. \end{aligned}$$

Now, Lemma 13.7 assures the existence of a constant $L > 0$ such that $\|\mathcal{U}(t)P\mathcal{U}^{-1}(t)\| \leqslant L$, for all $t \in [T_1, \infty)$. Proposition 13.3 shows that the linear generalized ODE (13.2) admits an exponential dichotomy on the interval $[T_1, \infty)$ and, finally, Lemma 13.2 concludes that (13.2) admits an exponential dichotomy on $[0, \infty)$. □

Remark 13.10: In Example 13.5, if $a(t) \geqslant a_0 > 0$, for all $t \geqslant 0$, then conditions (i) and (ii) from Lemma 13.7 are fulfilled with $P = I, K_1 = 1$, and $\alpha_1 = a_0$ (here K_2 and α_2 can be taken arbitrary). Consequently, by Lemma 13.7, item (a), given $\theta \in (0, 1)$, there is $T > 0$ such that for any solution $x: [0, \infty) \to \mathbb{R}$ of (13.5), we have

$$\|x(s)\| \leqslant \theta \sup_{|u-s| \leqslant T} \|x(u)\|, \text{ for all } s \geqslant T.$$

Moreover, since $P = I$, we conclude that

$$V_0 = \left\{x_0 \in \mathbb{R}: \|x_0\| = 1 \text{ and } x(\cdot, x_0) \text{is unbounded}\right\} = \emptyset.$$

Then, by Theorem 13.9, the generalized ODE (13.5) admits an exponential dichotomy on $[0, \infty)$.

13.2 Boundedness and Dichotomies

In the classical theory of ordinary differential equations, there exists a relation between exponential dichotomy and boundedness of solutions. In this section, we deal with this relation in the context of generalized ODEs, that is we investigate the relation between exponential dichotomy of the generalized ODE

$$\frac{dx}{d\tau} = D[A(t)x] \tag{13.10}$$

and the bounded solutions of

$$\frac{dx}{d\tau} = D\left[A(t)x + f(t)\right]. \tag{13.11}$$

Throughout this section, we will consider $J = \mathbb{R}$, the operator $A: \mathbb{R} \to L(X)$ satisfies conditions

(H_1') $A \in BV([a, b], L(X))$ for all interval $[a, b] \subset \mathbb{R}$,
(H_2') A satisfies condition (Δ) presented in Definition 6.2 with $J = \mathbb{R}$,

and $f: \mathbb{R} \to X$ is a function which will be specified later.

The definitions and the results presented in this section can be found in [29, Sec. 4].

Definition 13.11: We say that the linear generalized ODE (13.10) satisfies *condition* (D), whenever the operator A satisfies both conditions (H_1') and (H_2') and the generalized ODE (13.10) admits an exponential dichotomy, with positive constants K_1 and K_2, positive exponents α_1 and α_2 and a projection $P \in L(X)$, that is,

$$\begin{aligned} &\|\mathcal{U}(t)P\mathcal{U}^{-1}(s)\| \leqslant K_1 e^{-\alpha_1(t-s)}, \quad t \geqslant s, \quad \text{and} \\ &\|\mathcal{U}(t)(I-P)\mathcal{U}^{-1}(s)\| \leqslant K_2 e^{-\alpha_2(s-t)}, \quad s \geqslant t, \end{aligned} \tag{13.12}$$

where $\mathcal{U}: \mathbb{R} \to L(X)$ is given by $\mathcal{U}(t) = U(t,0)$, $t \in \mathbb{R}$, and $U: \mathbb{R} \times \mathbb{R} \to L(X)$ is the fundamental operator associated with the linear generalized ODE (13.10).

Proposition 13.12: *The unique bounded solution of the linear generalized ODE* (13.10), *satisfying condition* (D), *is the null solution.*

Proof. Let $x: \mathbb{R} \to X$ be a bounded solution of (13.10) and set $\xi = x(0)$. There exists $K > 0$ such that $\|x(t)\| \leqslant K$ for all $t \in \mathbb{R}$. Using Proposition 6.7, we conclude that the unique solution of the IVP

$$\begin{cases} \dfrac{dx}{d\tau} = D[A(t)x], \\ x(0) = \xi \end{cases}$$

is given by $x(t) = \mathcal{U}(t)\xi$, $t \in \mathbb{R}$. Note that

$$x(t) = \mathcal{U}(t)\xi = \mathcal{U}(t)P\xi + \mathcal{U}(t)(I-P)\xi, \quad t \in \mathbb{R}.$$

Set $x_1(t) = \mathcal{U}(t)P\xi$ and $x_2(t) = \mathcal{U}(t)(I-P)\xi$, $t \in \mathbb{R}$. Taking into account the inequalities in (13.12), we obtain

$$\|x_1(t)\| = \|\mathcal{U}(t)P\xi\| = \|\mathcal{U}(t)P\mathcal{U}^{-1}(0)\xi\| \leqslant K_1 e^{-\alpha_1 t} \|\xi\|, \quad t \geqslant 0, \quad \text{and} \tag{13.13}$$

$$\|x_2(t)\| = \|\mathcal{U}(t)(I-P)\xi\| = \|\mathcal{U}(t)(I-P)\mathcal{U}^{-1}(0)\xi\| \leqslant K_2 e^{\alpha_2 t} \|\xi\|, \quad t \leqslant 0. \tag{13.14}$$

Consequently,

$$\|x_1(t)\| = \|x(t) - x_2(t)\| \leqslant \|x(t)\| + \|x_2(t)\| \leqslant K + K_2\|\xi\|, \quad t \leqslant 0, \tag{13.15}$$

$$\|x_2(t)\| = \|x(t) - x_1(t)\| \leqslant \|x(t)\| + \|x_1(t)\| \leqslant K + K_1\|\xi\|, \quad t \geqslant 0. \tag{13.16}$$

The inequalities (13.13), (13.14), (13.15), and (13.16) say that x_1 and x_2 are bounded functions in $\mathbb{R}$. Thus, there exists $L > 0$ such that

$$\|x_1(t)\| + \|x_2(t)\| \leqslant L,$$

for all $t \in \mathbb{R}$. Using this boundedness of x_1 and x_2 and conditions (13.12), we obtain

$$\begin{aligned} \|P\xi\| &= \|x_1(0)\| = \|\mathcal{U}(0)P\xi\| = \|\mathcal{U}(0)P\mathcal{U}^{-1}(t)\mathcal{U}(t)P\xi\| \\ &\leqslant \|\mathcal{U}(0)P\mathcal{U}^{-1}(t)\| \, \|x_1(t)\| \leqslant K_1 e^{\alpha_1 t} L, \end{aligned}$$

for all $t \leqslant 0$, and

$$\begin{aligned}\|(I-P)\xi\| &= \|x_2(0)\| = \|\mathcal{U}(0)(I-P)\xi\| \\ &= \left\|\mathcal{U}(0)(I-P)\mathcal{U}^{-1}(t)\mathcal{U}(t)(I-P)\xi\right\| \\ &\leqslant \left\|\mathcal{U}(0)(I-P)\mathcal{U}^{-1}(t)\right\| \left\|x_2(t)\right\| \leqslant K_2 e^{-\alpha_2 t} L,\end{aligned}$$

for all $t \geqslant 0$. The previous estimates take us to $P\xi = 0$ and $(I-P)\xi = 0$, that is, $\xi = P\xi + (I-P)\xi = 0$. Therefore, $x(t) = 0$ for all $t \in \mathbb{R}$ as we have uniqueness of a solution. □

The next result gives us sufficient conditions for the nonhomogeneous linear generalized ODE (13.11) to not admit more than one bounded solution.

Corollary 13.13: *Assume that the linear generalized ODE* (13.10) *satisfies condition (D) and $f \in G(\mathbb{R}, X)$. Then the nonhomogeneous linear generalized ODE* (13.11) *admits at most one bounded solution.*

Proof. Suppose $x, y: \mathbb{R} \to X$ are two bounded solutions of the nonhomogeneous generalized ODE (13.11). Using Theorem 6.14 with $F(x,t) = f(t)$, these solutions can be written by

$$x(t) = \mathcal{U}(t)x(0) + f(t) - f(0) - \int_0^t d_s\left[\mathcal{U}(t)\mathcal{U}^{-1}(s)\right](f(s) - f(0)), \quad t \in \mathbb{R}, \text{ and}$$

$$y(t) = \mathcal{U}(t)y(0) + f(t) - f(0) - \int_0^t d_s\left[\mathcal{U}(t)\mathcal{U}^{-1}(s)\right](f(s) - f(0)), \quad t \in \mathbb{R}.$$

Recall that we are assuming in this section that $\mathcal{U}(t) = U(t,0)$, $t \in \mathbb{R}$.

Now, define $z: \mathbb{R} \to X$ by $z(t) = x(t) - y(t)$, $t \in \mathbb{R}$, which is clearly bounded. Since

$$z(t) = \mathcal{U}(t)x(0) - \mathcal{U}(t)y(0) = \mathcal{U}(t)z(0),$$

for all $t \in \mathbb{R}$, we have z is a bounded solution of the linear generalized ODE (13.10). By Proposition 13.12, $z(t) = x(t) - y(t) = 0$ for all $t \in \mathbb{R}$, that is, $x(t) = y(t)$ for all $t \in \mathbb{R}$, and the proof is complete. □

In the sequel, we present two results which give us sufficient conditions for the nonhomogeneous linear generalized ODE (13.11) to admit a unique bounded solution. These results are stated in Propositions 13.14 and 13.15.

Proposition 13.14: *Assume that the linear generalized ODE* (13.10) *satisfies condition (D), $f \in G(\mathbb{R}, X)$, the Perron–Stieltjes integrals*

$$\int_{-\infty}^t d_s\left[\mathcal{U}(t)P\mathcal{U}^{-1}(s)\right](f(s) - f(0)) \quad \textit{and}$$

$$\int_t^\infty d_s \left[\mathcal{U}(t)(I-P)\mathcal{U}^{-1}(s)\right](f(s)-f(0))$$

exist for each $t \in \mathbb{R}$, and the functions

$$t \in \mathbb{R} \mapsto \int_{-\infty}^t d_s \left[\mathcal{U}(t)P\mathcal{U}^{-1}(s)\right](f(s)-f(0)) \in X \quad \textit{and}$$
$$t \in \mathbb{R} \mapsto \int_t^\infty d_s \left[\mathcal{U}(t)(I-P)\mathcal{U}^{-1}(s)\right](f(s)-f(0)) \in X$$

are bounded, where $\mathcal{U}(t) = U(t,0)$, $t \in \mathbb{R}$. Then the nonhomogeneous linear generalized ODE (13.11) admits a unique bounded solution.

Proof. Denote by $x: \mathbb{R} \to X$ the solution of the initial value problem

$$\begin{cases} \dfrac{dx}{d\tau} = D\left[A(t)x + f(t)\right], \\ x(0) = -\displaystyle\int_{-\infty}^0 d_s \left[P\mathcal{U}^{-1}(s)\right](f(s)-f(0)) \\ \qquad + \displaystyle\int_0^\infty d_s \left[(I-P)\mathcal{U}^{-1}(s)\right](f(s)-f(0)). \end{cases}$$

According to Theorem 6.14, $x(t)$ is given by

$$x(t) = \mathcal{U}(t)x(0) + f(t) - f(0) - \int_0^t d_s \left[\mathcal{U}(t)\mathcal{U}^{-1}(s)\right](f(s)-f(0)), \quad t \in \mathbb{R}.$$

Since $\mathcal{U}(t)\mathcal{U}^{-1}(s) = \mathcal{U}(t)P\mathcal{U}^{-1}(s) + \mathcal{U}(t)(I-P)\mathcal{U}^{-1}(s)$, we have

$$\begin{aligned} x(t) &= \mathcal{U}(t)x(0) + f(t) - f(0) - \int_0^t d_s \left[\mathcal{U}(t)P\mathcal{U}^{-1}(s)\right](f(s)-f(0)) \\ &\quad - \int_0^t d_s \left[\mathcal{U}(t)(I-P)\mathcal{U}^{-1}(s)\right](f(s)-f(0)) \\ &= \mathcal{U}(t)x(0) + f(t) - f(0) \\ &\quad - \mathcal{U}(t)\left[\int_0^t d_s \left[P\mathcal{U}^{-1}(s)\right](f(s)-f(0)) \pm \int_{-\infty}^0 d_s \left[P\mathcal{U}^{-1}(s)\right](f(s)-f(0))\right. \\ &\quad + \int_0^t d_s \left[(I-P)\mathcal{U}^{-1}(s)\right](f(s)-f(0)) \\ &\quad \left. \pm \int_0^\infty d_s \left[(I-P)\mathcal{U}^{-1}(s)\right](f(s)-f(0))\right] \\ &= \mathcal{U}(t)x(0) + f(t) - f(0) - \mathcal{U}(t)\left[\int_{-\infty}^t d_s \left[P\mathcal{U}^{-1}(s)\right](f(s)-f(0))\right. \\ &\quad \left. - \int_t^\infty d_s \left[(I-P)\mathcal{U}^{-1}(s)\right](f(s)-f(0)) + x(0)\right] \end{aligned}$$

$$= f(t) - f(0) - \int_{-\infty}^{t} d_s \left[\mathcal{U}(t)P\mathcal{U}^{-1}(s)\right](f(s) - f(0))$$
$$+ \int_{t}^{\infty} d_s \left[\mathcal{U}(t)(I-P)\mathcal{U}^{-1}(s)\right](f(s) - f(0)).$$

The last equality shows that x is bounded, since by hypothesis the functions

$$t \in \mathbb{R} \mapsto \int_{-\infty}^{t} d_s \left[\mathcal{U}(t)P\mathcal{U}^{-1}(s)\right](f(s) - f(0)) \in X \quad \text{and}$$
$$t \in \mathbb{R} \mapsto \int_{t}^{\infty} d_s \left[\mathcal{U}(t)(I-P)\mathcal{U}^{-1}(s)\right](f(s) - f(0)) \in X$$

are bounded. In conclusion, Corollary 13.13 assures that x is the only bounded solution of (13.11). □

Proposition 13.15: *Assume that the linear generalized ODE (13.10) satisfies condition (D), $f \in G(\mathbb{R}, X)$, the Perron–Stieltjes integrals*

$$W_1(t) = \int_{-\infty}^{t} \mathcal{U}(t)P\mathcal{U}^{-1}(s)d[f(s)] \quad \textit{and}$$
$$W_2(t) = \int_{t}^{\infty} \mathcal{U}(t)(I-P)\mathcal{U}^{-1}(s)d[f(s)]$$

exist for all $t \in \mathbb{R}$, the functions $t \in \mathbb{R} \mapsto W_1(t) \in X$ and $t \in \mathbb{R} \mapsto W_2(t) \in X$ are bounded and the function $k: \mathbb{R} \to X$ given by

$$k(t) = \begin{cases} \mathcal{U}(t)\left(\sum\limits_{0 \leqslant \tau < t} \Delta^+\mathcal{U}^{-1}(\tau)\Delta^+ f(\tau) - \sum\limits_{0 < \tau \leqslant t} \Delta^-\mathcal{U}^{-1}(\tau)\Delta^- f(\tau)\right), & t > 0 \\ \mathcal{U}(t)\left(\sum\limits_{t \leqslant \tau < 0} \Delta^+\mathcal{U}^{-1}(\tau)\Delta^+ f(\tau) - \sum\limits_{t < \tau \leqslant 0} \Delta^-\mathcal{U}^{-1}(\tau)\Delta^- f(\tau)\right), & t < 0 \\ 0, \quad t = 0 \end{cases}$$

is bounded. Then the nonhomogeneous linear generalized ODE (13.11) admits a unique bounded solution.

Proof. Consider the initial value problem

$$\begin{cases} \dfrac{dx}{d\tau} = D\left[A(t)x + f(t)\right], \\ x(0) = \displaystyle\int_{-\infty}^{0} P\mathcal{U}^{-1}(s)d[f(s)] - \int_{0}^{\infty} (I-P)\mathcal{U}^{-1}(s)d[f(s)], \end{cases}$$

and let $x: \mathbb{R} \to X$ be its solution. According to Theorem 6.14, $x(t)$ can be represented by

$$x(t) = \mathcal{U}(t)x(0) + f(t) - f(0) - \int_{0}^{t} d_s \left[\mathcal{U}(t)\mathcal{U}^{-1}(s)\right](f(s) - f(0)), \ t \in \mathbb{R}.$$

On the other hand, by Proposition 1.70, for each $t \in \mathbb{R}$, we have

$$
\begin{aligned}
x(t) &= \mathcal{U}(t)x(0) + f(t) - f(0) \\
&\quad - \left(f(t) - f(0) - \int_0^t \mathcal{U}(t)\mathcal{U}^{-1}(s)d\left[f(s) - f(0)\right] - k(t) \right) \\
&= \mathcal{U}(t)x(0) + k(t) + \int_0^t \mathcal{U}(t)P\mathcal{U}^{-1}(s)d\left[f(s)\right] \\
&\quad + \int_0^t \mathcal{U}(t)(I-P)\mathcal{U}^{-1}(s)d\left[f(s)\right] \\
&= \mathcal{U}(t)x(0) + k(t) + \mathcal{U}(t)\left[\int_0^t P\mathcal{U}^{-1}(s)d\left[f(s)\right] \pm \int_{-\infty}^0 P\mathcal{U}^{-1}(s)d\left[f(s)\right] \right. \\
&\quad \left. + \int_0^t (I-P)\mathcal{U}^{-1}(s)d[f(s)] \pm \int_0^\infty (I-P)\mathcal{U}^{-1}(s)d\left[f(s)\right] \right] \\
&= \int_{-\infty}^t \mathcal{U}(t)P\mathcal{U}^{-1}(s)d\left[f(s)\right] - \int_t^\infty \mathcal{U}(t)(I-P)\mathcal{U}^{-1}(s)d\left[f(s)\right] + k(t) \\
&= W_1(t) - W_2(t) + k(t).
\end{aligned}
$$

By hypothesis, the functions W_1, W_2, and k are bounded in $\mathbb{R}$. Hence, x is bounded in $\mathbb{R}$. By Corollary 13.13, x is the only bounded solution of the nonhomogeneous linear generalized ODE (13.11). □

In the next result, we present sufficient conditions for the nonhomogeneous linear generalized ODE (13.11) to admit a unique τ-periodic solution.

Proposition 13.16: *Assume the same hypotheses of Proposition 13.14 hold. Assume, in addition, that the projection P in (13.12) is uniquely determined and A and f are τ-periodic mappings, $\tau > 0$. Then the nonhomogeneous linear generalized ODE (13.11) admits a unique τ-periodic solution.*

Proof. According to the proof of Proposition 13.14, the solution $x : \mathbb{R} \to X$ of the following initial value problem

$$
\begin{cases}
\dfrac{dx}{d\tau} = D\left[A(t)x + f(t)\right], \\
x(0) = -\displaystyle\int_{-\infty}^0 d_s\left[P\mathcal{U}^{-1}(s)\right](f(s) - f(0)) \\
\qquad + \displaystyle\int_0^\infty d_s\left[(I-P)\mathcal{U}^{-1}(s)\right](f(s) - f(0)),
\end{cases}
$$

can be rewritten as

$$x(t) = f(t) - f(0) - \int_{-\infty}^{t} d_s \left[\mathcal{U}(t)P\mathcal{U}^{-1}(s)\right] (f(s) - f(0))$$
$$+ \int_{t}^{\infty} d_s \left[\mathcal{U}(t)(I - P)\mathcal{U}^{-1}(s)\right] (f(s) - f(0)), \quad t \in \mathbb{R}. \tag{13.17}$$

Claim. The mappings $\mathbb{R} \ni t \mapsto \mathcal{U}(t)P\mathcal{U}^{-1}(t)$ and $\mathbb{R} \ni t \mapsto \mathcal{U}(t)(I - P)\mathcal{U}^{-1}(t)$ are τ-periodic.

We are going to show that the mapping $\mathbb{R} \ni t \mapsto \mathcal{U}(t)P\mathcal{U}^{-1}(t)$ is τ-periodic. We may proceed in the same way to conclude that the mapping $\mathbb{R} \ni t \mapsto \mathcal{U}(t)(I - P)\mathcal{U}^{-1}(t)$ is τ-periodic. In fact, define the operator $Y : \mathbb{R} \to L(X)$ by $Y(t) = \mathcal{U}(t + \tau)$. This operator is a solution of the linear generalized ODE

$$\begin{cases} \dfrac{dY}{d\tau} = D\left[A(t)Y\right], \\ Y(0) = \mathcal{U}(\tau), \end{cases}$$

since

$$Y(t) = I + \int_{0}^{t+\tau} d[A(s)]\mathcal{U}(s) = \mathcal{U}(\tau) + \int_{\tau}^{t+\tau} d[A(s)]\mathcal{U}(s)$$
$$= \mathcal{U}(\tau) + \int_{0}^{t} d[A(s+\tau)]\mathcal{U}(s+\tau) = \mathcal{U}(\tau) + \int_{0}^{t} d[A(s)]Y(s), \quad t \in \mathbb{R},$$

where in the third equality, we used Theorem 2.18, and in the last equality, we used the periodicity of A. Hence, we may write $Y(t) = \mathcal{U}(t)\mathcal{U}(\tau)$, $t \in \mathbb{R}$. In other words, $\mathcal{U}(t + \tau) = \mathcal{U}(t)\mathcal{U}(\tau)$, for all $t \in \mathbb{R}$.

Now, consider the projection $\tilde{P} = \mathcal{U}(\tau)P\mathcal{U}^{-1}(\tau)$. Using the estimates (13.12), we obtain

$$\begin{aligned} \left\|\mathcal{U}(t)\tilde{P}\mathcal{U}^{-1}(s)\right\| &= \left\|\mathcal{U}(t)\mathcal{U}(\tau)P\mathcal{U}^{-1}(\tau)\mathcal{U}^{-1}(s)\right\| \\ &= \left\|\mathcal{U}(t+\tau)P\mathcal{U}^{-1}(s+\tau)\right\| \\ &\leqslant K_1 e^{-\alpha_1(t-s)}, \end{aligned}$$

for all $t \geqslant s$, and

$$\left\|\mathcal{U}(t)(I - \tilde{P})\mathcal{U}^{-1}(s)\right\| = \left\|\mathcal{U}(t+\tau)(I - P)\mathcal{U}^{-1}(s+\tau)\right\| \leqslant K_2 e^{-\alpha_2(s-t)},$$

for all $s \geqslant t$. By the uniqueness of the projection, we obtain $P = \tilde{P} = \mathcal{U}(\tau)P\mathcal{U}^{-1}(\tau)$. Therefore,

$$\mathcal{U}(t+\tau)P\mathcal{U}^{-1}(t+\tau) = \mathcal{U}(t)\tilde{P}\mathcal{U}^{-1}(t) = \mathcal{U}(t)P\mathcal{U}^{-1}(t),$$

that is, $\mathcal{U}(t)P\mathcal{U}^{-1}(t)$ is τ-periodic.

Calculating $x(t+\tau)$ in the expression (13.17) and using the τ-periodicity of $\mathcal{U}(t)P\mathcal{U}^{-1}(t)$, we get

$$\begin{aligned}
x(t+\tau) &= f(t+\tau) - f(0) - \int_{-\infty}^{t+\tau} d_s\left[\mathcal{U}(t+\tau)P\mathcal{U}^{-1}(s)\right](f(s)-f(0)) \\
&\quad + \int_{t+\tau}^{\infty} d_s\left[\mathcal{U}(t+\tau)(I-P)\mathcal{U}^{-1}(s)\right](f(s)-f(0)) \\
&= f(t) - f(0) - \int_{-\infty}^{t} d_s\left[\mathcal{U}(t+\tau)P\mathcal{U}^{-1}(s+\tau)\right](f(s+\tau)-f(0)) \\
&\quad + \int_{t}^{\infty} d_s\left[\mathcal{U}(t+\tau)(I-P)\mathcal{U}^{-1}(s+\tau)\right](f(s+\tau)-f(0)) \\
&= f(t) - f(0) \\
&\quad - \int_{-\infty}^{t} d_s\left[\mathcal{U}(t+\tau)\mathcal{U}^{-1}(s+\tau)\mathcal{U}(s+\tau)P\mathcal{U}^{-1}(s+\tau)\right](f(s)-f(0)) \\
&\quad + \int_{t}^{\infty} d_s\left[\mathcal{U}(t+\tau)\mathcal{U}^{-1}(s+\tau)\mathcal{U}(s+\tau)(I-P)\mathcal{U}^{-1}(s+\tau)\right] \\
&\quad \times (f(s)-f(0)) \\
&= f(t) - f(0) - \int_{-\infty}^{t} d_s\left[\mathcal{U}(t)\mathcal{U}(\tau)\mathcal{U}^{-1}(\tau)\mathcal{U}^{-1}(s)\mathcal{U}(s)P\mathcal{U}^{-1}(s)\right] \\
&\quad \times (f(s)-f(0)) \\
&\quad + \int_{t}^{\infty} d_s\left[\mathcal{U}(t)\mathcal{U}(\tau)\mathcal{U}^{-1}(\tau)\mathcal{U}^{-1}(s)\mathcal{U}(s)(I-P)\mathcal{U}^{-1}(s)\right] \\
&\quad \times (f(s)-f(0)) \\
&= f(t) - f(0) - \int_{-\infty}^{t} d_s\left[\mathcal{U}(t)P\mathcal{U}^{-1}(s)\right](f(s)-f(0)) \\
&\quad + \int_{t}^{\infty} d_s\left[\mathcal{U}(t)(I-P)\mathcal{U}^{-1}(s)\right](f(s)-f(0)) = x(t),
\end{aligned}$$

which shows that x is τ-periodic. Therefore, by Corollary 13.13, x is the only solution τ-periodic of the nonhomogeneous linear generalized ODE (13.11). □

We may ask about the existence of the integrals presented in Propositions 13.14 and 13.15. In the next result, we present some sufficient conditions for the existence of such integrals.

Proposition 13.17: *Assume that the linear generalized ODE* (13.10) *satisfies condition* (D).

(i) *If* $f \in G(\mathbb{R}, X)$ *and*

$$\left\| \int_a^b Y(r)d[f(r)] \right\| \leqslant \int_a^b \|Y(r)\|\, \delta dr$$

holds for every operator $Y \in BV_{loc}(\mathbb{R}, L(X))$, *for some* $\delta > 0$ *and for all* $a, b \in \mathbb{R}$ *with* $a < b$, *then the hypotheses of Proposition 13.15 are satisfied.*

(ii) *If* $f \in G(\mathbb{R}, X)$ *is of bounded variation on* $\mathbb{R}$, *that is,*

$$V_f = \sup\{var_a^b(f): a, b \in \mathbb{R}, a < b\} < \infty,$$

then the hypotheses of Proposition 13.15 are satisfied.

(iii) *If* $f: \mathbb{R} \to X$ *is a bounded function,* $\|[I - (A(t) - A(t^-))]^{-1}\| \leqslant C$ *for all* $t \in \mathbb{R}$ *and*

$$V_A = \sup\{var_a^b(A): a, b \in \mathbb{R}, a < b\} < \infty,$$

then the hypotheses of Proposition 13.14 are satisfied.

Proof. (i) It is enough to prove that the integral $W_1(t)$ exists, and it is bounded. Analogously, one can prove that $W_2(t)$ exists, and it is bounded. Let $t \in \mathbb{R}$ be fixed and define the sequence

$$x_n = \int_{-n}^{t} \mathcal{U}(t)P\mathcal{U}^{-1}(r)d[f(r)], \quad n \in \mathbb{N}.$$

Let us show that $\{x_n\}_{n\in\mathbb{N}}$ is a Cauchy sequence. In fact, for $n \geqslant m > |t|$, we have

$$\begin{aligned}\|x_n - x_m\| &= \left\|\int_{-n}^{-m} \mathcal{U}(t)P\mathcal{U}^{-1}(r)d[f(r)]\right\| \leqslant \int_{-n}^{-m} K_1 e^{-\alpha_1(t-r)}\delta dr \\ &\leqslant K_1\delta e^{-\alpha_1 t}\frac{1}{\alpha_1}\left(e^{-\alpha_1 m} - e^{-\alpha_1 n}\right).\end{aligned}$$

Then,

$$\|x_n - x_m\| \to 0, \quad \text{as } n, m \to \infty.$$

Hence, $\{x_n\}_{n\in\mathbb{N}}$ is a Cauchy sequence. Consequently, the limit $\lim_{n\to\infty} x_n$ exists, that is, the Perron–Stieltjes integral

$$W_1(t) = \int_{-\infty}^{t} \mathcal{U}(t)P\mathcal{U}^{-1}(r)d[f(r)]$$

exists (the existence of this integral is due to the Hake-type theorem – see Theorem 2.12 and Remark 2.13; see also the comments after Lemma 2.2 concerning integration over unbounded intervals). Moreover, using the previous calculations, we obtain

$$\|x_n\| = \left\|\int_{-n}^{t} \mathcal{U}(t)P\mathcal{U}^{-1}(r)d[f(r)]\right\| \leqslant \frac{K_1}{\alpha_1}\delta,$$

for all $n \in \mathbb{N}$, and, consequently, $\|W_1(t)\| \leqslant \frac{K_1}{\alpha_1}\delta$.

By the arbitrariness of t and by the fact that the limiting factor $\frac{K_1}{\alpha_1}\delta$ does not depend on the choice of $t \in \mathbb{R}$, we have the existence of the Perron–Stieltjes integral

$$W_1(t) = \int_{-\infty}^{t} \mathcal{U}(t)P\mathcal{U}^{-1}(r)d[f(r)],$$

for every $t \in \mathbb{R}$, and the map $\mathbb{R} \ni t \mapsto W_1(t)$ is bounded.

(ii) We prove just that the integral $W_1(t)$ exists, and it is bounded. For a fixed $t \in \mathbb{R}$, we define $x_n = \int_{-n}^{t} \mathcal{U}(t)P\mathcal{U}^{-1}(r)d[f(r)]$, $n \in \mathbb{N}$. As in item (i), for $n \geqslant m > |t|$, and using Corollary 1.48, we obtain

$$\begin{aligned}
\|x_n - x_m\| &\leqslant \int_{-n}^{-m} \|\mathcal{U}(t)P\mathcal{U}^{-1}(r)\| \, d[\mathrm{var}_{-n}^{r}(f)] \\
&\leqslant (\mathrm{var}_{-n}^{-m}(f)) \sup_{r\in[-n,-m]} \|\mathcal{U}(t)P\mathcal{U}^{-1}(r)\| \\
&\leqslant K_1 V_f \sup_{r\in[-n,-m]} e^{-\alpha_1(t-r)} \leqslant K_1 V_f \, e^{-\alpha_1(t+m)}.
\end{aligned}$$

Then $\|x_n - x_m\| \to 0$ as $n, m \to \infty$, which shows that $\{x_n\}_{n\in\mathbb{N}}$ is a Cauchy sequence. Thus, the Perron–Stieltjes integral $W_1(t) = \int_{-\infty}^{t} \mathcal{U}(t)P\mathcal{U}^{-1}(r)d[f(r)]$ exists. Since

$$\|x_n\| = \left\| \int_{-n}^{t} \mathcal{U}(t)P\mathcal{U}^{-1}(r)d[f(r)] \right\| \leqslant K_1 V_f \sup_{r\in[-n,t]} e^{-\alpha_1(t-r)} \leqslant K_1 V_f,$$

for every $n \in \mathbb{N}$, we get $\|W_1(t)\| \leqslant K_1 V_f$.

By the arbitrariness of t and by the fact that the limiting factor $K_1 V_f$ does not depend on the choice of $t \in \mathbb{R}$, we have the existence of the Perron–Stieltjes integral $W_1(t)$ for every $t \in \mathbb{R}$ and the boundedness of the map $\mathbb{R} \ni t \mapsto W_1(t)$.

(iii) Looking at the proof of [209, Theorem 6.15], we also can obtain the following estimates:

$$\|U(t,s)\| \leqslant Ce^{C\,\mathrm{var}_a^b(A)} \quad \text{for all} \quad t, s \in [a,b],$$

$$\mathrm{var}_a^b(U(\cdot,s)) \leqslant Ce^{C\,\mathrm{var}_a^b(A)}\,\mathrm{var}_a^b(A) \quad \text{and} \quad \mathrm{var}_a^b(U(t,\cdot)) \leqslant C^2e^{2C\,\mathrm{var}_a^b(A)}\,\mathrm{var}_a^b(A).$$

Claim. $\mathrm{var}_\eta^t(\mathcal{U}(t)P\mathcal{U}^{-1}(\cdot)) \leqslant L_1 e^{-\beta_1(t-\eta)}\,\|P\|\,C^3e^{3CV_A}V_A$, for every $t \geqslant \eta$, where the constants L_1 and β_1 come from Proposition 13.3.

Indeed, let $d = s_0 \leqslant s_1 \leqslant \cdots \leqslant s_{|d|}$ be a division of $[\eta, t]$ and remember that $\mathcal{U}^{-1}(t) = U(0,t)$, $t \in \mathbb{R}$, then

$$\begin{aligned}
&\sum_{i=1}^{|d|} \|\mathcal{U}(t)P\mathcal{U}^{-1}(s_i) - \mathcal{U}(t)P\mathcal{U}^{-1}(s_{i-1})\| \leqslant \sum_{i=1}^{|d|} \|\mathcal{U}(t)P\|\,\|\mathcal{U}^{-1}(s_i) - \mathcal{U}^{-1}(s_{i-1})\| \\
&\leqslant L_1 e^{-\beta_1(t-\eta)}\,\|\mathcal{U}(\eta)P\|\,C^2e^{2C\,\mathrm{var}_\eta^t(A)}\,\mathrm{var}_\eta^t(A) \leqslant L_1 e^{-\beta_1(t-\eta)}\,\|P\|\,C^3e^{3CV_A}V_A,
\end{aligned}$$

and the claim is proved.

Now, for a fixed $t \in \mathbb{R}$, define $x_n = \int_{-n}^{t} d_s[\mathcal{U}(t)P\mathcal{U}^{-1}(s)](f(s) - f(0))$, $n \in \mathbb{N}$. Since f is bounded, there is $M > 0$ such that $\|f(t)\| \leqslant M$ for all $t \in \mathbb{R}$. As in item (i), we have $\{x_n\}_{n\in\mathbb{N}}$ is a Cauchy sequence because for $n \geqslant m > |t|$, we have

$$\begin{aligned}\|x_n - x_m\| &\leqslant \text{var}_{-n}^{-m}(\mathcal{U}(t)P\mathcal{U}^{-1}(\cdot)) \sup_{s\in[-n,-m]} \|f(s) - f(0)\| \\ &\leqslant 2M \, \text{var}_{-n}^{t}(\mathcal{U}(t)P\mathcal{U}^{-1}(\cdot)) \\ &\leqslant 2ML_1 e^{-\beta_1(t+n)} \|P\| \, C^3 e^{3CV_A} V_A,\end{aligned}$$

which shows that $\|x_n - x_m\| \to 0$ as $n, m \to \infty$. Hence, using the arbitrariness of t, we can conclude that the integral $\displaystyle\int_{-\infty}^{t} d_s[\mathcal{U}(t)P\mathcal{U}^{-1}(s)](f(s) - f(0))$ exists for every $t \in \mathbb{R}$. It is not difficult to see that

$$\sup_{t\in\mathbb{R}} \left\| \int_{-\infty}^{t} d_s[\mathcal{U}(t)P\mathcal{U}^{-1}(s)](f(s) - f(0)) \right\| \leqslant 2ML_1 \, \|P\| \, C^3 e^{3CV_A} V_A.$$

Similarly, we show the existence of the integral $\int_t^\infty d_s[\mathcal{U}(t)(I - P)\mathcal{U}^{-1}(s)](f(s) - f(0))$ and the boundedness

$$\sup_{t\in\mathbb{R}} \left\| \int_{t}^{\infty} d_s[\mathcal{U}(t)(I - P)\mathcal{U}^{-1}(s)](f(s) - f(0)) \right\| \leqslant 2ML_2(1 + \|P\|)C^3 e^{3CV_A} V_A.$$

□

13.3 Applications to MDEs

In this section, we apply the results on exponential dichotomies for a class of measure differential equations (MDEs). We shall use the correspondence between generalized ODEs and MDEs to obtain the results. The reader may find all the concepts and results presented in this section in [29, Section 5].

We consider $J \subset \mathbb{R}$ an interval and X a Banach space. We will study the following MDE in the integral form:

$$x(t) = x(t_0) + \int_{t_0}^{t} \mathcal{F}(s)x(s)ds + \int_{t_0}^{t} \mathcal{G}(s)x(s)du(s), \tag{13.18}$$

where $t_0 \in J$ and the functions $\mathcal{F}: J \to L(X)$, $\mathcal{G}: J \to L(X)$, and $u: J \to \mathbb{R}$ satisfy the following conditions:

(C1) $\mathcal{F}(\cdot)$ is Perron integrable over J;
(C2) u is of locally bounded variation in J and left-continuous on $J \setminus \{\inf J\}$;
(C3) $\mathcal{G}(\cdot)$ is Perron–Stieltjes integrable with respect to u over J;
(C4) there exists a locally Perron integrable function $m_1 : J \to \mathbb{R}$ such that for each $a, b \in J$, $a \leqslant b$, we have $\left\|\int_a^b \mathcal{F}(s)ds\right\| \leqslant \int_a^b m_1(s)ds$;

(C5) there is a locally Perron–Stieltjes integrable function $m_2: J \to \mathbb{R}$ with respect to u, such that for each $a, b \in J$, $a \leqslant b$, we have $\left\| \int_a^b \mathcal{G}(s)du(s) \right\| \leqslant \int_a^b m_2(s)du(s)$;

(C6) for all t such that t is a point of discontinuity of u, we have

$$\left(I + \lim_{r \to t+} \int_t^r \mathcal{G}(s)du(s) \right)^{-1} \in L(X).$$

Let $t_0 \in [\alpha, \beta]$. According to Definition 3.2, a function $x: [\alpha, \beta] \subset J \to X$ is a *solution* of the MDE (13.18) on $[\alpha, \beta]$, if the Perron integral $\int_\alpha^\beta \mathcal{F}(s)x(s)ds$ and the Perron–Stieltjes integral $\int_\alpha^\beta \mathcal{G}(s)x(s)du(s)$ exist and the equality (13.18) holds for all $t \in [\alpha, \beta]$.

Next, we present the correspondence result between the MDE (13.18) and its corresponding generalized ODE. See Theorem 4.14. The reader also may consult [209, Theorem 5.17].

Theorem 13.18: *Let $t_0 \in [\alpha, \beta]$. The function $x: [\alpha, \beta] \subset J \to X$ is a solution of (13.18), with initial condition $x(t_0) = x_0$, if and only if x is solution of*

$$\begin{cases} \dfrac{dx}{d\tau} = D[A(t)x + F(t)x], \\ x(t_0) = x_0, \end{cases}$$

where $A(t) = \displaystyle\int_{t_0}^t \mathcal{F}(s)ds$ *and* $F(t) = \displaystyle\int_{t_0}^t \mathcal{G}(s)du(s)$ *for all* $t \in [\alpha, \beta]$.

Theorem 13.19 concerns the existence and uniqueness of a solution of the MDE (13.18).

Theorem 13.19: *Let $\mathcal{F}$ and $\mathcal{G}$ satisfy conditions (C1)–(C6). Then the MDE (13.18) admits a unique solution defined in J.*

Proof. Consider the operators $A: J \to L(X)$ and $F: J \to L(X)$ defined in Theorem 13.18. We need to show that $A + F$ satisfies conditions (H_1)–(H_2).

At first, we show that $A + F$ satisfies condition H_1. Indeed, let $a, b \in J, a < b$, and $d = t_0 \leqslant t_1 \leqslant \ldots \leqslant t_{|d|} = b$ be a division of $[a, b]$. Since

$$\sum_{i=1}^{|d|} \|A(t_i) + F(t_i) - A(t_{i-1}) - F(t_{i-1})\|$$

$$\leqslant \sum_{i=1}^{|d|} \left\| \int_{t_{i-1}}^{t_i} \mathcal{F}(s)ds \right\| + \sum_{i=1}^{|d|} \left\| \int_{t_{i-1}}^{t_i} \mathcal{G}(s)du(s) \right\|$$

$$\leqslant \sum_{i=1}^{|d|} \int_{t_{i-1}}^{t_i} m_1(s)ds + \sum_{i=1}^{|d|} \int_{t_{i-1}}^{t_i} m_2(s)du(s) = \int_a^b m_1(s)ds + \int_a^b m_2(s)du(s),$$

we conclude that

$$\mathrm{var}_a^b(A+F) \leqslant \int_a^b m_1(s)ds + \int_a^b m_2(s)du(s) < \infty.$$

Hence, condition H_1 is satisfied.

Now, let us prove that $A + F$ satisfies condition H_2. At first, we note that $A \in C(J, L(X))$ as

$$\|A(t_2) - A(t_1)\| = \left\| \int_{t_1}^{t_2} \mathcal{F}(s)ds \right\| \leqslant \int_{t_1}^{t_2} m_1(s)ds,$$

for all $t_1, t_2 \in J$ with $t_1 \leqslant t_2$. On the other hand,

$$\|F(t_2) - F(t_1)\| = \left\| \int_{t_1}^{t_2} \mathcal{G}(s)du(s) \right\| \leqslant \int_{t_1}^{t_2} m_2(s)du(s),$$

for all $t_1, t_2 \in J$ with $t_1 \leqslant t_2$, which implies that F is left-continuous on $J \setminus \{\inf J\}$ since u has the same property. Thus, for $t \in J$, we have $A(t^-) = A(t^+) = A(t)$ and $F(t) = F(t^-)$, which shows that

$$(I - [(A+F)(t) - (A+F)(t^-)])^{-1} = I \in L(X)$$

and

$$(I + [(A+F)(t^+) - (A+F)(t)])^{-1} = (I + F(t^+) - F(t))^{-1}$$
$$= \left(I + \lim_{\alpha \to t^+} \int_t^{\alpha} \mathcal{G}(s)du(s) \right)^{-1} \in L(X),$$

where we used condition (C6). Hence, $A + F$ satisfies condition H_2.

According to Proposition 6.7 and Theorem 13.18 the result is proved. □

In the next result, we exhibit the concept of the fundamental operator associated with the MDE (13.18), see Theorem 13.20.

Theorem 13.20: *There exists a unique operator $V: J \times J \to L(X)$ such that*

$$V(t,s) = I + \int_s^t \mathcal{F}(r)V(r,s)dr + \int_s^t \mathcal{G}(r)V(r,s)du(r), \tag{13.19}$$

for all $t, s \in J$. Moreover, for each fixed $s \in J$, $V(\cdot, s)$ is an operator of locally bounded variation. For $t_0 \in J$, the function $x(t) = V(t, t_0)\tilde{x}$ is the unique solution of (13.18) satisfying the initial condition $x(t_0) = \tilde{x}$, with $\tilde{x} \in X$. This operator is called the fundamental operator of the MDE (13.18).

Proof. By Theorem 6.6, there exists a unique operator $U: J \times J \to L(X)$ such that

$$U(t,s) = I + \int_s^t d[A(r) + F(r)]U(r,s), \quad t,s \in J,$$

where $A: J \to L(X)$ and $F: J \to L(X)$ are the operators defined in Theorem 13.18. Note that $A + F$ satisfies conditions H_1–H_2 by the proof of Theorem 13.19. Moreover, for each $s \in J$, $U(\cdot, s)$ is an operator of locally bounded variation in J.

Using Proposition 1.67, we obtain

$$\int_s^t d[A(r)]U(r,s) = \int_s^t d[A(r) - A(s)]U(r,s) = \int_s^t \mathcal{F}(r)U(r,s)dr \quad \text{and}$$
$$\int_s^t d[F(r)]U(r,s) = \int_s^t d[F(r) - F(s)]U(r,s) = \int_s^t \mathcal{G}(r)U(r,s)du(r),$$

that is,

$$U(t,s) = I + \int_s^t \mathcal{F}(r)U(r,s)dr + \int_s^t \mathcal{G}(r)U(r,s)du(r), \quad t,s \in J.$$

Hence, $V: J \times J \to L(X)$ defined by $V(t,s) = U(t,s)$, $t,s \in J$, is the unique operator that satisfies (13.19).

Lastly, let $x(t) = V(t,t_0)\tilde{x}$ with $\tilde{x} \in X$ and $t \in J$. Since

$$\begin{aligned}
&\int_{t_0}^t \mathcal{F}(s)x(s)ds + \int_{t_0}^t \mathcal{G}(s)x(s)du(s) \\
&\quad = \int_{t_0}^t \mathcal{F}(s)V(s,t_0)\tilde{x}ds + \int_{t_0}^t \mathcal{G}(s)V(s,t_0)\tilde{x}du(s) \\
&\quad = (V(t,t_0) - I)\tilde{x} = x(t) - \tilde{x},
\end{aligned}$$

and we are assuming that conditions (C1)–(C5) hold, it follows by Theorem 13.19 that $x(t) = V(t,t_0)\tilde{x}$ is the unique solution of (13.18) such that $x(t_0) = \tilde{x}$. □

As a consequence of Theorem 13.20, we have the following result.

Corollary 13.21: *Let $V: J \times J \to L(X)$ be the fundamental operator of (13.18) and $U: J \times J \to L(X)$ be the fundamental operator of the corresponding linear generalized ODE*

$$\frac{dx}{d\tau} = D[A(t)x + F(t)x], \tag{13.20}$$

where $A(t) = \int_{t_0}^t \mathcal{F}(s)ds$ and $F(t) = \int_{t_0}^t \mathcal{G}(s)du(s)$, for all $t \in J$. Then $U(t,s) = V(t,s)$ for all $t,s \in J$.

Let V be the fundamental operator of (13.18). Define the operator $\mathcal{V}: J \to L(X)$ by

$$\mathcal{V}(t) = V(t, t_0), \quad t \in J,$$

where $t_0 \in J$. We denote $\mathcal{V}^{-1}(t) = V(t_0, t)$, $t, t_0 \in J$. In the next definition, we present the concept of exponential dichotomy for the MDE (13.18).

Definition 13.22: The MDE (13.18) admits an exponential dichotomy on J, if there exist positive constants K_1, K_2, α_1, and α_2 and a projection $P: X \to X$ such that

(i) $\|\mathcal{V}(t)P\mathcal{V}^{-1}(s)\| \leqslant K_1 e^{-\alpha_1(t-s)}$ for all $t, s \in J$, with $t \geqslant s$;
(ii) $\|\mathcal{V}(t)(I-P)\mathcal{V}^{-1}(s)\| \leqslant K_2 e^{-\alpha_2(s-t)}$ for all $t, s \in J$, with $s \geqslant t$.

Remark 13.23: The operator V satisfies conditions (i), (ii), (iii), (iv), and (v) from Proposition 6.8.

Remark 13.24: If $\mathcal{U}(t) = U(t, t_0)$, $t, t_0 \in J$, where $U: J \times J \to L(X)$ is the fundamental operator of (13.20), then $\mathcal{V}(t) = \mathcal{U}(t)$ for all $t \in J$.

Using Theorem 13.18 that gives the correspondence between the solutions of (13.18) and (13.20) and Remark 13.24, we can state the following result that gives us the equivalence between the dichotomies of the MDE (13.18) and its corresponding generalized ODE.

Proposition 13.25: *The MDE* (13.18) *admits an exponential dichotomy on J, if and only if the generalized ODE* (13.20) *admits an exponential dichotomy on J.*

By Propositions 13.3 and 13.25, we obtain the following result that gives a characterization of an exponential dichotomy for the MDE (13.18).

Proposition 13.26: *The MDE (13.18) admits an exponential dichotomy on J, if and only if there exist positive constants L_1, L_2, L, α_1,, and α_2, and a projection $P: X \to X$ such that, for all $\xi \in X$, the following estimates hold:*

(i) $\|\mathcal{V}(t)P\xi\| \leqslant L_1 e^{-\alpha_1(t-s)} \|\mathcal{V}(s)P\xi\|$, *for all $t, s \in J$, with $t \geqslant s$;*
(ii) $\|\mathcal{V}(t)(I-P)\xi\| \leqslant L_2 e^{-\alpha_2(s-t)} \|\mathcal{V}(s)(I-P)\xi\|$, *for all $t, s \in J$, with $s \geqslant t$;*
(iii) $\|\mathcal{V}(t)P\mathcal{V}^{-1}(t)\| \leqslant L$, *for all $t \in J$.*

Using Theorem 13.18, Remark 13.24, and Theorem 13.9, we state in the next result sufficient conditions to obtain exponential dichotomy on $[0, \infty)$.

Theorem 13.27: *Let $J = [0, \infty)$ and assume that*

$$V_0 = \{x_0 \in X : \|x_0\| = 1 \text{ and } x(t, x_0) \text{ is unbounded}\}$$

is a compact set in X, where $x(t, x_0)$ denotes the solution of (13.18) satisfying the condition $x(0) = x_0$. Assume that there exist constants $T > 0$, $C > 1$, and $0 < \theta < 1$ such that any solution $x : [0, \infty) \to X$ of (13.18) satisfies the conditions:

(i) $\|x(t)\| \leqslant C\,\|x(s)\|$ *for all* $0 \leqslant s \leqslant t \leqslant s + T$;
(ii) $\|x(t)\| \leqslant \theta \sup_{|u-t| \leqslant T} \|x(u)\|$ *for all* $t \geqslant T$.

Moreover, assume that for each $x_0 \in V_0$, there is an increasing sequence $\{t_n^{x_0}\}_{n\in\mathbb{N}} \subset \mathbb{R}_+$ with $t_{n+1}^{x_0} \leqslant t_n^{x_0} + T$, for all $n \in \mathbb{N}$, such that $\|x(t, x_0)\| < \theta^{-n} C$ for $t \in [0, t_n^{x_0})$ and $\left\|x(t_n^{x_0}, x_0)\right\| \geqslant \theta^{-n} C$, $n \in \mathbb{N}$. Then the MDE (13.18) admits an exponential dichotomy on J.

From now on, we study the relation between the exponential dichotomy of the MDE

$$x(t) = x_0 + \int_{t_0}^{t} \mathcal{F}(s)x(s)ds + \int_{t_0}^{t} \mathcal{G}(s)x(s)du(s),$$

and the bounded solutions of

$$x(t) = x_0 + \int_{t_0}^{t} \mathcal{F}(s)x(s)ds + \int_{t_0}^{t} \mathcal{G}(s)x(s)du(s) + \int_{t_0}^{t} h(s)du(s), \tag{13.21}$$

where $h : J \to X$ is a Perron–Stieltjes integrable function with respect to u over J and $t_0 \in J$. We assume that $\mathcal{F}$, $\mathcal{G}$ and u satisfy conditions (C1)–(C6).

Remark 13.28: Let $t_0 \in J$ and consider the functions

$$A(t) = \int_{t_0}^{t} \mathcal{F}(s)ds, \quad F(t) = \int_{t_0}^{t} \mathcal{G}(s)du(s), \quad \text{and} \quad g(t) = \int_{t_0}^{t} h(s)du(s) + g(t_0),$$

for $t \in J$. Consequently, the generalized ODE corresponding to the MDE (13.21) is given by

$$\frac{dx}{d\tau} = D[(A(t) + F(t))x + g(t)], \quad t \in J. \tag{13.22}$$

In addition,

$$\frac{dx}{d\tau} = D[(A(t) + F(t))x], \quad t \in J, \tag{13.23}$$

is the linear generalized ODE corresponding to the MDE (13.21).

Lemma 13.29: *$x: J \to X$ is a solution of the MDE (13.21) if and only if x is a solution of generalized ODE (13.22).*

Proof. According to Proposition 1.67, we have

$$\int_{t_0}^{t} \mathcal{F}(s)x(s)ds + \int_{t_0}^{t} \mathcal{G}(s)x(s)du(s) + \int_{t_0}^{t} h(s)du(s)$$
$$= \int_{t_0}^{t} d[A(s)]x(s) + \int_{t_0}^{t} d[F(s)]x(s) + g(t) - g(t_0),$$

for all $t \in J$. Consequently, we obtain the desired result. □

Lemma 13.30: *Let $J = \mathbb{R}$ and u be of locally bounded variation. Assume that the MDE (13.18) admits an exponential dichotomy on $\mathbb{R}$. Then the linear generalized ODE (13.23) satisfies the condition (D).*

Proof. Proposition 13.25 gives us the equivalence between the dichotomies of the MDE (13.18) and its corresponding generalized ODE (13.23). Since $A + F$ satisfies conditions H_1' and H_2' (see the proof of Theorem 13.19), then the linear generalized ODE (13.23) satisfies the condition (D) as presented in Definition 13.11. □

Let us suppose that the MDE (13.18) admits an exponential dichotomy on $\mathbb{R}$. Using Lemma 13.29, Lemma 13.30, and take into account that $g \in G(\mathbb{R}, X)$, we can conclude from Corollary 13.13 that the generalized ODE (13.22) admits at most one bounded solution. Using again the correspondence between the solutions of the MDE (13.21) and of the generalized ODE (13.22), see Lemma 13.29, we conclude that MDE (13.21) admits at most one bounded solution. This result is stated in the next proposition.

Proposition 13.31: *Let $J = \mathbb{R}$ and u be of locally bounded variation. Assume that the MDE (13.18) admits an exponential dichotomy on $\mathbb{R}$. Then the MDE (13.21) admits at most one bounded solution.*

In what follows, we give sufficient conditions for the perturbed MDE (13.21) to admit a unique bounded solution.

Proposition 13.32: *Let $J = \mathbb{R}$ and u be of locally bounded variation in $\mathbb{R}$. Assume that (13.18) admits an exponential dichotomy on $\mathbb{R}$, the Perron–Stieltjes integrals*

$$Z_1(t) = \int_{-\infty}^{t} \left(d_s\left[\mathcal{V}(t)P\mathcal{V}^{-1}(s)\right] \int_0^s h(r)du(r) \right) \quad \textit{and}$$
$$Z_2(t) = \int_{t}^{\infty} \left(d_s\left[\mathcal{V}(t)(I-P)\mathcal{V}^{-1}(s)\right] \int_0^s h(r)du(r) \right)$$

exist for each $t \in \mathbb{R}$ and the functions

$$t \in \mathbb{R} \mapsto Z_1(t) \in X \quad \text{and} \quad t \in \mathbb{R} \mapsto Z_2(t) \in X$$

are bounded, where $\mathcal{V}(t) = V(t, 0)$ for all $t \in \mathbb{R}$. Then the perturbed MDE (13.21) *admits a unique bounded solution.*

Proof. Since (13.18) admits an exponential dichotomy on $\mathbb{R}$, it follows by Lemma 13.30 that the generalized ODE (13.23) satisfies condition (D). Let $U: \mathbb{R} \times \mathbb{R} \to L(X)$ be the fundamental operator of (13.23) and $V: \mathbb{R} \times \mathbb{R} \to L(X)$ be the fundamental operator of (13.18). Define $\mathcal{U}(t) = U(t, 0)$ and $\mathcal{V}(t) = V(t, 0)$, $t \in \mathbb{R}$ $(t_0 = 0)$. Then by Remark 13.24, we have

$$\mathcal{V}(t) = \mathcal{U}(t)$$

for all $t \in \mathbb{R}$. Thus, we obtain

$$Z_1(t) = \int_{-\infty}^{t} d_s \left[\mathcal{U}(t) P \mathcal{U}^{-1}(s)\right] (g(s) - g(0)) \quad \text{and}$$

$$Z_2(t) = \int_{t}^{\infty} d_s \left[\mathcal{U}(t)(I - P)\mathcal{U}^{-1}(s)\right] (g(s) - g(0)),$$

for all $t \in \mathbb{R}$. Now, Proposition 13.14 assures that the nonhomogeneous linear generalized ODE (13.22) admits a unique bounded solution. Consequently, through the correspondence between the solutions of the MDE (13.21) and of the Eq. (13.22), we obtain the result. □

In Proposition 13.33, we present another result concerning sufficient conditions for the perturbed MDE (13.21) to admit a unique bounded solution.

Proposition 13.33: *Let $J = \mathbb{R}$ and u be of locally bounded variation in $\mathbb{R}$. Assume that* (13.18) *admits an exponential dichotomy on $\mathbb{R}$, the Perron–Stieltjes integrals*

$$Z_3(t) = \int_{-\infty}^{t} \mathcal{V}(t) P \mathcal{V}^{-1}(s) h(s) du(s) \quad \text{and}$$

$$Z_4(t) = \int_{t}^{\infty} \mathcal{V}(t)(I - P)\mathcal{V}^{-1}(s) h(s) du(s)$$

exist for all $t \in \mathbb{R}$, the functions $t \in \mathbb{R} \mapsto Z_3(t) \in X$ and $t \in \mathbb{R} \mapsto Z_4(t) \in X$ are bounded, and the function $K: \mathbb{R} \to X$ given by

$$K(t) = \begin{cases} \mathcal{V}(t)\left(\displaystyle\sum_{0 \leqslant \tau < t} \Delta^+ \mathcal{V}^{-1}(\tau) \Delta^+ H(\tau) - \sum_{0 < \tau \leqslant t} \Delta^- \mathcal{V}^{-1}(\tau) \Delta^- H(\tau)\right), & t > 0 \\ \mathcal{V}(t)\left(\displaystyle\sum_{t \leqslant \tau < 0} \Delta^+ \mathcal{V}^{-1}(\tau) \Delta^+ H(\tau) - \sum_{t < \tau \leqslant 0} \Delta^- \mathcal{V}^{-1}(\tau) \Delta^- H(\tau)\right), & t < 0 \\ 0, \quad t = 0 \end{cases}$$

is bounded, where $H(\tau) = \int_0^\tau h(r)du(r)$. *Then the perturbed MDE* (13.21) *admits a unique bounded solution.*

Proof. Since (13.18) admits an exponential dichotomy on $\mathbb{R}$, it follows by Lemma 13.30 that the generalized ODE (13.23) satisfies condition (D). As described in the proof of Proposition 13.32, we have $\mathcal{U}(t) = \mathcal{V}(t)$ for all $t \in \mathbb{R}$, where $U: \mathbb{R} \times \mathbb{R} \to L(X)$ is the fundamental operator of (13.23) and $V: \mathbb{R} \times \mathbb{R} \to L(X)$ is the fundamental operator of (13.18). In this way, for each $t \in \mathbb{R}$, we have

$$\begin{aligned} Z_3(t) &= \int_{-\infty}^{t} \mathcal{V}(t)P\mathcal{V}^{-1}(s)h(s)du(s) = \int_{-\infty}^{t} \mathcal{V}(t)P\mathcal{V}^{-1}(s)d\left[\int_0^s h(r)du(r)\right] \\ &= \int_{-\infty}^{t} \mathcal{U}(t)P\mathcal{U}^{-1}(s)d[g(s) - g(0)] = \int_{-\infty}^{t} \mathcal{U}(t)P\mathcal{U}^{-1}(s)dg(s), \\ Z_4(t) &= \int_{t}^{\infty} \mathcal{V}(t)(I-P)\mathcal{V}^{-1}(s)h(s)du(s) \\ &= \int_{t}^{\infty} \mathcal{V}(t)(I-P)\mathcal{V}^{-1}(s)d\left[\int_0^s h(r)du(r)\right] \\ &= \int_{t}^{\infty} \mathcal{U}(t)(I-P)\mathcal{U}^{-1}(s)d[g(s) - g(0)] = \int_{t}^{\infty} \mathcal{U}(t)(I-P)\mathcal{U}^{-1}(s)dg(s) \end{aligned}$$

and

$$H(t) = g(t) - g(0).$$

By Proposition 13.15, the nonhomogeneous linear generalized ODE (13.22) admits a unique bounded solution. Using Lemma 13.29, we conclude that the MDE (13.21) admits a unique bounded solution. □

Using Proposition 13.17, we can obtain sufficient conditions for the existence of the integrals presented in Propositions 13.32 and 13.33. See the next result.

Proposition 13.34: *Let* $J = \mathbb{R}$ *and* u *be of locally bounded variation in* $\mathbb{R}$. *Assume that the MDE* (13.18) *admits an exponential dichotomy on* $\mathbb{R}$.

(i) *If* $t \in \mathbb{R} \mapsto \int_0^t h(s)du(s) \in X$ *is a bounded function and the functions* m_1 *and* m_2 *given in the conditions (C4) and (C5) are such that*

$$\int_{-\infty}^{\infty} m_1(s)ds + \int_{-\infty}^{\infty} m_2(s)du(s) < \infty,$$

then the hypotheses of Proposition 13.32 are satisfied.

(ii) *If* $t \in \mathbb{R} \mapsto \int_0^t h(s)du(s) \in X$ *is of bounded variation on* $\mathbb{R}$, *then the hypotheses of Proposition 13.33 are satisfied.*

13.4 Applications to IDEs

This section concerns the study on exponential dichotomies for the following IDE

$$\begin{cases} \dot{x} = f(t)x, \ t \neq t_i, \\ \Delta(x(t_i)) = x(t_i^+) - x(t_i) = B_i x(t_i), \ i = \pm 1, \pm 2, \dots. \end{cases}$$

The definitions and the results presented here can be found in [29, Section 5].

We consider $J \subset \mathbb{R}$ an interval such that $[0, \infty) \subset J$ and X is a Banach space. The function $f\colon J \to L(X)$ satisfies the conditions:

(B1) f is locally Perron integrable in J;
(B2) there exists a locally Perron integrable function $m\colon J \to \mathbb{R}$ such that for each $a, b \in J$, $a \leqslant b$, we have $\left\| \int_a^b f(s)ds \right\| \leqslant \int_a^b m(s)ds$.

In relation to the impulses, we assume $\{\dots, t_{-k}, \dots, t_{-1}, t_1, \dots, t_k, \dots\} \subset \mathbb{R}$ is such that

$$\cdots < t_{-k} < \cdots < t_{-1} < t_1 < 0 < t_1 < t_2 < \cdots < t_k < \cdots$$

and $\lim_{k\to\pm\infty} t_k = \pm\infty$. Moreover, we assume that the operators $B_i \in L(X)$, with $i = \pm 1, \pm 2, \dots$, are such that $(I + B_i)^{-1} \in L(X)$ for all $i = \pm 1, \pm 2, \dots$. Now, define the auxiliary sets $\mathcal{I} = \{i \in \mathbb{Z} : t_i \in J\}$ and $\mathcal{I}_r^s = \{i \in \mathcal{I} : r \leqslant t_i \leqslant s\}$ for $r, s \in J$.

We shall study the exponential dichotomy for the following IDE:

$$\begin{cases} \dot{x} = f(t)x, \ t \neq t_i, \\ \Delta(x(t_i)) = x(t_i^+) - x(t_i) = B_i x(t_i), \ i \in \mathcal{I}. \end{cases} \tag{13.24}$$

Let $t_0 \in J$ and $x_0 \in X$. A function $x\colon J \to X$ is said to be a *solution* of the IVP

$$\begin{cases} \dot{x} = f(t)x, \ t \neq t_i, \\ \Delta(x(t_i)) = x(t_i^+) - x(t_i) = B_i x(t_i), \ i \in \mathcal{I}, \\ x(t_0) = x_0, \end{cases} \tag{13.25}$$

if $\dot{x}(t) = f(t)x(t)$ for almost all $t \in J \setminus \{t_i : i \in \mathcal{I}\}$, $x(t_0) = x_0$ and

$$x(t_i^+) = \lim_{t\to t_i^+} x(t) = x(t_i) + B_i(x(t_i)) \quad \text{for all} \quad i \in \mathcal{I}.$$

Let H_d be the Heaviside function at point $d \in J$.

The solution x of (13.25) satisfies the following integral equation:

$$x(t) = \begin{cases} x_0 + \displaystyle\int_{t_0}^t f(s)x(s)ds + \sum_{i\in\mathcal{I}_{t_0}^t} B_i x(\tau_i) H_{\tau_i}(t) & \text{for } \ t \geqslant t_0 \ \ (t \in J), \\ x_0 + \displaystyle\int_{t_0}^t f(s)x(s)ds - \sum_{i\in\mathcal{I}_t^{t_0}} B_i x(\tau_i)(1 - H_{\tau_i}(t)) & \text{for } \ t < t_0 \ \ (t \in J), \end{cases}$$

where the integral is in the sense of Perron.

The next result shows a correspondence between a solution of the IDE (13.24) and a solution of its corresponding generalized ODE. The proof follows similarly as the proof of [209, Theorem 5.20].

Theorem 13.35: *Let $t_0 \in J$. The function $x: J \to X$ is a solution of (13.24) if and only if x is a solution of the generalized ODE:*

$$\frac{dx}{d\tau} = D[A(t)x], \tag{13.26}$$

where A is given by

$$A(t) = \begin{cases} \int_{t_0}^{t} f(s)ds + \sum_{i\in \mathcal{I}_{t_0}^{t}} B_i H_{t_i}(t), & \text{for } t \geqslant t_0 \ (t \in J), \\ \int_{t_0}^{t} f(s)ds - \sum_{i\in \mathcal{I}_{t}^{t_0}} B_i(1 - H_{t_i}(t)), & \text{for } t < t_0 \ (t \in J). \end{cases} \tag{13.27}$$

Remark 13.36: The operator A defined in (13.27) is left-continuous. It is not difficult to see that $I - \Delta^{-}A(t) = I$ for all $t \in J$,

$$I + \Delta^{+}A(t) = I \quad \text{if} \quad t \neq t_i, \quad \text{and} \quad I + \Delta^{+}A(t) = I + B_i \quad \text{if} \quad t = t_i,$$

with $i \in \mathcal{I}$. In addition, since we are assuming that $(I + B_i)^{-1} \in L(X)$ for all $i = \pm 1, \pm 2, \ldots$, then the operator A also satisfies conditions H_1 and H_2 (in case $J = \mathbb{R}$, we have A satisfies H_1' and H_2').

In order to define the fundamental operator for the IDE (13.24), let us consider the fundamental operator of the ODE $\dot{x} = f(t)x$, which we denote by $\Phi: J \times J \to L(X)$. In this way, as done in [209], we define $V: J \times J \to L(X)$ by

$$V(t,s) = \Phi(t,t_i)\left(\prod_{k=i}^{j+1}[I + B_k]\Phi(t_k, t_{k-1})\right)[I + B_j]\Phi(t_j, s)$$

whenever $t \geqslant s$, $t \in (t_i, t_{i+1}]$ and $s \in (t_{j-1}, t_j]$ ($j \leqslant i$ and $i,j \in \mathcal{I}$), and

$$V(t,s) = [V(s,t)]^{-1} = \Phi(t,t_j)[I + B_j]^{-1} \cdots [I + B_i]^{-1}\Phi(t_i, s)$$

whenever $t < s$, $s \in (t_i, t_{i+1}]$ and $t \in (t_{j-1}, t_j]$ ($j \leqslant i$ and $i,j \in \mathcal{I}$).

Lemma 13.37: *The operator $V(t,s)$ is the fundamental operator of the IDE (13.24) and $x(t) = V(t,s)\tilde{x}$ is the solution of (13.24) with condition $x(s) = \tilde{x}$, for $t \geqslant s$, $t \in (t_i, t_{i+1}]$ and $s \in (t_{j-1}, t_j]$ ($j \leqslant i$ and $i,j \in \mathcal{I}$). Moreover, $U(t,s) = V(t,s)$ for all $t, s \in J$, where $U(t,s)$ is the fundamental operator of the generalized ODE (13.26).*

Proof. The proof follows by Example 6.20, page 194, from [209]. □

Let $V: J \times J \to L(X)$ be the fundamental operator of the IDE (13.24). Now, we define the operator $\mathcal{V}: J \to L(X)$ by

$$\mathcal{V}(t) = V(t, t_0), \quad t \in J,$$

where $t_0 \in J$. We denote $\mathcal{V}^{-1}(t) = V(t_0, t)$, $t, t_0 \in J$. Next, we give the concept of exponential dichotomy for the IDE (13.24).

Definition 13.38: The IDE (13.24) admits an exponential dichotomy on J, if there exist positive constants K_1, K_2, α_1 and α_2, and a projection $P: X \to X$ such that

(i) $\|\mathcal{V}(t)P\mathcal{V}^{-1}(s)\| \leqslant K_1 e^{-\alpha_1(t-s)}$, for all $t, s \in J$, with $t \geqslant s$;
(ii) $\|\mathcal{V}(t)(I - P)\mathcal{V}^{-1}(s)\| \leqslant K_2 e^{-\alpha_2(s-t)}$, for all $t, s \in J$, with $s \geqslant t$.

Using Lemma 13.37 and Theorem 13.35 that gives the correspondence between the solutions of (13.24) and (13.26), we can state the following equivalence between the dichotomies of the IDE (13.24) and its corresponding generalized ODE.

Proposition 13.39: *The IDE (13.24) admits an exponential dichotomy on J if and only if the corresponding generalized ODE (13.26) admits an exponential dichotomy on J.*

In the next result, we present a characterization of the exponential dichotomy for the IDE (13.24).

Proposition 13.40: *The IDE (13.24) admits an exponential dichotomy on J, if and only if there exist constants $L_1, L_2, L, \alpha_1, \alpha_2 > 0$ such that*

(i) $\|\mathcal{V}(t)P\xi\| \leqslant L_1 e^{-\alpha_1(t-s)} \|\mathcal{V}(s)P\xi\|$, *for all $t, s \in J$, with $t \geqslant s$;*
(ii) $\|\mathcal{V}(t)(I - P)\xi\| \leqslant L_2 e^{-\alpha_2(s-t)} \|\mathcal{V}(s)(I - P)\xi\|$, *for all $t, s \in J$, with $s \geqslant t$;*
(iii) $\|\mathcal{V}(t)P\mathcal{V}^{-1}(t)\| \leqslant L$, *for all $t \in J$.*

Proof. The proof is a consequence of Propositions 13.3 and 13.39. □

Theorem 13.41 exhibits sufficient conditions to obtain exponential dichotomy on $[0, \infty)$.

Theorem 13.41: *Suppose $V_0 = \{x_0 \in X : \|x_0\| = 1 \text{ and } x(t, x_0) \text{ is unbounded}\}$ is a compact subset of X, where $x(t, x_0)$ denotes the solution of (13.24) such that $x(0) = x_0$. Assume that there are constants $T > 0$, $C > 1$, and $0 < \theta < 1$ such that any solution $x: [0, \infty) \to X$ of (13.24) satisfies the conditions:*

(i) $\|x(t)\| \leqslant C\,\|x(s)\|$ *for* $0 \leqslant s \leqslant t \leqslant s+T$;
(ii) $\|x(t)\| \leqslant \theta \sup_{|u-t|\leqslant T} \|x(u)\|$ *for all* $t \geqslant T$.

Moreover, assume that for each $x_0 \in V_0$, *there is an increasing sequence* $\{t_n^{x_0}\}_{n\in\mathbb{N}} \subset \mathbb{R}_+$ *with* $t_{n+1}^{x_0} \leqslant t_n^{x_0} + T$, $n \in \mathbb{N}$, *such that* $\|x(t,x_0)\| < \theta^{-n}C$ *for* $t \in [0, t_n^{x_0})$ *and* $\left\|x(t_n^{x_0}, x_0)\right\| \geqslant \theta^{-n}C$, $n \in \mathbb{N}$. *Then the IDE* (13.24) *admits an exponential dichotomy on* J.

Proof. This result follows from Theorems 13.9 and 13.35. □

In what follows, we consider $J = \mathbb{R}$ and we study the relation between the exponential dichotomy of the IDE (13.24) and the bounded solutions of

$$\begin{cases} \dot{x} = f(t)x + h(t), \ t \neq t_i, \\ \Delta(x(t_i)) = x(t_i^+) - x(t_i) = B_i x(t_i), \ i = \pm 1, \pm 2, \dots, \end{cases} \tag{13.28}$$

where $h : \mathbb{R} \to L(X)$ is a locally Perron integrable function over $\mathbb{R}$.

Next, we give sufficient conditions for the nonhomogeneous IDE (13.28) to admit at most a unique bounded solution.

Proposition 13.42: *Let* $J = \mathbb{R}$. *Assume that the IDE* (13.24) *admits an exponential dichotomy on* $\mathbb{R}$ *and let* $h : \mathbb{R} \to L(X)$ *be a locally Perron integrable function over* $\mathbb{R}$. *Then the nonhomogeneous IDE* (13.28) *admits at most one bounded solution.*

Proof. Consider the linear generalized ODE

$$\frac{dx}{d\tau} = D[A(t)x], \tag{13.29}$$

where $A : \mathbb{R} \to L(X)$ is given by

$$A(t) = \begin{cases} \displaystyle\int_{t_0}^{t} f(s)ds + \sum_{i \in \mathcal{I}_0^t} B_i H_{t_i}(t), & \text{for } t \geqslant 0, \\ \displaystyle\int_{t_0}^{t} f(s)ds - \sum_{i \in \mathcal{I}_t^0} B_i (1 - H_{t_i}(t)), & \text{for } t < 0. \end{cases}$$

Remark 13.36 assures that A satisfies conditions H_1' and H_2'.

Since the IDE (13.24) admits an exponential dichotomy on $\mathbb{R}$, it follows by Proposition 13.39 that the linear generalized ODE (13.29) also admits an exponential dichotomy on $\mathbb{R}$.

Define a function $g : \mathbb{R} \to X$ by

$$g(t) = \int_0^t h(s)ds + g(0), \quad t \in \mathbb{R}. \tag{13.30}$$

Clearly, $g \in C(\mathbb{R}, X) \subset G(\mathbb{R}, X)$. Then, by Corollary 13.13, the generalized ODE

$$\frac{dx}{d\tau} = D[A(t)x + g(t)] \tag{13.31}$$

admits, at most, one bounded solution. Since $x : \mathbb{R} \to X$ is solution of (13.31) if and only if x is solution of the corresponding IDE

$$\begin{cases} \dot{x} = f(t)x + h(t), & t \neq t_i, \\ \Delta(x(t_i)) = x(t_i^+) - x(t_i) = B_i x(t_i), & i = \pm 1, \pm 2, \ldots, \end{cases} \tag{13.32}$$

we obtain the IDE (13.32) admits at most one bounded solution. □

Remark 13.43: We emphasize here, as mentioned in the proof of Proposition 13.42, that $x : \mathbb{R} \to X$ is a solution of (13.32) if and only if x is a solution of (13.31).

Proposition 13.44: *Let $J = \mathbb{R}$. Assume that (13.24) admits an exponential dichotomy on $\mathbb{R}$, the Perron–Stieltjes integrals*

$$I_1(t) = \int_{-\infty}^{t} \left(d_s \left[\mathcal{V}(t) P \mathcal{V}^{-1}(s) \right] \int_0^s h(r) dr \right)$$

and

$$I_2(t) = \int_{t}^{\infty} \left(d_s \left[\mathcal{V}(t)(I - P) \mathcal{V}^{-1}(s) \right] \int_0^s h(r) dr \right)$$

exist for each $t \in \mathbb{R}$ and the functions $t \in \mathbb{R} \mapsto I_1(t) \in X$ and $t \in \mathbb{R} \mapsto I_2(t) \in X$ are bounded, where $\mathcal{V}(t) = V(t, 0)$ for all $t \in \mathbb{R}$. Then the nonhomogeneous IDE (13.32) admits a unique bounded solution.

Proof. By Proposition 13.39, the corresponding generalized ODE (13.29) admits an exponential dichotomy on $\mathbb{R}$. Thus, (13.29) satisfies condition (D) (see Definition 13.11) as A satisfies conditions H_1' and H_2'. According to Lemma 13.37,

$$\mathcal{U}(t) = \mathcal{V}(t)$$

for all $t \in \mathbb{R}$, where $\mathcal{V}(t) = V(t, 0)$ and $\mathcal{U}(t) = U(t, 0)$ for all $t \in \mathbb{R}$ (V is the fundamental operator of (13.24) and U is the fundamental operator of (13.29)). Thus, we may rewrite $I_1(t)$ and $I_2(t)$ as

$$I_1(t) = \int_{-\infty}^{t} d_s \left[\mathcal{U}(t) P \mathcal{U}^{-1}(s) \right] (g(s) - g(0))$$

and

$$I_2(t) = \int_{t}^{\infty} d_s \left[\mathcal{U}(t)(I - P) \mathcal{U}^{-1}(s) \right] (g(s) - g(0)),$$

for all $t \in \mathbb{R}$, where g comes from (13.30). Since $g \in G(\mathbb{R}, X)$, it follows by Proposition 13.14 that the nonhomogeneous linear generalized ODE (13.31) admits a unique bounded solution. Lastly, by Remark 13.43, we conclude that the nonhomogeneous IDE (13.32) admits a unique bounded solution. □

Next, we establish sufficient conditions for the nonhomogeneous IDE (13.32) to admit a unique bounded solution.

Proposition 13.45: *Let $J = \mathbb{R}$. Assume that (13.24) admits an exponential dichotomy, the Perron integrals*

$$I_3(t) = \int_{-\infty}^{t} \mathcal{V}(t)P\mathcal{V}^{-1}(s)h(s)ds \quad \text{and} \quad I_4(t) = \int_{t}^{\infty} \mathcal{V}(t)(I-P)\mathcal{V}^{-1}(s)h(s)ds$$

exist for all $t \in \mathbb{R}$, and the functions $t \in \mathbb{R} \mapsto I_3(t) \in X$ and $t \in \mathbb{R} \mapsto I_4(t) \in X$ are bounded. Then the nonhomogeneous IDE (13.32) admits a unique bounded solution.

Proof. Looking at to the proof of Proposition 13.44, we have the generalized ODE (13.29) satisfies condition (D) and $\mathcal{U}(t) = \mathcal{V}(t)$ for all $t \in \mathbb{R}$. Consequently,

$$\begin{aligned} I_3(t) &= \int_{-\infty}^{t} \mathcal{V}(t)P\mathcal{V}^{-1}(s)h(s)ds = \int_{-\infty}^{t} \mathcal{V}(t)P\mathcal{V}^{-1}(s)d\left[\int_0^s h(r)dr\right] \\ &= \int_{-\infty}^{t} \mathcal{U}(t)P\mathcal{U}^{-1}(s)d[g(s)-g(0)] = \int_{-\infty}^{t} \mathcal{U}(t)P\mathcal{U}^{-1}(s)d[g(s)] \end{aligned}$$

and

$$\begin{aligned} I_4(t) &= \int_{t}^{\infty} \mathcal{V}(t)(I-P)\mathcal{V}^{-1}(s)h(s)ds = \int_{t}^{\infty} \mathcal{V}(t)(I-P)\mathcal{V}^{-1}(s)d\left[\int_0^s h(r)dr\right] \\ &= \int_{t}^{\infty} \mathcal{U}(t)(I-P)\mathcal{U}^{-1}(s)d[g(s)-g(0)] \\ &= \int_{t}^{\infty} \mathcal{U}(t)(I-P)\mathcal{U}^{-1}(s)d[g(s)], \end{aligned}$$

for all $t \in \mathbb{R}$.

Now, note that the function k defined in Proposition 13.15 is null since $\Delta^+ g(t) = \Delta^- g(t) = 0$ for all $t \in \mathbb{R}$. Therefore, the result follows by Proposition 13.15 and Remark 13.43. □

In the last result of this chapter, we give conditions for the existence of the integrals presented in Propositions 13.44 and 13.45. The result follows by Proposition 13.17.

Proposition 13.46: *Assume that the IDE (13.24) admits an exponential dichotomy on* $\mathbb{R}$.

(i) *If the mapping* $t \in \mathbb{R} \mapsto \int_0^t h(s)ds \in X$ *is a bounded function and* $\int_{-\infty}^{\infty} m(s)ds < \infty$, *where* m *comes from condition (B2), then the conditions of Proposition 13.44 hold.*

(ii) *If the mapping* $t \in \mathbb{R} \mapsto \int_0^t h(s)ds \in X$ *is of bounded variation on* $\mathbb{R}$, *then the hypotheses of Proposition 13.45 are satisfied.*

14

Topological Dynamics

Suzete M. Afonso[1], *Marielle Ap. Silva*[2], *Everaldo M. Bonotto*[3], *and Márcia Federson*[2]

[1] *Departamento de Matemática, Instituto de Geociências e Ciências Exatas, Universidade Estadual Paulista "Júlio de Mesquita Filho" (UNESP), Rio Claro, SP, Brazil*
[2] *Departamento de Matemática, Instituto de Ciências Matemáticas e de Computação (ICMC), Universidade de São Paulo, São Carlos, SP, Brazil*
[3] *Departamento de Matemática Aplicada e Estatística, Instituto de Ciências Matemáticas e de Computação (ICMC), Universidade de São Paulo, São Carlos, SP, Brazil*

This chapter is dedicated to the study of semidynamical systems generated by generalized ODEs. Besides the existence of a local semidynamical system, we also show the existence of an associated impulsive semidynamical system. As a consequence, we do not only present a version of LaSalle's invariance principle for impulsive semidynamical systems corresponding to an impulsive generalized ODE, but we also include some recursive properties such as minimality and recurrency.

A local semidynamical system related to a generalized ODE was first considered by Zvi Artstein in [9]. Motivated by [9], the authors of [4], constructed an impulsive semidynamical system related to a class of generalized ODEs which we explore here in more details (see Section 14.3). The version of LaSalle's invariance principle that will be presented in Section 14.4 is also borrowed from [4].

Recently, recursive properties for impulsive semidynamical systems were investigated by E. M. Bonotto and M. Z. Jimenez in [30, 31]. These works motivated the study of the mentioned qualitative properties for an impulsive semidynamical system in the sense of generalized ODEs (see also [28]). Such results are contained in Section 14.5.

It is worth mentioning that the treatment of generalized ODEs subject to impulses at variable times which we provide here is in consonance with the concepts and approaches of S. K. Kaul in [141] and of K. Ciesielski in [41, 42].

Generalized Ordinary Differential Equations in Abstract Spaces and Applications, First Edition.
Edited by Everaldo M. Bonotto, Márcia Federson, and Jaqueline G. Mesquita.

Nevertheless, our approach differs from those of V. Lakshmikantham, D. Bainov and P. S. Simeonov in [156] and of A. M. Samoilenko and N. A. Perestyuk in [201].

In [156, 201] and in some other papers (see [223] and [240], for instance), the study of properties of differential systems with impulses is somehow reduced to the preassigned case by the imposition of additional hypotheses, such as a limited number of times at which impulse surfaces are reached by some integral curve (usually, not more than once). Here, on the other hand, we construct an impulse operator acting on a closed surface M satisfying a certain condition (see Section 14.3 – Eq. (14.18)).

Throughout this chapter, we shall assume that $X = \mathbb{R}^n$ and $\Omega = \mathcal{O} \times [0, \infty)$, where $\mathcal{O} \subset \mathbb{R}^n$ is an open set. We denote by $\|\cdot\|$ the norm of $X = \mathbb{R}^n$ and by $|r|$ the absolute value of a real number r. We shall also consider a generalized ODE

$$\frac{dx}{d\tau} = DF(x, t),$$

where F belongs to a subclass of $\mathcal{F}(\Omega, h)$ (see Definition 4.3) given by

$$\mathcal{F}_0(\Omega, h) = \{F : \Omega \to \mathbb{R}^n : F \in \mathcal{F}(\Omega, h) \text{ and } F(x, 0) = 0 \text{ for all } x \in \Omega\},$$

with $h : [0, \infty) \to \mathbb{R}$ being a nondecreasing continuous function. Thus, a function $F : \Omega \to \mathbb{R}^n$ belongs to the *class* $\mathcal{F}_0(\Omega, h)$ if, for all $(x, s_2), (x, s_1), (y, s_2), (y, s_1) \in \Omega$, we have

$$F(x, 0) = 0, \tag{14.1}$$

$$\| F(x, s_2) - F(x, s_1) \| \leqslant |h(s_2) - h(s_1)| \quad \text{and} \tag{14.2}$$

$$\| F(x, s_2) - F(x, s_1) - F(y, s_2) + F(y, s_1) \| \leqslant \| x - y \| \, |h(s_2) - h(s_1)|. \tag{14.3}$$

14.1 The Compactness of the Class $\mathcal{F}_0(\Omega, h)$

In this section, we show that the class $\mathcal{F}_0(\Omega, h)$ is a compact set, provided h is a nondecreasing continuous function. The results are borrowed from [4], see Lemma 3.1 and Theorem 3.2 there.

We start by constructing a metric for the space $\mathcal{F}_0(\Omega, h)$.

Since $\Omega = \mathcal{O} \times [0, \infty)$ and $\mathcal{O} \subset \mathbb{R}^n$ is open, we may consider a sequence of compact sets $\{K_n\}_{n\in\mathbb{N}}$ in Ω such that $K_n \subset \operatorname{Int} K_{n+1}$ and $\Omega = \bigcup_{n=1}^{\infty} K_n$, where $\operatorname{Int} K_{n+1}$ denotes the interior of K_{n+1}. In this way, for each $n \in \mathbb{N}$, we can define a mapping $\|\cdot\|_n : \mathcal{F}_0(\Omega, h) \to [0, \infty)$ by

$$\| F - G \|_n = \sup \{ \| F(x, t) - G(x, t) \| : (x, t) \in K_n \},$$

for $F, G \in \mathcal{F}_0(\Omega, h)$, where $\|\cdot\|$ is any norm in $\mathbb{R}^n$. Thus, the pseudometric

$$\rho_n(F, G) = \frac{\| F - G \|_n}{1 + \| F - G \|_n}, \quad F, G \in \mathcal{F}_0(\Omega, h),$$

is well-defined. Now, setting

$$\rho(F,G) = \sum_{n=1}^{\infty} 2^{-n} \rho_n(F,G), \quad F, G \in \mathcal{F}_0(\Omega, h), \tag{14.4}$$

$(\mathcal{F}_0(\Omega, h), \rho)$ becomes a metric space. Although the metric ρ depends on the choice of the sequence $\{K_n\}_{n\in\mathbb{N}}$, any other sequence of compact sets generates an equivalent metric. See [214], for instance.

Let $\{F_k\}_{k\in\mathbb{N}} \subset \mathcal{F}_0(\Omega, h)$ be a sequence and $F \in \mathcal{F}_0(\Omega, h)$. Note that the convergence $\rho(F_k, F) \overset{k\to\infty}{\to} 0$ is equivalent to the uniform convergence $\| F_k(x,t) - F(x,t) \| \overset{k\to\infty}{\to} 0$ on every compact subset of Ω. To see that, it is enough to consider the isometric immersion $\gamma : (\mathcal{F}_0(\Omega, h), \rho) \to (\prod_{i=1}^{\infty} \mathcal{F}_0(K_i, h), D)$ given by

$$\gamma(F) = (F|_{K_1}, F|_{K_2}, F|_{K_3}, \dots) = (F_i)_{i\in\mathbb{N}},$$

where $D(f,g) = \sum_{i=1}^{\infty} 2^{-i} \frac{\|f_i - g_i\|}{1+\|f_i - g_i\|}$, $f = (f_i)_{i\in\mathbb{N}}, g = (g_i)_{i\in\mathbb{N}} \in \prod_{i=1}^{\infty} \mathcal{F}_0(K_i, h)$, with $\| f_i - g_i \| = \sup_{(x,t)\in K_i} \| f(x,t) - g(x,t) \|$. In addition, note that given a compact subset $K \subset \Omega$, there exists $i \in \mathbb{N}$ such that $K \subset K_i$.

The class $\mathcal{F}_0(\Omega, h)$, defined in the introduction, is equicontinuous as shown by the next lemma.

Lemma 14.1: *The class $\mathcal{F}_0(\Omega, h)$ is equicontinuous on compact subsets of $\Omega = \mathcal{O} \times [0, \infty)$, where $\mathcal{O} \subset \mathbb{R}^n$ is open.*

Proof. Assume that $A \subset \mathcal{O}$ and $C \subset [0,\infty)$ are compact sets and $F \in \mathcal{F}_0(\Omega, h)$. Let $x \in A$ and $t \in C$ be arbitrary. Since F satisfies conditions (14.1) and (14.3), we obtain

$$\begin{aligned} \| F(x,t) - F(y,t) \| &= \| F(x,t) - F(x,0) - F(y,t) + F(y,0) \| \\ &\leqslant \| x - y \| \, |h(t) - h(0)| \\ &\leqslant \| x - y \| \, (|h(t)| + |h(0)|), \end{aligned} \tag{14.5}$$

for each $y \in A$. In addition, by condition (14.2), for all $(y,s) \in A \times C$, we have

$$\| F(y,t) - F(y,s) \| \leqslant |h(t) - h(s)|. \tag{14.6}$$

Therefore, using relations (14.5) and (14.6), for all $(y,s) \in A \times C$, we get

$$\begin{aligned} \| F(x,t) - F(y,s) \| &\leqslant \| F(x,t) - F(y,t) \| + \| F(y,t) - F(y,s) \| \\ &\leqslant \| x - y \| \, (|h(t)| + |h(0)|) + |h(t) - h(s)|. \end{aligned}$$

On the other hand, the continuity of h (and, consequently, the uniform continuity of h on C), assures that, for every $\epsilon > 0$, there exists $\delta > 0$ such that

$$|h(t) - h(s)| < \frac{\epsilon}{2},$$

whenever $|t-s|<\delta$, $t,s\in C$. Furthermore, since C is compact, there exists $M>0$ such that $|h(s)|\leqslant M$ for all $s\in C$. Thus,

$$\| F(x,t)-F(y,s) \| < \epsilon,$$

whenever $\| x-y \| < \frac{\epsilon}{2}(M+|h(0)|)^{-1}$ and $|t-s|<\delta$, with $(y,s)\in A\times C$, which proves the result. □

Remark 14.2: Let $\{F_k\}_{k\in\mathbb{N}}\subset \mathcal{F}_0(\Omega,h)$ and $F\in\mathcal{F}_0(\Omega,h)$. As a consequence of Lemma 14.1, we may conclude that the convergence $\rho(F_k,F)\overset{k\to\infty}{\to}0$ is equivalent to the pointwise convergence $F_k\overset{k\to\infty}{\to}F$, that is, $\| F_k(x,t)-F(x,t) \|\overset{k\to\infty}{\to}0$ for each $(x,t)\in\Omega$. Indeed, as we presented before Lemma 14.1, the convergence $\rho(F_k,F)\overset{k\to\infty}{\to}0$ is equivalent to the uniform convergence $\| F_k(x,t)-F(x,t) \|\overset{k\to\infty}{\to}0$ on every compact subset of Ω. In this way, we need to verify that the pointwise convergence implies the uniform convergence. Let us assume that $\| F_k(x,t)-F(x,t) \|\overset{k\to\infty}{\to}0$ for each $(x,t)\in\Omega$. Let $\epsilon>0$ and $K\subset\Omega$ be a compact subset. For each $(x,t)\in K$, there exists $K_{x,t}\in\mathbb{N}$ such that

$$\| F_k(x,t)-F(x,t) \| < \frac{\epsilon}{3} \quad \text{whenever} \quad k>K_{x,t}.$$

By Lemma 14.1, there exists $\delta>0$ such that

$$\| F_k(x,t)-F_k(y,s) \| < \frac{\epsilon}{3} \quad \text{and} \quad \| F(x,t)-F(y,s) \| < \frac{\epsilon}{3},$$

whenever $\| x-y \|<\delta$ and $|t-s|<\delta$, with $(x,t),(y,s)\in K$.

Using the compactness of K, one can find $(x_1,t_1),\dots,(x_m,t_m)\in K$ such that

$$K\subset B((x_1,t_1),\delta)\cup\cdots\cup B((x_m,t_m),\delta),$$

where $B((x,t),\delta)=\{(y,s):\| y-x \|<\delta,\ |s-t|<\delta\}$. For a given arbitrary $(x,t)\in K$, there exists (x_p,t_p) with $1\leqslant p\leqslant m$ such that $(x,t)\in B((x_p,t_p),\delta)$. If $k>\max\{K_{x_1,t_1},\dots,K_{x_m,t_m}\}$, we obtain

$$\begin{aligned}
&\| F_k(x,t)-F(x,t) \| \\
&\leqslant \| F_k(x,t)-F_k(x_p,t_p) \| + \| F_k(x_p,t_p)-F(x_p,t_p) \| + \| F(x_p,t_p)-F(x,t) \| \\
&<\epsilon.
\end{aligned}$$

Hence, $\| F_k(x,t)-F(x,t) \|<\epsilon$ for all $(x,t)\in K$ and $k>\max\{K_{x_1,t_1},\dots,K_{x_m,t_m}\}$. This yields the uniform convergence $\| F_k(x,t)-F(x,t) \|\overset{k\to\infty}{\to}0$ on every compact subset of Ω.

Since the functions in the class $\mathcal{F}_0(\Omega,\ h)$ satisfy conditions (14.1), (14.2), and (14.3), we have the following straightforward lemma.

Lemma 14.3: *The class $\mathcal{F}_0(\Omega,\ h)$ is closed.*

Now, we present the result that guarantees the compactness of the class $\mathcal{F}_0(\Omega,\ h)$.

Theorem 14.4: *The class $\mathcal{F}_0(\Omega,\ h)$ is compact.*

Proof. By Lemma 14.1, we have the equicontinuity of $\mathcal{F}_0(\Omega,\ h)$ on compact subsets of $\Omega = \mathcal{O} \times [0, \infty)$. Furthermore, using (14.1) and (14.2),

$$\| F(x,t) \| = \| F(x,t) - F(x,0) \| \leqslant |h(t) - h(0)| \leqslant |h(t)| + |h(0)| \tag{14.7}$$

for every $t \in [0, \infty)$, $x \in \mathcal{O}$ and $F \in \mathcal{F}_0(\Omega,\ h)$.

We claim that $\mathcal{F}_0(\Omega,\ h)$ is uniformly bounded on compact sets. In fact, let $A \subset \mathcal{O}$ and $C \subset [0, \infty)$ be compact sets and take $(x, t) \in A \times C$. By the continuity of h and the compactness of C, there exists $M > 0$ such that $|h(s)| \leqslant M$ for all $s \in C$. Now, by the relation (14.7), we have

$$\| F(x,t) \| \leqslant M + |h(0)|$$

for all $(x,t) \in A \times C$ and all $F \in \mathcal{F}_0(\Omega,\ h)$, which shows the claim. Consequently, the Ascoli's Theorem (see [184]) guarantees that any sequence $\{F_n\}_{n\in\mathbb{N}}$ in $\mathcal{F}_0(\Omega,\ h)$ admits a subsequence $\{F_{n_k}\}_{k\in\mathbb{N}}$ which converges uniformly on compact subsets to some function F_0. Finally, since $\mathcal{F}_0(\Omega,\ h)$ is closed (see Lemma 14.3), we conclude that $F_0 \in \mathcal{F}_0(\Omega,\ h)$ and the proof is finished. □

14.2 Existence of a Local Semidynamical System

This section concerns the existence of a local semidynamical system generated by the following nonautonomous generalized ODE

$$\frac{dx}{d\tau} = DF(x,\ t), \tag{14.8}$$

where $F : \Omega \to \mathbb{R}^n$ belongs to the class $\mathcal{F}_0(\Omega,\ h)$. Recall that we are assuming that $\Omega = \mathcal{O} \times [0,\ \infty)$, where $\mathcal{O} \subset \mathbb{R}^n$ in an open set and h is a nondecreasing continuous real function. The results of this section are borrowed from [4].

Remark 14.5: Lemma 4.9 ensures us that the solutions of (14.8) are continuous, since h is a continuous function.

According to Remark 14.2, a sequence $\{F_k\}_{k\in\mathbb{N}} \subset \mathcal{F}_0(\Omega, h)$ converges to a function $F \in \mathcal{F}_0(\Omega, h)$ if

$$F_k(x,\ t) \overset{k\to\infty}{\to} F(x,\ t)$$

in $\mathbb{R}^n$ for each $(x, t) \in \Omega$, that is,

$$\| F_k(x, t) - F(x, t) \| \overset{k\to\infty}{\to} 0,$$

for every $(x, t) \in \Omega$. In this case, we write $F_k \overset{k\to\infty}{\to} F$. In addition, given a sequence $\{v_k\}_{k\in\mathbb{N}} \subset \mathbb{R}^n$ and $v \in \mathbb{R}^n$, we have

$$(v_k, F_k) \overset{k\to\infty}{\to} (v, F) \quad \text{in} \quad \mathbb{R}^n \times \mathcal{F}_0(\Omega, h),$$

if and only if $\| v_k - v \| \overset{k\to\infty}{\to} 0$ and $\| F_k(x, s) - F(x, s) \| \overset{k\to\infty}{\to} 0$ for every $(x, s) \in \Omega$.

In the next definition, we exhibit the concept of a general local semidynamical system. Our aim is to show that the generalized ODE (14.8) generates a local semidynamical system.

Given $(v, F) \in \mathcal{O} \times \mathcal{F}_0(\Omega, h)$, we denote by $I_{(v,F)}$ an interval of type $[0, b) \subset \mathbb{R}$, with $b \in \mathbb{R}_+$. Consider, also, the following set

$$S = \{(t, v, F) \in \mathbb{R}_+ \times \mathcal{O} \times \mathcal{F}_0(\Omega, h) : t \in I_{(v,F)}\}.$$

Definition 14.6: A *local semidynamical system* on the space $\mathcal{O} \times \mathcal{F}_0(\Omega, h)$ is a mapping $\pi : S \to \mathcal{O} \times \mathcal{F}_0(\Omega, h)$ that satisfies the following conditions:

(i) $\pi(0, v, F) = (v, F)$, for every $(v, F) \in \mathcal{O} \times \mathcal{F}_0(\Omega, h)$;

(ii) given $(v, F) \in \mathcal{O} \times \mathcal{F}_0(\Omega, h)$, if $t \in I_{(v,F)}$ and $s \in I_{\pi(t,v,F)}$, then $t + s \in I_{(v,F)}$ and $\pi(s, \pi(t, v, F)) = \pi(t + s, v, F)$;

(iii) $\pi : S \to \mathcal{O} \times \mathcal{F}_0(\Omega, h)$ is continuous;

(iv) $I_{(v,F)} = [0, b_{(v,F)})$ is maximal in the following sense: either $I_{(v,F)} = \mathbb{R}_+$ or, if $b_{(v,F)} \neq \infty$, then the positive orbit $\{\pi(t, v, F) : t \in [0, b_{(v,F)})\} \subset \mathcal{O} \times \mathcal{F}_0(\Omega, h)$ cannot be continued to a larger interval $[0, b_{(v,F)} + c)$, $c > 0$;

(v) if $(v_k, F_k) \overset{k\to\infty}{\to} (v, F)$, where (v, F) and $(v_k, F_k) \in \mathcal{O} \times \mathcal{F}_0(\Omega, h)$, $k \in \mathbb{N}$, then $I_{(v,F)} \subset \liminf I_{(v_k,F_k)}$.

We observe that the definition of a local semidynamical system presented in [17] consists of items (i), (ii), (iii), and (iv) from Definition 14.6. Item (v) from [17], known as *Kamke's axiom*, assures that the domain of π is open. In our case, we replace condition (v) from [17] by an equivalent property of lower semicontinuity. The reader may want to consult [17, p. 12, 13] for more details.

It is important to mention that if the domain of π is $\mathbb{R}_+ \times \mathcal{O} \times \mathcal{F}_0(\Omega, h)$, then conditions (iv) and (v) are satisfied straightforwardly. Taking this fact into account, we have the following definition.

Definition 14.7: If the domain of π is $\mathbb{R}_+ \times \mathcal{O} \times \mathcal{F}_0(\Omega, h)$, then π will be called a *global semidynamical system*.

Given $F \in \mathcal{F}_0(\Omega,\ h)$ and $t \geqslant 0$, we denote the *translate* F_t *of* F by

$$F_t(x,\ s) = F(x,\ t+s) - F(x,\ t), \quad (x,\ s) \in \Omega. \tag{14.9}$$

The next result exhibits some basic properties of the translates F_t, $t \geqslant 0$. The reader also may consult [9, p. 234].

Lemma 14.8: *Given* $F \in \mathcal{F}_0(\Omega,\ h)$ *and* $t \geqslant 0$*, the translate* F_t *of* F *satisfies the following properties:*

(i) $F_0 = F$ *(normalization of* F*);*
(ii) $F_{t+\tau} = (F_t)_\tau$ *for all* $t, \tau \geqslant 0$ *(semigroup property);*
(iii) *the mapping* $(t, F) \mapsto F_t$ *is continuous.*

Proof. Item (i) follows immediately from the definition of F_t given in (14.9). Let $t, \tau \geqslant 0$ and $(x,\ s) \in \Omega$, then

$$\begin{aligned}(F_t)_\tau(x,s) &= F_t(x,\tau+s) - F_t(x,\tau)\\ &= F(x,t+\tau+s) - F(x,t) - F(x,t+\tau) + F(x,t)\\ &= F(x,t+\tau+s) - F(x,t+\tau)\\ &= F_{t+\tau}(x,s),\end{aligned}$$

that is, condition (ii) holds.

In order to show that condition (iii) holds, let $t, t_k \geqslant 0$ and $F, F_k \in \mathcal{F}_0(\Omega,\ h)$, $k \in \mathbb{N}$, be such that $F_k \overset{k\to\infty}{\to} F$ and $t_k \overset{k\to\infty}{\to} t$. For $(x,\ s) \in \Omega$ and using (14.2), we have

$$\begin{aligned}\|\ (F_k)_{t_k}(x,s) - F_t(x,s)\ \| &\leqslant \|\ (F_k)_{t_k}(x,s) - (F_k)_t(x,s)\ \| + \|\ (F_k)_t(x,s) - F_t(x,s)\ \|\\ &\leqslant \|\ F_k(x,s+t_k) - F_k(x,s+t)\ \| + \|\ F_k(x,t_k) - F_k(x,t)\ \|\\ &\quad + \|\ F_k(x,t+s) - F(x,t+s)\ \| + \|\ F_k(x,t) - F(x,t)\ \|\\ &\leqslant |h(s+t_k) - h(s+t)| + |h(t_k) - h(t)|\\ &\quad + \|\ F_k(x,t+s) - F(x,t+s)\ \| + \|\ F_k(x,t) - F(x,t)\ \|.\end{aligned}$$

Since h is continuous, $F_k \overset{k\to\infty}{\to} F$ and $t_k \overset{k\to\infty}{\to} t$, we conclude that $(F_k)_{t_k} \overset{k\to\infty}{\to} F_t$. □

Given $t \geqslant 0$ and $F \in \mathcal{F}_0(\Omega, h)$, we cannot assure that $F_t \in \mathcal{F}_0(\Omega, h)$. In the next definition, we provide some restriction on the function h in order to obtain an invariant subset of $\mathcal{F}_0(\Omega, h)$ under the translate F_t.

Definition 14.9: For a given nondecreasing continuous function $h : [0, \infty) \to \mathbb{R}$, we say that a function $F : \Omega \to X$ belongs to the *class* $\mathcal{F}_0^*(\Omega, h)$, if F belongs to the class $\mathcal{F}_0(\Omega, h)$ and the function h satisfies

$$|h(t_1+s) - h(t_2+s)| \leqslant |h(t_1) - h(t_2)|, \quad t_1, t_2, s \in [0,\ \infty).$$

The next result follows immediately from Theorem 14.4.

Corollary 14.10: *If $h : [0, \infty) \to \mathbb{R}$ is a nondecreasing continuous function, then the space $\mathcal{F}_0^*(\Omega, h)$ is compact.*

Next, we show that the class $\mathcal{F}_0^*(\Omega, h)$ is invariant, that is, $F_t \in \mathcal{F}_0^*(\Omega, h)$ for all $t \geqslant 0$ provided $F \in \mathcal{F}_0^*(\Omega, h)$.

Lemma 14.11: *If $F \in \mathcal{F}_0^*(\Omega, h)$, then the translate F_t of F belongs to $\mathcal{F}_0^*(\Omega, h)$ for each $t \geqslant 0$.*

Proof. Let $F \in \mathcal{F}_0^*(\Omega, h)$ and $t \geqslant 0$. Note that $F_t(x, 0) = F(x, t) - F(x, t) = 0$ for all $x \in \mathcal{O}$. Also, for all $(x, s_2), (x, s_1), (y, s_2), (y, s_1) \in \Omega$, and taking into account conditions (14.2)–(14.3), we obtain

$$\begin{aligned} \| F_t(x, s_2) - F_t(x, s_1) \| &= \| F(x, t + s_2) - F(x, t + s_1) \| \\ &\leqslant |h(t + s_2) - h(t + s_1)| \leqslant |h(s_2) - h(s_1)| \end{aligned}$$

and

$$\begin{aligned} &\| F_t(x, s_2) - F_t(x, s_1) - [F_t(y, s_2) - F_t(y, s_1)] \| \\ &= \| F(x, t + s_2) - F(x, t + s_1) - F(y, t + s_2) + F(y, t + s_1) \| \\ &\leqslant \| x - y \| \, |h(t + s_2) - h(t + s_1)| \\ &\leqslant \| x - y \| \, |h(s_2) - h(s_1)|. \end{aligned}$$

Hence, $F_t \in \mathcal{F}_0^*(\Omega, h)$ for all $t \geqslant 0$. □

Remark 14.12: The function F_t, defined in (14.9), is continuous for each $t \geqslant 0$, since we are assuming that $F \in \mathcal{F}_0^*(\Omega, h)$, where h is nondecreasing and continuous.

We finish this section with the construction of a local semidynamical system related to the nonautonomous generalized ODE (14.8). Theorem 14.13 generalizes [9, Theorem 6.3] and [91, Theorem 4.1].

Theorem 14.13: *Assume that for each $u \in \mathcal{O}$ and each $F \in \mathcal{F}_0^*(\Omega, h)$, $x(t, u, F)$ is the unique maximal solution of the initial value problem*

$$\begin{cases} \dfrac{dx}{d\tau} = DF(x, t), \\ x(0) = u. \end{cases} \tag{14.10}$$

Let $[0, \omega(u, F))$, $\omega(u, F) > 0$, be the maximal interval of definition of the solution $x(\cdot, u, F)$ and define $\pi : S \to \mathcal{O} \times \mathcal{F}_0^*(\Omega, h)$ by

$$\pi(t, u, F) = (x(t, u, F), F_t), \tag{14.11}$$

where $S = \{(t, u, F) \in \mathbb{R}_+ \times \mathcal{O} \times \mathcal{F}_0^*(\Omega, h) : t \in I_{(u,F)}\}$. *Then,* π *is a local semidynamical system on* $\mathcal{O} \times \mathcal{F}_0^*(\Omega, h)$.

Proof. Since the second component F_t of π in (14.11) is defined for all $t \in [0, \infty)$, the maximal interval $I_{(u,F)}$ of the semidynamical system, given by (14.11), coincides with $[0, \omega(u, F))$. We need to verify that the five conditions of Definition 14.6 hold. We prove at first that conditions (i), (ii), (iv), and (v) hold, and, finally, we prove condition (iii). *Proof of* (i): Let $(u, F) \in \mathcal{O} \times \mathcal{F}_0^*(\Omega, h)$. Since $F_0(x, s) = F(x, s)$ for all $(x, s) \in \Omega$ (see Lemma 14.8(i)) and $x(0, u, F) = u$ for each (u, F), we obtain $\pi(0, u, F) = (u, F)$.

Proof of (ii): Let $t \in I_{(u, F)}$, $\sigma \in I_{\pi(t,u, F)}$ and $(u, F) \in \mathcal{O} \times \mathcal{F}_0^*(\Omega, h)$.

For $\tau \in I_{(u, F)}$, set

$$x(\tau) = x(\tau, u, F), \quad \psi(\tau) = x(\tau, x(t), F_t), \quad \text{and} \quad \xi(\tau) = x(\tau + t),$$

where x is the maximal solution of (14.10) and ψ is a solution of the following initial value problem:

$$\begin{cases} \dfrac{d\psi}{d\tau} = D[F_t(\psi, s)], \\ \psi(0) = x(t) = x(t, u, F). \end{cases} \tag{14.12}$$

We are going to show that ξ is a solution of problem (14.12). Initially, we point out that

$$\xi(\sigma) - \xi(0) = x(\sigma + t) - x(t) = \int_t^{\sigma+t} DF(x(\tau), s).$$

On the other hand, using the change of variable $\phi(s) = s + t$ and Theorem 2.18, we obtain

$$\begin{aligned} \int_t^{t+\sigma} DF(x(\tau), s) &= \int_{\phi(0)}^{\phi(\sigma)} DF(x(\tau), s) = \int_0^{\sigma} DF(x(\phi(\varsigma)), \phi(\mu)) \\ &= \int_0^{\sigma} DF(x(\varsigma + t), \mu + t), \end{aligned}$$

whence

$$\xi(\sigma) - \xi(0) = \int_0^{\sigma} DF(x(\tau + t), s + t) = \int_0^{\sigma} DF_t(\xi(\tau), s).$$

Furthermore, since $\xi(0) = x(t) = x(t, u, F)$, we use the uniqueness of the solution of (14.10) (see Theorem 5.1), and we conclude that $\psi(\sigma) = \xi(\sigma) = x(\sigma + t)$ for all $\sigma \in I_{\pi(t, u, F)} = [0, \omega(u, F))$. Using this fact, we obtain

$$\begin{aligned} \pi(\sigma, \pi(t, u, F)) &= \pi(\sigma, x(t, u, F), F_t) = \pi(\sigma, x(t), F_t) \\ &= (x(\sigma, x(t), F_t), (F_t)_\sigma) = (\xi(\sigma), (F_t)_\sigma) \end{aligned}$$

$$= (\xi(\sigma),\ F_{\sigma+t}) = (x(\sigma+t),\ F_{t+\sigma})$$
$$= (x(\sigma+t,\ u,\ F),\ F_{t+\sigma}) = \pi(\sigma+t,\ u,\ F).$$

Proof of (iv): Assume that $\omega = \omega(u, F) < \infty$. Since h is continuous, $\Omega = \Omega_F$. Thus, if $x(t, u, F) \to z$ as $t \to \omega^-$, then it follows by Corollary 5.15 that $z \notin \mathcal{O}$.

Proof of (v): Take $(y_0,\ F_0) \in \mathcal{O} \times \mathcal{F}_0^*(\Omega,\ h)$ and a sequence $\{(y_k,\ F_k)\}_{k\in\mathbb{N}} \subset \mathcal{O} \times \mathcal{F}_0^*(\Omega,\ h)$, such that $(y_k,\ F_k) \overset{k\to\infty}{\to} (y_0,\ F_0)$. We are going to show that $\omega(u,\ F)$ is lower semicontinuous on $\mathcal{O} \times \mathcal{F}_0^*(\Omega,\ h)$. The idea of the proof is based on [9, Thm. A.8]. Let $t_k \overset{k\to\infty}{\to} t_0$ and $x(s) = x(s,\ y_0,\ F_0)$ be the unique solution of the initial value problem:

$$\begin{cases} \dfrac{dx}{d\tau} = DF_0(x,\ s), \\ x(0) = y_0, \end{cases} \tag{14.13}$$

defined on the maximal interval $[0,\ \omega(y_0,\ F_0))$, with $\omega(y_0,\ F_0) > 0$.

According to [209, Theorem 8.6], we can assert that there exists $k_1 \in \mathbb{N}$ such that, for each $k \geqslant k_1$, one can obtain a solution $x_k(s,\ y_k,\ F_k)$ of the nonautonomous generalized ODE:

$$\begin{cases} \dfrac{dx}{d\tau} = DF_k(x,\ s), \\ x(0,\ y_k,\ F_k) = y_k, \end{cases} \tag{14.14}$$

defined on $[0,\ \gamma]$, $0 < \gamma < \omega(y_0,\ F_0)$, satisfying $\lim_{k\to\infty} x_k(s,\ y_k,\ F_k) = x(s,\ y_0,\ F_0)$ for every $s \in [0,\ \gamma]$. The result [209, Theorem 8.6] assures that γ does not depend on $k \geqslant k_1$.

Consider a set $A \subset [0, \infty)$ defined by

$$A = \{b \geqslant 0 :\ \text{for } k \geqslant k_1 \text{ the functions } x_k(s,\ y_k,\ F_k) \text{ are defined in } [0, b] \text{ and are equicontinuous on } [0, b]\}. \tag{14.15}$$

Claim 1. The functions $x_k(\cdot,\ y_k,\ F_k)$, $k \geqslant k_1$, are equicontinuous on $[0,\ \gamma]$.

Indeed, Lemma 4.9 provides the relation

$$\| x_k(s_2,\ y_k,\ F_k) - x_k(s_1,\ y_k,\ F_k) \| \leqslant |h(s_2) - h(s_1)|, \quad s_1,\ s_2 \in [0,\ \gamma],$$

consequently, the equicontinuity of $x_k(s,\ y_k,\ F_k)$ follows immediately, since h is continuous and does not depend on k. This yields $A \neq \emptyset$.

Set $\beta = \sup A$. In order to prove the lower semicontinuity of ω, we will show that $[0,\ \beta)$ is the maximal positive interval of definition of the solution $x(\cdot,\ y_0,\ F_0)$.

Claim 2. The sequence of functions $x_k(\cdot,\ y_k,\ F_k)$ is an equibounded sequence for $k > k_2$ with k_2 sufficiently larger than k_1.

In fact, let $0 \leqslant b < \beta$. Again, using Lemma 4.9, we obtain

$$\begin{aligned}\| x_k(s, y_k, F_k) \| &\leqslant \| y_k \| + \| x_k(s, y_k, F_k) - y_k \| \\ &= \| y_k \| + \| x_k(s, y_k, F_k) - x_k(0, y_k, F_k) \| \\ &\leqslant \| y_k \| + [h(s) - h(0)] \\ &\leqslant \| y_k \| + [h(b) - h(0)],\end{aligned}$$

for every $s \in [0, b]$, which proves *Claim 2*, since $y_k \overset{k\to\infty}{\to} y_0$.

Claims 1 and *2* allow us to conclude that $\{x_k(s, y_k, F_k)\}_{k>k_2}$ is a pointwise relatively compact family of uniformly bounded variation. Therefore, by a Helly's Choice Principle (see, e.g. [13]), we can infer that the sequence $x_k(\,\cdot\,, y_k, F_k)$ is relatively compact in $C([0, b], \mathbb{R}^n)$ for $k > k_2$. The result [209, Theorem 8.2] guarantees that every limiting point of this sequence is a solution of system (14.13) defined on $[0, b]$. Besides, from the uniqueness of solutions of this equation, we obtain exactly one limiting point of the sequence $\{x_k(s, y_k, F_k)\}_{k>k_2}$. Thus, the whole sequence converges uniformly to the solution $x(s, y_0, F_0)$ on $[0, b]$.

Suppose to the contrary that $x(\beta) = x(\beta, y_0, F_0)$ is defined. Consequently, $x(\beta) \in \mathcal{O}$ and, by Theorem 5.1, there exists $\Delta_\beta > 0$ such that $x(s, y_0, F_0)$ is defined for each $s \in [\beta, \beta + \Delta_\beta]$. Moreover, [209, Theorem 8.6] ensures the existence of an integer $\bar{k}$ such that the sequence $x_k(s, y_k, F_k)$ is defined and is equicontinuous on $[0, \beta + \Delta_\beta]$ for all $k \geqslant \bar{k}$. But this is a contradiction as $\beta = \sup A$. Therefore, $x(\cdot, y_0, F_0)$ is not defined at β and $\beta = \omega(y_0, F_0)$.

Proof of (iii): Note that for each fixed $(u, F) \in \mathcal{O} \times \mathcal{F}_0^*(\Omega, h)$, $\pi(t, u, F)$ is continuous at every $t \in I_{(u,F)}$. This fact follows from Remarks 14.12, and Lemma 14.8(iii).

Let $(t_0, u_0, F_0) \in I_{(u_0,F_0)} \times \mathcal{O} \times \mathcal{F}^*(\Omega, h)$ be arbitrary and $\{(t_k, u_k, F_k)\}_{k\in\mathbb{N}} \subset I_{(u_k,F_k)} \times \mathcal{O} \times \mathcal{F}^*(\Omega, h)$ be a sequence such that $(t_k, u_k, F_k) \overset{k\to\infty}{\to} (t_0, u_0, F_0)$. It follows from the proof of item (v) that

$$x(s, u_k, F_k) \overset{k\to\infty}{\to} x(s, u_0, F_0), \tag{14.16}$$

uniformly on compact intervals of $[0, \beta)$, where $\beta = \sup A$, with A defined in (14.15), $x(s, u_0, F_0)$ is the unique solution of the initial value problem

$$\begin{cases} \dfrac{dx}{d\tau} = DF_0(x, s), \\ x(0) = u_0, \end{cases}$$

and, for each k, $x(s, u_k, F_k)$ is the unique solution of the initial value problem

$$\begin{cases} \dfrac{dx}{d\tau} = DF_k(x, s), \\ x(0) = u_k. \end{cases}$$

Note that $x(t_k, u_k, F_k) \overset{k\to\infty}{\to} x(t_0, u_0, F_0)$. In fact, this follows from item (v), the continuity of $\pi(\cdot, u_0, F_0)$, relation (14.16), and the inequality

$$\| x(t_k, u_k, F_k) - x(t_0, u_0, F_0) \| \leqslant \| x(t_k, u_k, F_k) - x(t_k, u_0, F_0) \| + \| x(t_k, u_0, F_0) - x(t_0, u_0, F_0) \| .$$

Finally, the continuity of mapping π is guaranteed, because $(F_k)_{t_k}$ converges to $(F_0)_{t_0}$ by the properties of the translates of F. Therefore, the proof is complete. □

14.3 Existence of an Impulsive Semidynamical System

In this section, based on the paper [4], we investigate properties of the following impulsive generalized ODE associated with the initial value problem (14.10):

$$\begin{cases} \dfrac{dx}{d\tau} = DF(x, s) \\ \mathrm{I} : \mathrm{M} \to \mathbb{R}^n \\ x(0) = u, \end{cases} \tag{14.17}$$

where $F \in \mathcal{F}_0^*(\Omega,\ h)$ ($\Omega = \mathcal{O} \times [0,\ \infty)$ with $\mathcal{O} \subset \mathbb{R}^n$ an open set), $u \in \mathcal{O}$, $M \subset \mathbb{R}^n$ is a closed subset, and I is a continuous function such that $\mathrm{I}(\mathrm{M} \cap \mathcal{O}) \subset \mathcal{O} \setminus M$. We also assume that M satisfies the following condition: if for $(u, F) \in \mathcal{O} \times \mathcal{F}_0^*(\Omega,\ h)$, the solution of (14.10) is such that $x(t_0, u, F) \in M$ for some $t_0 > 0$, then there exists $\epsilon = \epsilon(u, F) > 0$ such that

$$x(t, u, F) \notin \mathrm{M} \quad \text{for} \quad t \in (t_0 - \epsilon, t_0) \cup (t_0, t_0 + \epsilon). \tag{14.18}$$

Now, we define a function which represents the least positive time at which the trajectory of (u, F) meets M. This function, which we denote by $\varphi : \mathcal{O} \times \mathcal{F}_0^*(\Omega, h) \to (0, \infty]$, is given by

$$\varphi(u, F) = \begin{cases} s, & \text{if } x(s, u, F) \in \mathrm{M} \text{ and } x(t, u, F) \notin \mathrm{M} \text{ for } 0 < t < s, \\ \infty, & \text{if } x(t, u, F) \notin \mathrm{M} \text{ for all } t > 0. \end{cases} \tag{14.19}$$

From now on, in this chapter, we shall assume that the function φ defined in (14.19) is continuous on $(\mathcal{O} \setminus \mathrm{M}) \times \mathcal{F}_0^*(\Omega,\ h)$. The reader may want to consult [41] to obtain sufficient conditions for φ to be continuous.

Using the function φ, we are now able to describe the impulsive trajectory of system (14.17). Let $(u, F) \in \mathcal{O} \times \mathcal{F}_0^*(\Omega,\ h)$ be fixed and arbitrary. Next, we characterize the solution $\widetilde{x}(t, u, F)$ of the impulsive generalized ODE (14.17).

(I) If $\varphi(u,F)=\infty$, then we define $\widetilde{x}(t,u,F)=x(t,u,F)$ for all $t\geqslant 0$, where $x(t,u,F)$ is the solution of the generalized ODE (14.10). But, if $\varphi(u,F)=s_0$, then $u_1=x(s_0,u,F)\in \mathrm{M}$. Thus, we define $\widetilde{x}(t,u,F)$ on $[0,s_0]$ by

$$\widetilde{x}(t,u,F)=\begin{cases} x(t,u,F), & 0\leqslant t<s_0,\\ u_1^+, & t=s_0,\end{cases}$$

where $u_1^+=\mathrm{I}(u_1)$. Let us consider $u=u_0^+$. The process now continues from u_1^+ on.

(II) If $\varphi(u_1^+,F)=\infty$,, then we define $\widetilde{x}(t,u,F)=x(t-s_0,u_1^+,F)$ for $t\geqslant s_0$, where $x(\cdot,u_1^+,F)$ is the solution of the generalized ODE

$$\begin{cases} \dfrac{dx}{d\tau}=DF(x,s),\\ x(0)=u_1^+.\end{cases}$$

However, if $\varphi(u_1^+,F)=s_1$, then $u_2=x(s_1,u_1^+,F)\in \mathrm{M}$. In this way, we define $\widetilde{x}(t,u,F)$ on $[s_0,\ s_0+s_1]$ by

$$\widetilde{x}(t,u,F)=\begin{cases} x(t-s_0,u_1^+,F), & s_0\leqslant t<s_0+s_1,\\ u_2^+, & t=s_0+s_1,\end{cases}$$

where $u_2^+=\mathrm{I}(u_2)$.

(III) Assume that $\widetilde{x}(t,u,F)$ is defined on the interval $[t_{n-1},t_n]$ with $\widetilde{x}(t_n,u,F)=u_n^+$, where $t_n=\sum_{i=0}^{n-1}s_i$, $n\in\mathbb{N}$. If $\varphi(u_n^+,F)=\infty$, then we define

$$\widetilde{x}(t,u,F)=x(t-t_n,u_n^+,F)$$

for all $t\geqslant t_n$. However, if $\varphi(u_n^+,F)=s_n$, then

$$\widetilde{x}(t,u,F)=\begin{cases} x(t-t_n,u_n^+,F), & t_n\leqslant t<t_{n+1},\\ u_{n+1}^+, & t=t_{n+1},\end{cases}$$

where $u_{n+1}^+=\mathrm{I}(u_{n+1})$ and $u_{n+1}=x(s_n,u_n^+,F)\in \mathrm{M}$.

The solution $\widetilde{x}(t,u,F)$ is defined on each interval $[t_n,\ t_{n+1}]$, where $t_0=0$ and $t_{n+1}=\sum_{i=0}^{n}s_i$, $n\in\mathbb{N}_0$. Consequently, $\widetilde{x}(t,u,F)$ is defined on the interval $[0,\ t_{n+1}]$. Furthermore, provided $\varphi(u_n^+,F)=\infty$ for some n, the process described above ends after a finite number of steps. However, if $\varphi(u_n^+,F)<\infty$ for all $n\in\mathbb{N}_0$, then the process continues indefinitely and, in this case, $\widetilde{x}(t,u,F)$ is defined on $[0,\ T(u,F))$, where $T(u,F)=\sum_{i=0}^{\infty}s_i$.

From now on, we assume that the solutions of Eqs. (14.10) and (14.17) are defined in the whole interval $[0,\infty)$. The reader may consult Chapter 5 to obtain sufficient conditions to prolongate a solution to the interval $[0,\infty)$.

By Theorem 14.13 and Definition 14.7, the mapping

$$\pi:\mathbb{R}_+\times\mathcal{O}\times\mathcal{F}_0^*(\Omega,\ h)\to\mathcal{O}\times\mathcal{F}_0^*(\Omega,\ h)$$

given by

$$\pi(t,\ u,\ F) = (x(t,\ u,\ F),\ F_t),$$

defines a global semidynamical system associated with the generalized ODE (14.10). Let us denote a global semidynamical system by $(\mathcal{O} \times \mathcal{F}_0^*(\Omega,\ h), \pi)$ and, for the sake of simplicity, we refer to such system as a semidynamical system.

Definition 14.14: Let $(u, F) \in \mathcal{O} \times \mathcal{F}_0^*(\Omega,\ h)$.

(i) The *motion of* (u, F) is the continuous function $\pi_{(u,F)} : \mathbb{R}_+ \to \mathcal{O} \times \mathcal{F}_0^*(\Omega,\ h)$ defined by $\pi_{(u,\ F)}(t) = \pi(t, u, F)$, $t \in \mathbb{R}_+$.
(ii) The *positive orbit* of (u, F) is given by $\pi^+(u, F) = \{\pi(t, u, F) : t \geqslant 0\}$.

The concept of impulsive semidynamical system related to a generalized ODE was introduced in [4] and follows below.

Definition 14.15: An *impulsive semidynamical system* on $\mathcal{O} \times \mathcal{F}_0^*(\Omega,\ h)$ is a mapping $\tilde{\pi} : \mathbb{R}_+ \times \mathcal{O} \times \mathcal{F}_0^*(\Omega,\ h) \to \mathcal{O} \times \mathcal{F}_0^*(\Omega,\ h)$ which satisfies the following conditions:

(i) $\tilde{\pi}(0, u, F) = (u, F)$ for all $(u, F) \in \mathcal{O} \times \mathcal{F}_0^*(\Omega,\ h)$;
(ii) $\tilde{\pi}(s, \tilde{\pi}(t, u, F)) = \tilde{\pi}(t + s, u, F)$ for all $(u, F) \in \mathcal{O} \times \mathcal{F}_0^*(\Omega,\ h)$ and $t, s \in [0,\ \infty)$;
(iii) for each $(u, F) \in \mathcal{O} \times \mathcal{F}_0^*(\Omega,\ h)$, the mapping $\tilde{\pi}(\cdot, u, F)$ is right-continuous at every point in $[0, \infty)$ and the left limits $\tilde{\pi}(t^-, u, F)$ exist for all $t > 0$.

The reader may consult [25, 27, 41] to obtain more details about the theory of impulsive semidynamical systems in the classic ordinary case.

Definition 14.16: Let $(u, F) \in \mathcal{O} \times \mathcal{F}_0^*(\Omega,\ h)$. The *positive impulsive orbit* of (u, F) is given by the set $\tilde{\pi}^+(u, F) = \{\tilde{\pi}(t, u, F) : t \geqslant 0\}$.

Consider a semidynamical system $(\mathcal{O} \times \mathcal{F}_0^*(\Omega,\ h), \pi)$ associated with the system (14.10) and let $\pi(t, u, F) = (x(t, u, F), F_t)$ be its motion, where $x(t, u, F)$ is the unique solution of (14.10) defined on the interval $[0, \infty)$. Associated with this motion, we define a mapping $\tilde{\pi} : \mathbb{R}_+ \times \mathcal{O} \times \mathcal{F}_0^*(\Omega,\ h) \to \mathcal{O} \times \mathcal{F}_0^*(\Omega,\ h)$ by

$$\tilde{\pi}(t, u, F) = \pi(t - t_n, u_n^+, F) \qquad \text{for } t_n \leqslant t < t_{n+1} \text{ and } n \in \mathbb{N}_0, \tag{14.20}$$

where $u = u_0^+$, $t_0 = 0$, $t_n = \sum_{i=0}^{n-1} s_i$, with $n \in \mathbb{N}$, and $s_n = \varphi(u_n^+, F)$, $n \in \mathbb{N}_0$ (recall the definition of φ in (14.19)).

It is worth noting that $\tilde{\pi}(t, u, F) = (\tilde{x}(t, u, F), F_{t-t_n})$ for $t_n \leqslant t < t_{n+1}$, $n \in \mathbb{N}_0$, where $\tilde{x}(t, u, F)$ is the solution of (14.17).

The next result, established in [4, Theorem 5.2], guarantees that $\tilde{\pi}$ given by (14.20) is an impulsive semidynamical system associated with the impulsive generalized ODE (14.17).

Theorem 14.17: *The mapping $\tilde{\pi}$ given by (14.20) is an impulsive semidynamical system associated with the impulsive generalized ODE (14.17). We denote such system by $(\mathcal{O} \times \mathcal{F}_0^*(\Omega, h), \tilde{\pi})$.*

Proof. Using the ideas of the proof of [25, Proposition 2.1], we obtain conditions (i) and (ii) of Definition 14.15. Moreover, condition (iii) of Definition 14.15 is easily verified, since $\tilde{x}(t, u, F)$ and F_t are right-continuous at every point $t \in [0, \infty)$ and the left limits $\tilde{x}(t-, u, F)$ and F_{t-} exist for all $t > 0$. This completes the proof. □

Although $\tilde{\pi}$ is not continuous, we have the following result concerning convergence. It is worth mentioning that its proof is analogous to the proof presented in [141, Lemma 2.3]. See also [4, Lemma 6.1].

Lemma 14.18: *Let $(\mathcal{O} \times \mathcal{F}_0^*(\Omega, h), \tilde{\pi})$ be an impulsive semidynamical system. Assume that $u \in \mathcal{O} \setminus M$ and $\{v_n\}_{n\in\mathbb{N}}$ is a sequence in $\mathcal{O}$ which converges to u. Let $\{F_n\}_{n\in\mathbb{N}}$ be a sequence in $\mathcal{F}_0^*(\Omega, h)$ such that $F_n \overset{n\to\infty}{\to} F$. Then, for every $t \geqslant 0$, there exists a sequence of real numbers $\{\epsilon_n\}_{n\in\mathbb{N}}$, with $\epsilon_n \overset{n\to\infty}{\to} 0$, such that*

$$\tilde{\pi}(t + \epsilon_n, v_n, F_n) \overset{n\to\infty}{\to} \tilde{\pi}(t, u, F).$$

Proof. For each $n \in \mathbb{N}$, let $x(t, v_n, F_n)$ denote the solution of the initial value problem

$$\begin{cases} \dfrac{dx}{d\tau} = DF_n(x, s) \\ x(0) = v_n, \end{cases} \tag{14.21}$$

defined for all $t \geqslant 0$. According to [209, Theorem 8.6], $x(t, v_n, F_n) \overset{n\to\infty}{\to} x(t, u, F)$, where $x(t, u, F)$ is the solution of the generalized ODE (14.10). Then

$$\pi(t, v_n, F_n) \overset{n\to\infty}{\to} \pi(t, u, F)$$

for each $t \geqslant 0$, since $(F_n)_t \overset{n\to\infty}{\to} F_t$.

Note that if $\varphi(u, F) = \infty$, then the statement follows taking $\epsilon_n = 0$, $n \in \mathbb{N}$, as $\varphi(v_n, F_n) \overset{n\to\infty}{\to} \varphi(u, F)$. However, if $\varphi(u, F) < \infty$, then we use the ideas presented in [141, Lemma 2.3] to conclude the result.

Case 1: $0 \leqslant t < s_0 = \varphi(u, F)$.

Let $\epsilon > 0$ be such that $\epsilon < s_0 - t$. Since φ is continuous on $(\mathcal{O} \setminus \mathrm{M}) \times \mathcal{F}_0^*(\Omega,\ h)$, there exists $n_0 \in \mathbb{N}$ such that $-\epsilon < \varphi(v_n, F_n) - \varphi(u, F)$ for all $n \geqslant n_0$. Consequently, $t < s_0 - \epsilon < \varphi(v_n, F_n)$ for $n \geqslant n_0$, and taking $\epsilon_n = 0$, yields

$$\tilde{\pi}(t + \epsilon_n, v_n, F_n) = \tilde{\pi}(t, v_n, F_n) = \pi(t, v_n, F_n) \overset{n\to\infty}{\to} \pi(t, u, F) = \tilde{\pi}(t, u, F).$$

Case 2: $t = s_0 = \varphi(u, F)$.

Choosing $\epsilon_n = \varphi(v_n, F_n) - \varphi(u, F)$, $n \in \mathbb{N}$, we have

$$\tilde{\pi}(t + \epsilon_n, v_n, F_n) = \tilde{\pi}(\varphi(v_n, F_n), v_n, F_n) = \pi(0, \mathrm{I}((v_n)_1), F_n),$$

where $(v_n)_1 = x(\varphi(v_n, F_n), v_n, F_n)$, $n \in \mathbb{N}$. However, since $\mathrm{I}((v_n)_1) \overset{n\to\infty}{\to} \mathrm{I}(u_1)$, we obtain

$$\tilde{\pi}(t + \epsilon_n, v_n, F_n) = \pi(0, \mathrm{I}((v_n)_1), F_n) \overset{n\to\infty}{\to} \pi(0, u_1^+, F) = \tilde{\pi}(t, u, F).$$

Case 3: $t > \varphi(u, F)$.

In this case, we may write

$$t = \sum_{i=0}^{m-1} s_i + t',$$

for some $m \in \mathbb{N}$ and $0 \leqslant t' < s_m$. Now, set $t_n = \sum_{i=0}^{m-1} \varphi((v_n)_i^+, F_n)$, where

$$\begin{aligned}
&(v_n)_0^+ = v_n,\\
&(v_n)_i = x(\varphi((v_n)_{i-1}^+, F_n), (v_n)_{i-1}^+, F_n) \quad \text{and}\\
&\mathrm{I}((v_n)_i) = (v_n)_i^+, \ \text{for } 1 \leqslant i \leqslant m-1.
\end{aligned}$$

Thus,

$$\tilde{\pi}(t_n, v_n, F_n) = ((v_n)_m^+, F_n) \overset{n\to\infty}{\to} (u_m^+, F).$$

Defining $\epsilon_n = t_n + t' - t$, $n \in \mathbb{N}$, and taking into account that $u_m^+ \notin M$ (since $\mathrm{I}(\mathrm{M} \cap \mathcal{O}) \subset \mathcal{O} \setminus M$ and $t' < s_m = \varphi(u_m^+, F)$), it follows from Case 1 that

$$\tilde{\pi}(t + \epsilon_n, v_n, F_n) = \tilde{\pi}(t', \tilde{\pi}(t_n, v_n, F_n)) \overset{n\to\infty}{\to} \tilde{\pi}(t', u_m^+, F) = \tilde{\pi}(t, u, F),$$

and the proof is finished. □

Definition 14.19: A subset Γ of $\mathcal{O} \times \mathcal{F}_0^*(\Omega,\ h)$ is called *positively invariant*, if for every $(v_0,\ F_0) \in \Gamma$ and every $t \in [0,\ \infty)$, we have $\tilde{\pi}(t,\ v_0,\ F_0) \in \Gamma$.

The positive orbit of a point $(v, H) \in \mathcal{O} \times \mathcal{F}_0^*(\Omega,\ h)$ is positively invariant. The closure of a positive orbit $\tilde{\pi}^+(v, H)$, with $(v, H) \in \mathcal{O} \times \mathcal{F}_0^*(\Omega,\ h)$, is not positively invariant in general. However, we have the following result.

Lemma 14.20: *For each $(v, H) \in \mathcal{O} \times \mathcal{F}_0^*(\Omega,\ h)$, the set $\overline{\tilde{\pi}^+(v, H)} \setminus (M \times \mathcal{F}_0^*(\Omega,\ h))$ is positively invariant.*

Proof. Let $(u, F) \in \overline{\tilde{\pi}^{+}(v, H)} \setminus (\mathrm{M} \times \mathcal{F}_0^*(\Omega,\ h))$ and $t \geqslant 0$ be arbitrary. Then, there exists a sequence $\{t_n\}_{n\in\mathbb{N}} \subset \mathbb{R}_+$ such that $\tilde{\pi}(t_n, v, H) \overset{n\to\infty}{\to} (u, F)$. Since $u \notin \mathrm{M}$, it follows from Lemma 14.18 that there exists a sequence of real numbers $\{\epsilon_n\}_{n\in\mathbb{N}}$, with $\epsilon_n \overset{n\to\infty}{\to} 0$, such that

$$\tilde{\pi}(t + t_n + \epsilon_n, v, H) = \tilde{\pi}(t + \epsilon_n, \pi(t_n, v, H)) \overset{n\to\infty}{\to} \tilde{\pi}(t, u, F).$$

Therefore, $\tilde{\pi}(t, u, F) \in \overline{\tilde{\pi}^{+}(v, H)} \setminus (\mathrm{M} \times \mathcal{F}_0^*(\Omega,\ h))$, and the proof is complete. □

14.4 LaSalle's Invariance Principle

In this section, we present a version of LaSalle's invariance principle in the context of generalized ODEs. The definitions and the results presented is this section can be found in [4]. The existence of an impulsive semidynamical system $(\mathcal{O} \times \mathcal{F}_0^*(\Omega,\ h),\ \tilde{\pi})$ (Theorem 14.17) will be crucial for obtaining such result.

Next, we exhibit the concept of a limit set on impulsive semidynamical systems in the frame of generalized ODEs.

Definition 14.21: Let $(\mathcal{O} \times \mathcal{F}_0^*(\Omega,\ h),\ \tilde{\pi})$ be an impulsive semidynamical system. The *set of all limiting points* of $\tilde{\pi}(t,\ u,\ F)$, when $t \to \infty$, is given by

$$\Omega^+(u,\ F) = \{(u^*,\ F^*) \in \mathcal{O} \times \mathcal{F}_0^*(\Omega,\ h)\ :\ \tilde{\pi}(\lambda_n,\ u,\ F) \overset{n\to\infty}{\to} (u^*,\ F^*)$$

$$\text{for some sequence of positive real numbers} \quad \lambda_n \overset{n\to\infty}{\to} \infty\}.$$

We call $\Omega^+(u,\ F)$ the *positive limit set* of $\tilde{\pi}(t,\ u,\ F)$.

The next result follows straightforwardly using Lemma 14.18.

Lemma 14.22: *The set $\Omega^+(u,\ F) \setminus (M \times \mathcal{F}_0^*(\Omega,\ h))$ is positively invariant. In particular, if $\Omega^+(u,\ F) \cap (M \times \mathcal{F}_0^*(\Omega,\ h)) = \emptyset$, then $\Omega^+(u,\ F)$ is positively invariant.*

In the sequel, we establish sufficient conditions for the limit set to be non-empty.

Proposition 14.23: *Let $(\mathcal{O} \times \mathcal{F}_0^*(\Omega,\ h),\ \tilde{\pi})$ be an impulsive semidynamical system. If $\tilde{x}(t,\ u,\ F)$ remains in a compact subset C of $\mathcal{O}$ for all $t \in [0,\ \infty)$, then $\Omega^+(u,\ F)$ is non-empty.*

Proof. Let $\{\lambda_n\}_{n\in\mathbb{N}} \subset \mathbb{R}_+$ be a sequence such that $\lambda_n \overset{n\to\infty}{\to} \infty$. Note that for each $n \in \mathbb{N}$, there exists $p(n) \in \mathbb{N}^*$ satisfying $t_{p(n)} \leqslant \lambda_n < t_{p(n)+1}$, where $t_{p(n)} = \sum_{i=0}^{p(n)-1} s_i$. By the definition of $\tilde{\pi}$, we may write

$$\tilde{\pi}(\lambda_n, u, F) = \pi(\lambda_n - t_{p(n)}, u^+_{p(n)}, F) = (x(\lambda_n - t_{p(n)}, u^+_{p(n)}, F), F_{\lambda_n - t_{p(n)}}).$$

Using the compactness of C and of $\mathcal{F}_0^*(\Omega,\ h)$ guaranteed by Corollary 14.10, we obtain a subsequence $\{n_k\}_{k\in\mathbb{N}}$, $u^* \in C$ and $F^* \in \mathcal{F}_0^*(\Omega,\ h)$ such that

$$\widetilde{x}(\lambda_{n_k}, u, F) = x(\lambda_{n_k} - t_{p(n_k)}, u^+_{p(n_k)}, F) \overset{k\to\infty}{\to} u^* \quad \text{and}$$

$$F_{\lambda_{n_k} - t_{p(n_k)}} \overset{k\to\infty}{\to} F^* \text{ in } \mathcal{F}_0^*(\Omega,\ h).$$

Therefore, $\tilde{\pi}(\lambda_{n_k},\ u,\ F) \overset{k\to\infty}{\to} (u^*,\ F^*)$ and, since $\lambda_{n_k} \overset{k\to\infty}{\to} \infty$, we conclude that $(u^*,\ F^*) \in \Omega^+(u,\ F)$. □

In the next definition, we present the concept of a Lyapunov functional associated with the impulsive semidynamical system $(\mathcal{O} \times \mathcal{F}_0^*(\Omega,\ h),\ \tilde{\pi})$.

Definition 14.24: A *Lyapunov functional associated to the impulsive semidynamical system* $(\mathcal{O} \times \mathcal{F}_0^*(\Omega, h),\ \tilde{\pi})$ is a nonnegative function $V : \mathcal{O} \times \mathcal{F}_0^*(\Omega, h) \to \mathbb{R}_+$ which satisfies the following conditions:

(i) V is continuous on $\mathcal{O} \times \mathcal{F}_0^*(\Omega, h)$;
(ii) $\dot{V}(u, F) \leqslant 0$ for $(u, F) \in \mathcal{O} \times \mathcal{F}_0^*(\Omega, h)$, where

$$\dot{V}(u,\ F) = \limsup_{h\to 0^+} \frac{V(\tilde{\pi}(h, u, F)) - V(u, F)}{h}.$$

Note that condition (ii) of Definition 14.24 ensures us that $V(\tilde{\pi}(t, u, F)) \leqslant V(u, F)$ for every $t \geqslant 0$.

Now, we are able to present a version of LaSalle's invariance principle for generalized ODEs.

Theorem 14.25 LaSalle's Invariance Principle: *Let* $(\mathcal{O} \times \mathcal{F}_0^*(\Omega,\ h),\ \tilde{\pi})$ *be an impulsive semidynamical system. Suppose* $\widetilde{x}(t,\ u,\ F)$ *remains in a compact subset* C *of* $\mathcal{O}$ *for all* $t \in [0,\ \infty)$. *Let* $V : \mathcal{O} \times \mathcal{F}_0^*(\Omega,\ h) \to \mathbb{R}_+$ *be a Lyapunov functional as defined in Definition 14.24. Define* $E = \{(u, F) \in \mathcal{O} \times \mathcal{F}_0^*(\Omega,\ h) : \dot{V}(u, F) = 0\}$. *Let* W *be the largest set in* E *which is positively invariant. If* $\Omega^+(u,\ F) \cap (M \times \mathcal{F}_0^*(\Omega,\ h)) = \emptyset$, *then* $\Omega^+(u,\ F)$ *is contained in* W.

Proof. The proof follows some ideas of [26, Theorem 3.1]. We know that $\Omega^+(u,\ F) \neq \emptyset$ and $\Omega^+(u,\ F)$ is positively invariant, see Lemma 14.22 and Proposition 14.23. Then, consider $(u^*, F^*) \in \Omega^+(u,\ F)$.

Case 1: $\Omega^+(u,\ F)$ is a singleton.

In this case, $\Omega^+(u, F) = \{(u^*, F^*)\}$. By the positive invariance of $\Omega^+(u,\ F)$, we have

$$\tilde{\pi}(t, u^*, F^*) = (u^*, F^*)$$

for all $t \geqslant 0$. Consequently, we obtain $\dot{V}(u^*, F^*) = 0$ and $\Omega^+(u,F) \subset E$. Since W is the largest positively invariant subset in E, we conclude that $\Omega^+(u, F) \subset W$.

Case 2: $\Omega^+(u, F)$ is not a singleton.

Let $(u_1, F_1), (u_2, F_2) \in \Omega^+(u, F)$. We claim that $V(u_1, F_1) = V(u_2, F_2)$. Indeed, by definition of a positive limit set, there exist sequences $\{\lambda_n\}_{n\in\mathbb{N}}$ and $\{\kappa_n\}_{n\in\mathbb{N}}$ in $\mathbb{R}_+$ with $\lambda_n \overset{n\to\infty}{\to} \infty$ and $\kappa_n \overset{n\to\infty}{\to} \infty$ such that

$$\tilde{\pi}(\lambda_n, u, F) \overset{n\to\infty}{\to} (u_1, F_1) \quad \text{and} \quad \tilde{\pi}(\kappa_n, u, F) \overset{n\to\infty}{\to} (u_2, F_2).$$

We may choose subsequences $\{\lambda_{n_k}\}_{k\in\mathbb{N}}$ and $\{\kappa_{n_k}\}_{k\in\mathbb{N}}$ such that $\lambda_{n_k} \leqslant \kappa_{n_k}$, $k \in \mathbb{N}$. Then, condition (ii) of Definition 14.24 implies that

$$V(\tilde{\pi}(\kappa_{n_k}, u, F)) \leqslant V(\tilde{\pi}(\lambda_{n_k}, u, F)). \tag{14.22}$$

By the continuity of V, $V(u_2, F_2) \leqslant V(u_1, F_1)$ as $k \to \infty$ in (14.22). Analogously, we may choose subsequences $\{\kappa_{n_m}\}_{m\in\mathbb{N}}$ and $\{\lambda_{n_m}\}_{m\in\mathbb{N}}$ such that $\kappa_{n_m} \leqslant \lambda_{n_m}$, $m \in \mathbb{N}$, and then $V(u_1, F_1) \leqslant V(u_2, F_2)$. This yields the claim and, hence, V is constant on $\Omega^+(u, F)$. By the positive invariance of $\Omega^+(u, F)$, the derivative of V satisfies $\dot{V}(u^*, F^*) = 0$ for every $(u^*, F^*) \in \Omega^+(u, F)$. Therefore, $\Omega^+(u, F) \subset W$ which completes the proof. □

14.5 Recursive Properties

This section brings out some topological properties of an impulsive semidynamical system $(\mathcal{O} \times \mathcal{F}_0^*(\Omega, h), \tilde{\pi})$ as presented in Section 14.3.

Consider the space $\mathcal{O} \times \mathcal{F}_0^*(\Omega, h)$ with the following metric

$$\varrho((x, F_1), (y, F_2)) = \| x - y \| + \rho(F_1, F_2),$$

for all $(x, F_1), (y, F_2) \in \mathcal{O} \times \mathcal{F}_0^*(\Omega, h)$, where ρ is defined in (14.4). In addition, we shall assume the following condition (T):

(T) If $(u, F) \in \mathrm{M} \times \mathcal{F}_0^*(\Omega, h)$, $(v, H) \in \mathcal{O} \times \mathcal{F}_0^*(\Omega, h)$, and $\{t_n\}_{n\in\mathbb{N}}$ is a sequence such that $\tilde{\pi}(t_n, v, H) \overset{n\to\infty}{\to} (u, F)$, then there exists a sequence $\{\alpha_n\}_{n\in\mathbb{N}} \subset \mathbb{R}_+$, $\alpha_n \overset{n\to\infty}{\to} 0$, such that $\pi(\alpha_n, \tilde{\pi}(t_n, v, H)) \in \mathrm{M} \times \mathcal{F}_0^*(\Omega, h)$ for n sufficiently large, that is, $\tilde{\pi}(t_n + \alpha_n, v, H) \overset{n\to\infty}{\to} (\mathrm{I}(u), F)$.

Next, we present the concepts of minimality and recurrence. The reader may consult these definitions in [18] for the case of continuous dynamical systems. For general impulsive systems, the reader can consult [30] for instance.

Definition 14.26: A subset $\Sigma \subset \mathcal{O} \times \mathcal{F}_0^*(\Omega, h)$ is called *minimal*, whenever $\Sigma \setminus (\mathrm{M} \times \mathcal{F}_0^*(\Omega, h)) \neq \emptyset$, Σ is closed, $\Sigma \setminus (\mathrm{M} \times \mathcal{F}_0^*(\Omega, h))$ is positively invariant, and Σ does not contain any proper subset satisfying the previous conditions.

Definition 14.27: A point $(u, F) \in \mathcal{O} \times \mathcal{F}_0^*(\Omega, h)$ is said to be *recurrent*, if for every $\epsilon > 0$, there exists a $T = T(\epsilon) > 0$, such that for every $t, s \geqslant 0$, the interval $[0, T]$ contains a number $\tau > 0$ such that

$$\rho(\tilde{\pi}(t, u, F), \tilde{\pi}(s + \tau, u, F)) < \epsilon.$$

A subset $\Sigma \subset \mathcal{O} \times \mathcal{F}_0^*(\Omega, h)$ is said to be *recurrent*, if each point $(u, F) \in \Sigma$ is recurrent.

Next, we characterize minimal sets of $\mathcal{O} \times \mathcal{F}_0^*(\Omega, h)$. The proof of this result is based on the proof of [30, Theorem 4.4].

Theorem 14.28: *A subset $\Sigma \subset \mathcal{O} \times \mathcal{F}_0^*(\Omega, h)$ is minimal if and only if $\Sigma = \overline{\tilde{\pi}^+(u, F)}$, for all $(u, F) \in \Sigma \setminus (M \times \mathcal{F}_0^*(\Omega, h))$.*

Proof. Assume that Σ is minimal and let $(u, F) \in \Sigma \setminus (\mathrm{M} \times \mathcal{F}_0^*(\Omega, h))$. By the positive invariance of $\Sigma \setminus (\mathrm{M} \times \mathcal{F}_0^*(\Omega, h))$ and by the fact that Σ is closed, we have

$$\overline{\tilde{\pi}^+(u, F)} \subset \overline{\Sigma} = \Sigma.$$

Since $\overline{\tilde{\pi}^+(u, F)} \setminus (\mathrm{M} \times \mathcal{F}_0^*(\Omega, h)) \neq \emptyset$, $\overline{\tilde{\pi}^+(u, F)}$ is closed and $\overline{\tilde{\pi}^+(u, F)} \setminus (\mathrm{M} \times \mathcal{F}_0^*(\Omega, h))$ is positively invariant (see Lemma 14.20), the minimality of Σ yields $\Sigma = \overline{\tilde{\pi}^+(u, F)}$.

Now, assume that $\Sigma = \overline{\tilde{\pi}^+(u, F)}$, for all $(u, F) \in \Sigma \setminus (\mathrm{M} \times \mathcal{F}_0^*(\Omega, h))$. Let $\Gamma \subset \Sigma$ be such that $\Gamma \setminus (\mathrm{M} \times \mathcal{F}_0^*(\Omega, h)) \neq \emptyset$, Γ be closed and $\Gamma \setminus (\mathrm{M} \times \mathcal{F}_0^*(\Omega, h))$ be positively invariant. Take $(v, F) \in \Gamma \setminus (\mathrm{M} \times \mathcal{F}_0^*(\Omega, h))$. Then $(v, F) \in \Sigma \setminus (\mathrm{M} \times \mathcal{F}_0^*(\Omega, h))$, which yields $\Sigma = \overline{\tilde{\pi}^+(v, F)}$. Finally, since $\Gamma \setminus (\mathrm{M} \times \mathcal{F}_0^*(\Omega, h))$ is positively invariant, we obtain

$$\Gamma \subset \Sigma = \overline{\tilde{\pi}^+(v, F)} \subset \Gamma,$$

that is, $\Sigma = \Gamma$. Therefore, Σ is minimal. □

By the proof of Theorem 14.28, we obtain the following result.

Theorem 14.29: *Let $\Sigma \subset \mathcal{O} \times \mathcal{F}_0^*(\Omega, h)$ and assume that, for all $(u, F) \in \Sigma$, $\Omega^+(u, F) \setminus (M \times \mathcal{F}_0^*(\Omega, h)) \neq \emptyset$. Then, Σ is minimal if and only if $\Sigma = \Omega^+(u, F)$ for all $(u, F) \in \Sigma \setminus (M \times \mathcal{F}_0^*(\Omega, h))$.*

In the sequel, we prove that compact minimal sets are recurrent, as in Birkhoff's Theorem, see [30, Theorem 4.17]. With slight modifications, the proof follows similarly.

Theorem 14.30: *If the set $\Sigma \subset \mathcal{O} \times \mathcal{F}_0^*(\Omega, h)$ is minimal and compact, then the set $\Sigma \setminus (M \times \mathcal{F}_0^*(\Omega, h))$ is recurrent.*

Proof. Suppose the contrary, that is, there exists $(u, F) \in \Sigma \setminus (\mathrm{M} \times \mathcal{F}_0^*(\Omega, h))$ which is not recurrent. Then, there are $\epsilon > 0$ and sequences $\{\lambda_n\}_{n\in\mathbb{N}}$, $\{s_n\}_{n\in\mathbb{N}}$, $\{t_n\}_{n\in\mathbb{N}}$ $\subset \mathbb{R}_+$ such that $\lambda_n \overset{n\to\infty}{\to} \infty$ and

$$\varrho(\tilde{\pi}(t_n, u, F), \tilde{\pi}(s_n + \tau, u, F)) \geqslant \epsilon, \text{ for every } \tau \in [0, \lambda_n] \text{ and } n \in \mathbb{N}. \tag{14.23}$$

Note that $\tilde{\pi}(t_n, u, F), \tilde{\pi}\left(s_n + \frac{\lambda_n}{2}, u, F\right) \in \Sigma$ for all $n \in \mathbb{N}$, since $\Sigma \setminus (\mathrm{M} \times \mathcal{F}_0^*(\Omega, h))$ is positively invariant. By the compactness of Σ, we can assume that there are (u_1, F_1), $(u_2, F_2) \in \Sigma$ such that

$$\varrho(\tilde{\pi}(t_n, u, F), (u_1, F_1)) \overset{n\to\infty}{\to} 0 \quad \text{and} \quad \varrho\left(\tilde{\pi}\left(s_n + \frac{\lambda_n}{2}, u, F\right), (u_2, F_2)\right) \overset{n\to\infty}{\to} 0.$$

We have two cases to consider: when $u_2 \notin \mathrm{M}$ and when $u_2 \in \mathrm{M}$.

Case 1: $u_2 \notin \mathrm{M}$.

Let $t \geqslant 0$ be fixed and arbitrary and assume that $t \neq \sum_{j=0}^{k} \varphi((u_2)_j^+, F)$, for all $k \in \mathbb{N}_0$, that is, t is not a jump time. Using the continuity of π and I, we obtain $\delta > 0$ such that, if $\varrho((w, I), (u_2, F_2)) < \delta$, then

$$\varrho(\tilde{\pi}(t, w, I), \tilde{\pi}(t, u_2, F_2)) < \frac{\epsilon}{3}. \tag{14.24}$$

Now, let $n_0 \in \mathbb{N}$ be such that $\frac{\lambda_{n_0}}{2} > t$,

$$\varrho(\tilde{\pi}(t_{n_0}, u, F), (u_1, F_1)) < \frac{\epsilon}{3} \quad \text{and} \quad \varrho\left(\tilde{\pi}\left(s_{n_0} + \frac{\lambda_{n_0}}{2}, u, F\right), (u_2, F_2)\right) < \delta. \tag{14.25}$$

Using (14.23), (14.24), and (14.25), we get

$$\begin{aligned}
\varrho(\tilde{\pi}(t, u_2, F_2), (u_1, F_1)) \geqslant & \varrho\left(\tilde{\pi}(t_{n_0}, u, F), \pi\left(s_{n_0} + \frac{\lambda_{n_0}}{2} + t, u, F\right)\right) \\
& - \varrho\left(\tilde{\pi}(t, u_2, F_2), \tilde{\pi}\left(t, \pi\left(s_{n_0} + \frac{\lambda_{n_0}}{2}, u, F\right)\right)\right) \\
& - \varrho(\tilde{\pi}(t_{n_0}, u, F), (u_1, F_1)) \\
> & \epsilon - \frac{\epsilon}{3} - \frac{\epsilon}{3} = \frac{\epsilon}{3}.
\end{aligned}$$

By the arbitrariness of t, we conclude that

$$\varrho(\tilde{\pi}(t, u_2, F_2), (u_1, F_1)) > \frac{\epsilon}{3},$$

for all $t \geqslant 0$, with $t \neq \sum_{j=0}^{k} \varphi((u_2)_j^+, F)$, $k \in \mathbb{N}_0$.

On the other hand, if there exists $k \in \mathbb{N}$ such that $t = \sum_{j=0}^{k} \varphi((u_2)_j^+, F)$, then we may choose a sequence $\{\beta_n\}_{n\geqslant 1} \subset \mathbb{R}_+$ such that

$$\beta_n \overset{n\to\infty}{\to} \sum_{j=0}^{k} \phi((u_2)_j^+, F) \quad \text{and} \quad \sum_{j=0}^{k} \varphi((u_2)_j^+, F) < \beta_n < \sum_{j=0}^{k+1} \varphi((u_2)_j^+, F).$$

By the proof of the previous case,

$$\varrho(\tilde{\pi}(\beta_n, u_2, F_2), (u_1, F_1)) > \frac{\epsilon}{3}, \quad \text{for } n \in \mathbb{N}.$$

Since $n \to \infty$, we obtain

$$\varrho\left(\tilde{\pi}\left(\sum_{j=0}^{k} \phi((u_2)_j^+, F), u_2, F_2\right), (u_1, F_1)\right) \geqslant \frac{\epsilon}{3},$$

because $\tilde{\pi}$ is right-continuous. Therefore,

$$\varrho(\tilde{\pi}(t, u_2, F_2), (u_1, F_1)) \geqslant \frac{\epsilon}{3}, \quad \text{for all } t \geqslant 0.$$

Thus $(u_1, F_1) \notin \overline{\tilde{\pi}^+(u_2, F_2)}$. Now, since Σ is minimal, we have $\Sigma = \overline{\tilde{\pi}^+(u_2, F_2)}$, that is, $(u_1, F_1) \notin \Sigma$, which is a contradiction.

Case 2: $u_2 \in \mathrm{M}$.

By condition (T), there exists a sequence $\{\alpha_n\}_{n\in\mathbb{N}} \subset \mathbb{R}_+$, $\alpha_n \overset{n\to\infty}{\to} 0$, such that

$$\varrho\left(\tilde{\pi}\left(\alpha_n + s_n + \frac{\lambda_n}{2}, u, F\right), (\mathrm{I}(u_2), F_2)\right) \overset{n\to\infty}{\to} 0.$$

Since $\Sigma \setminus (\mathrm{M} \times \mathcal{F}_0^*(\Omega, h))$ is positively invariant and $(u, F) \in \Sigma \setminus (\mathrm{M} \times \mathcal{F}_0^*(\Omega, h))$, we get $(\mathrm{I}(u_2), F_2) \in \overline{\Sigma} = \Sigma$. Now, we consider the motion $\tilde{\pi}(t, \mathrm{I}(u_2), F_2)$ for every $t \geqslant 0$. Since $\mathrm{I}(\mathrm{M}) \cap \mathrm{M} = \emptyset$ we have $\mathrm{I}(u_2) \notin \mathrm{M}$. By following the ideas of *Case 1*, we conclude that

$$\varrho(\tilde{\pi}(t, I(u_2), F_2), (u_1, F_1)) \geqslant \frac{\epsilon}{3}, \quad \text{for all } t \geqslant 0,$$

which yields $(u_1, F_1) \notin \overline{\tilde{\pi}^+(\mathrm{I}(u_2), F_2)}$, which, in turn, is a contradiction, once Σ is minimal.

Cases 1 and *2* imply $\Sigma \setminus (\mathrm{M} \times \mathcal{F}_0^*(\Omega, h))$ is recurrent. □

Corollary 14.31: *Let $(u, F) \in \mathcal{O} \times \mathcal{F}_0^*(\Omega, h)$ be given. If $\Omega^+(u, F)$ is compact and minimal, then $\Omega^+(u, F) \setminus (M \times \mathcal{F}_0^*(\Omega, h))$ is recurrent.*

15

Applications to Functional Differential Equations of Neutral Type

Fernando G. Andrade[1], *Miguel V. S. Frasson*[2], *and Patricia H. Tacuri*[3]

[1] *Colégio Técnico de Bom Jesus, Universidade Federal do Piauí, Bom Jesus, PI, Brazil*
[2] *Departamento de Matemática Aplicada e Estatística, Instituto de Ciências Matemáticas e de Computação (ICMC), Universidade de São Paulo, São Carlos, SP, Brazil*
[3] *Departamento de Matemática, Centro de Ciências Exatas, Universidade Estadual de Maringá, Maringá, PR, Brazil*

15.1 Drops of History

Present in Mathematics for a long time, *differential equations* appear concomitantly with the emergence of Calculus. On 29 October 1675, Gottfried W. Leibniz introduced both the notation $\int$, which resembled the letter "S" in italics in old texts, and also the notation $\int l$ which represented the sum of functions l's. Such Leibniz's manuscript can be found in [102]. In particular, on page 125, one finds the sentence

> *Utile erit scribi* $\int$ *pro omn., ut* $\int l$ *pro omn. l, id est summa ipsorum l.*

which, translated to English, means

> *It will be helpful to write* $\int$ *instead of omn., and* $\int l$ *instead of omn. l, that is the sum of l.*

This implies that, before the introduction of the notations $\int$ and $\int l$, one would read *omn.* and *omn.l*, respectively.

Much later, in 1744, the Bernoulli brothers, Johann and Jakob, suggested the terminology *"integral of l"* for the notation $\int l$ for the first time. Reference [16], called *"Opera,"* contains several articles due to Jakob Bernoulli, one of which contains the sentence below with the term *"integral."* Notice that [16] was published

Generalized Ordinary Differential Equations in Abstract Spaces and Applications, First Edition.
Edited by Everaldo M. Bonotto, Márcia Federson, and Jaqueline G. Mesquita.

later than the year of publication of the same article in *Acta Eruditorum*. Thus, on [16, p. 423], one reads

> *Ergo & horum Integralia æquantur…*

which is equivalent to

> *And then these integrals are equal to…*

Back to Leibniz's manuscript of 29 October 29 1675, it is possible to encounter the words

> *…si sit $\int l \sqcap ya$. Ponemus $l \sqcap \frac{ya}{d}$, nempe ut $\int$ augebit, ita d minuet dimensiones. $\int$ autem significat summan, d diffenrentiam.*

or, equivalently,

> *…if $\int l \sqcap ya$. I will put $l \sqcap \frac{ya}{d}$, namely that $\int$ increases by less than d dimensions. $\int$ means summation, d differentiation.*

where the symbol $\sqcap$ stands for the symbol = for equality, signifying that while the integral, $\int$, increases the dimensions, the derivative, d, decreases them.

Continuing in 1675, but now in a different manuscript, dated 11 November, Leibniz wrote that dx and $\frac{x}{d}$ denoted the same thing and he not only introduced the notations dx and dy to represent the differentials of x and y but also the symbol $\frac{dx}{dy}$ to represent the derivative of x with respect to y, after concluding that neither $d\overline{xy}$ and $dx\,dy$ are the same, nor $d\frac{x}{y}$ equals $\frac{dx}{dy}$. In this manuscript, which can be found in [101, Beilage II (Appendix II)], pages 32–40, one reads

> *Videndum an dx dy idem sit quod $d\,\overline{xy}$, et an $\frac{dx}{dy}$ idem quod $d\frac{x}{y}$…*

which means

> *Let us see if dx dy is the same as $d\,\overline{xy}$, and whether $\frac{dx}{dy}$ is the same as $d\frac{x}{y}$…*

In the following year, Leibniz used, for the first time, the term "*æquatio differentialis*" to denote a relationship between the differentials dx and dy of the variables x and y respectively. But differentials only appeared in a manuscript published a few years later, in 1684. As a matter of fact, the year 1684 is considered to be the formal date for the arising of Calculus, and this manuscript of Leibniz, entitled

Nova Methodus pro Maximis et Minimis, can be found in [164]. In particular, we refer to pages 467 or 469, where one can find the term *"œquatio differentialis"*.

According to the historian F. Cajori in [34, p. 213], Sir Isaac Newton wrote a coded message to Leibniz containing the information

> *Data œquetione quotcunque fluentes quantitates s involvente fluxiones invenire, et viceversa.*

or, equivalently,

> *The given equation to quantify the number of fluent involving the flows to find, and vice-versa.*

which essentially contains the idea of a derivative. In 1671, Newton wrote his work entitled *Methodus Fluxionum et Serierum infinitorum* whose first version was only published much later, in 1736, when John Colson translated it from Latin into English as *The Method of Flows and Infinite Series with it Applications to the Geometry of* CURVE-LINES (see [187]). In this work, Newton tried to solve differential equations, which he referred to as *"Fluxional Equations."* In such work, Newton considered that if x is a quantity, called by him *fluents*, that is indefinitely increasing, then the derivative $\dot{x}$ of x would be the speed at which the quantity x is increasing (called by him *fluxions*). See [34, p. 194].

The progress of the theory of differential equations in the final years of the seventeenth century was mainly due to Leibniz, Newton and the Bernoulli brothers. The development of such a theory continued throughout the eighteenth century and, among other authors, we can highlight Daniel Bernoulli, son of Johann Bernoulli, and his contemporary, Leonhard Euler. As a matter of fact, in 1755, Euler published a rigorous reformulation of Calculus in *Institutiones calculi differentialis* (see [64]) and *Institutionum calculi integralis* (see [65–67]).

At the same time, a branch of a special type of equations was being developed, that is, the branch of functional equations. Recall that a functional equation is any equation whose argument of the unknown function is also a function. Some functional equations are well known. For instance, in 1821 (see [36, p. 104–113]), Augustin Louis Cauchy solved the following four different functional equations

$$\begin{aligned}
&f(t+s) = f(t) + f(s),\\
&f(ts) = f(t) + f(s),\\
&f(t+s) = f(t) \times f(s),\\
&f(ts) = f(t) \times f(s).
\end{aligned}$$

The first equation, known as Cauchy additive functional, was already known by Adrien-Marie Legendre in 1794 in his study of areas (see [159, Note IV, p. 293]).

> *Considérons un rectangle cont les dimensions sont p et q, et sa surface qui est une fonction de p et q, représentons la par $\varphi(p, q)$. Si on considere un autre rectangle dont les dimensions sont $p + p'$ et q, il est clair que ce rectangle est composé de deux autres, l'un qui a pour dimension p et q, l'autre qui a pour dimensions p' et q, de sorte qu'on aura*
>
> $$\varphi : (p + p', q) = \varphi : (p, q) + \varphi : (p', q)$$

which, translated to English, means

> *Consider a rectangle with dimensions p and q, and its area, which is a function of p and q, represents it by $\varphi(p, q)$. If we consider another rectangle whose dimensions are $p + p'$ and q, it is clear that this rectangle is composed of two others, one with dimensions p and q, the other with dimensions p' and q, so that we have*
>
> $$\varphi(p + p', q) = \varphi(p, q) + \varphi(p', q).$$

Cauchy functionals were also investigated by Johan Ludwig William Valdemar Jensen in 1878 (see [137]) and in 1905 (see [138]). In fact, Jensen solved the Cauchy functional equations in 1878, and it was only in 1905 Jensen wrote about the functional which carries his name and is a modification of Cauchy additive functional. Later, it became known as Jensen functional. Thus, in [138] one reads

$$f\left(\frac{t+s}{2}\right) = \frac{f(t) + f(s)}{2}.$$

Note that, when $s = 0$ in Jensen functional, one obtains the Cauchy additive functional.

As a matter of fact, functional equations had been investigated since 1747, by Jean le Rond d'Alembert, for example (see [50–52]). In 1769, d'Alembert solved the problem of the parallelogram of forces, by reducing the proof for solving the following functional equation:

$$f(t + s) - f(t - s) = 2f(t)f(s).$$

Indeed, on [53, p. 279], one reads

> *Donc, substituant cette dernière de valeur de $z\varphi u$ dans l'équation précédente, on aura après les réductions $\varphi\alpha + \varphi(\alpha + 2m') = \varphi m' \times \varphi(\alpha + m')$, ou en faisant $\alpha + m' = \bigstar$, $\varphi(\bigstar - m') + \varphi(\bigstar + m') = \varphi m' \times \varphi\bigstar$*

which means

> *Thus, substituting the latter with the value of $z\varphi u$ in the previous equation, we will have after the reductions $\varphi\alpha + \varphi(\alpha + 2m') = \varphi m' \times \varphi(\alpha + m')$, or in making $\alpha + m' = \bigstar$, $\varphi(\bigstar - m') + \varphi(\bigstar + m') = \varphi m' \times \varphi\bigstar$.*

Nowadays, in the particular case, where the difference in the arguments of a functional equation is constant, as in

$$f(t) = f(t-1) + f(t+2),$$

for example, the functional equation is said to be a *difference equation* (see [61]).

When the ideas of differential equations are combined with those of functional equations, functional differential equations emerge. The latter appeared for the first time in a work of *Marquis de Condorcet* (a codename for Marie Jean Antoine Nicolas de Caritat Condorcet) entitled *Sur la détermination des fonctions arbitraires qui entrent dans les intégrales des Équations aux differences partielles*, where one reads, on [46, p. 52],

> *Lorsqu'on a F′, on en déduit F par une équation ou finie, ou aux différences infiniment petites.*

This sentence means

> *When we have F′, we deduce F by an equation either finite, or with infinitely small differences.*

According to Anatoliı̆ D. Myshkis (see [186]), the motivation for Condorcet to develop this type of equation came from geometric problems proposed by Euler in [63]. More specifically, Euler was looking for curves whose evolutes are similar to the curves themselves.

Differential equations have always been used in applications in sciences and engineering. We can mention, for example, one of the first models for application in population dynamics which is due to Thomas Robert Malthus [171]. The Malthusian model (see [185, p. 2]) is described as

$$\frac{d}{dt}y(t) = by(t),$$

where $b > 0$ can be understood as the reproduction rate. The model says that the growth of a population is proportional to the number of individuals in the current population. After reading Malthus' work, Pierre François Verhulst (see [231, 232]) proposed the following model in which the environment has a capacity K,

$$\frac{d}{dt}y(t) = by(t)\left(1 - \frac{y(t)}{K}\right). \tag{15.1}$$

The term $b(1 - y(t)/K)$ in Verhulst model is the per capita birth rate. Such model is also known as the logistic equation, and it suggests that the population starts to decrease immediately after reaching the maximum capacity of the environment K. However, it is known that this equation does not describe the real behavior of a single species population. For example, in [198], David M. Pratt stated that the fluctuation in the population dynamics of small planktonic crustacean called *Daphnia magna* is due to a delay in the impact of the effects of density. In this case, George Evelyn Hutchinson (see [133]) suggested that the model for describing population dynamics shall be expressed by

$$\frac{d}{dt}y(t) = by(t)\left(1 - \frac{y(t-r)}{K}\right).$$

Unlike the Verhulst model, in Hutchinson's equation, the reproductive process continues after the population reaches the capacity K of the environment, being interrupted after a time $r > 0$. Such number r represents a delay or a retardation in time.

While in Hutchinson's model, known as "*delayed logistic equation*," the delay is considered within the death rate, some approaches based on the experiments of the entomologist Alexander John Nicholson (see [189]) suggest that the delay should be set at the birth rate, since the reproduction rate may not only depend on the current population density but also on the population density in the past. Such a model can be described by the equation

$$\frac{d}{dt}y(t) = b(y(t-r))y(t-r) - \lambda(y(t))y(t).$$

where $b(y(t))$ is the birth rate per head, $\lambda(y(t))$ is the death rate per head, and r denotes the time needed for an egg to become an adult fly. This equation is known as "*Nicholson's blowflies equation*." See, for instance, [111, 195] and [112].

Frederick Edward Smith, on the other hand, in an investigation using *Daphnia magna* (see [222]) mentioned that the per capita rate in (15.1) should be replaced by $b(1 - (y(t) + c\, dy(t)/dt)/K)$, since a growing population consumes much more food than a population that has already reached maturity, with c being a positive or negative constant. In this case, the population dynamics model would take the form

$$\frac{d}{dt}y(t) = by(t)\left(1 - \frac{y(t) - c\, dy(t)/dt}{K}\right).$$

For Yang Kuang (see [146]), it was reasonable to think that $y(t)$ represents in a population of individuals that feed on a pasture which, in turn, needs time to regenerate. A better approach is therefore, a neutral logistic equation of the form

$$\frac{d}{dt}y(t) = by(t)\left(1 - \frac{y(t-r) - c\, dy(t-r)/dt}{K}\right).$$

This equation was investigated for the first time by Kondalsamy Gopalsamy and Bing Gen Zhang in 1988 (see [106]).

The theory behind functional differential equations of neutral type (we write FDE of neutral type, for short) is relatively new and is generally related to the space of continuous functions as the phase space. Here, we propose an expansion of this branch of equations by studying measure FDEs of neutral type within the framework of generalized ODEs. In order to do this, we present a relation, borrowed from [76], between the latter and measure FDEs of neutral type. We use the existing theory of generalized ODEs to obtain our results, namely, existence and uniqueness of solutions and continuous dependence on parameters. We end this chapter with an example, also borrowed from [76], which illustrates the theory.

15.2 FDEs of Neutral Type with Finite Delay

Let $t_0 \in \mathbb{R}$, $r > 0$ and $\Omega \subset C([-r,0],\mathbb{R}^n) \times \mathbb{R}$ be an open subset. The theory of *FDEs of neutral type* is usually concerned with equations of the form

$$\frac{d}{dt}M(y_t,t) = f(y_t,t), \quad t \geqslant t_0, \tag{15.2}$$

where $f\colon \Omega \to \mathbb{R}^n$ is continuous and y_t denotes the memory function $y_t(\theta) = y(t+\theta)$, $\theta \in [-r,0]$, for every $t \geqslant t_0$. In the sequel, we recall the concept of normalized function.

Definition 15.1: We say that a matrix-valued function $\zeta\colon [-r,0] \to \mathbb{R}^{n\times n}$ of bounded variation is *normalized*, if ζ is left-continuous on $(-r,0)$ and $\zeta(0) = 0$.

We assume that the operator M in (15.2) is linear continuous with respect to the first variable (in this case, we use the notation $M(\varphi,t) = M(t)\varphi$) and is given by

$$M(t)\varphi = \varphi(0) - \int_{-r}^{0} d_\theta[\xi(t,\theta)]\varphi(\theta), \tag{15.3}$$

for $\varphi \in C([-r,0],\mathbb{R}^n)$ and $t \geqslant t_0$, where the function $\xi\colon \mathbb{R}\times\mathbb{R} \to \mathbb{R}^{n\times n}$ is measurable and, for each $t \geqslant t_0$, $\xi(t,\cdot)$ is a $n \times n$ matrix-valued normalized in $[-r,0]$ with

$$\xi(t,\theta) = \xi(t,-r),\ \theta \leqslant -r, \quad \text{and} \quad \xi(t,\theta) = 0,\ \theta \geqslant 0,$$

such that

(N1) the variation of $\xi(t,\cdot)$ on $[s,0]$, $\operatorname{var}_s^0(\xi(t,\cdot))$, tends to zero as $s \to 0$.

Next, we define a solution of Eq. (15.2).

Definition 15.2: We say that a function $y\colon [t_0, t_0+\sigma] \to \mathbb{R}^n$, with $\sigma > 0$, is a *solution of the FDE of neutral type* (15.2), if $(y_t,t) \in \Omega$ for all $t \in [t_0,t_0+\sigma]$, and $dM(y_t,t)/dt = f(y_t,t)$ holds for almost every $t \in [t_0,t_0+\sigma]$.

The integral form of the FDE of neutral type (15.2) is, then, given by

$$y(t) = y(t_0) + \int_{t_0}^{t} f(y_s, s)ds + \int_{-r}^{0} d_\theta[\xi(t,\theta)]y(t+\theta) - \int_{-r}^{0} d_\theta[\xi(t_0,\theta)]y(t_0+\theta),$$

for all $t \in [t_0, t_0 + \sigma]$, where the integrals can be taken in the sense of Riemann–, Lebesgue– or Perron–Stieltjes. A systematic study of this class of equations can be found in [115].

We are interested in broadening the class of neutral equations, for instance weakening continuity hypotheses, and getting qualitative information on solutions that may be not continuous as well.

Given $t_0 \in \mathbb{R}$ and $r, \sigma > 0$, let $O \subset G([t_0 - r, t_0 + \sigma], \mathbb{R}^n)$ be an open subset and consider the set

$$P = \{y_t : y \in O \text{and } t_0 \leqslant t \leqslant t_0 + \sigma\} \subset G([-r, 0], \mathbb{R}^n). \tag{15.4}$$

Assume that $f: P \times [t_0, t_0 + \sigma] \to \mathbb{R}^n$ is a function such that $t \mapsto f(y_t, t)$ is integrable over $[t_0, t_0 + \sigma]$ in the sense of Perron–Stieltjes, for each $y \in O$, with respect to a nondecreasing function $g: [t_0, t_0 + \sigma] \to \mathbb{R}$. Here, we deal with a *measure FDE of neutral type* with integral form

$$y(t) = y(t_0) + \int_{t_0}^{t} f(y_s, s)dg(s) + \int_{-r}^{0} d_\theta[\xi(t,\theta)]y(t+\theta)$$
$$- \int_{-r}^{0} d_\theta[\xi(t_0,\theta)]y(t_0+\theta), \tag{15.5}$$

where $t \in [t_0, t_0 + \sigma]$.

Definition 15.3: We say that a function $y: [t_0, t_0 + \sigma] \to \mathbb{R}^n$ is a *solution of the measure FDE of neutral type* (15.5), whenever $(y_t, t) \in P \times [t_0, t_0 + \sigma]$, the Perron–Stieltjes $\int_{t_0}^{t} f(y_s, s)dg(s)$ exists and the equality (15.5) is satisfied for all $t \in [t_0, t_0 + \sigma]$.

Given an arbitrary element $\tilde{x} \in G([t_0 - r, t_0 + \sigma], \mathbb{R}^n)$ and $c \geqslant 1$, consider the sets

$$O = B_c = \{z \in G([t_0 - r, t_0 + \sigma], \mathbb{R}^n) : \|z - \tilde{x}\|_\infty < c\} \quad \text{and}$$
$$P = P_c = \{y_t : \ y \in B_c \text{and } t_0 \leqslant t \leqslant t_0 + \sigma\}.$$

Notice that B_c has the prolongation property (see Definition 4.15).

In this section, we want to investigate some qualitative properties of a measure FDE of neutral type using a biunivocal relation between the solution of Eq. (15.5), with initial condition $y_{t_0} = \varphi \in P_c$, and the solution of the generalized ODE

$$\frac{dx}{d\tau} = DF(x, t), \tag{15.6}$$

with initial condition $x(t_0) = x_0 \in B_c$, where $F: B_c \times [t_0, t_0 + \sigma] \to G([t_0 - r, t_0 + \sigma], \mathbb{R}^n)$ is defined by

$$F(x,t)(\vartheta) = \begin{cases} 0, & \vartheta \in [t_0 - r, t_0], \\ \displaystyle\int_{t_0}^{\vartheta} f(x_s, s)dg(s) + \int_{-r}^{0} d_\theta[\xi(\vartheta,\theta)]x(\vartheta+\theta) & \\ \displaystyle\qquad - \int_{-r}^{0} d_\theta[\xi(t_0,\theta)]x(t_0+\theta), & \vartheta \in [t_0, t], \\ \displaystyle\int_{t_0}^{t} f(x_s, s)dg(s) + \int_{-r}^{0} d_\theta[\xi(t,\theta)]x(t+\theta) & \\ \displaystyle\qquad - \int_{-r}^{0} d_\theta[\xi(t_0,\theta)]x(t_0+\theta), & \vartheta \in [t, t_0+\sigma]. \end{cases} \tag{15.7}$$

In order to fulfill this purpose, given $t_0 < a < b < t_0 + \sigma$ and $y, z \in B_c$, we shall assume that the function $f: P_c \times [t_0, t_0 + \sigma] \to \mathbb{R}^n$ and the nondecreasing function $g: [t_0, t_0 + \sigma] \to \mathbb{R}$ satisfy conditions (A), (B), and (C) (see p. 162). Consider, in addition, the following condition on the normalized function $\xi: \mathbb{R} \times \mathbb{R} \to \mathbb{R}^{n\times n}$:

(N2) there exists a Perron integrable function $C: [t_0, t_0 + \sigma] \to \mathbb{R}$ such that, for all $s_1, s_2 \in [t_0, t_0 + \sigma]$, with $s_1 \leqslant s_2$, and $z \in O$, we have

$$\left\| \int_{-r}^{0} d_\theta \xi(s_2,\theta) z(s_2+\theta) - \int_{-r}^{0} d_\theta \xi(s_1,\theta) z(s_1+\theta) \right\| \leqslant \int_{s_1}^{s_2} C(s) \left(\int_{-r}^{0} d_\theta \xi(s,\theta) \, \|z(s+\theta)\| \right) ds.$$

Next, we present a lemma which gives sufficient conditions for the function F to belong to the class $\mathcal{F}(B_c \times [t_0, t_0 + \sigma], h)$ (see Definition 4.3), where h is a nondecreasing function. Then, we present a pair of theorems which form a bridge between generalized ODEs and measure FDEs of neutral type with finite delay. This bridge relates the solution $y: [t_0 - r, t_0 + \sigma] \to \mathbb{R}^n$ of the measure FDE of neutral type (15.5) with initial condition $y_{t_0} = \varphi \in P_c$ and the solution $x: [t_0, t_0 + \sigma] \to B_c$ of the generalized ODE (15.6) with initial condition $x(t_0) = x_0$. Such relation is described by Theorems 15.6 and 15.7 and based on

$$x(t)(\vartheta) = \begin{cases} y(\vartheta), & \vartheta \in [t_0 - r, t], \\ y(t), & \vartheta \in [t, t_0 + \sigma]. \end{cases} \tag{15.8}$$

Both theorems are borrowed from [76], as well as Lemma 15.4.

Lemma 15.4: *Let $c \geqslant 1$. Assume that $g: [t_0, t_0 + \sigma] \to \mathbb{R}$ is a nondecreasing function and $f: P_c \times [t_0, t_0 + \sigma] \to \mathbb{R}^n$ satisfies conditions (A)–(C). Moreover, suppose the* *normalized function* $\xi\colon \mathbb{R}\times\mathbb{R}\to\mathbb{R}^{n\times n}$ *satisfies conditions* (N1) and (N2). *Then, the function* $F\colon B_c\times[t_0,t_0+\sigma]\to G([t_0-r,t_0+\sigma],\mathbb{R}^n)$ *given by* (15.7) *belongs to the class* $\mathcal{F}(B_c\times[t_0,t_0+\sigma],h)$*, where* $h\colon[t_0,t_0+\sigma]\to\mathbb{R}$ *is given by*

$$h(t)=\int_{t_0}^{t}[L(s)+M(s)]dg(s)+\int_{t_0}^{t}C(s)\,var_{-r}^{0}(\xi(s,\cdot))ds\left(\|\tilde{x}\|_\infty+c\right). \tag{15.9}$$

Proof. At first, note that condition (A) implies that the integrals that appear in (15.7) exist. Given $x\in B_c$ and $t_0\leqslant t_1<t_2\leqslant t_0+\sigma$, we have

$$[F(x,t_2)-F(x,t_1)](\vartheta)= \tag{15.10}$$

$$\begin{cases}0, & \vartheta\in[t_0-r,t_1],\\ \displaystyle\int_{t_1}^{\vartheta}f(x_s,s)dg(s)+\int_{-r}^{0}d_\theta[\xi(\vartheta,\theta)]x(\vartheta+\theta) & \\ \qquad\displaystyle-\int_{-r}^{0}d_\theta[\xi(t_1,\theta)]x(t_1+\theta), & \vartheta\in[t_1,t_2],\\ \displaystyle\int_{t_1}^{t_2}f(x_s,s)dg(s)+\int_{-r}^{0}d_\theta[\xi(t_2,\theta)]x(t_2+\theta) & \\ \qquad\displaystyle-\int_{-r}^{0}d_\theta[\xi(t_1,\theta)]x(t_1+\theta), & \vartheta\in[t_2,t_0+\sigma],\end{cases} \tag{15.11}$$

and, using conditions (B) and (N2), we obtain

$$\begin{aligned}\|F(x,t_2)-F(x,t_1)\|_\infty &= \sup_{\vartheta\in[t_1,t_2]}\|[F(x,t_2)-F(x,t_1)](\vartheta)\|\\ &\sup_{\vartheta\in[t_1,t_2]}\left\|\int_{t_1}^{\vartheta}f(x_s,s)dg(s)+\int_{-r}^{0}d_\theta[\xi(\vartheta,\theta)]x(\vartheta+\theta)\right.\\ &\quad\left.-\int_{-r}^{0}d_\theta[\xi(t_1,\theta)]x(t_1+\theta)\right\|\\ &\leqslant \sup_{\vartheta\in[t_1,t_2]}\left(\int_{t_1}^{\vartheta}M(s)dg(s)+\int_{t_1}^{\vartheta}C(s)\left(\int_{-r}^{0}d_\theta[\xi(s,\theta)]\,\|x(s+\theta)\|\right)ds\right)\\ &\leqslant \int_{t_1}^{t_2}M(s)dg(s)+\int_{t_1}^{t_2}C(s)\left(\int_{-r}^{0}d_\theta[\xi(s,\theta)]\,\|x(s+\theta)\|\right)ds\\ &\leqslant \int_{t_1}^{t_2}M(s)dg(s)+\int_{t_1}^{t_2}C(s)\left(\int_{-r}^{0}d_\theta[\xi(s,\theta)]\right)ds\|x\|_\infty\\ &\leqslant \int_{t_1}^{t_2}M(s)dg(s)+\int_{t_1}^{t_2}C(s)\left(\int_{-r}^{0}d_\theta[\xi(s,\theta)]\right)ds\left(\|\tilde{x}\|_\infty+c\right)\\ &\leqslant h(t_2)-h(t_1).\end{aligned}$$

Now, for $x, y \in B_c$ and $t_0 \leqslant t_1 \leqslant t_2 \leqslant t_0 + \sigma$, we have

$$[F(x,t_2) - F(x,t_1) - F(y,t_2) + F(y,t_1)](\vartheta) =$$
$$= \begin{cases} 0, & \vartheta \in [t_0 - r, t_1], \\ \int_{t_1}^{\vartheta} [f(x_s,s) - f(y_s,s)]dg(s) & \\ \quad + \int_{-r}^{0} d_\theta[\xi(\vartheta,\theta)][x(\vartheta+\theta) - y(\vartheta+\theta)] & \\ \quad - \int_{-r}^{0} d_\theta[\xi(t_1,\theta)][x(t_1+\theta) - y(t_1+\theta)], & \vartheta \in [t_1, t_2], \\ \int_{t_1}^{t_2} [f(x_s,s) - f(y_s,s)]dg(s) & \\ \quad + \int_{-r}^{0} d_\theta[\xi(t_2,\theta)][x(t_2+\theta) - y(t_2+\theta)] & \\ \quad - \int_{-r}^{0} d_\theta[\xi(t_1,\theta)][x(t_1+\theta) - y(t_1+\theta)], & \vartheta \in [t_2, t_0+\sigma], \end{cases}$$

and, according to hypotheses (C) and (N2), we have

$$\begin{aligned} &\|F(x,t_2) - F(x,t_1) - F(y,t_2) + F(y,t_1)\|_\infty \\ &\quad = \sup_{\vartheta\in[t_1,t_2]} \|[F(x,t_2) - F(x,t_1) - F(y,t_2) + F(y,t_1)](\vartheta)\| \\ &\quad \leqslant \sup_{\vartheta\in[t_1,t_2]} \left\| \int_{t_1}^{\vartheta} [f(x_s,s) - f(y_s,s)]dg(s) \right. \\ &\quad + \int_{-r}^{0} d_\theta[\xi(\vartheta,\theta)][x(\vartheta+\theta) - y(\vartheta+\theta)] \\ &\quad \left. - \int_{-r}^{0} d_\theta[\xi(t_1,\theta)][x(t_1+\theta) - y(t_1+\theta)] \right\| \\ &\quad \leqslant \sup_{\vartheta\in[t_1,t_2]} \Big(\int_{t_1}^{\vartheta} L(s)\|x_s - y_s\|_\infty \, dg(s) \\ &\quad + \int_{t_1}^{\vartheta} C(s) \left(\int_{-r}^{0} d_\theta[\xi(s,\theta)] \, \|x(s+\theta) - y(s+\theta)\| \right) ds \Big) \\ &\quad \leqslant \left(\int_{t_1}^{t_2} L(s)dg(s) + \int_{t_1}^{t_2} C(s) \left(\int_{-r}^{0} d_\theta[\xi(s,\theta)] \right) ds \right) \|x - y\|_\infty \\ &\quad \leqslant (h(t_2) - h(t_1))\|x - y\|_\infty . \end{aligned}$$

For every $\vartheta \in [t_0, t_0 + \sigma]$, the existence of the integral $\int_{t_1}^{\vartheta} L(s)\|y_s - z_s\|_\infty \, dg(s)$ is guaranteed by the Lemma 3.3, since $s \mapsto \|z_s - y_s\|_\infty$ is a regulated function.

The above calculations show that $F \in \mathcal{F}(\Omega, h)$ (see Definition 4.3), for $\Omega = B_c \times [t_0, t_0 + \sigma]$ and the function h given by (15.9). □

The next result can be found in [76]. As it is a modified version of the Lemma 4.17, the proof can be adapted with no difficulty. Thus, we omit it here.

Lemma 15.5: *Let $c \geqslant 1$ and assume that $\varphi \in P_c$, $g\colon [t_0, t_0 + \sigma] \to \mathbb{R}$ is a nondecreasing function and $f\colon P_c \times [t_0, t_0 + \sigma] \to \mathbb{R}^n$ is such that the integral $\int_{t_0}^{t_0+\sigma} f(y_t, t)dg(t)$ exists for every $y \in P_c$. Moreover, suppose $\xi\colon \mathbb{R} \times \mathbb{R} \to \mathbb{R}^{n\times n}$ is a normalized function which satisfies conditions (N1) and (N2). Consider $F\colon B_c \times [t_0, t_0 + \sigma] \to G([t_0 - r, t_0 + \sigma], \mathbb{R}^n)$ given by (15.7) and assume that $x\colon [t_0, t_0 + \sigma] \to B_c$ is a solution of the generalized ODE (15.6) with initial condition $x(t_0)(\vartheta) = \varphi(\vartheta)$ for $\vartheta \in [t_0 - r, t_0]$, and $x(t_0)(\vartheta) = x(t_0)(t_0)$ for $\vartheta \in [t_0, t_0 + \sigma]$. If $v \in [t_0, t_0 + \sigma]$ and $\vartheta \in [t_0, t_0 + \sigma]$, then*

$$x(v)(\vartheta) = x(v)(v), \quad \vartheta \geqslant v,$$
$$x(v)(\vartheta) = x(\vartheta)(\vartheta), \quad v \geqslant \vartheta.$$

Theorem 15.6: *Suppose $\varphi \in P_c$, $c \geqslant 1$, $g\colon [t_0, t_0 + \sigma] \to \mathbb{R}$ is a nondecreasing function, $f\colon P_c \times [t_0, t_0 + \sigma] \to \mathbb{R}^n$ satisfies conditions (A)–(C). Moreover, suppose the normalized function $\xi\colon \mathbb{R} \times \mathbb{R} \to \mathbb{R}^{n\times n}$ satisfies conditions (N1) and (N2). Let $F\colon B_c \times [t_0, t_0 + \sigma] \to G([t_0 - r, t_0 + \sigma], \mathbb{R}^n)$ be given by (15.7) and $y\colon [t_0 - r, t_0 + \sigma] \to \mathbb{R}^n$ be a solution of the measure FDE of neutral type*

$$\begin{aligned} y(t) = y(t_0) &+ \int_{t_0}^{t} f(y_s, s)dg(s) + \int_{-r}^{0} d_\theta[\xi(t, \theta)]y(t + \theta) \\ &- \int_{-r}^{0} d_\theta[\xi(t_0, \theta)]y(t_0 + \theta), \end{aligned} \tag{15.12}$$

where $t \in [t_0, t_0 + \sigma]$, subject to the initial condition $y_{t_0} = \varphi$. For every $t \in [t_0, t_0 + \sigma]$, let $x(t)\colon [t_0 - r, t_0 + \sigma] \to \mathbb{R}^n$ be given by (15.8). Then, the function $x\colon [t_0, t_0 + \sigma] \to B_c$ is a solution of the generalized ODE (15.6).

Proof. At first, we prove that, for every $v \in [t_0, t_0 + \sigma]$, the Kurzweil integral $\int_{t_0}^{v} DF(x(\tau), t)$ exists and

$$x(v) - x(t_0) = \int_{t_0}^{v} DF(x(\tau), t).$$

Fix an arbitrary $\epsilon > 0$. By hypothesis, the function g is nondecreasing. Then, it admits only a finite number of points $t \in [t_0, v]$ such that $\Delta^+ g(t) \geqslant \epsilon$. We denote these points by $t_1, \dots, t_m$.

Let $\delta\colon [t_0, t_0 + \sigma] \to (0, \infty)$ be a gauge such that

$$\delta(\tau) < \min\left\{\frac{t_k - t_{k-1}}{2} : k = 2, \dots, m\right\}, \tau \in [t_0, t_0 + \sigma], \text{ and}$$
$$\delta(\tau) < \min\left\{|\tau - t_k|, |\tau - t_{k-1}| : \tau \in (t_{k-1}, t_k),\ k = 1, \dots, m\right\}.$$

These conditions assure that in a δ-fine point-interval pair, we must have at most one of the points $t_1, \dots, t_m$, for example t_k, and when this happens, the tag of the interval is necessarily equal to t_k.

Note that, for all $\theta \in [-r, 0]$, by Lemma 15.5, we have $x(t_k)(t_k + \theta) = x(t_k + \theta)(t_k + \theta)$, and, using the relation (15.8), $x(t_k + \theta)(t_k + \theta) = y(t_k + \theta)$. Thus, according to the Corollary 2.14, we have

$$\lim_{s \to t_k^+} \int_{t_k}^{s} L(s)\|y_s - x(t_k)_s\|_\infty \, dg(s) = L(t_k)\left\|y_{t_k} - x(t_k)_{t_k}\right\|_\infty \Delta^+ g(t_k) = 0,$$

for every $k \in \{1, \dots, m\}$. As a consequence of this last equality, we can choose the gauge δ such that

$$\int_{t_k}^{t_k+\delta(t_k)} L(s)\|y_s - x(t_k)_s\|_\infty dg(s) < \frac{\epsilon}{4m+1},\ k \in \{1, \dots, m\}, \text{ and}$$
$$\int_{t_k}^{t_k+\delta(t_k)} C(s)\mathrm{var}_{-r}^{0}(\xi(s, \cdot))\|y_s - x(t_k)_s\|_\infty \, ds < \frac{\epsilon}{4m+1},\ k \in \{1, \dots, m\}.$$

Furthermore, in view of conditions (B) and (N2), we obtain

$$\begin{aligned}
&\|y(\tau + t) - y(\tau)\| \\
&\leqslant \left\|\int_{\tau}^{\tau+t} f(y_s, s) dg(s) + \int_{-r}^{0} d_\theta \xi(t + \tau, \theta) y(t + \tau + \theta) \right. \\
&\quad \left. - \int_{-r}^{0} d_\theta \xi(\tau, \theta) y(\tau + \theta)\right\| \\
&\leqslant \int_{\tau}^{\tau+t} M(s) dg(s) + \int_{\tau}^{\tau+t} C(s)\mathrm{var}_{-r}^{0}(\xi(s, \cdot))\, \|y(s + \theta)\| \, ds \\
&\leqslant h(\tau + t) - h(\tau).
\end{aligned}$$

Thus, Corollary 2.14 yields

$$\left\|y(\tau^+) - y(\tau)\right\| \leqslant \Delta^+ h(\tau) < \epsilon, \text{ for all } \tau \in [t_0, t_0 + \sigma] \setminus \{t_1, \dots, t_m\}.$$

Therefore, the gauge δ can be adjusted so that

$$\|y(\rho) - y(\tau)\| \leqslant \epsilon,$$

for every $\tau \in [t_0, t_0 + \sigma] \setminus \{t_1, \dots, t_m\}$ and $\rho \in [\tau, \tau + \delta(\tau))$.

Now, let $d = (\tau_j, [s_{j-1}, s_j])$ be a δ-fine tagged division of the interval $[t_0, v]$. For every $j \in \{1, \dots, |d|\}$, it is not difficult to see that, directly from the definition of the function x given by (15.8), we have

$$[x(s_j) - x(s_{j-1})](\vartheta) =$$

$$\begin{cases} 0, & \vartheta \in [t_0 - r, s_{j-1}], \\ \displaystyle\int_{s_{j-1}}^{\vartheta} f(y_s, s)dg(s) + \int_{-r}^{0} d_\theta[\xi(\vartheta, \theta)]y(\vartheta + \theta) & \\ \qquad \displaystyle - \int_{-r}^{0} d_\theta[\xi(s_{j-1}, \theta)]y(s_{j-1} + \theta), & \vartheta \in [s_{j-1}, s_j], \\ \displaystyle\int_{s_{j-1}}^{s_j} f(y_s, s)dg(s) + \int_{-r}^{0} d_\theta[\xi(s_j, \theta)]y(s_j + \theta) & \\ \qquad \displaystyle - \int_{-r}^{0} d_\theta[\xi(s_{j-1}, \theta)]y(s_{j-1} + \theta), & \vartheta \in [s_j, t_0 + \sigma]. \end{cases}$$

Similarly,

$$[F(x(\tau_j), s_j) - F(x(\tau_j), s_{j-1})](\vartheta) =$$

$$\begin{cases} 0, & \vartheta \in [t_0 - r, s_{j-1}], \\ \displaystyle\int_{s_{j-1}}^{\vartheta} f(x(\tau_j)_s, s)dg(s) + \int_{-r}^{0} d_\theta[\xi(\vartheta, \theta)]x(\tau_j)(\vartheta + \theta) & \\ \qquad \displaystyle - \int_{-r}^{0} d_\theta[\xi(s_{j-1}, \theta)]x(\tau_j)(s_{j-1} + \theta), & \vartheta \in [s_{j-1}, s_j], \\ \displaystyle\int_{s_{j-1}}^{s_j} f(x(\tau_j)_s, s)dg(s) + \int_{-r}^{0} d_\theta[\xi(s_j, \theta)]x(\tau_j)(s_j + \theta) & \\ \qquad \displaystyle - \int_{-r}^{0} d_\theta[\xi(s_{j-1}, \theta)]x(\tau_j)(s_{j-1} + \theta), & \vartheta \in [s_j, t_0 + \sigma]. \end{cases}$$

If we combine the previous calculations, we obtain

$$[x(s_j) - x(s_{j-1})](\vartheta) - [F(x(\tau_j), s_j) - F(x(\tau_j), s_{j-1})](\vartheta) =$$

$$\begin{cases} 0, & \vartheta \in [t_0 - r, s_{j-1}], \\ \displaystyle\int_{s_{j-1}}^{\vartheta} [f(y_s, s) - f(x(\tau_j)_s, s)]dg(s) \\ \quad + \displaystyle\int_{-r}^{0} d_\theta[\xi(\vartheta, \theta)][y(\vartheta + \theta) - x(\tau_j)(\vartheta + \theta)] \\ \quad - \displaystyle\int_{-r}^{0} d_\theta[\xi(s_{j-1}, \theta)][y(s_{j-1} + \theta) - x(\tau_j)(s_{j-1} + \theta)], & \vartheta \in [s_{j-1}, s_j], \\ \displaystyle\int_{s_{j-1}}^{s_j} [f(y_s, s) - f(x(\tau_j)_s, s)]dg(s) \\ \quad + \displaystyle\int_{-r}^{0} d_\theta[\xi(s_j, \theta)][y(s_j + \theta) - x(\tau_j)(s_j + \theta)] \\ \quad - \displaystyle\int_{-r}^{0} d_\theta[\xi(s_{j-1}, \theta)][y(s_{j-1} + \theta) - x(\tau_j)(s_{j-1} + \theta)], & \vartheta \in [s_j, t_0 + \sigma]. \end{cases}$$

Thus,

$$\begin{aligned} &\|x(s_j) - x(s_{j-1}) - [F(x(\tau_j), s_j) - F(x(\tau_j), s_{j-1})]\|_\infty \\ &\quad = \sup_{\vartheta \in [t_0 - r, t_0 + \sigma]} \|[x(s_j) - x(s_{j-1})](\vartheta) - [F(x(\tau_j), s_j) - F(x(\tau_j), s_{j-1})](\vartheta)\| \\ &\quad = \sup_{\vartheta \in [s_{j-1}, s_j]} \left\| \int_{s_{j-1}}^{\vartheta} [f(y_s, s) - f(x(\tau_j)_s, s)]dg(s) \right. \\ &\qquad + \int_{-r}^{0} d_\theta[\xi(\vartheta, \theta)][y(\vartheta + \theta) - x(\tau_j)(\vartheta + \theta)] \\ &\qquad \left. - \int_{-r}^{0} d_\theta[\xi(s_{j-1}, \theta)][y(s_{j-1} + \theta) - x(\tau_j)(s_{j-1} + \theta)] \right\| \\ &\quad \leqslant \sup_{\vartheta \in [s_{j-1}, s_j]} \left\| \int_{s_{j-1}}^{\vartheta} [f(y_s, s) - f(x(\tau_j)_s, s)]dg(s) \right\| \\ &\qquad + \sup_{\vartheta \in [s_{j-1}, s_j]} \left\| \int_{-r}^{0} d_\theta[\xi(\vartheta, \theta)][y(\vartheta + \theta) - x(\tau_j)(\vartheta + \theta)] \right. \\ &\qquad \left. - \int_{-r}^{0} d_\theta[\xi(s_{j-1}, \theta)][y(s_{j-1} + \theta) - x(\tau_j)(s_{j-1} + \theta)] \right\|. \end{aligned}$$

By the definition of x, when $s \leqslant \tau_i$, we have $x(\tau_j)_s = y_s$. Thus,

$$\int_{s_{j-1}}^{\vartheta} [f(y_s, s) - f(x(\tau_j)_s, s)]dg(s)$$

$$= \begin{cases} 0, & \vartheta \in [s_{j-1}, \tau_j], \\ \displaystyle\int_{\tau_j}^{\vartheta} [f(y_s, s) - f(x(\tau_j)_s, s)] dg(s), & \vartheta \in [\tau_j, s_j]. \end{cases}$$

Then, according to hypothesis (C), for every $v \in [s_{j-1}, s_j]$, we have

$$\left\| \int_{s_{j-1}}^{\vartheta} [f(y_s, s) - f(x(\tau_j)_s, s)] dg(s) \right\| \leqslant \int_{\tau_j}^{s_j} L(s) \left\| y_s - x(\tau_j)_s \right\|_\infty dg(s).$$

Likewise, we can use condition (N2) to write

$$\begin{aligned} &\Bigg\| \int_{-r}^{0} d_\theta [\xi(\vartheta, \theta)](y(\vartheta + \theta) - x(\tau_j)(\vartheta + \theta)) \\ &\quad - \int_{-r}^{0} d_\theta [\xi(s_{j-1}, \theta)](y(s_{j-1} + \theta) - x(\tau_j)(s_{j-1} + \theta)) \Bigg\| \\ &\leqslant \int_{s_{j-1}}^{\vartheta} C(s) \operatorname{var}_{-r}^{0}(\xi(s, \cdot)) \parallel y_s - x(\tau_j)_s \parallel_\infty ds \\ &= \begin{cases} 0, & \vartheta \in [s_{j-1}, \tau_j], \\ \displaystyle\int_{\tau_j}^{\vartheta} C(s) \operatorname{var}_{-r}^{0}(\xi(s, \cdot)) \parallel y_s - x(\tau_j)_s \parallel_\infty ds, & \vartheta \in [\tau_j, s_j]. \end{cases} \end{aligned}$$

If we take $(\tau_j, [s_{j-1}, s_j])$ particular, then we need consider two cases:

(i) $[s_{j-1}, s_j] \cap \{t_1, \dots, t_m\} = t_k$ (in this case $t_k = \tau_j$);
(ii) $[s_{j-1}, s_j] \cap \{t_1, \dots, t_m\} = \emptyset$.

Considering case (i), as it was explained before, it follows from the choice of the gauge δ that

$$\int_{\tau_j}^{s_j} L(s) \parallel y_s - x(\tau_j)_s \parallel_\infty dg(s) \leqslant \frac{\epsilon}{4m+1},$$

$$\int_{\tau_j}^{s_j} C(s) \operatorname{var}_{-r}^{0}(\xi(s, \cdot)) \parallel y_s - x(\tau_j)_s \parallel_\infty ds < \frac{\epsilon}{4m+1}.$$

Note that this case occurs at most $2m$ times.

In case (ii), again, by the definition of the gauge δ, for all $s \in [\tau_j, s_j]$, we have

$$\|y_s - x(\tau_j)_s\| \infty = \sup_{\rho \in [\tau_j, s]} \|y(\rho) - y(\tau_j)\| \leqslant \epsilon.$$

It follows from the cases (i) and (ii) that

$$\begin{aligned} &\left\| x(v) - x(t_0) - \sum_{j=1}^{|d|} \left[F(x(\tau_j), s_j) - F(x(\tau_j), s_{j-1}) \right] \right\|_\infty \\ &\quad < \epsilon \int_{t_0}^{t_0+\sigma} L(s) dg(s) + \epsilon \int_{t_0}^{t_0+\sigma} C(s) \operatorname{var}_{-r}^{0}(\xi(s, \cdot)) ds + \frac{4m\epsilon}{4m+1} \end{aligned}$$

$$< \epsilon \left(\int_{t_0}^{t_0+\sigma} L(s)dg(s) + \int_{t_0}^{t_0+\sigma} C(s)\text{var}_{-r}^{0}(\xi(s,\cdot))ds + 1 \right),$$

for every $v \in [t_0, t_0 + \sigma]$, that is,

$$x(v) - x(t_0) = \int_{t_0}^{v} DF(x(\tau), t), \quad v \in [t_0, t_0 + \sigma],$$

since $\epsilon > 0$ is arbitrary, which concludes the proof. □

The next result is a reciprocal for the previous theorem.

Theorem 15.7: *Let $c \geqslant 1$, $\varphi \in P_c$, $g: [t_0, t_0 + \sigma] \to \mathbb{R}$ be a nondecreasing function and let $f: P_c \times [t_0, t_0 + \sigma] \to \mathbb{R}^n$ satisfy conditions (A)–(C). Assume that the normalized function $\xi: \mathbb{R} \times \mathbb{R} \to \mathbb{R}^{n\times n}$ satisfies conditions (N1) and (N2). Let $F: B_c \times [t_0, t_0 + \sigma] \to G([t_0 - r, t_0 + \sigma], \mathbb{R}^n)$ be given by (15.7) and assume that $F(x, t) \in G([t_0 - r, t_0 + \sigma], \mathbb{R}^n)$ for every $x \in B_c$ and every $t \in [t_0, t_0 + \sigma]$. Let $x: [t_0, t_0 + \sigma] \to B_c$ be a solution of the generalized ODE (15.6) with initial condition*

$$x(t_0)(\vartheta) = \begin{cases} \varphi(\vartheta - t_0), & \vartheta \in [t_0 - r, t_0], \\ x(t_0)(t_0), & \vartheta \in [t_0, t_0 + \sigma]. \end{cases}$$

Then, the function $y \in B_c$ defined by

$$y(\vartheta) = \begin{cases} x(t_0)(\vartheta), & \vartheta \in [t_0 - r, t_0], \\ x(\vartheta)(\vartheta), & \vartheta \in [t_0, t_0 + \sigma], \end{cases}$$

is a solution of the measure FDE of neutral type

$$\begin{cases} y(t) = y(t_0) + \displaystyle\int_{t_0}^{t} f(y_s, s)dg(s) + \int_{-r}^{0} d_\theta \xi(t, \theta) y(t + \theta) \\ \qquad - \displaystyle\int_{-r}^{0} d_\theta \xi(t_0, \theta) y(t_0 + \theta), \\ y_{t_0} = \varphi, \end{cases}$$

on $[t_0 - r, t_0 + \sigma]$.

Proof. First, notice that, under the hypotheses of this theorem, $y_{t_0} = \varphi$, because, for all $\vartheta \in [-r, 0]$, we have

$$y_{t_0}(\vartheta) = y(t_0 + \vartheta) = x(t_0)(t_0 + \vartheta) = \varphi(t_0 + \vartheta - t_0) = \varphi(\vartheta)$$

and, using Lemma 15.5, we obtain

$$y(v) - y(t_0) = x(v)(v) - x(t_0)(v) = \left(\int_{t_0}^{v} DF(x(\tau), t) \right)(v),$$

that is,

$$y(v) - y(t_0) - \int_{t_0}^{v} f(y_s, s)dg(s) - \int_{-r}^{0} d_\theta \xi(v,\theta)y(v+\theta)$$
$$+ \int_{-r}^{0} d_\theta \xi(t_0,\theta)y(t_0+\theta)$$
$$= \left(\int_{t_0}^{v} DF(x(\tau),t)\right)(v) - \int_{t_0}^{v} f(y_s, s)dg(s) - \int_{-r}^{0} d_\theta \xi(v,\theta)y(v+\theta)$$
$$+ \int_{-r}^{0} d_\theta \xi(t_0,\theta)y(t_0+\theta), \tag{15.13}$$

for every $v \in [t_0, t_0+\sigma]$.

Let an arbitrary $\epsilon > 0$ be given. We already know that there is only a finite number of points $t \in [t_0, v]$ such that $\Delta^+ g(t) \geqslant \epsilon$, because the function g is nondecreasing. Again, we denote these points by $t_1, \dots, t_m$.

As in the proof of previous theorem, consider a gauge $\delta\colon [t_0, t_0+\sigma] \to (0,\infty)$ fulfilling

$$\delta(\tau) < \min\left\{\frac{t_k - t_{k-1}}{2} : k = 2,\dots,m\right\},\ \ \tau \in [t_0, t_0+\sigma],$$
$$\delta(\tau) < \min\left\{|\tau - t_k|,\ |\tau - t_{k-1}| \colon\ \tau \in (t_{k-1}, t_k),\ k = 1,\dots,m\right\},$$
$$\int_{t_k}^{t_k+\delta(t_k)} L(s)\|y_s - x(t_k)_s\|_\infty\, dg(s) < \frac{\epsilon}{4m+1}, \tag{15.14}$$
$$\int_{t_k}^{t_k+\delta(t_k)} C(s)\mathrm{var}_{-r}^{0}(\xi(s,\cdot))\|y_s - x(t_k)_s\|_\infty\, ds < \frac{\epsilon}{4m+1}, \tag{15.15}$$

where $k \in \{1,\dots,m\}$.

Recall that the first two inequalities assure that if a point-interval pair, $(\tau,[\alpha,\beta])$, is δ-fine, then $[\alpha,\beta]$ contains at most one of the points $t_1,\dots,t_m$, and, moreover, $\tau = t_k$, whenever $t_k \in [\alpha,\beta]$.

By Lemma 15.4, the function F, defined in (15.7), satisfies the conditions to belong to the class $\mathcal{F}(B_c \times [t_0, t_0+\sigma], h)$, with

$$h(t) = \int_{t_0}^{t} [L(s)+M(s)]dg(s) + \int_{t_0}^{t} C(s)\,\mathrm{var}_{-r}^{0}(\xi(s,\cdot))ds\left(\|\tilde{x}\|_\infty + c\right), \tag{15.16}$$

where $t \in [t_0, t_0+\sigma]$. Thus, we can assume that the gauge δ satisfies

$$\|h(\rho) - h(\tau)\| \leqslant \epsilon, \quad \text{for all } \rho \in [\tau, \tau+\delta(\tau)).$$

For that adjusted gauge δ, by the definition of the Kurzweil integral, take a particular δ-fine division $d = \{(\tau_j,[s_{j-1},s_j])\colon j = 1,\dots,|d|\}$ of $[t_0, v]$, such that

$$\left\|\int_{t_0}^{v} DF(x(\tau),t) - \sum_{j=1}^{|d|}[F(x(\tau_j),s_j) - F(x(\tau_j),s_{j-1})]\right\|_\infty < \epsilon, \tag{15.17}$$

for every $v \in [t_0, t_0+\sigma]$. Then, considering (15.13) and (15.17), we obtain

$$\begin{aligned}&\left\|y(v) - y(t_0) - \int_{t_0}^{v} f(y_s, s)dg(s)\right.\\&\left.- \int_{-r}^{0} d_\theta \xi(v,\theta)y(v+\theta) + \int_{-r}^{0} d_\theta \xi(t_0,\theta)y(t_0+\theta)\right\|\\&< \epsilon + \sum_{j=1}^{|d|}\left\|[F(x(\tau_j), s_j) - F(x(\tau_j), s_{j-1})](v) - \int_{t_0}^{v} f(y_s, s)dg(s)\right.\\&\left.- \int_{-r}^{0} d_\theta \xi(s_i,\theta)y(s_i+\theta) + \int_{-r}^{0} d_\theta \xi(s_{j-1},\theta)y(s_{j-1}+\theta)\right\|,\end{aligned}$$

for every $v \in [t_0, t_0+\sigma]$.

Let us look at this last inequality with a little more attention. The definition of F in (15.7) yields

$$\begin{aligned}[F(x(\tau_j), s_j) - F(x(\tau_j), s_{j-1})](v) &= \int_{s_{j-1}}^{s_j} f(x(\tau_j)_s, s)dg(s)\\&+ \int_{-r}^{0} d_\theta \xi(s_j,\theta)x(\tau_j)(s_j+\theta)\\&- \int_{-r}^{0} d_\theta \xi(s_{j-1},\theta)x(\tau_j)(s_{j-1}+\theta),\end{aligned}$$

which implies

$$\begin{aligned}&\left\|[F(x(\tau_j), s_j) - F(x(\tau_j), s_{j-1})](v) - \int_{s_{j-1}}^{s_j} f(y_s, s)dg(s)\right.\\&\left.- \int_{-r}^{0} d_\theta \xi(s_j,\theta)y(s_j+\theta) + \int_{-r}^{0} d_\theta \xi(s_{j-1},\theta)y(s_{j-1}+\theta)\right\|\\&\leqslant \left\|\int_{s_{j-1}}^{s_j} f(x(\tau_j)_s, s)dg(s) - \int_{s_{j-1}}^{s_j} f(y_s, s)dg(s)\right\|\\&+ \left\|\int_{-r}^{0} d_\theta \xi(s_j,\theta)[x(\tau_j)(s_j+\theta) - y(s_j+\theta)]\right.\\&\left.- \int_{-r}^{0} d_\theta \xi(s_{j-1},\theta)[x(\tau_j)(s_{j-1}+\theta) - y(s_{j-1}+\theta)]\right\|,\end{aligned}$$

for every point-interval pair $(\tau_j, [s_{j-1}, s_j])$ of δ-fine division d.

With a little bit of calculation, using Lemma 15.5, it is possible to verify that for every $j \in \{1, \dots, |d|\}$, we have

$$\begin{aligned}&x(\tau_j)_s = x(s)_s = y_s, \quad \text{for } s \in [s_{j-1}, \tau_j] \text{ and}\\&y_s = x(s)_s = x(s_j)_s, \quad \text{for } s \in [\tau_j, s_j].\end{aligned}$$

Thus,

$$\begin{aligned}\left\|\int_{s_{j-1}}^{s_j}[f(x(\tau_j)_s,s)-f(y_s,s)]dg(s)\right\| &= \left\|\int_{\tau_j}^{s_j}[f(x(\tau_j)_s,s)-f(y_s,s)]dg(s)\right\| \\ &= \left\|\int_{\tau_j}^{s_j}[f(x(\tau_j)_s,s)-f(x(s_j)_s,s)]dg(s)\right\| \\ &\leqslant \int_{\tau_j}^{s_j} L(s)\left\|x(\tau_j)_s - x(s_j)_s\right\|_\infty dg(s),\end{aligned}$$

with the inequality being true by condition (C).

On the other hand, hypothesis (N2) ensures that

$$\begin{aligned}&\left\|\int_{-r}^{0} d_\theta\xi(s_j,\theta)[x(\tau_j)(s_j+\theta)-y(s_j+\theta)]\right.\\ &\quad\left.-\int_{-r}^{0} d_\theta\xi(s_{j-1},\theta)[x(\tau_j)(s_{j-1}+\theta)-y(s_{j-1}+\theta)]\right\|\\ &\leqslant \int_{s_{j-1}}^{s_j} C(s)\left(\int_{-r}^{0} d_\theta\xi(s,\theta)\|x(\tau_j)(s+\theta)-y(s+\theta)\|\right)ds\\ &\leqslant \int_{s_{j-1}}^{s_j} C(s)\mathrm{var}_{-r}^{0}(\xi(s,\cdot))\|x(\tau_j)_s - y_s\|_\infty\, ds\\ &= \int_{\tau_j}^{s_j} C(s)\mathrm{var}_{-r}^{0}(\xi(s,\cdot))\|x(\tau_j)_s - y_s\|_\infty\, ds.\end{aligned}$$

In the same way as in Theorem 15.6, we have two cases to consider:

(i) $[s_{j-1},s_j]\cap\{t_1,\dots,t_m\}=t_k$ (in this case $t_k=\tau_j$);
(ii) $[s_{j-1},s_j]\cap\{t_1,\dots,t_m\}=\emptyset$.

In case (i), if we use (15.14) and (15.15), then

$$\begin{aligned}&\left\|[F(x(\tau_j),s_j)-F(x(\tau_j),s_{j-1})](v)-\int_{s_{j-1}}^{s_j} f(y_s,s)dg(s)\right.\\ &\quad\left.-\int_{-r}^{0} d_\theta\xi(s_j,\theta)y(s_j+\theta)+\int_{-r}^{0} d_\theta\xi(s_{j-1},\theta)y(s_{j-1}+\theta)\right\| \leqslant \frac{2\epsilon}{4m+1}.\end{aligned}$$

In case (ii), we first need to note that

$$\|x(s_j)_s - x(\tau_j)_s\|\ \infty \leqslant \|x(s_j)-x(\tau_j)\|\ \infty \leqslant h(s_j)-h(\tau_j)\leqslant \epsilon$$

holds as a consequence of Lemma 4.9, whenever $s\in[\tau_i,s_i]$. Thus,

$$\begin{aligned}&\left\|[F(x(\tau_j),s_j)-F(x(\tau_j),s_{j-1})](v)-\int_{s_{j-1}}^{s_j} f(y_s,s)dg(s)\right.\\ &\quad\left.-\int_{-r}^{0} d_\theta\xi(s_j,\theta)y(s_j+\theta)+\int_{-r}^{0} d_\theta\xi(s_{j-1},\theta)y(s_{j-1}+\theta)\right\|\end{aligned}$$

$$\leqslant \epsilon \int_{\tau_j}^{s_j} L(s)dg(s) + \epsilon \int_{\tau_j}^{s_j} C(s)\text{var}_{-r}^{0}(\xi(s,\cdot))ds.$$

From the cases (i) and (ii), remembering that case (i) occurs at most $2m$ times, we have

$$\sum_{j=1}^{|d|} \left\| \left[F(x(\tau_j), s_j) - F(x(\tau_j), s_{j-1})\right](v) - \int_{s_{j-1}}^{s_j} f(y_s, s)dg(s) \right.$$
$$\left. - \int_{-r}^{0} d_\theta \xi(s_j, \theta) y(s_j + \theta) + \int_{-r}^{0} d_\theta \xi(s_{j-1}, \theta) y(s_{j-1} + \theta) \right\|$$
$$\leqslant \epsilon \int_{t_0}^{t_0+\sigma} L(s)\ dg(s) + \epsilon \int_{t_0}^{t_0+\sigma} C(s)\text{var}_{-r}^{0}(\xi(s,\cdot))ds + \frac{4m\epsilon}{4m+1},$$

for every point-interval pair $(\tau_j, [s_{j-1}, s_j])$ of the division d. Therefore, for every $v \in [t_0, t_0 + \sigma]$, we have

$$\left\| y(v) - y(t_0) - \int_{t_0}^{v} f(y_s, s)dg(s) \right.$$
$$\left. - \int_{-r}^{0} d_\theta \xi(v, \theta) y(v + \theta) + \int_{-r}^{0} d_\theta \xi(t_0, \theta) y(t_0 + \theta) \right\|$$
$$< \epsilon(\int_{t_0}^{t_0+\sigma} L(s)dg(s) + \int_{t_0}^{t_0+\sigma} C(s)\text{var}_{-r}^{0}(\xi(s,\cdot))ds + 1),$$

which completes the proof. □

Now, we present an existence-uniqueness theorem for measure FDEs of neutral type. Such result can be found in [76]. The idea of the proof is to employ the previous correspondence between measure FDEs of neutral type and generalized ODEs (applying Theorem 5.1) in order to get the result.

Theorem 15.8: *Let $c \geqslant 1$, $g\colon [t_0, t_0 + \sigma] \to \mathbb{R}$ be a left-continuous and nondecreasing function and let $f\colon P_c \times [t_0, t_0 + \sigma] \to \mathbb{R}^n$ satisfy conditions* (A)–(C). *Assume that the normalized function $\xi\colon \mathbb{R} \times \mathbb{R} \to \mathbb{R}^{n\times n}$ satisfies conditions* (N1) *and* (N2). *If $\varphi \in P_c$ is such that the function*

$$z(t) = \begin{cases} \varphi(t - t_0), & t \in [t_0 - r, t_0], \\ \varphi(0) + f(\varphi, t_0)\Delta^+ g(t_0), & t \in (t_0, t_0 + \sigma], \end{cases}$$

belongs to B_c, then there exist $\omega > 0$ and a function $y\colon [t_0 - r, t_0 + \omega] \to \mathbb{R}^n$ which is the unique solution of the initial value problem

$$\begin{cases} y(t) = y(t_0) + \int_{t_0}^{t} f(y_s, s)dg(s) + \int_{-r}^{0} d_\theta \xi(t, \theta) y(t + \theta) - \int_{-r}^{0} d_\theta \xi(t_0, \theta) y(t_0 + \theta), \\ y_{t_0} = \varphi. \end{cases} \tag{15.18}$$

Proof. Let F be given by (15.7). We know that this function belongs to the class $\mathcal{F}(B_c \times [t_0, t_0 + \sigma], h)$ (see Lemma 15.4), where

$$h(t) = \int_{t_0}^{t} [M(s) + L(s)]dg(s) + \int_{t_0}^{t} C(s)\,\mathrm{var}_{-r}^{0}(\xi(s,\cdot))ds(\|\tilde{x}\|_\infty + c)$$

is a nondecreasing function, for every $t \in [t_0, t_0 + \sigma]$. The function $x_0 \colon [t_0 - r, t_0 + \sigma] \to \mathbb{R}^n$ defined by

$$x_0(\vartheta) = \begin{cases} \varphi(\vartheta - t_0), & \vartheta \in [t_0 - r, t_0], \\ \varphi(0), & \vartheta \in [t_0, t_0 + \sigma], \end{cases}$$

is an element of B_c.

We need to fit the hypotheses of Theorem 5.1 on existence and uniqueness for generalized ODEs. In order to do this, we look at the limit $F(x_0, t_0^+)$. Since $F \in \mathcal{F}(B_c \times [t_0, t_0 + \sigma], h)$, F is a regulated function with respect to the variable t. Thus, the limit $F(x_0, t_0^+)$ must exist, and it is taken with respect to the supremum norm. Furthermore, according to Corollary 2.14, we have

$$F(x_0, t_0^+)(\vartheta) = \begin{cases} 0, & t \in [t_0 - r, t_0], \\ f(\varphi, t_0)\Delta^+ g(t_0), & t \in (t_0, t_0 + \sigma], \end{cases}$$

and, consequently,

$$x_0 + F(x_0, t_0^+) - F(x_0, t_0) = x_0 + F(x_0, t_0^+) = z \in B_c.$$

With the above observations, all hypotheses of the Theorem 5.1 are satisfied and, therefore, there exist $\omega > 0$ and a unique solution $x \colon [t_0, t_0 + \omega] \to G([t_0 - r, t_0 + \sigma], \mathbb{R}^n)$ of the initial value problem

$$\begin{cases} \dfrac{dx}{d\tau} = DF(x, t), \\ x(t_0) = x_0. \end{cases} \tag{15.19}$$

The function $y \colon [t_0 - r, t_0 + \omega] \to \mathbb{R}^n$ given by

$$y(\vartheta) = \begin{cases} x(t_0)(\vartheta), & \vartheta \in [t_0 - r, t_0], \\ x(\vartheta)(\vartheta), & \vartheta \in [t_0, t_0 + \omega], \end{cases}$$

is a solution of the measure FDE of neutral type (15.18), see Theorem 15.7. This solution must be unique. Indeed, if we have another solution for (15.18) then, in view of Theorem 15.6, we must have another solution for the associated IVP (15.19). Remember that the pair Theorems 15.6 and 15.7 form a biunivocal relation. But, this contradicts Theorem 5.1. Thus, the result follows. □

In the remaining of this section, we present a result on the continuous dependence on parameters for measure FDEs of neutral type. Such result is presented next, and it was borrowed from [76]. We proceed in the same way as we did previously, that is, we use the existing theory for generalized ODEs in search of our result.

Theorem 15.9: *Let* $c \geqslant 1$, $P = \{y_t\colon\ y \in G([t_0 - r, t_0 + \sigma], \mathbb{R}^n)$ *and* $t_0 \leqslant t \leqslant t_0 + \sigma\}$, $g\colon [t_0, t_0 + \sigma] \to \mathbb{R}$ *be a nondecreasing and left-continuous function and* $f_k\colon P \times [t_0, t_0 + \sigma] \to \mathbb{R}^n$, $k \in \mathbb{N}_0$, *be a sequence of functions which satisfy conditions* (A)–(C) *for the same functions* $L, M\colon [t_0, t_0 + \sigma] \to \mathbb{R}$ *for every* $k \in \mathbb{N}_0$. *Suppose the normalized function* $\xi_k\colon \mathbb{R} \times \mathbb{R} \to \mathbb{R}^{n\times n}$ *satisfies conditions* (N1) *and* (N2) *for the same function* $C\colon [t_0, t_0 + \sigma] \to \mathbb{R}$ *for every* $k \in \mathbb{N}_0$. *Moreover, assume*

(i) *For every* $y \in G([t_0 - r, t_0 + \sigma], \mathbb{R}^n)$,

$$\lim_{k\to\infty} \int_{t_0}^{t} f_k(y_s, s)dg(s) = \int_{t_0}^{t} f_0(y_s, s)dg(s)$$

uniformly with respect to $t \in [t_0, t_0 + \sigma]$.

(ii) *For every* $y \in G([t_0 - r, t_0 + \sigma], \mathbb{R}^n)$,

$$\lim_{k\to\infty} \int_{-r}^{0} d_\theta \xi_k(t, \theta) y(t + \theta) = \int_{-r}^{0} d_\theta \xi_0(t, \theta) y(t + \theta)$$

uniformly with respect to $t \in [t_0, t_0 + \sigma]$.

Consider a sequence of functions $\varphi_k \in P$, $k \in \mathbb{N}_0$, *such that* $\lim\limits_{k\to\infty} \varphi_k = \varphi_0$ *uniformly on* $[-r, 0]$. *If, for each* $k \in \mathbb{N}$, $y_k \in G([t_0 - r, t_0 + \sigma], \mathbb{R}^n)$ *is the solution of*

$$\begin{cases} y_k(t) = & y_k(t_0) + \displaystyle\int_{t_0}^{t} f_k((y_k)_s, s)dg(s) + \int_{-r}^{0} d_\theta \xi_k(t, \theta) y_k(t + \theta) \\ & - \displaystyle\int_{-r}^{0} d_\theta \xi_k(t_0, \theta) y_k(t_0 + \theta), \\ (y_k)_{t_0} = \varphi_k, & \end{cases}$$

on $[t_0 - r, t_0 + \sigma]$, *then there exists a function* $y_0 \in G([t_0 - r, t_0 + \sigma], \mathbb{R}^n)$ *such that* $\lim\limits_{k\to\infty} y_k = y_0$ *on* $[t_0, t_0 + \sigma]$ *and* $y_0\colon [t_0 - r, t_0 + \sigma] \to \mathbb{R}^n$ *is a solution of*

$$\begin{cases} y_0(t) = & y_0(t_0) + \displaystyle\int_{t_0}^{t} f_0((y_0)_s, s)dg(s) + \int_{-r}^{0} d_\theta \xi_0(t, \theta) y_0(t + \theta) \\ & - \displaystyle\int_{-r}^{0} d_\theta \xi_0(t_0, \theta) y_0(t_0 + \theta), \\ (y_0)_{t_0} = \varphi_0. & \end{cases} \tag{15.20}$$

Proof. For every $k \in \mathbb{N}$, we define the function $F_k : G([t_0 - r, t_0 + \sigma], \mathbb{R}^n) \times [t_0, t_0 + \sigma] \to G([t_0 - r, t_0 + \sigma], \mathbb{R}^n)$ by

$$F_k(x,t)(\vartheta) = \begin{cases} 0, & \vartheta \in [t_0 - r, t_0], \\ \displaystyle\int_{t_0}^{\vartheta} f_k(x_s, s)dg(s) + \int_{-r}^{0} d_\theta[\xi_k(\vartheta, \theta)]x(\vartheta + \theta) & \\ \qquad\displaystyle - \int_{-r}^{0} d_\theta[\xi_k(t_0, \theta)]x(t_0 + \theta), & \vartheta \in [t_0, t], \\ \displaystyle\int_{t_0}^{t} f_k(x_s, s)dg(s) + \int_{-r}^{0} d_\theta[\xi_k(t, \theta)]x(t + \theta) & \\ \qquad\displaystyle - \int_{-r}^{0} d_\theta[\xi_k(t_0, 0)]x(t_0 + \theta), & \vartheta \in [t, t_0 + \sigma]. \end{cases}$$

Note that, according to Lemma 15.4, $F_k \in \mathcal{F}(G([t_0 - r, t_0 + \sigma], \mathbb{R}^n) \times [t_0, t_0 + \sigma], h)$, for every $k \in \mathbb{N}$, where

$$h(t) = \int_{t_0}^{t} (M(s) + L(s))dg(s) + \int_{t_0}^{t} C(s)\operatorname{var}_{-r}^{0}(\xi(s, \cdot))ds(\|\tilde{x}\|_\infty + c),$$

for all $t \in [t_0, t_0 + \sigma]$ and, under the conditions of this theorem, for every $x \in G([t_0 - r, t_0 + \sigma], \mathbb{R}^n)$, we have

$$\lim_{k\to\infty} F_k(x, t) = F_0(x, t),$$

uniformly with respect to $t \in [t_0, t_0 + \sigma]$. Hence, $F_0(x, t) \in G([t_0 - r, t_0 + \sigma], \mathbb{R}^n)$ and $F_0 \in \mathcal{F}(G([t_0 - r, t_0 + \sigma], \mathbb{R}^n) \times [t_0, t_0 + \sigma], h)$. Furthermore, by the Moore–Osgood theorem,

$$\lim_{k\to\infty} F_k(x, t^+) = F_0(x, t^+),$$

for every $x \in G([t_0 - r, t_0 + \sigma], \mathbb{R}^n)$ and $t \in [t_0, t_0 + \sigma)$.

Given $k \in \mathbb{N}$ and $t_0 \leqslant t_1 \leqslant t_2 \leqslant t_0 + \sigma$, we have

$$\begin{aligned} \|y_k(t_2) - y_k(t_1)\| &= \left\| \int_{t_1}^{t_2} f_k((y_k)_s, s)dg(s) + \int_{-r}^{0} d_\theta \xi_k(t_2, \theta)y_k(t_2 + \theta) \right. \\ &\qquad \left. - \int_{-r}^{0} d_\theta \xi_k(t_1, \theta)y_k(t_1 + \theta) \right\| \\ &\leqslant h(t_2) - h(t_1) = \eta(h(t_2)) - \eta(h(t_1)), \end{aligned}$$

where $\eta(t) = t$, for every $t \in [0, \infty)$.

Since the sequence $\{y_k(t_0)\}_{k\in\mathbb{N}}$ is bounded and the function h is nondecreasing, the sequence $\{y_k\}_{k\in\mathbb{N}}$ contains a subsequence which is uniformly convergent on $[t_0, t_0 + \sigma]$ (see the remark after Corollary 1.19). Without loss of generality, we can denote this subsequence, again, by $\{y_k\}_{k\in\mathbb{N}}$. Furthermore, since $(y_k)_{t_0} = \varphi_k$, for each $k \in \mathbb{N}$, $\{y_k\}_{k\in\mathbb{N}}$ converges to y_0 uniformly on $[t_0 - r, t_0 + \sigma]$.

On the other hand, for every $k \in \mathbb{N}$, by Theorem 15.6, the function x_k, defined by

$$x_k(t)(\vartheta) = \begin{cases} y_k(\vartheta), & \vartheta \in [t_0 - r, t], \\ y_k(t), & \vartheta \in [t, t_0 + \sigma], \end{cases}$$

is the solution of the generalized ODE

$$\frac{dx}{d\tau} = DF_k(x, t),$$

with initial condition

$$x_k(t_0)(\vartheta) = \begin{cases} \varphi_k(\vartheta - t_0), & \vartheta \in [t_0 - r, t_0], \\ x_k(t_0)(t_0), & \vartheta \in [t_0, t_0 + \sigma], \end{cases}$$

and $\lim_{k\to\infty} x_k(t) = x_0(t)$ uniformly with respect to $t \in [t_0, t_0 + \sigma]$. Moreover, note that $x_k\colon [t_0, t_0 + \sigma] \to G([t_0 - r, t_0 + \sigma], \mathbb{R}^n)$ for every $k \in \mathbb{N}$. Then, Theorem 7.4 implies that x_0 is a solution of

$$\frac{dx}{d\tau} = DF_0(x, t)$$

on $[t_0, t_0 + \sigma]$, with initial condition

$$x_0(t_0)(\vartheta) = \begin{cases} \varphi_0(\vartheta - t_0), & \vartheta \in [t_0 - r, t_0], \\ x_0(t_0)(t_0), & \vartheta \in [t_0, t_0 + \sigma]. \end{cases}$$

Thus, applying Theorem 15.7, the function

$$y_0(\vartheta) = \begin{cases} x_0(t_0)(\vartheta), & \vartheta \in [t_0 - r, t_0], \\ x_0(\vartheta)(\vartheta), & \vartheta \in [t_0, t_0 + \sigma], \end{cases}$$

satisfies (15.20) on $[t_0 - r, t_0 + \sigma]$ and therefore, the proof is complete. □

References

1 Adamec, L. A note on continuous dependence of solutions of dynamic equations on time scales. *Journal of Difference Equations and Applications 17*, 5 (2011), 647–656.

2 Afonso, S. M., Bonotto, E. M., Federson, M., and Gimenes, L. P. Boundedness of solutions of retarded functional differential equations with variable impulses via generalized ordinary differential equations. *Mathematische Nachrichten 285*, 5–6 (2012), 545–561.

3 Afonso, S. M., Bonotto, E. M., Federson, M., and Gimenes, L. P. Stability of functional differential equations with variable impulsive perturbations via generalized ordinary differential equations. *Bulletin des Sciences Mathématiques 137*, 2 (2013), 189–214.

4 Afonso, S. M., Bonotto, E. M., Federson, M., and Schwabik, Š. Discontinuous local semiflows for Kurzweil equations leading to LaSalle's invariance principle for differential systems with impulses at variable times. *Journal of Differential Equations 250*, 7 (2011), 2969–3001.

5 Alexiewicz, A. Linear functionals on Denjoy-integrable functions. *Colloquium Mathematicum 1* (1948), 289–293.

6 Alvarez, E., Grau, R., Lizama, C., and Mesquita, J. G. Volterra–stieltjes integral equations and impulsive volterra–stieltjes integral equations. *Electronic Journal of Qualitative Theory of Differential Equations*, 2021, Paper No. 5, 20 pp.

7 Andrade da Silva, F., Federson, M., Grau, R., and Toon, E. Converse Lyapunov theorems for measure functional differential equations. *Journal of Differential Equations* 286 (2021), 1–46.

8 Artstein, Z. Continuous dependence on parameters: on the best possible results. *Journal of Differential Equations 19*, 2 (1975), 214–225.

9 Artstein, Z. Topological dynamics of ordinary differential equations and Kurzweil equations. *Journal of Differential Equations 23*, 2 (1977), 224–243.

Generalized Ordinary Differential Equations in Abstract Spaces and Applications, First Edition.
Edited by Everaldo M. Bonotto, Márcia Federson, and Jaqueline G. Mesquita.

10 Bainov, D. D., Kostadinov, S. I., and Van Minh, N. *Dichotomies and Integral Manifolds of Impulsive Differential Equations.* Science Culture and Technology Publishing, Singapore, 1994.

11 Bainov, D. D., and Milusheva, S. D. Justification of the averaging method for a system of functional-differential equations with variable structure and impulses. *Applied Mathematics and Optimization 16*, 1 (1987), 19–36.

12 Bartle, R. G. *A Modern Theory of Integration*, vol. 32 of *Graduate Studies in Mathematics*. American Mathematical Society, Providence, RI, 2001.

13 Belov, S. A., and Chistyakov, V. V. A selection principle for mappings of bounded variation. *Journal of Mathematical Analysis and Applications 249*, 2 (2000), 351–366.

14 Benchohra, M., Henderson, J., Ntouyas, S. K., and Ouahab, A. On first order impulsive dynamic equations on time scales. *Journal of Difference Equations and Applications 10*, 6 (2004), 541–548.

15 Benchohra, M., Henderson, J., Ntouyas, S. K., and Ouahab, A. Impulsive functional dynamic equations on time scales. *Dynamic Systems and Applications 15*, 1 (2006), 43–52.

16 Bernoulli, J., Battier, J. J., and Cramer, G. *Jacobi Bernoulli, Basileensis, Opera*, vol. 1. Cramer & Philibert, Geneve, 1744.

17 Bhatia, N. P., and Hajek,O. *Local Semi-Dynamical Systems, Lecture Notes in Mathematics.* Springer-Verlag, Berlin Heidelberg, 2006.

18 Bhatia, N. P., and Szegö, G. P. *Stability Theory of Dynamical Systems, Classics in Mathematics.* Springer-Verlag, Berlin Heidelberg, 1970.

19 Blake, A. A Boolean derivation of the Moore-Osgood theorem. *The Journal of Symbolic Logic 11* (1946), 65–70.

20 Bogolyubov, N. N., and Mitropol'skiĭ, Y. A. *Asimptotičeskie metody v teorii nelineĭnyh kolebaniĭ (Russian) [Asymptotic methods in the theory of nonlinear oscillations].* Gosudarstvennoe Izdatel'stvo Tehniko-Teoreticeskoi Literatury, Moscow, 1955.

21 Bohner, M., Federson, M., and Mesquita, J. G. Continuous dependence for impulsive functional dynamic equations involving variable time scales. *Applied Mathematics and Computation 221* (2013), 383–393.

22 Bohner, M., and Peterson, A. *Dynamic Equations on Time Scales: An Introduction with Applications.* Birkhäuser, Boston, MA, 2001.

23 Bohner, M., and Peterson, A. *Advances in Dynamic Equations on Time Scales.* Birkhäuser, Boston, MA, 2002.

24 Bongiorno, B. Relatively weakly compact sets in the Denjoy space. *Journal of Mathematical Study 27*, 1 (1994), 37–44.

25 Bonotto, E. M. Flows of characteristic 0+ in impulsive semidynamical systems. *Journal of Mathematical Analysis and Applications 332*, 1 (2007), 81–96.

26 Bonotto, E. M. Lasalle's theorems in impulsive semidynamical systems. *Nonlinear Analysis: Theory Methods & Applications 71*, 5 (2009), 2291–2297.

27 Bonotto, E. M., and Federson, M. Topological conjugation and asymptotic stability in impulsive semidynamical systems. *Journal of Mathematical Analysis and Applications 326*, 2 (2007), 869–881.

28 Bonotto, E. M., Federson, M., and Gadotti, M. C. Recursive properties on generalized ordinary differential equations and applications. Manuscript submitted for publication, 2020.

29 Bonotto, E. M., Federson, M., and Santos, F. L. Dichotomies for generalized ordinary differential equations and applications. *Journal of Differential Equations 264*, 5 (2018), 3131–3173.

30 Bonotto, E. M., and Jimenez, M. Z. On impulsive semidynamical systems: minimal, recurrent and almost periodic motions. *Topological Methods in Nonlinear Analysis 44*, 1 (2014), 121–141.

31 Bonotto, E. M., and Jimenez, M. Z. Weak almost periodic motions, minimality and stability in impulsive semidynamical systems. *Journal of Differential Equations 256*, 4 (2014), 1683–1701.

32 Bourbaki, N. *Fonctions d'une variable réelle: Théorie élémentaire, Eléments de mathématique*. Springer–Verlag, Berlin Heidelberg, 2007.

33 Burton, T. A. Stability theory for delay equations. *Funkcialaj Ekvacioj-Serio Internacia 22*, 1 (1979), 67–76.

34 Cajori, F. *A History of Mathematics*. The Macmillan Company, New York, 1919.

35 Cao, S. S. The Henstock integral for Banach–valued functions. *Southeast Asian Bulletin of Mathematics 16*, 1 (1992), 35–40.

36 Cauchy, A. L. *Cours d'analyse de l'École Royale Polytechnique. 1. Analyse algébrique*, vol. 1. de Bure, Paris, 1821.

37 Chang, Y.-K., and Li, W.-T. Existence results for impulsive dynamic equations on time scales with nonlocal initial conditions. *Mathematical and Computer Modelling 43*, 3–4 (2006), 377–384.

38 Chicone, C. *Ordinary Differential Equations with Applications*, 2nd ed., vol. 34 of *Texts in Applied Mathematics*. Springer–Verlag, New York, 2006.

39 Chow, S.-N., and Yorke, J. A. Lyapunov theory and perturbation of stable and asymptotically stable systems. *Journal of Differential Equations 15* (1974), 308–321.

40 Cichoń, K., Cichoń, M., and Satco, B. On regulated functions. *Polytechnica Posnaniensis. Institutum Mathematicum. Fasciculi Mathematici 60* (2018), 37–57.

41 Ciesielski, K. On semicontinuity in impulsive dynamical systems. *Bulletin of the Polish Academy of Sciences Mathematics 52*, 1 (2004), 71–80.

42 Ciesielski, K. On time reparametrizations and isomorphisms of impulsive dynamical systems. *Annales Polonici Mathematici 84*, 1 (2004), 1–25.

43 Coffman, C. V., and Schäffer, J. J. Dichotomies for linear difference equations. *Mathematische Annalen 172* (1967), 139–166.

44 Collegari, R. *Equações diferenciais ordinárias generalizadas lineares e aplicações às equações diferenciais funcionais lineares*. PhD thesis, Universidade de São Paulo, 2014.

45 Collegari, R., Federson, M., and Frasson, M. Linear FDEs in the frame of generalized ODEs: variation–of–constants formula. *Czechoslovak Mathematical Journal 68*, 4 (2018), 889–920.

46 Condorcet, Marie Jean Antoine Nicolas Caritat, Marquis de Sur la détermination des fonctions arbitraires qui entrent dans les intégrales des équations aux différences partielles. *Mémoires de l'Académie Royale des Sciences 1774* (1771), 49–74.

47 Congxin, W., and Xiaobo, Y. A Riemann–type definition of the Bochner integral. *Journal of Mathematical Study 27*, 3 (1994), 32–36.

48 Coppel, W. A. *Dichotomies in Stability Theory, Lecture Notes in Mathematics.* Springer–Verlag, Berlin Heidelberg, New York, 1978.

49 Daleckiĭ, J. L., and Kreĭn, M. G. *Stability of Solutions of Differential Equations in Banach Space*, vol. 43 of *Translations of Mathematical Monographs*. American Mathematical Society, Providence, RI, 1974.

50 D'Alembert, J. Récherches sur la courbe que forme une corde tenduë mise en vibration. I. *Memoires de l'Academie royale des sciences et belles lettres 3* (1747), 214–219.

51 D'Alembert, J. Récherches sur la courbe que forme une corde tenduë mise en vibration. II. *Memoires de l'Academie royale des sciences et belles lettres 3* (1747), 220–249.

52 D'Alembert, J. Récherches sur la courbe que forme une corde tenduë mise en vibration. *Memoires de l'Academie royale des sciences et belles lettres 6* (1750), 355–360.

53 D'Alembert, J. Mémoire sur les principes de la mécanique. *Mémoires de l'Académie Royale des Sciences Paris 6* (1769), 278–286.

54 Das, P. C., and Sharma, R. R. Existence and stability of measure differential equations. *Czechoslovak Mathematical Journal 22*, 1 (1972), 145–158.

55 Di Piazza, L., and Musial, K. A characterization of variationally Mcshane integrable Banach–space valued functions. *Illinois Journal of Mathematics 45*, 1 (2001), 279–289.

56 Diekmann, O., van Gils, S. A., Verduyn Lunel, S. M., and Walther, H.-O. *Delay Equations. Functional-, Complex-, and Nonlinear Analysis*, vol. 110 of *Applied Mathematical Sciences*. Springer–Verlag, New York, 1995.

57 Diestel, J. *Sequences and Series in Banach Spaces*, vol. 92 of *Graduate Texts in Mathematics*. Springer–Verlag, New York, 1984.

58 Dieudonné, J. *Foundations of Modern Analysis*, vol. 10 of *Pure and Applied Mathematics*. Academic Press, New York, London, 1960.

59 Duke, E. R., Hall, K. J., and Oberste-Vorth, R. Changing time scales I: the continuous case as a limit. In *Proceedings of the Sixth WSEAS International Conference on Applied Mathematics* (2004), N. Mastoraki, Ed., 1–8. http://www.wseas.us/e-li-brary/conferences/cor-fu2004/papers/488-477.pdf.

60 Dvoretzky, A., and Rogers, C. A. Absolute and unconditional convergence in normed linear spaces. *Proceedings of the National Academy of Sciences of the United States of America 36*, 3 (1950), 192–197.

61 Elaydi, S. *An Introduction to Difference Equations*, 3rd ed., *Undergraduate Texts in Mathematics*. Springer–Verlag, New York, 2005.

62 Esty, N., and Hilger, S. Convergence of time scales under the Fell topology. *Journal of Difference Equations and Applications 15*, 10 (2009), 1011–1020.

63 Euler, L. Investigatio curvarum quae evolutae sui similes producunt. *Commentarii academiae scientiarum Petropolitanae 12* (1750), 3–52.

64 Euler, L. Institutiones calculi differentialis cum eius usu in analysi finitorum ac doctrina serierum, Volume I. *Euler Archive – All Works 212* (1755).

65 Euler, L. Institutionum calculi integralis volumen primum. *Euler Archive – All Works 342* (1768).

66 Euler, L. Institutionum calculi integralis volumen secundum. *Euler Archive – All Works 366* (1769).

67 Euler, L. Institutionum calculi integralis volumen tertium. *Euler Archive – All Works 385* (1770).

68 Evans, L., and Gariepy, R. *Measure Theory and Fine Properties of Functions, Revised Edition, Textbooks in Mathematics*. CRC Press, Boca Raton, FL, 2015.

69 Fan, M., Dishen, J., Wan, Q., and Wang, K. Stability and boundedness of solutions of neutral functional differential equations with finite delay. *Journal of Mathematical Analysis and Applications 276*, 2 (2002), 545–560.

70 Federson, M. The fundamental theorem of calculus for multidimensional Banach space–valued Henstock vector integrals. *Real Analysis Exchange 25*, 1 (1999), 469–480.

71 Federson, M. The Monotone Convergence Theorem for multidimensional abstract Kurzweil vector integrals. *Czechoslovak Mathematical Journal 52* (2002), 429–437.

72 Federson, M. Substitution formulas for the Kurzweil and Henstock vector integrals. *Mathematica Bohemica 127*, 1 (2002), 15–26.

73 Federson, M. Some peculiarities of the Henstock and Kurzweil integrals of Banach space–valued functions. *Real Analysis Exchange 29*, 1 (2003), 439–460.

74 Federson, M., and Bianconi, R. Linear integral equations of Volterra concerning Henstock integrals. *Real Analysis Exchange 25*, 1 (1999), 389–418.
75 Federson, M., and Bianconi, R. Linear Volterra–Stieltjes integral equations in the sense of the Kurzweil–Henstock integral. *Archivum Mathematicum 37*, 4 (2001), 307–328.
76 Federson, M., Frasson, M., Mesquita, J. G., and Tacuri, P. Measure neutral functional differential equations as generalized ODEs. *Journal of Dynamics and Differential Equations 31*, 1 (2019), 207–236.
77 Federson, M., Grau, R., and Mesquita, C. Existence of affine–periodic solutions for generalized ODEs. Manuscript submitted for publication, 2020.
78 Federson, M., Grau, R., and Mesquita, J. G. Prolongation of solutions of measure differential equations and dynamic equations on time scales. *Mathematische Nachrichten 292*, 1 (2019), 22–55.
79 Federson, M., Grau, R., Mesquita, J. G., and Toon, E. Boundedness of solutions of measure differential equations and dynamic equations on time scales. *Journal of Differential Equations 263*, 1 (2017), 26–56.
80 Federson, M., Grau, R., Mesquita, J. G., and Toon, E. Lyapunov stability for measure differential equations and dynamic equations on time scales. *Journal of Differential Equations 267*, 7 (2019), 4192–4223.
81 Federson, M., Mawhin, J., and Mesquita, C. Existence of periodic solutions and bifurcation points for generalized ordinary differential equations. Bulletin des Sciences Mathématiques 169 (2021), 102991.
82 Federson, M., and Mesquita, J. G. Averaging for retarded functional differential equations. *Journal of Mathematical Analysis and Applications 382*, 1 (2011), 77–85.
83 Federson, M., and Mesquita, J. G. Averaging principle for functional differential equations with impulses at variable times via Kurzweil equations. *Differential and Integral Equations 26*, 11–12 (2013), 1287–1320.
84 Federson, M., and Mesquita, J. G. Non-periodic averaging principles for measure functional differential equations and functional dynamic equations on time scales involving impulses. *Journal of Differential Equations 255*, 10 (2013), 3098–3126.
85 Federson, M., Mesquita, J. G., and Slavík, A. Measure functional differential equations and functional dynamic equations on time scales. *Journal of Differential Equations 252*, 6 (2012), 3816–3847.
86 Federson, M., Mesquita, J. G., and Slavík, A. Basic results for functional differential and dynamic equations involving impulses. *Mathematische Nachrichten 286*, 2–3 (2013), 181–204.
87 Federson, M., Mesquita, J. G., and Toon, E. Lyapunov theorems for measure functional differential equations via Kurzweil-equations. *Mathematische Nachrichten 288*, 13 (2015), 1487–1511.

88 Federson, M., and Schwabik, Š. Generalized ODE approach to impulsive retarded functional differential equations. *Differential and Integral Equations 19*, 11 (2006), 1201–1234.

89 Federson, M., and Schwabik, Š. Stability for retarded functional differential equations. *Ukrainian Mathematical Journal 60*, 1 (2008), 121–140.

90 Federson, M., and Schwabik, Š. A new approach to impulsive retarded differential equations: stability results. *Functional Differential Equations 16*, 4 (2009), 583–607.

91 Federson, M., and Táboas, P. Impulsive retarded differential equations in banach spaces via Bochner–Lebesgue and Henstock integrals. *Nonlinear Analysis: Theory Methods & Applications 50* (2002), 389–407.

92 Federson, M., and Táboas, P. Topological dynamics of retarded functional differential equations. *Journal of Differential Equations 195*, 2 (2003), 313–331.

93 Fleury, M., Mesquita, J. G., and Slavík, A. Massera's theorems for various types of equations with discontinuous solutions. *Journal of Differential Equations 269*, 12 (2020), 11667–11693.

94 Fodčuk, V. I. The method of averaging for differential–difference equations of neutral type. *Akademiya Nauk Ukrainskoĭ SSR. Institut Matematiki. Ukrainskiĭ Matematicheskiĭ Zhurnal 20* (1968), 203–209.

95 Franková, D. Continuous dependence on a parameter of solutions of generalized differential equations. *Časopis pro pěstování matematiky 114*, 3 (1989), 230–261.

96 Franková, D. Regulated functions. *Mathematica Bohemica 116*, 1 (1991), 20–59.

97 Franková, D. Regulated functions with values in Banach space. *Mathematica Bohemica 144*, 4 (2019), 437–456.

98 Fu, X., and Zhang, L. On boundedness of solutions of impulsive integro–differential systems with fixed moments of impulse effects. *Acta Mathematica Scientia 17*, 2 (1997), 219–229.

99 Gaines, R. E., and Mawhin, J. L. *Coincidence Degree and Nonlinear Differential Equations, Lecture Notes in Mathematics*, vol. 568. Springer–Verlag, Berlin, New York, 1977.

100 Gallegos, C., Henríquez, H., and Mesquita, J. G. Measure functional differential equations with infinite time–dependent delay. *Mathematische Nachrichten* (accepted) (2020).

101 Gerhardt, C. I. *Die entdeckung der differentialrechnung durch Leibniz: mit benutzung der Leibnizischen manuscripte auf der Königlichen Bibliothek zu Hannover*. H. W. Schmidt, Halle, 1848.

102 Gerhardt, C. I. *Die Geschichte der höheren Analysis*. No. v. 1 in Die Geschichte der höheren Analysis. H. W. Schmidt, Halle, 1855.

103 Gichman, I. I. On the origins of a theorem of N. N. Bogoljubov. *Ukrainski Matematicheski Zurnal IV (Russian)* (1952), 215–219.

104 Gilary, I., Moiseyev, N., Rahav, S., and Fishman, S. Trapping of particles by lasers: the quantum Kapitza pendulum. *Journal of Physics A: Mathematical and General 36*, 25 (2003), 409–415.

105 Gilioli, A., Floret, K., and Hönig, C. S. Natural ultrabornological, non-complete, normed function spaces. *Archiv der Mathematik 61*, 5 (1993), 465–477.

106 Gopalsamy, K., and Zhang, B. On a neutral delay logistic equation. *Dynamics and Stability of Systems 2*, 3–4 (1988), 183–195.

107 Gordon, R. A. The McShane integral of Banach-valued functions. *Illinois Journal of Mathematics 34*, 3 (1990), 557–567.

108 Gordon, R. A. *The Integrals of Lebesgue, Denjoy, Perron, and Henstock*, vol. 4 of *Graduate Studies in Mathematics*. American Mathematical Society, Providence, RI, 1994.

109 Graef, J. R., and Ouahab, A. Nonresonance impulsive functional dynamic equations on times scales. *International Journal of Applied Mathematical Sciences 2*, 1 (2005), 65–80.

110 Grothendieck, A. *Topological Vector Spaces*. Gordon and Breach Science Publishers, New York, London, Paris, 1973. Translated from the French by Orlando Chaljub, Notes on Mathematics and its Applications.

111 Gurney, W., Blythe, S., and Nisbet, R. Nicholson's blowflies revisited. *Nature 287*, 5777 (1980), 17–21.

112 Hadeler, K. P., and Bocharov, G. Where to put delays in population models, in particular in the neutral case. *Canadian Applied Mathematics Quarterly 11*, 2 (2003), 159–173.

113 Halanay, A. On the method of averaging for differential equations with retarded argument. *Journal of Mathematical Analysis and Applications 14* (1966), 70–76.

114 Hale, J. K. Averaging methods for differential equations with retarded arguments and a small parameter. *Journal of Differential Equations 2* (1966), 57–73.

115 Hale, J. K., and Lunel, S. M. V. *Introduction to Functional Differential Equations, Applied Mathematical Sciences*. Springer-Verlag, New York, 1993.

116 Henríquez, H., Mesquita, J. G., and dos Reis, H. C. Measure functional differential equations with state-dependent delays and other types of equations. *Preprint* (2020).

117 Henry, D. *Geometric Theory of Semilinear Parabolic Equations, Lecture Notes in Mathematics*. Springer-Verlag, Berlin Heidelberg, 2006.

118 Henstock, R. Definitions of Riemann type of the variational integrals. *Proceedings of the London Mathematical Society. Third Series 11* (1961), 402–418.

119 Henstock, R. A Riemann–Type Integral of Lebesgue Power. *Canadian Journal of Mathematics 20* (1968), 79–87.

120 Henstock, R. *Lectures on the Theory of Integration, Series in Real Analysis.* World Scientific, Singapore, 1988.

121 Henstock, R. The construction of path integrals. *Mathematica Japonica 39*, 1 (1994), 15–18.

122 Henstock, R., Muldowney, P., and Skvortsov, V. Partitioning infinite-dimensional spaces for generalized Riemann integration. *Bulletin of The London Mathematical Society 38*, 5 (2006), 795–803.

123 Hilger, S. *Ein Maßkettenkalkül mit Anwendung auf Zentrumsmanningfaltigkeiten.* PhD thesis, Universität Würzburg, 1988.

124 Hino, Y., Murakami, S., and Naito, T. *Functional–Differential Equations with Infinite Delay*, vol. 1473 of *Lecture Notes in Mathematics.* Springer–Verlag, Berlin, 1991.

125 Holyst, J. A., and Wojciechowski, W. The effect of Kapitza pendulum and price equilibrium. *Physica A: Statistical Mechanics and its Applications 324*, 1 (2003), 388–395. Proceedings of the International Econophysics Conference.

126 Hönig, C. S. The abstract Riemann–Stieltjes integral and its applications to linear differential equations with generalized boundary conditions. *Notas do Instituto de Matemática e Estatística da Universidade de São Paulo Série Matemática 1* 1973.

127 Hönig, C. S. *Volterra Stieltjes–Integral Equations.* North-Holland Publishing Co., Amsterdam, 1975. Functional analytic methods; linear constraints, Mathematics Studies, No. 16, Notas de Matemática, No. 56. [Notes on Mathematics, No. 56].

128 Hönig, C. S. *Análise funcional e o problema de Sturm–Liouville.* Editora Blucher, São Paulo, 1978.

129 Hönig, C. S. There is no natural Banach space norm on the space of Kurzweil–Henstock–Denjoy–Perron integrable functions. *Seminário Brasileiro de Análise 30* (1989), 387–397.

130 Hönig, C. S. On a remarkable differential characterization of the functions that are Kurzweil–Henstock integrals. *Seminário Brasileiro de Análise 33* (1991), 331–341.

131 Hönig, C. S. A Riemannian characterization of the Bochner–Lebesgue integral. *Seminário Brasileiro de Análise 35* (1992), 351–358.

132 Hönig, C. S. The gauge integrals (in Portuguese). *Seminário Brasileiro de Análise 37* (1993), 1–60.

133 Hutchinson, G. E. Circular causal systems in ecology. *Annals of the New York Academy of Sciences 50*, 4 (1948), 221–246.

134 Jarník, J. Dependence of solutions of a class of differential equations of the second order on a parameter. *Czechoslovak Mathematical Journal 15*, 90 (1965), 124–160.

135 Jarník, J. On differential equations analogous to the equation of Kapica's pendulum. *Abhandlungen der Deutschen Akademie der Wissenschaften zu Berlin. Klasse für Mathematik, Physik und Technik* 1965, 1 (1965), 260–263.

136 Jarník, J., Schwabik, Š., Tvrdý, M., and Vrkoč, I. Eighty years of Jaroslav Kurzweil. *Mathematica Bohemica 131*, 2 (2006), 113–143.

137 Jensen, J. L. W. V. Om fundamentalligningers "opløsning" ved elementære midler. *Tidsskrift for mathematik 2* (1878), 149–155.

138 Jensen, J. L. W. V. Om konvekse funktioner og uligheder imellem middelværdier. *Nyt tidsskrift for matematik 16* (1905), 49–68.

139 Kapitza, P. L. Dynamic stability of a pendulum with an oscilatting point of suspension (in russian). *Zhurnal eksperimental'noi i teoreticheskoi fiziki 21*, 5 (1965), 588–597, 1951. Translation in: Collected Papers by P. L. Kapitza, vol. 2, 714–726, Pergamon Press, London.

140 Kapitza, P. L. A pendulum with vibrating point of suspension (in russian). *Uspekhi Fizicheskikh Nauk 44*, 1 (1965), XLIV, 7–20, 1951. Translation in: Collected Papers by P. L. Kapitza, vol. 2, 726–737, Pergamon Press, London.

141 Kaul, S. K. On impulsive semidynamical systems. III. Lyapunov stability. In *Recent Trends in Differential Equations*, vol. 1 of *World Scientific Series in Applicable Analysis*, R. P. Agarwal, Ed. World Scientific Publishing, River Edge, NJ, 1992, pp. 335–345.

142 Köthe, G. *Topological Vector Spaces II, Grundlehren der mathematischen Wissenschaften*. Springer–Verlag, New York, 1969.

143 Krasnoselskii, M. A., and Kreĭn, S. G. On the principle of averaging in nonlinear mechanics. *Akademiya Nauk SSSR i Moskovskoe Matematicheskoe Obshchestvo. Uspekhi Matematicheskikh Nauk 10 3*, 65 (1955), 147–152.

144 Kryloff, N., and Bogoliuboff, N. *Introduction to Non-Linear Mechanics*, vol. 11 of *Annals of Mathematics Studies*. Princeton University Press, Princeton, NJ, 1943.

145 Krylov, N. N., and Bogolyubov, N. N. *New Methods in Linear Mechanics*. GTTs, Kiev (Russian), 1934.

146 Kuang, Y. *Delay Differential Equations with Applications in Population Dynamics*, vol. 191 of *Mathematics in Science and Engineering*. Academic Press, Inc., Boston, MA, 1993.

147 Kurzweil, J. Generalized ordinary differential equations and continuous dependence on a parameter. *Czechoslovak Mathematical Journal 07*, 3 (1957), 418–449.

148 Kurzweil, J. Generalized ordinary differential equations. *Czechoslovak Mathematical Journal 08*, 3 (1958), 360–388.

149 Kurzweil, J. Unicity of solutions of generalized differential equations. *Czechoslovak Mathematical Journal 08*, 4 (1958), 502–509.

150 Kurzweil, J. Addition to my paper "Generalized ordinary differential equations and continuous dependence on a parameter". *Czechoslovak Mathematical Journal 09*, 4 (1959), 564–573.

151 Kurzweil, J. Problems which lead to a generalization of the concept of an ordinary nonlinear differential equation. In *Differential Equations and Their Applications. Proceedings of the Conference held in Prague in September 1962* Edited by Ivo Babuška. (Prague, 1963), Publishing House of the Czechoslovak Academy of Sciences, 1962, pp. 65–76.

152 Kurzweil, J. *Nichtabsolut Konvergente Integrale*. Teubner, Leipzig, 1980.

153 Kurzweil, J. *Generalized Ordinary Differential Equations. Not Absolutely Continuous Solutions*, vol. 11. of *Series in Real Analysis*. World Scientific Publishing Co. Pte. Ltd., Hackensack, NJ, 2012.

154 Kurzweil, Y., and Vorel, Z. Continuous dependence of solutions of differential equations on a parameter. *Czechoslovak Mathematical Journal 7* (82) (1957), 568–583.

155 Lakrib, M., and Sari, T. Averaging results for functional differential equations. *Siberian Mathematical Journal 45* (2004), 311–320.

156 Lakshmikantham, V., Bainov, D., and Simeonov, P. S. *Theory of Impulsive Differential Equations*, vol. 6 of *Series in Modern Applied Mathematics*. World Scientific, Singapore, 1989.

157 Lawrence, B. A., and Oberste-Vorth, R. W. Solutions of dynamic equations with varying time scales. In *Difference Equations, Special Functions and Orthogonal Polynomials*, S. Elaydi, J. Cushing, R. Lasser, A. Ruffing, V. Papageorgiou, and W, Van Assche, Ed. World Scientific Publishing, Hackensack, NJ, 2007, pp. 452–461.

158 Lee, P. Y. *Lanzhou Lectures on Henstock Integration, Series in Real Analysis*. World Scientific, Singapore, 1989.

159 Legendre, A.-M. *Éléments de géométrie, avec des notes*. Firmin Didot, Paris, 1794.

160 Lehman, B. The influence of delays when averaging slow and fast oscillating systems: overview. *IMA Journal of Mathematical Control and Information 19*, 03 (2002), 201–215.

161 Lehman, B., and Weibel, S. P. Averaging theory for delay difference equations with time-varying delays. *SIAM Journal on Applied Mathematics 59*, 4 (1999), 1487–1506.

162 Lehman, B., and Weibel, S. P. Averaging theory for functional differential equations. In *Proceedings of the 37th IEEE Conference on Decision and Control* (1999), vol. 2, pp. 1352–1357.

163 Lehman, B., and Weibel, S. P. Fundamental theorems of averaging for functional-differential equations. *Journal of Differential Equations 152*, 1 (1999), 160–190.

164 von Leibniz, G. W. Nova methodus pro maximis et minimis, itemque tangentibus, quae nec fractas, nec irrationales quantitates moratur, & singulare pro illis calculi genus. In *Acta Eruditorum Anno MDCLXXXIV [-Anno MDCLXXXV]*. Christopher Günther for Johann Gross and J.F. Gleditsch, 1684–1685, Leipzig, 1646–1716, 467–473.

165 Li, X., Bohner, M., and Wang, C.-K. Impulsive differential equations: periodic solutions and applications. *Automatica 52* (2015), 173–178.

166 Liapounoff, A. M. Problème général de la stabilité du mouvement. *Annales de la Faculté des sciences de Toulouse : Mathématiques 9* (1907), 203–474.

167 Liapounoff, A. M. *Problème Géneral de la Stabilité du Mouvement. (AM-17).* Princeton University Press, Princeton, NJ, 1947.

168 Lin, Z., and Lin, Y. X. *Linear Systems Exponential Dichotomy and Structure of Sets of Hyperbolic Points.* World Scientific, Singapore, 2000.

169 Luo, Z., and Shen, J. Stability and boundedness for impulsive differential equations with infinite delays. *Nonlinear Analysis: Theory Methods & Applications 46*, 4 (2001), 475–493.

170 Lyapunov, A. M. *The general problem of the stability of motion (In Russian).* PhD thesis, University of Kharkov, 1892.

171 Malthus, T. R. *First Essay on Population 1798.* Palgrave Macmillan, London, 1966.

172 McLeod, R. M. *The Generalized Riemann Integral, Carus Mathematical Monographs.* Mathematical Association of America, Gambier, OH, 1980.

173 McShane, E. J. *A Riemann-type Integral that Includes Lebesgue–Stieltjes, Bochner and Stochastic Integrals, American Mathematical Society, Memoirs.* American Mathematical Society, Providence, RI, 1969.

174 McShane, E. J. A unified theory of integration. *The American Mathematical Monthly 80*, 4 (1973), 349–359.

175 Medvedev, G. N. The asymptotic solution of certain systems of differential equations with deviating argument. *Doklady Akademii Nauk SSSR 178* (1968), 293–295.

176 Medvedev, G. N., Morgunov, B. I., and Volosov, V. M. Application of the averaging method to the solution of certain systems of differential equations with deviating argument. *Vestnik Moskovskogo Universiteta. Seriya III. Fizika, Astronomiya 1965 6* (1965), 89–91.

177 Mesquita, J. G. *Measure functional differential equations and impulse functional dynamic equations on time scales.* PhD thesis, Universidade de São Paulo, 2012.

178 Mesquita, J. G., and Slavík, A. Periodic averaging theorems for various types of equations. *Journal of Mathematical Analysis and Applications 387*, 2 (2012), 862–877.

179 Monteiro, G. A., Slavík, A., and Tvrdý, M. *Kurzweil-Stieltjes Integral: Theory and Applications*, vol. 15 of *Series in Real Analysis*. World Scientific Publishing Co. Pte. Ltd., Hackensack, NJ, 2019.

180 Monteiro, G. A., and Tvrdý, M. Generalized linear differential equations in a Banach space: continuous dependence on a parameter. *Discrete and Continuous Dynamical Systems. Series A 33* (2013), 283.

181 Muldowney, P. *General Theory of Integration in Function Spaces, Including Wiener and Feynman Integration, Pitman Research Notes in Mathematics Series*. London Scientific & Technical, New York, 1987.

182 Muldowney, P. Feynman's path integrals and Henstock's non-absolute integration. *Journal of Applied Analysis 6* (2000), 1–23.

183 Muldowney, P. *A Modern Theory of Random Variation: With Applications in Stochastic Calculus, Financial Mathematics, and Feynman Integration*. John Wiley & Sons, Inc., Hoboken, NJ, 2012.

184 Munkres, J. R. *Topology*. Prentice–Hall, Upper Saddle River, NJ, 2000.

185 Murray, J. D. *Mathematical Biology I. An Introduction*, 3rd ed., vol. 17 of *Interdisciplinary Applied Mathematics*. Springer–Verlag, New York, 2002.

186 Myškis, A. D. General theory of differential equations with retarded arguments. *Akademiya Nauk SSSR i Moskovskoe Matematicheskoe Obshchestvo. Uspekhi Matematicheskikh Nauk 4* 5, 33 (1949), 99–141.

187 Newton, I. *The Method of Fluxions and Infinite Series: With its Application to the Geometry of Curve-Lines*. Henry Woodfall, London, 1736.

188 Ngiamsunthorn, P. S. Existence of periodic solutions for differential equations with multiple delays under dichotomy condition. *Advances in Difference Equations 2015, 1* (2015), 259.

189 Nicholson, A. J. An outline of the dynamics of animal populations. *Australian Journal of Zoology 2*, 1 (1954), 9–65.

190 Oberste-Vorth, R. W. The Fell topology on the space of time scales for dynamic equations. *Advances in Dynamical Systems and Applications 3*, 1 (2008), 177–184.

191 Okamura, H. On the strong stability of the stationary point in the current. *Funcional Equations (in Japanese) 40* (1943), 6–19.

192 Palmer, K. J. A generalization of Hartman's linearization theorem. *Journal of Mathematical Analysis and Applications 41* (1973), 753–758.

193 Palmer, K. J. Exponential dichotomies, the shadowing lemma and transversal homoclinic points. In *Dynamics Reported*, Vol. 1 of *Dynamics Reported. A Series in Dynamical Systems and their Applications*, U. Kirchgraber and H.O. Walther, Ed. John Wiley & Sons, Ltd., Chichester, 1988, pp. 265–306.

194 Pandit, S. G., and Deo, S. G. *Differential Systems Involving Impulses, Lecture Notes in Mathematics*. Springer–Verlag, Berlin Heidelberg, 1982.

195 Perez, J., Malta, C., and Coutinho, F. Qualitative analysis of oscillations in isolated populations of flies. *Journal of Theoretical Biology 71*, 4 (1978), 505–514.

196 Peterson, A., and Thompson, B. Henstock-Kurzweil delta and nabla integrals. *Journal of Mathematical Analysis and Applications 323*, 1 (2006), 162–178.

197 Popescu, L. H. Exponential dichotomy roughness on Banach spaces. *Journal of Mathematical Analysis and Applications 314*, 2 (2006), 436–454.

198 Pratt, D. M. Analysis of population development in daphnia at different temperatures. *The Biological Bulletin 85*, 2 (1943), 116–140.

199 Qian, D., and Li, X. Periodic solutions for ordinary differential equations with sublinear impulsive effects. *Journal of Mathematical Analysis and Applications 303*, 1 (2005), 288–303.

200 Sakamoto, K. Estimates on the strength of exponential dichotomies and application to integral manifolds. *Journal of Differential Equations 107*, 2 (1994), 259–279.

201 Samoilenko, A. M., and Perestyuk, N. A. *Impulsive Differential Equations*, vol. 14 of *World Scientific Series on Nonlinear Science. Series A: Monographs and Treatises*. World Scientific Publishing Co. Inc., River Edge, NJ, 1995.

202 Sanders, J. A., and Verhulst, F. *Averaging Methods in Nonlinear Dynamical Systems*, vol. 59 of *Applied Mathematical Sciences*. Springer–Verlag, New York, 1985.

203 Sanders, J. A., and Verhulst, F. *Averaging Methods in Nonlinear Dynamical Systems*. Springer–Verlag, New York, 2014.

204 Sansone, G., and Conti, R. *Non-Linear Differential Equations*. Translated from the Italian by Ainsley H. Diamond. Rev. ed. *International Series of Monographs in Pure and Applied Mathematics*, vol. 67. Pergamon Press, Oxford, London, New York, Paris, Frankfurt. XIII, 533 p., 1964.

205 Santos, F. L. *Dicotomias em equações diferenciais ordinárias generalizadas e aplicações*. PhD thesis, Universidade de São Paulo, 2016.

206 Schmaedeke, W. W. Optimal control theory for nonlinear vector differential equations containing measures. *SIAM Journal on Control and Optimization 3* (1965), 231–280.

207 Schwabik, Š. Floquetova teorie pro zobecněné diferenciální rovnice. *Časopis pro pěstování matematiky 98*, 4 (1973), 416–418.

208 Schwabik, Š. Variational stability for generalized ordinary differential equations. *Československá Akademie Věd. Časopis Pro Pěstování Matematiky 109*, 4 (1984), 389–420.

209 Schwabik, Š. *Generalized Ordinary Differential Equations*, vol. 5 of *Series in Real Analysis*. World Scientific Publishing Co. Inc., River Edge, NJ, 1992.

210 Schwabik, Š. Abstract Perron–Stieltjes integral. *Mathematica Bohemica 121*, 4 (1996), 425–447.

211 Schwabik, Š. Linear Stieltjes integral equations in Banach spaces. *Mathematica Bohemica 124*, 4 (1999), 433–457.

212 Schwabik, Š. A note on integration by parts for abstract Perron–Stieltjes integrals. *Mathematica Bohemica 126*, 3 (2001), 613–629.

213 Schwabik, Š. General integration and extensions I. *Czechoslovak Mathematical Journal 60*, 4 (2010), 961–981.

214 Sell, G. R. Nonautonomous differential equations and topological dynamics. I. The basic theory. *Transactions of the American Mathematical Society 127* (1967), 241–262.

215 Sharma, R. An abstract measure differential equation. *Proceedings of the American Mathematical Society 32* (1972), 503–510.

216 Sharma, R. A measure differential inequality with applications. *Proceedings of the American Mathematical Society 48* (1975), 87–87.

217 Shen, J. H. Razumikhin techniques in impulsive functional differential equations. *Nonlinear Analysis. Theory, Methods & Applications. An International Multidisciplinary Journal 36*, 1 (1999), 119–130.

218 Slavík, A. Averaging dynamic equations on time scales. *Journal of Mathematical Analysis and Applications 388*, 2 (2012), 996–1012.

219 Slavík, A. Dynamic equations on time scales and generalized ordinary differential equations. *Journal of Mathematical Analysis and Applications 385*, 1 (2012), 534–550.

220 Slavík, A. Measure functional differential equations with infinite delay. *Nonlinear Analysis: Theory Methods & Applications 79* (2013), 140–155.

221 Slavík, A., Stehlík, P., and Volek, J. Well-posedness and maximum principles for lattice reaction-diffusion equations. *Advances in Nonlinear Analysis 8*, 1 (2019), 303–322.

222 Smith, F. E. Population dynamics in daphnia magna and a new model for population growth. *Ecology 44*, 4 (1963), 651–663.

223 Stamova, I. M. Boundedness of impulsive functional differential equations with variable impulsive perturbations. *Bulletin of the Australian Mathematical Society 77*, 2 (2008), 331–345.

224 Stephenson, A. On a new type of dynamic stability. *Memoirs and Proceedings of the Manchester Literary and Philosophical Society 52*, 8 (1908), 1–10.

225 Stephenson, A. XX. On induced stability. *The London, Edinburgh, and Dublin Philosophical Magazine and Journal of Science 15*, 86 (1908), 233–236.

226 Strygin, V. V. The averaging principle for equations with heredity. *Akademiya Nauk Ukrainskoĭ SSR. Institut Matematiki. Ukrainskiĭ Matematicheskiĭ Zhurnal 22* (1970), 503–513.

227 Swartz, C., and DePree, J. *Introduction to Real Analysis*. John Wiley and Sons, Inc., New York, 1988.

228 Teixeira, J., and Borges, M. J. Existence of periodic solutions of ordinary differential equations. *Journal of Mathematical Analysis and Applications 385*, 1 (2012), 414–422.

229 Tvrdý, M. Generalized differential equations in the space of regulated functions (boundary value problems and controllability). *Mathematica Bohemica 116*, 3 (1991), 225–244.

230 Tvrdý, M. Linear boundary value problems for generalized differential equations. *Annales Mathematicae Silesiane 14*, 2 (2000), 51–80.

231 Verhulst, P.-F. Notice sur la loi que la population suit dans son accroissement. *Correspondance Mathématique et Physique 10* (1838), 113–126.

232 Verhulst, P.-F. Recherches mathématiques sur la loi d'accroissement de la population. *Académie royale de Belgique 18* (1845), 39. Mémoires des Membres.

233 Verhulst, F. *Nonlinear Differential Equations and Dynamical Systems*, Universitext. Springer–Verlag, Berlin Heidelberg, 2006.

234 Vrkoč, I. Integral stability. *Czechoslovak Mathematical Journal (in Russian) 9*, 84 (1959), 71–128.

235 Wang, J. R., Ren,L., and Zhou, Y. (ω, c)–periodic solutions for time varying impulsive differential equations. *Advances in Difference Equations* 2019, 1 (2019), 259.

236 Ward Jr.,, J. R. Periodic solutions of ordinary differential equations with bounded nonlinearities. *Topological Methods in Nonlinear Analysis 19*, 2 (2002), 275–282.

237 Weiss, L. The concepts of differential controllability and differential observability. *Journal of Mathematical Analysis and Applications 10*, 2 (1965), 442–449.

238 Yoshizawa, T. On the stability of solutions of a system of differential equations. *Memoirs of the College of Science, University of Kyoto. Series A: Mathematics 29*, 1 (1955), 27–33.

239 Yosida, K. *Functional Analysis*, Classics in Mathematics. Springer–Verlag, Berlin Heidelberg, 1995.

240 Zhang, Y., and Sun, J. Stability of impulsive delay differential equations with impulses at variable times. *Dynamical Systems 20*, 3 (2005), 323–331.

List of Symbols

Generalized Ordinary Differential Equations in Abstract Spaces and Applications, First Edition.
Edited by Everaldo M. Bonotto, Márcia Federson, and Jaqueline G. Mesquita.

Index

Generalized Ordinary Differential Equations in Abstract Spaces and Applications, First Edition.
Edited by Everaldo M. Bonotto, Márcia Federson, and Jaqueline G. Mesquita.

L

M

N

O

P

Printed and bound by CPI Group (UK) Ltd, Croydon, CR0 4YY

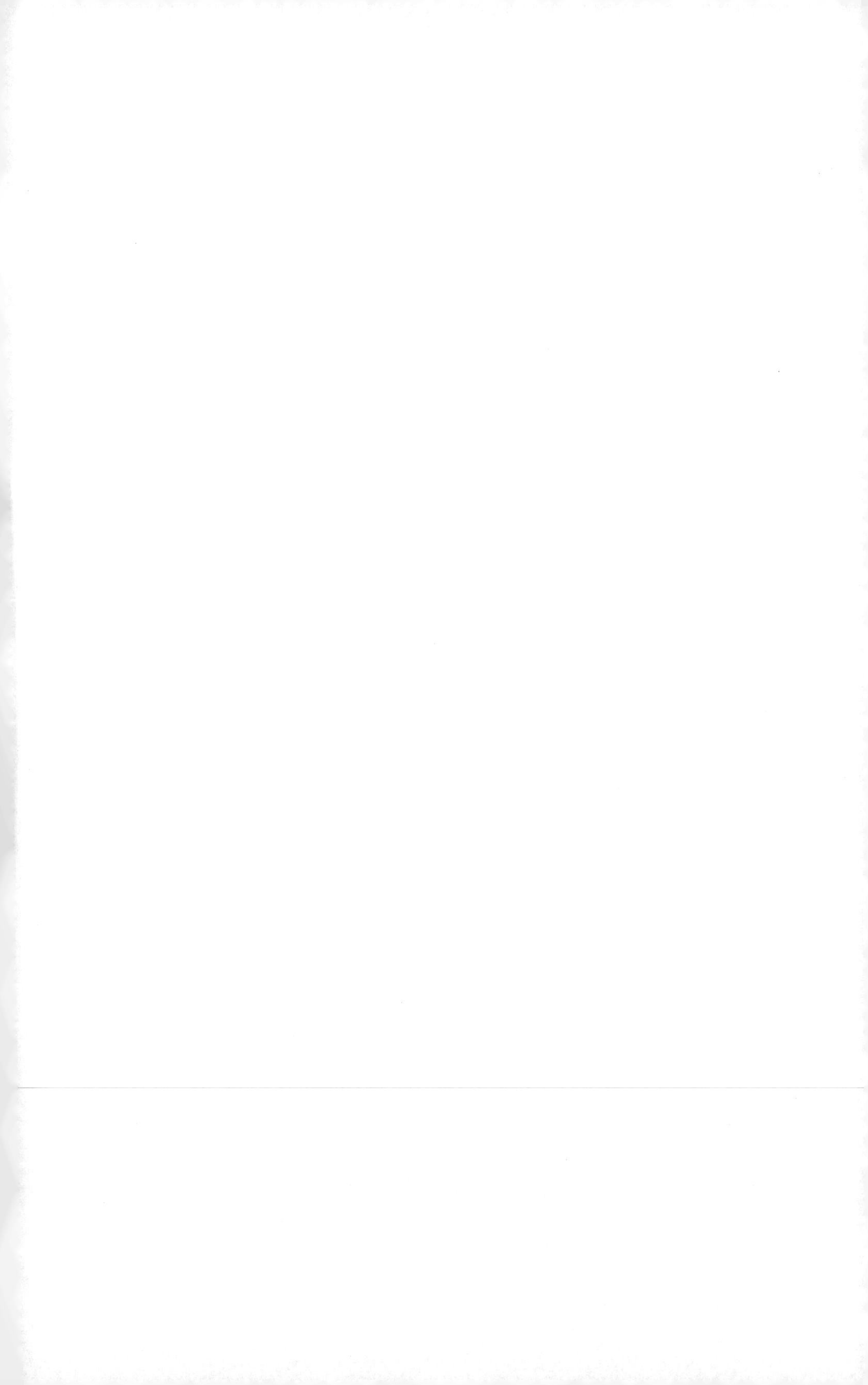